AF794785

Science and Engineering of Polyphenols

Science and Engineering of Polyphenols

Fundamentals and Industrial Scale Applications

Edited by
Chandrabhan Verma
Department of Chemical Engineering
Khalifa University of Science and Technology
Abu Dhabi,
United Arab Emirates

WILEY

Copyright © 2024 by John Wiley & Sons Inc. All rights reserved.

Published by John Wiley & Sons, Inc., Hoboken, New Jersey.
Published simultaneously in Canada.

No part of this publication may be reproduced, stored in a retrieval system, or transmitted in any form or by any means, electronic, mechanical, photocopying, recording, scanning, or otherwise, except as permitted under Section 107 or 108 of the 1976 United States Copyright Act, without either the prior written permission of the Publisher, or authorization through payment of the appropriate per-copy fee to the Copyright Clearance Center, Inc., 222 Rosewood Drive, Danvers, MA 01923, (978) 750-8400, fax (978) 750-4470, or on the web at www.copyright.com. Requests to the Publisher for permission should be addressed to the Permissions Department, John Wiley & Sons, Inc., 111 River Street, Hoboken, NJ 07030, (201) 748-6011, fax (201) 748-6008, or online at http://www.wiley.com/go/permission.

Trademarks: Wiley and the Wiley logo are trademarks or registered trademarks of John Wiley & Sons, Inc. and/or its affiliates in the United States and other countries and may not be used without written permission. All other trademarks are the property of their respective owners. John Wiley & Sons, Inc. is not associated with any product or vendor mentioned in this book.

Limit of Liability/Disclaimer of Warranty: While the publisher and author have used their best efforts in preparing this book, they make no representations or warranties with respect to the accuracy or completeness of the contents of this book and specifically disclaim any implied warranties of merchantability or fitness for a particular purpose. No warranty may be created or extended by sales representatives or written sales materials. The advice and strategies contained herein may not be suitable for your situation. You should consult with a professional where appropriate. Further, readers should be aware that websites listed in this work may have changed or disappeared between when this work was written and when it is read. Neither the publisher nor authors shall be liable for any loss of profit or any other commercial damages, including but not limited to special, incidental, consequential, or other damages.

For general information on our other products and services or for technical support, please contact our Customer Care Department within the United States at (800) 762-2974, outside the United States at (317) 572-3993 or fax (317) 572-4002.

Wiley also publishes its books in a variety of electronic formats. Some content that appears in print may not be available in electronic formats. For more information about Wiley products, visit our web site at www.wiley.com.

Library of Congress Cataloging-in-Publication Data:

Names: Verma, Chandrabhan, editor.
Title: Science and engineering of polyphenols : fundamentals and industrial scale applications / edited by Chandrabhan Verma, Khalifa University of Science and Technology, Abu Dhabi, United Arab Emirates.
Description: Hoboken, New Jersey : Wiley, [2024] | Includes bibliographical references and index.
Identifiers: LCCN 2023053751 (print) | LCCN 2023053752 (ebook) | ISBN 9781394203901 (hardback) | ISBN 9781394203918 (adobe pdf) | ISBN 9781394203925 (epub)
Subjects: LCSH: Polyphenols. | Polyphenols—Industrial applications. | Polyphenols—Therapeutic use.
Classification: LCC QK898.P764 S35 2024 (print) | LCC QK898.P764 (ebook) | DDC 547/.632—dc23/eng/20231221
LC record available at https://lccn.loc.gov/2023053751
LC ebook record available at https://lccn.loc.gov/2023053752

Cover Design: Wiley
Cover Image: © zhengshun tang/Getty Images

Set in 9.5/12.5pt STIXTwoText by Straive, Chennai, India

Contents

List of Contributors

Akshima
Department of Applied Chemistry
(CBFS–ASAS)
Amity University
Gurugram
Manesar, Haryana
India

Humira Assad
Department of Chemistry
School of Chemical Engineering and
Physical Sciences
Lovely Professional University
Punjab
India

Department of Chemistry
Faculty of Technology and Science
Lovely Professional University
Phagwara
Punjab
India

Liya Babu
Department of Biochemistry and
Biotechnology
CSIR-Central Leather Research
Institute (CLRI)
Adyar
Chennai
Tamil Nadu
India

Academy of Scientific and Innovative
Research (AcSIR)
Ghaziabad
India

Vincent Ball
Institut National de la Santé et de la
Recherche Médicale
Unité Mixte de Recherche
Strasbourg
Cedex
France

Université de Strasbourg
Faculté de Chirurgie Dentaire
Strasbourg
France

Manoj K. Banjare
MATS School of Sciences
MATS University
Raipur
Chhattisgarh
India

Kamalakanta Behera
Department of Chemistry
Faculty of Science
University of Allahabad
Prayagraj
Uttar Pradesh
India

Elyor Berdimurodov
Chemical & Materials Engineering
New Uzbekistan University
Tashkent
Uzbekistan

Medical School
Central Asian University
Tashkent
Uzbekistan

Faculty of Chemistry
National University of Uzbekistan
Tashkent
Uzbekistan

Khasan Berdimuradov
Faculty of Industrial Viticulture and
Food Production Technology
Shahrisabz Branch of Tashkent
Institute of Chemical Technology
Shahrisabz
Uzbekistan

Azimjon Choriev
Faculty of Chemistry-Biology
Karshi State University
Karshi
Uzbekistan

Omar Dagdag
Department of Mechanical
Engineering
Gachon University
Seongnam
Republic of Korea

Ilyos Eliboev
Faculty of Chemistry
National University of Uzbekistan
Tashkent
Uzbekistan

Salima El Yakhlifi
Institut National de la Santé et de la
Recherche Médicale
Unité Mixte de Recherche
Strasbourg
Cedex
France

Akbar Esmaeili
Department of Chemical Engineering
North Tehran Branch
Islamic Azad University
Tehran
Iran

Muhammad U. Farooq
Department of Chemical Engineering
Ghulam Ishaq Khan Institute of
Engineering Sciences & Technology
Topi
Swabi
KP
Pakistan

Youyou Feng
The Key Laboratory of Biomedical
Information Engineering of Ministry
of Education
School of Life Science and Technology
Xi'an Jiaotong University
Xi'an
P. R. China

Richika Ganjoo
Department of Chemistry
School of Chemical Engineering and
Physical Sciences
Lovely Professional University
Punjab
India

Rosa M. Giráldez-Pérez
University of Córdoba
Departments of Cellular Biology
Physiology and Immunology
Córdoba
Spain

Elia Grueso
University of Seville
Department of Physical Chemistry
Seville
Spain

Rajesh Haldhar
School of Chemical Engineering
Yeungnam University
Gyeongsan
Republic of Korea

Nastaran Hashemzadeh
Pharmaceutical Analysis Research
Center and Faculty of Pharmacy
Tabriz University of Medical Sciences
Tabriz
Iran

P. Indra Arulselvi
Department of Biotechnology
Periyar University
Salem
Tamil Nadu
India

Bhawana Jain
SIDDHACHALAM LABORATORY
Raipur
India

Kirubakaran Baskaran Jayalakshmi
Department of Biochemistry and
Biotechnology
CSIR-Central Leather Research
Institute (CLRI)
Adyar
Chennai
Tamil Nadu
India

Vikrant Jayant
Department of Natural and Applied
Sciences
School of Science and Technology
Glocal University
Mirzapur Pole
Saharanpur
Uttar Pradesh
India

Abolghasem Jouyban
Pharmaceutical Analysis Research
Center and Faculty of Pharmacy
Tabriz University of Medical Sciences
Tabriz
Iran

K.R. Kavya
Department of Biotechnology
Periyar University
Salem
Tamil Nadu
India

M. Kesavapriya
Department of Biotechnology
Periyar University
Salem
Tamil Nadu
India

Palraj Kasinathan
Department of Chemistry
Faculty of Science
University of Allahabad
Prayagraj
Uttar Pradesh
India

Anvar Khamidov
Faculty of Chemistry
National University of Uzbekistan
Tashkent
Uzbekistan

Wasim Khan
Department of Petroleum Engineering
School of Science and Technology
Glocal University
Mirzapur Pole
Saharanpur
Uttar Pradesh
India

Hansang Kim
Department of Mechanical Engineering
Gachon University
Seongnam
Republic of Korea

Alok Kumar
Nalanda College of Engineering
Bihar Engineering University
Department of Science,
Technology and Technical Education
Government of Bihar
India

Ashish Kumar
Nalanda College of Engineering
Bihar Engineering University
Department of Science, Technology and Technical Education
Government of Bihar
India

Deena Praveena Kumar
Department of Biochemistry and Biotechnology
CSIR-Central Leather Research Institute (CLRI)
Adyar
Chennai
Tamil Nadu
India

Academy of Scientific and Innovative Research (AcSIR)
Ghaziabad
India

Kamlesh Kumari
Department of Zoology
University of Delhi
New Delhi
India

Chandan K. Mandal
Department of Applied Chemistry
(CBFS–ASAS)
Amity University
Gurugram
Manesar, Haryana
India

Mangaiyarkarasi Manivel
Department of Fish Processing
Technology
Dr. M.G.R. Fisheries College and
Research Institute
Tamil Nadu Dr. J. Jayalalithaa
Fisheries University
Ponneri
Tamil Nadu
India

Garima Narang
Department of Chemistry
Atma Ram Sanatan Dharma College
University of Delhi
New Delhi
India

Srinivasan Narasimhan
Department of Biochemistry
Asthagiri Herbal Research Foundation
Chennai
Tamil Nadu
India

Garima Pandey
Department of Chemistry
SRM Institute of Science & Technology
Modinagar
India

Palmiro Poltronieri
Department of Agriculture
Institute of Sciences of Food
Productions
National Research Council of Italy
CNR-ISPA
Lecce
Italy

Ganesan Ponesakki
Department of Biochemistry and
Biotechnology
CSIR-Central Leather Research
Institute (CLRI)
Adyar
Chennai
Tamil Nadu
India

Academy of Scientific and Innovative
Research (AcSIR)
Ghaziabad
India

Elaheh Rahimpour
Pharmaceutical Analysis Research
Center and Faculty of Pharmacy
Tabriz University of Medical Sciences
Tabriz
Iran

Atta Ur. Rahman
Institute of Chemical Sciences
University of Peshawar
Peshawar
KP
Pakistan

Kamini Numbi Ramudu
Department of Biochemistry and Biotechnology
CSIR-Central Leather Research Institute (CLRI)
Adyar
Chennai
Tamil Nadu
India

Academy of Scientific and Innovative Research (AcSIR)
Ghaziabad
India

Jayasri Ravichandran
Department of Fish Processing Technology
Dr. M.G.R. Fisheries College and Research Institute
Tamil Nadu Dr. J. Jayalalithaa Fisheries University
Ponneri
Tamil Nadu
India

Mohamed Rbaa
The Higher Institute of Nursing Professions and Health Techniques of Casablanca
Morocco

Laboratory of Materials, Nanotechnology, and Environment
Faculty of Sciences
Mohammed V University in Rabat
Agdal-Rabat
Morocco

Garima Sachdeva
Department of Applied Chemistry (CBFS–ASAS)
Amity University
Gurugram
Manesar
Haryana
India

Srikanta Sahu
Department of Chemistry
School of Applied Sciences
Centurian University
Jatni
Odisha
India

Sanaz Sajedi-Amin
Pharmaceutical Analysis Research Center and Faculty of Pharmacy
Tabriz University of Medical Sciences
Tabriz
Iran

Muhammad Salman
Institute of Chemical Sciences
University of Peshawar
Peshawar
KP
Pakistan

Sekar Sasikala
Department of Biochemistry
Asthagiri Herbal Research Foundation
Chennai
Tamil Nadu
India

Manu Sharma
Department of Orthopaedics
Maharishi Markandeshwar Medical College and Hospital
Solan
HP
India

Praveen Kumar Sharma
Department of Chemistry
School of Chemical Engineering and Physical Sciences
Lovely Professional University
Punjab
India

Sanju Sharma
Department of Applied Chemistry (CBFS–ASAS)
Amity University
Gurugram
Manesar
Haryana
India

Shveta Sharma
Department of Chemistry
School of Chemical Engineering and Physical Sciences
Lovely Professional University
Punjab
India

Department of Chemistry
Government College Una
Affiliated to Himachal Pradesh University
Una
India

Prashant Singh
Department of Chemistry
Atma Ram Sanatan Dharma College
University of Delhi
New Delhi
India

Nimish Mol Stephen
Department of Fish Processing Technology
Dr. M.G.R. Fisheries College and Research Institute
Tamil Nadu Dr. J. Jayalalithaa Fisheries University
Ponneri
Tamil Nadu
India

Sidra Subhan
Department of Chemistry
Bacha Khan University
Charsadda
Pakistan

Abhinay Thakur
Department of Chemistry
School of Chemical Engineering and Physical Sciences
Lovely Professional University
Phagwara
Punjab
India

Gianluca Viscusi
Department of Industrial Engineering
University of Salerno
Fisciano
SA
Italy

Giuliana Gorrasi
Department of Industrial Engineering
University of Salerno
Fisciano
SA
Italy

Kalaiselvi Vinayagamurthy
Department of Biochemistry and Biotechnology
CSIR-Central Leather Research Institute (CLRI)
Adyar
Chennai
Tamil Nadu
India

Academy of Scientific and Innovative Research (AcSIR)
Ghaziabad
India

Vrinda
Department of Chemistry
Atma Ram Sanatan Dharma College
University of Delhi
New Delhi
India

Gen Wang
School of Environmental and Municipal Engineering
Xi'an University of Architecture and Technology
Xi'an
P. R. China

Jing Wei
The Key Laboratory of Biomedical Information Engineering of Ministry of Education
School of Life Science and Technology
Xi'an Jiaotong University
Xi'an
P. R. China

Mansi Yadav
Department of Applied Chemistry (CBFS–ASAS)
Amity University
Gurugram
Manesar, Haryana
India

Karthika Yalavarthi
Department of Biochemistry and Biotechnology
CSIR-Central Leather Research Institute (CLRI)
Adyar
Chennai
Tamil Nadu
India

Academy of Scientific and Innovative Research (AcSIR)
Ghaziabad
India

Muhmmad Yaseen
Institute of Chemical Sciences
University of Peshawar
Peshawar
KP
Pakistan

Chinmurot Yodgorov
Faculty of Chemistry
National University of Uzbekistan
Tashkent
Uzbekistan

Mohd Yusuf
Department of Natural and Applied Sciences
School of Science and Technology
Glocal University
Mirzapur Pole
Saharanpur
Uttar Pradesh
India

Mukhriddin Yusufov
Faculty of Chemistry
National University of Uzbekistan
Tashkent
Uzbekistan

Muhammad Zeeshan
Department of Chemistry
Bacha Khan University
Charsadda
Pakistan

Preface

Polyphenols are a class of micronutrients that are naturally present in plants, and they are of interest due to their potential for technological advancement and therapeutic effects. They are present in many supplements and readily available in various foods, including fruits, vegetables, teas, and spices. There are more than 8000 different varieties of polyphenols. The majority of the antioxidants in our diets are polyphenols. The two primary types of polyphenols, accounting for one-third and two-thirds, are phenolic acids (mostly caffeic acid) and flavonoids, respectively. The chemical structure of polyphenols is related to their biological characteristics, bioavailability, antioxidant activity, and specific interactions with cell receptors and enzymes. Polyphenols substantially impact diagnosing and preventing many diseases, including diabetes, obesity, Parkinson's, Alzheimer's, and others. They are connected to several significant properties, making them candidates for pharmacological, biological, and industrial applications.

The book, *Science and Engineering of Polyphenols: Fundamentals and Industrial Scale Applications*, describes polyphenols' science and engineering, including their fundamentals and industrial-scale applications. The book is divided into three sections. Part I is related to the overview of polyphenols and contains four chapters. Chapters 1 and 2 describe the fundamentals of polyphenols and their nomenclature, classification, properties, synthesis, characterization, and chemical reactivities. Chapters 3 and 4 account for the extraction, purification, and characterization of plant-derived polyphenols and the degradation and fermentation of the polyphenols, respectively. Part II describes the different industrial applications of polyphenols. Part II has eight chapters (Chapters 5–12), each reporting a specific industrial application of polyphenols. Chapters 5–10 describe wastewater treatment, corrosion protection, dyes, packaging, textile, and sensing/biosensing applications of polyphenols. Chapter 11 reports the fundamental properties and industrial applications of polyphenol-based hydrogels and nanocomplexes. Lastly, Chapter 12 describes the chemo- and bio-sensing applications of polyphenol-based materials,

Part III discusses the biological uses of polyphenols. It is divided into 11 chapters (Chapters 13–23), each discussing a different biological use. Chapters 13 and 14 discuss the applications of polyphenols in COVID-19 control and dietary applications, respectively. Chapter 15 reports the use of polyphenols in health management and disease control. Chapters 16 and 17 report polyphenols' antimicrobial and antioxidant uses, respectively. Chapter 18 describes the polyphenols derived from marine organisms and their biofunctional properties. Chapters 19–22 describe the different applications of polyphenols, including dietary, food additives, and pharmaceuticals (Diagnosis, Therapies, and Prevention of Diseases). The last chapter (Chapter 23) discusses the phenolic compounds (polyphenols) isolated from marine organisms and their biofunctional properties.

I, Dr. Chandrabhan Verma (the editor), thank all contributors for their outstanding efforts. On behalf of John Wiley & Sons, Inc. (Wiley Publications), I am very thankful to the authors of all chapters for their outstanding and passionate efforts in making this book. Special thanks to Dr. Michael Leventhal (commissioning editor) and Dr. Dorairaj Vijayan (Editorial Assistant) for their dedicated support and help during this project. In the end, all thanks to John Wiley & Sons, Inc. (Wiley Publications) for publishing the book.

15 January
Abu Dhabi, United Arab Emirates

Chandrabhan Verma, PhD
(Editor)

Editor Biography

Name: Chandrabhan Verma
Affiliation: Department of Chemical Engineering, Khalifa University of Science and Technology, Abu Dhabi, United Arab Emirates
Email: chandraverma.rs.apc@itbhu.ac.in

Chandrabhan Verma works at the Department of Chemical Engineering, Khalifa University of Science and Technology, Abu Dhabi, United Arab Emirates. Dr. Verma obtained B.Sc. and M.Sc. degrees from Udai Pratap Autonomous College, Varanasi (UP), India. He received his PhD from the Department of Chemistry, Indian Institute of Technology (Banaras Hindu University) Varanasi, under the supervision of Prof. Mumtaz A. Quraishi in Corrosion Science and Engineering. He has served as a member of the American Chemical Society (ACS) and a lifetime member of the World Research Council (WRC). He is a reviewer and editorial board member of various internationally recognized ACS, RSC, Elsevier, Wiley, and Springer platforms. Dr. Verma published numerous research and review articles in different areas of science and engineering at ACS, Elsevier, RSC, Wiley and Springer, etc. He has a total citation of more than 12000, with an

H-index of 61 and an i-10 index of 155. His current research focuses on designing and developing industrially applicable corrosion inhibitors. Dr. Verma edited a few books for the ACS, Elsevier, RSC, Springer, and Wiley. Dr. Verma received several awards for his academic achievements, including a Gold medal in M.Sc. (Organic Chemistry, 2010) and Best Publication awards from the Global Alumni Association of IIT-BHU (Second Prize 2013).

Part I

Overview of Polyphenols

1

Fundamentals of Polyphenols: Nomenclature, Classification and Properties

Sidra Subhan[2], Muhammad Zeeshan[2], Atta Ur. Rahman[1], and Muhmmad Yaseen[1,]*

[1] *Institute of Chemical Sciences, University of Peshawar, Peshawar, KP, Pakistan*
[2] *Department of Chemistry, Bacha Khan University, Charsadda, Pakistan*

1.1 Introduction

Polyphenols are a class of substances that are found in plants and are renowned for having antioxidant qualities [1]. The basic structure of polyphenols consists of one or more aromatic rings with hydroxyl (—OH) groups attached to them as shown in Figure 1.1, which helps to neutralize free radicals (which are highly reactive molecules that can cause damage to cells and contribute to various diseases, including cancer, heart disease, and Alzheimer's disease, etc.) [2]. In order to protect themselves against environmental factors including ultraviolet (UV) radiation, diseases, and pests, plants can produce them. Flavonoids, phenolic acids, lignans, and stilbenes are just a few of the numerous chemically different molecules that comprise the complex group of substances called polyphenols [3]. Fruits, vegetables, tea, coffee, red wine, and chocolate are a few major dietary sources of polyphenols. As per research, polyphenols may provide a wide range of health benefits [4]. For instance, it has been revealed that they possess antioxidant activity anti-inflammatory, anti-cancer, neuroprotective activity, antidiabetic activity, and cardiovascular health-promoting properties. They also reduce the heart disease risk. They may service people manage their blood sugar levels and have neuroprotection [5]. While phenols are generally safe and well-tolerated, it is

* Corresponding author

Science and Engineering of Polyphenols: Fundamentals and Industrial Scale Applications,
First Edition. Edited by Chandrabhan Verma.
© 2024 John Wiley & Sons Inc. Published 2024 by John Wiley & Sons Inc.

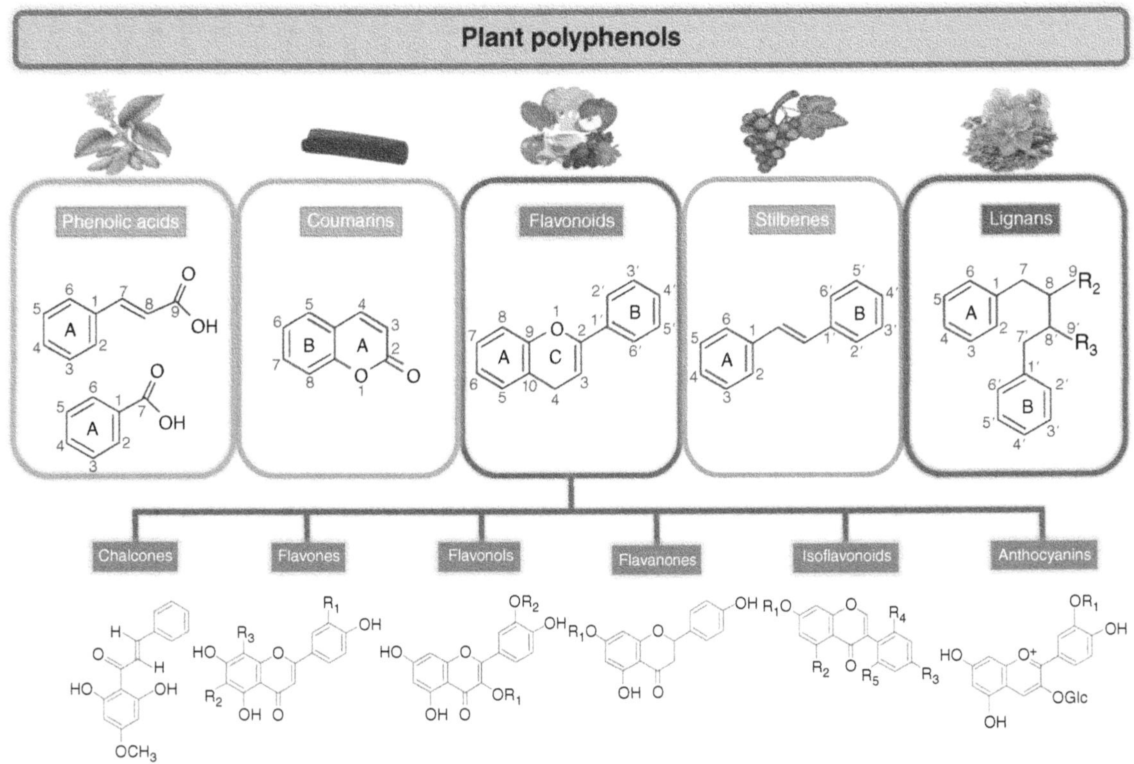

Figure 1.1 Basic structure of polyphenol. Source: Oscar Vidal-Casanella et al. [2]/Reproduced from MDPI/CC BY 4.0.

important to note that they can also interact with certain medications and affect their absorption and effectiveness. Therefore, before taking high doses of polyphenol supplements, it is essential to speak with a healthcare expert or making significant dietary changes to increase polyphenol consumption [6].

1.2 A Short History of Polyphenols

The name derives from the Ancient Greek word (*polus*, meaning "many, much") and the word "phenol," which refers to a chemical structure formed by attaching to an aromatic benzenoid (phenyl) ring to —OH group as is found in alcohols (hence the *-ol* suffix). The term "polyphenol" used since 1894 is a Greek-rooted word polus meaning many and phenol refereeing to the chemical structure formed by attaching phenyl ring to —OH group. The history of polyphenols can be traced back to ancient times when they were used for medicinal purposes and as food preservatives. One of the earliest recorded uses of polyphenols was by the ancient Egyptians, who used tannin-rich plant extracts to preserve mummies. In ancient Greece, Hippocrates, the father of modern medicine, recognized the medicinal properties of wine, which contains high levels of polyphenols. During the Middle Ages, polyphenols were used as natural remedies for a variety of ailments. For example, extracts of grape seeds were used to treat skin conditions and digestive disorders. In the 16th century, the Spanish physician Andres Laguna wrote about the medicinal properties of tea, which contains high levels of polyphenols. In the 19th and early 20th centuries, the study of polyphenols became more scientific. In 1853, the French chemist Henri Braconnot isolated tannins, which are a type of polyphenol, from oak bark. In the early 20th century, the German chemist Richard Willstätter discovered the chemical structure of flavonoids, which are another type of polyphenol. Today, research continues into the potential health benefits of polyphenols, which include reducing inflammation, improving cardiovascular health, and protecting against certain types of cancer. Polyphenols are found in a wide range of plant-based foods, including fruits, vegetables, tea, coffee, and wine and are an important part of a healthy diet as shown in Table 1.1.

Table 1.1 List of some polyphenol compounds along with their natural sources.

Polyphenols	Natural sources
Resveratrol	Grapes, red wine, peanuts, and berries
Catechins	Green tea, black tea, cocoa, and apples
Quercetin	Onions, apples, grapes, berries, broccoli, and tea
Flavonols	Onions, kale, broccoli, apples, berries, and tea
Anthocyanins	Berries, cherries, grapes, red cabbage, and eggplant

(Continued)

Table 1.1 (Continued)

Polyphenols	Natural sources
Proanthocyanidins	Grapes, red wine, cocoa, and apples
Ellagic acid	Strawberries, raspberries, pomegranates, and nuts
Curcumin	Turmeric
Chlorogenic acid	Coffee, apples, pears, and blueberries
Luteolin	Celery, parsley, thyme, and peppers
Apigenin	Parsley, celery, chamomile, and peppermint
Isoflavones	Soybeans, lentils, and chickpeas
Myricetin	Berries, grapes, and red wine
Oleuropein	Olives and olive oil
Rutin	Buckwheat, citrus fruits, and asparagus
Stilbenes	Grapes, blueberries, and red wine
Tannin	Tea, wine, and berry
Caffeic acid	Coffee, fruits, and vegetables such as tomatoes and carrot
Ferulic acid	Brown rice, oats, barley, and wheat bra
Hesperidin	Citrus fruits such as oranges and lemons
Delphinidin	Blueberries, cranberries, and grape
Epicatechin	Green tea, cocoa, and apple
Gallocatechin	Green tea, black tea, and cocoa
Isorhamnetin	Onions, grapes, and tomato sauce
Secoisolariciresinol	Flaxseed, sesame seeds, and whole grain
Prodelphinidin	Cocoa and dark chocolate
Ursolic acid	Apples, basil, and cranberries
Chrysin	Honey, propolis, and passionflower
Epigallocatechin	Green tea, black tea, and cocoa
Gallic acid	Grapes, blueberries, and green tea
Hydroxytyrosol	Olives, olive oil, and red win
Malvidin	Red grapes, blackberries, and blueberries
Piceatannol	Blueberries, and passionfruit
Theaflavins	Black tea
Hesperidin	Oranges and lemons
Lignans	Flaxseed, sesame seeds, and whole grains
Syringic acid	Walnuts, and berries such as raspberries and blackberries

1.3 Fundamental of Polyphenols

A class of naturally occurring substances called polyphenols are present in plants and are distinguished by having many phenolic rings. They may be found in a variety of fruits, vegetables, cereals, tea, coffee, wine, and other plant-based meals [7]. They are also extensively spread throughout the plant world. One or more aromatic rings with —OH groups attached compose up the fundamental structure of polyphenols [8]. These —OH groups give polyphenols their antioxidant capabilities, which aid in the body's ability to combat free radicals and oxidative stress. Based on their chemical structure, polyphenols may be divided into a number of distinct classes, including flavonoids, phenolic acids, lignans, and stilbenes. The most prevalent and most researched class of polyphenols are flavonoids, which are further divided into subcategories including flavones, flavanones, and flavonol. Deep research has been done on polyphenols' possible on health advantages, which are mostly related to their anti-inflammatory and antioxidant characteristics [9, 10]. Many chronic diseases, such as cancer, heart disease, and neurological illnesses have been proven to be protected against by them. Polyphenols have a variety of different biological effects that have been discovered in addition to their antioxidant capabilities [11]. For instance, modify the activity of enzymes involved in cell signaling, control gene expression, and interact with the body's proteins and lipids. The bioavailability of polyphenols might change based on the dietary source and processing techniques. it is important to note. For example, certain polyphenols are better absorbed when taken cooked as opposed to raw, while others are more accessible when consumed along with other nutrients like fat. You may improve your general health and well-being by including a range of foods high in polyphenols in your diet.

1.4 Nomenclature of Polyphenols

The nomenclature of polyphenols can be complex and varies depending on a specific type of polyphenol. However, there are some general rules that are commonly used to name these compounds. Polyphenols are frequently named based on their chemical structure, which usually contains multiple phenolic rings. The simplest polyphenols, such as catechins and flavanols, are named according to the number and position of the —OH groups on the phenolic rings. For example, catechins have two —OH groups on adjacent carbon atoms, while flavanols have —OH groups on different carbon atoms. The number and position of the —OH groups on the phenolic rings of more complex polyphenols, such as flavonoids and phenolic acids, are used to designate them. For instance, the flavonoid quercetin is so called because of the —OH groups that are located at positions 3, 5, 7, and 4′ on its flavone ring (numbering starts at the top of the ring and proceeds clockwise).

Polyphenols may be given names based on their biological function or place of origin. For example, curcumin, a phenolic molecule included in turmeric, has anti-inflammatory effects, whereas resveratrol is a stilbene polyphenol found in red wine and grapes. Moreover, shorthand acronyms or chemical symbols can be used to identify polyphenols, such as epigallocatechin gallate (EGCG) (a catechin present in green tea) or gallic acid equivalents (a measure of total polyphenol content in a food or beverage) [12–14]. The nomenclature of polyphenols can be complex, and naming conventions can vary depending on the specific type of polyphenol and the context in which it is being used. However, understanding the basic principles of polyphenol naming can help to clarify the structure and properties of these important plant-based compounds.

1.5 Classification of Polyphenols

Polyphenols can be classified based on their chemical structure and properties. Figure 1.2 shows some of the common classifications of polyphenols [15]:

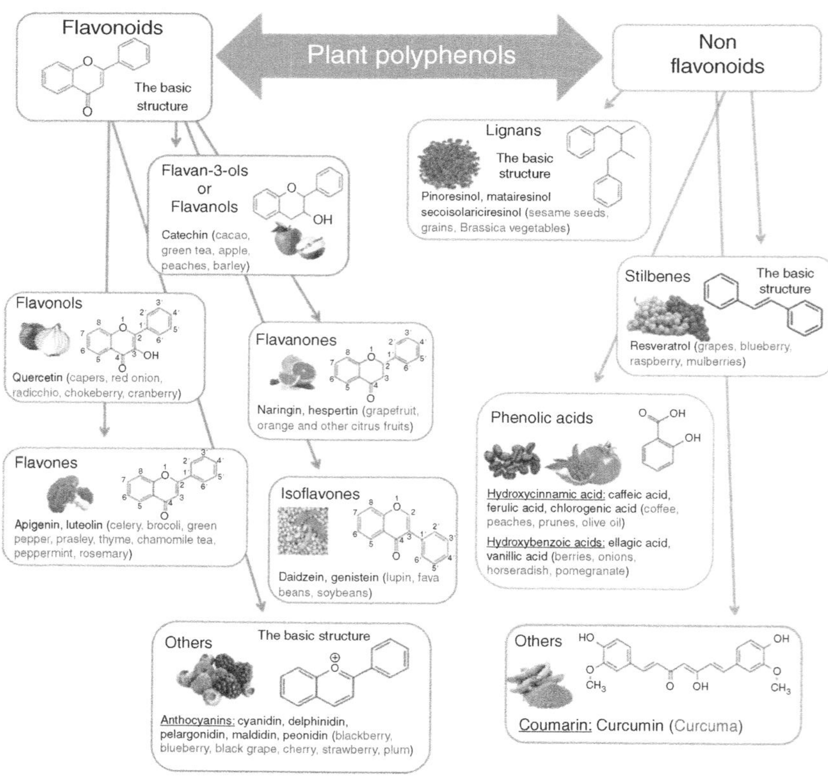

Figure 1.2 Classification of polyphenols. Source: Rosaria Meccariello et al. [15]/ Reproduced from MDPI/CC BY 4.0.

1.5.1 Flavonoids

The most well-known and well-researched class of polyphenols is flavonoids. They may be identified by their two aromatic rings joined by a three-carbon chain, which makes up their fundamental chemical structure. Depending on their chemical structure, flavonoids may be further divided into a number of subgroups, including flavones, flavonols, flavanones, flavan-3-ols (catechins), isoflavones, anthocyanidins, and anthocyanins. Moreover, sources from plants have given the identification of over 4000 flavonoids. Flavonols, between C3 and C2, a double bond could be observed, and a OH group is linked at C3. The majority of flavonoids found in various dietary sources are flavonols. These compounds are mostly found in onions however, leeks Kale, lettuce, tomatoes and broccoli also contain flavonols [16]. Flavanones due to their distinct patterns of substitution, flavanones exhibit a wide range of substituted derivatives (such as prenylated and benzylated flavanones). The most typical type of glycosylated flavanones has a disaccharide substitution at the C7 position. Most flavanones are present in aglycone forms and are primarily present in citrus fruits, such as oranges and lemons (e.g., hesperidin) [16, 17].

Flavones have a double bond between the C2 and C3 atoms, a carbonyl group at the C4 position, and ring B linked to the heterocyclic ring at the C2 location. Lutein and apigenin are the two key flavones. Moreover, ring B (in isoflavones) is attached to the heterocyclic ring (C3 position) rather than the C2 as in other classes, and an oxygen atom is placed in the C4 position [16, 18]. Flavanols (flavan-3-ols) have a saturated heterocyclic ring, a —OH group at position C3, and no double bond between C2 and C3. catechins' characterization derives from the ability to explain two chiral centers, unlike other types of flavonoids. Only aglycones, a kind of flavanol, are present in food (the glycosylated state is excluded). Moreover, they can be found in polymeric forms known as tannins and in monomeric forms known as catechins and epicatechins. epicatechins are diastereoisomers and have a —OH group linked to the C3 position. Epicatechin has the cis configuration of the stereoisomers (+)-epicatechin and (−)-epicatechin, whereas the catechin isomer has the Trans configuration of two stereoisomers, (+)-catechin and (−)-catechin. These structures, which are referred to as tannins in connection to polymeric flavanols, have excellent water solubility and a relatively high molecular weight. With the help of the B ring, ring C is joined to its C3 edge in isoflavones. Legumes are a rich source of these substances as well. While soybeans are the major supply for food. isoflavones have a greater impact on health. Together with glycitein, two more important isoflavones present in soy are daidzein and genistein. Such compounds can also be found in red clovers.

The most common types of isoflavone-aglycones are 7-*O*-glucosides and 6″-*O*-malonyl-7-*O*-glucosides. Dalbergin is the main neo-flavonoid found in plant-based meals [19–21]. The presence of two double bonds in the heterocyclic rings of anthocyanidins and anthocyanins sets them apart from other flavonoids. The hydroxylation and methoxylation patterns on ring B define anthocyanins, which are anthocyanidins in their glycosylated form. A range of anthocyanins are produced by differences in the number of hydroxylated groups, the kind, and the quantity of linked sugar units to their structures. Often, monosaccharides like glucose, galactose, and arabinose make up bonded sugar units. The majority of the colors seen are attributed to molecules called glycosylated anthocyanins, which are water-soluble pigments found in vibrant flowers and fruits [17, 21].

1.5.2 Nonflavonoids

Nonflavonoids consist mostly of a single aromatic ring. Stilbenes, lignans, and phenolic acids are examples of nonflavonoid chemicals. Phenolic acids, mostly derivatives of benzoic acid and cinnamic acid, make up the bulk of this group's chemical composition.

Phenolic acids: One or more —OH groups are connected to a single aromatic ring, which distinguishes phenolic acids from other polyphenols. These may be found in a variety of plant-based foods, such as grains, fruits, and vegetables. Gallic acid and hydroxycinnamic acids are the two most prevalent forms of phenolic acids (e.g. caffeic acid) [22].

Stilbenes: A class of polyphenols known as stilbenes is distinguished by two aromatic rings joined by a double bond. The most popular stilbene is resveratrol, which may be found in red wine, peanuts, and grapes. Many possible health advantages, such as anti-inflammatory and antioxidant capabilities, have been associated with resveratrol [23].

Lignans: Lignans are polyphenols that resemble phenolic acids structurally but are joined by a carbon–carbon bond. Foods including flaxseed, sesame seeds, whole grains, and some vegetables contain them. Bacteria in the stomach break down lignans to create enterolignans, which have been found to improve health [24].

1.6 Extraction Methods of Polyphenols

1.6.1 Ultrasonic-Assisted Extraction of Polyphenols

A popular and effective method for extracting polyphenols from plant sources is ultrasonic-assisted extraction (UAE) shown in Figure 1.3 [25]. It is a simple, quick, non-destructive, and energy-saving technique that can increase the polyphenols' extraction yield as well as quality. The frequency, power, duration, and solvent type used in UAE all have a significant effect on the quantity and quality of the polyphenols that are extracted. First, a dry, ground-up sample of plant materials is collected and combined with an appropriate extraction solvent in a process known as UAE. For polyphenol extraction, ethanol, methanol, and water are frequently used as solvents. The most used solvent for polyphenol extraction is ethanol. Put the

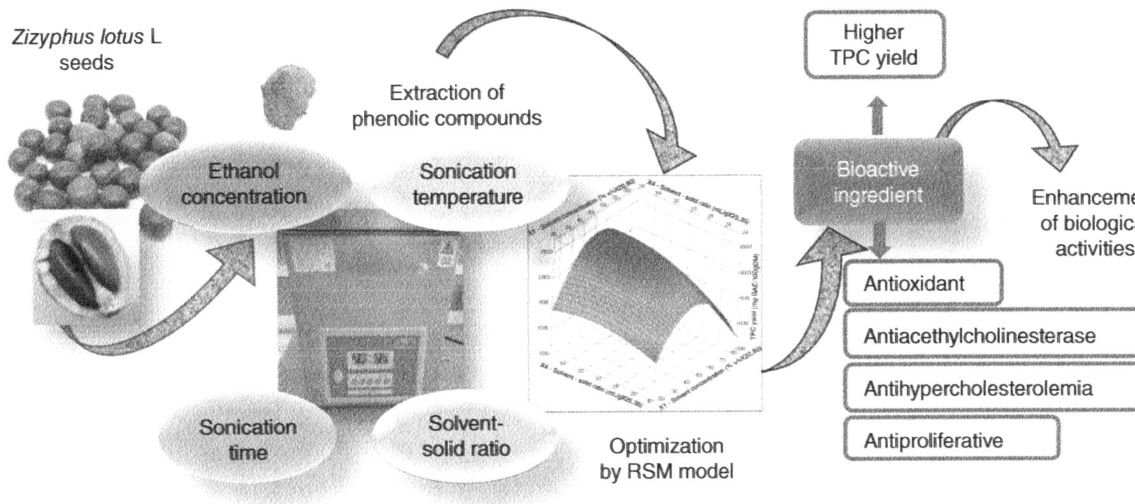

Figure 1.3 Ultrasonic assisted extraction of polyphenols. Source: Farida Berkani et al. [25]/Reproduced from MDPI/CC BY 4.0.

sample in an ultrasonic bath and sonicate it. The ultrasonic waves help to break down the cell walls and facilitate the release of polyphenols from the plant material. The extract is then filtered via filter paper or a syringe filter to remove any insoluble material. Make the extract more concentrated by using a vacuum concentrator or a rotary evaporator. UAE functions on the basic premise that high-frequency sound waves cause cavitation bubbles to form in the solvent, which then violently collapses, producing strong shear forces that rupture the cell walls of the plant material and release the desired chemicals. The mass transfer and solubility of the polyphenols in the solvent are also increased by the ultrasonic vibrations, leading to a faster and more efficient extraction. For example, compared to traditional solvent extraction, ultrasonic-assisted polyphenol extraction from mulberry leaves and grape skins and seeds produced considerably greater yields and total polyphenol contents. Yet, it was shown that a frequency of 40 kHz produced the best extraction yield and antioxidant activity when polyphenols were extracted from bamboo leaves. Methanol was the most efficient solvent for extraction of polyphenols from pomegranate peels when compared to conventional solvent extraction methods. Polyphenols from different plant materials such as EGCG from green tea leaves, quercetin from onion peel, anthocyanins from purple sweet potato, chlorogenic acid from *Eucommia ulmoides* olive leaves, catechins from grape seeds and Rutin from *Sophora japonica* L. flowers can be extracted.

1.6.2 Microwave-Assisted Extraction of Polyphenols

In the late 1980s, microwave-assisted extraction of polyphenols from natural products was introduced. As a result of technical advancements, today it is one of the most common and economical extraction techniques. Microwave-assisted extraction is a widely used technique for the extraction of polyphenols from plant materials. Microwave energy is used in microwave-assisted extraction to heat the solvent and the sample, speeding up the extraction process. In order to increase the extraction yield and quality of polyphenols, microwave-assisted extraction has been recommended as a simpler, quicker, non-destructive, less time-consuming, and energy-efficient method. Microwave-assisted extraction involves heating the solvent and the sample in a microwave oven, where the microwave energy causes the solvent to heat and release its polyphenols as shown in Figure 1.4 [26]. The extraction is usually done at a temperature of 50–100 °C and a pressure of 1–2 bar. The extraction time might vary based on the sample and the solvent employed. Microwave-assisted extraction factors like power, duration, and solvent type may all be optimized to increase extraction efficiency and aid in the production of new functional foods and nutraceuticals. Microwave-assisted extraction is based on the use of microwaves to produce an electromagnetic field that causes the polar molecules in the solvent to spin and produce heat. The target chemicals are released into the solvent as a result of the plant material's cell walls rupturing due

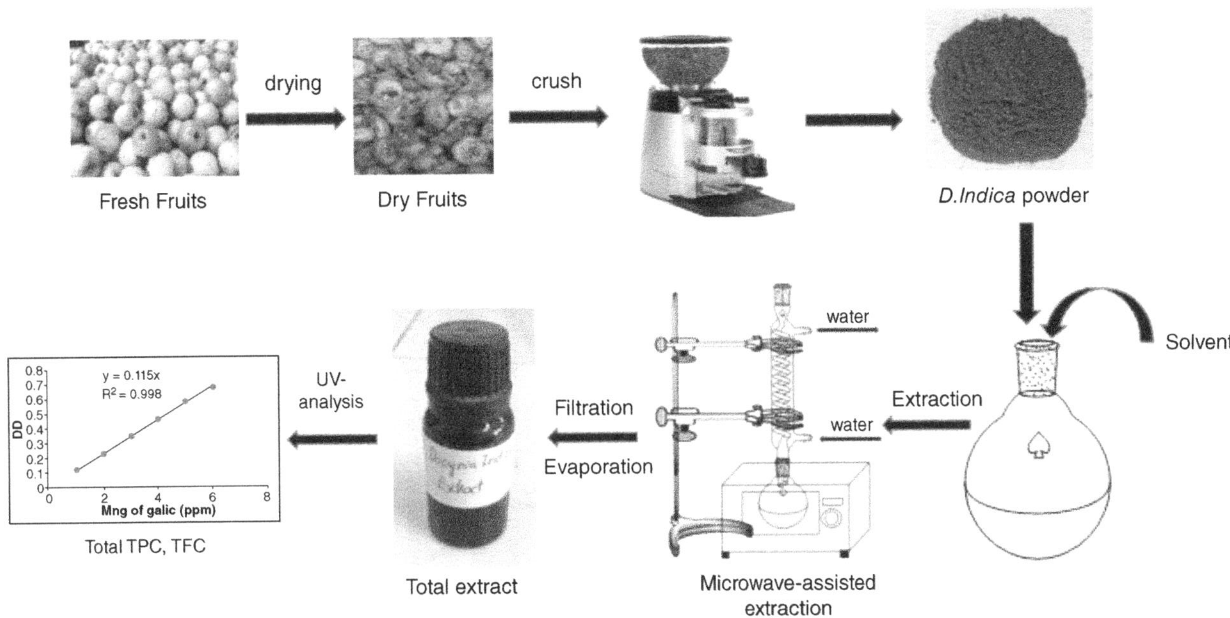

Figure 1.4 Microwave assisted extraction polyphenols. Source: Xuan Duy Le et al. [26]/Reproduced from MDPI/CC BY 4.0.

to heat. Also, microwaves promote the solubility and mass transfer of the polyphenols in the solvent, leading to a more rapid and efficient extraction. For extracting polyphenols from various plant sources, the method of microwave-assisted extraction may be used to extract a number of different polyphenols. For instance, polyphenols from Brazilian red propolis may be extracted via microwave-assisted extraction, which breaks down the cell walls and releases the desired chemicals. Another method of extracting polyphenols from black rice used microwave radiation to make cell membranes more permeable, which facilitated the ability for the polyphenols to permeate into the solvent for the conformation of the influence of microwave-assisted extraction parameters (power, duration, and solvent type), power of 500 W resulted in the maximum extraction quantity and antioxidant properties of polyphenols extract from grape pomace and comparing extraction efficiency of polyphenols. A wider range of polyphenols from blackcurrant leaves and from rose hips can be extracted by using microwave-assisted extraction producing higher yields of polyphenols as compared to other methods.

1.6.3 Supercritical Fluid Extraction of Polyphenols

Supercritical fluid extraction uses supercritical fluids as solvents to extract bioactive substances from a variety of plant materials in a generally safe, effective, and eco-friendly way. Due to their potential health advantages, polyphenols are one of the most investigated groups of bioactive compounds extracted by supercritical fluid extraction. Gases that have been compressed above their critical point to the point where they behave like both a gas and a liquid are known as supercritical fluids. In the process of extracting polyphenols, a number of supercritical fluids are applied, including supercritical carbon dioxide ($SC{-}CO_2$), supercritical water, supercritical ethanol, and supercritical propane. Due to its low toxicity, non-flammability, and availability, $SC{-}CO_2$ is the most often utilized supercritical fluid in supercritical fluid extraction. As per the literature, the plant material is prepared by drying and crushing into a fine powder to improve the extraction surface area as shown in Figure 1.5 [27]. Loading the extraction vessel: The ground plant material is placed into the extraction vessel, which is commonly a stainless-steel column or an extraction cell. Purging the vessel: Any air or moisture that could be present in the system is removed by purging the vessel with the supercritical fluid. Extraction: To reach the supercritical state, the supercritical fluid is heated and pressured until it can behave as both a liquid and a gas. It is then injected into the extraction tank, where it dissolves the polyphenols from the plant material. Supercritical fluid extraction has the benefit of producing significant yields of pure extracts with low solvent residues. Tea leaves, grape seeds, rosemary leaves, and red grape pomace have all been demonstrated to be excellent sources of polyphenols that may be extracted by supercritical fluid extraction. Several parameters, including the extraction pressure, temperature, and time, affect the yield and

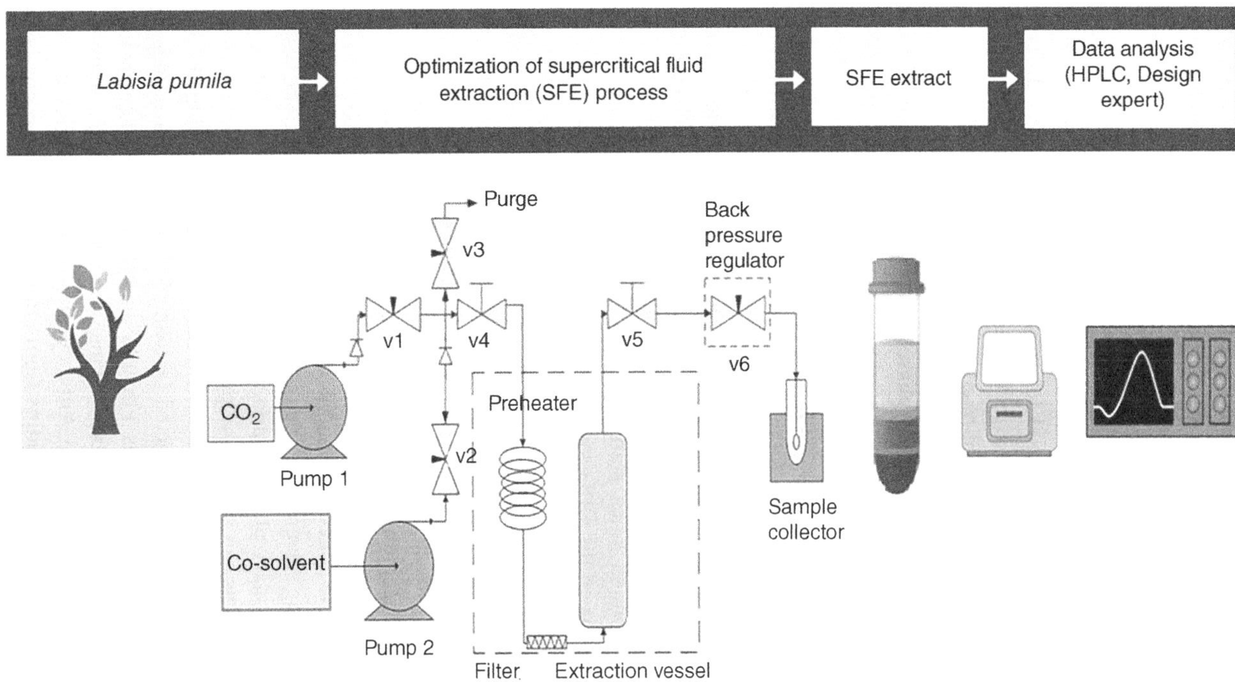

Figure 1.5 Supercritical fluid extraction of polyphenols. Source: (Reproduced from CCBY 4.0; No permission Needed) Radzali et al. [27].

composition of the extracts as well as the type of plant material used. Additionally, specific polyphenols may be specifically extracted via supercritical fluid extraction, which might be helpful for manufacturing. The obtained polyphenols can be purified and concentrated further using methods such as solid-phase extraction (SPE), chromatography, or evaporation.

1.6.4 SPE of Polyphenols

One popular and effective method for separating and purifying different compounds from complex mixtures, including polyphenols, is SPE as shown in Figure 1.6 [28]. Polyphenols from plant extracts can be selectively extracted and concentrated through SPE. The SPE procedure involves the use of a SPE cartridge or column packed with a stationary phase material that is selective for the target component, often a silica-based material(s). When the sample is put into the column, the undesired components are removed by washing with a solvent that is only susceptible to the desired compound(s). Finally, the target compound(s) are eluted from the column with a solvent that is appropriate for downstream analysis or application. A frequently utilized stationary phase material for SPE in the case of polyphenols is C18-bonded silica, which is hydrophobic in nature and selective for substances with polar functional groups, such as phenolic —OH groups. In the SPE, the polyphenol-containing sample is placed into the SPE column, which contains C18-bonded silica stationary phase material. Through hydrophobic interactions between the C18 groups and the phenolic —OH groups of the polyphenols, the sample's polyphenols interact with the stationary phase material. When the sample has been loaded, the undesirable components in the sample are removed by washing the column with a solvent that is selective for the target compound(s). The solvent used for washing depends on the physicochemical properties of the sample and the stationary phase material. Normally, washing is done with water, acetonitrile, or methanol. And thereafter, the hydrophobic contacts between the target polyphenols and the stationary phase material are broken by a solvent and the target polyphenols are eluting from the column. Elution can be carried out using solvents like acetone, methanol, or ethanol. The type of polyphenols and stationary phase component must be considered while selecting a solvent. Polyphenols can isolate and purify from different extracts such as blackcurrant pomace and hawthorn extract with a C18 column by using SPE methods. Several polyphenols can be extracted by using SPE such as flavonoids from tea leaves grape seeds and citrus peels. Phenolic acids from olive oil coffee beans and blackberries. Stilbenes from grape skins and peanut skins, tannins from pomegranate, green tea, and wine. Anthocyanins from blueberries, blackcurrants and black carrots, proanthocyanidins from cranberry pomace cocoa beans, and grape seeds

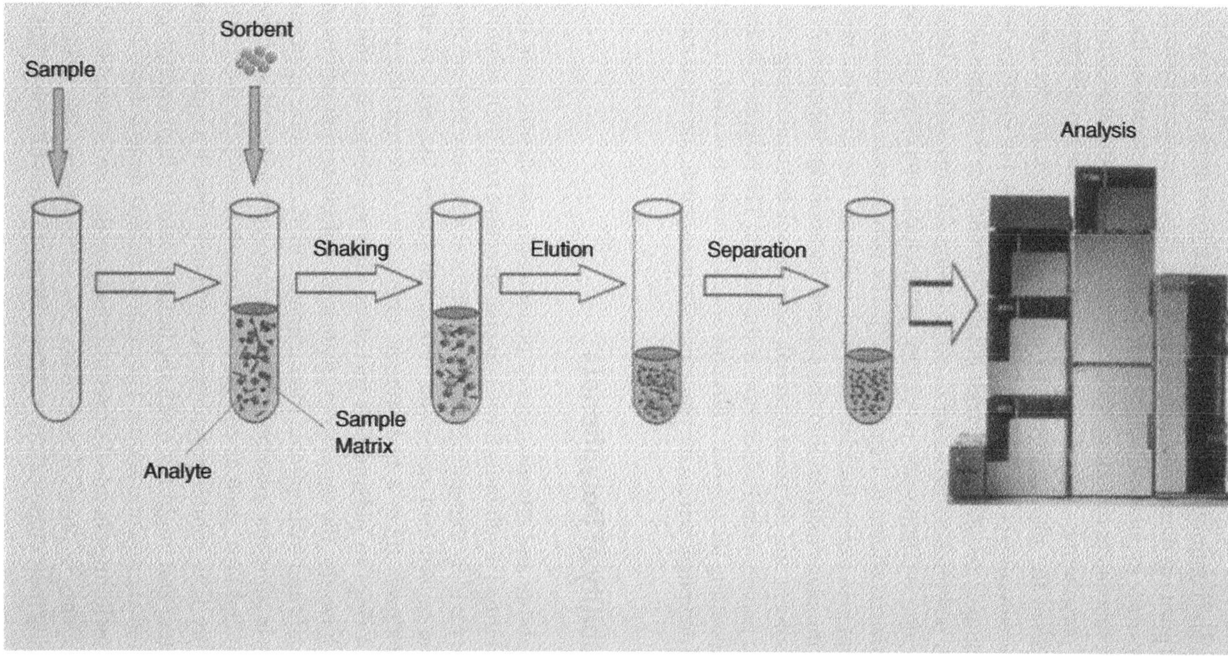

Figure 1.6 Solid phase extraction. Source: Piotr Ścigalski et al. [28]/Reproduced from MDPI/CC BY 4.0.

1.6.5 Solvent Extraction of Polyphenols

Solvent extraction persists as a common approach for extracting polyphenols from plant sources, and continuing studies are taking place to optimize the process for optimal yield and purity of these beneficial components. Polyphenols are hydrophilic substances that dissolve in polar solvents like methanol and water. However, the presence of other plant substances including lipids, pigments, and carbohydrates may limit the extraction of polyphenols. Therefore, organic solvents are used to selectively extract polyphenols from the plant matrix. The efficiency of the extraction process may depend on the solvent type and plant material composition. For instance, polar polyphenols are simpler to isolate using high-polarity solvents like methanol and ethanol, but non-polar polyphenols are easier to extract using solvents such as ethyl acetate. Solvent extraction is a multi-step procedure that includes processes like organic solvent penetration as shown in Figure 1.7 [29]: The polyphenols are dissolved by the organic solvent penetrating the plant material. Diffusion: The dissolved polyphenols diffuse from the plant material into the solvent. Partitioning: Polyphenols partition between the solvent and the plant material, with the solvent containing a larger proportion of the polyphenols. Evaporation of the solvent: The polyphenols are concentrated by evaporating the solvent. Filtration or centrifugation is used to separate the concentrated polyphenols from the plant material in a solid–liquid separation process.

1.7 Solubility of Polyphenols

The bioavailability and effectiveness of polyphenols are significantly influenced by their solubility. Polyphenols have a diverse structure and their solubility properties can vary widely depending on their chemical structure and environmental variables. The presence of non-polar functional groups like methyl, ethyl, and propyl groups gives them a hydrophobic character. Moreover, the presence of many phenol rings in polyphenols increases their molecular weight and size., polyphenols have also the capacity to create intermolecular interactions including hydrogen bonds and hydrophobic interactions, which can cause aggregation and precipitation in aqueous solutions. Some such polyphenols can also form complexes with metal ions, such as complex as, a result many polyphenols have low water solubility. Low-soluble polyphenols could only absorb a little amount in the digestive system and might not reach the target tissues in high enough quantities to activate their therapeutic action. They can be turned more soluble by a variety of methods, including chemical structural changes, the use of suitable solvents or solubilizing agents, or physical treatments. The solubility can also be enhanced by either changing their chemical structure or using any suitable solvent like adding

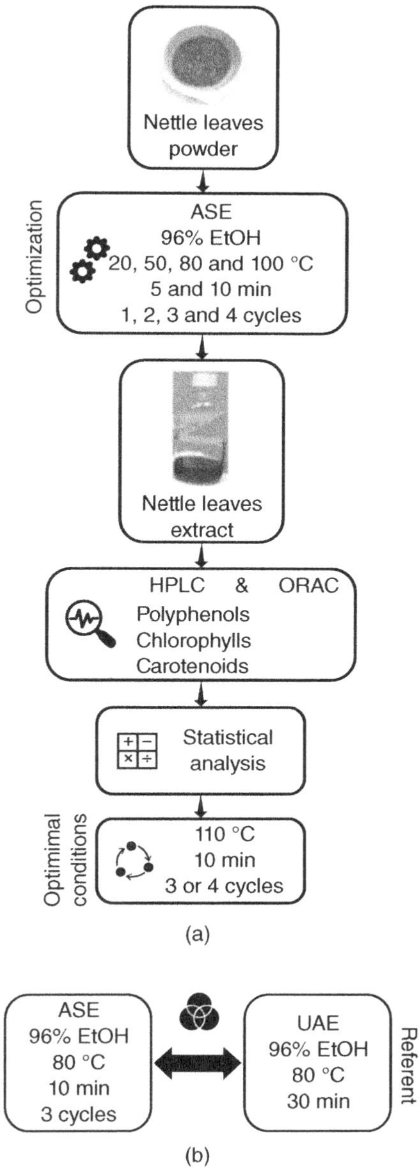

Figure 1.7 Solvent extraction. Source: (Reproduced from CCBY 4.0; No permission Needed) Repajić et al. [29].

sulfonic acid or carboxylic acid, polyphenols can be changed into more water-soluble forms. Similarly, by using solubilizing substances like cyclodextrins (CDs), surfactants, or polymers or by using physical processes like sonication or high-pressure homogenization, many environmental factors, including pH, temperature, and the presence of other matrix constituents, can also have an impact on the solubility of polyphenols. For instance, some polyphenols' solubility might be increased in acid conditions, while other polyphenols might be more soluble in alkaline pH ranges.

1.7.1 Chemical Modification

The solubility of polyphenols can be increased by chemical modification, which can increase their use in a variety of applications. The chemical structure of polyphenols is changed by chemical treatment. The polyphenol molecule is modified chemically by the addition of polar functional groups. To introduce negatively charged groups, for example, sulfonation, carboxylation, or phosphorylation can be used, while amino or —OH groups can be added to improve the polarity of the molecule. These additional functional groups can improve the interactions between the polyphenols and the water molecules, to enhance the solubility of polyphenols in water. Another strategy is to add a sugar molecule to the polyphenol and change its structure through a process called glycosylation. By adding a polar sugar moiety to the hydrophobic polyphenol, which enhances the molecule's solubility and stability, this modification increases the solubility of polyphenols in water. The solubility of polyphenols including quercetin, EGCG, and resveratrol has been shown to be enhanced by glycosylation. The bioavailability and stability of polyphenols can both be improved by improving their solubility by chemical modification. Similarly, galloylation, the addition of gallic acid to polyphenols like catechins, can increase their stability and antioxidant activity by increasing the amount of —OH groups available for hydrogen bonding.

1.7.2 Solubilizing Agents

To increase the solubility of polyphenols, solubilizing agents such CDs, surfactants, and polymers are used. CDs are cyclic oligosaccharides composed of glucose units. They possess a hydrophilic outer surface and a hydrophobic cavity that may trap hydrophobic compounds like polyphenols, to increase their water solubility. The solubility and stability of polyphenols including curcumin, EGCG, and resveratrol have been improved by CDs in a variety of experiments. Surfactants are compounds that possess both hydrophilic and hydrophobic characteristics. They have the capacity to form micelles, which are tiny aggregates capable of dissolving hydrophobic substances in water. It has been found that surfactants like sodium dodecyl sulfate and Tween 80 can increase the solubility of polyphenols including

EGCG, quercetin, and genistein. Polymers are large, repetitive units that can bind to polyphenols to make them more soluble in water. In order to solubilize polyphenols, two most common polymers polyethylene glycol (PEG) and polyvinylpyrrolidone are used. For example, PEGylation can enhance the solubility and stability of curcumin. It is important to note that the use of solubilizing agents can affect the stability and bioactivity of polyphenols, and the choice of solubilizing agent depends on the specific polyphenol and the intended application.

1.7.3 Physical Treatment

Sonication, high-pressure processing (HPP), and grinding are referring to physical treatment of polyphenols to improve the solubility of polyphenols. Sonication is a method that creates cavitation by using high-frequency sound waves, which can break down polyphenol aggregates and improve their solubility in water. Whereas grinding is a physical treatment technique that includes decreasing the particle size of polyphenols to increase their surface area and enhance their solubility in water, HPP is a non-thermal process technique that employs high pressure to modify the structure of polyphenols and increase their solubility in water. HPP and sonication can be used to increase the solubility and stability of polyphenols such EGCG, resveratrol, and quercetin, while grape seeds can be ground to increase the solubility of their polyphenols. Physical processing techniques can affect the chemical and physical properties of polyphenols, which may have an effect on their bioactivity and efficiency. To obtain the required solubility improvement while limiting degradation or other negative consequences, the physical treatment parameters, including temperature, pressure, and time, must also be optimized.

1.7.4 Solubility of Polyphenols by pH Adjustment

The solubility of polyphenols can be increased by pH change, which ionizes the compounds and makes them more soluble in water. At various pH levels, polyphenols' —OH groups can ionize and improve their solubility. In general, polyphenols are more soluble at higher pH levels because the —OH groups are deprotonated and become negatively charged. Thus, increasing pH levels the not only enhances the solubility but also contribute well to the effectiveness of substances. This can be seen by increasing the pH of quercetin, whose solubility is increased from 0.45 mg/mL at pH 3.0 to 3.16 mg/mL at pH 9.0. The solubility of resveratrol has also been found to rise at pH levels higher than its pK_a. Adjusting the pH can also help to minimize the interactions in the presence of other compounds that affect solubility, such as proteins or carbohydrates. To increase their solubility without compromising the polyphenols' bioactivity, it is important to determine the optimal pH settings for each given polyphenol to achieve the required solubility.

1.7.5 Co-solvents

Co-solvents increase the solubility of poorly soluble compounds in a solvent by lowering intermolecular interactions and increasing the molecular gap that is accessible to solute molecules. They can also act as solubilizing agents for compounds that are insoluble in water. Polyphenols have a low solubility in aqueous solutions because of their hydrophobic nature; co-solvents have been commonly used to increase the solubility. Due to its great solubility in water and capacity to create hydrogen bonds with both the solute and water molecules, ethanol is one of the most often used co-solvents. Methanol, propylene glycol, glycerol, and PEG are also used as co-solvents. It has been proven that co-solvent systems considerably improve the solubility of different polyphenols. For instance, a 50% (v/v) ethanol/water combination may improve quercetin's solubility from 1.7 g/mL in water to 90 g/mL. Similar to this, curcumin, a polyphenol included in turmeric, becomes more soluble when mixed with 10% (v/v) PEG 400 and water, rising from 11 g/mL in water to 26 mg/mL. The stability and bioactivity of polyphenols may be potentially suffered from the use of co-solvents. In order to optimize solubility, the kind and quantity of co-solvent should be carefully selected. Co-solvents that produce reactive oxygen species (ROS) might potentially reduce polyphenols' antioxidant activity.

1.8 Flavor and Color of Polyphenols

1.8.1 Flavor

Several plant-based meals and beverages benefit from the bitterness, astringency, sweetness, and scent that polyphenols provide. Understanding the role of polyphenols in food flavor can aid in the development of new foods and the improvement of current products. For instance, the polyphenol catechin is mostly responsible for the bitterness of green tea, whereas tannins, another kind of polyphenol, are predominantly responsible for the astringency of red wine. In addition to flavonoids like rutin, quercetin, and kaempferol, steviol glycosides, a kind of polyphenol found in the leaves of the stevia plant, are frequently utilized as natural sweeteners while Resveratrol and anthocyanins, present in red wine, are mostly responsible for the aroma. while in black tea, the aroma is attributed to theaflavins, a type of polyphenol that is formed during the oxidation process. In addition to their direct influence on flavor, polyphenols have the potential to interact with other flavor components in foods and beverages, influencing overall flavor perception. For instance, it has been observed that the presence of polyphenols in red wine interacts with the aroma compounds in the wine, changing the overall flavor profile as shown Table 1.2.

Table 1.2 Some common polyphenols and their complementary taste.

Polyphenols	Taste
Catechins	Bitter, astringent taste
Flavanols	Bitter or astringent taste
Tannins	Puckering taste and can be quite bitter
Quercetin	Slightly bitter, quite astringent
Anthocyanins	Slightly sweet or tart flavor

1.8.2 Color of Polyphenols

Several plant-based foods and beverages have different colors because of polyphenols. They can range in color from yellow to red to blue, depending on the specific type of polyphenol present as shown in Table 1.3. Anthocyanins are responsible for the red, blue, and purple colors in many fruits and vegetables, including berries, grapes, and eggplants. However, several fruits and vegetables, including onions, apples, and citrus fruits, have a yellow color because of flavonols. It is revealed that a number of factors, including processing methods and storage environments, can affect the color of polyphenols. The color of tea polyphenols, for example, can be influenced by the type of tea, the temperature and time of boiling, and the presence of other compounds in the tea. Depending on their chemical composition, they can be any hue from yellow to red to purple to even black.

Table 1.3 Some of the common polyphenols with their complementary colors.

Polyphenol	Color
Flavonols	Yellow
Anthocyanins	Red, purple, and blue
Hydroxycinnamic acids	Often yellow in color
Flavan-3-ols (catechins)	Usually colorless, but can appear yellow or brown in some cases
Stilbenes	Colorless or pale yellow
Hydroxycinnamic acids	Yellow
Carotenoids	Yellow to orange to red
Tannins	Range in color from yellow to brown
Lignans	Usually colorless or pale yellow
Ellagitannins	Often red or purple in color
Proanthocyanidins	Range in color from yellow to red to blue

1.9 Health-Promoting Properties of Polyphenols

Polyphenols have a variety of health-promoting qualities. Below are a few of polyphenols' main characteristics and some of the common properties are tabulated in Table 1.4.

1.9.1 Antioxidative Properties

Polyphenols have been extensively studied for their antioxidant properties, which are believed to contribute to their potential health benefits [30]. Antioxidants are compounds that protect cells from damage caused by free radicals, which are unstable molecules that can damage cells and contribute to the development of various diseases. Among the most valued characteristics of polyphenols is their

Table 1.4 Polyphenols with health-promoting activity.

Polyphenols	Properties
Resveratrol	Antioxidative, anti-inflammatory, and anti-cancer
Quercetin	Antioxidative, anti-inflammatory, and anti-cancer
Catechins	Cardiovascular, antioxidative, anti-inflammatory, and anti-cancer
Curcumin	Neuroprotective, anti-inflammatory, antioxidant, and anti-cancer
Anthocyanins	Cardiovascular, antioxidative, anti-inflammatory, and anti-cancer
Epigallocatechin gallate	Neuroprotective, cardiovascular, antioxidative, anti-inflammatory, and anti-cancer
Ellagic acid	Anti-inflammatory, antioxidative, and anti-cancer
Proanthocyanidins	Neuroprotective, cardiovascular, antioxidative, anti-inflammatory, and anti-cancer
Kaempferol	Cardiovascular, antioxidative, anti-inflammatory, and anti-cancer
Genistein	Neuroprotective, cardiovascular, antioxidative, anti-inflammatory, and anti-cancer, reduces risk of chronic diseases
Chlorogenic acid	Cardiovascular, anti-inflammatory, antioxidative, anti-cancer, reduces risk of chronic diseases
Myricetin	Cardiovascular, anti-inflammatory, antioxidative, anti-cancer, reduces risk of chronic diseases
Luteolin	Cardiovascular, anti-inflammatory, antioxidative, anti-cancer, reduces risk of chronic diseases
Apigenin	Cardiovascular, anti-inflammatory, antioxidative, anti-cancer
Isoflavones	Cardiovascular, anti-inflammatory, antioxidative, anti-cancer
Procyanidins	Cardiovascular, anti-inflammatory, antioxidative, anti-cancer, reduces risk of chronic diseases

potential antioxidative activity. These substances exhibit a variety of biological activities associated with the human body, which is mostly due to their antioxidative properties. Free radicals in the body are often produced by reactive forms of oxygen (ROS). They are produced when molecules of molecular oxygen progressively decay in one-electron processes. Nitric acid, nitrogen dioxide, and nitrogen oxide are all reactive forms of nitrogen (reactive nitrogen species (RNS)) that are harmful to human health. Equally damaging the organic radicals produced when ROS and RNS interact with organic compounds [31, 32]. The imbalance between the generation and deactivation of ROS can cause a variety of illnesses [33]. Cellular lipids, nucleic acids, and proteins can all be oxidized by free radicals. They aid in the mutagenesis, carcinogenesis, and onset of disorders like Parkinson's and Alzheimer's as well as the aging of cell proteins. Moreover, they raise the risk of cardiovascular problems, possibly through weakening cellular membranes and oxidizing LDL lipoprotein [32].

Many in vitro, animal, and human research have been done to look into the antioxidant properties of polyphenols. As an illustration, it has been discovered that blueberry polyphenols showed high antioxidant activity, as determined by their capacity to scavenge free radicals and prevent lipid peroxidation [34]. Similarly, green tea polyphenols have strong antioxidant activity, as determined by their capacity to scavenge free radicals and lessen oxidative stress in rats [35]. From the conclusion of these studies, we can say that one of the primary processes through which polyphenols may boost health is by acting as antioxidants, which lower the risk of chronic diseases including cancer, cardiovascular disease, and neurological disorder.

1.9.2 Anti-inflammatory Properties

As the body's initial reflex action to an infection or damage, inflammation is important for both the immune system's innate and adaptive responses. The intricate biological reaction of vascular tissues to adverse stimuli, such as infections, injured or tumoral cells, or allergens, is what distinguishes it [36]. Many health advantages of polyphenols are well-known, including their ability to reduce inflammation. However, chronic inflammation has been linked to a number of diseases, including autoimmune disorders, cancer, and cardiovascular disease etc. [36, 37]. Numerous inflammatory mechanisms in the body have been revealed to be regulated by polyphenols. The synthesis of pro-inflammatory cytokines, chemokines, and enzymes can be inhibited by these substances. Nuclear factor kappa B (NF-B), a transcription factor important in the control of inflammation, can also be inhibited by polyphenols. Moreover, polyphenols have the ability to activate Nrf2, a transcription factor that controls the body's anti-inflammatory and antioxidant pathways [38, 39].

The anti-inflammatory properties of various polyphenols have been studied in several research. The production of pro-inflammatory cytokines including interleukin-1 beta (IL-1), interleukin-6 (IL-6), and tumor necrosis factor-alpha (TNF-alpha) has been demonstrated to be inhibited by resveratrol which is present in grapes, berries, and peanuts [40]. The formation of prostaglandin E2, a pro-inflammatory mediator, has been demonstrated to be inhibited by another polyphenol called quercetin, which is present in onions, apples, and berries [41].

In vitro and in vivo studies have revealed that the polyphenol curcumin, which is present in turmeric, reduces NF-B activation and the production of pro-inflammatory cytokines [42]. Also, it has been revealed that the flavonoid a class of polyphenolic compound, which is present in several fruits, vegetables, and beverages, has anti-inflammatory properties. For instance, apigenin, a flavonoid present in parsley, celery, and chamomile, has been demonstrated to reduce the synthesis of pro-inflammatory cytokines including IL-1, IL-6, and TNF- in both in vitro and in vivo [43]. All these studies indicate that polyphenols also have anti-inflammatory properties through a number of different mechanisms and also these substances may be used to both prevent and treat chronic inflammatory disorders.

1.9.3 Anti-cancer Properties

By helping cell senescence and death, regulating autophagy, and preventing cancer cell growth and migrations, polyphenols show their promise as an anti-cancer treatment. By an increase in ROS and a decrease in cellular antioxidants [44]. Like the interaction glutathione with chemotherapeutic drugs, and the maintenance or reverse of multi-drug resistance, they can cause cellular stress and catabolism. Their potential for health benefits, including anti-cancer action, has been thoroughly investigated. A variety of polyphenols present in natural sources such as resveratrol can be found in berries, peanuts, and grapes [45], curcumin and Epigallocatechin-3-gallate (EGCG) which is found in turmeric and green tea [46, 47]. Quercetin is found in various plant sources, including apples, onions, and green tea. Kaempferol is found in various fruits, vegetables, and herbs [48, 49]. Certain polyphenols have been proven to have anti-cancer efficacy in a number of cancer types, including colon, breast, lung, colorectal, and prostate cancer [50]. Through controlling many signaling pathways, such as NF-kB, PI3K/AKT, MAPK/ERK, and Wnt/β-catenin, these polyphenols promote cell cycle arrest and death in cancer cells and also increase the efficiency of chemotherapy medications and radiation therapy in the treatment of cancer [51]. Most of the polyphenols have shown promising anti-cancer activity in various types of cancer.

1.9.4 Cardiovascular Health

Both the oxidative stress and ROS have been strongly linked to endothelial damage, the development of atherosclerosis, injury in prolonged myocardial infarction, and ischemic recovery in the treatment of cardiovascular diseases [44].

An increasing number of studies indicate that polyphenols may help in cardiovascular health improvement. A major contributing element to the development of cardiovascular disease is endothelial dysfunction [52]. By improving the generation of nitric oxide, lowering oxidative stress, and decreasing inflammation, polyphenols have been proven to enhance endothelial function and also lowering blood pressure, Cardiovascular disease is significantly increased by having high blood pressure [53]. It has been shown that polyphenols, particularly flavonoids, reduce blood pressure by relaxing blood vessels and enhancing endothelial function [54]. A polyphenol-rich extract from grape seeds and beverages has been shown in a randomized controlled experiment to significantly enhance endothelial function and decrease oxidative stress in individuals with metabolic syndrome [55]. Also, it has been shown that polyphenols contain antioxidant capabilities, which can aid in lowering oxidative stress and preventing cardiovascular system damage. Inflammation is thought to be a major factor in the development of cardiovascular disease. The anti-inflammatory properties of certain polyphenols have been discovered to help decrease inflammation and guard against cardiovascular system harm [56]. While elevated lipid levels, or dyslipidemia, are a significant risk factor for cardiovascular disease. It has been discovered that polyphenols particularly flavonoids, can enhance lipid profiles by lowering cholesterol and triglycerides and raising HDL cholesterol such as flavonoid-rich cocoa products can significantly improve the lipid profile in people with hypertension [57]. Results showed that polyphenols can also be used to improve the endothelial function, lower the blood pressure, reduce the oxidative stress, lessen the inflammation, and improve the lipid profiles, thus help to improve the overall cardiovascular health. Consuming foods and drinks high in polyphenols may help lower your chance of getting cardiovascular disease.

1.9.5 Neuroprotective Properties

The brain is more vulnerable to oxidative stress than other organs due to the low activity of the antioxidant defense system. To activate their neuroprotective properties, polyphenols interact with a variety of brain targets. The term "neuroprotection" describes a substance's ability to shield neurons against harm or death brought on by a number of stresses, including oxidative stress, inflammation, and toxicity. With promising findings, polyphenols' neuroprotective activity has been thoroughly investigated in both in vitro and in vivo models.

One of the main processes is their ability to neutralize free radicals and decrease oxidative stress, which is a significant cause of neurodegenerative disorders including Alzheimer's and Parkinson's disease. Moreover, polyphenols have been demonstrated to modify a number of signaling pathways involved in apoptosis and inflammation, which strengthens their neuroprotective effects. In animal models of neurodegenerative disorders, polyphenols have been demonstrated in several studies to improve cognition and memory. A polyphenol resveratrol has been proven, for instance, to improve cognition and lower amyloid-beta levels in a mouse model of Alzheimer's disease [58]. Similarly, a rat model of Parkinson's disease has demonstrated that the polyphenol epigallocatechin-3-gallate (EGCG) enhances spatial memory and lowers oxidative stress [59].

Polyphenols have been found to protect against neuronal damage brought on by ischemia, traumatic brain injury, and other stressors in addition to their benefits on cognitive performance. In a rat model of traumatic brain injury, quercetin, a polyphenol present in fruits and vegetables, has been demonstrated to decrease neuronal damage and increase functional recovery [60]. Similarly, in animal models of ischemic stroke, curcumin, a polyphenol present in turmeric, has been demonstrated to lessen brain damage and enhance functional results [61].

Additionally, Rutin (flavonoid) [62], Luteolin (flavone) [63], and Caffeic acid (hydroxycinnamic) possess antioxidative, anti-inflammatory, and neuroprotective properties. It can prevent neuroinflammation and neurodegeneration by inhibiting the activation of microglia and improving cognitive function. and reducing oxidative stress. From the above evidences, polyphenols are a prospective challenger for the prevention and treatment of neurodegenerative disorders due to their neuroprotective properties [64].

1.9.6 Anti-diabetic Activity

The ability of a drug or molecule to reduce blood sugar levels and control diabetes symptoms is referred to as anti-diabetic activity. Diabetes is a long-term metabolic condition defined by increased blood glucose levels brought on by the body's inefficiency in producing or using insulin. Anti-diabetic medications can function in a number of different ways, including increasing insulin secretion, enhancing insulin sensitivity, preventing glucose absorption, or lowering hepatic glucose synthesis. Polyphenols exhibit a range of anti-diabetic activities [65]. The majority of polyphenols have anti-diabetic properties by increasing insulin sensitivity, decreasing oxidative stress, and improving glucose metabolism [66]. Increasing insulin levels is one of the ways that polyphenols work to prevent diabetes. The insulin signaling pathway may be activated by polyphenols, which enable the body's cells to use glucose more effectively. The expression of several genes involved in glucose metabolism can also be altered by polyphenols, which enhance cells' ability to absorb glucose. It has been shown that polyphenols such as flavonoids, phenolic acids, and stilbenes increase the absorption and utilization of glucose in peripheral tissues including skeletal muscle and adipose tissue [67, 68]. These substances also help pancreatic beta-cells produce more insulin, which improves glucose homeostasis. By preventing the digestion and absorption of carbohydrates, polyphenols also have anti-diabetic effects. Alpha-amylase and alpha-glucosidase, two enzymes involved in the breakdown of carbohydrates, have been demonstrated to be inhibited by polyphenols such catechins, procyanidins, and anthocyanins. Polyphenols slow down the rate of sugar absorption in the small intestine by inhibiting these enzymes, resulting in lower postprandial glucose levels [69].

In both animal models as well as human clinical testing, several researchers have looked at whether polyphenols have any anti-diabetic properties. For instance, one study found that supplementing obese rats with green tea extract high in catechins increased their ability to tolerate glucose and maintained their insulin sensitivity [70]. Another study found that type 2 diabetic individuals with resveratrol supplementation had better glycemic control and insulin sensitivity [71]. Polyphenols also have anti-inflammatory and antioxidant qualities, which can help with diabetes control and prevention in addition to these processes [38]. Persistent low-grade inflammation and oxidative stress have been shown to advance type 2 diabetes by causing beta-cell dysfunction and insulin resistance [72]. It has been demonstrated that polyphenols like curcumin, resveratrol, and quercetin can lower inflammation and oxidative stress by altering several signaling pathways and gene expression patterns. These activities make polyphenols promising candidates for the prevention and management of diabetes.

1.9.7 Anti-microbial Properties

A substance's capacity to destroy or slow down the growth of microorganisms, such as bacteria, viruses, fungi, and parasites, is referred to as antimicrobial activity [73]. To prevent and control microbial infections and diseases, antimicrobial compounds are widely used in a wide range of industries, including medicine, agriculture, and the food industry. Many polyphenols exhibit significant antimicrobial activity, including disrupting the cell membrane and enzymatic activity of microorganisms, which can be attributed to their ability to interfere with various cellular processes in microorganisms [74, 75]. The disruption of microorganisms' cell membranes is one mechanism by that polyphenols exert their antimicrobial properties [75]. Reports suggested that polyphenols like tannins and flavonoids react with the lipid bilayer of bacterial cell membranes, thus damaging the membrane and allowing the internal contents to leak thus interferingwith the cell's regular functions andcausingbacterial death. Moreover, polyphenols have the ability to damage fungal cell walls by affecting their functionality [76, 77].

Polyphenols also exhibit antimicrobial activity by interfering with the enzymatic activity of microorganisms. Like DNA gyrase and topoisomerase IV, two enzymes involved in bacterial DNA replication and transcription, have been revealed to be inhibited by polyphenols such as catechins and resveratrol. This stops the growth and reproduction of bacteria. In addition, chitinase, an enzyme necessary for the production of fungal cell walls, is reduced by polyphenols [78]. Many researchers have proven in vivo and in vitro antibacterial properties of polyphenols. For example, resveratrol exhibited powerful antibacterial action against several bacterial strains, including methicillin-resistant *Staphylococcus aureus* and vancomycin-resistant *Enterococcus faecalis* [79]. EGCG, a polyphenol included in green tea, was shown to reduce the development and virulence of the fungus *Candida albicans* [80].

1.10 Skin Benefits of Polyphenols

The biggest organ in the human body, skin protects internal organs, controls body temperature, and serves as a barrier to potentially harmful environmental factors. Polyphenols have numerous skin benefits, including their photoprotective, skin whitening, wound healing, anti-inflammatory and antioxidant effects. Consuming foods high in polyphenols like green tea, chocolate, and berries as well as skincare products containing polyphenols may help maintain healthy skin and prevent skin problems. Photoprotective effects: Skin damage from UV light from the sun can contribute to premature aging, sunburns, and skin cancer. It has been demonstrated that polyphenols have photoprotective properties, which means they can help in shielding the skin from UV damage. EGCG, a polyphenol present in green tea, can prevent UV-induced DNA damage and shield the body from the damaging effects of UV radiation. Skin whitening effects: Common skin conditions like hyperpigmentation can be caused by a number of different factors, including hormone fluctuations, UV exposure, and inflammation. Polyphenols, such as flavonoids and catechins, have skin-whitening effects by inhibiting the activity of tyrosinase, an enzyme involved in melanin production. Wound healing effects: meaning they can contribute to promoting skin damage healing. Epicatechin, a polyphenol contained in cocoa, can accelerate the healing of wounds by increasing angiogenesis (the formation of new blood vessels), collagen production, and skin cell proliferation. Anti-inflammatory effects: Acne, rosacea, and psoriasis are just a few of the skin conditions that are frequently brought on by inflammation. Due to their anti-inflammatory properties, polyphenols can help to reduce skin irritation. Resveratrol and quercetin have the ability to block inflammatory pathways and lower the generation of cytokines that are pro-inflammatory and Antioxidant effects: The powerful antioxidant properties of polyphenols are known to help protect the skin from oxidative stress brought on by free radicals. Flavonoids and phenolic acids are examples of polyphenols that can scavenge free radicals and stop oxidative damage to the skin.

1.11 Conclusions

The discussion in this chapter demonstrates that polyphenols are a group of chemical compounds that are prevalent in plants and that, because of their antioxidant capabilities, have a significant impact on human health. They are divided into categories according to how many phenol rings they contain, including simple phenols, flavonoids, and tannins. Polyphenols, which are extensively distributed across the plant kingdom, are what give fruits, vegetables, and alcoholic beverages their distinct hues, smells, and scents. The capacity of polyphenols to scavenge free radicals, which are highly reactive entities that can harm cells and tissues, is thought to be the cause of their antioxidant capabilities. Numerous health advantages of polyphenols have been demonstrated, including a decreased risk of chronic diseases like cancer, cardiovascular disease, and neurological disorders. Fruits, vegetables, nuts, tea, coffee, wine, and other common dietary sources of polyphenols include tea, coffee, and wine. Polyphenol bioavailability varies based on the particular component and the dietary matrix it is found in. The basic concepts of polyphenols will be introduced in this chapter, followed by a thorough discussion of their nomenclature, classification, and properties with an emphasis on their potential as dietary antioxidants and their significance to human health.

References

1. Zamora-Ros, R., Knaze, V., Rothwell, J.A. et al. (2016). Dietary polyphenol intake in Europe: the European Prospective Investigation into Cancer and Nutrition (EPIC) study. *European Journal of Nutrition* 55: 1359–1375.
2. Vidal-Casanella, O., Núñez, O., Granados, M. et al. (2021). Analytical methods for exploring nutraceuticals based on phenolic acids and polyphenols. *Applied Sciences* 11 (18): 8276.
3. Zamora-Ros, R., Touillaud, M., Rothwell, J.A. et al. (2014). Measuring exposure to the polyphenol metabolome in observational epidemiologic studies: current tools and applications and their limits. *The American Journal of Clinical Nutrition* 100 (1): 11–26.
4. Arts, I.C. and Hollman, P.C. (2005). Polyphenols and disease risk in epidemiologic studies. *The American Journal of Clinical Nutrition* 81 (1): 317S–325S.
5. Buck, K., Zaineddin, A.K., Vrieling, A. et al. (2010). Meta-analyses of lignans and enterolignans in relation to breast cancer risk. *The American Journal of Clinical Nutrition* 92 (1): 141–153.
6. Davinelli, S. and Scapagnini, G. (2022). Interactions between dietary polyphenols and aging gut microbiota: a review. *BioFactors* 48 (2): 274–284.
7. Rasouli, H., Farzaei, M.H., and Khodarahmi, R. (2017). Polyphenols and their benefits: a review. *International Journal of Food Properties* 20 (suppl 2): 1700–1741.
8. Brito, J., Hlushko, H., Abbott, A. et al. (2021). Integrating antioxidant functionality into polymer materials: fundamentals, strategies, and applications. *ACS Applied Materials & Interfaces* 13 (35): 41372–41395.
9. Belščak-Cvitanović, A., Durgo, K., Huđek, A. et al. (2018). Overview of polyphenols and their properties. *Polyphenols: Properties, Recovery, and Applications: Elsevier* 3–44.

10 Durazzo, A., Lucarini, M., Souto, E.B. et al. (2019). Polyphenols: a concise overview on the chemistry, occurrence, and human health. *Phytotherapy Research* 33 (9): 2221–2243.

11 Di Donato, P., Taurisano, V., Tommonaro, G. et al. (2018). Biological properties of polyphenols extracts from agro industry's wastes. *Waste and Biomass Valorization* 9: 1567–1578.

12 Lim, D.Y., Jeong, Y., Tyner, A.L., and Park, J.H. (2007). Induction of cell cycle arrest and apoptosis in HT-29 human colon cancer cells by the dietary compound luteolin. *American Journal of Physiology-Gastrointestinal and Liver Physiology* 292 (1): G66–G75.

13 Lee, Y., Howard, L., and Villalon, B. (1995). Flavonoids and antioxidant activity of fresh pepper (Capsicum annuum) cultivars. *Journal of Food Science* 60 (3): 473–476.

14 Beecher, G.R. (2003). Overview of dietary flavonoids: nomenclature, occurrence and intake. *The Journal of Nutrition* 133 (10): 3248S–3254S.

15 Meccariello, R. and D'Angelo, S. (2021). Impact of polyphenolic-food on longevity: an elixir of life. An overview. *Antioxidants* 10 (4): 507.

16 Oliveira, L.L., Carvalho, M.V., and Melo, L. (2014). Health promoting and sensory properties of phenolic compounds in food. *Revista Ceres* 61: 764–779.

17 Giada, M. (2013). Food phenolic compounds: main classes, sources and their antioxidant power. *Oxidative Stress and Chronic Degenerative Diseases-A Role for Antioxidants* 2013: 87–112.

18 Tsao, R. (2010). Chemistry and biochemistry of dietary polyphenols. *Nutrients* 2 (12): 1231–1246.

19 Lattanzio, V. (2013). Phenolic compounds: introduction 50. *Nature Production* 1543–1580.

20 Nakamoto, K. (2009). *Infrared and Raman Spectra of Inorganic and Coordination Compounds, Applications in Coordination, Organometallic, and Bioinorganic Chemistry*. Hoboken, New Jersey: Wiley.

21 Ozcan, T., Akpinar-Bayizit, A., Yilmaz-Ersan, L., and Delikanli, B. (2014). Phenolics in human health. *International Journal of Chemical Engineering and Applications* 5 (5): 393.

22 Khalil, M., Alam, N., Moniruzzaman, M. et al. (2011). Phenolic acid composition and antioxidant properties of Malaysian honeys. *Journal of Food Science* 76 (6): C921–C928.

23 Dubrovina, A. and Kiselev, K. (2017). Regulation of stilbene biosynthesis in plants. *Planta* 246: 597–623.

24 Smeds, A.I., Eklund, P.C., and Willför, S.M. (2012). Content, composition, and stereochemical characterisation of lignans in berries and seeds. *Food Chemistry* 134 (4): 1991–1998.

25 Berkani, F., Serralheiro, M.L., Dahmoune, F. et al. (2020). Ultrasound assisted extraction of phenolic compounds from a jujube by-product with valuable bioactivities. *Processes* 8 (11): 1441.

26 Le, X.D., Nguyen, M.C., Vu, D.H. et al. (2019). Optimization of microwave-assisted extraction of total phenolic and total flavonoid contents from fruits of Docynia indica (Wall.) Decne. using response surface methodology. *Processes* 7 (8): 485.

27 Radzali, S.A., Markom, M., and Md, S.N. (2022). Parameter effects and optimisation in supercritical fluid extraction of phenolic compounds from Labisia pumila. *Separations* 9 (12): 385.

28 Ścigalski, P. and Kosobucki, P. (2020). Recent materials developed for dispersive solid phase extraction. *Molecules* 25 (21): 4869.

29 Repajić, M., Cegledi, E., Kruk, V. et al. (2020). Accelerated solvent extraction as a green tool for the recovery of polyphenols and pigments from wild nettle leaves. *Processes* 8 (7): 803.

30 Budryn, G. and Nebesny, E. (2005). Structure and antioxidant properties of coffee bean polyphenols. *Bromatologia i Chemia Toksykologiczna* 38 (3): 203.

31 Silva, M.M., Santos, M.R., Caroço, G. et al. (2002). Structure-antioxidant activity relationships of flavonoids: a re-examination. *Free Radical Research* 36 (11): 1219–1227.

32 Heim, K.E., Tagliaferro, A.R., and Bobilya, D.J. (2002). Flavonoid antioxidants: chemistry, metabolism and structure-activity relationships. *The Journal of Nutritional Biochemistry* 13 (10): 572–584.

33 Czeczot, H. (2000). Flawonoidy-naturalne antyoksydanty w naszej diecie. *Żywienie Człowieka i Metabolizm* 27 (4): 372–382.

34 Solomon, A., Golubowicz, S., Yablowicz, Z. et al. (2006). Antioxidant activities and anthocyanin content of fresh fruits of common fig (Ficus carica L.). *Journal of Agricultural and Food Chemistry* 54 (20): 7717–7723.

35 Zhou, Y., Xu, X.-Y., Gan, R.-Y. et al. (2019). Optimization of ultrasound-assisted extraction of antioxidant polyphenols from the seed coats of red sword bean (Canavalia gladiate (Jacq.) DC.). *Antioxidants* 8 (7): 200.

36 Penedo, F.J. and Dahn, J.R. (2005). Exercise and well-being: a review of mental and physical health benefits associated with physical activity. *Current Opinion in Psychiatry* 18 (2): 189–193.

37 Group CCLoORS (2009). Survival after laparoscopic surgery versus open surgery for colon cancer: long-term outcome of a randomised clinical trial. *The Lancet Oncology* 10 (1): 44–52.

38 Seo, E.-J., Fischer, N., and Efferth, T. (2018). Phytochemicals as inhibitors of NF-κB for treatment of Alzheimer's disease. *Pharmacological Research* 129: 262–273.

39 Carluccio, M.A., Siculella, L., Ancora, M.A. et al. (2003). Olive oil and red wine antioxidant polyphenols inhibit endothelial activation: antiatherogenic properties of Mediterranean diet phytochemicals. *Arteriosclerosis, Thrombosis, and Vascular Biology* 23 (4): 622–629.

40 Qin, N., Wei, L., Li, W. et al. (2017). Local intra-articular injection of resveratrol delays cartilage degeneration in C57BL/6 mice by inducing autophagy via AMPK/mTOR pathway. *Journal of Pharmacological Sciences* 134 (3): 166–174.

41 Kinger, M., Kumar, S., and Kumar, V. (2018). Some important dietary polyphenolic compounds: an anti-inflammatory and immunoregulatory perspective. *Mini Reviews in Medicinal Chemistry* 18 (15): 1270–1282.

42 Aggarwal, B.B. and Shishodia, S. (2006). Molecular targets of dietary agents for prevention and therapy of cancer. *Biochemical Pharmacology* 71 (10): 1397–1421.

43 Koo, H.-J., Lim, K.-H., Jung, H.-J., and Park, E.-H. (2006). Anti-inflammatory evaluation of gardenia extract, geniposide and genipin. *Journal of Ethnopharmacology* 103 (3): 496–500.

44 Ray, P.D., Huang, B.-W., and Tsuji, Y. (2012). Reactive oxygen species (ROS) homeostasis and redox regulation in cellular signaling. *Cellular Signalling* 24 (5): 981–990.

45 Ahmad, M. and Gani, A. (2021). Development of novel functional snacks containing nano-encapsulated resveratrol with anti-diabetic, anti-obesity and antioxidant properties. *Food Chemistry* 352: 129323.

46 Wang, S., Chen, R., Zhong, Z. et al. (2014). Epigallocatechin-3-gallate potentiates the effect of curcumin in inducing growth inhibition and apoptosis of resistant breast cancer cells. *The American Journal of Chinese Medicine* 42 (05): 1279–1300.

47 Jena, A.B., Dash, U.C., and Duttaroy, A.K. (2022). An in silico investigation on the interactions of curcumin and epigallocatechin-3-gallate with NLRP3 Inflammasome complex. *Biomedicine & Pharmacotherapy* 156: 113890.

48 Ezzati, M., Yousefi, B., Velaei, K., and Safa, A. (2020). A review on anti-cancer properties of Quercetin in breast cancer. *Life Sciences* 248: 117463.

49 Bhosale, P.B., Ha, S.E., Vetrivel, P. et al. (2020). Functions of polyphenols and its anticancer properties in biomedical research: a narrative review. *Translational Cancer Research* 9 (12): 7619.

50 Stoner, G.D. and Mukhtar, H. (1995). Polyphenols as cancer chemopreventive agents. *Journal of Cellular Biochemistry* 59 (S22): 169–180.

51 Malar, D.S., Prasanth, M.I., Jeyakumar, M. et al. (2021). Vitexin prevents Aβ proteotoxicity in transgenic Caenorhabditis elegans model of Alzheimer's disease by modulating unfolded protein response. *Journal of Biochemical and Molecular Toxicology* 35 (1): e22632.

52 Jay Widmer, R. and Lerman, A. (2014). Endothelial dysfunction and cardiovascular disease. *Global Cardiology Science and Practice* 2014 (3): 43.

53 Allen, J.D., Cobb, F.R., Kraus, W.E., and Gow, A.J. (2006). Total nitrogen oxide following exercise testing reflects endothelial function and discriminates health status. *Free Radical Biology and Medicine* 41 (5): 740–747.

54 Geleijnse, J.M. and Hollman, P.C. (2008). *Flavonoids and Cardiovascular Health: Which Compounds, What Mechanisms?* 12–13. Oxford University Press.

55 Kar, P., Laight, D., Rooprai, H. et al. (2009). Effects of grape seed extract in Type 2 diabetic subjects at high cardiovascular risk: a double blind randomized placebo controlled trial examining metabolic markers, vascular tone, inflammation, oxidative stress and insulin sensitivity. *Diabetic Medicine* 26 (5): 526–531.

56 Miles, E.A., Zoubouli, P., and Calder, P.C. (2005). Differential anti-inflammatory effects of phenolic compounds from extra virgin olive oil identified in human whole blood cultures. *Nutrition* 21 (3): 389–394.

57 Taubert, D., Roesen, R., and Schömig, E. (2007). Effect of cocoa and tea intake on blood pressure: a meta-analysis. *Archives of Internal Medicine* 167 (7): 626–634.

58 Agis-Torres, A., Sollhuber, M., Fernandez, M., and Sanchez-Montero, J. (2014). Multi-target-directed ligands and other therapeutic strategies in the search of a real solution for Alzheimer's disease. *Current Neuropharmacology* 12 (1): 2–36.

59 Mazzio, E.A., Close, F., and Soliman, K.F. (2011). The biochemical and cellular basis for nutraceutical strategies to attenuate neurodegeneration in Parkinson's disease. *International Journal of Molecular Sciences* 12 (1): 506–569.

60 Ülker, S., McKeown, P.P., and Bayraktutan, U. (2003). Vitamins reverse endothelial dysfunction through regulation of eNOS and NAD (P) H oxidase activities. *Hypertension* 41 (3): 534–539.

61 Rajakrishnan, V., Viswanathan, P., Rajasekharan, K., and Menon, V.P. (1999). Neuroprotective role of curcumin from Curcuma longa on ethanol-induced brain damage. *Phytotherapy Research: An International Journal Devoted to Pharmacological and Toxicological Evaluation of Natural Product Derivatives* 13 (7): 571–574.

62 Rodríguez-Arce, E. and Saldías, M. (2021). Antioxidant properties of flavonoid metal complexes and their potential inclusion in the development of novel strategies for the treatment against neurodegenerative diseases. *Biomedicine & Pharmacotherapy* 143: 112236.

63 Harikumar, K.B. and Aggarwal, B.B. (2008). Resveratrol: a multitargeted agent for age-associated chronic diseases. *Cell Cycle* 7 (8): 1020–1035.

64 Shi, H., Xie, D., Yang, R., and Cheng, Y. (2014). Synthesis of caffeic acid phenethyl ester derivatives, and their cytoprotective and neuritogenic activities in PC12 cells. *Journal of Agricultural and Food Chemistry* 62 (22): 5046–5053.

65 Hafizur, R.M., Kabir, N., and Chishti, S. (2012). Asparagus officinalis extract controls blood glucose by improving insulin secretion and β-cell function in streptozotocin-induced type 2 diabetic rats. *British Journal of Nutrition* 108 (9): 1586–1595.

66 Saravanan, S. and Parimelazhagan, T. (2014). In vitro antioxidant, antimicrobial and anti-diabetic properties of polyphenols of Passiflora ligularis Juss. fruit pulp. *Food Science and Human Wellness* 3 (2): 56–64.

67 Zhang, X., Du, L., Zhang, W. et al. (2022). Pomegranate peel polyphenols alleviate insulin resistance through the promotion of insulin signaling pathway in skeletal muscle of metabolic syndrome rats. *Food Science and Human Wellness* 11 (4): 1076–1085.

68 Qin, B., Dawson, H.D., Schoene, N.W. et al. (2012). Cinnamon polyphenols regulate multiple metabolic pathways involved in insulin signaling and intestinal lipoprotein metabolism of small intestinal enterocytes. *Nutrition* 28 (11–12): 1172–1179.

69 Nair, S.S., Kavrekar, V., and Mishra, A. (2013). In vitro studies on alpha amylase and alpha glucosidase inhibitory activities of selected plant extracts. *European Journal of Experimental Biology* 3 (1): 128–132.

70 Yoo, S.-H., Lee, Y.-E., Chung, J.-O. et al. (2020). Addition of flavonols and polysaccharides as excipient ingredients into epicatechin rich green tea extract inhibited free radical formation and glucose uptake. *Food & Function* 11 (4): 3105–3111.

71 Delpino, F.M. and Figueiredo, L.M. (2022). Resveratrol supplementation and type 2 diabetes: a systematic review and meta-analysis. *Critical Reviews in Food Science and Nutrition* 62 (16): 4465–4480.

72 Khan, M.S., Ikram, M., Park, T.J., and Kim, M.O. (2021). Pathology, risk factors, and oxidative damage related to type 2 diabetes-mediated Alzheimer's disease and the rescuing effects of the potent antioxidant anthocyanin. *Oxidative Medicine and Cellular Longevity* 2021.

73 Khezerlou, A., Alizadeh-Sani, M., Azizi-Lalabadi, M., and Ehsani, A. (2018). Nanoparticles and their antimicrobial properties against pathogens including bacteria, fungi, parasites and viruses. *Microbial Pathogenesis* 123: 505–526.

74 Allaker, R.P. and Douglas, C.I. (2009). Novel anti-microbial therapies for dental plaque-related diseases. *International Journal of Antimicrobial Agents* 33 (1): 8–13.

75 Ferrazzano, G.F., Amato, I., Ingenito, A. et al. (2011). Plant polyphenols and their anti-cariogenic properties: a review. *Molecules* 16 (2): 1486–1507.

76 Olchowik-Grabarek, E., Sekowski, S., Mies, F. et al. (2023). Electrophysiological and spectroscopic investigation of hydrolysable tannins interaction with α-hemolysin of S. aureus. *Bioelectrochemistry* 150: 108318.

77 Tarahovsky, Y.S., Kim, Y.A., Yagolnik, E.A., and Muzafarov, E.N. (2014). Flavonoid–membrane interactions: involvement of flavonoid–metal complexes in raft signaling. *Biochimica et Biophysica Acta (BBA)-Biomembranes* 1838 (5): 1235–1246.

78 Yi, S.-m., Zhu, J.-l., Fu, L.-l., and Li, J.-r. (2010). Tea polyphenols inhibit Pseudomonas aeruginosa through damage to the cell membrane. *International Journal of Food Microbiology* 144 (1): 111–117.

79 Sakagami, Y., Sawabe, A., Komemushi, S. et al. (2007). Antibacterial activity of stilbene oligomers against vancomycin-resistant Enterococci (VRE) and methicillin-resistant Staphylococcus aureus (MRSA) and their synergism with antibiotics. *Biocontrol Science* 12 (1): 7–14.

80 Rahman, L., Asif, S., Ullah, A. et al. (2022). Biofunctionalized nano-antimicrobials-progress, prospects and challenges. *Current Topics in Medicinal Chemistry* 22 (13): 1046–1067.

2

Chemistry of Polyphenols: Synthesis, Characterization, and Structure (Electronic and Molecular) and Reactivity

Omar Dagdag[1], Rajesh Haldhar[2], and Hansang Kim[1]

[1] *Department of Mechanical Engineering, Gachon University, Seongnam, Republic of Korea*
[2] *School of Chemical Engineering, Yeungnam University, Gyeongsan, Republic of Korea*

2.1 Introduction

Polyphenols are a compound widely present in nature and exhibit a high degree of diversity. In their most fundamental structural manifestation, they are distinguished by possessing more than one phenolic ring and devoid of nitrogen-based functional groups [1]. Various plant species have been found to contain over 8000 distinct phenolic compounds [2]. Polyphenolic compounds are commonly found in a variety of foods, including fruits, vegetables, grains, tea, coffee, legumes, wine, chocolate, and more [3]. As secondary metabolites, plants create polyphenols that are important to their defenses against UV radiation as well as pathogenic invasion. The sensory qualities of food, including color, taste, bitterness, and astringency, are enhanced by polyphenols. Polyphenols may be divided into four primary types based on the number of phenol rings and the components used in their conjugation. These categories include lignans, stilbenes, flavonoids, and phenolic acids (Figure 2.1) [5]. The two categories of polyphenols that are most frequently detected are phenolic acids and flavonoids [6].

Science and Engineering of Polyphenols: Fundamentals and Industrial Scale Applications,
First Edition. Edited by Chandrabhan Verma.
© 2024 John Wiley & Sons Inc. Published 2024 by John Wiley & Sons Inc.

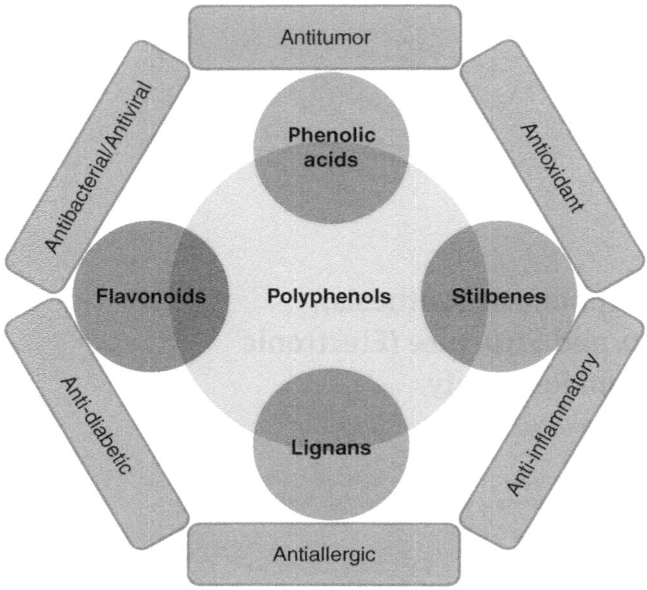

Figure 2.1 Polyphenol classification and biological functions. Source: Reproduced with permission from Xu et al. [4]/Elsevier.

2.2 Synthesis and Characterization

When dealing with naturally occurring polyphenols, it is necessary to analyze the natural product in order to identify its monomeric building blocks. This process may or may not lead to the discovery of a simpler phenolic counterpart. The complexity of biological processes can result in original simple phenolic components being transformed into other compounds through multiple or single-step transformations, and these transformed components may appear as repeated building blocks. In the case of synthetic polyphenolic compounds, complications can be avoided as the synthetic protocol for the monomeric unit is known, and the synthetic process can be controlled. Representative images of polyphenolic compounds with repeated phenolic units and linkers are shown in Scheme 2.1. Therefore, in synthetic chemistry, the construction of polyphenolic compounds typically starts with the primary units, whereas in the biological field, correlating a polyphenol with its simple building blocks is often retrospective. Finding an appropriate reagent that may serve as a foundation for a polymerization reaction in the correct direction is crucial in synthetic chemistry. The following subsections will cover how polyphenolic compounds are synthesized from basic building blocks, their possible uses, and physicochemical characteristics.

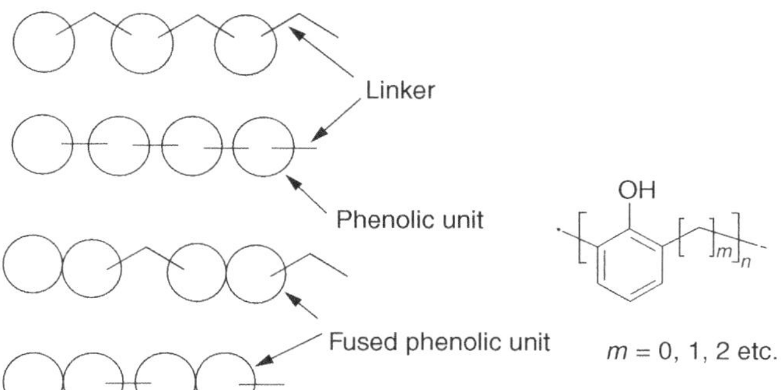

[IV]

Scheme 2.1 Photographs of Polyphenolic compounds often represent repeated phenolic units connected by linkers. Source: Reproduced with permission from [7]/Elsevier.

2.3 Polyphenolics Without Intervening Atoms

These most fundamental polyphenols are produced when a direct C—C bond between two aromatic regions of different phenolic units is used to join the units together. They can be found in cyclic or open chain forms.

2.3.1 Open Chain

Although it is straightforward to picture phenolic oligomers connected by direct C—C bonds, there are really very few of these systems because of their oxidative instability and the lack of chemical techniques available to build a uniform chain. The synthesis of these oligomers often involves oxidative radical coupling reactions. For example, dimer or trimer compounds can be produced when phenols that have one or two of their reactive sites blocked by alkyl groups undergo oxidation with a Cu(II) salt. However, this reaction requires harsh conditions that may degrade the oligomers [8]. When naphthols undergo oxidative reactions, they typically yield low molecular weight compounds linked by C—C bonds. Binaphthols comprise the bulk of the compounds produced by such reactions [9]. The palladium-catalyzed linking reaction involving aromatic boric esters and halobenzene, as indicated in Scheme 2.2, is a potential approach for synthesizing C—C bonded polyphenolics [10].

2.3.2 Cyclic

Spherands are cyclic oligomers of aromatic molecules, with a cyclohexametaphenylene system serving as the first example of this type of chemical to be reported. This substance was created [10] by oxidatively coupling *p*-cresol with $FeCl_3$ to produce a trimer. The equivalent spherand is produced by brominating the trimer, then methylating it, and finishing the process by adding butyl lithium (Scheme 2.3).

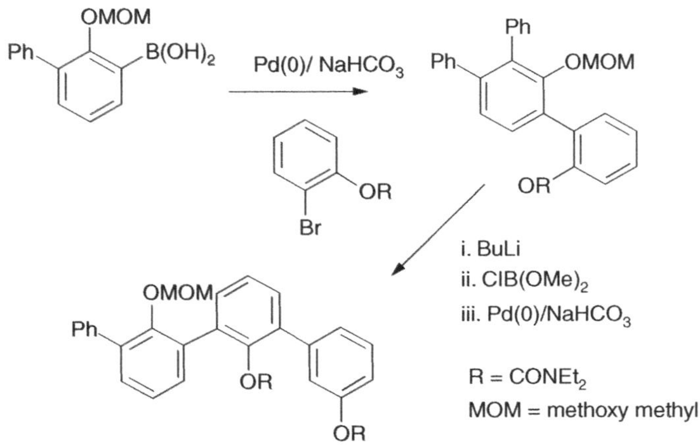

Scheme 2.2 The palladium-catalyzed coupling of aromatic boric esters with a halobenzene compound. Source: Reproduced with permission from [7]/Elsevier.

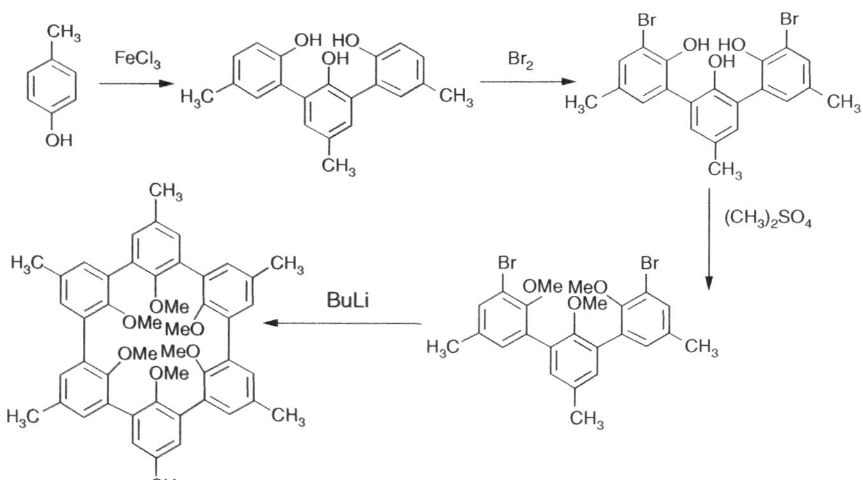

Scheme 2.3 Bromination of the trimer. Source: Reproduced with permission from[7]/ Elsevier.

2.4 Polyphenols with Intervening Carbon Atoms

2.4.1 Phenol–Aldehyde Oligomers

When phenol and an aldehyde, like formaldehyde, are combined under regulated conditions, cyclic and polyphenolic molecules are produced. Condensation can result in various products depending on the circumstances of the reaction.

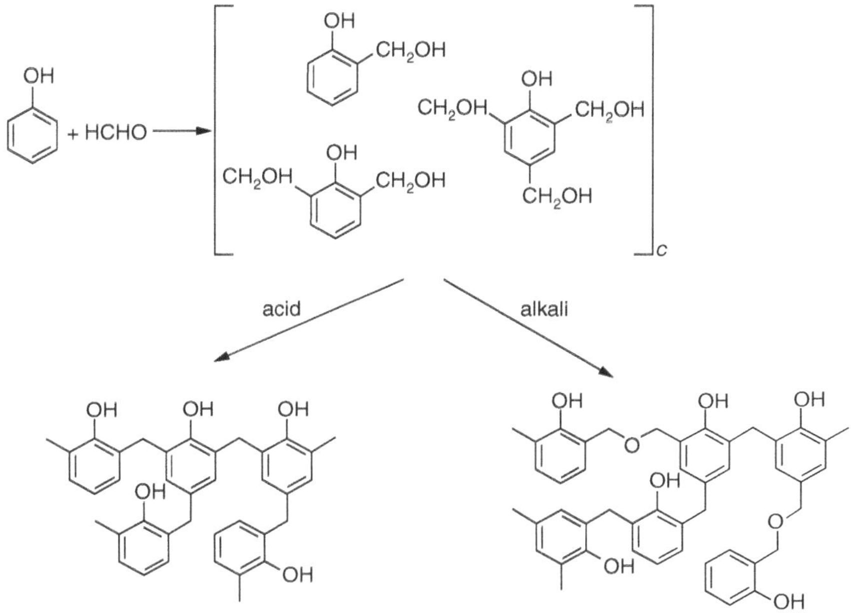

Scheme 2.4 Methylol group condensation with phenols in the *ortho*- and *para*-positions. Source: Reproduced with permission from[7]/Elsevier.

Baeyer initially described the condensation process of phenols and aldehydes in 1872 [11]. Aldehyde and phenol react in both basic and acidic conditions, forming methylol groups (CH_2OH) at positions 2 and 6 (*ortho*) and/or 4 (*para*) about the phenolic –OH. Two different condensations can cause further chain growth: (i) condensation involving two methylol phenols, which produces CH_2OCH_2 bridges, and/or (ii) additional condensation of the methylol group both *ortho* as well as the *para* position of phenols, which produces CH_2 bridges among phenolic rings (Scheme 2.4).

2.5 Polyphenols with Intervening Heteroatoms

2.5.1 Open Chain

Open-chain polyphenols of various kinds with interstitial heteroatoms are conceivable. These are some examples:

2.5.1.1 Poly-Phenylene-Ether

These substances, which include intervening oxygen bridges, are generated from phenolic substances. In Scheme 2.5, one instance is displayed [7]. This family of

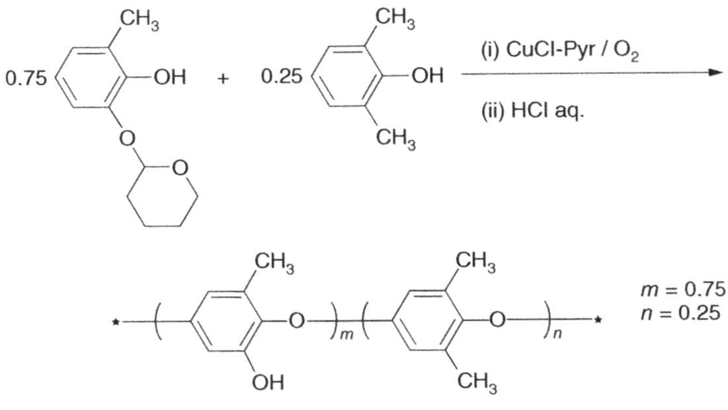

Scheme 2.5 Poly-phenylene-ether. Source: Reproduced with permission from [7]/ Elsevier.

oligomers has useful uses as engineering polymers and intriguing optical characteristics [11]. This type of oligomer is made by oxidative coupling processes [12].

2.5.1.2 Oxacalixarenes

These are analogs of calixarenes in which CH_2OCH_2 bridges entirely (in homooxacalixarenes) or partially (in heterooxacalixarenes) replace CH_2 bridges between the aromatic units [13]. In Scheme 2.6, Gutsche and colleagues [14] present a broad and straightforward procedure for the thermal dehydration of bishydroxymethylated phenols to yield oxacalixarenes.

Tannic acid terephthalate (TAT) was created through interfacial polycondensation, according to Xia et al. [15]. Figure 2.2 depicts the reaction plan for TA's straightforward one-step interfacial polycondensation. Because of the complexity of the structure of tannic acid treated with terephthaloyl chloride, direct structural elucidation employing Fourier-transform infrared spectroscopy (FTIR) as well as nuclear magnetic resonance spectroscopy (NMR) was not possible. As a result, three model compounds were created and studied using FTIR as well as ^{1}H NMR spectroscopy: methyl gallate modified together with benzoyl chloride (MGB), methyl gallate modified along with terephthaloyl chloride (MGT), and tannic acid modified along with benzoyl chloride (TAB). The peak assignments from these model compounds assisted in the precise elucidation of the structure of TAT. Figure 2.3 displays the reaction plans for creating the model compounds. Compared to TA, TAT has a 30% greater char production, a deficient heat release capacity (80 J/g-K), and is thermally stable up to 230 °C (less than 3% weight loss).

In order to broaden the uses of lignin-based phenols, Zhao et al. [17] synthesized a variety of triphenylmethane-type polyphenols (TPs) from guaiacols (methylguaiacol as well as propylguaiacol) and aldehydes (4-hydroxybenzaldehyde,

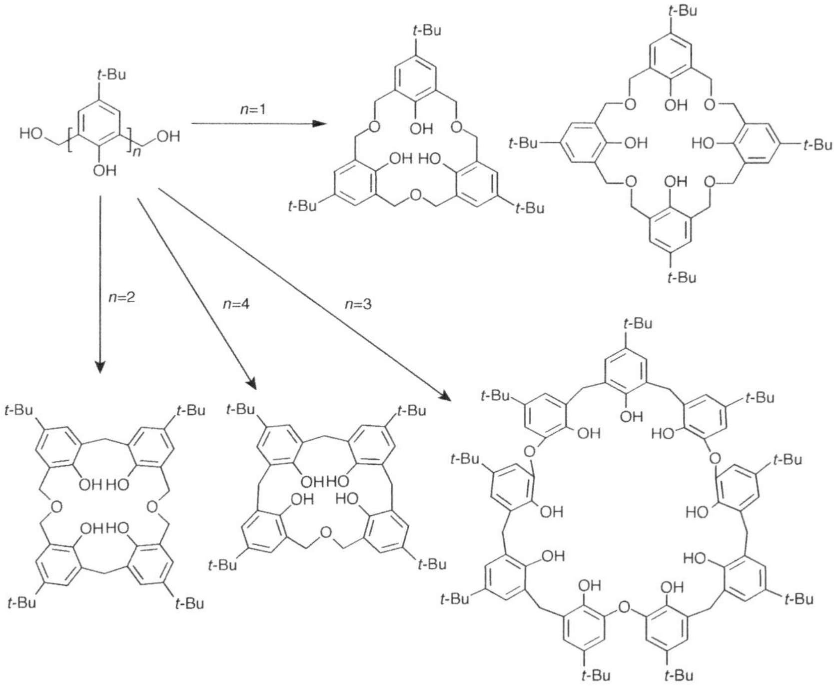

Scheme 2.6 Oxacalixarenes are made by thermally dehydrating bishydroxymethylated phenols. Source: Reproduced with permission from [7]/Elsevier.

vanillin, with syringaldehyde). Creating a newly reactive phenolic para site from the ortho-demethylation of guaiacols to catechols dramatically increased the reactivity and conversion of TPs. This improvement was made possible by the phenols generated from lignin's ortho-demethylation of the ortho methoxy group. Additional para and ortho sites are created when the conventional lignin-based guaiacols are ortho-demethylated. A typical TP (VAN-M-CAT), generated from the condensation of vanillin with methylcatechol) Then, epichlorohydrin is reacted to form glycidyl ether prepolymer (GE-VAN-M-CAT) to investigate their potential as precursors to epoxy resin. VAN-M-CAT was chosen because it was simple to remove from the reaction mixture and had the best yield (88%) of all TPs.

In a different research, Zhao et al. [18] applied a number of modifications to enhance the molecular weight, orientation, or quantity of functional groups in 2-methoxy-4-propylphenol-based biobased epoxy networks (dihydroeugenol, DHE). O-demethylation and phenol-formaldehyde processes were used to produce these changes. The architectures of cured networks and DHE-based monomers were determined using NMR alongside FTIR spectroscopy.

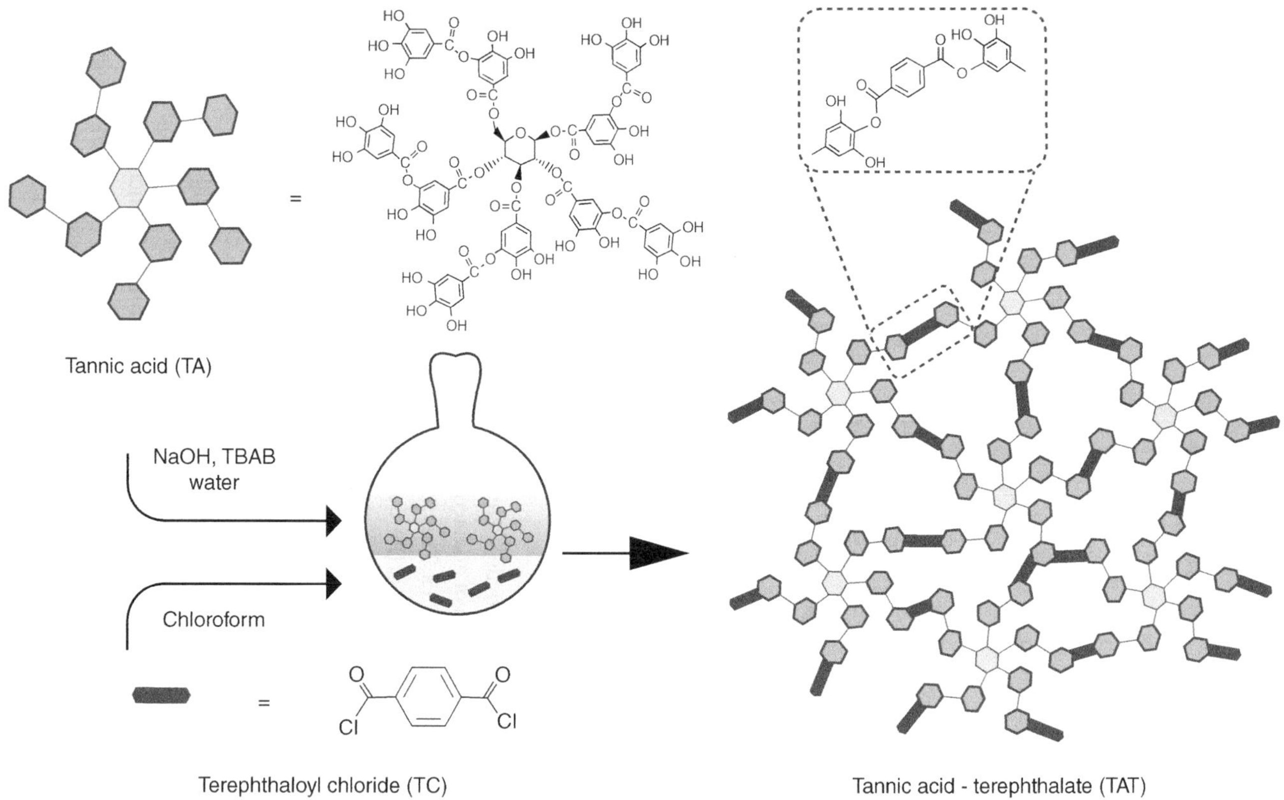

Figure 2.2 Proposed tannic acid with terephthaloyl chloride reaction pathway. Source: Reproduced with permission from Xia et al. [16]/Elsevier.

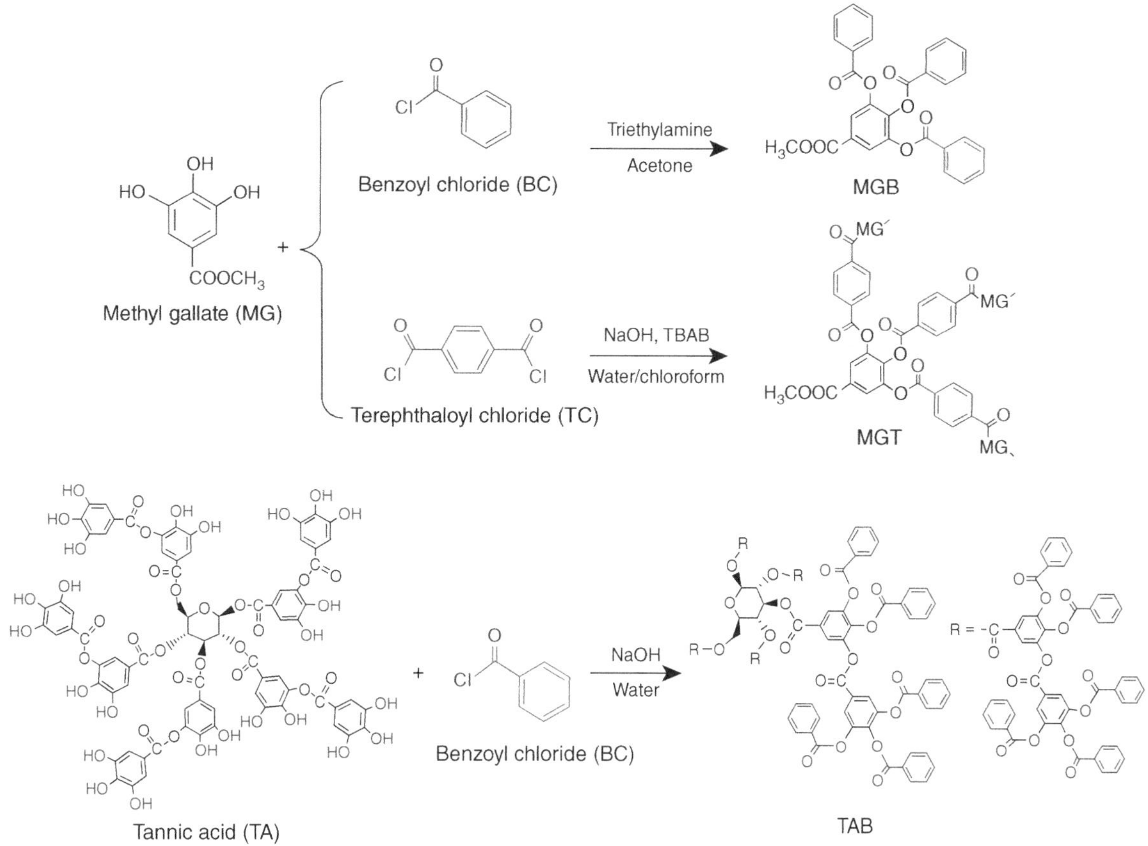

Figure 2.3 Reaction plans suggested for model compounds. Source: Reproduced with permission from Xia et al. [16]/ Elsevier.

2.6 Characterization

Over the last several years, there has been a growing interest among researchers, food manufacturers, and the general public in the quality of food products (vitamins, lipids, proteins, organic acids, carbohydrates) and others generally derived from technological processes, agro-chemical treatments, or packaging materials. Our society places a high priority on the study of food product quality, not only in terms of the vital nutrients and bioactive substances they contain that have a direct positive impact on health but also in terms of the presence of undesirable substances (such as contaminants) that are frequently hazardous to human health even at deficient concentrations. Although organoleptic (such as color, taste, and aroma) and socioeconomic (such as ecological production, guaranteed origin, and quality) factors frequently affect consumer preferences regarding food products, people are becoming more and more interested in the presence of some specific compounds with health-beneficial properties, which has even led to the production of functionalized food products.

A family of bioactive food ingredients known as polyphenols has gained public interest recently. The plant world is home to around 8000 different polyphenols and aromatic secondary metabolites with widely varying structural makeup. Molecules with masses ranging from small (100 Da) ones like phenolic acids to large (>30 000 Da) ones like those found in highly polymerized compounds. The primary drivers of interest in polyphenols are the discovery of their antioxidant qualities, the high quantity of these compounds in our food, and their potential to prevent several diseases [19]. Polyphenols, the main components of many medicinal plants'active ingredients, also control the activity of various enzymes and cell receptors [14]. Further, polyphenols' importance in food products stems from their contribution to sensory qualities. Concerning organoleptic issues, it has been noted that the presence of substances like anthocyanins and proanthocyanidins has a significant impact on color characteristics [15]. For instance, specific descriptors of the color of wines have been discovered as glycosides of anthocyanins (such as malvidin, petunidin, and peonidin) [17]. Additionally, additional substances including phenolic acids, catechins, and certain flavonoids significantly impact the flavor and color of food [17]. Evidence shows that other sensory qualities like bitterness and astringency depend on tannin molecules [18]. The main dietary sources of polyphenols for humans have been identified as foods like berries, chocolate, tea, wine, and fresh fruits, with concentrations varying from a few mg/kg to hundreds of mg/kg, depending on the constituent. Additionally, it has been noted that they are present in significant amounts in processed goods, dietary supplements, and pharmaceutical preparations [20].

The analysis of polyphenols in food samples is relatively challenging due to the wide range of compounds that can be present, which vary in size and polarity (from simple phenolic acids to tannins), along with the fact that a lot of these compounds can be found in food products at low concentration levels. The chemical variety of polyphenols has made it difficult to extract and treat samples and separate, quantify, and identify them. Liquid chromatography coupled with tandem mass spectrometry (LC–MS/MS) is the most effective technology for structural characterization and identification of high and low-molecular-weight polyphenols in food samples [21]. High-resolution and lengthy analysis periods are required for the LC determination of such compounds in complex matrices; the latter is occasionally a considerable restriction when high-throughput analysis is needed. Using either sub-2-m particle-packed columns or ultra-high performance liquid chromatography (UHPLC) [22] or porous-shell columns (with sub-3 μm superficially porous particles) [23], achieving 5- to 10-fold quicker separations than with conventional LC, while retaining or even enhancing resolution, has opened up new opportunities for upgrading the analytical procedures for complicated sample matrices. One of the most popular methods used today for food analysis is UHPLC coupled to MS (UHPLC–MS(/MS)), and there are an increasing number of studies focusing on the identification of polyphenols [13].

Accurate mass measurements and high-resolution mass spectrometry (HRMS) have grown in popularity because of their outstanding capacity to deliver more thorough information regarding a specific chemical's precise molecular mass, elemental makeup, and intricate molecular structure [24]. Among the several advantages of HRMS over conventional unit-mass-resolution tandem mass spectrometry are the following: (i) differentiation of isobaric compounds, which are substances with the same nominal mass but different elemental compositions; (ii) simplification of sample-preparation procedures, leading to faster methodologies requiring less sample manipulation; (iii) information gathered by a single injection that can be used for quantification as well as screening purposes, including targeted, suspect, and nontargeted analyses; and (iv) gathering full-scan spectra that can be saved and used [24].

Notably, a rising tendency relies on examining compositional profiles and fingerprints as a data source for categorization and authentication purposes in addition to qualitative and quantitative studies of polyphenols [25]. Chemometric data analysis has been used in a growing variety of applications during the past few years. LC–MS and LC-HRMS generate extremely high-quality data that can be further examined using chemometric techniques like PCA and partial least square-discriminant analysis (PLS-DA). The information that needs to be examined includes concentrations of relevant polyphenols (profiling approach) or instrumental signals made up of intensity counts that depend on m/z and retention time (fingerprinting approach). Additional data treatments have proven to be quite effective in making it easier to extract pertinent data on the descriptive and functional properties of food products to be used for characterization, classification, and authentication [26].

2.7 Chemical Structure and Reactivity

2.7.1 Functional Groups and Classification

The existence of a benzene ring (C_6H_6) containing a hydroxyl group (−OH) distinguishes the primary backbone in phenolic structures [27]. Due to the solitary OH group in their cyclic structure, which confers low polarity, phenols have a modest solubility in polar liquids. The likelihood of forming hydrogen bonds depending on the quantity and kind of substituents in the chemical conformation determines the solubility of polyphenols in water [28]. The range of chemical substituents in polyphenolic structures is seen in Figure 2.4.

The conformation of polyphenols can vary depending on the presence of substituents in their chemical structure. Polyphenols can contain various functional groups, such as cetonic and carboxylic groups, as well as mono- or disaccharides [30]. Most polyphenols are found in nature as glucosides attached to either glucose or a galactose molecule. But there are other polyphenols known as aglycones, which are devoid of carbohydrate rings. Stilbenes, lignans, flavonoids,

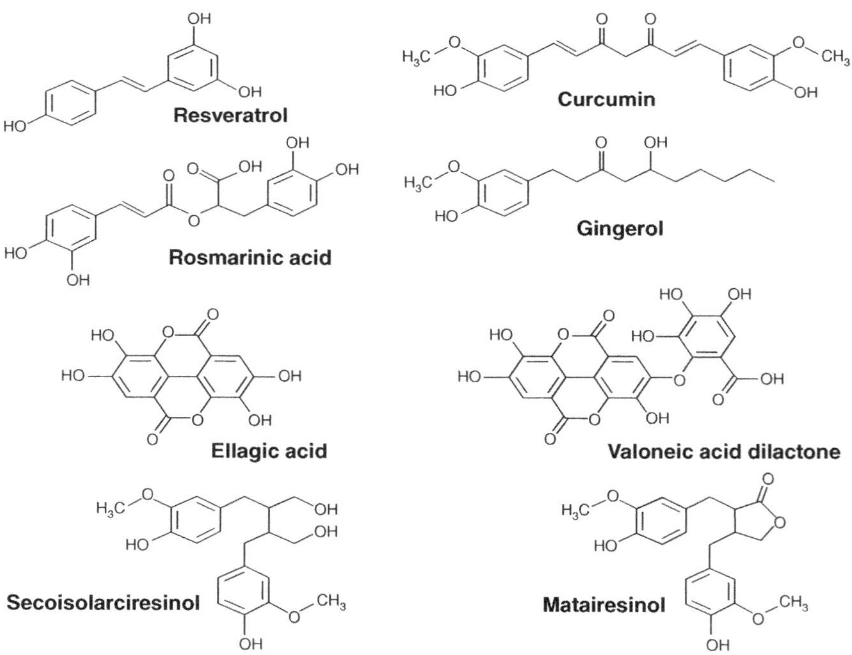

Figure 2.4 Chemical structure of chemicals in polyphenols. Source: Carmona-Hernandez and coworkers [29].

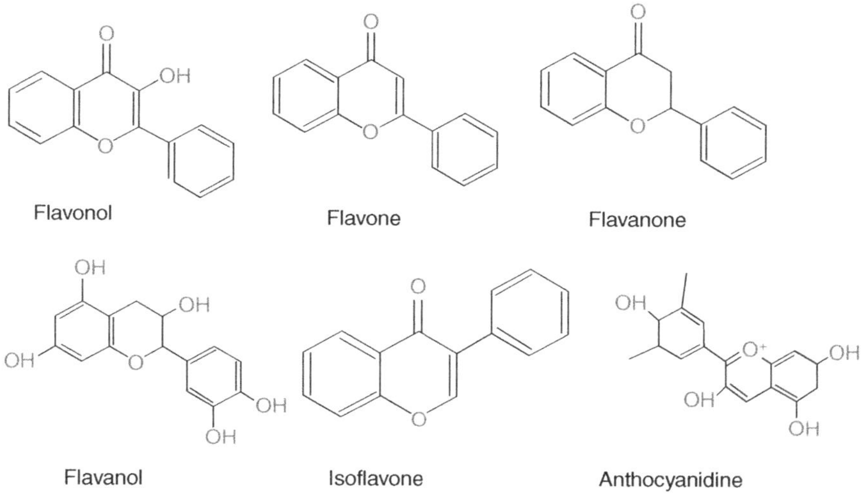

Figure 2.5 Typical flavonoid-type polyphenols' cyclic structures. Source: Carmona-Hernandez and coworkers [29].

hydroxybenzoic acids, and hydrocinnamic acids are the principal groups of polyphenols. The most prevalent and diversified of these are flavonoids, which include the subclasses flavanols, anthocyanidins, flavanols, isoflavones, flavanones, and flavones [31]. Examples of different flavonoid structures are shown in Figure 2.5. Currently, 4000 flavonoids have been discovered [30]. The existence of phenolic rings, as well as numerous three-dimensional changes, defines flavonoids. They have a wide spectrum of chemical reactivity, and the many plant flavonoid subgroups allow for various reactions and functions [30].

2.7.2 Chemical Arrangement of Polyphenols

Numerous studies describe polyphenols as natural antioxidants, emphasizing their structure is crucial for controlling free radicals, actively participating in oxido-reducing processes, suppressing cycloxygenases, and controlling cell aging [32]. Polyphenols are effective and active antioxidants due to a number of variables, including the presence of several hydroxyl groups, the frequency of single and double covalent bonds, and electronic resonance [32]. In pharmaceutical synthesis, polyphenols take part in polymeric processes and can block some portions of the arachidonic route by directly interacting with enzymes like cycloxygenase type I [32]. Functionality of polyphenols is influenced by electron mobility and free orbital availability, as well as the position and amount of hydroxyl ring substitutions in ortho, meta, and para orientations [32]. Its structural dynamic and mobility give the cyclic structure extra stability and resistance [33]. By allowing oxygen to acquire a greater electronic density along with free p orbitals, deprotonation of the –OH group creates places for electron relocation. This approach, ring alteration, and the addition of additional substituents might boost polyphenol reactivity [34]. Consequently, it is possible to produce reactions like addition, substitution, and rearrangement [35].

In addition, polymerization and higher molecular weight brought on by additional phenolic rings alter the structural makeup of polyphenols. Numerous researches have reported on polyphenol tannins' antioxidant properties. These compounds are made up of many rings and numerous hydroxyl substituents [36]. Because they can form hydrogen bonds, transport electrons quickly, and rearrange hydroxyl groups, tannins are very effective in oxido-reducing processes because of their fast electronic transfer [37].

2.7.3 Presence of Multiple Rings

Polyphenols have several hydroxyl groups and multiple benzene rings. Their general structure is shown as C6–C3–C6′, with the rings denoted by the letters A, B, and C [37]. Ring C is connected to rings A and B by a heterocyclic pyrane [38]. The typical cyclic polyphenol structure is shown in Figure 2.6. Ring C alterations affect polyphenol diversity, whereas ring A and B variations affect several functions [38]. Polyphenols with a single structure denoted as C6 include tannins, gallic acids, and hydroxybenzoic acid derivatives such as hydroxytyrosol [37]. Double-ringed polyphenols (C6–C3) connected to hydroxycinnamic acid include caffeine and cumaric acids. Some polyphenols, including resveratrol, stilbenes, and polydatin, have three rings (C6–C2–C6). Secoisolariciresinol is one of the four-ringed (C6–C4–C6) polyphenols known as lignans [39].

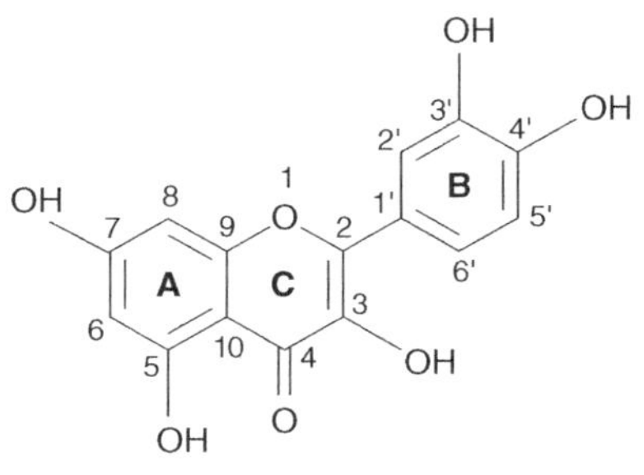

Figure 2.6 Numerology and depiction of polyphenol cycles in general. Source: Carmona-Hernandez and coworkers [29].

2.8 Conclusion

Polyphenols are among the most abundant phytochemicals known, with over 8000 currently identified, and flavonoids being the most numerous subgroup. A diet rich in polyphenols can benefit human health, with specific actions dependent on their structural characteristics, such as covalent bond configuration, cyclic structure, and numerous chemical functional groups. Polyphenols behave similarly to enzymes and vitamins in their ability to control reactive oxygen species and regulate oxidative stress.

Author Contribution Statement: Every author contributed equally.

References

1 Quideau, S., Deffieux, D., Douat-Casassus, C., and Pouységu, L. (2011). Plant polyphenols: chemical properties, biological activities, and synthesis. *Angewandte Chemie International Edition* 50: 586–621.

2 Oliver, S., Vittorio, O., Cirillo, G., and Boyer, C. (2016). Enhancing the therapeutic effects of polyphenols with macromolecules. *Polymer Chemistry* 7: 1529–1544.

3 Hu, Q. and Luo, Y. (2016). Polyphenol-chitosan conjugates: synthesis, characterization, and applications. *Carbohydrate Polymers* 151: 624–639.

4 Xu, L.Q., Neoh, K.-G., and Kang, E.-T. (2018). Natural polyphenols as versatile platforms for material engineering and surface functionalization. *Progress in Polymer Science* 87: 165–196.

5 Manach, C., Scalbert, A., Morand, C. et al. (2004). Polyphenols: food sources and bioavailability. *The American Journal of Clinical Nutrition* 79: 727–747.

6 Dykes, L. and Rooney, L. (2007). Phenolic compounds in cereal grains and their health benefits. *Cereal Foods World* 52: 105–111.

7 Handique, J. and Baruah, J. (2002). Polyphenolic compounds: an overview. *Reactive and Functional Polymers* 52: 163–188.

8 Bacon, R. and Hill, H. (1965). Copper complexes in organic chemistry. *Quarterly Review* 19: 95.

9 Sakamoto, T., Yonehara, H., and Pac, C. (1997). Catalytic activities of $CuSO_4$/Al_2O_3 in dehydrogenation of arenes by dioxygen. *The Journal of Organic Chemistry* 62: 3194–3199.

10 Sharp, M., Cheng, W., and Snieckus, V. (1987). Synthetic connections to the aromatic directed metalation reaction. Functionalized aryl boronic acids by ipso borodesilylation. General syntheses of unsymmetrical biphenyls and m-terphenyls. *Tetrahedron Letters* 28: 5093–5096.

11 Higashimura, H., Fujisawa, K., Moro-oka, Y. et al. (1998). Highly regioselective oxidative polymerization of 4-phenoxyphenol to poly (1,4-phenylene oxide) catalyzed by tyrosinase model complexes. *Journal of the American Chemical Society* 120: 8529–8530.

12 Mitsunobu, O., Trost, B., and Fleming, I. (1991). *Comprehensive Organic Synthesis*, vol. 6 (ed. B.M. Trost and I. Fleming), 22–28. Oxford: Pergamon Press.

13 Masci, B. (2001). New parent homooxacalixarenes through thermal dehydration of a bishydroxymethylated triphenol. *The Journal of Organic Chemistry* 66: 1497–1499.

14 Middleton, E., Kandaswami, C., and Theoharides, T.C. (2000). The effects of plant flavonoids on mammalian cells: implications for inflammation, heart disease, and cancer. *Pharmacological Reviews* 52: 673–751.

15 Bueno, J.M., Sáez-Plaza, P., Ramos-Escudero, F. et al. (2012). Analysis and antioxidant capacity of anthocyanin pigments. Part II: chemical structure, color, and intake of anthocyanins. *Critical Reviews in Analytical Chemistry* 42: 126–151.

16 Xia, Z., Kiratitanavit, W., Facendola, P. et al. (2018). Fire resistant polyphenols based on chemical modification of bio-derived tannic acid. *Polymer Degradation and Stability* 153: 227–243.

17 Boulton, R. (2001). The copigmentation of anthocyanins and its role in the color of red wine: a critical review. *American Journal of Enology and Viticulture* 52: 67–87.

18 Lesschaeve, I. and Noble, A.C. (2005). Polyphenols: factors influencing their sensory properties and their effects on food and beverage preferences. *The American Journal of Clinical Nutrition* 81: 330S–335S.

19 Bach-Faig, A., Berry, E.M., Lairon, D. et al. (2011). Mediterranean diet pyramid today. Science and cultural updates. *Public Health Nutrition* 14: 2274–2284.

20 Puigventós, L., Navarro, M., Alechaga, É. et al. (2015). Determination of polyphenolic profiles by liquid chromatography-electrospray-tandem mass spectrometry for the authentication of fruit extracts. *Analytical and Bioanalytical Chemistry* 407: 597–608.

21 Motilva, M.-J., Serra, A., and Macià, A. (2013). Analysis of food polyphenols by ultra high-performance liquid chromatography coupled to mass spectrometry: an overview. *Journal of Chromatography A* 1292: 66–82.

22 Fekete, S., Schappler, J., Veuthey, J.-L., and Guillarme, D. (2014). Current and future trends in UHPLC. *TrAC Trends in Analytical Chemistry* 63: 2–13.

23 Fekete, S., Fekete, J., and Ganzler, K. (2009). Shell and small particles; evaluation of new column technology. *Journal of Pharmaceutical and Biomedical Analysis* 49: 64–71.

24 Lucci, P., Martins, C.P., Núñez, O., and Gallart-Ayala, H. (2015). Liquid chromatography–high-resolution mass. spectrometry in environmental and food analysis. In: *Fast Liquid Chromatography–Mass Spectrometry Methods in Food and Environmental Analysis* (ed. O. Nuñez, H. Gallart-Ayala, C.P.B. Martins, and P. Lucci), 325–345. World Scientific Publishing Co Pte Ltd.

25 Ibáñez, C., García-Cañas, V., Valdés, A., and Simó, C. (2013). Novel MS-based approaches and applications in food metabolomics. *TrAC Trends in Analytical Chemistry* 52: 100–111.

26 Saurina, J. (2010). Characterization of wines using compositional profiles and chemometrics. *TrAC Trends in Analytical Chemistry* 29: 234–245.

27 Kurikawa, Y., Togo, M., Murata, M. et al. (2020). Mechanistic insights into visible light-induced direct hydroxylation of benzene to phenol with air and water over pt-modified wo_3 photocatalyst. *Catalysts* 10: 557.

28 Sas, O.G., Domínguez, I., González, B., and Domínguez, Á. (2018). Liquid–liquid extraction of phenolic compounds from water using ionic liquids: literature review and new experimental data using [C_2mim] FSI. *Journal of Environmental Management* 228: 475–482.

29 Cano-Avendaño, B.A., Carmona-Hernandez, J.C., Rodriguez, R.E. et al. (2021). Chemical properties of polyphenols: a review focused on anti-inflammatory and anti-viral medical application. *Biomedicine* 41: 3–8.

30 Durazzo, A., Lucarini, M., Souto, E.B. et al. (2019). Polyphenols: a concise overview on the chemistry, occurrence, and human health. *Phytotherapy Research* 33: 2221–2243.

31 Rothwell, J.A., Knaze, V., and Zamora-Ros, R. (2017). Polyphenols: dietary assessment and role in the prevention of cancers. *Current Opinion in Clinical Nutrition and Metabolic Care* 20: 512–521.

32 Zhang, H. and Tsao, R. (2016). Dietary polyphenols, oxidative stress and antioxidant and anti-inflammatory effects. *Current Opinion in Food Science* 8: 33–42.

33 Siddiqi, R., Naz, S., Sayeed, S.A. et al. (2013). Antioxidant potential of the polyphenolics in *Grewia asiatica, Eugenia jambolana* and *Carissa carandas. Journal of Agricultural Science* 5: 217.

34 Subramanya, M., Pai, S.R., Upadhya, V. et al. (2015). Total polyphenolic contents and *in vitro* antioxidant properties of eight Sida species from Western Ghats, India. *Journal of Ayurveda and Integrative Medicine* 6: 24.

35 Lucarini, M. and Pedulli, G.F. (2010). Free radical intermediates in the inhibition of the autoxidation reaction. *Chemical Society Reviews* 39: 2106–2119.

36 Furia, E., Marino, T., and Russo, N. (2014). Insights into the coordination mode of quercetin with the Al (III) ion from a combined experimental and theoretical study. *Dalton Transactions* 43: 7269–7274.

37 M'hiri, N., Ioannou, I., Ghoul, M., and Boudhrioua, N.M. (2017). Phytochemical characteristics of citrus peel and effect of conventional and nonconventional processing on phenolic compounds: a review. *Food Reviews International* 33: 587–619.

38 Xiao, J., Ni, X., Kai, G., and Chen, X. (2015). Advance in dietary polyphenols as aldose reductases inhibitors: structure–activity relationship aspect. *Critical Reviews in Food Science and Nutrition* 55: 16–31.

39 Anand, U., Jacobo-Herrera, N., Altemimi, A., and Lakhssassi, N. (2019). A comprehensive review on medicinal plants as antimicrobial therapeutics: potential avenues of biocompatible drug discovery. *Metabolites* 9: 258.

3

Plant-Derived Polyphenols: Extraction, Purification, and Characterization

Humira Assad[1,3], Alok Kumar[2], and Ashish Kumar[2]

[1] *Department of Chemistry, Faculty of Technology and Science, Lovely Professional University, Phagwara, Punjab, India*
[2] *Nalanda College of Engineering, Bihar Engineering University, Department of Science, Technology and Technical Education, Government of Bihar, India*
[3] *Department of Chemistry, School of Chemical Engineering and Physical Sciences, Lovely Professional University, Punjab, India*

3.1 Introduction

The enormous economic and societal costs associated with chronic diseases are worrying and are growing daily, necessitating the immediate implementation of preventive measures. Without a doubt, poor diets are a major contributor to many ailments. Diet and nutrition-related interest is rising, which has led to a rise in the significance of and faith in nutritional products and nutritional supplements [1]. Polyphenols (PP), peptides, vitamins, and polyunsaturated fatty acids are a few bioactive complexes that have been extracted from plant food sources and are said to provide health benefits against several ailments [2]. The evidence for the potential function of PP in the fight against chronic illnesses is rapidly growing. A large class of chemicals known as PP are present in plants and are abundant dietary micronutrients in our diet. The number of phenolic frameworks currently stands at over 8×10^3, making PP among the most pervasive and common types of phytochemicals in the plant kingdom [3]. This heterogeneous category of substances can range in architecture from meek molecules, including phenolic acids (PA), to extensively polymerized substances, such as tannins. They are all chemically distinguished by an aromatic ring containing a minimum of one hydroxyl group [4].

Science and Engineering of Polyphenols: Fundamentals and Industrial Scale Applications,
First Edition. Edited by Chandrabhan Verma.
© 2024 John Wiley & Sons Inc. Published 2024 by John Wiley & Sons Inc.

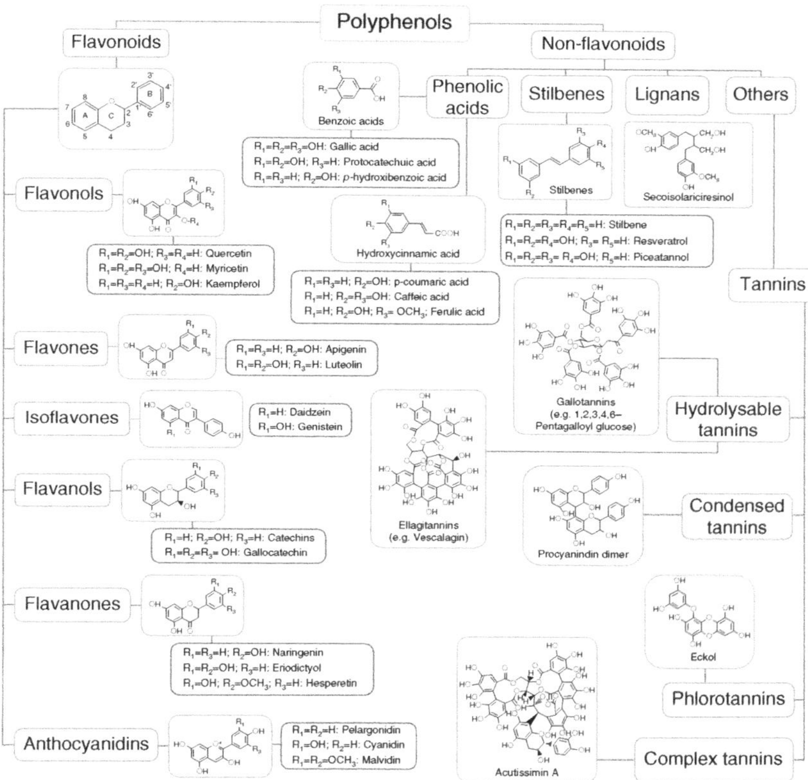

Figure 3.1 Overview of different types of polyphenols. Source: Reproduced from CCBY 4.0; No permission needed [7].

Plants synthesize PP as secondary metabolites, mostly through the acetate/polyketide and shikimate primary fabrication processes. Acetic acid and shikimic acid are both products of the metabolism of glucose [5]. PP can be categorized in many ways, including their biological role, chemical makeup, and place of origin [6]. Nevertheless, the structural makeup is undoubtedly one of the most often used classifications and separates PP into flavonoids and non-flavonoids as their two primary groupings (Figure 3.1).

Flavonoids are extremely prevalent in plant-based meals, including fruits, vegetables, tea, wine, and other beverages. Too far, several hundred distinct flavonoids have been recognized, and the number is continuing to rise [8, 9]. In general, these compounds have the same basic structure known as diphenyl-propane (C6–C3–C6). There are two benzene rings (Rings A and B) in this structure, joined by a 3-C connection, which typically materializes an oxygenated heterocycle (Ring C) [10]. According to changes in the heterocyclic ring and patterns of hydroxylation, flavonoids can be categorized into six main subclasses, as illustrated in Figure 3.1. In terms of non-flavonoids, this category consists of tannins, PA, stilbenes, and lignans. Because of their extensive distribution in edible plants and extremely variable chemical makeup, PP are, therefore, quite attractive as a component in functional meals or nutraceuticals [11].

Although PP share a phenolic characteristic, they exhibit structural variety, which causes their physicochemical characteristics to vary. They are regarded as effective chronic and degenerative disease preventive agents because of their complex and unique chemical structures, which give them a broad array of biological functions. Numerous favorable impacts on human well-being, comprising those as an anticancer, a neuro-protector, an antihypertensive, etc., have also been supported by studies, as shown in Figure 3.2 [13].

Because of their supposed safety and possible nutritional and therapeutic usefulness, natural antioxidants like polyphenolics are receiving more attention. Due to rising consumer demand and market size, "Transparency Market Research, a global market intelligence firm," recently forecasted an explosion in the PP industry. According to this survey, in 2018, the worldwide marketplace for PP is forecasted to surpass USD 873.7 million. For the years 2012–2018, depending on income and volume (tons), with a yearly average growth rate of 6.1%, consumer interest is projected [14]. Therefore, it is apparent that researchers are extremely interested in PP, particularly their in vitro and in vivo physiological functions, their extraction methods (EM), and the creation of products on a commercial basis. Research on the extraction as well as the analysis of polyphenolics is primarily concerned with a variety of factors, including the chemical makeup of the materials, the method of extraction, and additional analytical techniques. No global common methodology has been devised that will provide an overarching framework for the analytical analyses of PP compounds because of the enormous variety of phenolic substances and their sources of origin. The analytical methods and the type of solvents utilized, which can differ concerning the sources and their sequence of magnitudes, are the key determinants of polyphenolic compound extraction [15]. Therefore, to gain greater accuracy in the results, it is required to establish an ideal and suitable procedure for the correct removal, sanctification, and identification of polyphenolic compounds. Numerous articles on various elements of extraction and analytical procedures have been published; however, the research is still dispersed. To characterize these substances uniformly, this chapter addresses and recapitulates some elements of various categories of PP, PP EM, purification, and associated identification methods. It is crucial to understand the many kinds of polyphenolic compounds to choose the precise approach to be used for extraction and characterization.

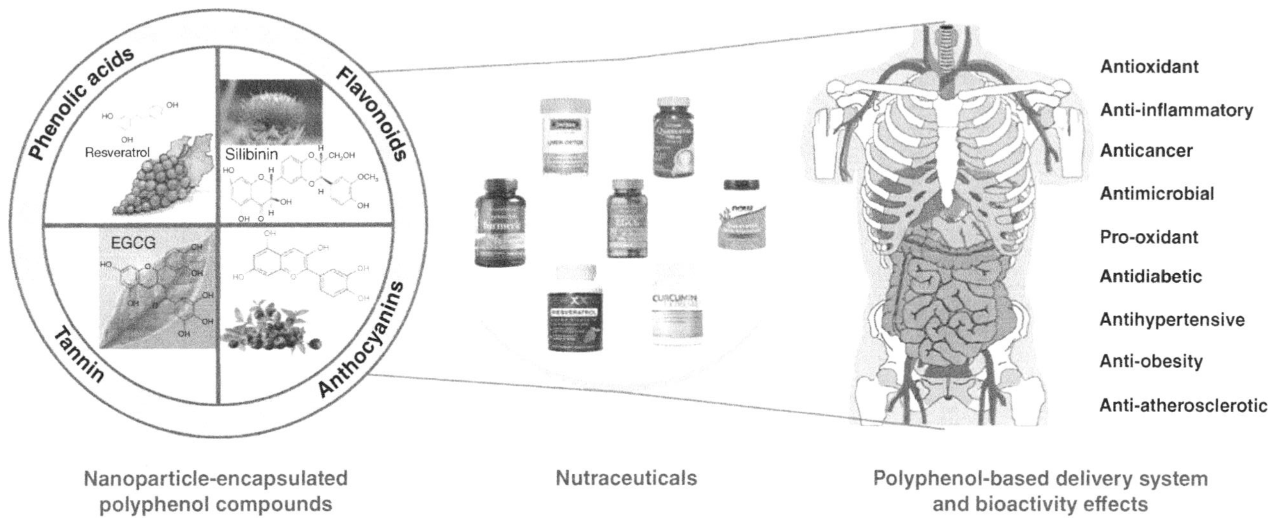

Figure 3.2 Potential application of plant-based polyphenols in nutraceutical formulations. Source: Ref. [12]/Reproduced from MDPI/CC BY 4.0.

3.2 Extraction of Polyphenols

A growing number of plant components are being used to separate and purify bioactive chemicals, and recent research has revealed the antioxidant potential of fruit leftovers, as shown in Figure 3.3 [17]. Traditional methods apply solid–liquid extraction (SLE) solely by the usage of solvent and leaching. Everyone is pretty familiar with the domestic use of conventional solvent extraction (CSE) from brewing coffee or tea at home. Maceration extraction (ME), Soxhlet extraction (SE), and percolation are examples of traditional techniques that are included in SLE.

These methods have been used for the separation of PP for more than a century. However, some drawbacks of CSE make its use relatively uneconomical because of the excessive use of time, energy, and harmful solvents [18]. These intrinsic disadvantages have spurred research into more efficient, environmentally friendly methods of extracting PP from a variety of plants and their byproducts. New extraction technologies that are quicker, more effective, and more environmentally benign than conventional processes are now in greater demand. Numerous researchers have examined the use of innovative extraction technologies thus far (see illustration below) [19, 20]. According to published data, the following

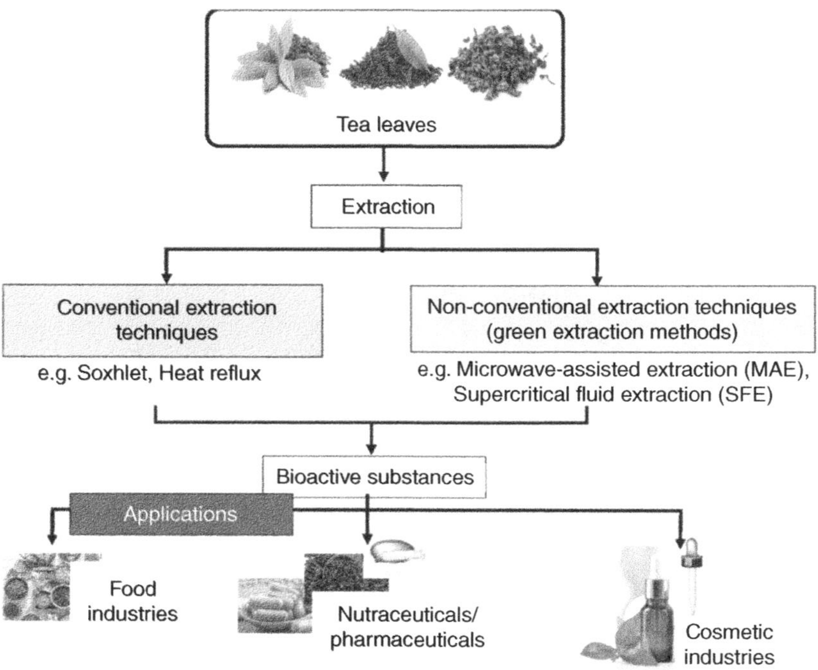

Figure 3.3 Extraction approaches and application of bioactive substances. Source: Ref. [16]/Reproduced from MDPI/CC BY 4.0.

methods offer several advantages, including a larger quantity of PP, a reduced extraction duration or speedier kinetics, a decrease in the utilization of organic solvents, and a favorable effect on the natural world. Additionally, the commercial method of implementing green extraction involves three stages that are associated with the optimization of processing variables: energy, solvent usage, and raw material. The following level-based development and customization of current procedures as well as technologies make up the three-tier approach to implementation [21, 22]:

- Improving through creative design and guaranteeing adherence to optimization methodologies;
- Making use of equipment that is intrinsically unsuited; and
- Investigating fresh substitutes for traditional solvents while creating novel methods.

The chapter's next section offers updated technical details on EM, benefits, and drawbacks. It also shows a contrasting examination of the uses of contemporary green extraction techniques, as shown in Figure 3.4.

3.2.1 Ultrasound-Assisted Extraction (UAE)

- Using an ultrasonic probe or ultrasound bath, extraction can be done with ultrasound assistance. The process of ultrasonic-assisted extraction (UAE) involves the annihilation of plant tissue by ultrasonic waves, which flow through plant cells as mechanical vibrations and induce cycles of enlargement and contraction as they pass through the extraction media. These mechanical and thermal processes result in the breakdown of cell walls, the liberation of cell contents, increased solvent diffusion of plant matter, a boost in the transmission of mass, and ultimately an upsurge in the extracted amount of PP [23]. The amount produced of PP is not greatly affected by the dimension of the particles of the plant material because ultrasound waves cause a decline in the particle dimensions (of about 5%), destruction of the cells, and alteration of the microstructure, all of which boost solvent access into the herbal matrices and improve the

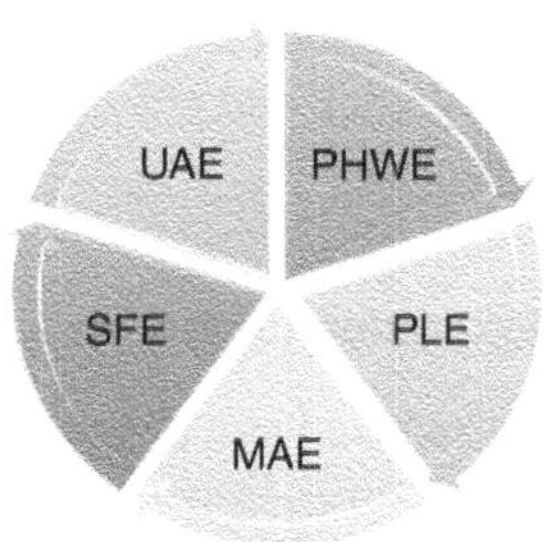

Figure 3.4 Green extraction techniques.

expulsion of PP, irrespective of the initial dimensions of the plant particulates [24]. The most effective solvent for recovering the most PP from a variety of species of plants during UAE was an alcohol/water mixture, while pure water or alcohol could not entirely extricate PP [25]. Results from earlier studies show that as the solid/solvent ratio boosted, the number of PP in botanical extracts produced through UAE improved [24]. This could be attributed to the way ultrasonic waves work, which alter plant tissue and speed up mass transfer, causing the liquid to saturate more quickly. Therefore, a boost in the solid/solvent ratio thwarts the extraction media from becoming saturated and upsurges the number of PP in the extracts [25]. However, the ultrasonic waves are attenuated by the inclusion of a lot of plant matrix, and the active component is only present in a tiny area. In the UAE of stinging nettle leaf and wild thyme herb, no statistically significant difference was seen in the generation of PP during the extraction phase [26]. Additionally, because of the greater temperature and free radicals that the ultrasonic waves produce, the prolonged period can harm extracted natural antioxidants and decrease the quality of the extract. UAE has several benefits, including

- An improvement in extraction yield;
- An increase in extract quality;
- Rapid kinetics, a crucial characteristic for industrial needs;
- In comparison to other contemporary extraction techniques, the ultrasonic device is less expensive and easier to operate; and
- A broad array of solvents can be employed to extract various phytochemicals that are pharmacologically and physiologically active.

The following are some drawbacks of UAE:

- Deterioration of antioxidant PP chemicals;
- Dependence of the influence of ultrasound waves on the amount produced and EM of PP from plant material; and
- Limitations of the active component of ultrasound to an area near the ultrasonic emitter due to the existence of a greater number of plant particulates.

Taking into consideration the aforementioned discussion, several studies demonstrate that, in contrast to CE methods, extracts prepared to utilize UAE had an exceptionally high amount of PP and took less time for extraction [27]. Therefore, it is possible to suggest UAE as a quick and effective method for removing bioactive PP from a variety of plant sources.

3.2.2 Supercritical Fluid Extraction (SFE)

SFE has become extensively recognized as a prevalent green approach, especially in the context of solid matrix structures, because of benefits like producing purer extracts under the right circumstances than those generated by a CSE with little to no organic solvents. SFE is frequently utilized to add value to plant byproducts that are produced throughout the preparation. These chemicals appear to have no economic worth, and extraction makes it possible to effectively utilize waste materials to extract the desired phenolic substances [28]. A fluid that is beyond its typical critical pressure and temperature and exhibits characteristics of both liquids and gases is known as a supercritical (SC) fluid. The critical pressure, or PC, is the least amount of pressure thought to be necessary for a gas to liquefy at its particular critical temperature. A gas's critical temperature (Tc) is the temperature at which it will remain a gas until further pressure is applied before it turns into a liquid. Due to the combined features of two distinct phases—liquid and gaseous—at once, SC fluids have become the preferred solvents. As a result, there is an increase in the rate of solute mass transfer during extraction.

In light of this, SC fluid offers special density, solvation, and viscosity characteristics at the transitional phase between gases and liquids. Temperature and pressure changes can alter these qualities [29]. Raw material from plants from the extraction procedure is introduced into an extraction container. A pressure release valve and thermostats linked to the extraction container are in service to preserve the desired extraction parameters. CO_2 has been classified as generally acknowledged as secure (GRAS) by the "European Food Safety Authority" and the "US Food and Drug Administration." Due to its unique characteristics, most notably its economic affordability and GRAS status, CO_2 is the most commonly utilized SC solvent. The extraction vessel receives CO_2 that has been injected as a fluid at $Tc < 278.15\,K$ and $Pc = 5.7\,MPa$. The CO_2 at the CO_2 pump's inlet and exit is cooled using heat exchangers. By adjusting temperature and pressure simultaneously in the CO_2-extract centrifuge, the SC fluid's solvation potential is reduced. An exit gate on the bottom portion of the separator is used to divide the fluid containing phenolic chemicals from the solution in the CO_2-extract separator. When maximal PP recovery rates are achieved from the selected plant specimen composites, the cyclic SFE process is repeated [30].

3.2.3 Microwave-Assisted Extraction

Modern methods, such as MAE, have been employed in place of traditional techniques for extracting PP to increase extraction proficiency while also reducing extraction time, using less solvent, and yielding more PP. A microwave reactor or microwave oven can be used for microwave-assisted extraction (MAE). The quick energy delivery provided by microwaves to the whole volume of the extraction media and plant particles results in the efficient and uniform heating of the sample. In addition, microwave energy is absorbed by water inside plant cells, which leads to internal superheating and cell rupture, which improves the recovery of PP [30]. Concerning Chupin et al. the maximum proportion of PP was achieved utilizing the smallest particulates, <0.1 mm, in the MAE of *Pinus pinaster*. However, there was no statistically substantial variation between the particle sizes of 0.4–1 mm [31].

Additionally, with an extent of fragmentation smaller than 0.5 mm, the highest effectiveness of MAE of PP from *Achillea melanocarpa* (*AM*) and *Hippophae rhamnoides* was accomplished [32, 33]. According to recent studies, the polar dispersion and dielectric constant may clarify why the concentration of ethanol in the water-based medium increased, which in turn led the PP output from *AM* and *Myrtus communis* to rise as well [19].

The incorporation of water into ethanol results in a rise in the dielectric constant, which enables the uptake of microwave energy. Nevertheless, more water results in less heat spreading into the extraction medium, which also lowers the effectiveness of MAE [34]. Further studies have demonstrated that in MAE, a boost in the solid/solvent ratio culminated in a boost in PP content from *AM* because the plant matter was wetted more effectively, leading to a greater amount of PP from *AM* when utilizing 50% C_2H_5OH, whereas additional increases in C_2H_5OH concentration were associated with a decline in PP production. The quick saturation of the extraction medium is produced by an increase in the diffusion coefficient and a vigorous discharge of PP from plant cells that have been damaged by microwaves. As a result, the retrieval of PP from plant material is improved by increasing the solvent volume. Data from the literature indicate that using strong microwaves for a brief period during an extraction process can guarantee cell wall breakdown and improve PP dispersion into the extraction media.

In addition, because microwaves trigger the liberation of a lot of ballast materials, including polysaccharides, lipids, and proteins, longer extraction times and improved yields do not necessarily translate into a lot of target chemicals [35]. The primary benefits of MAE include:

- A substantial reduction in extraction duration;
- A decline in solvent use;
- A spike in extraction generate; and
- An approach that is simpler and more cost-effective than other novel extraction techniques, like supercritical fluid extraction.

The MAE has certain drawbacks, including:

- The breakdown of thermosensitive PP components;
- The reduced amount of volatile and non-polar target substances;
- The dependence of microwave impacts on PP generation on solvent polarity; and
- The potential extraction of significant amounts of ballast materials.

Chokeberry berries, yarrow, myrtle, and sea buckthorn leaves are only a few examples of diverse plant materials from which PP extracts can be effectively and commercially produced [32, 33, 35]. MAE is a recommended technique to generate excellent herbal extracts from a variety of plant resources, with elevated PP production, from the perspective of safeguarding the environment and the alimentary, medicinal products, and aesthetic sectors, which demand secure and excellent quality goods.

3.2.4 Pressurized Liquid Extraction (PLE)

An EM recognized as "pressurized liquid extraction" (PLE), additionally referred to as "accelerated solvent extraction," uses organic and aqueous solvents to keep them liquid past their typical boiling points while under pressure and at a higher temperature. This temperature rise enhances the kinetics of the substance and makes extractions more effective. The EM is simplified by the use of high pressure. The solvents employed in the pressurized liquid extraction are typically water, methanol, acetone, and hexane, with temperatures fluctuating from 75 to 150 °C and a pressure of typically 10.4 MPa. The high temperature helps break the bindings of the PP constituents with the matrix while the imposed pressure increases contact between the solvent and the sample. The amount of solvent that enters the cells rises with heating, creating an augmented internal pressure that forces the dissolved components out via the pores in the cell wall. It was suggested to use this method of extraction to obtain phytochemicals from sources such as herbal plants and vegetables [36, 37], as shown in Figure 3.5. The method previously described is called subcritical water extraction (SWE) when H_2O is the solvent. With its distinct properties and high dielectric constant, water that is still liquid at temperatures between 100 and 374 °C under high pressure (typically ~ 10 bar) effectively dissolves substances that are only moderately or barely solvable in H_2O at room temperature.

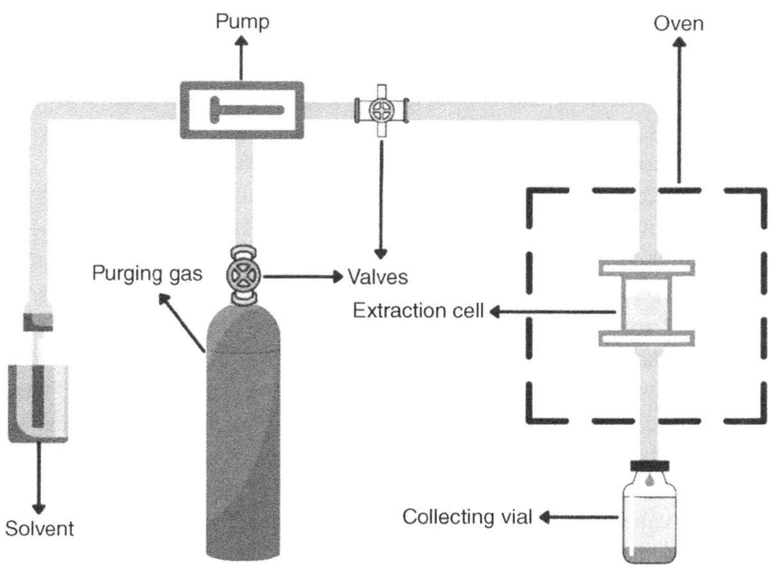

Figure 3.5 Diagrammatic illustration of polyphenol extraction employing PLE. Source: Reproduced from CCBY 4.0; No permission needed [38].

In contrast, getting the water into its subcritical state does not affect the solubility of chemicals that are readily solvable in H_2O at ambient temperature. Numerous studies have provided systematic ways to improve the PLE and SWE processes for the gathering of PP elements and to assess the productivity of the aforementioned procedures with more traditional ones [39, 40]. In general, using these methods has many benefits over traditional EM, including quick extraction duration, better-quality extracts, cheaper solvent extraction expenses, an environmentally friendly method, and exclusivity that is easily modifiable by adjusting the extraction temperature. Some outcomes of using PLE to recover PP from aromatic plants or to compare it to other techniques were explored by Machado et al. [41].

3.2.5 Pressurized Hot Water Extraction (PHWE)

Among the different green EM, PHWE is another appreciated methodology. The scientific literature has praised this water-based extraction method for being attractive due to its prepared and cost-free accessibility, non-toxic nature, non-flammability, and recognition as an environmentally acceptable solvent. However, the application of this method as a commonly used approach has been constrained by the dissolution of organic molecules in water, such as PP [42]. But as temperatures rise, physicochemical properties like viscosity, surface tension, and dissociation constant undergo considerable modifications. Regarding preference for phenolic ingredients and a decrease in extraction instances, these consequences increase PHWE performance as a green extraction technique [43]. With the groundbreaking research of Hawthorne and colleagues addressing natural specimens in the 1990s, hot water was first used in the PHWE approach. It is also known as SWE, superheated WE, or ASE using H_2O as an extractant [44].

Using pressurized water requires temperatures that are above and below the water's critical points, which at a pressure of 0.1 and 22.1 MPa are 273 and 647 K, respectively. Practically speaking, the equipment used for PHWE using pressurized hot water (PHW) as the extractant is similar to that employed for PLE. For the abstraction of PP from different plant matrices, PHWE has established a solid reputation as a practical substitute to CEM. Temperature-dependent choosiness, low operating costs, and quick operation are the most attractive qualities that make this method a popular option. The main benefit of using PHW as an extractant rather than harmful and expensive chemical solvents is that it is harmless. By using PHWE and conventional methods, Budrat and Shotipruk have examined the withdrawal capabilities of various PP (genistic acid, catechin, etc.) from *Momordica charantia* (bitter melon) [45]. In this instance of PHWE gathers, a favorable correlation between increasing temperatures and the amount of phenolic compounds was discovered, whereas an opposite relationship was found between rising temperature and the effectiveness of antioxidants. When contrasted to the yields seen with MeOH separation (6.6 mg GAE/g DW) and SE with aqueous methanol, specifically at 473.15 K, PHWE has yielded isolates with significantly higher overall phenolic concentrations (52 mg GAE/g DW). Furthermore, PHWE extracts were shown to have three times the amount of antioxidant activity of other extracts generated through more traditional methods. Comparing the amount of solvent used for PHWE extraction to that needed for MeOH extraction, and SE extraction, PHWE has been shown to require the least amount of solvent (2 mL/min).

The enhanced extraction yield of hesperidin and narirutin (PP) from *Citrus unshiu* (citrus) peel, a derivative of the citrus trade, employing PHWE has also been reported in a similar publication. When contrasted to SE and ME employing C_2H_5OH and CH_3OH as the extractants, respectively, PHWE was shown to be proficient and appealing in relation to flavanone recovery and antioxidant properties of PHWE residues. When utilizing hot water, methanol, and ethanol, the scientists found extract yields from PHWE that were, respectively, 34.21, 3.2, and 9-fold higher than those from CSE [46]. In a further investigation, RSM was used to optimize the PHWE approach to isolate PP moieties from apple pomace. The authors have optimized the extraction conditions using RSM to get the highest yields of flavonols from the byproducts of cider production. They observed that, utilizing PHWE, an optimal extraction amount of antioxidant chemicals occurred at a temperature of 393.15 K and an extraction period of 3 minutes. The primary benefits of PHWE generally include the following:

- Using water as an environmentally friendly extractant with polarity comparable to that of organic solvents (such as alcohol) while posing the fewest disposal problems;
- Improves selectivity while facilitating the diffusion of several solutes over a spectrum of low to medium polarity;
- A larger decrease in the utilization of organic solvents;
- An affordable and safe extraction method for organic substances from a variety of plant materials and byproducts;
- Less operating and maintenance requirements than those of more advanced techniques;
- Better support for bioremediation treatments as a green extraction technique.

The main disadvantages of PHWE include the following:

- The physicochemical characteristics of superheated water may change as a result of the incorporation of various modifiers and additives;
- Need for micelle-mediated PHWE with Triton X-100 for moieties that are thermosensitive.

3.3 Purification of Polyphenols

The PP discovered in plant extracts (PE) may possess a diversity of pollutants and interference-causing agents. Therefore, it is essential to eliminate the contaminants before identifying and measuring the phenolic compounds. The lipids and other undesirable chemicals can be removed with ethyl acetate, or diethyl ether, petroleum ether, and samples of PP may be purified for additional research under vacuum [47]. It should be noted that, in some phenolic complexes, like anthocyanins, excessive concentrations may result in the deficit of sugar and labile acyl remnants; it is therefore preferable to utilize boiling water, weak organic acids, or neutral organic solvents, for preventing this degradation. In general, solid phase or liquid–liquid techniques can be used to purify and fractionate phenolic substances.

One of the often-used methods for isolating, purifying, and pre-concentrating phytochemicals, particularly phenolics, is solid phase purification [48]. Alkylated silica gels (mostly C8, C18) are frequently employed as sorbents for SPE, and for increased isolation specificity, two cartridges with C18 and quaternary amine (sorbents) are combined. However, it was also shown that polymeric solvents, like RP 105 and OASIS™ HLB, were the best solvents for phenolic ingredient separation. In comparison to the traditional liquid–liquid extraction method, solid phase purification is quicker, more repeatable, and results in cleaner extracts. For the most part, phenolics were purified using graphitized carbon black, styrene-divinylbenzene copolymers, or C18-bonded silica reversed-phase sorbents. For the isolation of plant components, combination SPE methods with C18 and anion exchange columns have occasionally been utilized. Because of the hydrophobicity of their substrates, co-polymers can encounter difficulties throughout the conditioning phase, which causes the prepared sorbent to dry out. The recuperation and extraction effectiveness of polar organic molecules employing novel polymeric sorbents with chemical modifications have been found to increase. These substances make the polymeric material more wettable, and the sorbents are considerably better able to hold onto the polar phenolic compounds. A relatively new sample extraction and purification method called solid-phase microextraction (SPME) gives a chromatographic examination of the sample's special capabilities. It can cut down on solvent use as well as any problems with how to properly dispose of solvents used to prepare samples.

SPME is a quick, low-cost approach for extracting organic chemicals like PP from gaseous and aqueous specimens without the use of solvents [49]. In this extraction technique, analytes are drawn out of the sample mixture using sorbent-coated silica fibers. The fiber is typically immersed to extract the analytes for sample extraction from an aqueous medium, and the fibers are then immediately shipped to perform additional chromatographic examination, where the specimens are ultimately sorted and investigated. Two important coating fibers utilized for the study of PP by SPE (micro) techniques (SPMET) are polydimethylsiloxane and polyacrylate [49]. For obtaining and determining the presence of phenolic chemicals in polluted wastewater, a SPE (micro) approach has been devised employing a polyacrylate-coated fiber [50]. For the measurement of PP in grapes and wine employing gas chromatography-mass spectrometry, a newly invented specimen pre-treatment technique centered around SPME has been produced [51]. Because it has merits over other procedures, like liquid–liquid phase extraction (LLPE) approaches, SPE is widely employed. With other technologies, including the online coupling of analytical columns, it is simpler to automate the solid extraction processes. The SPE approach, in contrast to further procedures, produces cleaner extracts. The automated SPE approach can shorten the duration needed for chemical extraction and purification. When contrasted to LLP purification, it can manage minor sample volumes and uses a fraction of the solvent volume. The method's primary drawback is that it restricts sample sizes, making it impossible to process huge numbers of samples.

The process of purification of phenolic chemicals from PE often involves liquid–liquid extraction. According to the type of phenolic components, the PE were typically dried using this process at lower pressure and temperature [15]. After being immersed in boiling water and allowed to cool for 12–24 hours, the dried material was strained. Water was used to clean the resinous residue precipitate, and petroleum ether was used to separate the filtrate. Diethyl ether was used to extract the aqueous solution, and 5% Na_2CO_3 was used to further treat the PA in the $(CH_3CH_2)_2O$ layer so that they would become sodium salts that could be dissolved in water. The bi-carbonate layer was then extracted with diethyl ether after being acidified with HCl. The resulting $(CH_3CH_2)_2O$ extract was carefully rinsed with H_2O to eliminate the acid, and then Na_2CO_3 was introduced to the extract to eliminate the remaining water [15].

Phenolics' LLP separation was primarily predicated on their segregating between two immiscible liquids, and these moveable phases typically have access to a miscible auxiliary solvent that can help with this process [52]. In comparison to SPE, LLP purification has many drawbacks, including the necessity of using numerous solvents, labor-intensive processes, the need for evaporation phases to confiscate excess solvent, the difficulty of determining the right solvent for each specimen, the possibility of problems with emulsified samples in LLE, and the possibility of contamination. In comparison to previous procedures, this method's main benefit was its capacity to handle a huge number of samples.

3.4 Characterization of Polyphenols

The primary factors affecting the quantification as well as the recognition of phenolic materials are the chemical makeup of the substances, the extraction technique, the storage space and circumstances, the particle size, the choice of standards, interfering elements and contaminants, and the test techniques. Many different spectrometric techniques have been employed to quantify phenolic chemicals in plant material. There are now many different contemporary tools available thanks to the development of analytical science, including high-performance liquid chromatography (HPLC), gas chromatography (GC), Fourier-transform infrared spectroscopy (FT-IR), and nuclear magnetic resonance (NMR), among others [15]. However, because different approaches have different properties and drawbacks, it is challenging to suggest a strategy for confirming the presence of PP in PE. In this regard, only one investigation has been successful in offering a technique for recognizing and defining the characteristics of PP that is typically suitable. For this goal, several techniques have been created, and how they are used relies on different parameters, such as the characteristics of the analyte and the target chemical.

Some analytical techniques, such as total phenols content (TPC) and total flavonoid content (TFC), are generic and unable to distinguish between specific chemicals in extracts. The Folin Ciocalteu spectrophotometric test and the $AlCl_3$ assay, respectively, are frequently used to measure the extract's TFC and TPC [53]. This method's limited specificity, meanwhile, is a drawback. One significant issue is that it reacts with numerous other non-phenolic chemicals, which greatly distorts TPC readings. Everette et al. demonstrated in their investigation that proteins, thiols, numerous vitamin compounds, and the inorganic ions SO_3^{2-}, I, Mn^{+4-}, and Fe^{+2} were all reactive toward the Folin reagent in addition to phenols [54]. Other methods have been created to offer improved precision in determining the presence of PP in light of this difficulty. In comparison to TPC assay, HPLC is an improved version of liquid chromatography (LC) and can be employed to more accurately recognize PP, as displayed in Figure 3.6.

Retention time (RT) of isolated compounds is compared with benchmarks to accomplish this. It is important to remember that the effectiveness of this method depends on the solubility of the target complexes and the mobile phase solutions utilized. Elution solution factors like pH and polarity, as well as other setup variables like column temperature, are crucial. However, in contrast to GC, it has lesser precision, extremely long downtimes, and significant procurement and operational expenses. Additionally, the HPLC method lacks a universally sensitive detector, which causes findings to vary. HPLC can be employed with mass spectrophotometry (MS) to increase the sensitivity of the procedure and generate more precise findings centered on the unique mass/charge characteristics of atoms. When juxtaposed with LC and HPLC, GC is more accurate. It provides higher accuracy when paired with MS (GCMS). A significant difficulty is poor chemical recognition with a substantial amount of unconfirmed peaks [56].

PP characterization is additionally made challenging by restrictions in MS data collection speeds brought on by insufficient metabolomes. A variety of techniques

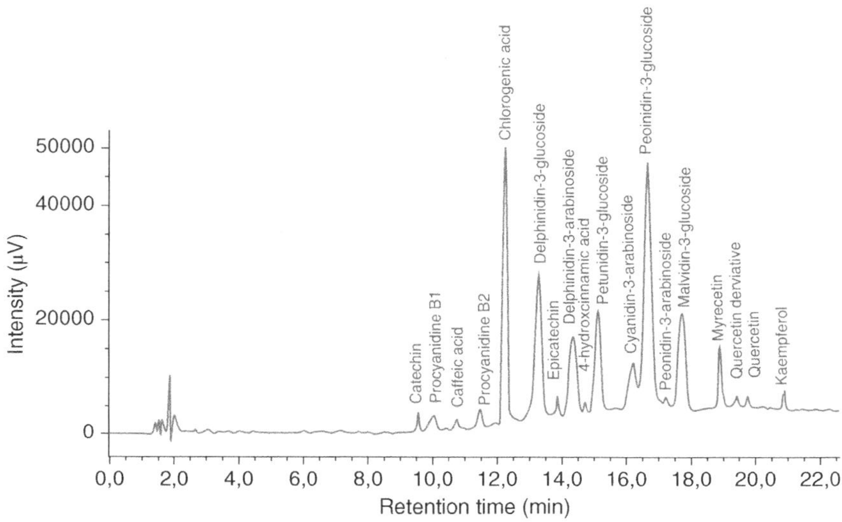

Figure 3.6 HPLC analysis for the identification of polyphenols obtained for a blueberry pomace sample. Source: Reproduced from CCBY 4.0; No permission needed [55].

are utilized to address the difficulties in PP identification. The chromatographic method known as supercritical fluid chromatography (SFC) is very new. The utilization of a SF as the mobile phase distinguishes SFC from other chromatographic methods, including GC and HPLC. SFC has some merits over other traditional chromatographic methods (GC and HPLC), including better compound segregation with less solvent use, speed, and the ability to analyze thermolabile substances. A SFC approach for the segregation and characterization of various PP from grape seeds has also been established [57]. An approach involving thermal desorption, like laser diode thermal desorption (LDTD), has just been invented in light of the recent tendency to include green chemistry as a key procedure among eco-friendly procedures. An additional ion generator for mass spectrometry is LDTD. The ion source operates without any LC apparatus, which will minimize solvent use. It is primarily based on a shotgun technique. Without a mobile phase, external matrix, or solvent, the specimens can be ionized. The LDTD method can analyze 90 samples simultaneously and performs well with samples that are under 2 L in volume. The use of this approach has been covered in several research studies, including the study of hormones from wastewater, contaminants, and drug metabolites [58].

In the meantime, PP extraction can also be done using this technique. The three main chromatographic techniques used in investigations are HPLC, LC–MS/MS, and GC–MS/MS, with supplementary approaches such as FTIR and NMR coming in second and third, respectively. The different functional groups in extracts can be distinguished using FTIR [59, 60]. However, it has a narrow reach and cannot distinguish between specific compounds belonging to the same category. A potent supplemental spectroscopic method utilized for the structural clarification of complicated phenolic substances separated from various plant-based sources is NMR. Because a reference molecule is required and the approach uses fairly substantial quantities of the substances, phenol NMR spectra are typically complicated. Nevertheless, it has a narrow range and cannot distinguish between particular compounds belonging to the same category. Complex phenolic compounds isolated from various plant sources can be structurally clarified using the potent complementary spectroscopic method known as NMR. The intricacy of the NMR spectra of phenols is typically due to the need for a reference chemical and the approach's requirement for comparatively hefty quantities of the compounds. A reference standard is not necessary when using 2D-NMR spectrometry for structural investigation.

The primary NMR techniques used to analyze substances are ^{13}C and ^{1}H NMR, as well as 2D^{1}H-^{1}H correlated NMR spectroscopy (COSY), rotating frame of reference (ROESY), nuclear Overhauser effect in the laboratory frame (NOESY), heteronuclear chemical shift correlation NMR (C-H HECTOR), and totally correlated NMR spectroscopy (TOCSY). Despite being a costly, new, and infrequent approach for phenolic moieties, NMR spectroscopy has numerous benefits over other approaches. NMR analysis frequently offers greater accuracy and precision than conventional HPLC techniques. The NMR analysis of chemicals does not require expensive standards. It also provides details on the compounds' impurities and isomeric structures. NMR has the potential to be more precise, quicker to perform (no equilibration time), simpler, and more reproducible. NMR is optimal for nondestructive purposes because it may infer structural details from unaltered molecules' oscillations in their natural surroundings. NMR collections can be utilized for recognizing molecules, but their application is only as broad as the library's collection of chemicals. Results can be more accurately interpreted when complementary data from many useful techniques are combined. The problems with PP identification and characterization haven't been fully solved by a single technique up to this point.

3.5 Conclusion and Outlook

Due to their widespread availability in plants and positive biological activity, polyphenolic chemicals are attracting more attention. PP extraction from plant sources is gaining popularity as a natural, affordable, and reliable substitute for synthetic substances, some of which have harmful and carcinogenic consequences. Owing to the extensive use of time, samples, organic solvents, and energy, traditional procedures have been demonstrated to be less efficient, time-consuming, and environmentally friendly than modernized alternatives. As a result, scientists from all over the globe are extremely engaged in researching more sustainable and ecological contemporary "green" methods for effective extraction of PP from plant matrix structures that comply with recommendations being laid out by international regulatory organizations to mitigate the issues connected with the phenomenon of global warming. Concerning PP extraction efficiency, consistency, and overall beneficial influence on environmental change by lowering CO_2 release and harmful solvent treatment, modern green methods of extraction are likely to be quick, simple, cost-effective, environmentally sound, and eco-friendly. Given their significant health benefits, analytical science should pay close attention to the discovery, characterization, and structural understanding of polyphenolic substances.

This chapter covered several varieties of PP as well as diverse approaches to their extraction, purification, and structural characterization. Owing to the enormous variety of PP and their sources, significant extraction production, purification, and accurate recognition of PP moieties remain a mystery to researchers. These investigations show that each particular polyphenolic chemical executes a unique set of extraction settings, which are then carried out using distinct approaches. Each approach has limitations, notwithstanding the specificity with which it analyzes the polyphenolic chemicals. Therefore, mechanistic investigations are necessary to fully comprehend the mechanisms driving the kinetics of extraction using green extraction techniques. Centrifugal ultrafiltration might be a common technique in this context. Additionally, with some additional breakthroughs in analytical sciences, EM employing moderate working settings with enhanced retrieval are sought. The development of several purification methods and characterization instruments will lead to pledging exploration in the field of polyphenolic chemicals when these approaches are used correctly. Since each analyte type exerts different process requirements, no single method can be used to characterize and analyze polyphenolic substances on a global scale; instead, a collection of approaches is needed.

List of Abbreviations

AM	*Achillea melanocarpa*
CE	Conventional extraction
CSE	Conventional solvent extraction
FT-IR	Fourier-transform infrared spectroscopy
GC	Gas chromatography
HPLC	High-performance liquid chromatography
LLPE	Liquid-liquid phase extraction
MAE	Microwave-assisted extraction
ME	Maceration extraction
NMR	Nuclear magnetic resonance
PA	Phenolic acids

PE	Plant extracts
PHWE	Pressurized hot water extraction
PLE	Pressurized liquid extraction
PP	Polyphenols
RT	Retention time
SC	Supercritical
SE	Soxhlet extraction
SFE	Supercritical fluid extraction
SLE	Solid–liquid extraction
SPE	Solid-phase microextraction
SWE	Subcritical water extraction
Tc	Critical temperature
UAE	Ultrasound-assisted extraction

References

1 Wansink, B., Westgren, R.E., and Cheney, M.M. (2005). Hierarchy of nutritional knowledge that relates to the consumption of a functional food. *Nutrition* 21 (2): 264–268.

2 Siriwardhana, N., Kalupahana, N.S., and Moustaid-Moussa, N. (2012). Health benefits of n−3 polyunsaturated fatty acids: eicosapentaenoic acid and docosahexaenoic acid. *Advances in Food and Nutrition Research* 65: 211–222.

3 Tsao, R. (2010). Chemistry and biochemistry of dietary polyphenols. *Nutrients* 2 (12): 1231–1246.

4 Durazzo, A., Lucarini, M., Souto, E.B. et al. (2019). Polyphenols: a concise overview on the chemistry, occurrence, and human health. *Phytotherapy Research* 33 (9): 2221–2243.

5 Rajbhar, K., Dawda, H., and Mukundan, U. (2015). Polyphenols: methods of extraction. *Scientific Reviews & Chemical Communications* 5 (1): 1–6.

6 Belščak-Cvitanović, A., Durgo, K., Huđek, A. et al. (2018). Overview of polyphenols and their properties. In: *Polyphenols: Properties, Recovery, and Applications* (ed. C.M. Galanakis), 3–44. Elsevier.

7 Câmara, J.S., Albuquerque, B.R., Aguiar, J. et al. (2020). Food bioactive compounds and emerging techniques for their extraction: polyphenols as a case study. *Food* 10 (1): 37.

8 Aguiar, J., Gonçalves, J.L., Alves, V.L., and Câmara, J.S. (2020). Chemical fingerprint of free polyphenols and antioxidant activity in dietary fruits and vegetables using a non-targeted approach based on QuEChERS ultrasound-assisted extraction combined with UHPLC-PDA. *Antioxidants* 9 (4): 305.

9 Amawi, H., Ashby, C.R., and Tiwari, A.K. (2017). Cancer chemoprevention through dietary flavonoids: what's limiting? *Chinese Journal of Cancer* 36 (1): 1–13.

10 Pietta, P., Minoggio, M., and Bramati, L. (2003). Plant polyphenols: structure, occurrence and bioactivity. *Studies in Natural Products Chemistry* 28: 257–312.

11 Manach, C., Scalbert, A., Morand, C. et al. (2004). Polyphenols: food sources and bioavailability. *The American Journal of Clinical Nutrition* 79 (5): 727–747.

12 Zhang, Z., Li, X., Sang, S. et al. (2022). Polyphenols as plant-based nutraceuticals: health effects, encapsulation, nano-delivery, and application. *Food* 11 (15): 2189.

13 Scalbert, A., Johnson, I.T., and Saltmarsh, M. (2005). Polyphenols: antioxidants and beyond. *The American Journal of Clinical Nutrition* 81 (1): 215S–217S.

14 Ameer, K., Shahbaz, H.M., and Kwon, J.H. (2017). Green extraction methods for polyphenols from plant matrices and their byproducts: a review. *Comprehensive Reviews in Food Science and Food Safety* 16 (2): 295–315.

15 Ajila, C., Brar, S., Verma, M. et al. (2011). Extraction and analysis of polyphenols: recent trends. *Critical Reviews in Biotechnology* 31 (3): 227–249.

16 Koina, I.M., Sarigiannis, Y., and Hapeshi, E. (2023). Green extraction techniques for the determination of active ingredients in tea: current state, challenges, and future perspectives. *Separations* 10 (2): 121.

17 Shahbaz, H.M., Akram, K., Ahn, J.-J., and Kwon, J.-H. (2016). Worldwide status of fresh fruits irradiation and concerns about quality, safety, and consumer acceptance. *Critical Reviews in Food Science and Nutrition* 56 (11): 1790–1807.

18 Mustafa, A. and Turner, C. (2011). Pressurized liquid extraction as a green approach in food and herbal plants extraction: a review. *Analytica Chimica Acta* 703 (1): 8–18.

19 Dahmoune, F., Nayak, B., Moussi, K. et al. (2015). Optimization of microwave-assisted extraction of polyphenols from *Myrtus communis* L. leaves. *Food Chemistry* 166: 585–595.

20 Oniszczuk, A. and Podgórski, R. (2015). Influence of different extraction methods on the quantification of selected flavonoids and phenolic acids from Tilia cordata inflorescence. *Industrial Crops and Products* 76: 509–514.

21 Chemat, F., Vian, M.A., and Cravotto, G. (2012). Green extraction of natural products: concept and principles. *International Journal of Molecular Sciences* 13 (7): 8615–8627.

22 Cravotto, G., Binello, A., and Orio, L. (2011). Green extraction techniques. *Agro Food Industry Hi-Tech* 22 (6): 57–59.

23 Deng, Y., Zhao, Y., Padilla-Zakour, O., and Yang, G. (2015). Polyphenols, antioxidant and antimicrobial activities of leaf and bark extracts of *Solidago canadensis* L. *Industrial Crops and Products* 74: 803–809.

24 Jovanović, A.A., Đorđević, V.B., Zdunić, G.M. et al. (2017). Optimization of the extraction process of polyphenols from *Thymus serpyllum* L. herb using maceration, heat-and ultrasound-assisted techniques. *Separation and Purification Technology* 179: 369–380.

25 Paz, J.E.W., Márquez, D.B.M., Ávila, G.C.M. et al. (2015). Ultrasound-assisted extraction of polyphenols from native plants in the Mexican desert. *Ultrasonics Sonochemistry* 22: 474–481.

26 Vajić, U.-J., Grujić-Milanović, J., Živković, J. et al. (2015). Optimization of extraction of stinging nettle leaf phenolic compounds using response surface methodology. *Industrial Crops and Products* 74: 912–917.

27 Horžić, D., Jambrak, A.R., Belščak-Cvitanović, A. et al. (2012). Comparison of conventional and ultrasound assisted extraction techniques of yellow tea and bioactive composition of obtained extracts. *Food and Bioprocess Technology* 5: 2858–2870.

28 Herrero, M., Mendiola, J.A., Cifuentes, A., and Ibáñez, E. (2010). Supercritical fluid extraction: recent advances and applications. *Journal of Chromatography A* 1217 (16): 2495–2511.

29 Pereira, C.G. and Meireles, M.A.A. (2010). Supercritical fluid extraction of bioactive compounds: fundamentals, applications and economic perspectives. *Food and Bioprocess Technology* 3: 340–372.

30 Wang, L. and Weller, C.L. (2006). Recent advances in extraction of nutraceuticals from plants. *Trends in Food Science & Technology* 17 (6): 300–312.

31 Chupin, L., Maunu, S., Reynaud, S. et al. (2015). Microwave assisted extraction of maritime pine (*Pinus pinaster*) bark: impact of particle size and characterization. *Industrial Crops and Products* 65: 142–149.

32 Asofiei, I., Calinescu, I., Trifan, A. et al. (2016). Microwave-assisted batch extraction of polyphenols from sea buckthorn leaves. *Chemical Engineering Communications* 203 (12): 1547–1553.

33 Dandena A, Zuka L, Kostova M (2014). Natural antioxidants in black chokeberry marc extracts depending on the extraction method. (ed. Evita Straumite). 9th Baltic Conference on Food Science and Technology "Food for Consumer Well-Being", FOODBALT 2014 Conference Proceedings. Jelgava, LLU, 2014, 346 pages.

34 Milutinović, M., Radovanović, N., Ćorović, M. et al. (2015). Optimisation of microwave-assisted extraction parameters for antioxidants from waste *Achillea millefolium* dust. *Industrial Crops and Products* 77: 333–341.

35 Simić, V.M., Rajković, K.M., Stojičević, S.S. et al. (2016). Optimization of microwave-assisted extraction of total polyphenolic compounds from chokeberries by response surface methodology and artificial neural network. *Separation and Purification Technology* 160: 89–97.

36 Hossain, M.B., Brunton, N.P., Patras, A. et al. (2012). Optimization of ultrasound assisted extraction of antioxidant compounds from marjoram (*Origanum majorana* L.) using response surface methodology. *Ultrasonics Sonochemistry* 19 (3): 582–590.

37 Wijngaard, H.H., Ballay, M., and Brunton, N. (2012). The optimisation of extraction of antioxidants from potato peel by pressurised liquids. *Food Chemistry* 133 (4): 1123–1130.

38 Ebrahimi, P. and Lante, A. (2022). Environmentally friendly techniques for the recovery of polyphenols from food by-products and their impact on polyphenol oxidase: a critical review. *Applied Sciences* 12 (4): 1923.

39 Kronholm, J., Revilla-Ruiz, P., Porras, S.P. et al. (2004). Comparison of gas chromatography–mass spectrometry and capillary electrophoresis in analysis of phenolic compounds extracted from solid matrices with pressurized hot water. *Journal of Chromatography A* 1022 (1–2): 9–16.

40 Hartonen, K., Parshintsev, J., Sandberg, K. et al. (2007). Isolation of flavonoids from aspen knotwood by pressurized hot water extraction and comparison with other extraction techniques. *Talanta* 74 (1): 32–38.

41 Machado, A.P.D.F., Pasquel-Reátegui, J.L., Barbero, G.F., and Martínez, J. (2015). Pressurized liquid extraction of bioactive compounds from blackberry (*Rubus fruticosus* L.) residues: a comparison with conventional methods. *Food Research International* 77: 675–683.

42 Plaza, M. and Turner, C. (2015). Pressurized hot water extraction of bioactives. *TrAC Trends in Analytical Chemistry* 71: 39–54.

43 Vergara-Salinas, J.R., Cuevas-Valenzuela, J., and Pérez-Correa, J.R. (2015). Pressurized hot water extraction of polyphenols from plant material. *Biotechnology of Bioactive Compounds: Sources and Applications* 63–101.

44 Hawthorne, S.B., Grabanski, C.B., Martin, E., and Miller, D.J. (2000). Comparisons of Soxhlet extraction, pressurized liquid extraction, supercritical fluid extraction and subcritical water extraction for environmental solids: recovery, selectivity and effects on sample matrix. *Journal of Chromatography A* 892 (1–2): 421–433.

45 Budrat, P. and Shotipruk, A. (2009). Enhanced recovery of phenolic compounds from bitter melon (*Momordica charantia*) by subcritical water extraction. *Separation and Purification Technology* 66 (1): 125–129.

46 Cheigh, C.-I., Chung, E.-Y., and Chung, M.-S. (2012). Enhanced extraction of flavanones hesperidin and narirutin from citrus unshiu peel using subcritical water. *Journal of Food Engineering* 110 (3): 472–477.

47 Naczk, M. and Shahidi, F. (2004). Extraction and analysis of phenolics in food. *Journal of Chromatography A* 1054 (1–2): 95–111.

48 Bi, K. and Kubáň V. (2000). High performance liquid chromatographic determination of phenolic compounds in seed exudates of *Festuca arundinacea* and *F. pratense*. *Phytochemical Analysis: An International Journal of Plant Chemical and Biochemical Techniques* 11 (6): 375–379.

49 Potter, D.W. and Pawliszyn, J. (1994). Rapid determination of polyaromatic hydrocarbons and polychlorinated biphenyls in water using solid-phase microextraction and GC/MS. *Environmental Science & Technology* 28 (2): 298–305.

50 Möder, M., Schrader, S., Franck, U., and Popp, P. (1997). Determination of phenolic compounds in waste water by solid-phase micro extraction. *Fresenius' Journal of Analytical Chemistry* 357: 326–332.

51 Vinas, P., Campillo, N., Martínez-Castillo, N., and Hernández-Córdoba, M. (2009). Solid-phase microextraction on-fiber derivatization for the analysis of some polyphenols in wine and grapes using gas chromatography–mass spectrometry. *Journal of Chromatography A* 1216 (9): 1279–1284.

52 Degenhardt, A., Hofmann, S., Knapp, H., and Winterhalter, P. (2000). Preparative isolation of anthocyanins by high-speed countercurrent chromatography and application of the color activity concept to red wine. *Journal of Agricultural and Food Chemistry* 48 (12): 5812–5818.

53 Dzah, C.S., Duan, Y., Zhang, H. et al. (2020). Latest developments in polyphenol recovery and purification from plant by-products: a review. *Trends in Food Science & Technology* 99: 375–388.

54 Everette, J.D., Bryant, Q.M., Green, A.M. et al. (2010). Thorough study of reactivity of various compound classes toward the Folin–Ciocalteu reagent. *Journal of Agricultural and Food Chemistry* 58 (14): 8139–8144.

55 Lončarić, A., Celeiro, M., Jozinović, A. et al. (2020). Green extraction methods for extraction of polyphenolic compounds from blueberry pomace. *Food* 9 (11): 1521.

56 Cui, L., Lu, H., and Lee, Y.H. (2018). Challenges and emergent solutions for LC-MS/MS based untargeted metabolomics in diseases. *Mass Spectrometry Reviews* 37 (6): 772–792.

57 Kamangerpour, A., Ashraf-Khorassani, M., Taylor, L. et al. (2002). Supercritical fluid chromatography of polyphenolic compounds in grape seed extract. *Chromatographia* 55: 417–421.

58 Fayad, P.B., Prevost, M., and Sauve, S. (2010). Laser diode thermal desorption/atmospheric pressure chemical ionization tandem mass spectrometry analysis of selected steroid hormones in wastewater: method optimization and application. *Analytical Chemistry* 82 (2): 639–645.

59 Forino, M., Tartaglione, L., Dell, Aversano, C., and Ciminiello, P. (2016). NMR-based identification of the phenolic profile of fruits of Lycium barbarum (goji berries). Isolation and structural determination of a novel N-feruloyl tyramine dimer as the most abundant antioxidant polyphenol of goji berries. *Food Chemistry* 194: 1254–1259.

60 Said, I.H., Truex, J.D., Heidorn, C. et al. (2020). LC-MS/MS based molecular networking approach for the identification of cocoa phenolic metabolites in human urine. *Food Research International* 132: 109119.

4

Fermentation and Degradation of Polyphenols

Chandan K. Mandal[1], Palraj Kasinathan[2], Srikanta Sahu[3], Akshima[1], Mansi Yadav[1], Garima Sachdeva[1], Sanju Sharma[1], Manoj K. Banjare[4], and Kamalakanta Behera[2]

[1] *Department of Applied Chemistry (CBFS–ASAS), Amity University, Gurugram, Manesar, Haryana, India*
[2] *Department of Chemistry, Faculty of Science, University of Allahabad, Prayagraj, Uttar Pradesh, India*
[3] *Department of Chemistry, School of Applied Sciences, Centurian University, Jatni, Odisha, India*
[4] *MATS School of Sciences, MATS University, Raipur, Chhattisgarh, India*

4.1 Introduction

Since ancient times, humans have consumed fermented foods as a main diet component. Various items, including vegetables and fruits, have been fermented historically because it was a successful method of food preservation [1]. Nutritionists, biologists, and consumers all have a restored in consideration and creating unusual foods that ferment; meanwhile, the abundance of distinctive bioactive compounds in these meals might be responsible for their better health-cooperative properties [2]. Foods that have undergone fermentation often exhibit several distinctive physicochemical and biological changes, including extended shelf-life, improved organoleptic qualities, and higher bioavailability of minerals and vitamins. Additionally, investigations in both clinical and preclinical settings have shown that fermented foods help treat long-term human diseases like cancer [3, 4], gut microbial dysbiosis [5], obesity [6], inflammatory aggravation [7], as well as immunosenescence [8].

In the process of fermentation, bacteria break down the food's nutrients into simpler, frequently distinctively different components. For instance, bioactive peptides have several health benefits and are frequently created because of protein degradation by microbes through generated [9]. Exopolysaccharides, which contribute to the flavor and pharmacological effects of fermented foods, are also

Science and Engineering of Polyphenols: Fundamentals and Industrial Scale Applications,
First Edition. Edited by Chandrabhan Verma.
© 2024 John Wiley & Sons Inc. Published 2024 by John Wiley & Sons Inc.

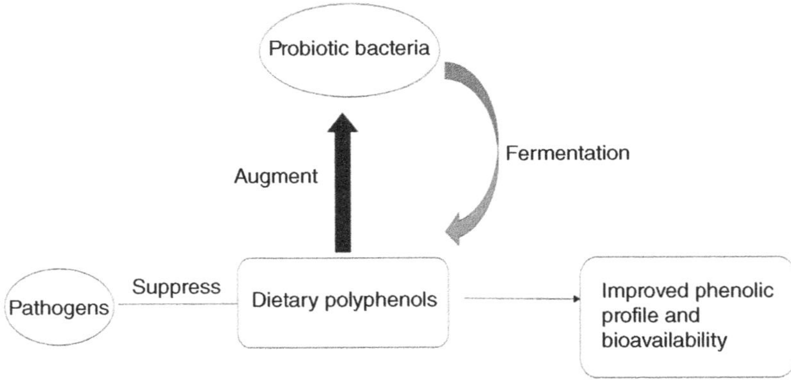

Figure 4.1 Innovative functional diets.

created by microbes during the fermentation of carbohydrates (Figure 4.1) [10]. The fact that main nutrients are not the only substrates for microbial fermentation in food must be taken into consideration. There is mounting evidence that suggests probiotic bacteria may have a deeper connection to secondary plant compounds like polyphenols [11]. The purpose of this chapter is to inform the audience of recent advances in the fermentation of foods high in polyphenols by probiotics. We first give a general introduction of the fermentation of macronutrients before reviewing the current research on foods containing polyphenols. After that, with a focus on the development of innovative functional food products, the relationship between probiotics and polyphenols bacteria, and the effectiveness and application of fermented polyphenolic meals, are examined.

The term "phenolic compound" was used to designate hundreds of compounds that can be found in edible plants and have a minimum of one hydroxyl group in the benzene ring. These chemicals are of different classes depending on the number of phenolic rings and the elements of structure that keep those rings together. There are distinctions among flavonoids, phenolic acids, and lignans (which are benzoic or hydroxycinnamic acid derivatives). Catechins and proanthocyanidins, flavanols, flavones, isoflavones, flavanones, and flavanols are some more subgroups of flavonoids. Other organic acids and carbohydrates may be associated with polyphenols in addition to this kind [12].

4.2 Plant-Based Fermented Food Products with Lactic Acid Bacteria

Due to their large antioxidant, vitamin, dietary fiber, and mineral content, vegetables should be a staple of any food. Most fruits and veggies eaten in the diet of humans are fresh, seldom analyzed, pasteurized, and often cooked in a water bath as well as the microwave. Due to rapid bacterial destruction, fresh vegetables have a relatively short shelf life. Additionally, cooking techniques would adversely impact the vegetables' physical characteristics and chemical composition. One of the various technical options for preserving and/or improving the safety, nutritional, sensory, and durability properties of vegetables is fermentation by lactic acid bacteria (LAB). Fermentation and the quality of the final food products are significantly influenced by the microbiota's composition and development. LAB make up a relatively minor portion of the native microbiota found in raw vegetables. The intrinsic qualities of the raw materials must be altered by LAB to obtain the desired features in fermented vegetable food items. Several autochthonous and contaminating microorganisms fight with one another to produce spontaneous fermentations. Microorganisms that have acclimated to the conditions during fermentation will eventually take control. This spontaneous process takes a while to begin, and there is a significant chance that it will fail. When fermentation processes fail, it can cause food to spoil and/or allow infections to survive, posing unforeseen health hazards. Hence, a product of consistent quality and quick acidity, which would prevent the formation of degradation and hazardous bacteria, are two benefits of employing starter cultures from a hygiene and safety standpoint. Even while a significant amount of LAB starters is regularly utilized in the fermentation of meat, dairy, and baked goods, only a limited number of cultured ones are used to ferment vegetables. *Lactobacillus plantarum* is the commercial starter that is often used in the fermentation process of cabbages, olives, and cucumbers [13, 14].

The chemical, physical, and microbiological changes brought about by the fermentation of cabbage or sauerkraut impact the products' safety and quality. An initially heterofermentative stage is followed by a homofermentative phase, which is a generic description of this fermentation's phases. The bacteria that are present in the fermented cabbage fermentations have historically been recognized as *Lactobacillus brevis, Leuconostoc mesenteroides, L. plantarum,* and *Pediococcus pentosaceus*; however, recent research has revealed that the varieties of LAB available are more diverse than previously observed and include, among others, *Leuconostoc citreum* and *Lactobacillus paraplantarum* varieties (Table 4.1) [15]. *L. plantarum, Lactobacillus pentosus, L. brevis, Leuconostoc* spp., and other similar LAB species have been reported to mostly occur in spontaneous cucumber fermentation [16]. The bacterium Le and *Lactobacillus* predominated in the earliest phases of the fermentation process, followed by *Lactobacillus* and *Pediococcus*, and then *Pediococcus* developed into the controlling genus in the latter stages of fermentation, according to other reports [17]. The spontaneous growth of LAB is required for the fermentation of capers and caper berries, whereas isolated straining of *L. paraplantarum, L. pentosus, pediococci,* and *Lactobacillus fermentum* as well as *L. brevis* correspondingly have been found [17].

Table 4.1 Isolated from certain vegetable fermentation was the major LAB species.

Herbal source	Laboratory varieties	Reference
Cabbage	*Leuconostoc mesenteroides*	[15]
	Leuconostoc citreum	
	Lactobacillus paraplantarum	
	Lactobacillus brevis	
	Pediococcus pentosaceus	
Cucumbers	*Lactobacillus pentosus*	[16]
	Leuconostoc sp.	
	Lactobacillus plantarum	
	Pediococcus sp.	
	L. brevis	
Caper berries	*L. brevis*	[17]
	L. paraplantarum	
	Lactobacillus fermentum	
	L. pentosus	
	L. plantarum	
Olives	*P. pentosaceus*	[18–20]
	L. pentosus	[21]
	Leuconostoc pseudomesenteroides	[13]
	Leuconostoc mesenteroides	[22]
	L. plantarum	[14, 23]

4.3 Microbial Fermentation of Food Macronutrients

Based on different types of fermenter microorganisms, fermentation can be roughly divided into two types: starter culture-dependent fermentation and spontaneous fermentation. When microbes are organically present in uncooked food or in the environment during processing, as in kimchi or dahi, spontaneous fermentation takes place [24]. As starter cultures are frequently linked to distinctive sensory, nutritional, and health-beneficial alterations, they are extensively used in the food industry [25]. Despite the source, microbial fermentation significantly alters the chemical nature of the fermented foodstuff product due to the food's numerous macronutrients being broken down during the process. To perform this, microorganisms have various distinctive enzymes that the host may lack, including ureases, peptidases, proteases, polysaccharide degrading enzymes, amylases, esterase's, phenol oxidases, and lipases [26]. Probiotic bacteria use the various carbohydrates found in raw dietary material, including polysaccharides, disaccharides, and monosaccharides as substrate.

Various probiotic bacterial species can metabolize various types of carbohydrates, making them capable for a wide range of food-related applications [27]. Probiotic bacteria primarily need carbohydrates to satisfy their energy requirements, and they then convert them into organic acids like short-chain fatty acids (SCFAs). Oligosaccharides, which are nondigestible like pectins and hemicellulose behave as prebiotics for the growth of probiotic bacteria [28, 29]. The decomposition of proteins, in addition to the amount of carbohydrates present, has an important role in describing the nutritional content and quality of fermented foods, particularly dairy products. Large proteins are broken down into tiny pieces during fermentation, frequently producing bioactive peptides that are found to have anti-thrombogenic, antioxidant, anti-hypertensive, and anti-inflammatory effects in various in vitro and preclinical investigations [30–32].

Generally, probiotic microorganisms enhance the overall digestibility and nutritional content of proteins in the diet by dissolving complicated storage proteins into more soluble, absorbable forms [33]. Moreover, probiotic bacteria could metabolize amino acids through procedures like deamination and decarboxylation, which help control the synthesis of biogenic amines in meals that are otherwise thought to be harmful [34]. Lipids play a significant role in determining the bioactivity and favor of fermented foods. According to studies, the composition of both short-chain and long-chain fatty acids, including butyric acid, palmitic acid, oleic acid, and conjugated linoleic acid, can be impacted by fermentation. This is mostly related to certain probiotic bacteria's lipolytic and proteolytic enzyme activity [35]. As an example, the fermentation of milk with the probiotic bacteria *Lactobacillus casei* and *Lactobacillus acidophilus* produced higher amounts of butyric and linoleic acids [36].

4.4 Metabolism of Food Phenolics by Lactic Acid Bacteria

4.4.1 Tannase

The ester linkages in the hydrolysable tannins and gallic acid ester are hydrolyzed by tannase, also generally recognized as tannin acyl hydrolase (EC 3.1.1.20). The creation of instant tea or acorn liquor as well as the manufacturing of gallic acid, which is used in the food sector as a substratum for the making of propylgallate, a powerful antioxidant, are currently the main commercial uses of tannase. Moreover, tannase is employed as a clarifying agent in various fruit juices, wines, and refreshing beverages with coffee flavors, as well as during the production of beer, to prevent the unfavorable turbidity caused by the protein-tannin synthesis [37–39]. Tannase may be used to clean food waste that includes high concentrations of potentially harmful contaminants called polyphenols and tannins, mostly tannins. The elimination of these chemicals using tannase on these residues is inexpensive and saves a lot of money.

Considering that this enzymatic capability is frequently linked to fermentations of plant materials, it has been hypothesized that it has an ecological advantage for this species. *L. plantarum* tannase was utilized in all of these investigations to hydrolyze methyl gallate, a straightforward galloylester of methanol. Gallic acid made from methyl gallate was then subjected to oxidation to give a color that was sufficient for visual or spectrometer detection [40]. It has been observed that the presence of Hg^{2+} and Mg^{2+} ions partially inactivate tannase. Although, the activity of *L. plantarum* tannase was unaffected by metal ions such as K^+, Ca^{2+}, and Zn^{2+}. Certain chelators (e.g. EDTA), surfactants (e.g. Tween 80), denaturing agents (e.g. urea), and inhibitors (DMSO) have no effect on the activity of *L. plantarum* tannase at the examined concentration [41]. Concentrated cell-free extracts were separated by nondenaturing SDS-PAGE and their enzyme activity was measured using a zymogram to identify *L. plantarum* tannase. Tannase activity was contained to a single band [41].

4.4.2 Phenolic Acid Decarboxylase (PAD)

When the microorganisms grow on plants, PAD activity may give them a selection advantage. The stress response brought on by phenolic acid may be represented by PAD expression. The *L. plantarum* phenolic acid decarboxylase (PAD) and its transcriptional repressor (PadR) have been successfully introduced into *Escherichia coli* (*E. coli*) and thoroughly examined at the molecular level [42, 43]. When ferulic, p-coumaric, or caffeic acid is introduced to an *L. plantarum* culture that is still growing, PadR is inactivated by a mechanism that is still being fully understood but may require a particular mediator. The stress brought on by the phenolic acid is eliminated by the subsequent creation of the PAD enzyme, which quickly breaks down the acid. When the padA gene is deleted in *L. plantarum*, *p*-coumaric acid concentrations beyond 3 mM, particularly at low pH, restrict growth [44].

4.4.3 Other Lactic Acid Bacteria Species

According to Whiting and Carr (1957), chlorogenic acid frequently vanished during the fermentation of cider. Studies using *Lactobacillus paracollinoides* (previously known as *L. pastorianus* var. *quinicus*) cell extracts showed that the hydrolysis of chlorogenic acid to caffeic acid and quinic acids was the first step in the metabolism of the compound. The products were further metabolized. Subsequent studies demonstrated that ethyl catechol and dihydrocaffeic acid are produced during the metabolism of caffeic acid [45]. *L. paracollinoides* also had the ability to decrease the side chains of 3, 4, and 3 hydroxycinnamic acids. Once their byproducts have been decarboxylated, they may yield ethyl catechol and ethyl phenol.

Gallic acid and catechin have been described as degraded by *Lactobacillus hilgardii* [45]. Gallic acid or catechin was added to a complex medium in which *L. hilgardii* was cultivated. Pyrogallol, gallic acid, protocatechuic acid, *p*-hydroxybenzoic acid, catechol, *p*-hydroxybenzldehyde, and *p*-hydroxybenzyl alcohol were found in acid gallic-grown cultures. Gallic acid, homovanillic acid, pyrogallol, acetovanillone, *p*-hydroxybenzoic acid, and catechol were all found in cultures that had grown catechin. According to the scientist LoCascio et al., *P. pentosaceus* cultures remove 21.3% of the quercetin that is found in culture media; although, it is still unclear how the quercetin is lost (absorbed or transferred into the cell). Moreover, it has been reported that *Oenococcus oeni* strains metabolize phenolics such as anthocyanins and other phenolics to produce significant wine aroma components [46–51]. Bloem et al. [52] investigated the generation of vanillin by wine-associated LAB from simple phenols. They discovered that neither eugenol nor isoeugenol nor vanillic acid could be converted into vanillin by LAB. Though with a small yield, *O. oeni* or *Lactobacillus* sp. could produce vanillin from ferulic acid. Vanillin was additionally converted by LAB to the appropriate vanillin alcohol [52], Cavin et al. [53] demonstrated that *L. brevis*, *L. plantarum*, and *P. pentosaceus* decarboxylated *p*-coumaric and ferulic acids to produce volatile phenols [54].

4.5 Polyphenols Differentially Modulate Bacterial Growth

Polyphenols are historically valued for their antioxidant and antimicrobial properties; they are also thought to be helpful in the organic preservation of food [55]. A previous study showed that the existence of polyphenol-rich traces of various medicinal plants can specifically suppress the growth of pathogens like *Listeria monocytogenes*, *Enterococcus faecalis*, *Escherichia coli*, *Pseudomonas aeruginosa*, *Candida albicans*, and *Staphylococcus aureus* in vitro while directly stimulating the growth of probiotic *L. plantarum* and yeast Sacchar [56]. Like this, several bioactive polyphenols found in fruits and vegetables, including catechin, ferulic, gallic, protocatechuic, and vanillic acid, have been shown to promote probiotic development and inhibit pathogens [57]. A mouse in vivo study utilizing polyphenols from *Pueraria lobata* root showed a rise in the number of helpful microbes in the gut, including Lactobacillaceae and Bacteroidetes, while simultaneously suppressing the affluence of Ruminococcaceae, Prevotellaceae, and Burkholderiaceae [58]. Additionally, we discovered that the green tea catechin epigallocatechin gallate (EGCG) differentially affects the growth of microbes in vitro, strongly suppressing the growth of pathogenic microbes (*Escherichia coli*, *Salmonella typhi*, *Staphylococcus aureus*, *Shigella fexneri*, and *Vibrio cholerae*) while having a lesser to no impact on the growth of probiotic LAB (*Lactobacillus rhamnosus*, *Lactobacillus plantarum* and *L. fermentum*) [59]. In short, the use of polyphenols is likely to alter the composition of the entire microbial community by preventing potentially dangerous species, allowing probiotic bacteria to flourish, and maybe maintaining the eubiosis of the gut microbiota. Understanding this polyphenolic feature is thought to have significant implications for the food business. In the latter scenario, polyphenols appear to behave as prebiotics [60].

It is logical to conclude that polyphenols hold a unique place among various bioactive compounds used for probiotic fermentation because they protect the probiotic bacteria from pathogens, permitting their viability and proliferation, and they may also lead to excellent functional food formulations.

4.6 Functional Food Products Based on Polyphenolic fermentation by Probiotics

The probiotic bacteria ability lies not only in surviving in polyphenol-rich settings but also in generating fermentation in a range of foods, resulting in the biotransformation of the naturally occurring polyphenolic content. This has created opportunities for the creation of various innovative functional foods. For this reason, we discovered a collegial mixture of probiotic *L. fermentum* and green tea EGCG that, when consumed, resulted in increased antioxidant and anti-immunosenescence activities in aged mice compared to solo treatments with EGCG or probiotic bacteria [59]. Similar findings were made in another study, which found that a novel symbiotic formulation containing Triphala's polyphenols and three different probiotic bacterial strains (*L. plantarum* NCIMB 8826, *L. fermentum* NCIMB 5221, and *Bifidobacterium longum* subsp. *infantis* NCIMB 702255) was better than individual treatments at reducing multiple markers of diet-induced diabetes and obesity in *Drosophila melanoga* [61]. A collegial relationship between *Cudrania tricuspidata* leaf extract and the probiotic *Lactobacillus gasseri* has been reported. This mixture, when added to milk, caused the phenolic acids to be broken down, resulting in the production of a new metabolite known as bioactive peptides and 3,4-dihydroxyhydrocinnamic acid, which enhanced the in vivo efficacy and free radical scavenging capabilities of fermented milk [62–64]. These data give proof of the collegial properties of polyphenol with certain probiotic bacteria, which are to be utilized for producing innovative functional foods. Like this, many studies claim that by fermenting mixtures of medicinal plants or polyphenol-rich foods with probiotic bacteria, new and enhanced functional food products can be created. For instance, green tea made from *Houttuynia cordata* leaves fermented by *Lactobacillus paracasei* subsp. paracasei produced larger quantities of chlorogenic acid, EGC, and EGCG than unfermented tea leaves, which also had better in vitro cytoprotective effects on fat cells [65]. It was recently discovered that fermenting Jerusalem artichokes, apples, cucumber, pumpkin, pineapple, and spinach, with probiotic LAB enhanced the survivability of the LAB and produced a tolerable sensory response [66].

4.7 Fermentation of Whole Grain Foods

The conversion of food crops into edible forms depends on food processing. An ancient method of food processing, fermentation has been used for millennia, particularly in developing countries. It entails the purposeful conversion or modification of a substrate by microbe actions to produce a desired result. Usually, microbiological processes are used to do this, which improve the foods' look, flavor, functionality, nutritional content, color, and texture. Direct microbial action during the fermentation process as well as the creation of metabolites and other complex chemicals, result in positive effects [67–69]. The three conventional approaches to fermentation are (i) naturally occurring, (ii) back slopping, and (iii) controlled fermentation. Because of the preservation effect of the fermentation process, the resultant products are shelf stable, have improved bioavailability and palatability, as well as favorable organoleptic features that affect scent, texture, and flavor and enhance the nutritional value of food [70, 71]. Some fermented food products from whole grain (WL) are presented in Table 4.2 and Figure 4.2.

Table 4.2 Some fermented food products from whole grain (WG).

Whole grain's	Food	Categories of fermentation	Reference
Sorghum	Fermented balls	SSF	[72]
Sorghum	Ogi	SmF	[73]
Sorghum	Injera	SSF	[74]
Sorghum	Fermented porridge	SSF	[75]
Sorghum	Uji	SmF	[74]
Sorghum	Ting	SSF	[76–78]
Wheat	Sour-dough bread	SSF	[79]
Wheat	Bread	SSF	[80, 81]
Wheat	Tempe	SSF	[82, 83]
Quinoa, wheat	Fermented product	SSF	[84]
Rye, wheat	Sour-dough bread	SSF	[85]
Rye, wheat, oat	Bread	SSF	[86]
Millet	Fermented drink	SmF	[87]
Millet	Sour-dough bread	SSF	[88]
Millet	Probiotic drink	SmF	[89]
Maize	Akamu/Ogi	SSF	[90, 91]
Oat, wheat	Bread	SSF	[92]
Quinoa	yogurt	SmF	[93]

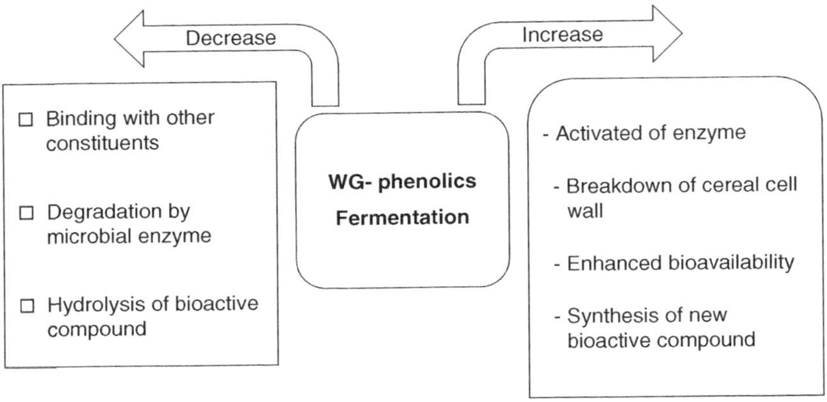

Figure 4.2 Different ways of fermentation of whole grain (WG) phenolic compounds.

Lactic acid fermentation (LAF) is an important option that can preserve nutritional and sensory features and improve the shelf-life of fruits and vegetables. LAF can occur instinctively with anaerobic conditions, right temperature and salt concentration, proper moisture and water activity, that are favorable for autochthonous lactic acid bacteria growth [94]. Due to the intrinsic chemical and physical parameters of some raw vegetables and fruits, they are highly susceptible to microorganisms. Phenolic compounds are secondary plant metabolites found throughout the plant kingdom [95]. These nutrition constituents have fascinated substantial curiosity in the food industry because of their extensive fitness paybacks. Because differing fruit dimensions and compositions can impact the microbial dynamics, causing fermentation and influencing consumer perceptions of the product, grower types also have a significant impact on the development of the microbiota [96].

There are significant physical modifications happening during the microbial fermentation phase. Organic acids, which are necessary for the stability and continuation of the fermenting process, are provided by fermentable sugars, which serve as the primary source of carbon for microorganisms. Lactic acid bacteria-based fermentation is one of the economical and eco-friendly approaches [97]. During fermentation, microorganism induces a chemical change in the organic substrate, which in turn aid in the formation of bioactive compounds, either by hydrolysis of large polymer to simple molecules or transformation of substrate [98]. Lactic acid fermentation, which is efficient, adds value to the food, reduces waste and environmental pollution, and ultimately contributes to the positive impact on the economy. The phenolic compounds were aerobically biodegraded via 4-hydroxy-3-methylbenzoic acid and 4-hydroxybenzoic acid, which were retained by electro-migration in the anode chamber. Oxygen, iron(III), nitrate, and sulfate at the electrode were electron acceptors for biodegradation [97]. Electricity is produced by the action of the electro-active bacteria devoted to the anode, utilizing acetate as an electron donor. During the electricity generation alongside iron reduction at the anode takes pace to give thermodynamically favorable iron(III).

Since the 1990s, phenolic compounds in food have gained considerable interest due to the increasing proof of their favorable impact on human health. A diet rich in plant-based items has been reported for decreased risk of certain diseases, including cardiovascular diseases and cancer [99]. Because of their numerous health advantages as antioxidants and for reducing long-term inflammation, malignancy, heart problems, and diabetes100, phenolic compounds from plant food products may play a substantial part in the positive effect [100]. The next-most significant class of phytochemicals comprise phenolic acids, most of the dietary polyphenols, present in fruit in a bound form (Table 4.3) [101]. The hydroxybenzoic and hydroxycinnamic compounds are two readily distinct constitutive carbon frameworks found in these compounds. The only variation in these frameworks is brought about by the quantity and placement of the hydroxyl groups on the rings that are aromatic. These compounds, hydroxybenzoic and hydroxycinnamic acid, exhibit an acidic property because they contain one carboxylic group. Most hydroxycinnamic acid compounds are derivatives with a (C6–C1) skeleton [101].

The perfect structural makeup for the activity of scavenging free radicals is found in phenolic compounds, giving them antioxidant properties exhibiting a wide range of beneficial health effects. The health benefits of these substances rely on their bioavailability; the majority of phytochemicals in veggies and fruits are in natural or conjugated forms [102]. Phenolic in the insoluble form is covalently bound to cell

Table 4.3 Principal groups of global fermented food product.

Food origin	Food source	Fermented product
Plants	Cereals and grains	Beer, bread, rice wine, malt, and grain whiskey, tamari, miso, idli, dosa, vodka, vinegar
	Honey	Mead, metheglin
	Vegetables	Asinan, kimchi, sauerkraut
	Fruits	Atchara, brandy, cider, natadepina, vinegar, wine
	Tea	Kombucha, miang
Animal	Meat	Beef sticks, Icelandic slatur, pepperoni, sausages, salami
	Fish	Bagoong, jeotgal, garum, shiokara

Table 4.4 The enzymatic activities of lactic acid bacteria.

Species	Enzymatic activities
Lactobacillus acidophilus	Amylase, lactate dehydrogenase, peptidase, proteinase
Lactobacillus casei	Amylase, lactate dehydrogenase, peptidase, proteinase
Lactobacillus fermentum	Amylase, lactate dehydrogenase, peptidase, proteinase
Lactobacillus lactis	Amylase, aminotransferase, decarboxylase, esterase, lactate, proteinase
Lactobacillus rhamnosus	Amylase, cellulose, esterase, lactate dehydrogenase, peptidase, proteinase

wall structural components such as cellulose hemicellulose [103]. This inexpensive technique using fermented food for efficient preservation and increased availability of nutrients makes this production technique an important aspect of the human diet globally [104]. Due to the nutritious and medicinal importance of fermented foods, region-specific origin predominates; thus, ongoing efforts are made to advance technology to enable their mass production.

The use of fermented foods, once restricted to foods like bread and wine, has now spread to a variety of other substrates, including vegetables, fish, meat, milk, and cereals [105]. The type of substrate or fermenting will determine the type of food that has been fermented. Prebiotic oligosaccharides, bioactive proteins, and metabolites from microbes are among the advantageous end products produced during fermentation. For preserving the security and longevity features of food, fermenting may be thought of as a straightforward and beneficial biotechnological procedure [106]. The fermentative microbiota also changes some phenolic foods when the food is fermenting. For those dietary components to be absorbed and to control their biological action, this transformation is frequently necessary. The modification of phenolic compounds by certain fermentative microbiota components via glucosidase and decarboxylase activities determines their action in a significant way [107]. Table 4.4 represents the enzymatic activities of lactic acid bacteria.

4.8 Degradation of Polyphenol

The degradation of aromatic compounds by microbes is a significant part of the planets carbon cycle. Pyrolysis of organic material from anthropogenic or plant and animal-based sources generates a variety of aromatic compounds that are relatively recalcitrant to natural degradation. If these aromatic products are not degraded, the carbon contents become locked and are unavailable for the biosphere. Phloroglucinol and associated polyphenolics have a benzene ring that is highly inert to degradation because of its resonance structure stability. It shows very less reactivity because of its meta-substituted hydroxyl groups.

4.8.1 Aerobic Degradation

The biodegradation of polyphenols could be achieved by aerobic microorganisms (Figures 4.3–4.5) [108–112]. Isolated microorganism naturally occurring in soil and feces uses PG as the carbon source. At first, this PG compound is transformed into dihydrophloroglucinol, which instantly dehydrates, resulting in a resorcinol molecule. Resorcinol is consequently hydroxylated into 1,2,4-trihydroxybenzene, undergoing ortho and meta cleavage by enzymes to maleylacetate and 2,4,6-trioxohexanoate. Orcinol is hydroxylated to 1,2,4-trihydroxytoluene, which undergometa-cleavage to produce 2,4,6-trihydroxyheptanoate and acetylpyruvate, with product acetate and pyruvate produced [112].

4.8.2 Thermal Degradation

The behavior of polyphenols at high temperatures could be explained basically by three proposed mechanisms. At higher temperatures, the bonds between lignin and phenolic acids are broken. As a result, the phenolic compounds, which are insoluble, could be liberated. It has been shown that the quality of bound phenol acids is almost double that of free phenolic compound [113]. Second, the bonds in

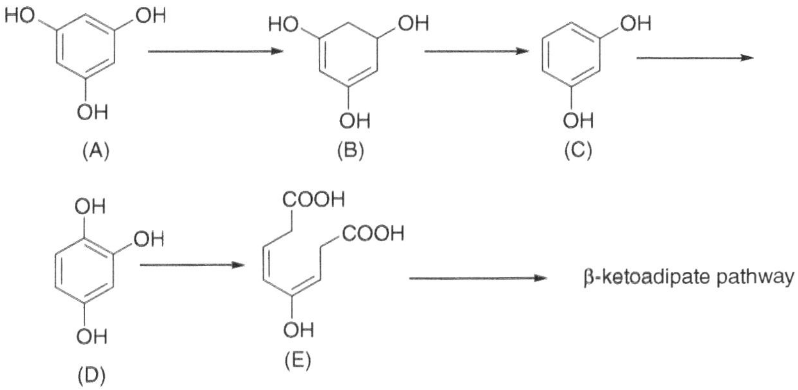

Figure 4.3 Aerobic degradation of phloroglucinol by *Pseudomonas* sp. MAC. 451 and *Penicillium* sp. MAC. M-47. The pathway includes: (A), phloroglucinol; (B), dihydrophloroglucinol; (C), resorcinol; (D), 1,2,4-trihydroxybenzene; (E),3-hydroxymuconic acid; fl-ketoadipate pathway. Source: [108]/S.M. Armstrong and T.R. Patel.

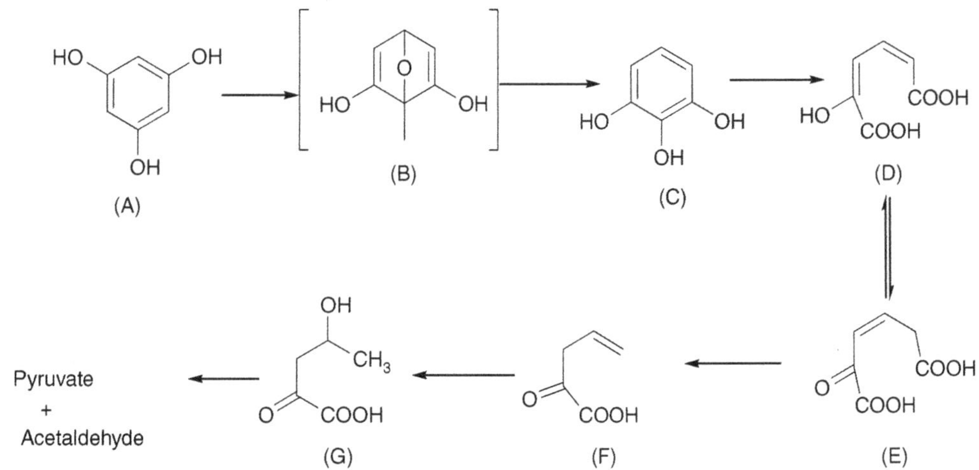

Figure 4.4 Phloroglucinol catabolism proposed pathway by busariumsolmi. (A) phloroglucinol; (B) resorcinol epoxide; (C) pyrogallol; (D) 2-hydroxymuconatc; (E) oxalocrotonate; (F) vinylpyruvate; (G) 4- hydroxy-2-oxovalcrate. Source: [109]/Springer Nature.

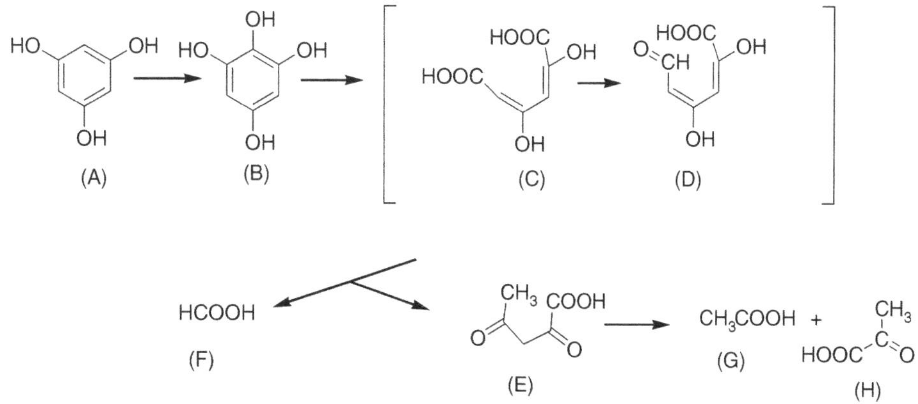

Figure 4.5 Proposed pathway for phloroct a, M glucinol catabolism by *Rhodococcus* sp. BPG-8. (A) phloroglucinol; (B) 1,2,3,5-tetrahydroxybenzene; (C) 2,4-dihyroxy-2,4-hexadiene-1,6-dioic acid; (D) 2,4-dihydroxy-6-0~0- acetopyruvate; (F) formic acid; (G) acetic acid. Source: [110]/S.M. Armstrong and T.R. Patel.

lignin may be activated and broken at elevated temperatures, causing phenolic acid yield to increase. Thermal degradation is the prevailing reason to justify the decrease in polyphenol yield when extracted at higher temperatures. But, thermal degradation is not the only reason to justify the decrease in yield of polyphenol. Larrauri et al. [114] noted a drastic decline in total phenolic content when the plant material was dried at temperature above 100 °C, leading to a decline in antioxidant activity and in total phenolic content (TPC) which suggests that thermal degradation of the polyphenyl might have taken place. In another study, the same authors investigated the outcome of drying temperatures between 20 and 120 °C where they suggested that the oxidation of polyphenol at high temperature might be the reason for the loss in antioxidant property [115].

Roses et al. [115] studied the extraction of polyphenol with ethanol at various temperatures from grape seed flour. In the study they noticed that the TPC yield dropped when the temperature was beyond 180 °C, which suggests that the total flavonoids are sensitive to temperatures. The study also showed that the TPC yield increased at 150 °C, which might be due to the liberation of bonded phenolic compound or the breakdown of lignin itself. At higher temperatures, beyond 180 °C, Larrauri et al. [112] also suggested that the main mechanism behind the reduction in phenolic yield is thermal degradation. It is observed that catechin drops at above 150 °C but gallo catechin and gallic acid increases massively with time and temperature [115]. Other studies have also reported that flavonoids are more heat sensitive and the degradation of flavones and flavanols are observed at lower temperatures [124]. Anthocyanins are one of the most abundant flavonoids constituents found in fruits and vegetables [125–126]. The thermal effect was studied between 20–60 °C and an increase in temperature causes an increase in anthocyanins extraction. A drastic decrease in anthocyanins extraction was observed at 45 °C. Beyond that temperature, the anthocyanin content in the TPC dropped from 70% to 54%. The initial rise in yield of anthocyanins with extraction temperature could be due to its higher solubility in ethanol at higher temperature. While, TPC yield increase with temperature and showed the trend opposite to anthocynin [127]. The reverse trend of anthocyanins compared to TPC could be due to high sensitivity of anthocyanins to higher temperature. Hence, the impact of temperature on both anthocyanins and TPC is significant. They also found that the optimal extraction temperature for anthrocyanins occurred at lower temperature compared with TPC. Volden et al. [123] found that blanching and boiling red cabbage increased the total polyphenol while the anthocyanin content decreased by 59% and 41% for the blanched and boiled red cabbage, the anthocyanin was degraded by heat. Another study showed that the anthocyanin content of elderberry, black carrot and strawberry reduced with heating at 95 °C for 1h. After 3 hr of heating, the anthocyanin content in elderberry decreased by 50% [98]. Table 4.5 provides a summary of the findings of the above research on Thermal Degradation.

Table 4.5 Impact on individual phenolic compound.

Source	Extraction	Temperature tested	Effect on polyphenol	Reference
Red grape pomace peel	TSE	Dried at 60, 100, and 140 °C	TPC at 100 °C	[113]
Grape seed flour	TSE	Heated at 120, 150180, 210, and 240 °C	TPC AND TFC above 120–180 °C	[116]
Black rice	TSE	Dried at 20, 40, 60, 80, and 100 °C	TFC AND TPC above 40–80 °C	[117]
Spinach	TSE	Extraction between 50 and 190 °C	No decrease in TPC	[118]
Black currants	TSE	Extraction between 20 and 60 °C	Anthocyanin above 45 °C	[119]
Hemp, Flax, canola seed cake	TSE	Extraction at 40, 50, 60, 70 °C	TFC above 70 °C in flax and canola seed cake	[120]
Peach	TSE	Extraction between 25 and 70 °C	TFC remains same between 25 and 70 °C	[121]
Mango peel and seed	TSE	Extraction at 25, 50, and 75 °C	TFC at 50 and 75 °C	[122]
Red grape skin	TSE	Extraction between 20 and 140 °C	Anthocyanins above 100 C TPC above 120 °C	[123]
Elderberry, Strawberry and black carrot	TSE	Heated at 95 °C	Anthocyanins	[124]
Red cabbage	TSE	Blanched at 94–96 °C	TPC at 94–96 °C	[125]

4.9 Conclusions and Future Perspectives

Plant foods that are abundant in polyphenols offer numerous health advantages. When these polyphenol-rich foods undergo microbial fermentation with the help of specific carbohydrates and active biocatalysts, it leads to an increase in polyphenol levels during the fermentation process and facilitates their recovery. Application of bioengineered microorganisms for the production of polyphenols is currently of immense importance in the field of genetic engineering. Polyphenols help promote the growth of health by promoting microorganisms in the gut. In order to produce important bioactive compounds from natural resources, the fermentation method is one of those suitable methods showing great potential. Various novel bioactive compounds are produced through fermentation, which are important for human health. Formation of different bioactive compounds via fermentation is yet fully unexplored that can be done by explication of the various metabolic pathways of fermentative microorganisms. Further, microbial degradation of aromatic compounds including polyphenols is of immense importance attracting many researchers worldwide. This book chapter highlights the fermentation and degradation pathways of polyphenols and their importance on human health.

Acknowledgments

Dr. K. Behera acknowledges the University of Allahabad for the facilities and funding.

List of Abbreviations

EGC	Epigallocatechin
EGCG	Epigallocatechin-3-gallate
LAB	Lactic acid bacteria
LAF	Lactic acid fermentation
NTIMB	National collection of industrial, food, and marine bacteria
PG	Phloroglucinol
SmS	Submerded/liquid fermentation
SSF	Solid state fermentation
TFC	Total flavonoid content
TPC	Total phenolic content
TSE	Traditional solvent extraction
WG	Whole grain

References

1 Tamang, J.P., Cotter, P.D., Endo, A. et al. (2020). Fermented foods in a global age: east meets west. *Comprehensive Reviews in Food Science and Food Safety* 19 (1): 184–217. https://doi.org/10.1111/1541-4337.12520.
2 Şanlier, N., Gökcen, B.B., and Sezgin, A.C. (2019). Health benefits of fermented foods. *Critical Reviews in Food Science and Nutrition* 59 (3): 506–527. https://doi.org/10.1080/10408398.2017.1383355.
3 Bermejo, L.M., López-Plaza, B., Santurino, C. et al. (2019). Milk and dairy product consumption and bladder cancer risk: a systematic review and meta-analysis of observational studies. *Advances in Nutrition* 10 (5): S224–S280. https://doi.org/10.1093/advances/nmy119.
4 Zhang, K., Dai, H., Liang, W. et al. (2019). Fermented dairy foods intake and risk of cancer. *International Journal of Cancer* 144 (9): 2099–2108. https://doi.org/10.1002/ijc.31959.

5 Stiemsma, L.T., Nakamura, R.E., Nguyen, J.G., and Michels, K.B. (2020). Does consumption of fermented foods modify the human gut microbiota? *The Journal of Nutrition* 150 (7): 1680–1692. https://doi.org/10.1093/jn/nxaa077.

6 Mohammadi, H., Ghavami, A., Faghihimani, Z. et al. (2021). Effects of probiotics fermented milk products on obesity measure among adults: a systematic review and meta-analysis of clinical trials. *Journal of Functional Foods* 82 (April): 104494. https://doi.org/10.1016/j.jff.2021.104494.

7 Wastyk, H.C., Fragiadakis, G.K., Perelman, D. et al. (2021). Gut-microbiota-targeted diets modulate human immune status. *Cell* 184 (16): 4137–4153.e14. https://doi.org/10.1016/j.cell.2021.06.019.

8 Sharma, R., Kapila, R., Dass, G., and Kapila, S. (2014). Improvement in Th1/Th2 immune homeostasis, antioxidative status and resistance to pathogenic *E. coli* on consumption of probiotic *Lactobacillus rhamnosus* fermented milk in aging mice. *Age (Omaha) 36* (4): 1–17. 10.1007/s11357-014-9686-4.

9 Chakrabarti, S., Guha, S., and Majumder, K. (2018). Food-derived bioactive peptides in human health: challenges and opportunities. *Nutrients* 10 (11): 1–17. https://doi.org/10.3390/nu10111738.

10 Lynch, K.M., Zannini, E., Coffey, A., and Arendt, E.K. (2018). Lactic acid bacteria exopolysaccharides in foods and beverages: isolation, properties, characterization, and health benefits. *Annual Review of Food Science and Technology* 9: 155–176. https://doi.org/10.1146/annurev-food-030117-012537.

11 Piekarska-Radzik, L. and Klewicka, E. (2021). Mutual influence of polyphenols and *Lactobacillus* spp. bacteria in food: a review. *European Food Research and Technology* 247 (1): 9–24. https://doi.org/10.1007/s00217-020-03603-y.

12 Manach, C., Scalbert, A., Morand, C. et al. (2004). Polyphenols: food sources and bioavailability. *The American Journal of Clinical Nutrition* 79 (5): 727–747. https://doi.org/10.1093/ajcn/79.5.727.

13 Leal-Sánchez, M.V., Ruiz-Barba, J.L., Sánchez, A.H. et al. (2003). Fermentation profile and optimization of green olive fermentation using *Lactobacillus plantarum* LPCO10 as a starter culture. *Food Microbiology* 20 (4): 421–430. https://doi.org/10.1016/S0740-0020(02)00147-8.

14 Ruiz-Barba, J.L., Cathcart, D.P., Warner, P.J., and Jimenez-Diaz, R. (1994). Use of *Lactobacillus plantarum* LPCO10, a bacteriocin producer, as a starter culture in Spanish-style green olive fermentations. *Applied and Environmental Microbiology* 60 (6): 2059–2064. https://doi.org/10.1128/aem.60.6.2059-2064.1994.

15 Plengvidhya, V., Breidt, F., Lu, Z., and Fleming, H.P. (2007). DNA fingerprinting of lactic acid bacteria in sauerkraut fermentations. *Applied and Environmental Microbiology* 73 (23): 7697–7702. https://doi.org/10.1128/AEM.01342-07.

16 Tamminen, M., Joutsjoki, T., Sjöblom, M. et al. (2004). Screening of lactic acid bacteria from fermented vegetables by carbohydrate profiling and PCR-ELISA. *Letters in Applied Microbiology* 39 (5): 439–444. https://doi.org/10.1111/j.1472-765X.2004.01607.x.

17 Singh, A.K. and Ramesh, A. (2008). Succession of dominant and antagonistic lactic acid bacteria in fermented cucumber: insights from a PCR-based approach. *Food Microbiology* 25 (2): 278–287. https://doi.org/10.1016/j.fm.2007.10.010.

18 Ercolini, D., Villani, F., Aponte, M., and Mauriello, G. (2006). Fluorescence in situ hybridisation detection of *Lactobacillus plantarum* group on olives to be used in natural fermentations. *International Journal of Food Microbiology* 112 (3): 291–296. https://doi.org/10.1016/j.ijfoodmicro.2006.09.003.

19 Chamkha, M., Sayadi, S., Bru, V., and Godon, J.J. (2008). Microbial diversity in *Tunisian olive* fermentation brine as evaluated by small subunit RRNA - single strand conformation polymorphism analysis. *International Journal of Food Microbiology* 122 (1–2): 211–215. https://doi.org/10.1016/j.ijfoodmicro.2007.11.052.

20 Hurtado, A., Reguant, C., Esteve-Zarzoso, B. et al. (2008). Microbial population dynamics during the processing of arbequina table olives. *Food Research International* 41 (7): 738–744. https://doi.org/10.1016/j.foodres.2008.05.007.

21 Nychas, G.J.E., Panagou, E.Z., Parker, M.L. et al. (2002). Microbial colonization of naturally black olives during fermentation and associated biochemical activities in the cover brine. *Letters in Applied Microbiology* 34 (3): 173–177. https://doi.org/10.1046/j.1472-765x.2002.01077.x.

22 Panagou, E.Z., Tassou, C.C., and Katsaboxakis, C.Z. (2003). Induced lactic acid fermentation of untreated green olives of the Conservolea cultivar by *Lactobacillus pentosus*. *Journal of the Science of Food and Agriculture* 83 (7): 667–674. https://doi.org/10.1002/jsfa.1336.

23 Ruiz-Barba, J.L. and Jiménez-Díaz, R. (1994). Vitamin and amino acid requirements of *Lactobacillus plantarum* strains isolated from green olive fermentations. *Journal of Applied Bacteriology* 76 (4): 350–355.

24 Dimidi, E., Cox, S., Rossi, M., and Whelan, K. (1806). Fermented foods: definitions and characteristics, gastrointestinal health and disease. *Nutrients* 2019 (11): 1–26.

25 García-Díez, J. and Saraiva, C. (2021). Use of starter cultures in foods from animal origin to improve their safety. *International Journal of Environmental Research and Public Health* 18 (5): 1–25. https://doi.org/10.3390/ijerph18052544.

26 Chugh, B. and Kamal-Eldin, A. (2020). Bioactive compounds produced by probiotics in food products. *Current Opinion in Food Science* 32: 76–82. https://doi.org/10.1016/j.cofs.2020.02.003.

27 Hedberg, M., Hasslöf, P., Sjöström, I. et al. (2008). Sugar fermentation in probiotic bacteria - an in vitro study. *Oral Microbiology and Immunology* 23 (6): 482–485. https://doi.org/10.1111/j.1399-302X.2008.00457.x.

28 Wang, Y., Wu, J., Lv, M. et al. (2021). Metabolism characteristics of lactic acid bacteria and the expanding applications in food industry. *Frontiers in Bioengineering and Biotechnology* 9 (May): 1–19. https://doi.org/10.3389/fbioe.2021.612285.

29 Kelly, S.M., Munoz-Munoz, J., and van Sinderen, D. (2021). Plant glycan metabolism by bifidobacteria. *Frontiers in Microbiology* 12 (February): 1–18. https://doi.org/10.3389/fmicb.2021.609418.

30 Daliri, E.B.M., Oh, D.H., and Lee, B.H. (2017). Bioactive peptides. *Food* 6 (5): 1–21. https://doi.org/10.3390/foods6050032.

31 Daroit, D.J. and Brandelli, A. (2021). In vivo bioactivities of food protein-derived peptides – a current review. *Current Opinion in Food Science* 39: 120–129. https://doi.org/10.1016/j.cofs.2021.01.002.

32 Karami, Z. and Akbari-adergani, B. (2019). Bioactive food derived peptides: a review on correlation between structure of bioactive peptides and their functional properties. *Journal of Food Science and Technology* 56 (2): 535–547. https://doi.org/10.1007/s13197-018-3549-4.

33 Skalickova, S., Ridoskova, A., Slama, P. et al. (2022). Effect of lactic fermentation and cooking on nutrient and mineral digestibility of peas. *Frontiers in Nutrition* 9 (February): 1–12. https://doi.org/10.3389/fnut.2022.838963.

34 Fong, F.L.Y., Lam, K.Y., Lau, C.S. et al. (2020). Reduction in biogenic amines in Douchi fermented by probiotic bacteria. *PLoS One* 15 (3): 1–10. https://doi.org/10.1371/journal.pone.0230916.

35 Feng, L., Tang, N., Liu, R. et al. (2021). The relationship between flavor formation, lipid metabolism, and microorganisms in fermented fish products. *Food & Function* 12 (13): 5685–5702. https://doi.org/10.1039/d1fo00692d.

36 Yadav, H., Jain, S., and Sinha, P.R. (2007). Production of free fatty acids and conjugated linoleic acid in probiotic Dahi containing *Lactobacillus acidophilus* and *Lactobacillus casei* during fermentation and storage. *International Dairy Journal* 17 (8): 1006–1010. https://doi.org/10.1016/j.idairyj.2006.12.003.

37 Lekha, P.K. and Lonsane, B.K. (1997). *Production and Application of Tannin Acyl Hydrolase: State of the Art*, vol. 44. Elsevier Masson SAS https://doi.org/10.1016/S0065-2164(08)70463-5.

38 Belmares, R., Contreras-Esquivel, J.C., Rodríguez-Herrera, R. et al. (2004). Microbial production of tannase: an enzyme with potential use in food industry. *LWT* 37 (8): 857–864. https://doi.org/10.1016/j.lwt.2004.04.002.

39 Aguilar, C.N., Rodríguez, R., Gutiérrez-Sánchez, G. et al. (2007). Microbial tannases: advances and perspectives. *Applied Microbiology and Biotechnology* 76 (1): 47–59. https://doi.org/10.1007/s00253-007-1000-2.

40 Inoue, K.H. and Hagerman, A.E. (1988). Determination of gallotannin with rhodanine. *Analytical Biochemistry* 169 (2): 363–369. https://doi.org/10.1016/0003-2697(88)90296-5.

41 Rodríguez, H., de las Rivas, B., Gómez-Cordovés, C., and Muñoz, R. (2008). Characterization of tannase activity in cell-free extracts of *Lactobacillus plantarum* CECT 748T. *International Journal of Food Microbiology* 121 (1): 92–98. https://doi.org/10.1016/j.ijfoodmicro.2007.11.002.

42 Gury (2016). *Sector de la Automoción en España* 70 (4): 2146–2153. https://doi.org/10.1128/AEM.70.4.2146.

43 Cavin, J.F., Barthelmebs, L., and Diviès, C. (1997). Molecular characterization of an inducible P-Coumaric acid decarboxylase from *Lactobacillus plantarum*: gene cloning, transcriptional analysis, overexpression in *Escherichia coli*, purification, and characterization. *Applied and Environmental Microbiology* 63 (5): 1939–1944. https://doi.org/10.1128/aem.63.5.1939-1944.1997.

44 Barthelmebs, L., Divies, C., and Cavin, J.F. (2000). Knockout of the P-coumarate decarboxylase gene from *Lactobacillus plantarum* reveals the existence of two other inducible enzymatic activities involved in phenolic acid metabolism. *Applied and Environmental Microbiology* 66 (8): 3368–3375. https://doi.org/10.1128/AEM.66.8.3368-3375.2000.

45 Alberto, M.R., Gómez-Cordovés, C., and Manca De Nadra, M.C. (2004). Metabolism of gallic acid and catechin by *Lactobacillus hilgardii* from wine. *Journal of Agricultural and Food Chemistry* 52 (21): 6465–6469. https://doi.org/10.1021/jf049239f.

46 Vivas, N., Lonvaud-Funel, A., and Glories, Y. (1997). Effect of phenolic acids and anthocyanins on growth, viability and malolactic activity of a lactic acid bacterium. *Food Microbiology* 14 (3): 291–299. https://doi.org/10.1006/fmic.1996.0086.

47 Boido, E., Lloret, A., Medina, K. et al. (2002). Effect of β-glycosidase activity of *Oenococcus oeni* on the glycosylated flavor precursors of tannat wine during malolactic fermentation. *Journal of Agricultural and Food Chemistry* 50 (8): 2344–2349. https://doi.org/10.1021/jf0109367.

48 Ugliano, M., Genovese, A., and Moio, L. (2003). Hydrolysis of wine aroma precursors during malolactic fermentation with four commercial starter cultures of *Oenococcus oeni*. *Journal of Agricultural and Food Chemistry* 51 (17): 5073–5078. https://doi.org/10.1021/jf0342019.

49 D'Incecco, N., Bartowsky, E., Kassara, S. et al. (2004). Release of glycosidically bound flavour compounds of chardonnay by *Oenococcus oeni* during malolactic fermentation. *Food Microbiology* 21 (3): 257–265. https://doi.org/10.1016/j.fm.2003.09.003.

50 De Revel, G., Bloem, A., Augustin, M. et al. (2005). Interaction of Œnococcus Œni and oak wood compounds. *Food Microbiology* 22 (6): 569–575. https://doi.org/10.1016/j.fm.2004.11.006.

51 Bloem, A., Lonvaud-Funel, A., and de Revel, G. (2008). Hydrolysis of glycosidically bound flavour compounds from oak wood by *Oenococcus oeni*. *Food Microbiology* 25 (1): 99–104. https://doi.org/10.1016/j.fm.2007.07.009.

52 Bloem, A., Bertrand, A., Lonvaud-Funel, A., and De Revel, G. (2007). Vanillin production from simple phenols by wine-associated lactic acid bacteria. *Letters in Applied Microbiology* 44 (1): 62–67. https://doi.org/10.1111/j.1472-765X.2006.02037.x.

53 Cavin, J.F., Andioc, V., Etievant, P.X., and Divies, C. (1993). Ability of wine lactic acid bacteria to metabolize phenol carboxylic acids. *American Journal of Enology and Viticulture* 44: 76–80. https://doi.org/10.5344/ajev.1993.44.1.76.

54 Rodríguez, H., Curiel, J.A., Landete, J.M. et al. (2009). Food phenolics and lactic acid bacteria. *International Journal of Food Microbiology* 132 (2–3): 79–90. https://doi.org/10.1016/j.ijfoodmicro.2009.03.025.

55 Martinengo, P., Arunachalam, K., and Shi, C. (2021). Polyphenolic antibacterials for food preservation: review, challenges, and current applications. *Food* 10 (10): https://doi.org/10.3390/foods10102469.

56 Milutinović, M., Dimitrijević-Branković, S., and Rajilić-Stojanović, M. (2021). Plant extracts rich in polyphenols as potent modulators in the growth of probiotic and pathogenic intestinal microorganisms. *Frontiers in Nutrition* 8 (July): 1–11. https://doi.org/10.3389/fnut.2021.688843.

57 Palareti, G., Legnani, C., Cosmi, B. et al. (2016). Comparison between different D-dimer cutoff values to assess the individual risk of recurrent venous thromboembolism: analysis of results obtained in the DULCIS study. *International Journal of Laboratory Hematology* 38 (1): 42–49. https://doi.org/10.1111/ijlh.12426.

58 Xu, X., Guo, Y., Chen, S. et al. (2022). The positive influence of polyphenols extracted from *Pueraria lobata* root on the gut microbiota and its antioxidant capability. *Frontiers in Nutrition* 9 (March): 1–11. https://doi.org/10.3389/fnut.2022.868188.

59 Sharma, R., Kumari, M., Kumari, A. et al. (2019). Diet supplemented with phytochemical epigallocatechin gallate and probiotic *Lactobacillus fermentum* confers second generation synbiotic effects by modulating cellular immune responses and antioxidant capacity in aging mice. *European Journal of Nutrition* 58 (7): 2943–2957. https://doi.org/10.1007/s00394-018-01890-6.

60 Nazzaro, F., Fratianni, F., De Feo, V. et al. (2020). *Polyphenols, the New Frontiers of Prebiotics*, 1e, vol. 94. Elsevier Inc. https://doi.org/10.1016/bs.afnr.2020.06.002.

61 Westfall, S., Lomis, N., and Prakash, S. (2017). A polyphenol-rich prebiotic in combination with a novel probiotic formulation alleviates markers of obesity and diabetes in *Drosophila*. *Journal of Functional Foods* 2018 (48): 374–386. https://doi.org/10.1016/j.jff.2018.07.012.

62 Ha, J., Oh, H., Oh, N.S. et al. (2020). Anti-inflammatory effect of a peptide derived from the synbiotics, fermented *Cudrania tricuspidata* with *Lactobacillus gasseri*, on inflammatory bowel disease. *Mediators of Inflammation* 2020: https://doi.org/10.1155/2020/3572809.

63 Oh, N.S., Lee, J.Y., Oh, S. et al. (2016). Improved functionality of fermented milk is mediated by the synbiotic interaction between *Cudrania tricuspidata* leaf extract and *Lactobacillus gasseri* strains. *Applied Microbiology and Biotechnology* 100 (13): 5919–5932. https://doi.org/10.1007/s00253-016-7414-y.

64 Oh, N.S., Joung, J.Y., Lee, J.Y. et al. (2020). A synbiotic combination of *Lactobacillus gasseri* 505 and *Cudrania tricuspidata* leaf extract prevents hepatic toxicity induced by colorectal cancer in mice. *Journal of Dairy Science* 103 (4): 2947–2955. https://doi.org/10.3168/jds.2019-17411.

65 Wang, L.C., Pan, T.M., and Tsai, T.Y. (2018). Lactic acid bacteria-fermented product of green tea and *Houttuynia cordata* leaves exerts anti-adipogenic and anti-obesity effects. *Journal of Food and Drug Analysis* 26 (3): 973–984. https://doi.org/10.1016/j.jfda.2017.11.009.

66 Güney, D. and Güngörmüşler, M. (2021). Development and comparative evaluation of a novel fermented juice mixture with probiotic strains of lactic acid bacteria and bifidobacteria. *Probiotics and Antimicrobial Proteins* 13 (2): 495–505. https://doi.org/10.1007/s12602-020-09710-2.

67 Masebe, K.M. and Adebo, O.A. (2019). *Production and Quality Characteristics of a Probiotic Beverage from Watermelon (Citrullus Lanatus)*. doi: 10.17758/eares8.eap1119243.

68 Marsh, A.J., Hill, C., Ross, R.P., and Cotter, P.D. (2014). Fermented beverages with health-promoting potential: past and future perspectives. *Trends in Food Science and Technology* 38 (2): 113–124. https://doi.org/10.1016/j.tifs.2014.05.002.

69 Adebo, O.A., Njobeh, P.B., Adebiyi, J.A. et al. (2017). Fermented pulse-based food products in developing nations as functional foods and ingredients. *Functional Food - Improve Health through Adequate Food* https://doi.org/10.5772/intechopen.69170.

70 Adebo, O.A., Kayitesi, E., Tugizimana, F., and Njobeh, P.B. (2019). Differential metabolic signatures in naturally and lactic acid bacteria (LAB) fermented ting (a southern African food) with different tannin content, as revealed by gas chromatography Mass spectrometry (GC–MS)-based metabolomics. *Food Research International* 121 (March): 326–335. https://doi.org/10.1016/j.foodres.2019.03.050.

71 Adebiyi, J.A., Kayitesi, E., Adebo, O.A. et al. (2019). Food fermentation and mycotoxin detoxification: an African perspective. *Food Control* 106. https://doi.org/10.1016/j.foodcont.2019.106731.

72 Ragaee, S. and Abdel-Aal, E.S.M. (2006). Pasting properties of starch and protein in selected cereals and quality of their food products. *Food Chemistry* 95 (1): 9–18. https://doi.org/10.1016/j.foodchem.2004.12.012.

73 Akingbala, J.O., Rooney, L.W., and Faubion, J.M. (1981). Physical, chemical, and sensory evaluation of Ogi from sorghum of differing kernel characteristics. *Journal of Food Science* 46 (5): 1532–1536. https://doi.org/10.1111/j.1365-2621.1981.tb04214.x.

74 Taylor, J. and Taylor, J.R.N. (2011). Protein biofortified sorghum: effect of processing into traditional African foods on their protein quality. *Journal of Agricultural and Food Chemistry* 59 (6): 2386–2392. https://doi.org/10.1021/jf104006v.

75 Dlamini, N.R., Taylor, J.R.N., and Rooney, L.W. (2007). The effect of sorghum type and processing on the antioxidant properties of African sorghum-based foods. *Food Chemistry* 105 (4): 1412–1419. https://doi.org/10.1016/j.foodchem.2007.05.017.

76 Kruger, J., Taylor, J.R.N., and Oelofse, A. (2012). Effects of reducing phytate content in sorghum through genetic modification and fermentation on in vitro iron availability in whole grain porridges. *Food Chemistry* 131 (1): 220–224. https://doi.org/10.1016/j.foodchem.2011.08.063.

77 Adebo, O.A., Njobeh, P.B., Adebiyi, J.A., and Kayitesi, E. (2018). Co-influence of fermentation time and temperature on physicochemical properties, bioactive components and microstructure of ting (a southern African food) from whole grain sorghum. *Food Bioscience* 25: 118–127. https://doi.org/10.1016/j.fbio.2018.08.007.

78 Adebo, O.A., Njobeh, P.B., and Kayitesi, E. (2018). Fermentation by *Lactobacillus fermentum* strains (singly and in combination) enhances the properties of ting from two whole grain sorghum types. *Journal of Cereal Science* 82: 49–56. https://doi.org/10.1016/j.jcs.2018.05.008.

79 García-Mantrana, I., Yebra, M.J., Haros, M., and Monedero, V. (2016). Expression of bifidobacterial phytases in *Lactobacillus casei* and their application in a food model of whole-grain sourdough bread. *International Journal of Food Microbiology* 216: 18–24. https://doi.org/10.1016/j.ijfoodmicro.2015.09.003.

80 Demir, M.K. and Elgün, A. (2014). Comparison of autoclave, microwave, IR and UV-C stabilization of whole wheat flour branny fractions upon the nutritional properties of whole wheat bread. *Journal of Food Science and Technology* 51 (1): 59–66. https://doi.org/10.1007/s13197-011-0475-0.

81 Struyf, N., Laurent, J., Verspreet, J. et al. (2017). Saccharomyces cerevisiae and luyveromyces marxianus cocultures allow reduction of fermentable oligo-, di-, and monosaccharides and polyols levels in whole wheat bread. *Journal of Agricultural and Food Chemistry* 65 (39): 8704–8713. https://doi.org/10.1021/acs.jafc.7b02793.

82 Dey, T.B. and Kuhad, R.C. (2014). Enhanced production and extraction of phenolic compounds from wheat by solid-state fermentation with *Rhizopus oryzae* RCK2012. *Biotechnology Reports* 4 (1): 120–127. https://doi.org/10.1016/j.btre.2014.09.006.

83 Starzyńska-Janiszewska, A., Stodolak, B., Socha, R. et al. (2019). Spelt wheat tempe as a value-added whole-grain food product. *LWT* 113: 108250. https://doi.org/10.1016/j.lwt.2019.108250.

84 Ayyash, M., Johnson, S.K., Liu, S.Q. et al. (2018). Cytotoxicity, antihypertensive, antidiabetic and antioxidant activities of solid-state fermented lupin, quinoa and wheat by *Bifidobacterium* species: in-vitro investigations. *LWT* 95: 295–302. https://doi.org/10.1016/j.lwt.2018.04.099.

85 Koistinen, V.M., Mattila, O., Katina, K. et al. (2018). Metabolic profiling of sourdough fermented wheat and rye bread. *Scientific Reports* 8 (1): 1–11. https://doi.org/10.1038/s41598-018-24149-w.

86 Buddrick, O., Jones, O.A.H., Cornell, H.J., and Small, D.M. (2014). The influence of fermentation processes and cereal grains in wholegrain bread on reducing phytate content. *Journal of Cereal Science* 59 (1): 3–8. https://doi.org/10.1016/j.jcs.2013.11.006.

87 Sheela, P., UmaMaheswari, T., Kanchana, S. et al. (2018). Development and evaluation of fermented millet milk based curd. *Journal of Pharmacognosy and Phytochemistry* 7 (4): 714–717.

88 Wang, Y., Compaoré-Sérémé, D., Sawadogo-Lingani, H. et al. (2019). Influence of dextran synthesized in situ on the rheological, technological and nutritional properties of whole grain pearl millet bread. *Food Chemistry* 285: 221–230. https://doi.org/10.1016/j.foodchem.2019.01.126.

89 Di Stefano, E., White, J., Seney, S. et al. (2017). A novel millet-based probiotic fermented food for the developing world. *Nutrients* 9 (5): 1–18. https://doi.org/10.3390/nu9050529.

90 Muui, C.W., Muasya, R.M., and Kirubi, D.T. (2013). *African Journal of Food, Agriculture, Nutrition and Development 13* (1): 7339–7357. ISSN 16845374.

91 Obinna-Echem, P.C., Beal, J., and Kuri, V. (2015). Effect of processing method on the mineral content of Nigerian fermented maize infant complementary food - Akamu. *LWT* 61 (1): 145–151. https://doi.org/10.1016/j.lwt.2014.11.001.

92 Gamel, T.H., Abdel-Aal, E.S.M., and Tosh, S.M. (2015). Effect of yeast-fermented and sour-dough making processes on physicochemical characteristics of β-glucan in whole wheat/oat bread. *LWT - Food Science and Technology* 60 (1): 78–85. https://doi.org/10.1016/j.lwt.2014.07.030.

93 Zannini, E., Jeske, S., Lynch, K., and Arendt, E.K. (2018). Development of novel quinoa-based yoghurt fermented with dextran producer *Weissella cibaria* MG1. *International Journal of Food Microbiology* 268 (January): 19–26. https://doi.org/10.1016/j.ijfoodmicro.2018.01.001.

94 Osawa, R., Kuroiso, K., Goto, S., and Shimizu, A. (2000). Isolation of tannin-degrading lactobacilli from humans and fermented foods. *Applied and Environmental Microbiology* 66 (7): 3093–3097. https://doi.org/10.1128/AEM.66.7.3093-3097.2000.

95 Reverón, I., de las Rivas, B., Muñoz, R., and López de Felipe, F. (2012). Genome-wide transcriptomic responses of a human isolate of *Lactobacillus plantarum* exposed to *p*-coumaric acid stress. *Molecular Nutrition & Food Research* 56 (12): 1848–1859. https://doi.org/10.1002/mnfr.201200384.

96 Yokotsuka, T. (1986). Soy sauce biochemistry. *Advances in Food Research* 30 (C): 195–329. https://doi.org/10.1016/S0065-2628(08)60350-X.

97 Zhao, F., Slade, R.C.T., and Varcoe, J.R. (2009). Techniques for the study and development of microbial fuel cells: an electrochemical perspective. *Chemical Society Reviews* 38 (7): 1926–1939. https://doi.org/10.1039/b819866g.

98 Jiménez, N., Esteban-Torres, M., Mancheño, J.M. et al. (2014). Tannin degradation by a novel tannase enzyme present in some *Lactobacillus plantarum* strains. *Applied and Environmental Microbiology* 80 (10): 2991–2997. https://doi.org/10.1128/AEM.00324-14.

99 Ávila, M., Hidalgo, M., Sánchez-Moreno, C. et al. (2009). Bioconversion of anthocyanin glycosides by *Bifidobacteria* and *Lactobacillus*. *Food Research International* 42 (10): 1453–1461. https://doi.org/10.1016/j.foodres.2009.07.026.

100 Landete, J.M. (2012). Updated knowledge about polyphenols: functions, bioavailability, metabolism, and health. *Critical Reviews in Food Science and Nutrition* 52 (10): 936–948. https://doi.org/10.1080/10408398.2010.513779.

101 Haminiuk, C.W.I., Maciel, G.M., Plata-Oviedo, M.S.V., and Peralta, R.M. (2012). Phenolic compounds in fruits - an overview. *International Journal of Food Science and Technology* 47 (10): 2023–2044. https://doi.org/10.1111/j.1365-2621.2012.03067.x.

102 Acosta-Estrada, B.A., Gutiérrez-Uribe, J.A., and Serna-Saldívar, S.O. (2014). Bound phenolics in foods, a review. *Food Chemistry* 152: 46–55. https://doi.org/10.1016/j.foodchem.2013.11.093.

103 Wong, D.W.S. (2006). Feruloyl esterase: a key enzyme in biomass degradation. *Applied Biochemistry and Biotechnology* 133 (2): 87–112. https://doi.org/10.1385/ABAB:133:2:87.

104 Gadaga, T.H., Mutukumira, A.N., Narvhus, J.A., and Feresu, S.B. (1999). A review of traditional fermented foods and beverages of Zimbabwe. *International Journal of Food Microbiology* 53 (1): 1–11. https://doi.org/10.1016/S0168-1605(99)00154-3.

105 Selhub, E.M., Logan, A.C., and Bested, A.C. (2014). Fermented foods, microbiota, and mental health: ancient practice meets nutritional psychiatry. *Journal of Physiological Anthropology* 33 (1): 1–12. https://doi.org/10.1186/1880-6805-33-2.

106 Ohta, Y. and Douglas, W. (1976). Bacterial metabolism of resorcinylic compound. *European Journal of Biochemistry* 61: 259–269.

107 Selma, M.V., Espín, J.C., and Tomás-Barberán, F.A. (2009). Interaction between phenolics and gut microbiota: role in human health. *Journal of Agricultural and Food Chemistry* 57 (15): 6485–6501. https://doi.org/10.1021/jf902107d.

108 Armstrong, S.M. and Patel, T.R. (1994). Microbial degradation of phloroglucinol and other polyphenolic compounds. *Journal of Basic Microbiology* 34 (2): 123–135. https://doi.org/10.1002/jobm.3620340208.

109 Walker, J.R.L. and Taylor, B.G. (1983). Metabolism of phloroglucinol by *Fusarium solani*. *Archives of Microbiology* 134 (2): 123–126. https://doi.org/10.1007/BF00407944.

110 Armstrong, S., Patel, T.R., and Whalen, M. (1993). *Hicroldolegy* 8: 136–140.

111 Hofmann, K. and Hammer, E. (1999). Anaerobic formation and degradation of toxic aromatic compounds in agricultural and communal sewage deposits. *Chemosphere* 38 (11): 2561–2568. https://doi.org/10.1016/S0045-6535(98)00463-9.

112 Chapman, P.J. and Ribbons, D.W. (1976). Alternative Pathways for Resorcinol Catabolism In Pseudomonas putida. *Journal of Bacteriology* 125 (3): 985–998.

113 Maillard, M.N. and Berset, C. (1995). Evolution of antioxidant activity during kilning: role of insoluble bound phenolic acids of barley and malt. *Journal of Agricultural and Food Chemistry* 43 (7): 1789–1793. https://doi.org/10.1021/jf00055a008.

114 Larrauri, J.A., Rupérez, P., and Saura-Calixto, F. (1997). Effect of drying temperature on the stability of polyphenols and antioxidant activity of red grape pomace peels. *Journal of Agricultural and Food Chemistry* 45 (4): 1390–1393. https://doi.org/10.1021/jf960282f.

115 Zhu, Y., Hu, J., He, W. et al. (2019). Gallic acid disruption of Aβ1–42 aggregation rescues cognitive decline of APP/PS1 double transgenic mouse. *Neurobiology of Disease* July 2018 (247): 67–80.

116 Ross, C.F., Hoye, C., and Fernandez-Plotka, V.C. (2011). Influence of heating on the polyphenolic content and antioxidant activity of grape seed flour. *Journal of Food Science* 76 (6): 884–890. https://doi.org/10.1111/j.1750-3841.2011.02280.x.

117 Lang, G.H., Lindemann, I., Ferreira, C.D. et al. (2019). Effects of drying temperature and long-term storage conditions on black rice phenolic compounds. *Food Chemistry* 287 (September 2018): 197–204. https://doi.org/10.1016/j.foodchem.2019.02.028.

118 Howard, L. and Pandjaitan, N. (2008). Pressurized liquid extraction of flavonoids from spinach. *Journal of Food Science* 73 (3). https://doi.org/10.1111/j.1750-3841.2007.00658.x.

119 Cacace, J.E. and Mazza, G. (2003). Optimization of extraction of anthocyanins from black currants with aqueous ethanol. *Food Science* 68 (1): 240–248.

120 Teh, S.S. and Birch, E.J. (2014). Effect of ultrasonic treatment on the polyphenol content and antioxidant capacity of extract from defatted hemp, flax and canola seed cakes. *Ultrasonics Sonochemistry* 21 (1): 346–353. https://doi.org/10.1016/j.ultsonch.2013.08.002.

121 Mokrani, A. and Madani, K. (2016). *Effect of Solvent, Time and Temperature on the Extraction of Phenolic Compounds and Antioxidant Capacity of Peach (Prunus persica L.) Fruit*, vol. 162. Elsevier BV https://doi.org/10.1016/j.seppur.2016.01.043.

122 Dorta, E., Lobo, M.G., and Gonzalez, M. (2012). Reutilization of mango byproducts: study of the effect of extraction solvent and temperature on their antioxidant properties. *Journal of Food Science* 77 (1): 80–88. https://doi.org/10.1111/j.1750-3841.2011.02477.x.

123 Ju, Z.Y. and Howard, L.R. (2003). Effects of solvent and temperature on pressurized liquid extraction of anthocyanins and total phenolics from dried red grape skin. *Journal of Agricultural and Food Chemistry* 51 (18): 5207–5213. https://doi.org/10.1021/jf0302106.

124 Sadilova, E., Stintzing, F.C., and Carle, R. (2006). Thermal degradation of acylated and nonacylated anthocyanins. *Journal of Food Science* 71 (8): https://doi.org/10.1111/j.1750-3841.2006.00148.x.

125 Casazza, A.A., Aliakbarian, B., Sannita, E., and Perego, P. (2012). High-pressure high-temperature extraction of phenolic compounds from grape skins. *International Journal of Food Science Technology* 47 (2): 399–405. https://doi.org/10.1111/j.1365-2621.2011.02853.x.

126 Patras, A., Brunton, N.P., O'Donnell, C., and Tiwari, B.K. (2010). Effect of thermal processing on anthocyanin stability in foods; mechanisms and kinetics of degradation. *Trends in Food Science and Technology* 21 (1): 3–11. https://doi.org/10.1016/j.tifs.2009.07.004.

127 Alvarez-Suarez, J.M., Cuadrado, C., Redondo, I.B., et al. (2021). Novel approaches in anthocyanin research – plant fortification and bioavailability issues. *Trends in Food Science and Technology* 117: 92–105. https://doi.org/10.1016/j.tifs.2021.01.049.

Part II

Industrial Applications of Polyphenols

5

Polyphenols for Wastewater and Industrial Influents' Treatment

Muhammad Salman[1], Muhammad U. Farooq[2], Atta Ur. Rahman[1], and Muhammad Yaseen[1,]*

[1] *Institute of Chemical Sciences, University of Peshawar, Peshawar, KP, Pakistan*
[2] *Department of Chemical Engineering, Ghulam Ishaq Khan Institute of Engineering Sciences & Technology, Topi, Swabi, KP, Pakistan*

5.1 Introduction

Freshwater is a crucial resource for sustaining life, supporting food production, promoting public health and safety, and managing energy resources on Earth [1]. However, rapid industrialization and urbanization have fronted a marginal rise in demand for freshwater, resulting in a significant decrease in availability [2]. Freshwater scarcity threatens human and aquatic life and exacerbates global pollution [3]. More than two billion people are currently experiencing water scarcity, which is expected to gradually increase [4]. Consequently, developing sustainable technologies for freshwater preservation and wastewater purification has become a critical research focus requiring careful attention and technical expertise. To address this challenge, various methods for the treatment of wastewater and industrial influents, such as membrane-based processes [5], solar desalination [6], electrochemical oxidation [7], adsorption [8], ultraviolet disinfection [9], advanced oxidation [10], ozonation [11], and advanced biological processes [12], have emerged, offering promising solutions for the treatment of wastewater and industrial influents. The design and fabrication of resilient and versatile water remediation composite materials and devices, including disinfectants, adsorbents, separation membranes, catalysts, and photothermal materials, are critical determinants for successfully implementing these techniques [13].

* Corresponding author

Science and Engineering of Polyphenols: Fundamentals and Industrial Scale Applications, First Edition. Edited by Chandrabhan Verma.
© 2024 John Wiley & Sons Inc. Published 2024 by John Wiley & Sons Inc.

A water remediation system that can effectively treat wastewater and industrial influents must possess specific characteristics. These include excellent machinability, multifunctionalities such as photothermal conversion, hydrophilicity/hydrophobicity, and bactericidal activity, as well as cost-effectiveness, accessibility, and abundant resources for large-scale application. The global energy crisis and concerns about secondary environmental pollution have led to a growing interest in using natural resources as alternatives to petrochemical substances for developing water remediation materials. Chitosan, cellulose, and carbonized biomass substances have been studied and documented for their perspective on water remediation through various methods, including adsorption [14].

Natural polyphenols are plant-derived (e.g., fruits, vegetables, tea, plant seeds, and medicinal herbs) molecules containing two or more phenolic hydroxyl groups. Over 8000 polyphenolic molecules have been identified, and they are known for their abundant presence in foods and beverages and their role as leading antioxidants in human eats [15]. Additionally, their natural mucilaginous property and biological compatibility make them suitable candidates for preparing biomaterials with various properties, i.e., antiviral, antioxidant, antimicrobial, immunoregulatory ability, and myocardial protection [16]. These materials have diverse applications in tissue engineering [17], cancer therapy [17], theranostics [18], sunscreen [19], and hair dye [20]. Natural polyphenols can easily conjugate with various building blocks, allowing for the development of multifunctional composite materials in different dimensions [21]. For instance, researchers have successfully fabricated one-dimensional (1D) polyphenolic fibers, two-dimensional (2D) polyphenolic membranes, and three-dimensional (3D) polyphenolic hydrogels/particles for producing clean and safe freshwater, using various techniques such as solar distillation, membrane filtration, adsorption, advanced oxidation processes, and disinfection. With these promising advancements in water purification applications, there is a growing need for an overview of natural polyphenolic composites' fabrication strategies, structures, functions, and applications in water remediation.

This chapter aims to present a detailed analysis of natural polyphenols' structural and functional attributes, focusing on their application in creating composite materials for water remediation. It also commences with a brief historical account of natural polyphenols and outlines their different types. The chapter highlights polyphenols' exceptional structural and functional properties, making them ideal for constructing composite materials for water purification. It then examines recent research that has investigated the use of natural polyphenolic compounds in various water remediation methods, such as solar distillation, membrane filtration, coagulation, adsorption, advanced oxidation processes, disinfection, and flocculation, to produce superior quality water. Lastly, this chapter concludes by identifying emerging trends and prospects for developing multifunctional natural polyphenolic composites to enhance the efficiency of water remediation processes.

5.2 History of Natural Polyphenols

Natural polyphenols have a long and rich history dating back thousands of years. These compounds, found in various plants, have been used by various cultures worldwide for medicinal and dietary purposes. The first recorded use of polyphenols dates back to the ancient Egyptians, who used tannins, a type of polyphenol, to preserve animal hides and protect them from decay [22]. Polyphenols were employed in traditional Chinese medicine to treat diverse health conditions in ancient times, including digestive problems, infections, and inflammation. Green tea, rich in polyphenols, has been a part of Chinese culture for centuries and is still popular today [23].

In the Western world, natural polyphenols became more widespread during the Middle Ages when herbal medicine was widely practiced. In particular, polyphenols found in plants such as St. John's worth, elderflower, and chamomile were used to treat various ailments, including wounds, infections, and anxiety [24]. The scientific study of polyphenols began in the late 19th century when researchers discovered that these compounds were responsible for the astringent taste in tea and wine. Since then, the study of polyphenols has expanded to include their role in plant growth and development and their possible human health benefits [25]. Polyphenols are acknowledged as crucial dietary constituents that keep a wide spectrum of possible health advantages such as anticancer, anti-inflammatory, and antioxidant properties. They are acquired in a collection of foods, including nuts, fruits, vegetables, and grains, and are also available in supplement form [26].

5.3 Types of Natural Polyphenols

Polyphenols are a diverse class of natural composites categorized based on their chemical structure into four major classes: flavonoids, phenolic acids, stilbenes, and lignans [27].

5.3.1 Flavonoids

Flavonoids are the most commonly found natural polyphenols extensively distributed in herbs, vegetables, and fruits. The flavonoid class includes several subclasses such as flavones, flavanones, flavanols, isoflavones, and anthocyanidins. From a chemical perspective, flavonoids possess a distinctive 15-carbon skeleton composed of 2-phenyl (A and B) and a heterocyclic ring (C) that contains an embedded oxygen molecule. The carbon framework of flavonoids can be denoted as C6–C3–C6, representing the prearrangement of the carbon atoms in the 2-phenyl and the heterocyclic ring, respectively (Figure 5.1a). These compounds

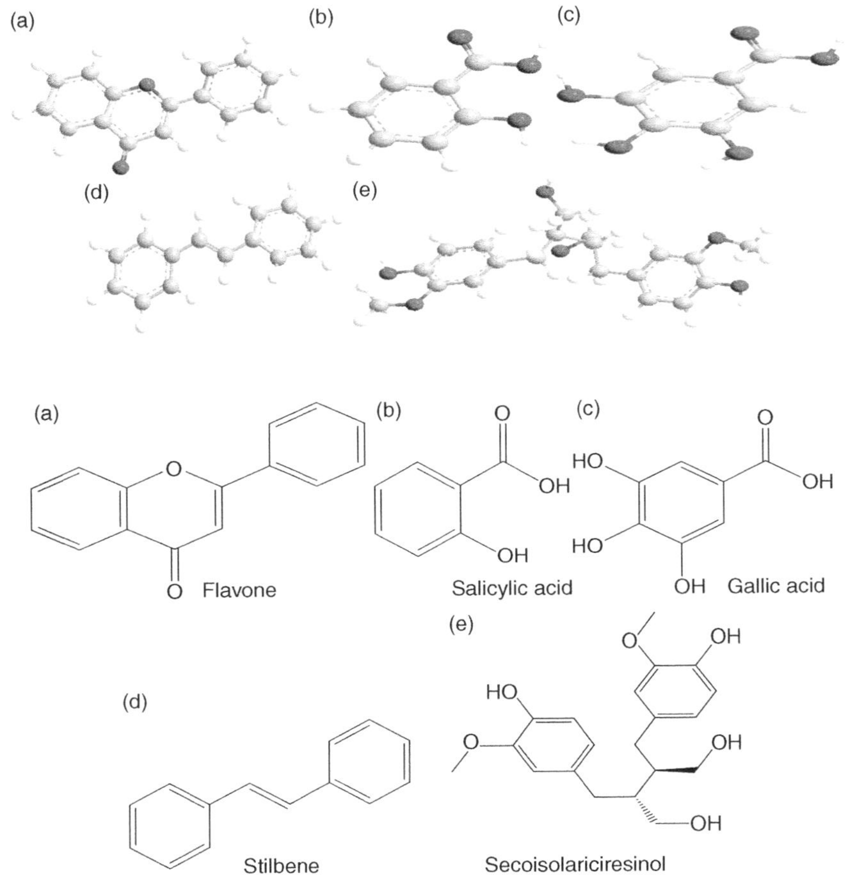

Figure 5.1 Optimized (above) and chemical structure of (a) flavonoid (flavone), (b) salicylic acid, (c) gallic acid, (d) stilbene, and (e) secoisolariciresinol.

are renowned for demonstrating various biological actions, encompassing anti-cancer, anti-inflammatory, antiviral, and antioxidant effects [28].

5.3.2 Phenolic Acids

Phenolic acids are another class of natural polyphenols broadly spread in the plant kingdom. These compounds are distinguished by a simple phenolic structure with one or more carboxylic acid groups attached as shown in Figure 5.1b. Polyphenols are categorized into two groups, which include hydroxycinnamic acids (HCAs) and hydroxybenzoic acids (HBAs). The HBAs group includes gallic acid, protocatechuic acid, and vanillic acid, while the HCAs group comprises ferulic acid, caffeic acid, and p-coumaric acid. These compounds are renowned for their anti-inflammatory, antimicrobial, and antioxidant properties [29].

5.3.3 Tannins

Tannins are naturally occurring polyphenols in vegetables, fruits, and other plant sources. They are a large and diverse group of polyphenols with a complex structure (Figure 5.1c), typically consisting of multiple phenol units linked to each other through several bonds such as ester, ether, and carbon–carbon bonds [30]. Tannins can be further categorized into two separate groups based on their structure: condensed tannins and hydrolyzable tannins. Condensed tannins are characterized by multiple phenol units linked together by carbon–carbon bonds. In contrast, hydrolyzable tannins are characterized by an ester bond between a sugar molecule and a polyphenol unit [31]. Tannins can be simple phenolic compounds, such as catechins, or complex polymeric structures, such as proanthocyanidins. Tannins are recognized for their characteristic astringent taste and have been employed in traditional medicine for their anti-inflammatory and antimicrobial properties [32].

5.3.4 Stilbenes

Stilbenes are another class of natural polyphenols found in several plants, including grapes, peanuts, and berries. They are characterized by their two phenyl rings connected by an ethylene bridge as depicted in Figure 5.1d. Resveratrol is the most well-known stilbene and has undergone extensive research for its potential benefits, including antioxidant, anti-inflammatory, and anticancer properties [33].

5.3.5 Lignans

Lignans are natural polyphenols abundant in various plant sources, including flaxseeds, sesame seeds, and berries. These polyphenols are formed by polymerizing two or more phenolic units, as shown in Figure 5.1e. Examples include secoisolariciresinol and enterolactone. They exhibit a broad list of biological functions, containing anti-inflammatory, anticancer, and antioxidant effects [34].

5.4 Polyphenols in Wastewater Remediation

Natural polyphenols are becoming increasingly popular for their potential to serve as effective components in multifunctional water treatment processes, with a particular focus on treating wastewater and industrial effluents. A diverse range of naturally occurring polyphenols has been utilized effectively for water treatment. Several naturally occurring polyphenols have been utilized for the purification of water, including ellagic acid (EA) [35], epicatechin gallate (ECG) [36], gallic acid (GA) [37], anthocyanin (ACN) [38], curcumin (CUR) [39], protocatechualdehyde (PA) [40], tannic acid (TA) [41], lignin (LIG) [42], quercetin (QN) [43], epigallocatechin (EGC) [44], epigallocatechin gallate (EGCG) [45], tea polyphenols (TP) [36], caffeic acid (CA) [46], and others. Figure 5.2 provides the chemical structures of the most common natural polyphenols [47, 48].

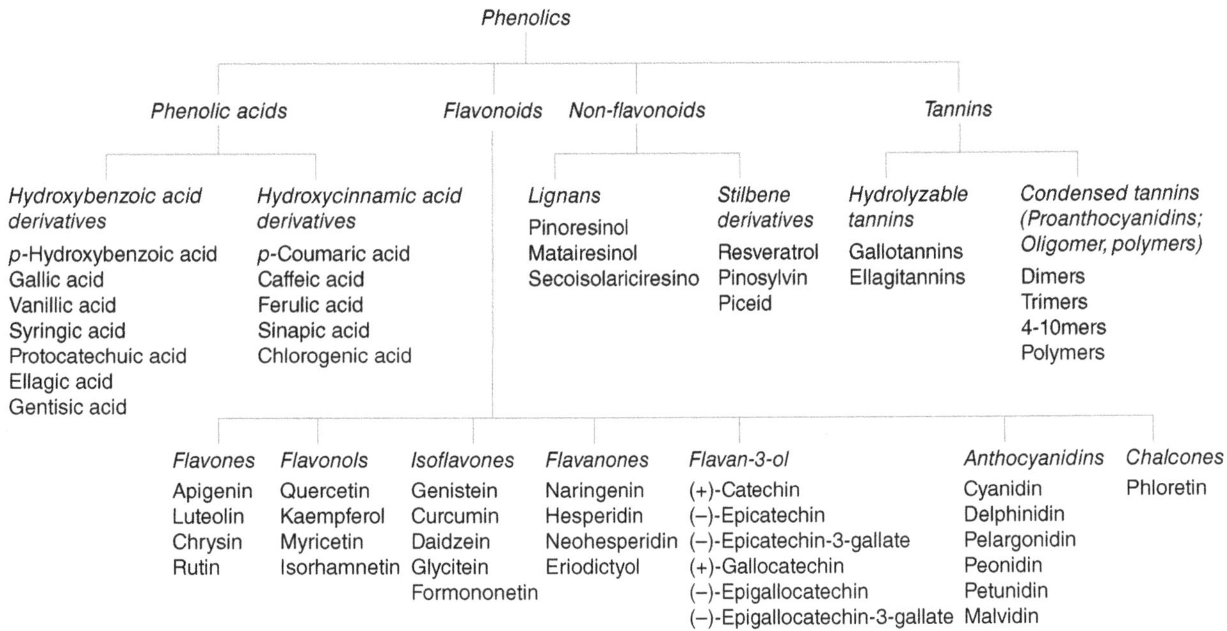

Figure 5.2 Different types of natural polyphenols for wastewater treatment.

5.5 Structural and Functional Features of Polyphenols

5.5.1 Natural Polyphenols' Structural Features

The sole structural characteristic of naturally occurring polyphenols is their abundant presence of phenolic hydroxyl functional groups. This unique trait enables natural polyphenols to form covalent interactions through several means, such as Schiff base reactions, polyphenol–metal coordination, Michael addition, radical coupling reactions, and polyphenol–boronate complexation. Moreover, natural polyphenols can also partake in non-covalent interactions, including hydrophobic interactions, π interactions, electrostatic interactions, and hydrogen bonds, as shown in Figure 5.3. This remarkable ability to interact with other structures through covalent and non-covalent means makes natural polyphenols such versatile and valuable compounds [49]. The versatility of natural polyphenols as building blocks stems from their unique structural features, making them excellent candidates for developing and designing functional substances for water remediation purposes. For instance, multifunctional coatings on a variety of substrate surfaces can be formed solely from natural polyphenol. TA, EGCG, and ECG from the crude extract of green tea, red wine, and cocoa bean have been incorporated as transparent coatings, improving nanoparticles' intrinsic properties [50].

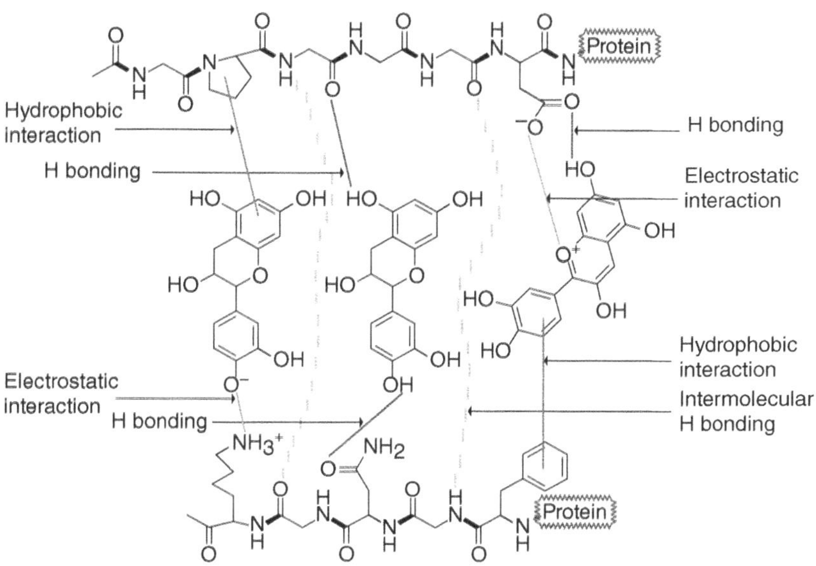

Figure 5.3 A schematic illustration demonstrating the standard covalent and non-covalent interactions observed in natural polyphenolic compounds (Reproduced from CCBY 4.0; No permission needed) [47].

Similarly, metal phenolic networks (MPNs) were used as polyphenols coating on various substrates. Caruso et al. developed a universal coating using Fe^{3+} ions and natural polyphenols, resulting in efficient crosslinking and rapid coating formation on various surfaces. The popularity of metal–phenolic interactions in functional materials has increased due to their presence in nature [51]. In addition to MPNs, other metal–polyphenol interactions have been studied. The catechol/pyrogallol moieties in polyphenols possess aromaticity and p-electron content, which results in the formation of binding sites carrying negative charge for alkaline-earth metal ions or cationic alkali, thus leading to the formation of cation-p interactions [52]. Polyphenols can prompt the reduction of noble metal ions into stable metallic forms by accepting electrons in redox reactions. As a result, polyphenols have been extensively utilized to make the in-situ synthesis of metal oxide nanoparticles (NPs) and metals, thereby eliminating the need for harmful reducing reagents [21]. Thus, metal–polyphenol composites hold potential for biomedical and environmental applications [53].

Additionally, the extremely reactive structure of naturally occurring polyphenols allows them to serve as versatile structural building blocks for composites used in water purification technologies. For instance, they can be utilized to make adaptable and more stable primers or anchors for the immobilization of functional molecules, thereby facilitating added effective functional integration [54]. Natural polyphenols possess multiple binding sites and can form hydrogels by cross-linking or complexing with other available hydrophilic macromolecules. This integration of polyphenols and hydrogels combines their respective advantages, resulting in great potential for application in wastewater treatment. Furthermore, the abundant presence of catechol or pyrogallol groups in natural polyphenols allows for easy coordination with transition metal ions, facilitating the construction of metal–phenolic networks. These networks can serve as precursors for creating metal-organic frameworks [55].

To summarize, the distinctive structural characteristics of natural polyphenols—including chelating sites, effective crosslinking, and multibinding sites—allow them to network with other available functional materials, forming various novel functional substances for water remediation.

5.5.2 Natural Polyphenols Functional Features

Polyphenols have surfaced as futuristic materials for various water remediation applications because of their unique functional features, including adhesion [56], hydrophilicity [57], photothermal performance [58], and antibacterial properties [59] as depicted in Figure 5.4. These compounds have catechol/pyrogallol moieties that allow them to strongly adhere to various substrates, similar to the way marine mussels do. This adhesion is due to bonding mechanisms between the moieties and the substrate surfaces [60]. Polyphenols can be used as coatings on

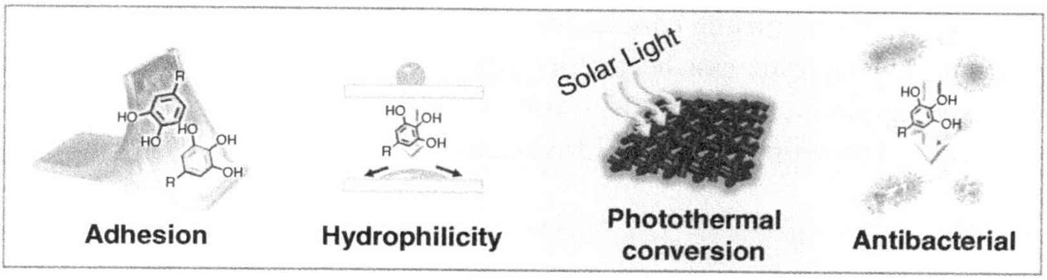

Figure 5.4 Functional features of polyphenols for wastewater and industrial influents' treatment [55].

different surfaces, such as metal, glass, and polymer, and can be functionalized with antibacterial, pH-responsive, and smart drug-release properties [21]. Researchers have utilized polyphenols to create robust underwater adhesives with adhesive strengths on metal surpassing 10 MPa, a level higher than all reported and commercially available underwater adhesives. The adhesion properties of polyphenols make them helpful in creating multifunctional composite materials for water treatment and other applications [55]. Additionally, the presence of functional groups having hydrophilic properties, i.e., hydroxyl and carboxylic groups, within polyphenols enhances their water solubility and promotes their dispersibility in aqueous media. This makes them highly suitable for wastewater treatment applications [61].

Furthermore, the photothermal transformation capability of polyphenols is an intriguing feature of its applications in water treatment technologies. By introducing metal ions, the cross-section area of absorption can be expanded to increase their ability to separation through absorption. The chelation of polyphenols and metal ions could cause a band gap decrease in polyphenolic compounds, thereby enhancing their total photothermal effect and light harvesting capacity. This feature can be handy for solar-driven seawater desalination, as polyphenols have demonstrated efficient photothermal conversion ability in previous studies. For example, coconut husk-based photothermal materials coordinated with phenolic compounds and Fe^{3+} have an evaporation rate of 1.83 kg m^2/h, approximately four times greater than saltwater [55].

Another significant functional feature of polyphenols is their antibacterial property; it can be attributed to their capacity to impede the biosynthesis of cell components, harm bacteria cell walls, and disrupt bacterial cell membranes. This property has been employed in developing composite materials with antibacterial properties, where polyphenols act as building blocks to produce nano-thin coatings with enhanced antibacterial performance [62]. Polyphenols can also help produce silver nanoparticles via their interaction with Ag^+, demonstrating exceptional antibacterial action against gram-positive and gram-negative bacteria. In summary, the multifunctional properties of polyphenols, including adhesion, hydrophilicity, photothermal performance, and antibacterial activity, make them highly versatile and promising candidates for wastewater treatment applications. These functional features can be further optimized by carefully selecting and modifying polyphenolic compounds to suit specific wastewater treatment needs, thus opening up new opportunities for sustainable and eco-friendly wastewater treatment processes.

5.6 Overview of Polyphenol-Based Materials

Natural polyphenols possess unique structural and functional properties that make them attractive to act as structure units for the construction of composite materials for water remediation. These compounds have been shown to exhibit strong adhesive properties, high antioxidant activity, and the ability to complex with metals and other pollutants. Additionally, their complex structures allow facile integration with other functional units, such as nanoparticles, via various techniques such as surface coating, chemical grafting, and physical blending. This makes them versatile and promising for developing innovative composite materials with improved performance for water remediation applications. Several polyphenolic composites have been used for water remediation, including nanofibers, synthetic membranes, phenolic nanoparticles, adsorbent, and phenolic hydrogels.

5.6.1 Nanofibers

Various techniques, including self-assembly, electrospinning, and phase separation, can generate nanofibers with diameters in the nanometer range. Among these methods, electrospinning has become the predominant approach for producing nanofibers on a large scale. Electrospun nanofibers are particularly interesting for environmental remediation applications because of their high porosity, widespread specific surface area, and adaptable surface functionality. In the context of wastewater treatment, natural polyphenols are an attractive option. Polyphenols containing catechol/pyrogallol moieties can act as crosslinking agents, enhancing the mechanical performance of nanofibers by forming chemically and physically cross-linked networks with the electro-spun polymer matrix. Moreover, polyphenols can enable versatility to modify electro-spun nanofibers, facilitating efficient wastewater purification through photocatalytic degradation or adsorption. Recent studies have demonstrated the potential of electrospun nanofibers modified with polyphenols for wastewater treatment. For instance, incorporating tannic acid (TA) into sodium alginate nanofibers has enhanced their mechanical strength and enabled the on-site production of Photo-Fenton catalyst b-FeOOH nanoparticles on the surface of nanofibers. The resulting nanofibers exhibited underwater superoleophobic properties, making them highly effective for separating crude oil-in-water emulsions, and degrading contaminants via the photo-Fenton reaction, consequently enabling self-cleaning performance [63].

Significantly, these one-dimensional (1D) nanofibers can be employed to create materials with two-dimensional (2D) nanofiber membranes and three-dimensional (3D) nanofiber aerogels, with promising prospects for wastewater treatment applications. As an example, superhydrophilic and underwater superoleophobic polysulfone nanofibrous membranes have been fabricated via a two-step MPN coating process, exhibiting outstanding separation efficacy and ultrahigh permeation flux for oil-in-water mixtures and emulsions [64]. Similarly, electro-spun polyimide nanofibrous membranes modified with TA-Al^{3+} MPNs have been shown to have effective antibacterial and UV-shielding properties [65]. Lastly, hybrid coatings on cellulose nanofiber-based aerogels, produced by TA-induced siloxane grafting, have exhibited effective adsorption capabilities for oil droplets and Cu^{2+} [66].

5.6.2 Synthetic Membranes

Membranes, one of the most common two-dimensional materials, have become popular for treating wastewater and producing clean water for reuse. Membranes can be grouped into categories based on their separation principles and intrinsic properties. These types include microfiltration (MF) having pore sizes ranging from 0.1 to 5 μm, ultrafiltration (UF) having pore sizes ranging from 0.5 to 100 nm, nanofiltration (NF) with pore sizes ranging from 0.5 to 10 nm, and membranes for reverse osmosis (RO) with pore sizes less than 0.5 nm [5]. These diverse membrane types can remove various impurities from effluent water, including suspended solids, organic matter, oil emulsions, colloidal haze, fats, proteins, viruses, bacteria, and even metal ions. These membranes are typically composed of synthetic materials such as polymers, ceramics, or metals, and their pore size, shape, and surface properties can be precisely engineered to allow selective passage of specific molecules or ions [67]. Interestingly, natural polyphenols such as tannic acid, catechol, pyrogallol, and dopamine are widely utilized as surface coating agents for membranes to attain desirable characteristics such as desired functionality and improved antibacterial and antifouling performance during membrane filtration. Natural polyphenols can also improve water purification performance by changing the microstructure and surface chemistry of 2D nanomaterial and membranes [68]. For instance, TA-based graphene hydrogel revealed enhanced adsorption capacity for organic dyes due to an improved porous system and increased surface area [69]. Moreover, polyphenols, for example, EGC, TA, EGCG, and tea catechins, can be effectively employed as exfoliating agents for ultrathin transition metal dichalcogenides, i.e., MoS_2 that have diverse environmental remediation applications. Furthermore, the exfoliated MoS_2 nanosheets demonstrated outstanding photothermal properties and improved antibacterial performance [70].

5.6.3 Phenolic Nanoparticles

Phenolic chemistry that mimics nature's reactivity has played a significant role in advancing materials science. This chemistry has made assembling and stabilizing phenolic nanoparticles with precise sizes and uniform properties easier. Natural phenolic nanoparticles from propolis, gallic acid, green tea, and grape seed have been investigated for their potential in UV protection and radical scavenging. In the meantime, researchers have produced functionalized nanoparticles using tea and fruit polyphenols. These nanoparticles have efficiently eliminated antibiotics, heavy metal ions, and bacteria from effluent water. For instance, Fe_3O_4@GTPs NPs, superparamagnetic nanoparticles with a Green Tea Polyphenols (GTP) coating, were produced using an environmentally safe and eco-friendly wet-chemical method. Due to their hydrophilic GTP coating, the resulting NPs exhibited excellent stability and dispersion in aqueous media. The tested NPs showed an impressive tendency to separate methylene blue (MB) dye from effluent via absorption, with a separation capacity of 7.25 mg/g, making them an effective tool for wastewater treatment [71].

Besides, catechol and pyrogallol have been applied as coating materials for nanoparticles in wastewater treatment. In this study, researchers developed a method for creating robust superhydrophilic surfaces to facilitate the separation of an oil–water mixture inspired by natural polyphenols. By taking inspiration from the reducing capacity of free catechol/pyrogallol groups as well as the adhesion of polyphenol, investigators first immersed commercially available materials, including nonwoven cotton fabrics, PET, and melamine sponge, in aqueous tannic acid (TA) solution to create a multifunctional coating layer on the surface of the fabric or sponge. This coating acted as a reducing agent, leading to the creation of silver nanoparticles (Ag-NPs). Next, the investigators added 1*H*, 1*H*, 2*H*, 2*H*-perfluorodecanethiol (PFDT) molecules to create superhydrophobic surfaces. The resulting sponge demonstrated higher absorption capacity (66–150 g/g) and oil/water selectivity for a wide extend of oils and organic solvents. It was also tough against abrasion, chemically stable, and demonstrated long-term endurance in severe environments. An innovative approach created a strong superhydrophobic surface to separate the oil/water mixture. This nature-inspired polyphenol technique was inspired by the ability of polyphenols to adhere and the reducing capability of available catechol/pyrogallol groups in polyphenols. Materials like PET, nonwoven cotton fabrics, and melamine sponges were immersed in an aqueous tannic acid (TA) solution to start the process. This allowed for the creation of a multifunctional thin layer on the surface of the fabric or sponge, which acted as a reducing reagent to produce silver nanoparticles (Ag-NPs). Next, the surface was coated with PFDT molecules to achieve superhydrophobic surfaces. The resulting sponge demonstrated high selectivity for separating oil and water, with a high absorption capacity of 66–150 g/g for a wide range of organic solvents and oils. Additionally, the sponge was highly tough to chemical damage, and mechanical abrasion, and could withstand harsh environmental conditions, making it a highly durable and reliable option [72].

5.6.4 Adsorbent

Adsorbents are materials that are broadly exploited in the treatment of effluent water. They are designed to remove contaminants and pollutants such as fertilizers, heavy metals, organic compounds, pesticides, dyes, pharmaceutical products, and other harmful substances from wastewater. There are numerous types of adsorbents available for the treatment of wastewater, such as activated carbon, zeolites, silica gel, bentonite clay, alumina, magnesia, carbon nanotubes, iron oxide nanoparticles, copper-based adsorbents, biochar, chitosan, fly ash, graphene oxide, polymeric adsorbents, and lignite-based activated carbon [73]. The polyphenol-based adsorbent is a type of adsorbent material that utilizes polyphenols derived from natural sources such as tea, coffee, and grape pomace. The use of polyphenols in adsorbent materials has been extensively studied for its potential to remove a broad range of contaminants from wastewater which includes but is not limited to organic compounds, heavy metal ions, and dyes [74]. Several studies have demonstrated the effectiveness of polyphenol-based adsorbent materials in removing contaminants from wastewater. A work done by Gao et al. illustrated that a museel-inspired triple bionic adsorbent, i.e., hydroxide@polydopamine@metal-polyphenol networks (LDH@PDA@MPNs) was highly effective in removing organic dyes, i.e., from wastewater, with a removal efficiency of up to 89.608 and 40.481 mg/g [75]. Similarly, Inoue et al. developed a cost-effective and environment-friendly process for separating Cs^+ from water samples using biopolymer-based adsorbents equipped with polyphenols. The adsorbents exhibited high adsorption efficiency and selectivity for Cs^+ over Na^+ ions, with a fast adsorption rate that achieved complete adsorption within 5 minutes of contact time [76]. These studies highlight the potential of polyphenol-based adsorbent materials in wastewater treatment.

5.6.5 Phenolic Hydrogels

Hydrogel, a 3D structure that tends to absorb significant quantities of water, has emerged as a versatile material platform with the potential to address global water security challenges. This is due to its unique physicochemical properties such as its ability to swell, porous structure, and affinity for water [77]. For instance, hydrogels with a large surface area can rapidly adsorb different pollutants. They can be regenerated to reuse their active sites, making them a sustainable solution for water treatment and purification [78]. A highly absorbent hydrogel possessing bactericidal properties has been developed to address the issue of microorganism-contaminated water [79]. This hydrogel can absorb contaminated water and subsequently bring out disinfected water via extrusion. Another innovative approach is a photothermal hydrogel that transforms seawater vapor into fresh water that can be collected [80]. The chemistry for forming a polyphenolic network has enabled the development of various fabrication approaches for producing nanoparticles, nanofibers, hydrogel materials, and membranes through crosslinking with other substances. Incorporating naturally occurring polyphenols has improved hydrogel systems' functional and structural properties, including increased toughness, strength, and strong adhesion, enabling hydrogels to effectively perform in complex wastewater environments [81].

In static adsorption experiments, Zhang and his fellows created a double network porous hydrogel adsorbent by combining tea residue and acrylic acid. This adsorbent has demonstrated exceptional selectivity and adsorption capacity towards heavy metal ions, including Cr^{3+}, Pb^{2+}, and Fe^{3+}. The researchers then tested the tea residue-derived hydrogel adsorbent in fixed-bed column experiments to demonstrate its ability to separate several heavy metals from simulated wastewater in a dynamic environment. These findings suggest that polyphenol-based hydrogels, particularly those derived from tea residue, have significant applications in effective and sustainable wastewater treatment technologies, particularly for separating heavy metal ions [81].

5.7 Applications of Natural Polyphenol-Based Material in Water Remediation

The composite materials based on natural polyphenols have shown significant potential in treating wastewater and industrial effluents due to their separation ability towards pollutants and heavy metal ions from water. They can be used in various treatment processes such as adsorption, solar-powered distillation, membrane filtration, advanced oxidation processes as well as water sterilization.

5.7.1 Membrane Filtration

Membrane technology involves selectively permeable membranes to separate and purify liquids or gases. The membranes are designed to allow certain molecules to pass through while rejecting others, based on size, charge, or other properties. This technology has become increasingly important in wastewater and industrial influent treatment, where it can remove suspended solids, dissolved contaminants, and microorganisms from water. Membrane processes have become popular in water treatment because of their high efficiency, energy efficiency, ease of large-scale application, and minimal environmental impact. Ultrafiltration, microfiltration, reverse osmosis, and nanofiltration are good examples of these processes [5]. Benefiting from the exceptional adhesive capability and promising hydrophilicity, coatings based on natural polyphenols have emerged as a potent approach to enhance the structure and properties of membranes. Specifically, these coatings have been utilized to optimize membranes' wettability and antifouling properties, thereby significantly improving membrane filtration performance [82]. Natural polyphenol-based composite membranes have been utilized in several membrane filtration applications such as produced water [83], desalination [84], and contaminant interception [85].

Produced water is a significant environmental concern, and its safe disposal or treatment is necessary to protect human health and the environment [83]. Applying natural polyphenol materials in membrane-based separation is an attractive approach to separating and purifying oily wastewater. For instance, an overall single-step strategy has been developed to transform membranes having highly hydrophobic properties such as polypropylene (PP), polyvinylidene fluoride (PVDF), polytetrafluoroethylene (PTFE), copper mesh, stainless steel wire, and nylon mesh into superhydrophilic and underwater superoleophobic ones by co-depositing aqueous tannic acid (TA) and aminopropyltriethoxysilane (APTES). This approach has been demonstrated to efficiently separate numerous oil-in-water emulsions having high oil rejection, elevated permeation flux, exceptional stability, and outstanding antifouling performance [86].

The TA-APTES coating has been utilized to create various materials with different properties such as superhydrophobic, adsorbent, and catalytic materials, through different secondary reactions. One such application is the development of a membrane that can remove high-concentration dyes and catalyze the integration of 4-nitrophenol in effluent water. The coupling of this coating with a silane coupling agent could also transform it into a superhydrophobic material capable of effectively separating oil and intercepting water from mixtures containing oil and water [54]. In addition, polyphenols can be utilized in creating multilayer membranes that display superior selectivity for separating various inorganic salts and charged organic dyes. MOFs have the potential to be applied in (MOFs) in advanced membrane separation applications. Using natural polyphenols, such as tannic acid, in the construction of MOF nanofiltration membranes, improved water permeability and rejection rates for salts [87]. Tannic acid, the source of natural polyphenols, has also been used to improve the separation efficiency of membranes [88].

5.7.2 Interfacial Solar Distillation

Interfacial solar distillation is a technique used to obtain freshwater using solar energy. Unlike traditional solar distillation, it heats only the evaporative quantity of water at the interface, achieving highly efficient water purification. The described method utilizes a photothermal material as a crucial element in its functional apparatus. When solar light strikes the absorber layer of the photothermal material, it transforms into thermal energy, which heats the water at the interface. The thermal insulation layer prevents heat dissipation to the surrounding water. Additionally, the channels within the absorber and insulation layers help direct the water towards the heated interface for the distillation process [89].

Natural polyphenols can enhance the efficiency of interfacial solar distillation. Composite polyphenols can offer properties with exceptional photothermal characteristics and hydrophilicity, which can improve interfacial solar distillation and water transport [90]. Indeed, the robust adhesive properties of natural polyphenols render them valuable in constructing functional devices for efficient solar distillation. For example, polyphenol-based adhesives have been used to attach photothermal materials to various substrates to create efficient solar distillation devices. Polyphenols have been incorporated into membranes and coatings to improve their adhesion to surfaces and enhance their performance in solar distillation applications [91]. Tannic acid (TA) is a widely used substance for creating functional coatings on numerous materials having photothermal properties [92].

Li et al. explained a versatile but simple approach for preparing photothermal woods. The process involved submersing the wood sequentially in a tannic acid and ferric sulfate aqueous solution. The resulting TA-Fe^{3+} coordination compounds assisted wide and high light absorption, resulting in photothermal woods having high solar-to-evaporation efficiency of approximately equal to 90% [93]. To further improve the efficiency of interfacial solar distillation, a common approach is to create an added accessible water/air interface. The TA/APTES/Fe^{3+} composite material coating was applied to a porous sponge to create a confinement capillarity effect of a very thin coating photothermal layer. This effect promoted water transportation through the numerous micro channels among black/hydrophilic nanoparticles. It exposed as much evaporation area as possible by enhancing the vapor escape channel, resulting in a great solar evaporation rate [94]. Natural polyphenols, such as lignin-derived graphene (LIG), have also been used to fabricate photothermal composite materials for efficient solar distillation.

5.7.3 Adsorption

Adsorption is a technique in which an adsorbate is removed from the gas or liquid phase by adhering to a solid surface [95]. Natural polyphenolic composites have shown great potential for adsorption applications because of many advantages, such as enhanced surface area, tunable surface chemistry, and chemical stability [74]. Natural polyphenolic composites can be modified or functionalized to selectively adsorb specific types of pollutants in wastewater, making them a versatile and effective tool for wastewater treatment. This can be achieved by adjusting the chemical composition and structure of the polyphenolic composite, such as modifying the type and amount of functional groups or incorporating other materials with specific properties. For example, LIG-based adsorbents have been prepared with plentiful phenolic and carboxylic groups for the adsorption of Co^{2+} ions [96]. Similarly, a mesopore-dominated adsorbent was prepared by blending TA, lignosulfonate, and cationic polyacrylamide for the adsorption of boron [97].

In addition to their effectiveness in adsorbing specific contaminants, natural polyphenolic composites demonstrate low cost, easy operation, and high efficiency. The availability of active sites for adsorption often limits the capability of adsorbents. However, the active sites can be easily regenerated through a desorption process, enabling reuse and ensuring economic feasibility. Furthermore, natural polyphenolic composites can be manufactured using the carbonization of different carbon foundations, making them a low-cost, renewable option for adsorption applications. Carbon materials synthesized from natural polyphenols have unique features, such as strong bonding interactions with polymeric pores, making them highly porous and suitable for use as adsorbents in environmental remediation. In conclusion, natural polyphenolic composites offer a favorable avenue for effectively and efficiently separating toxins from wastewater. With their ability to be customized for specific types of contaminants, their low cost, and their renewable nature, they are a highly desirable option for adsorption applications [98].

5.7.4 Advanced Oxidation Process

Advanced oxidation processes (AOPs) are highly effective for degrading toxic organic contaminants into non-toxic mineral salts, CO_2, and H_2O. These processes produce highly reactive free radicals from different superoxides, which possess potent oxidation capabilities [99]. One promising approach to catalyze AOPs is utilizing natural polyphenolic composites [100]. Polyphenols can act as metal catalysts and photocatalysts for pollutant degradation due to their tunable light absorption and ability to interact with various metals [101]. Additionally, natural polyphenolic composites can adsorb pollutants and facilitate their contact with catalysts, thereby accelerating AOPs. A critical aspect of AOPs is visible light-driven photocatalysis, which harnesses inexhaustible sunlight to degrade pollutants. Graphitic carbon nitride (g-C_3N_4) is a commonly utilized photocatalyst for environmental remediation, although its effectiveness can be limited due to inactive sites and the suppression of photo-induced carriers [102]. Recent studies have shown that polydopamine (PTA) nanorods, formed by self-polymerization of tannic acid (TA), can be utilized as both an adhesive as well as carbonization template to fabricate a novel g-C_3N_4-based nanorod photocatalyst. The PTA/g-C_3N_4 composite nanorods that were formed showed a considerable improvement in photocatalytic performance, breaking down 97% of rhodamine B dyes into 80 minutes and 78% of tetracycline antibiotics in 3 hour under exposure to visible light [103]. Metal oxides carrying mesoporous structure are also crucial in photocatalysis, as they possess a substantial specific surface area and promote improved interfacial transport.

Peroxymonosulfate (PMS)-based advanced oxidation processes (AOPs) rely on producing OH. and SO_4 radicals to break down organic pollutants. Copper ions (Cu^{2+}) can activate PMS and transform into Cu^+, enhancing organic pollutants' integration. However, this transformation is typically constrained by a limited pH range, particularly under acidic or neutral conditions. Recent studies have shown that adding gallic acid (GA) to the PMS/Cu^{2+} system at a pH value of 4.3 can enhance the degradation of tetrabromobisphenol A (TBBPA). The GA/PMS/Cu^{2+} structure can decay 84.6% of TBBPA within 10 minutes, which is more than twice faster than the PMS/Cu^{2+} system. The adjoining phenolic hydroxyl groups in gallic acid have a strong electron-donating and chelation ability, accelerating the reduction of Cu^{2+} to Cu^+ ions. In the peroxymonosulfate (PMS) structural system, Cu^+ is rapidly oxidized to Cu^{3+}, which leads to the prompt generation of extremely reactive OH radicals with strong oxidation capability [104].

Recent developments have led to the design of new catalysts for PMS activation, for example, biochar derived from lignin to support a mesoporous carbon-supported Cu–Fe oxides catalyst ($CuO/CuFe_2O_4$) and cobalt ferrite ($CoFe_2O_4$). The Cu–Fe oxides were evenly distributed throughout the structure of the mesoporous carbon framework, increasing the adsorption tendency of bisphenol A as well as improving its efficiency to transport electron, ultimately increasing the catalytic productivity in PMS activation. The $CoFe_2O_4$@BC composite NPs showed excellent catalytic performance in PMS activation, degrading 95.4% of sulfamethazine within 30 minutes [105]. These developments highlight the potential of natural polyphenolic composites as effective catalysts for AOPs, offering new opportunities for environmental remediation. The polyphenolic composite compounds have also been applied in advanced oxidation processes (AOPs), a promising field of study with numerous potential uses in the purification of polluted water and air.

5.7.5 Disinfection and Purification of Water

Polyphenols have surfaced as propitious agents for water disinfection and purification because of their strong antibacterial, antiviral, and antifungal properties. These compounds can effectively inhibit the growth and proliferation of various pathogenic microorganisms in water, making them useful for water treatment [106]. One of the fundamental mechanisms of action of polyphenols is their ability to disrupt the cell membrane of microorganisms, leading to their death. This disruption can occur through various mechanisms, including the development of hydrogen bonds between the membrane components and polyphenols and the oxidation of cellular components by reactive oxygen species generated by the polyphenols [107]. In addition to their direct antimicrobial effects, polyphenols can function as natural coagulants and flocculants. This means they can help remove suspended particles, turbidity, and other impurities from water, making it clearer and safer for consumption [108].

Numerous classes of polyphenols have been reviewed for their potential use in water disinfection and purification, including tannins, flavonoids, and lignin. Tannins, for instance, effectively reduce bacterial growth in water by up to 99%. On the other hand, Flavonoids have strong antiviral properties and can inhibit the replication of viruses in water. The use of polyphenols for water disinfection and purification is beautiful because it is a natural, non-toxic, and environmentally friendly approach. Unlike many chemical disinfectants, polyphenols are biodegradable and do not generate harmful byproducts that can harm the environment [109]. Overall, the use of polyphenols for water disinfection and purification shows great promise and has the potential to provide a safe, effective, and sustainable solution for ensuring access to clean water for all.

5.8 Conclusions, Challenges, and Prospects

Polyphenols have shown great perspective in the treatment of wastewater and industrial influents. Nevertheless, despite notable progress, several challenges must still be addressed. Firstly, most studies have focused on purified single-component polyphenols, which can be costly. To address this, future research should explore mixtures containing polyphenols precisely extracted from organic plants. An in-depth investigation of the relationships between the components, processing methods, and properties of naturally occurring extractives having multiple components can lead to the optimum functionalization of polyphenolic composites derived from natural sources. Secondly, achieving precise control over the functional and structural features of polyphenolic compounds is essential for the development of highly efficient water remediation devices. To achieve this, emerging processing technologies such as microfluidics and 3D printing, which offer higher accuracy, may become the future trend for fabricating naturally occurring polyphenolic compounds. Thirdly, the recycling of materials is crucial for cost-effectiveness and sustainability.

Although polyphenolic composites have shown good reusability, their decontamination efficiency may decrease during the multiple-use process. To enable the reuse of functional devices and leverage the plentiful source of polyphenols present in nature, it is mandatory to synthesize bio-degradable materials that can act as functional portions of water purification devices, such as functional coatings. A promising approach for future developments is to peel off the functional portions of natural polyphenols and further functionalize them in an environmentally friendly manner. Additionally, the oxidation of natural polyphenols at ease may offer a more stable performance of compounds, and their robustness under extreme conditions requires further investigation. In conclusion, natural polyphenols hold enormous potential for water treatment, and another study of their structure–function relationships and the use of evolving composite materials synthesis techniques will lead to their widespread use in the future. This review provides valuable insights for the optimal design of natural polyphenolic compounds that achieve highly efficient, eco-friendly, and long-lasting water remediation applications.

References

1 Liu, J., Yang, H., Gosling, S.N. et al. (2017). Water scarcity assessments in the past, present, and future. *Earth's Future* 5 (6): 545–559.

2 Kavitha, V. and Kandasubramanian, B. (2020). Tannins for wastewater treatment. *SN Applied Sciences* 2: 1–21.

3 Hoekstra, A.Y. (2014). Water scarcity challenges to business. *Nature Climate Change* 4 (5): 318–320.

4 Zhongming, Z., Linong, L., Xiaona, Y. et al. (2021). *UN World Water Development Report 2021 'Valuing Water'*. Paris, France: United Nations Educational, Scientific and Cultural Organization (UNESCO).

5 Salman, M., Shakir, M., and Yaseen, M. (2022). Recent developments in membrane filtration for wastewater treatment. *Industrial Wastewater Treatment: Emerging Technologies for Sustainability* 2: 1–25.

6 Antar, M.A., Bilton, A., Blanco, J., and Zaragoza, G. (2012). Solar desalination. *Annual Review of Heat Transfer* 15.

7 Weinberg, N. and Weinberg, H. (1968). Electrochemical oxidation of organic compounds. *Chemical Reviews* 68 (4): 449–523.

8 Ulhaq, I., Ahmad, W., Ahmad, I. et al. (2021). Engineering TiO_2 supported CTAB modified bentonite for treatment of refinery wastewater through simultaneous photocatalytic oxidation and adsorption. *Journal of Water Process Engineering* 43: 102239.

9 Das, T.K. (2001). Ultraviolet disinfection application to a wastewater treatment plant. *Clean Products and Processes* 3 (2): 69–80.

10 Ma, D., Yi, H., Lai, C. et al. (2021). Critical review of advanced oxidation processes in organic wastewater treatment. *Chemosphere* 275: 130104.

11 Wei, C., Zhang, F., Hu, Y. et al. (2017). Ozonation in water treatment: the generation, basic properties of ozone and its practical application. *Reviews in Chemical Engineering* 33 (1): 49–89.

12 Dezotti, M., Lippel, G., and Bassin, J.P. (2018). *Advanced Biological Processes for Wastewater Treatment*, vol. 10, 978–973. İsviçre: Springer.

13 Bhattacharya, S., Gupta, A.B., Gupta, A., and Pandey, A. (2018). Introduction to water remediation: importance and methods. *Water Remediation* 3–8.

14 Fernandes, S.C., Freire, C.S., Silvestre, A.J. et al. (2011). Novel materials based on chitosan and cellulose. *Polymer International* 60 (6): 875–882.

15 Abbas, M., Saeed, F., Anjum, F.M. et al. (2017). Natural polyphenols: an overview. *International Journal of Food Properties* 20 (8): 1689–1699.

16 Wang, H., Wang, C., Zou, Y. et al. (2020). Natural polyphenols in drug delivery systems: current status and future challenges. *Giant* 3: 100022.

17 Gao, X., Xu, Z., Liu, G., and Wu, J. (2021). Polyphenols as a versatile component in tissue engineering. *Acta Biomaterialia* 119: 57–74.

18 Zhang, Z., Xie, L., Ju, Y., and Dai, Y. (2021). Recent advances in metal-phenolic networks for cancer theranostics. *Small* 17 (43): 2100314.

19 Wang, T., Zhao, J., Yang, Z. et al. (2022). Polyphenolic sunscreens for photoprotection. *Green Chemistry* 24 (9): 3605–3622.

20 Sun, M., Deng, Y., Cao, X. et al. (2022). Effects of natural polyphenols on skin and hair health: a review. *Molecules* 27 (22): 7832.

21 Xu, L.Q., Neoh, K.-G., and Kang, E.-T. (2018). Natural polyphenols as versatile platforms for material engineering and surface functionalization. *Progress in Polymer Science* 87: 165–196.

22 Elnaggar, A., Leona, M., Nevin, A., and Heywood, A. (2017). The characterization of vegetable tannins and colouring agents in ancient Egyptian leather from the collection of the metropolitan museum of art. *Archaeometry* 59 (1): 133–147.

23 Liu, J. and Henkel, T. (2002). Traditional Chinese medicine (TCM): are polyphenols and saponins the key ingredients triggering biological activities? *Current Medicinal Chemistry* 9 (15): 1483–1485.

24 Ponmurugan, P., Kavitha, S., Suganya, M., and Gnanamangai, B.M. (2019). Tea polyphenols chemistry for pharmaceutical applications. *Tea Chemistry and Pharmacology* 1: 1.

25 Haslam, E. (1989). *Plant Polyphenols: Vegetable Tannins Revisited*. Cambridge, UK: CUP Archive.

26 Gorzynik-Debicka, M., Przychodzen, P., Cappello, F. et al. (2018). Potential health benefits of olive oil and plant polyphenols. *International Journal of Molecular Sciences* 19 (3): 686.

27 Ahmad, A., Ahmad, V., Zamzami, M.A. et al. (2021). Introduction and classification of natural polyphenols. Polyphenols-based nanotherapeutics for cancer. *Management* 1–16.

28 Panche, A.N., Diwan, A.D., and Chandra, S.R. (2016). Flavonoids: an overview. *Journal of Nutritional Science* 5: e47.

29 Kumar, N. and Goel, N. (2019). Phenolic acids: natural versatile molecules with promising therapeutic applications. *Biotechnology Reports* 24: e00370.

30 Khanbabaee, K. and Van Ree, T. (2001). Tannins: classification and definition. *Natural Product Reports* 18 (6): 641–649.

31 Koleckar, V., Kubikova, K., Rehakova, Z. et al. (2008). Condensed and hydrolysable tannins as antioxidants influencing the health. *Mini Reviews in Medicinal Chemistry* 8 (5): 436–447.

32 Chung, K.-T., Wong, T.Y., Wei, C.-I. et al. (1998). Tannins and human health: a review. *Critical Reviews in Food Science and Nutrition* 38 (6): 421–464.

33 Shen, T., Wang, X.-N., and Lou, H.-X. (2009). Natural stilbenes: an overview. *Natural Product Reports* 26 (7): 916–935.

34 Adlercreutz, H. (2007). Lignans and human health. *Critical Reviews in Clinical Laboratory Sciences* 44 (5-6): 483–525.

35 Siddiqui, M.F., Oh, H.-S., Rzechowicz, M. et al. (2015). Biofouling control potential of tannic acid, ellagic acid, and epigallocatechin against Pseudomonas aeruginosa and reverse osmosis membrane multispecies community. *Journal of Industrial and Engineering Chemistry* 30: 204–211.

36 Raman, C.D. and Kanmani, S. (2018). Epigallocatechin gallate supported iron particles on dye decolorization and textile wastewater treatment. *Desalination and Water Treatment* 113: 288–295.

37 Nazim, V.S., El-Sayed, G.M., Amer, S.M., and Nadim, A.H. (2023). Functionalized SnO_2 nanoparticles with gallic acid via green chemical approach for enhanced photocatalytic degradation of citalopram: synthesis, characterization and application to pharmaceutical wastewater treatment. *Environmental Science and Pollution Research* 30 (2): 4346–4358.

38 Karkoosh, H., Vithanage, M., and Sarmah, A.K. (2023). The role of anthocyanin and kaolinite in modifying cabbage leaves biochar for removal of potentially toxic elements and pharmaceutical from aqueous solution. *Environmental Pollution* 121435.

39 Shajari, M., Zamani, M., Ahmadi, N. et al. (2022). Improving the antibacterial activity of curcumin loaded nanoparticles in wastewater treatment by enhancing permeability and sustained release. *Journal of Polymers and the Environment* 30 (6): 2658–2668.

40 Li, M., Wang, H., Hu, J. et al. (2019). Smart hydrogels with antibacterial properties built from all natural building blocks. *Chemistry of Materials* 31 (18): 7678–7685.

41 Yang, W., Hu, W., Zhang, J. et al. (2021). Tannic acid/Fe^{3+} functionalized magnetic graphene oxide nanocomposite with high loading of silver nanoparticles as ultra-efficient catalyst and disinfectant for wastewater treatment. *Chemical Engineering Journal* 405: 126629.

42 Sun, Y., Wang, T., Sun, X. et al. (2021). The potential of biochar and lignin-based adsorbents for wastewater treatment: comparison, mechanism, and application—a review. *Industrial Crops and Products* 166: 113473.

43 Zhao, R., Li, Y., Mao, Y. et al. (2022). Recycling the high-salinity textile wastewater by quercetin-based nanofiltration membranes with minimal water and energy consumption. *Environmental Science & Technology* 56 (24): 17998–18007.

44 Zhou, Y. and Li, X. (2022). Green synthesis of modified polyethylene packing supported tea polyphenols-NZVI for nitrate removal from wastewater: characterization and mechanisms. *Science of The Total Environment* 806: 150596.

45 Lade, H., Paul, D., and Kweon, J.H. (2015). Combined effects of curcumin and (−)-epigallocatechin gallate on inhibition of N-acylhomoserine lactone-mediated biofilm formation in wastewater bacteria from membrane bioreactor. *Journal of Microbiology and Biotechnology* 25 (11): 1908–1919.

46 Krishnamoorthi, R., Anbazhagan, R., Tsai, H.-C. et al. (2022). Biodegradable, superwettable caffeic acid/chitosan polymer coated cotton fibers for the simultaneous removal of oils, dyes, and metal ions from water. *Chemical Engineering Journal* 427: 131920.

47 Shahidi, F. and Dissanayaka, C.S. (2023). Phenolic-protein interactions: insight from in-silico analyses–a review. *Food Production, Processing and Nutrition* 5 (1): 1–21.

48 Serra, V., Salvatori, G., and Pastorelli, G. (2021). Dietary polyphenol supplementation in food producing animals: effects on the quality of derived products. *Animals* 11 (2): 401.

49 Chen, C., Yang, H., Yang, X., and Ma, Q. (2022). Tannic acid: a crosslinker leading to versatile functional polymeric networks: a review. *RSC Advances* 12 (13): 7689–7711.

50 Sileika, T.S., Barrett, D.G., Zhang, R. et al. (2013). Colorless multifunctional coatings inspired by polyphenols found in tea, chocolate, and wine. *Angewandte Chemie* 125 (41): 10966–10970.

51 Ejima, H., Richardson, J.J., Liang, K. et al. (2013). One-step assembly of coordination complexes for versatile film and particle engineering. *Science* 341 (6142): 154–157.

52 Geng, H., Zhang, P., Peng, Q. et al. (2022). Principles of cation– π interactions for engineering mussel-inspired functional materials. *Accounts of Chemical Research* 55 (8): 1171–1182.

53 Geng, H., Zhong, Q.-Z., Li, J. et al. (2022). Metal ion-directed functional metal–phenolic materials. *Chemical Reviews* 122 (13): 11432–11473.

54 Wang, Z., Han, M., Zhang, J. et al. (2019). Designing preferable functional materials based on the secondary reactions of the hierarchical tannic acid (TA)-aminopropyltriethoxysilane (APTES) coating. *Chemical Engineering Journal* 360: 299–312.

55 Xu, Y., Hu, J., Zhang, X. et al. (2022). Robust and multifunctional natural polyphenolic composites for water remediation. *Materials Horizons*.

56 Rahim, M.A., Kristufek, S.L., Pan, S. et al. (2019). Phenolic building blocks for the assembly of functional materials. *Angewandte Chemie International Edition* 58 (7): 1904–1927.

57 Wu, J., Wang, Z., Yan, W. et al. (2015). Improving the hydrophilicity and fouling resistance of RO membranes by surface immobilization of PVP based on a metal-polyphenol precursor layer. *Journal of Membrane Science* 496: 58–69.

58 Qin, J., Liang, G., Cheng, D. et al. (2021). Controllable synthesis of iron-polyphenol colloidal nanoparticles with composition-dependent photothermal performance. *Journal of Colloid and Interface Science* 593: 172–181.

59 Rani, P., Yu, X., Liu, H. et al. (2021). Material, antibacterial and anticancer properties of natural polyphenols incorporated soy protein isolate: a review. *European Polymer Journal* 152: 110494.

60 Zhang, H., Xiao, Y., Chen, P. et al. (2022). Robust natural polyphenolic adhesives against various harsh environments. *Biomacromolecules* 23 (8): 3493–3504.

61 Shi, L., Zhang, Y., Tong, Y. et al. (2023). Plant-inspired biomimetic hybrid PVDF membrane co-deposited by tea polyphenols and 3-amino-propyl-triethoxysilane for high-efficiency oil-in-water emulsion separation. *Chinese Journal of Chemical Engineering* 53: 170–180.

62 Yu, R., Wang, H., Wang, R. et al. (2022). Polyphenol modified natural collagen fibrous network towards sustainable and antibacterial microfiltration membrane for efficient water disinfection. *Water Research* 218: 118469.

63 Wang, Y., He, Y., Yu, J. et al. (2022). Alginate-based nanofibrous membrane with robust photo-Fenton self-cleaning property for efficient crude oil/water emulsion separation. *Separation and Purification Technology* 287: 120569.

64 Obaid, M., Mohamed, H.O., Alayande, A.B. et al. (2021). Facile fabrication of superhydrophilic and underwater superoleophobic nanofiber membranes for highly efficient separation of oil-in-water emulsion. *Separation and Purification Technology* 272: 118954.

65 Ma, W., Ding, Y., Zhang, M. et al. (2020). Nature-inspired chemistry toward hierarchical superhydrophobic, antibacterial and biocompatible nanofibrous membranes for effective UV-shielding, self-cleaning and oil-water separation. *Journal of Hazardous Materials* 384: 121476.

66 Ji, Y., Wen, Y., Wang, Z. et al. (2020). Eco-friendly fabrication of a cost-effective cellulose nanofiber-based aerogel for multifunctional applications in Cu (II) and organic pollutants removal. *Journal of Cleaner Production* 255: 120276.

67 Xiang, H., Min, X., Tang, C.-J. et al. (2022). Recent advances in membrane filtration for heavy metal removal from wastewater: a mini review. *Journal of Water Process Engineering* 49: 103023.

68 Cassano, A., De Luca, G., Conidi, C., and Drioli, E. (2017). Effect of polyphenols-membrane interactions on the performance of membrane-based processes. A review. *Coordination Chemistry Reviews* 351: 45–75.

69 Yao, G., Liu, X., Zhang, G. et al. (2021). Green synthesis of tannic acid functionalized graphene hydrogel to efficiently adsorb methylene blue. *Colloids and Surfaces A: Physicochemical and Engineering Aspects* 625: 126972.

70 Zhang, C., Hu, D.-F., Xu, J.-W. et al. (2018). Polyphenol-assisted exfoliation of transition metal dichalcogenides into nanosheets as photothermal nanocarriers for enhanced antibiofilm activity. *ACS Nano* 12 (12): 12347–12356.

71 Singh, K.K., Senapati, K.K., and Sarma, K.C. (2017). Synthesis of superparamagnetic Fe_3O_4 nanoparticles coated with green tea polyphenols and their use for removal of dye pollutant from aqueous solution. *Journal of Environmental Chemical Engineering* 5 (3): 2214–2221.

72 Bu, Y., Huang, J., Zhang, S. et al. (2018). Robust superhydrophobic surface by nature-inspired polyphenol chemistry for effective oil-water separation. *Applied Surface Science* 440: 535–546.

73 Crini, G., Lichtfouse, E., Wilson, L.D., and Morin-Crini, N. (2019). Conventional and non-conventional adsorbents for wastewater treatment. *Environmental Chemistry Letters* 17: 195–213.

74 Sadegh, N., Haddadi, H., Sadegh, F., and Asfaram, A. (2022). Recent advances and perspectives of tannin-based adsorbents for wastewater pollutants elimination: a review. *Environmental Nanotechnology, Monitoring & Management* 100763.

75 Gao, M., Xu, D., Gao, Y. et al. (2021). Mussel-inspired triple bionic adsorbent: facile preparation of layered double hydroxide@ polydopamine@ metal-polyphenol networks and their selective adsorption of dyes in single and binary systems. *Journal of Hazardous Materials* 420: 126609.

76 Gurung, M., Adhikari, B.B., Alam, S. et al. (2013). Adsorptive removal of Cs(I) from aqueous solution using polyphenols enriched biomass-based adsorbents. *Chemical Engineering Journal* 231: 113–120.

77 Sinha, V. and Chakma, S. (2019). Advances in the preparation of hydrogel for wastewater treatment: a concise review. *Journal of Environmental Chemical Engineering* 7 (5): 103295.

78 Van Tran, V., Park, D., and Lee, Y.-C. (2018). Hydrogel applications for adsorption of contaminants in water and wastewater treatment. *Environmental Science and Pollution Research* 25: 24569–24599.

79 Li, S., Dong, S., Xu, W. et al. (2018). Antibacterial hydrogels. *Advanced Science* 5 (5): 1700527.

80 Sharshir, S.W., Algazzar, A.M., Elmaadawy, K. et al. (2020). New hydrogel materials for improving solar water evaporation, desalination and wastewater treatment: a review. *Desalination* 491: 114564.

81 Zhang, S., Liu, C., Yuan, Y. et al. (2020). Selective, highly efficient extraction of Cr(III), Pb(II) and Fe(III) from complex water environment with a tea residue derived porous gel adsorbent. *Bioresource Technology* 311: 123520.

82 Chen, X., He, Y., Fan, Y. et al. (2019). Nature-inspired polyphenol chemistry to fabricate halloysite nanotubes decorated PVDF membrane for the removal of wastewater. *Separation and Purification Technology* 212: 326–336.

83 Zhang, J., Wang, L., Zhang, H. et al. (2022). Biomimetic modified polypropylene membranes based on tea polyphenols for efficient oil/water separation. *Progress in Organic Coatings* 164: 106723.

84 Xu, D., Liang, H., Zhu, X. et al. (2020). Metal-polyphenol dual crosslinked graphene oxide membrane for desalination of textile wastewater. *Desalination* 487: 114503.

85 Tong, X., Liu, S., Qu, D. et al. (2021). Tannic acid-metal complex modified MXene membrane for contaminants removal from water. *Journal of Membrane Science* 622: 119042.

86 Wang, Z., Ji, S., He, F. et al. (2018). One-step transformation of highly hydrophobic membranes into superhydrophilic and underwater superoleophobic ones for high-efficiency separation of oil-in-water emulsions. *Journal of Materials Chemistry A* 6 (8): 3391–3396.

87 Xiao, Y., Zhang, W., Jiao, Y. et al. (2021). Metal-phenolic network as precursor for fabrication of metal-organic framework (MOF) nanofiltration membrane for efficient desalination. *Journal of Membrane Science* 624: 119101.

88 Xu, Y., Xiao, Y., Zhang, W. et al. (2021). Plant polyphenol intermediated metal-organic framework (MOF) membranes for efficient desalination. *Journal of Membrane Science* 618: 118726.

89 Chen, R., Zhang, T., Kim, J. et al. (2021). Interfacial solar distillation for freshwater production: fate of volatile and semivolatile organic contaminants. *Environmental Science & Technology* 55 (9): 6248–6256.

90 Liu, F., Liang, W., Wang, C. et al. (2020). Superhydrophilic and mechanically robust phenolic resin as double layered photothermal materials for efficient solar steam generation. *Materials Today Energy* 16: 100375.

91 Behboodi-Sadabad, F., Li, S., Lei, W. et al. (2021). High-throughput screening of multifunctional nanocoatings based on combinations of polyphenols and catecholamines. *Materials Today Bio* 10: 100108.

92 Zhang, C., Chen, Z., Xia, Z. et al. (2020). Ferric tannate photothermal material for efficient water distillation. *Environmental Science: Water Research & Technology* 6 (4): 911–915.

93 He, F., Han, M., Zhang, J. et al. (2020). A simple, mild and versatile method for preparation of photothermal woods toward highly efficient solar steam generation. *Nano Energy* 71: 104650.

94 Wang, Z., Han, M., He, F. et al. (2020). Versatile coating with multifunctional performance for solar steam generation. *Nano Energy* 74: 104886.

95 Rashid, R., Shafiq, I., Akhter, P. et al. (2021). A state-of-the-art review on wastewater treatment techniques: the effectiveness of adsorption method. *Environmental Science and Pollution Research* 28: 9050–9066.

96 Budnyak, T.M., Modersitzki, S., Pylypchuk, I.V. et al. (2020). Tailored hydrophobic/hydrophilic lignin coatings on mesoporous silica for sustainable cobalt (II) recycling. *ACS Sustainable Chemistry & Engineering* 8 (43): 16262–16273.

97 Xia, N.N., Zhang, H.Y., Hu, Z.H. et al. (2021). A functionalized bio-based material with abundant mesopores and catechol groups for efficient removal of boron. *Chemosphere* 263: 128202.

98 Li, M., Wang, X., Gong, G. et al. (2022). Natural polyphenol-based nanoengineering of collagen-constructed hemoperfusion adsorbent for the excretion of heavy metals. *Journal of Hazardous Materials* 428: 128145.

99 Munter, R. (2001). Advanced oxidation processes–current status and prospects. *Proceedings of the Estonian Academy of Sciences. Chemistry* 50 (2): 59–80.

100 Yang, Z., Hou, J., Pan, Z. et al. (2022). A innovative stepwise strategy using magnetic Fe_3O_4-co-graft tannin/polyethyleneimine composites in a coupled process of sulfate radical-advanced oxidation processes to control harmful algal blooms. *Journal of Hazardous Materials* 439: 129485.

101 Zheng, A.L.T., Boonyuen, S., and Andou, Y. (2021). Green synthesis of Ag/Cu_2O using tannic acid as highly efficient photocatalyst for tetracycline degradation. *SPAST Abstracts* 1 (01).

102 Zhao, D., Wang, Y., Dong, C.-L. et al. (2021). Boron-doped nitrogen-deficient carbon nitride-based Z-scheme heterostructures for photocatalytic overall water splitting. *Nature Energy* 6 (4): 388–397.

103 Yang, D., Wang, W., An, K. et al. (2021). Bioinspired construction of carbonized poly(tannic acid)/g-C_3N_4 nanorod photocatalysts for organics degradation. *Applied Surface Science* 562: 150256.

104 Wang, Y., Wu, Y., Yu, Y. et al. (2020). Natural polyphenols enhanced the Cu (II)/peroxymonosulfate (PMS) oxidation: the contribution of Cu (III) and HO•. *Water Research* 186: 116326.

105 Wang, G., An, W., Zhang, Y. et al. (2022). Mesoporous carbon framework supported Cu-Fe oxides as efficient peroxymonosulfate catalyst for sustained water remediation. *Chemical Engineering Journal* 430: 133060.

106 Li, J., Feng, C., Jin, J. et al. (2022). Current understanding on antibacterial mechanisms and research progress of tea polyphenols as a supplementary disinfectant for drinking water. *Journal of Water and Health* 20 (11): 1611–1628.

107 Lobiuc, A., Pavăl, N.-E., Mangalagiu, I.I. et al. (2023). Future antimicrobials: natural and functionalized phenolics. *Molecules* 28 (3): 1114.

108 Cuesta-Parra, D.M., Correa-Mahecha, F., Rodríguez-Miranda, J.P. et al. (2022). Extraction of polyphenols from unripened coffee (Coffea arabica) residues and use as a natural coagulant for removing turbidity. *Processes* 10 (6): 1105.

109 Vijayaraghavan, G., Sivakumar, T., and Kumar, A.V. (2011). Application of plant based coagulants for waste water treatment. *International Journal of Advanced Engineering Research and Studies* 1 (1): 88–92.

6

Polyphenols for Anticorrosion Application

Shveta Sharma[1,2], *Richika Ganjoo*[2], *and Ashish Kumar*[3]

[1] *Department of Chemistry, Government College Una, Affiliated to Himachal Pradesh University, Una, India*
[2] *Department of Chemistry, School of Chemical Engineering and Physical Sciences, Lovely Professional University, Punjab, India*
[3] *Nalanda College of Engineering, Bihar Engineering University, Department of Science, Technology and Technical Education, Government of Bihar, India*

6.1 Introduction

Metals are the most prevalent material used in modern civilization and in a wide variety of sectors; yet, because of how they react to the surroundings, the quality of metals degrades gradually but unavoidably and irreversibly due to their vulnerability [1–3]. Before this study, it was thought that the primary method for preventing the corrosion of metals was to separate the metals from the corrosive agents. Inhibitors are today one of the most prominent practical strategies for protecting against corrosion. In order to inhibit or slow the corrosion of metals, many efforts have been made to identify compounds that are suitable for use as corrosion inhibitors in a variety of corrosive situations [4–6]. In recent years, a lot of studies have been carried out on the suppression of metal corrosion by synthetic organic chemicals, which are often considered to be substances that are harmful to the environment. As a preventive measure against corrosion, organic chemicals are frequently employed. Despite this, the cumulative toxicity of organic compounds has a deleterious impact on the surrounding ecology. In addition, organic compounds need to be replaced with biodegradable inhibitors in order for environmental protection to be effective. These inhibitors are harmonious with nature and do not have any adverse effects on living beings [7]. In the realm of corrosion research, sustainable or "green" corrosion inhibitors are attracting a lot of attention mainly because they are non-hazardous and biodegradable, in addition to having the capability to be reprocessed.

Science and Engineering of Polyphenols: Fundamentals and Industrial Scale Applications,
First Edition. Edited by Chandrabhan Verma.
© 2024 John Wiley & Sons Inc. Published 2024 by John Wiley & Sons Inc.

The field of green inhibitors is still considerably unexplored, despite the fact that a major proportion of ground-breaking studies have been carried out and that a huge number of research articles have been produced. Because of the exceptionally high degree of interest in the concept, the number of chemicals that were investigated significantly increased. The researchers who carried out this study refer to the chosen inhibitors as being environmentally friendly, environmentally sustainable, or appropriate. Research has been done on several different classes of chemicals derived from natural sources to investigate their potential as corrosion inhibitors for aluminum, mild steel, and carbon steel materials. An overwhelming number of these investigations have focused their attention on extracts that were taken from natural sources. The outcomes of the investigations that were carried out indicate that the primary components of these extracts consist of an extensive number of organic compounds. As a consequence of this, practically all of these compounds have a large number of bonds in addition to atoms of oxygen (O), nitrogen (N), sulfur (S), and phosphorus (P), which act as bonding sites for the adsorption of the compounds onto the surface of the metal [8–11]. For the purpose of reducing the effects of corrosion, the various chemicals extracted from plants like amino acids, alkaloids, and polyphenols are utilized, even though commercially, these are not beneficial. In spite of this, there is a potential that this method will one day be utilized for the purpose of protecting against corrosion, particularly if there is a greater demand for the aforementioned service. Because of this, there seems to be a lot of enthusiasm for researching the impacts of utilizing these naturally occurring compounds that have strong inhibitor qualities in the field of metal corrosion protection, particularly in light of the current environmental condition.

In addition to being less toxic than most other plant extracts, including alkaloid extracts, polyphenol-based extracts appear to fulfill maximum conditions for inhibition and biological degradation of materials [11–13]. In addition to having a phenolic quality, the structural diversity of polyphenols causes them to exhibit a number of different physicochemical characteristics. Even more impressive are the processes of extracting, isolating, characterizing, and measuring polyphenols. On the other hand, recent technological improvements have resulted in an increased quantity of polyphenols in plants and a greater structural variety among them. The development of a standardized method for dealing with polyphenols is hampered by the complexity of glycosylation and polymerization processes. It has also been discovered that the displacement of phenolic compounds in plants varies at the tissue, cellular, and sub-cellular levels [14–17]. Based on the chemical structure of polyphenols, one can place them into a wide variety of different families, which leads to different physicochemical properties that ultimately have an influence on polyphenol extraction. Therefore, they are a promising option for protecting metals from corrosion.

Phenolic compounds, often known as polyphenols, are members of one of the chemical families found in plants with the greatest diversity and distribution, identifiable by a phenyl ring as well as numerous connected hydroxyl groups (−OH) [18–20]. Polyphenols provide defense against parasitic organisms, pathogenic organisms, free radicals, and ultraviolet light as a part of a plant's secondary metabolism. Polyphenols are a broad and diverse class of chemicals that include at least one phenyl ring along with one or even more OH groups in their chemical makeup. There are over 10,000 distinct polyphenol compounds that are known to exist. In their natural environments, polyphenols can exist in a wide variety of forms, from molecules as straightforward as phenolic acids to molecules as complex and heavily polymerized as tannins. They are found conjugated the vast majority of the time, with one or more sugar by-products connected with OH groups [21–24]. The carbon atom of the phenyl, however, has strong bonds with the sugar group. Glucose is the most common sugar residue, although other mono-, dis-, and oligosaccharides are possible as well, including galactose, rhamnose, xylose, glucuronic acid, and galacturonic acid. There are still plant products like exotic fruits and cereals whose polyphenol composition has not been determined. Not only that, but polyphenols regularly link up with other molecules, including carboxylic or organic acids, amines, fatty acids, and even other phenols [25, 26]. Approximately 8,000 distinct forms of polyphenols have been identified so far, with the vast majority occurring in common foods like fruits, vegetables, and beverages. The primary structural variations among compounds are found in their varying degrees of oxidation, hydroxylation, methylation, and glycosylation [27–29]. Figure 6.1 provides a basic structure for the four main types of polyphenols (which consist of non-flavonoids, flavonoids, phenolic acids, and coumarins).

In general, every subgroup of polyphenols has a one-of-a-kind structure, and their biological activities set them apart from other potential candidates for use in the pharmaceutical and nutraceutical industries. In addition to the possibility that they could have a therapeutic effect, it has been demonstrated that polyphenols are intriguing options for the inhibition of corrosion. Polyphenols, by virtue of their antioxidant capabilities, are capable of preventing the oxidation of metal's outermost layer, which is one of the most common causes of corrosion. In recent years, polyphenols have attracted a lot of attention due to the fact that they have the potential to act as environmentally friendly corrosion inhibitors.

They not only have strong corrosion reduction characteristics, but they are also substances that occur naturally with desirable basic features, which makes them an appealing choice for usage in the industry because of the potential benefits that may be realized through their application. As a result of this, the use of polyphenols as corrosion inhibitors for various metallic substances such as steel, copper, and aluminum has been the focus of a number of research studies. The findings of this research are quite encouraging since they suggest that polyphenols can

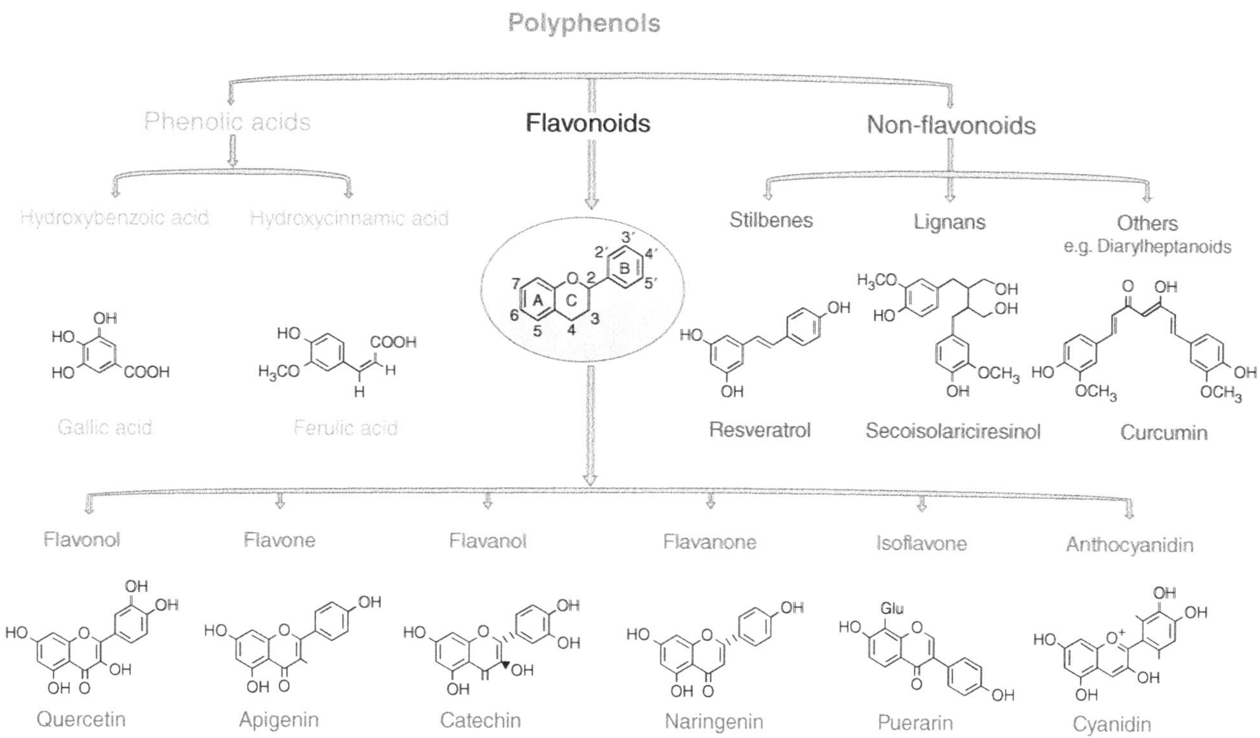

Figure 6.1 Classification of polyphenols. Source: Adapted from Rambaran [30] CCBY 4.0.

successfully prevent corrosion and protect metal structures. Additionally, obtaining polyphenols from natural plant extracts may be able to provide an alternative to conventional corrosion inhibitors (which may have hazardous chemicals) that are less harmful to the environment and more sustainable as well, which often contain harmful chemicals.

6.2 Anticorrosive Properties of Polyphenols: Literature Survey

Polyphenols are now recognized as being new types of important naturally occurring compounds derived from plants. The majority of the appearance of aromatic OH groups can be attributed to the anticorrosion characteristics of these compounds. In addition to this, this is the case due to the fact that heteroatoms can transfer electrons to metals in order to fill empty d-orbitals, resulting in the formation of coordination bonds. The presence of ring structure (because of conjugated bonds or electrons) makes it easier for molecules to adsorb onto metal surfaces. The leaf extract (MIL) of the very abundantly present tropical plant *Mangifera indica* L. was prepared, and it has been found that it mitigated the corrosion of commercial steel in 3.5 wt% NaCl by Veedu et al. [31]. An excellent anticorrosive property was provided by it because of its high polyphenol content. At the metal–electrolyte interface, corrosion can be stopped by the formation of an undissolved organometallic complex between selected metal functional groups present in the selected inhibitor. This complex prevents corrosion from occurring. A novel and straightforward method for producing bio-extract with epoxy coating on steel has been implemented here. The epoxy coating includes a MIL-precipitated amorphous silica hybrid as an ingredient. Electrochemical testing showed that the newly created coating has a corrosion mitigation capacity of 99%, which is higher to that of other coatings. Therefore, the current findings provide evidence of a new corrosion inhibitor that is safe for the environment and effective for protecting steel from saline environments [31].

Ramirez et al. carried out research with the goal of enhancing the corrosion resistance of metal frames and formed two layers formed by organic corrosion inhibitors (tannins) and anodic protection through the assimilation of zinc oxide (size in nanoscale). In this investigation, two layers were studied by using at 0, 720 hours at lab conditions, and almost –four months of atmospheric conditions. According to the findings, the combination of tannins derived from pine bark and contained zinc oxide nanoparticles is an effective substitute for commercial coatings that contain a greater proportion of synthetic substances. These coatings are now available for purchase. Despite the fact that the properties of thin layer formed suffer a little reduction, investigations conducted by changing the time of exposure revealed they are still able to be classified as high-performance coatings [32].

Two primers containing organic corrosion inhibitors were developed by L.F. Montoya et al. using bark from *Pinus radiata* trees. The bark was obtained from the waste produced by the paper and wood industries [33]. Spectroscopy techniques, along with electrochemical analysis were used for the extract and coating characterizations [33]. Leaves of *Cucumis anguria* L. were examined for phytochemicals and their capabilities to mitigate corrosion in steel samples in acidic solution by Chung et al. [34]. First, various phytochemicals, such as phenols (TPC) and total flavonoids (TFC) 89.32 and 35.88 mg/g, were determined, respectively. Further, phenolic component was minutely investigated, and myricetin, quercetin, salicylic acid, chlorogenic acid, and kaempferol were quantified. This intriguing identification led the way for the identification of the corrosion-reducing capacity of the extract using electrochemical and weight loss experiments. The weight loss study proved good efficiency with an inhibitor concentration of 250 ppm. Raman spectroscopy demonstrated the existence of a defensive coating of material on the surface of the metal, and the maximum inhibition efficiency was 92.39% at 250 ppm of the concentration. The potentiodynamic polarization (PDP) investigation for metal under study by using *C. anguria* L. leaf extract as inhibitor with varying concentrations was studied and demonstrated that a rise in the concentration of the selected inhibitor results in an increase in the corrosion potential while simultaneously resulting in a decrease in the corrosion current an inhibitor reduce both cathodic and anodic reactions. As a result of the inhibitor's ability to adhere to and adsorb onto the surface of the metal, the metal is protected from the corrosive medium, and therefore the rise in the corrosion potential values was there when compared to the acidic medium without an inhibitor. The plant extract will adsorb onto the surface of the metal once the inhibitor has been introduced, obstructing the pores of the metal. As a result, a defensive coating is formed on the metallic sample, increasing its resistance to corrosion. Various electrochemical parameters were calculated from the electrochemical analysis. In the Nyquist plot, the formation of a protective layer on the metallic sample specimen is represented by semicircles whose diameters increase as the leaf extract concentration rises. The presence of a thin film (defensive in nature) on the surface of sample was initially suspected based on the weight loss study and then confirmed by surface morphological techniques. Overall, the research supports employing *C. anguria* L. as a green inhibitor against acid corrosion, and its potential use as a commercial inhibitor should be considered.

In order to better protect steel samples from corrosion, 3-aminopropyltriethoxysilane (APTES)-functionalized zinc oxide nanoparticles and an epoxy resin containing tannin as an inhibitor were investigated by Jaramillo et al. [35]. Standardized methods such as electrochemical techniques and SKP were employed to evaluate the properties of the film. With the help of polarization curves, tannin extract significantly reduced the corrosion I_{corr} via altering the cathodic behavior but had no effect on the active anodic behavior. Similar to the protective layer created by the tannin-iron complex, electrochemical impedance spectroscopy (EIS) study demonstrated that coatings containing surface-modified nanoparticles offer superior protection against corrosion. The impedance module graphs were used to identify the coating that offered the best protection from corrosion, and the best results were obtained from test items coated with a 1% solution of funct-ZnO-NPs [35]. To conclude, it was shown that using zinc oxide nanoparticles in the coating formula resulted in a small improvement to the surface properties and partially increased the hydrophobicity. Tannins successfully produce an iron tannate protective layer, as shown by the results.

The *Cryptostegia grandiflora* leaf extract with phenolic and flavonoid was examined by Prabakaran et al. [36]. Extract was tested for its ability to prevent corrosion on mild steel in an acidic medium. When used at a concentration of 500 ppm, the inhibitor was able to achieve an inhibitory efficiency of up to 87.54%. According to the results of the polarization tests, the extract has a mixed inhibitory effect. SEM-EDS and atomic force microscopy (AFM) analysis both backed up the experimentation that an inhibitor causes a productive layer to form on mild steel. The extract was adsorbed onto the mild steel in a manner that was consistent with the Temkin adsorption isotherm [36].

Magnesium (Mg) and the alloys made from it have garnered a lot of attention in the past 10 years because they have the potential to be biodegradable metallic materials. However, there are still obstacles to overcome because of the material's high rate of corrosion and inadequate biocompatibility after implantation. In this study, we produce a green tea phenol–metal induced multilayer conversion coating that is both simple to use and highly adaptable, and the surface morphology techniques revealed that, with the incorporation of Mg^{2+}, in coating, it showed more homogeneous and dense morphology, more planar structure with very less pits than the pure EGCG coating. Corrosion resistance was investigated by Zhang et al. [37] using electrochemical analysis, which validated the effectiveness of EGCG/Mg coating. These findings pointed to the possibility of employing a green tea polyphenol-induced Mg^{2+}-rich multilayer conversion coating for the purpose of achieving the desired level of biocompatibility in biodegradable cardiovascular implants and increasing the level of protection against corrosion.

Since ancient times, people have recognized the medicinal benefits of *Panax ginseng C.A.* (*P. ginseng*) Meyer and have used it for this purpose. Although a number of studies have been done that look into the capacities that phytoconstituents serve in human and animal health, there is still a lot that is unknown. Phenols such as chlorogenic acid, gentisic acid, and rutin were present in large quantities in fruit; on the other hand, leaf exhibited noteworthy quantities of *m*-coumaric acid and *p*-coumaric acid. The corrosion-inhibiting capacity of the fruit extract with methanol was studied for a copper specimen in an acidic medium, and around 96% of the corrosion-mitigating capacity at a 1000-ppm concentration by Malathy et al. [38]. When the surface of the specimen was in contact with the test

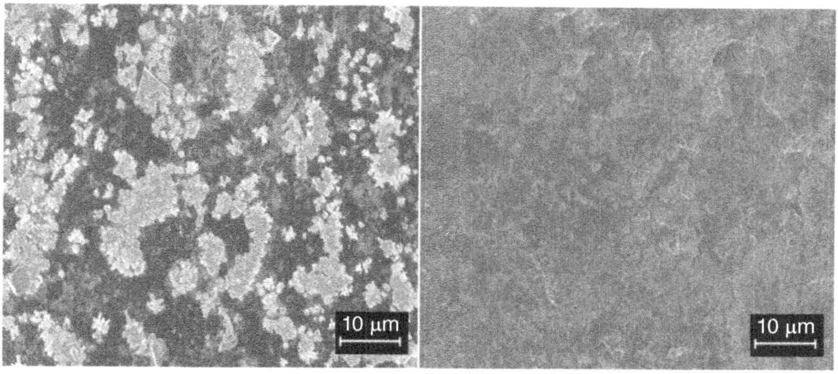

(a) (b)

Figure 6.2 Pictorial presentation of FESEM micrographs of copper dipped in (a) blank and (b) acidic solution with 1000 ppm of inhibitor. Source: Ramalingam Malathy et al. [38]/Reproduced from MDPI/CC BY 4.0.

solution, its surface was studied by field emission scanning electron microscope (FESEM) in image (Figure 6.2), and also it validated the corrosion-reducing tendency of ginseng plant extract with the help of AFM, as AFM confirmed the reduced roughness of the surface after using ginseng extract as an inhibitor.

Further, Figure 6.3 is a representation of the EDX spectra that display the elements that are present in copper that have been dipped in acidic medium as well as copper that has been dipped in acidic medium with 1000 ppm of inhibitor (extracted from the plant under study).

EDX also proved that the inhibitor has been successfully protecting the copper metal from corrosion. As there was a significant decrease in the wt% of carbon, oxygen and chlorine and copper wt% increased.

Polyphenol coating with phosphate ions solutions was developed to reduce corrosion on iron surfaces by Jain et al. [39]. Moreover, the reactivity of coating at the surface was evaluated by various surface morphological techniques.

Polymerization was used for preparing poly(*m*-aminophenol) (PmAP), and further characterization was performed with SEM, FTIR spectra, etc., by Quan et al. [40]. According to the findings, the coatings that included PmAP showed a significant capacity to prevent biofilm formation, which is the single most effective method of delaying biofouling's onset. Further, EIS was used to get an idea about the capacity of coating in corrosion protection, and it was validated that PmAP behaved as an excellent inhibitor. The experiment's findings may lead to significant advances in our understanding of how to best design and implement materials with multiple purposes for use in coatings manufacture.

Bee pollen extract (BPE) was investigated for its capacity to inhibit the corrosion of copper in acidic media by Ahmed et al. [41]. BP powder can have a wide variety

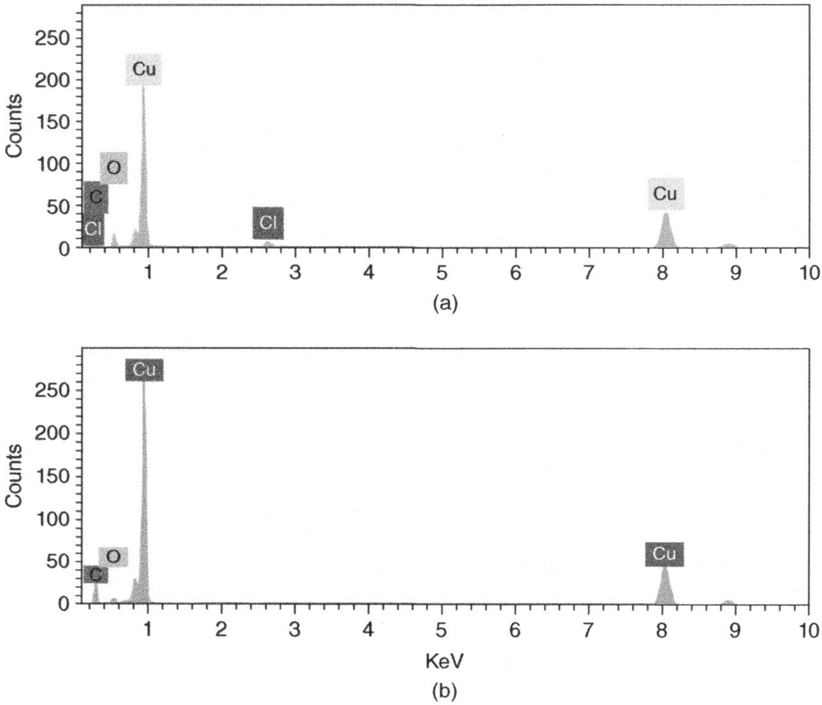

Figure 6.3 EDX spectra of copper submerged in (a) pure acid solution and (b) acid with inhibitor. Source: Adapted form from Malathy et al. [38]; CCBY 4.0.

of component mixes, which are determined by the botanical and geographical origin of the product. The nutritional value of BP is quite high, and it contains a significant amount of antioxidants, vitamins, quantities of natural antibiotics, etc. It is also rich in carotenoids, flavonoids, polyphenols, and other healthy compounds. The main BPE components were isolated and identified using gas chromatography-mass spectrometry. Various other techniques like electrochemical analysis including EIS, potentiodynamic polarization (PDP), weight loss measurement, and other multiscale testing techniques, were employed to investigate the corrosion reduction behavior for the pure copper metal in an acidic medium with a specific quantity of BPE (0–7 g/l). Adsorption was in accordance with Langmuir-type isotherm with the control on both the anodic and cathodic reactions and showed extraordinary inhibiting properties for controlling the corrosion of copper in acidic medium. With the increased concentration of BPE, an increased inhibition efficacy, and at 7 g/l of concentration, an inhibition efficacy of 94.5% was observed. Various spectroscopic techniques, such as Fourier-transform infrared spectroscopy (FTIR), and surface morphological techniques, such as AFM, were used to validate the corrosion reduction of the metallic sample. The adsorption on the surface of the copper is thought to be the mechanism behind the corrosion-preventing abilities of BPE when applied to pure copper. The results of this investigation shed light on how effective BPE could be as a potential corrosion inhibitor of copper.

Prabakaran et al. [42] worked to identify the phenolic content present and antioxidative capacity of *Epipremnum aureum* leaves. The amount calculated for phenol and flavonoid contents was 125.21 and 52.22 mg/g in methanolic *E. aureum* extract (EA-MeOH). Various assays were performed depending upon the antioxidative potency of phenol, where with 100 μg/ml EA-MeOH, it showed the greatest suppression of free radicals. In light of the growing interest in the anticorrosive properties of plant-based antioxidants, the effect of 100% EA-MeOH on low-carbon steel in acidic medium was tested. The inhibition efficiency was calculated based on many techniques such as weight loss, atomic absorption spectroscopy analyses and surface analysis techniques (FTIR, UV–Vis, and XRD). The results showed after FTIR analysis that absorption bands appear on 3611.86, 3018.72, 1043.53, and 756.12 cm^{-1}, and for metallic samples when EA-MeOH was utilized as inhibitor, UV–Vis bands appeared at absorbance bands at 221 and 251 nm. The inhibition efficacy of 92.37% was given with 600 ppm of *E. aureum* extract in acidic medium at specific test conditions. All of these show that the inhibitor forms a layer of protection around the specimen being studied.

Hlushko et al. [43] synthesized polymeric antioxidants and investigated for their potential as a potent metal corrosion inhibitor. The polymeric units were coated by using epoxy resin on alloy of aluminum in the form of a very thin layer of the order of sub-micrometer. Significant protection against corrosion was provided to alloy by thin coatings of antioxidant polymer. Antioxidant polymers, in contrast to poly(hexyl methacrylamide), which lacked polyphenol functional units, were compatible with normal epoxy coatings and provided remarkable protection of surfaces against extreme corrosion for more than 100 days in saline solution of particular concentration. EIS, SEM, and optical microscopy were utilized in order to characterize the anticorrosion performance of coatings that contained antioxidants and additionally analyzed using the comparable electrical circuit analysis using the data from the EIS. Importantly, unlike small-molecule analogs like tannic acid, antioxidant copolymers could not escape from epoxy coatings into water or ethanol. This was a key difference. When all of these results are considered together, it becomes clear that antioxidant copolymers are effective in preventing metal corrosion.

Coal tar epoxy, also known as CTE coatings, provides efficient protection for steel buildings in hazardous environments, such as immersion and atmosphere. But because of its hazardous nature, its use has been banned in countries like the United States and Japan. The usage of Kraft lignin, a bio-polyphenol that is regarded safe and acquired as waste from the pulp and paper industry, in order to prepare the epoxy coating, which will be free from coal tar content. The goal was to replace the use of CTE coatings, and it does so by proposing the use of Kraft lignin.

Diogenes et al. [44] put lignin through an acetylation procedure so that it would be more compatible with bisphenol, a diglycidyl ether (DGEBA). This was done so that the results of their experiment would be more positive. The epoxy-lignin resin was produced by integrating acetylated lignin into DGEBA, which resulted in the resin. There were three distinct formulations of epoxy-lignin resin that were manufactured. The structural analysis was performed by FTIR. Isophorone diamine, abbreviated as IPDA, was the chemical that was used to cure the resin during the coating preparation. The DGEBA and CTE coatings were used as a point of reference, while the resulting thermosetting material was investigated in terms of the chemical, thermal, anticorrosive, and UV-blocking capabilities it possessed. EIS was utilized in order to appraise the anticorrosive qualities of the coatings. The pull-off method (ASTM D4541) was used to determine how well the coatings adhered to the substrate. A 100% absorption of ultraviolet light in the UV-A and UV-B areas was seen in the coatings supplemented with acetylated lignin, which is a quality that is highly valuable in order to prevent ultraviolet damage and the decline in durability of epoxy coatings. According to the results of the EIS, the epoxy-lignin coatings (with 7.5% and 15% lignin content) provided modulus of impedance values that were comparable to the commercially available coating as well as superior to the CTE coating. This indicates that this coating may be useful as a corrosion protector.

In a wide variety of cutting-edge applications, polybenzoxazine has proven to be an excellent and extremely promising resin. These coatings were produced using paraformaldehyde by Deng et al. [45]. Curcumin is a yellowish polyphenolic compound that can be taken out from rhizomes of turmeric. Due to its biochemical actions, such as antioxidant, antibacterial, antiviral, anti-inflammatory, and antiproliferative, curcumin possesses the potential to have a pharmacological effect and great effectiveness in therapy. In addition to this, the functional phenolic component in curcumin makes it possible for polymer synthesis to be carried out at a high level of efficiency. To the best of our knowledge, no recent research has ever used curcumin as a phenolic resource to synthesize benzoxazine. It is anticipated that the incorporation of curcumin as a chemical feedstock for benzoxazine production will result in improvements to the resins' anticorrosion property, physical and chemical resilience, and antibacterial and antiviral activity, as well as their sustainability. Incorporating the chemical moieties or polymer chains of benzoxazine units is a good approach to take if you want to inherit other functionalities that benzoxazine possesses and broaden the applications that it can be used for. This curcumin-based polybenzoxazine resin was experimentally proved for the first time to be a successful protective layer. It was shown to be able to demonstrably reduce the corrosion rate of the substrate metal and greatly resist the fouling attachment. The antifouling performance of the coating was found to be further enhanced by an increased content of PEG in the resin, as this was discovered to be a contributing factor. Various characterization techniques were utilized such as spectroscopic methods, calorimetric techniques, and thermogravimetric technique for characterizing benzoxazine monomer and polymer. The anticorrosion and antifouling of the developed polybenzoxazine resins were evaluated by PDP, and EIS. Even after prolonged use in hostile conditions, the polybenzoxazine coatings demonstrated outstanding performance in terms of their ability to prevent corrosion, as demonstrated by the results, and few polybenzoxazine coatings, like PCBG-10 and PCBG-15, had an excellent antifouling capacity. The I_{corr} values decreased when the sample was coated with the coating under investigation as compared to the bar metallic sample (I_{corr} for the bare sample was 4.721 μA/cm^2, 3.456×10^{-3} to 3.864×10^{-6} μA/cm^2) and the values of E_{corr} for coated metallic sample increased as compared to the bare sample in comparison to the uncoated steel, it was abundantly clear that every coating offered significantly improved levels of protection. However, a higher concentration of PEG in the polybenzoxazine coating made it easier for the corrosive medium to make contact with the metal surface. As a result, the coating exhibited a higher I_{corr} value and a lower PE value. It was hypothesized that the hydrophilicity of the PEG in the coating helped to maximize the coating's capacity to absorb water and sped up the process by which water moved throughout the coating. A double-layer coating system was devised with the intention of simultaneously improving the anticorrosion and antifouling properties. The PCB resin was used as the sub-layer, and either the PCBG-10 or PCBG-15 film was used as the up-layer. Electrochemical tests conducted in salt immersion and salt-spray environments provided conclusive evidence of the coatings' exceptional ability to resist corrosion over the long term. It is anticipated that this newly designed version of the promising polybenzoxazine resin coating system will discover widespread uses in the maritime industry and other related fields.

Purple rice bran extract, by farming waste products high in antioxidants, was used to create a green corrosion inhibitor that is safe for the environment, which was investigated by Pal and Das [46]. This purple rice bran extract (PRBE) was used to evaluate the corrosion efficacy in two types of acidic media, one in hydrochloric acid and second in sulfuric acid medium. The corrosion mitigation capacity was confirmed by some electrochemical techniques. Maximum efficacy came out to be 91.13% and 87.43% in HCl and H_2SO_4 media with inhibitor, respectively. Langmuir adsorption was obeyed and inhibitor mitigated both cathodic and anodic reactions. Despite being highly resistant to corrosion, polybenzoxazine (PBZ) has proven difficult to use due to its brittleness and poor adhesion strength. In this work, Cao et al. added ether amine molecules into benzoxazine monomers, and then copolymerization was done with tannic acid (TA) to form flexible PB-TA coating, to enhance corrosion mitigation property [47]. Compared to polyurethane and epoxy resin coatings, the PB-6TA coating has been shown to have significantly higher corrosion resistance in additional trials.

Oil palm (*Elaeis guineensis* Jacq.) fronds, which consist of cellulose, lignin, and hemicelluloses, can be collected as extra material when oil palm fruits are harvested. Lignins, along with other polyphenols, are powerful antioxidants that eliminate free radical, and they are regarded as an important source of antioxidant phenolic compounds. The researchers Hussin et al. wanted to determine the extent to which lignins that were removed from oil palm biomass through the processes of Kraft, soda, and organosolv pulping had antioxidant characteristics [48]. Additionally, the lignins that were extracted were evaluated for their potential as inhibitors of corrosion in mild steel. Corrosive properties of lignins were verified by employing electrochemical analysis and various surface morphological activities. Lignins extracted using soda showed the highest capacity to save the metal from corrosion. According to a study, the fronds of the oil palm are a possible source of lignins, which can be used as an environmentally friendly antioxidant for the suppression of corrosion in mild steel.

Izadi et al. studied first by preparing Fe_3O_4 nanoparticles (Fe_3O_4-NS) by using ultrasound-assisted route [49]. The incorporation of Nettle extract (NE) as a sustainable corrosion inhibitor resulted in the production of an active nanoreservoir. A multiple electrolyte layer of polyaniline (PANI) was used in order to load Fe_3O_4 nanoparticles (Fe_3O_4-NP) as inhibiting molecule, and it also kept the inhibitor safe from release from Fe_3O_4-NS prior to application. Several distinct analytical methods were utilized in order to characterize Fe_3O_4-NS. The obtained results demonstrated that the PANI and NE layers had been successfully deposited on the surface of the Fe_3O_4-NP. Corrosive properties were confirmed by electrochemical measurements.

In another research, boron nitride nanosheets (BNNSs) were prepared and then modified by TA-M-BNNSs by Pan et al. [50]. After the addition of M-BNNS, there was a significant increase in the thermal conductivity and mechanical tensile strength of epoxy resin. Most notably, the EP composite coating with the proper amount of M-BNNS not only has a great corrosion-reducing tendency on the metal surfaces but forms a stable layer on the metal. It was confirmed I_{corr} value of 10%M-BNNS@EP which came out to be 8.053×10^{-7} A/cm^2, in saline solution (with dipping time 30 minutes) and even when compared (for 120 hours of dipping) the value of the impedance modulus ($|Z|0.01$) of the 10%M-BNNS@EP coating and epoxy and other composites, the value of synthesized coating came out to be higher. After being submerged in an aqueous solution containing saline solution for a period of 120 hours, electrochemical studies revealed the extraordinary protection given to the metallic surface by coating.

Anticorrosive coatings made of thermosetting polyurethane (PU) that are sustainable and based on renewable biomass resources are in high demand because they provide enduring protection against corrosion. Coatings made of thermosetting lignin-based polyurethanes (LPU) that have excellent resistance to corrosion are presented here, as well as Cao et al. [51] prepared lignin by mixing lignin-based polyol (OH-EL) with hexamethylene diisocyanate (HDI). In addition, the phenolic hydroxyls that were produced as a by-product of enzymatic hydrolysis of lignin were sequentially alkylated and subjected to a thiol-ene reaction in order to transform them preferentially into primary aliphatic hydroxyls. This was done in order to boost the reactivity and solubility of OH-EL when it reacted with isocyanates. The remarkable thermal stability and outstanding mechanical qualities exhibited by the resultant LPUs can be attributed to the efficient dispersion of OH-EL inside the polyurethane networks as well as the crosslinking reactivity of OH-EL. Remarkably, LPU-3, with the constituents of OH-EL with a weight percentage of 54.8, has high values of Tg modulus and high tensile strength of the order of 81.6 MPa and 1.4 GPa, respectively. Also, the LPU-3 coating, when used for carbon steel, showed a low value of I_{corr} and corrosion voltage E_{corr} of 7.58×10^{-11} A/cm^2 and 118 mV, respectively, exhibiting a greater resistance to corrosion than any other bio-based polymer coatings that have been described. These findings indicate that the synthesis of very reactive response lignin-based polyols using thiol-ene chemistry is an efficient and straightforward strategy. This approach is essential for the development of high-performance bio-based polyurethane coatings that are resistant to corrosion. The PDP measurements were carried out to evaluate the anticorrosive properties of LPU coatings. The I_{corr} and E_{corr} of bare CS, LPU-3, and LPU-5 are calculated and came out to be 3.08×10^{-6} A/cm^2 and −663 mV for bare CS, 3.36×10^{-11} A/cm^2 and 118 mV for LPU-3, 1.20×10^{-10} A/cm^2 and −380 mV for LPU-5, respectively, and according to the calculated values, the corrosion resistance of the LPU-3 and LPU-5 coatings is significantly better than that of the bare CS. Values confirmed that the I_{corr} of LPU-3 is lower than that of LPU-5, and the E_{corr} of LPU-3 is much more positive than that of LPU-5, and it turns out that LPU-3 has substantially superior anticorrosive performances than LPU-5, and the reason for this should be credited with the fact that LPU-3 has more uniform polymer networks. In addition, a passivation area was seen in both the anodic polarization curves of LPU-3 and LPU-5. This shows that the LPU coatings' resistance to corrosion is a mixed form of corrosion inhibition. EIS was utilized in order to research the anticorrosive properties of LPU coatings. When comparing the Nyquist plots of the bare CS, LPU-3, and LPU-5, it was a single depressed semicircle for CS (time considered was 0.5 hour and 6 hours), was close to a straight line for LPU-3 (after six hours immersion), and second-time constant was generated for LPU-5. Excellent corrosion resistance over a short period of time can be found in LPU-3. After six hours of immersion, the radius of the first capacitive arc has significantly shrunk, which is evidence that the corrosion process has begun at the metal's surface. As a result, the LPU-5's resistance to corrosion has drastically decreased within six hours [51]. The phytochemical, antioxidant, antidiarrheal, and antibacterial capabilities of mangrove plants are the subject of a significant amount of research because of

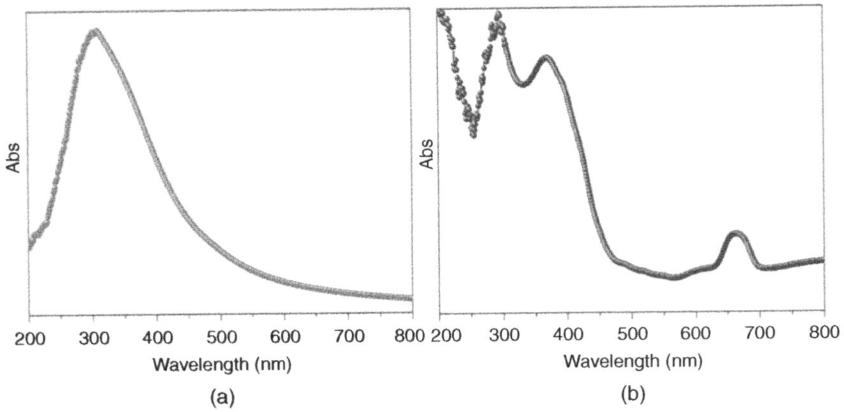

Figure 6.4 UV–Vis presentation of *C. tagal* extract without inhibitor and with inhibitor. Source: Adapted form from Shamsuzzaman et al. [52]; CCBY 4.0.

the widespread recognition of the economic and medical benefits associated with these plants. Shamsuzzaman et al. [52] worked on the *Ceriops tagal*, for its corrosion-mitigating capacity. A good amount of phenolic content (TPC) and total flavonoid content (TFC) were found. At the concentration of 600 ppm of *C. tagal* extract corrosion inhibition efficiency was 95% in acidic medium for the mild steel sample, and a dip was shown over other concentrations and temperatures. An investigation using UV–Visible spectroscopy demonstrated the production of an inhibitor metal complex given in Figure 6.4.

The elemental analysis provided the high percentage of presence of Fe, O, and C, in inhibited mild steel, as compared to uninhibited specimen, confirming the defensive layer on the metallic surface. Other surface morphological techniques were also used to confirm the formation of the protective layer. The *C. tagal* plant is a superior natural alternative to any of the available synthetic inhibitors, as shown by the SEM, which found that both uninhibited and inhibited mild steel underwent some notable alterations. The presence of 600 ppm of *C. tagal* extract greatly reduced the amount of corrosion that was present on the surface of mild steel, as determined by AFM surface topography analysis. On comparing the roughness values with the optimized concentration of *C. tagal* extract, it was found that roughness reduced to 65 nm from 251 nm, for mild steel and this significant reduction in roughness confirmed the formation of defensive layer (AFM shown in Figure 6.5).

At a concentration of 600 ppm, the level of inhibition was found to be at its highest (81.74%), whereas it was only 28% at a concentration of 100 ppm. These findings demonstrate the existence of a protective layer on mild steel. The concentration of the Fe^{+} ions in the solution determined the severity of the corrosion. The presence of a protective coating, which limited the corrosion to approximately 82%,

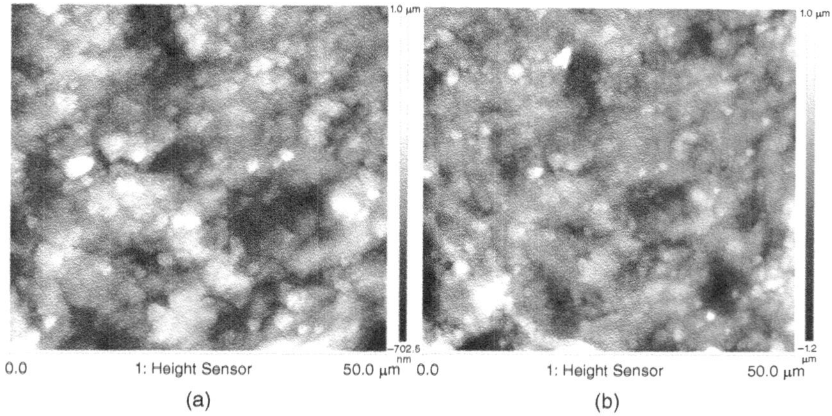

Figure 6.5 2D images of mild steel: (a) 1 M HCl; (b) 1 M HCl with 600 ppm of inhibitor Source: Md Shamsuzzaman et al. [52]/Reproduced from MDPI / CC BY 4.0.

was the cause of variations in the amounts of ions found in both of the solutions (without and with plant extract). The inhibition was dependent on the concentration, as a drop in η% was detected at lower concentrations, ranging from 28% to 72% for 200–500 ppm, indicating that the inhibition was being caused by the higher concentration. At low concentrations of inhibitors, the extent of acid attack increased, and rate of deterioration of metal increased. In most cases, organic compounds that contain S, N, and O atoms are attached to aromatic rings in order to form a conjugate system, which provides the chances for the bond formation with metal, so reducing the possibility of oxidation and minimizing corrosion.

Bao et al. [53] investigated steel samples anodes in phenolic solution (of concentration 0.1 mol/l), and the electrolyte used was sodium sulfate for anodic oxidation. Because of this oxidation, a yellow-brown polyphenol coating was produced on the surface of the steel anode. These methods of linear scanning, cyclic voltammetry, and constant current oxidation were used to investigate the reaction condition. The proper anodic electrode potential and bath voltage were 1.45 and 2.5 V, respectively. Further, FTIR and SEM were also employed in the study for structural determination and microstructure analysis of polyphenol coating, respectively. Gel permeation chromatography helps to find out the weight of soluble coating, and the number average molecular weight came out as 741 and 1200 was the weighted average, and further viscosity average weight was 1117, but GPC was not able to find out the weight of insoluble polyphenol [53].

In this study by Wang et al. [54] an effort to improve the interfacial adhesion with soy protein isolate (SPI) matrix, an innovative and cost-effective surface-modifying technique for montmorillonite (MMT) nanosheets, a biodegradable coupling crosslinking agent, was created, and TA was utilized to modify the surface and was prepared by the method used for natural polyphenol, the "green dip-coating." The casting procedure was used to make SPI nanocomposite films that had either MMT or TA@MMT included in them, in addition to the control films. Through involving ferric ions, the TA layer was successfully coated on the MMT surface and the characterization of prepared samples was done with various spectroscopic methods along with surface morphological techniques. Because of the TA-Fe^{III} coating that was applied to the MMT surface, interaction and interface contacts between the improved MMT and the SPI substrate were significantly improved. The generated biofilm has increased mechanical characteristics, increased water resistance, and increased thermal resilience as a direct consequence of this. The tensile strength of the nanocomposite films that were modified by the green dip-coating was raised by 113.3% when compared to that of the SPI film before any modifications were made to it. Due to the effective barriers that these SPI-based nanocomposite films provide against water vapor as well as ultraviolet and/or visible light, there is a promising future for the use of these films in food packaging applications.

The incorporation of synthesized smart nanocontainers into a new protective coating system enabled the development of an advanced anticorrosion coating with outstanding impermeable and active regeneration capabilities by Liu et al. [55]. In this case corrosion-mitigating compounds (benzotriazole, BTA) are confined in graphene knotted with the tannic acid-ferric coordination complex. EIS and immersion measurements were used for the detection of localized corrosion reactions. It was proved through the enhanced impedance modulus that this coating has superior barrier performance. In furtherance of this, the molecules of deconstructed TA could potentially combine with the rust that was produced to form layers that are relatively dense. Based on these findings, it was determined that the coating has triple-action protective properties thanks to the BTA inhibitor, graphene barrier, and TA rust convertor.

Again, the EIS technique is used for checking any failure of these coatings and how coating mitigates the corrosion on long exposures, and for that impedance modulus is one of the most useful parameters. During the long-term immersion period, a decreasing trend was observed in the variation of impedance modulus exhibited by all of the samples. It is abundantly clear that the impedance modulus for pure EP has gradually reduced, which indicates that the coating protection function has been rendered ineffective. The self-healing coating that was based on containers was also designed to have triple-action protective qualities. In this study, the I_{corr} around the artificial scratch decreased and finally was not there after ultimately 50 hours of dipping, which validated the self-healing function. In addition to this, the composite coatings were given an excellent barrier property thanks to the impermeable graphene nanosheets and innovative nanocontainer and therefore successfully mitigate the corrosion.

The extract of *Cinnamomum tamala* by Prasad et al. [56] was prepared, and its corrosion inhibition capacity for low carbon steel (LCS) in acidic solution was measured by employing electrochemical analysis techniques and gravimetric loss studies. A much higher efficiency of 96.76% was confirmed at a very low 100 mg/l critical concentration. Other techniques, like theoretical studies and UV–Vis, were also performed, and that also validated corrosion inhibition and helped to get an idea of electronic adsorption. The findings that were obtained suggested that an extract of *C. tamala* could be utilized as an extremely effective corrosion inhibitor that is required by industry.

Li et al. [57] prepared waterborne polyurethane coating. They studied its corrosion-mitigating properties for steel samples. They proved the long-term capacity of coating to save the metal from corrosion and the synergistic action of corrosion inhibitors Ce^{3+} and TA, which were also responsible for the same. After 50 days of immersion, the impedance modulus was still greater than $10^8\,\Omega\,cm^2$, which was about three orders of magnitude higher than that of the pure PU coating. Also, the disulfide bond exchange was responsible for the self-healing of the coating. Importantly, the healed coating demonstrated long-lasting corrosion resistance after healing. In the realm of long-term anticorrosion protection, this type of double-functional anticorrosive coating demonstrates a great deal of potential.

By adding innovative UV-resistant fillers into an epoxy matrix, a durable coating with improved corrosion-resistant and weathering-resistant qualities was produced by Wang et al. [58]. Mesoporous silica nanoparticles encapsulated the tea polyphenol (TP) to form the filler. The barrier property of the coating can be strengthened as a result of the uniform distribution of MSN-TP nanoparticles, which have the ability to prevent the penetration of corrosive fluids. In addition, the ability of TP to eliminate free radicals and the UV shielding effect of MSN worked together to stabilize the surface qualities of the epoxy coating, which was protected from the deterioration caused by weathering thanks to both of these properties. Such coating was able to provide long-term protection to the sample.

6.3 Conclusion

Metallic materials are often used in industrial applications; nevertheless, these materials are susceptible to corrosion damage when interacting with the acid medium that is present in a variety of engineering activities. The use of inhibitors might potentially prevent corrosion, which will in turn restrict damage. The hunt for a corrosion inhibitor that is easy to recognize, can be derived from natural sources, and can be referred to as "green" is undeniably one of the study fields in which new discoveries are made on a regular basis. Following an exhaustive analysis of the relevant published research in the field, it was discovered that virtually all of the polyphenolic compounds that were being researched held significant potential for application in a variety of media as environmentally friendly metal corrosion inhibitors. In addition, there is a lack of a polyphenolic component that is capable of being applied to the majority of metals or alloys by solution procedures. Because of this, the capacity of these compounds to prevent corrosion varied depending on their chemical composition, concentration, the type of corrosion medium, the type of metal surface, and a number of other factors. In spite of the fact that some polyphenols have the potential to be effective corrosion inhibitors, the manner in which they are used in operation might determine whether or not they have the opposite effect, which is to hasten the corrosion response. Recent research on the use of polyphenols as corrosion inhibitors has focused on the synthesis of new derivatives that have a stronger effectiveness as corrosion inhibitors than their equivalent compounds (i.e., phenolic acid) and using quantum chemistry computing combined with molecular mechanical modeling, describe the inhibitory mechanism that they use. Studying these substances, such as their interaction with metal surfaces, by employing simulations of molecular mechanics (based on classical mechanics), the density functional theory (DFT) has been widely employed in this subject due to its accessibility and accuracy. When taking into consideration the "inhibitor molecule-metal surface" combination, DFT techniques face additional limitations due to the fact that they are tedious and have a size limit on the structures they can analyze. Because of this same reason, we strongly encourage the researchers to concentrate even more of their attention on this subject.

References

1 Hau, N.N. and Huong, D.Q. (2023). Effect of aromatic rings on mild steel corrosion inhibition ability of nitrogen heteroatom-containing compounds: experimental and theoretical investigation. *Journal of Molecular Structure* 1277: 134884.

2 Zheng, H., Zhang, B., Wang, X. et al. (2023). Improved corrosion resistance of carbon steel in soft water with dendritic-polymer corrosion inhibitors. *Chemical Engineering Journal* 452: 139043.

3 Daoudi, W., Guo, L., Azzouzi, M. et al. (2023). Evaluation of the corrosion inhibition of mild steel by newly synthesized imidazo [1,2-*a*]pyridine derivatives: experimental and theoretical investigation. *Journal of Adhesion Science and Technology* 1–24.

4 Allah, M.A.A.H., Balakit, A.A., Salman, H.I. et al. (2023). New heterocyclic compound as carbon steel corrosion inhibitor in 1 M H_2SO_4, high efficiency at low concentration: experimental and theoretical studies. *Journal of Adhesion Science and Technology* 37: 525–547.

5 Wang, Y., Qiang, Y., Zhi, H. et al. (2023). Evaluating the synergistic effect of maple leaves extract and iodide ions on corrosion inhibition of Q235 steel in H_2SO_4 solution. *Journal of Industrial and Engineering Chemistry* 117: 422–433.

6 Wang, J., Zhao, J., Tabish, M. et al. (2023). Long-term corrosion inhibition for AA5052 aluminum alloy by an eco-friendly hybrid inhibitor: Synergism inhibition between rosemary extract and zinc chloride in 0.05 M NaCl solution. *Journal of Industrial and Engineering Chemistry* 120: 302–315.

7 Alahiane, M., Oukhrib, R., Albrimi, Y.A. et al. (2023). Corrosion inhibition of SS 316L by organic compounds: experimental, molecular dynamics, and conceptualization of molecules–surface bonding in H_2SO_4 solution. *Applied Surface Science* 612: 155755.

8 Liu, H., Jin, Z., Wang, Z. et al. (2023). Corrosion inhibition of deposit-covered X80 pipeline steel in seawater containing *Pseudomonas stutzeri*. *Bioelectrochemistry* 149: 108279.

9 Kaya, F., Solmaz, R., and Geçibesler, İ.H. (2023). The use of methanol extract of Rheum ribes (Işgın) flower as a natural and promising corrosion inhibitor for mild steel protection in 1 M HCl solution. *Journal of Industrial and Engineering Chemistry* 122: 102–117.

10 Guo, L., Zhu, M., Shi, W. et al. (2023). Evaluation of the corrosion inhibition behavior of N, S co-doped memory-type carbon dots for mild steel in HCl solution. *Materials Letters* 330: 133403.

11 Kokalj, A. (2023). On the estimation of standard adsorption free energy from corrosion inhibition efficiencies. *Corrosion Science* 217: 111139.

12 Wisetkomolmat, J., Arjin, C., Hongsibsong, S. et al. (2023). Antioxidant activities and characterization of polyphenols from selected northern thai rice husks: relation with seed attributes. *Rice Science* 30: 148–159.

13 Shah, D., Gandhi, M., Kumar, A. et al. (2023). Current insights into epigenetics, noncoding RNA interactome and clinical pharmacokinetics of dietary polyphenols in cancer chemoprevention. *Critical Reviews in Food Science and Nutrition* 63: 1755–1791.

14 Islam, T., Albracht-Schulte, K., Ramalingam, L. et al. (2023). Anti-inflammatory mechanisms of polyphenols in adipose tissue: role of gut microbiota, intestinal barrier integrity and zinc homeostasis. *The Journal of Nutritional Biochemistry* 115: 109242.

15 Meng, W., Sun, H., Mu, T., and Garcia-Vaquero, M. (2023). Pickering emulsions with chitosan and macroalgal polyphenols stabilized by layer-by-layer electrostatic deposition. *Carbohydrate Polymers* 300: 120256.

16 Xu, K., Cui, X., Ren, X. et al. (2023). Discovery of natural polyphenols from the wild vegetable *Suaeda salsa* L. with potential cardioprotective functions. *Food Chemistry* 405: 134968.

17 Li, S., Yin, S., Ding, H. et al. (2023). Polyphenols as potential metabolism mechanisms regulators in liver protection and liver cancer prevention. *Cell Proliferation* 56: e13346.

18 Spiegel, M., Cel, K., and Sroka, Z. (2023). The mechanistic insights into the role of pH and solvent on antiradical and prooxidant properties of polyphenols — nine compounds case study. *Food Chemistry* 407: 134677.

19 Zhang, L., García-Pérez, P., Martinelli, E. et al. (2023). Different fractions from wheat flour provide distinctive phenolic profiles and different bioaccessibility of polyphenols following in vitro digestion. *Food Chemistry* 404: 134540.

20 Huang, W., Tian, F., Wang, H. et al. (2023). Comparative assessment of extraction, composition, and in vitro antioxidative properties of wheat bran polyphenols. *LWT* 180: 114706.

21 Gabsi, M., Ferkous, H., Delimi, A. et al. (2023). The curious case of polyphenols as green corrosion inhibitors: a review on their extraction, design, and applications. *Environmental Science and Pollution Research* 30: 59081.

22 Liu, K., Chen, Q., Luo, H. et al. (2023). An in vitro catalysis of tea polyphenols by polyphenol oxidase. *Molecules* 28: 1722.

23 Athanasiadis, V., Voulgaris, A., Katsoulis, K. et al. (2023). Development of enriched oil with polyphenols extracted from olive mill wastewater. *Foods* 12: 497.

24 Fecka, I., Bednarska, K., and Kowalczyk, A. (2023). In vitro antiglycation and methylglyoxal trapping effect of peppermint leaf (*Mentha* × *piperita* L.) and its polyphenols. *Molecules* 28: 2865.

25 Rodríguez-Daza, M.C. and de Vos, W.M. (2023). Polyphenols as drivers of a homeostatic gut microecology and immuno-metabolic traits of *Akkermansia muciniphila*: from mouse to man. *International Journal of Molecular Sciences* 24: 45.

26 Pan, L., Chen, J., Fu, H. et al. (2023). Effects of fabrication of conjugates between different polyphenols and bovine bone proteins on their structural and functional properties. *Food Bioscience* 52: 102375.

27 Li, Y., Yang, Y., Zhu, S. et al. (2023). Tea polyphenols-OSA starch interaction and its impact on interface properties and oxidative stability of O/W emulsion. *Food Hydrocolloids* 135: 108187.

28 Poojary, M.M., Hellwig, M., Henle, T., and Lund, M.N. (2023). Covalent bonding between polyphenols and proteins: synthesis of caffeic acid-cysteine and chlorogenic acid-cysteine adducts and their quantification in dairy beverages. *Food Chemistry* 403: 134406.

29 Wang, Z., Ma, L., Liu, P. et al. (2023). Transcription factor NtWRKY33a modulates the biosynthesis of polyphenols by targeting NtMYB4 and NtHCT genes in tobacco. *Plant Science* 326: 111522.

30 Rambaran, T.F. (2020). Nanopolyphenols: a review of their encapsulation and anti-diabetic effects. *SN Applied Sciences* 2: 1335.

31 Karattu Veedu, K., Peringattu Kalarikkal, T., Jayakumar, N., and Gopalan, N.K. (2019). Anticorrosive performance of *Mangifera indica* L. leaf extract-based hybrid coating on steel. *ACS Omega* 4: 10176–10184.

32 Ramírez, J., Díaz-Gómez, A., Montoya, L.F. et al. (2023). Evaluation in real conditions of new anticorrosive formulations based on polyphenols from natural sources and encapsulated nanoparticles. *Coatings* 13: 8.

33 Montoya, L.F., Contreras, D., Jaramillo, A.F. et al. (2019). Study of anticorrosive coatings based on high and low molecular weight polyphenols extracted from the *Pine radiata* bark. *Progress in Organic Coatings* 127: 100–109.

34 Ill-Min, C., Kim, S.-H., and Prabakaran, M. (2020). Evaluation of phytochemical, polyphenol composition and anti-corrosion capacity of *Cucumis anguria* L. leaf extract on metal surface in sulfuric acid medium. *Protection of Metals and Physical Chemistry of Surfaces* 56: 214–224.

35 Jaramillo, A.F., Montoya, L.F., Prabhakar, J.M. et al. (2019). Formulation of a multifunctional coating based on polyphenols extracted from the *Pine radiata* bark and functionalized zinc oxide nanoparticles: evaluation of hydrophobic and anticorrosive properties. *Progress in Organic Coatings* 135: 191–204.

36 Prabakaran, M., Kim, S.-H., Hemapriya, V., and Chung, I.-M. (2016). Evaluation of polyphenol composition and anti-corrosion properties of *Cryptostegia grandiflora* plant extract on mild steel in acidic medium. *Journal of Industrial and Engineering Chemistry* 37: 47–56.

37 Zhang, B., Yao, R., Li, L. et al. (2019). Green tea polyphenol induced Mg^{2+}-rich multilayer conversion coating: toward enhanced corrosion resistance and promoted in situ endothelialization of AZ31 for potential cardiovascular applications. *ACS Applied Materials & Interfaces* 11: 41165–41177.

38 Malathy, R., Prabakaran, M., Kalaiselvi, K. et al. (2021). Comparative polyphenol composition, antioxidant and anticorrosion properties in various parts of *Panax ginseng* extracted in different solvents. *Applied Sciences* 11: 93.

39 Jain, R., Bhagawati, B., Khandagiri, P. et al. (2017). Anticorrosive and lubricating polyphenol coatings on galvanneal steel. *Surface Engineering* 33: 410–427.

40 Quan, X., Wang, J., Zhao, S. et al. (2018). Improved antibacterial, antifouling and corrosion protective performance of epoxy coatings with poly(*m*-aminophenol). *Progress in Organic Coatings* 115: 9–17.

41 Ahmed, R.K. and Zhang, S. (2020). Bee pollen extract as an eco-friendly corrosion inhibitor for pure copper in hydrochloric acid. *Journal of Molecular Liquids* 316: 113849.

42 Prabakaran, M., Hemapriya, V., Kim, S.-H., and Chung, I.-M. (2019). Evaluation of antioxidant and anticorrosion properties of *Epipremnum aureum*. *Arabian Journal for Science and Engineering* 44: 169–178.

43 Hlushko, H., Cubides, Y., Hlushko, R. et al. (2018). Hydrophobic antioxidant polymers for corrosion protection of an aluminum alloy. *ACS Sustainable Chemistry & Engineering* 6: 14302–14313.

44 Diógenes, O.B.F., de Oliveira, D.R., da Silva, L.R.R. et al. (2021). Development of coal tar-free coatings: acetylated lignin as a bio-additive for anticorrosive and UV-blocking epoxy resins. *Progress in Organic Coatings* 161: 106533.

45 Deng, Y., Xia, L., Song, G.-L. et al. (2021). Development of a curcumin-based antifouling and anticorrosion sustainable polybenzoxazine resin composite coating. *Composites Part B: Engineering* 225: 109263.

46 Pal, A. and Das, C. (2022). New eco-friendly anti-corrosion inhibitor of purple rice bran extract for boiler quality steel: experimental and theoretical investigations. *Journal of Molecular Structure* 1251: 131988.
47 Cao, X., Cai, G., Wang, Y. et al. (2023). Bio-inspired polybenzoxazine coating: anti-corrosion and anti-abrasion performance enhancement through monomer design. *Progress in Organic Coatings* 174: 107283.
48 Hussin, M.H., Shah, A.M., Rahim, A.A. et al. (2015). Antioxidant and anticorrosive properties of oil palm frond lignins extracted with different techniques. *Annals of Forest Science* 72: 17–26.
49 Izadi, M., Shahrabi, T., and Ramezanzadeh, B. (2018). Synthesis and characterization of an advanced layer-by-layer assembled Fe_3O_4/polyaniline nanoreservoir filled with Nettle extract as a green corrosion protective system. *Journal of Industrial and Engineering Chemistry* 57: 263–274.
50 Pan, D., Zhang, X., Yang, G. et al. (2020). Thermally conductive anticorrosive epoxy nanocomposites with tannic acid-modified boron nitride nanosheets. *Industrial & Engineering Chemistry Research* 59: 20371–20381.
51 Cao, Y., Liu, Z., Zheng, B. et al. (2020). Synthesis of lignin-based polyols via thiol-ene chemistry for high-performance polyurethane anticorrosive coating. *Composites Part B: Engineering* 200: 108295.
52 Shamsuzzaman, M., Kalaiselvi, K., and Prabakaran, M. (2021). Evaluation of antioxidant and anticorrosive activities of *Ceriops tagal* plant extract. *Applied Sciences* 11: 10150.
53 Bao, L., Xiong, R., and Wei, G. (2010). Electrochemical polymerization of phenol on 304 stainless steel anodes and subsequent coating structure analysis. *Electrochimica Acta* 55: 4030–4038.
54 Wang, Z., Kang, H., Zhang, W. et al. (2017). Improvement of interfacial interactions using natural polyphenol-inspired tannic acid-coated nanoclay enhancement of soy protein isolate biofilms. *Applied Surface Science* 401: 271–282.
55 Liu, C., Jin, Z., Zhao, H., and Wang, L. (2020). Triple-action smart nanocontainers enhanced protective coatings with superior active and passive properties. *Progress in Organic Coatings* 148: 105887.
56 Prasad, D., Dagdag, O., Safi, Z. et al. (2022). *Cinnamoum tamala* leaves extract highly efficient corrosion bio-inhibitor for low carbon steel: applying computational and experimental studies. *Journal of Molecular Liquids* 347: 118218.
57 Li, S., Xu, Y., Xiang, F. et al. (2023). Enhanced corrosion resistance of self-healing waterborne polyurethane coating based on tannic acid modified cerium-montmorillonites composite fillers. *Progress in Organic Coatings* 178: 107454.
58 Wang, J., Ma, L., Ding, X. et al. (2023). Tea polyphenol radical scavenger loaded UV absorber for corrosion resistant and weathering resistant epoxy coating fabrication. *Progress in Organic Coatings* 180: 107553.

7

Polyphenols for Dyes Application

Abhinay Thakur[1] and Ashish Kumar[2]

[1] *Department of Chemistry, School of Chemical Engineering and Physical Sciences, Lovely Professional University, Phagwara, Punjab, India*
[2] *Nalanda College of Engineering, Bihar Engineering University, Department of Science, Technology and Technical Education, Government of Bihar, India*

7.1 Introduction

The negative environmental consequences of synthetic dyes have become increasingly apparent today. The textile industry is one of the largest consumers of dyes globally. Artificial dyes have been widely used in the textile industry since the mid-19th century, offering various colors and shades for different fabrics. The global demand for textile dyes has increased significantly over the years due to population growth, urbanization, and fashion trends. According to a report by Grand View Research, the global textile dyes market was valued at USD 8.2 billion in 2020. The report projects a compound annual growth rate (CAGR) of 5.6% from 2021 to 2028 [1, 2]. The market is driven by the growing textile industry, especially in emerging economies, and the increasing demand for colored and patterned textiles. However, the environmental impact of synthetic dyes has raised concerns globally. Releasing toxic chemicals and heavy metals from dyeing processes into water bodies has led to water pollution and ecological damage. In response to these concerns, a shift has been made toward more sustainable and eco-friendly dyeing practices, including natural dyes.

The global market for natural dyes is also experiencing growth as consumers and textile manufacturers seek more environmentally friendly alternatives. Natural dyes, including those derived from polyphenols, offer a sustainable and biodegradable option compared to synthetic dyes. Using natural dyes aligns with sustainability principles and supports the development of a circular economy.

Science and Engineering of Polyphenols: Fundamentals and Industrial Scale Applications, First Edition. Edited by Chandrabhan Verma.
© 2024 John Wiley & Sons Inc. Published 2024 by John Wiley & Sons Inc.

Various initiatives and regulations have recently been implemented to promote sustainable dyeing practices [3, 4]. For instance, organizations like the Global Organic Textile Standard (GOTS) and the Sustainable Apparel Coalition (SAC) have set standards and guidelines for the use of eco-friendly dyes and processes in the textile industry. Natural dyes have gained traction as a viable solution due to their lower environmental impact and potential for biodegradability. Natural dyes are derived from renewable sources such as plants, insects, and minerals, and their use dates back thousands of years in various cultures around the world. However, the resurgence of interest in natural dyes is driven by the urgent need to address the environmental challenges associated with synthetic dyes. Polyphenols have emerged as promising candidates among the various natural dye sources. Polyphenols are a diverse group of compounds found abundantly in plants, including fruits, vegetables, leaves, barks, and seeds. They are known for their antioxidant properties and contribute to the vibrant colors observed in many plant-based foods and natural materials [5, 6].

Polyphenols exhibit a wide range of chemical structures, including flavonoids, tannins, lignans, and stilbenes [7–10]. These structures contain multiple phenolic rings and hydroxyl groups, which contribute to their dyeing properties. The presence of aromatic rings allows polyphenols to absorb light in specific wavelength ranges, resulting in the perception of color. The hydroxyl groups enable interaction with the substrate during the dyeing process, leading to attachment and color fixation. The abundance and diversity of polyphenols in plants make them an attractive resource for natural dye production. By utilizing plant waste or by-products, such as fruit peels or tree bark, the extraction of polyphenols for dyeing purposes contributes to the valorization of agricultural and forestry waste, promoting a circular economy approach [11–13]. Furthermore, the use of polyphenols as natural dyes aligns with the principles of sustainable agriculture and reduces dependence on petrochemical-based synthetic dyes. Polyphenols offer significant advantages over synthetic dyes in terms of their environmental impact. They are generally nontoxic, biodegradable, and do not contain harmful chemicals or heavy metals. This eliminates concerns regarding the release of toxic substances into the environment during the dyeing process or at the end of a product's life cycle. Furthermore, polyphenols have the potential to be compatible with various dyeing techniques and can be applied to a wide range of materials, including textiles, leather, and even natural fibers for cosmetic applications. Figure 7.1 shows the classification of polyphenols [14].

The use of polyphenols as natural dyes extends beyond their environmental benefits. Their diverse chemical structures provide a rich palette of colors, ranging from vibrant reds, oranges, and yellows to soothing blues, greens, and browns. This versatility allows for a wide range of color options in dyeing applications, catering to the preferences and needs of different industries and consumers. The exploration and utilization of polyphenols as natural dyes offer significant opportunities

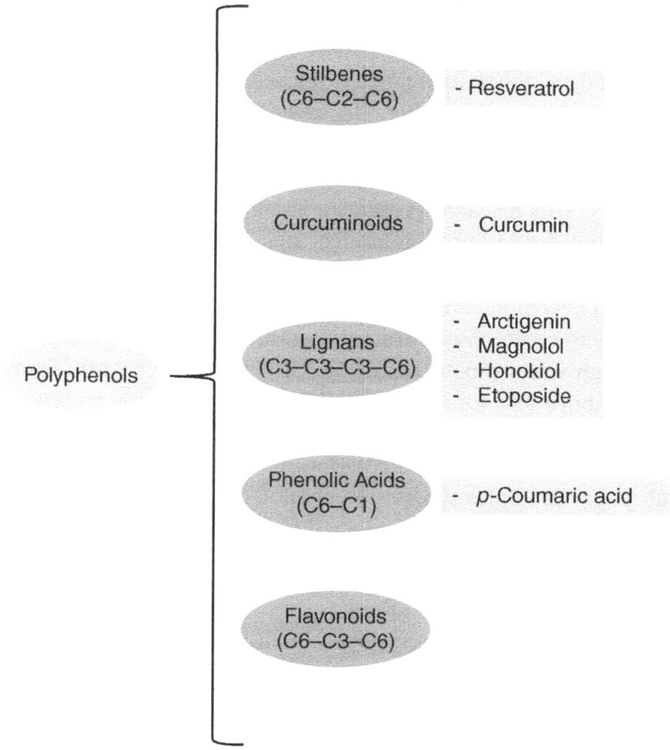

Figure 7.1 Classification of polyphenols. Source: Adapted with permission from Montané et al. [14] under CCBY 4.0.

for innovation and development. Researchers and industry professionals are actively investigating various extraction methods to optimize the yield and quality of polyphenols from different plant sources [15, 16]. Solvent extraction, solid-phase extraction, and ultrasound-assisted extraction are among the techniques being employed to extract polyphenols efficiently. These methods aim to maximize the recovery of polyphenols while minimizing the use of harmful solvents and energy consumption.

Chen et al. [1] present a simple method for surface modification, resulting in the creation of a new type of membrane. The modification process involved the simultaneous deposition of tannic acid (TA), polyethyleneimine (PEI), and halloysite nanotubes (HNTs) onto a polyvinylidene fluoride (PVDF) membrane. The modified membrane's surface morphology and chemical composition were examined using scanning electron microscopy (SEM) and X-ray photoelectron spectroscopy (XPS), respectively. The analyses confirmed the successful modification of the membrane with TA/PEI/HNTs. Compared to the original membrane, the coated membrane exhibited enhanced hydrophilicity and pure water flux (PWF). Furthermore, the TA/PEI/HNTs-modified membrane demonstrated excellent performance in rejecting dyes, achieving rejection rates of 90.7% for direct yellow, 92.8% for direct red, and 96.0% for direct blue. The treated membrane also demonstrated remarkable resistance to fouling after multiple cycles. Additionally, the introduction of the TA/PEI/HNTs coating improved the membrane's ability to remove heavy metal ions. This research presents a straightforward approach for achieving hydrophilic modification of membranes, and it is anticipated that the prepared membrane will find practical applications in wastewater treatment. You et al. [3] developed a new bio-coating comprising a Ni^{2+}-polyphenol network using a one-step strategy to create nanofiltration membranes. These membranes exhibited remarkably high-water flux, reaching $56.1\,L\,m^{-2}\,h^{-1}\,bar^{-1}$ while maintaining a rejection rate above 95% for rose bengal (RB). This research presents a straightforward method for producing highly efficient nanofiltration membranes, which could be valuable for wastewater remediation.

There are several global examples where polyphenols have been successfully utilized for dye applications, highlighting their potential as sustainable alternatives to synthetic dyes.

- *India*
  India has a rich tradition of using natural dyes, including those derived from polyphenols, in textile dyeing. For example, the use of plant-based dyes containing polyphenols, such as indigo derived from the *Indigofera tinctoria* plant, has been practiced for centuries [17, 18]. The revival of natural dyeing techniques and the promotion of eco-friendly practices have gained momentum in recent years, with efforts to sustainably cultivate dye-yielding plants and preserve traditional dyeing knowledge.
- *Japan*
  In Japan, traditional dyeing techniques using polyphenol-rich plant extracts have been an integral part of their cultural heritage. For instance, the dyeing technique called "Kusakizome" involves using plant materials, including those containing polyphenols, to achieve a wide range of colors [19–21]. The use of natural dyes in Japanese textiles promotes sustainability and fosters a connection with nature.
- *Europe*
  In Europe, there is a growing interest in utilizing polyphenols as natural dyes to reduce the environmental impact of textile production. In countries like Italy, Spain, and France, natural dye workshops and research initiatives are emerging to explore the potential of polyphenols for sustainable dyeing. Local plant sources, such as grape pomace and chestnut shells, rich in polyphenols, are being investigated for their dyeing properties, offering opportunities for waste valorization and circular economy practices.

- *United States*
  In the United States, the demand for sustainable and natural dyeing practices has led to the establishment of artisanal and small-scale natural dye businesses. These businesses often focus on using locally sourced plant materials, including polyphenol-rich sources, to create unique and environmentally friendly dyes. The use of polyphenols derived from plants like black walnut, avocado pits, and marigold flowers has gained popularity in the natural dyeing community.
- *South America*
  South American countries, such as Peru and Bolivia, have a rich history of utilizing natural dyes for textile production. The use of polyphenol-containing plant materials, such as cochineal insects and various tree bark, has been practiced for centuries. These natural dyes provide vibrant and long-lasting colors while promoting sustainable and traditional dyeing practices. Furthermore, understanding the dyeing mechanisms of polyphenols is essential for achieving the desired color fastness and stability [22]. Polyphenols can interact with substrates through adsorption, covalent bonding, or metal complexation. Factors such as pH, temperature, and the use of mordants play a crucial role in the dyeing process and can influence the color intensity and durability of the dyed material. Research in this area seeks to optimize dyeing conditions and improve the overall performance of polyphenol-based dyes. The chapter will delve into the dyeing mechanisms of polyphenols, which include adsorption, covalent bonding, and metal complexation. Understanding these mechanisms is essential for optimizing dyeing processes and achieving desired color fastness and stability. Factors such as pH, temperature, and the use of mordants will be examined, as they play crucial roles in the dyeing process. Moreover, the chapter will showcase the diverse applications of polyphenols as natural dyes. From fabric dyeing to leather tanning, and from hair colorants to natural food coloring, polyphenols offer versatile and sustainable solutions. Case studies and examples of successful applications will be presented to highlight the practical viability of using polyphenols as dyes.

7.2 Natural Dyes and the Need for Sustainable Alternatives

7.2.1 Synthetic Dyes vs. Natural Dyes

The textile industry has long relied on synthetic dyes as the primary choice for coloring fabrics. Synthetic dyes are chemically derived compounds, often derived from petrochemicals, and offer a vast spectrum of colors, excellent color fastness, and reproducibility. These dyes have revolutionized the industry by providing a wide range of vibrant and long-lasting hues, allowing for intricate and complex designs [23, 24]. Moreover, synthetic dyes are highly predictable, ensuring consistent color results, which is crucial for large-scale textile production. However, the environmental impact of synthetic dyes has raised significant concerns in recent years. During the dyeing process, synthetic dyes release toxic chemicals and heavy metals into the environment. These pollutants find their way into water bodies, contributing to water pollution and posing a threat to aquatic ecosystems. Moreover, the runoff from dyeing processes contaminates soil, affecting its fertility and potentially harming plants and other organisms. The detrimental effects of synthetic dyes on the environment and human health have become increasingly evident, prompting a demand for sustainable and environmentally friendly alternatives [23].

In contrast, natural dyes offer a more sustainable option for coloring textiles. Derived from renewable sources such as plants, animals, and minerals, natural dyes have a long history of use in various cultures worldwide. They are often obtained from parts of plants, such as roots, leaves, flowers, or barks, which are rich in pigments. Natural dyes offer several advantages over their synthetic counterparts. One of the primary advantages of natural dyes is their lower environmental impact. Unlike synthetic dyes, natural dyes are biodegradable and do not accumulate in the environment [25]. They break down naturally, minimizing the pollution of water bodies and soil. Additionally, the extraction processes for natural dyes typically involve less energy and fewer chemical inputs compared to the production of synthetic dyes, further reducing their ecological footprint. Furthermore, natural dyes can offer potential health benefits. Some natural dyes contain compounds with antimicrobial or antioxidant properties, which may have positive effects on human health. For example, turmeric, a commonly used natural dye, contains curcumin, a compound known for its antioxidant and anti-inflammatory properties. When used in textiles, natural dyes derived from such plants may contribute to the overall well-being of the wearer. Figure 7.2 shows the major categories of natural dyes divided by chemical structure.

However, natural dyes also come with certain challenges. Achieving consistent color and excellent color fastness can be more challenging with natural dyes compared to synthetic ones. Natural dyes often exhibit variations in color due to factors such as the growing conditions of the source plant, seasonal variations, and the extraction process. Achieving color consistency in large-scale production can take more work, requiring careful quality control and standardization. Color fastness, which refers to the ability of a dye to retain its color over time and under various conditions, is another crucial aspect. Natural dyes may exhibit a different level of color fastness than synthetic dyes, particularly in the face of exposure to sunlight, washing, and other environmental factors [27, 28]. However, research and development efforts are underway to address these challenges and enhance the color fastness of natural dyes through improved dyeing techniques and the use of mordants or natural fixatives. Despite these challenges, the demand for

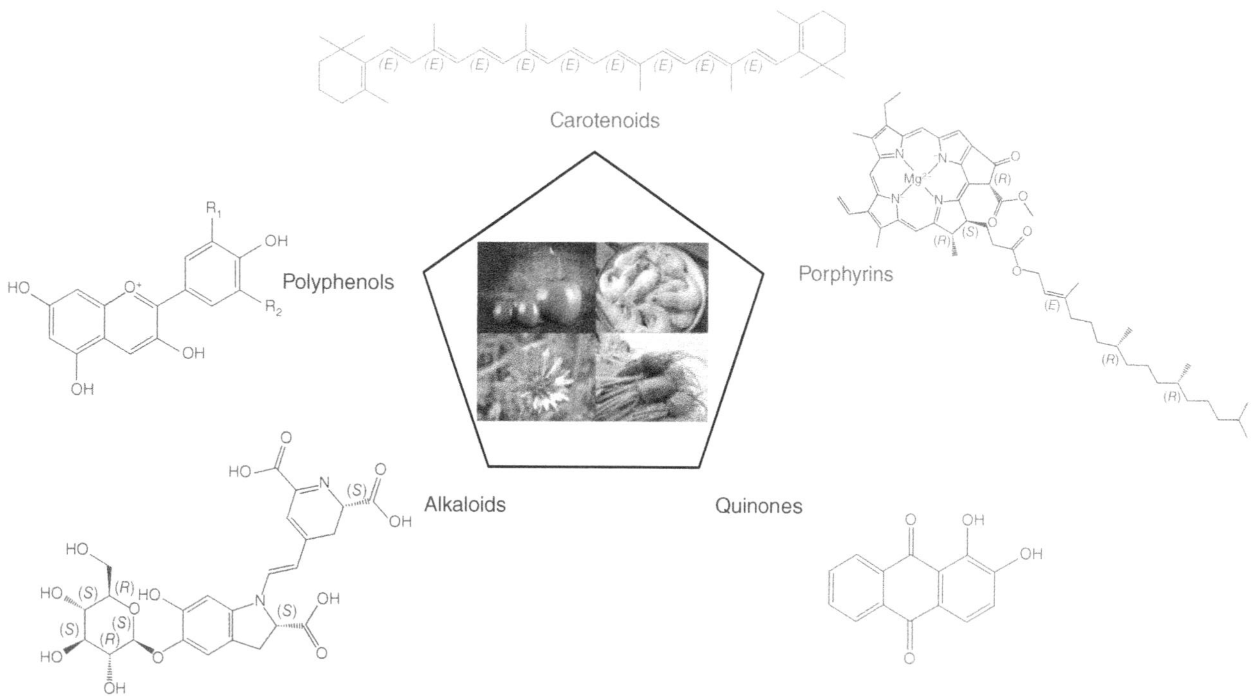

Figure 7.2 The major categories of natural dyes divided by chemical structure. Source: Li et al. [26]/Reproduced from MDPI/ CC BY 4.0.

natural dyes has been steadily growing. Consumers and industries are increasingly seeking sustainable and environmentally friendly alternatives, driven by a desire to minimize their ecological footprint and support ethical practices. The use of natural dyes aligns with these values, offering a more sustainable and socially conscious approach to textile coloring [29–42].

Ouyang et al. [15] successfully synthesized a heterogeneous Fenton-like catalyst, known as GFe0.5, using tea polyphenols (TP) and ferric trichloride, and applied it for the degradation of lincomycin (LCM). The results demonstrated that a dosage of 0.01 g/L of GFe0.5 could completely degrade a 20 mg/L concentration of LCM within 90 minutes. Furthermore, toxicity assessment using luminous bacteria revealed that the GFe0.5/H_2O_2 system effectively mitigated the toxicity risk associated with this pollutant. Characterization techniques such as transmission electron microscopy, XRD, Fourier transform infrared spectroscopy, and XPS confirmed the triangular structure of the synthesized material, resembling a combination of nonacarbonyldiiron and ferrous chloride, with Fe^{2+}, Fe^{3+}, and organic groups attached to the surface. The degradation mechanism of the iron-based material was determined through free-radical capture and electron spin resonance experiments, indicating that the main active free radical involved was –OH. Cyclic experiments, as well as inductively coupled plasma and XRD analysis of the materials after the reaction, revealed that the loss of free iron was the primary factor contributing to the decrease in activity. The Fe-based material exhibited the ability to bind free iron during the reaction, thereby preventing excessive consumption of Fe^{2+} during the initial stage of the Fenton-like reaction. This binding mechanism played a crucial role in the efficient Fenton-like catalytic performance of GFe0.5 and the prevention of iron sludge formation. Overall, this environmentally friendly synthesis approach yielded a Fe-based material resembling a mixture of nonacarbonyldiiron and ferrous chloride, demonstrating the promising potential for effective Fenton-like catalysis.

In response to the growing demand, efforts are being made to expand the availability and accessibility of natural dyes. Research focuses on exploring new sources of natural dyes, developing efficient extraction methods, and optimizing dyeing techniques to improve color consistency and fastness. Additionally, collaborations between textile manufacturers, researchers, and artisans are fostering innovation and knowledge exchange in the field of natural dyeing. Overall, the shift toward natural dyes signifies a broader movement toward sustainable practices in the textile industry. While synthetic dyes have long been the preferred choice due to their color range and reproducibility, the environmental concerns associated with them have catalyzed the search for alternative options. Natural dyes offer a more environmentally friendly and socially responsible approach to textile coloring, showcasing the potential for a harmonious coexistence between the textile industry and the natural world.

7.2.2 Growing Demand for Sustainable Dyeing Solutions

The growing demand for sustainable dyeing solutions is driven by several key factors. First and foremost, the adverse environmental effects of synthetic dyes have become more widely known. Synthetic dyes are typically derived from petrochemicals and involve complex chemical processes, resulting in the release of toxic substances and heavy metals during dyeing [43, 44]. These pollutants contaminate water bodies, contribute to soil degradation, and harm ecosystems. The awareness of these environmental issues has prompted consumers to seek alternatives that have a lesser impact on the planet. Secondly, consumers are becoming more conscious of the social and ethical aspects of the products they purchase. The textile industry, particularly in developing countries, has often been associated with poor working conditions, low wages, and exploitative labor practices. The demand for sustainable dyeing solutions is closely linked to the broader movement toward ethical and fair-trade practices. Consumers are increasingly interested in knowing the origins of their garments and whether they have been produced using sustainable and socially responsible methods. Natural dyes, derived from renewable sources and often produced through traditional and artisanal techniques, provide an opportunity to promote sustainable practices and support local communities [45].

Additionally, the influence of regulations and sustainability initiatives must be considered. Governments and regulatory bodies are enacting stricter regulations regarding the use of hazardous chemicals in textile production. For instance, the European Union has implemented the REACH regulation, which restricts the use of certain chemicals in textile manufacturing [14, 46]. Compliance with such regulations necessitates the exploration of alternative dyeing methods that are safer for both the environment and human health. Furthermore, many textile manufacturers and brands have recognized the importance of sustainability in their business strategies. They are proactively adopting environmentally friendly practices and seeking partnerships with suppliers and dye manufacturers who share their commitment to sustainability. This shift is driven not only by the desire to reduce the industry's environmental impact but also by the recognition that sustainability can be a source of competitive advantage. Companies that prioritize sustainability and offer products with eco-friendly dyeing solutions can appeal to a growing segment of environmentally conscious consumers, enhancing their brand reputation and market positioning. The demand for sustainable dyeing solutions has also been influenced by advancements in technology and innovation. Researchers and scientists are constantly exploring new methods and materials that can offer improved sustainability in the dyeing process. This includes the development of eco-friendly dyeing techniques, such as low-water and low-energy dyeing processes, as well as the utilization of natural dyes [26, 47]. Technological advancements are driving the optimization of dyeing processes, improving color fastness, and enhancing the overall performance of natural dyes.

Moreover, the awareness of climate change and its impact on the environment has further fueled the demand for sustainable dyeing solutions. The textile industry is a significant contributor to greenhouse gas emissions, primarily through its energy consumption and chemical usage. By adopting sustainable dyeing practices, such as using natural dyes and reducing energy consumption, the industry can make substantial progress in mitigating its environmental footprint and contributing to global efforts to combat climate change. In response to this growing demand, various initiatives and collaborations have emerged to promote sustainable dyeing solutions. Organizations, industry associations, and research institutions are working together to develop standards and guidelines for sustainable dyeing practices, share knowledge and best practices, and provide resources and support to textile manufacturers seeking to transition to more sustainable dyeing methods. These initiatives foster innovation, collaboration, and knowledge exchange, driving the adoption of sustainable dyeing solutions across the industry.

7.2.3 Role of Polyphenols as Natural Dyes

Polyphenols, a diverse group of compounds found abundantly in plants, have emerged as attractive candidates for natural dyes due to their unique properties and potential applications in the field of dyeing. With their aromatic rings and hydroxyl groups, polyphenols possess chemical structures that contribute to their dyeing properties. The utilization of polyphenols as natural dyes offers several advantages, including their sustainability, biodegradability, and potential health benefits. One of the key advantages of using polyphenols as natural dyes is their sustainability. Polyphenols are derived from renewable plant sources, making them an environmentally friendly alternative to synthetic dyes [48–51]. The extraction of polyphenols can be performed from various plant materials such as fruits, vegetables, and tree bark, allowing for the utilization of by-products or waste materials from agricultural processes. This not only provides opportunities for waste valorization but also supports the principles of a circular economy. Furthermore, polyphenols offer biodegradability, meaning that they can be naturally broken down by microorganisms and environmental factors without leaving persistent pollutants. This characteristic is particularly important as it reduces the environmental impact associated with the disposal of dyed textiles at the end of their life cycle. Compared to synthetic dyes that often persist in the environment and contribute to water pollution and soil contamination, the use of polyphenols as natural dyes aligns with sustainable and environmentally conscious practices.

Another intriguing aspect of polyphenols is their potential health benefits. Some polyphenols possess antioxidant and antimicrobial properties, which can provide additional functionality to dyed textiles. Antioxidants can help protect the fabric from degradation caused by exposure to environmental factors such as sunlight and air pollution. Antimicrobial properties can inhibit the growth of bacteria, fungi, and other microorganisms, reducing the risk of unpleasant odors and improving the hygiene of textiles [52, 53]. These properties make polyphenols a desirable choice for applications where health and well-being are of importance, such as in the medical, sportswear, or baby clothing industries. To harness the potential of polyphenols as natural dyes, research and development efforts have focused on optimizing the extraction methods from different plant sources. Various techniques have been explored, each with its advantages and limitations. Solvent extraction is a commonly used method that involves using organic solvents such as ethanol or methanol to dissolve and extract polyphenols from plant materials. Solid-phase extraction (SPE) is an alternative method that utilizes solid-phase adsorbents to selectively capture and retain polyphenols from plant extracts.

Ultrasound-assisted extraction (UAE) is an emerging technique that employs ultrasound waves to enhance the extraction process. Each method has its own specific requirements and considerations, including the solubility of polyphenols, extraction efficiency, and environmental impact [54]. Once the polyphenols are extracted, they can be used as natural dyes in the textile dyeing process. The dyeing properties of polyphenols are influenced by factors such as concentration, pH, temperature, and dyeing time. Optimization of these parameters is crucial to achieve the desired color intensity, fastness, and reproducibility. The use of mordants, which are substances that enhance the dyeing process by forming complexes with both polyphenols and fibers, can also improve the dyeing performance. Common mordants include metallic salts such as aluminum, iron, or copper salts. These mordants aid in the fixation of polyphenols onto the textile fibers and enhance color fastness.

Polyphenols exhibit different dyeing mechanisms depending on their chemical structure and the fiber substrate. One of the primary mechanisms is adsorption, where polyphenols adhere to the surface of textile fibers through noncovalent interactions such as hydrogen bonding and van der Waals forces. Covalent bonding can also occur, particularly with fibers that contain reactive functional groups, such as cellulose or protein fibers. This bonding leads to a more permanent attachment of the polyphenols to the fibers [55]. Metal complexation is another mechanism where polyphenols form complexes with metal ions, resulting in enhanced dyeing properties and color stability. The color range achievable with polyphenols as natural dyes is extensive. Different polyphenols exhibit various colors, ranging from yellow and orange to red, purple, and blue. However, achieving consistent and reproducible colors with natural dyes can be challenging due to the complexity and variability of plant-derived polyphenols. Factors such as plant species, growing conditions, and extraction methods can influence the composition and concentration of polyphenols, resulting in variations in color outcomes. Nevertheless, the uniqueness and unpredictability of natural dyes contribute to their appeal and aesthetic value in textile applications.

In recent years, the role of polyphenols as natural dyes has gained attention from both the industry and academia. Researchers, textile manufacturers, and fashion designers have been exploring the potential of polyphenols to create innovative and sustainable textile products. Efforts are underway to develop standardized extraction methods, optimize dyeing processes, and improve color fastness and durability [56, 57]. Additionally, collaborations between academia and industry are fostering knowledge exchange and innovation, leading to advancements in the field of polyphenol-based natural dyes.

7.3 Chemical Structure of Polyphenols and their Dyeing Properties

7.3.1 Overview of Polyphenols

Polyphenols are a diverse group of naturally occurring compounds that are widely distributed in plants. They are characterized by the presence of multiple phenolic ring structures and are known for their antioxidant properties and potential health benefits. Polyphenols play essential roles in plant growth, development, defense against pathogens, and protection from environmental stresses. They have also gained significant attention in the field of nutrition and medicine due to their potential therapeutic effects. The chemical structure of polyphenols is complex and varied, with different subclasses and subclasses within subclasses [58]. The main subclasses of polyphenols include flavonoids, phenolic acids, stilbenes, lignans, and tannins. Flavonoids are the most abundant and well-studied class of polyphenols, and they can be further classified into subclasses such as flavones, flavonols, flavanones, and anthocyanins. Phenolic acids are another important class of polyphenols, with examples including hydroxybenzoic acids and hydroxycinnamic acids [59].

One of the distinguishing features of polyphenols is their antioxidant activity. They have the ability to scavenge free radicals and reactive oxygen species, thereby protecting cells and tissues from oxidative damage. This antioxidant capacity is attributed to the presence of hydroxyl groups on the phenolic rings, which allow for the donation of hydrogen atoms or electrons to neutralize free radicals. The antioxidant properties of polyphenols have been linked to various health benefits, including the prevention of chronic diseases such as cardiovascular diseases, cancer, and neurodegenerative disorders [60, 61]. Polyphenols also exhibit a wide range of biological activities beyond their antioxidant effects. They have been shown to possess anti-inflammatory, antimicrobial, anti-cancer, and anti-diabetic properties, among others. These activities are attributed to their interactions with cellular signaling pathways, enzyme inhibition, modulation of gene expression, and regulation of oxidative stress. The dietary intake of polyphenols has been associated with numerous health benefits. Fruits, vegetables, whole grains, legumes, nuts, and beverages such as tea, coffee, and red wine are rich sources of polyphenols. The bioavailability of polyphenols, however, can be influenced by various factors, including food processing, digestion, absorption, metabolism, and the presence of other dietary components. Studies have shown that the gut microbiota also plays a crucial role in the metabolism of polyphenols, as they can transform and metabolize these compounds into bioactive metabolites with enhanced health benefits [62, 63].

The health effects of polyphenols have been extensively studied in epidemiological studies, animal models, and in vitro experiments. These studies have provided evidence for their potential preventive and therapeutic effects against various diseases. For example, epidemiological studies have shown an inverse association between polyphenol-rich diets and the risk of chronic diseases, including cardiovascular diseases, certain types of cancer, and neurodegenerative disorders. Animal studies have demonstrated the protective effects of polyphenols in various experimental models of diseases, such as inflammation, oxidative stress, and carcinogenesis [64]. In vitro studies have further elucidated the mechanisms underlying the biological activities of polyphenols, including their interactions with cellular targets and signaling pathways. In addition to their health benefits, polyphenols have also found applications in various industries. Their antioxidant properties make them suitable for use as natural food preservatives, as they can help prevent lipid oxidation and delay the spoilage of food products. Polyphenols are also used in the cosmetic and skincare industry due to their potential anti-aging and skin-protective effects. Furthermore, polyphenols have been explored as natural dyes, as mentioned in the previous sections, offering a sustainable and environmentally friendly alternative to synthetic dyes in the textile industry.

Despite the extensive research conducted on polyphenols, there are still challenges and gaps in knowledge that need to be addressed [65]. The complexity of polyphenol chemistry, their interactions with other dietary components, and variations in their bioavailability make it difficult to establish clear cause-and-effect relationships between polyphenol intake and health outcomes. Standardization of analytical methods for polyphenol identification and quantification is also necessary to ensure accurate and reliable data across studies. Additionally, more research is needed to understand the optimal doses, timing, and combinations of polyphenols for specific health benefits and to explore potential interactions with medications.

7.3.2 Aromatic Rings and Hydroxyl Groups in Polyphenols

Aromatic rings and hydroxyl groups are fundamental structural features of polyphenols, which contribute to their unique properties and diverse applications. Polyphenols are a class of organic compounds that are characterized by the presence of multiple phenolic rings and hydroxyl (–OH) groups [66, 67]. These structural elements play crucial roles in defining the chemical, physical, and biological characteristics of polyphenols. Aromatic rings are cyclic structures composed of carbon atoms arranged in a planar ring with alternating single and double bonds. Polyphenols contain one or more aromatic rings, which confer several important properties. Firstly, aromatic rings contribute to the stability of polyphenols by delocalizing electrons along the ring structure. This electron delocalization, known as aromaticity, imparts stability to the compound, making polyphenols less prone to degradation and more resistant to chemical reactions. Additionally, the presence of aromatic rings allows polyphenols to absorb ultraviolet (UV) light, making them potentially useful as UV protectants in various applications such as sunscreens or photoprotective coatings. Hydroxyl groups (–OH) are functional groups composed of an oxygen atom bonded to a hydrogen atom. Hydroxyl groups are highly polar and can form hydrogen bonds with other molecules or functional groups. In polyphenols, hydroxyl groups are attached to the aromatic rings and are responsible for many of their unique properties. Firstly, hydroxyl groups enhance the water solubility of polyphenols, allowing them to dissolve in aqueous solutions and facilitating their extraction from plant materials [68].

This solubility is crucial for various applications, such as the production of natural dyes or the formulation of polyphenol-rich beverages or nutraceuticals. Hydroxyl groups also contribute to the antioxidant properties of polyphenols. The presence of hydroxyl groups in the structure of polyphenols enables them to act as hydrogen donors, effectively neutralizing free radicals and reactive oxygen species. Through this mechanism, polyphenols can scavenge and neutralize harmful oxidative species, protecting cells and tissues from oxidative damage. The antioxidant capacity of polyphenols has been associated with potential health benefits, including the prevention of chronic diseases such as cardiovascular diseases, cancer, and neurodegenerative disorders [69, 70].

The hydroxyl groups in polyphenols also participate in chemical reactions, such as oxidation, esterification, or glycosylation. These reactions can modify the chemical structure and properties of polyphenols, leading to the formation of new compounds with different activities or functionalities. For example, the oxidation of polyphenols can generate quinones, which are reactive intermediates involved in the antimicrobial and anticancer activities of certain polyphenols. Esterification reactions involving hydroxyl groups can lead to the formation of esters, which can alter the solubility, stability, or bioavailability of polyphenols. The combination of aromatic rings and hydroxyl groups in polyphenols also influences their interactions with other molecules or biological systems [71]. Aromatic rings can participate in π–π stacking interactions with other aromatic compounds, leading to the formation of complexes or aggregates. These interactions are important in various applications, such as the binding of polyphenols to receptors or enzymes in biological systems. Hydroxyl groups, on the other hand, can form hydrogen bonds with functional groups in proteins, carbohydrates, or other molecules, influencing their binding affinity or biological activity. The presence of aromatic rings and hydroxyl groups in polyphenols offers a wide range of opportunities for their application in various fields. In an experiment, Liu et al. [72] developed a loose nanofiltration membrane (LNM) by grafting tannic acid (TA) and polyethyleneimine (PEI) onto a PES/Fe ultrafiltration membrane using coordination assembly and Michael addition strategies. The influence of PEI concentration on the membrane's morphology and properties was systematically investigated as shown in Figure 7.3.

The grafting of PEI resulted in increased hydrophilicity and a positive surface charge of the PES/Fe-TA-PEI membrane. The optimized membrane exhibited high pure water flux (124.6 $L\,m^{-2}\,h^{-1}$) and excellent dye rejections (98.5%, 99.8%, 98.4%, and 86.4% for Congo red, Eriochrome black T, Alcian blue 8GX, and Bromophenol blue, respectively) under an operation pressure of 2 bar. Furthermore, the LNM demonstrated effective separation of dye/salt mixed solutions, with high rejection (>98.4%) for Alcian blue 8GX and low rejection (<5.3%) for NaCl. Additionally, the PES/Fe-TA-PEI LNM exhibited excellent antifouling performance and long-term stability. SEM images in Figure 7.4 depict the top surface and cross-sectional morphology of PES/Fe-TA-PEI membranes cross-linked with varying concentrations of PEI (0, 0.5, 1.0, 1.5, and 2.0 g/L). As shown in Figure 7.4a–e, the pores on the membrane surface diminish and eventually disappear as the PEI cross-linking concentration increases. Consequently, the surface of the membrane becomes rougher and denser, leading to the development of a new separation layer. Cross-sectional images illustrate the presence of double layers on the supporting layer (Figure 7.4[a1–e1]). The thickness of the cross-linked separation layer on the membrane surface gradually increases with higher PEI concentrations, ranging from approximately 122 to 405 nm. This observation confirms the successful cross-linking of PEI on the membrane surface, resulting in the formation of a compact active layer and modifying the filtration performance.

In addition to their antioxidant and health-promoting properties, polyphenols have been explored for their potential as natural dyes, functional ingredients in food and beverages, pharmaceuticals, cosmetics, and materials science. The aromaticity and hydrophilicity conferred by the aromatic rings and hydroxyl groups make polyphenols versatile compounds with unique chemical and physical properties [73, 74].

7.3.3 Contribution of Chemical Structure to Dyeing Properties

The chemical structure of a dye molecule plays a crucial role in its dyeing properties. It determines the color, affinity, and stability of the dye on the substrate, as well as the mechanism of dye-substrate interaction. Understanding the contribution of

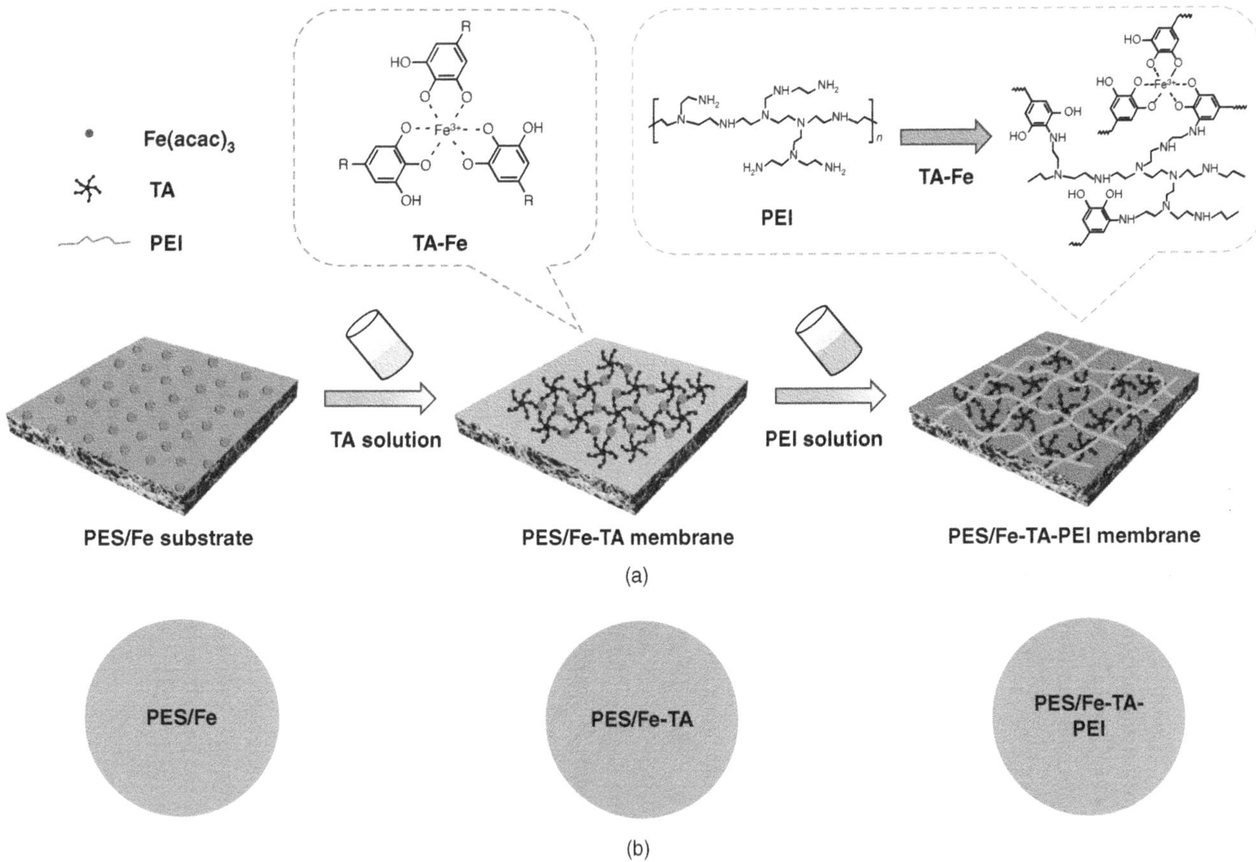

Figure 7.3 (a) Depiction of preparing the PES/Fe-TA-PEI membranes and (b) digital pictures of the PES/Fe, PES/Fe-TA, and PES/Fe-TA-PEI membranes. Source: Liu et al. [72] under CCBY 4.0.

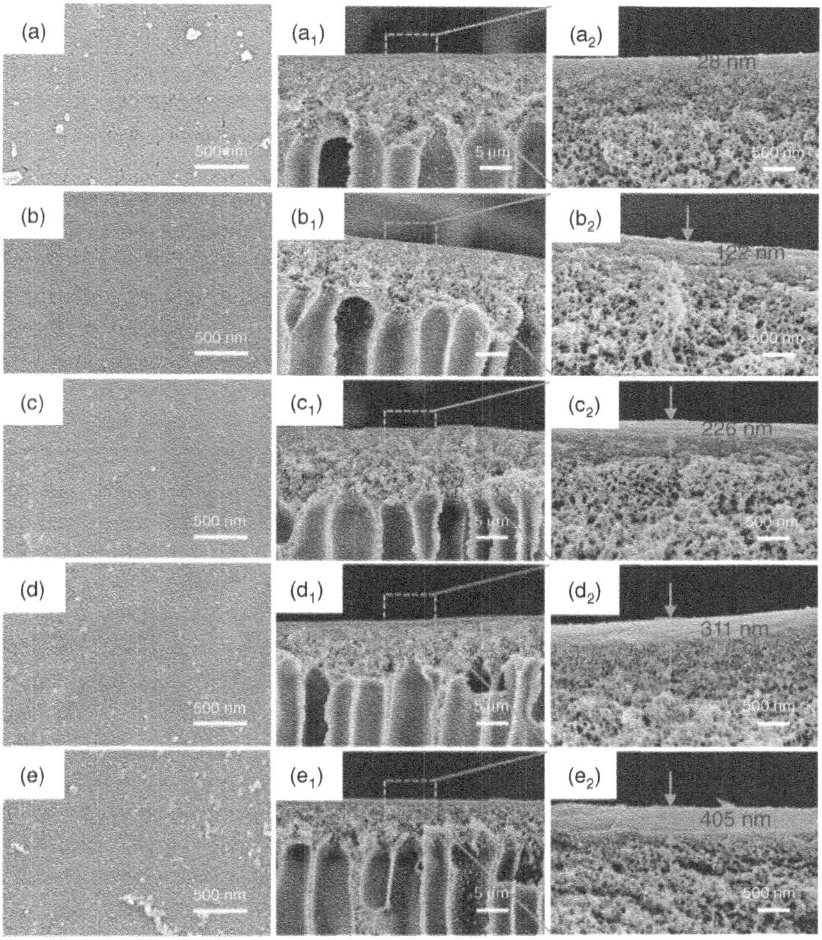

Figure 7.4 SEM scans of the top substrate and cross-section topologies of PES/Fe-TA-PEI membranes having distinctive PEI quantities ((a–a2) 0, (b–b2) 0.5, (c–c2) 1.0, (d–d2) 1.5, and (e–e2) 2.0 g/L). Source: Liu et al. [72]/Reproduced from MDPI/CC BY 4.0.

chemical structure to dyeing properties is essential for the development of efficient and sustainable dyeing processes. In this discussion, we will explore the key aspects of the chemical structure that influence dyeing properties, including chromophores, auxochromes, solubility, and reactive groups. Chromophores are the functional groups responsible for the color of a dye molecule. These groups absorb light in the visible spectrum, resulting in the perception of color. Common chromophores in dyes include azo (–N=N–), nitro (–NO_2), carbonyl (C=O), and conjugated systems such as aromatic rings.

The type and arrangement of chromophores in a dye molecule determine its absorption spectrum and, thus, its color [75]. For example, azo dyes exhibit a wide range of colors due to the variation in the substituents in the azo group. Conjugated systems, such as those found in anthraquinone or phthalocyanine dyes, contribute to intense and vibrant colors [76]. Auxochromes are functional groups that modify the color of a dye molecule by altering its absorption characteristics. These groups enhance the intensity and stability of the color, as well as the affinity of the dye for the substrate. Common auxochromes include hydroxyl (–OH), amino ($-NH_2$), and sulfonic acid ($-SO_3H$) groups. Hydroxyl and amino groups, for instance, can increase the solubility of the dye molecule in water and enhance its interaction with the substrate through hydrogen bonding or electrostatic forces. Sulfonic acid groups, on the other hand, provide water solubility and increase the dye's affinity for substrates with positive charges, such as wool or silk [77].

Solubility is another important aspect of dyeing properties influenced by the chemical structure of a dye molecule. Dye molecules must be soluble in the dyeing medium (usually water) to ensure uniform distribution and penetration into the substrate. The presence of hydrophilic groups, such as hydroxyl or sulfonic acid groups, enhances the water solubility of the dye. Additionally, the molecular size and shape of the dye molecule influence its solubility. Smaller molecules tend to have better solubility due to the reduced intermolecular interactions. Reactive groups in the dye molecule enable the formation of covalent bonds with the substrate, resulting in a durable and wash-fast coloration. Reactive dyes, for example, contain functional groups (e.g. vinyl sulfone and chlorotriazine) that can react with nucleophilic sites on the substrate, such as amino or hydroxyl groups.

The formation of covalent bonds between the dye and substrate ensures excellent color fastness properties, as the dye becomes an integral part of the textile material. Reactive groups also play a crucial role in the fixation and anchoring of the dye onto the substrate, preventing its easy release during subsequent washing or exposure to environmental factors. Furthermore, the molecular weight and structure of the dye molecule influence its migration and diffusion during the dyeing process. Large and complex dye molecules tend to have slower diffusion rates, requiring longer dyeing times or higher dye concentrations to achieve desired color depths. The molecular weight also affects the substantively of the dye, which refers to its ability to be attracted to and retained by the substrate. Dyes with appropriate molecular weights and structures can achieve optimal substantively, resulting in efficient and uniform dye uptake by the substrate. In addition to the aforementioned factors, the overall charge of the dye molecule also impacts its dyeing properties. Cationic dyes carry a positive charge and are attracted to negatively charged substrates, such as wool or cotton. On the other hand, anionic dyes bear a negative charge and are suitable for dyeing positively charged substrates like acrylic or nylon. Understanding the charge characteristics of the dye molecule is crucial for selecting the appropriate dye-substrate combination and optimizing the dyeing process.

7.4 Extraction Methods for Polyphenols from Plant Sources

7.4.1 Plant Sources of Polyphenols

Polyphenols are a diverse group of compounds that are widely distributed in the plant kingdom. They are synthesized by plants as secondary metabolites and serve various functions, including defense against pathogens, UV radiation protection, and pigmentation. Polyphenols have gained significant attention due to their potential health benefits and applications in various industries, including food, pharmaceuticals, and cosmetics. In this discussion, we will explore the plant sources of polyphenols, their classification, and the factors that influence their content and composition. Plant sources of polyphenols encompass a wide range of botanical species, including fruits, vegetables, cereals, nuts, seeds, herbs, and spices. The content and composition of polyphenols vary among different plant sources and even within different parts of the same plant. Understanding the diversity of plant sources is crucial for harnessing the potential of polyphenols in various applications. Fruits are one of the richest sources of polyphenols. Berries, such as strawberries, blueberries, raspberries, and blackberries, are particularly abundant in polyphenols, including anthocyanins, flavonols, and ellagitannins. Citrus fruits, such as oranges, lemons, and grapefruits, contain flavonoids, especially flavanones like hesperidin and naringin. Other fruits like apples, pears, and grapes are also notable sources of polyphenols, including flavanols (catechins and epicatechins) and procyanidins [78].

Vegetables are another significant source of polyphenols. Dark leafy greens, such as spinach, kale, and Swiss chard, are rich in flavonoids like quercetin and kaempferol. Cruciferous vegetables like broccoli, Brussels sprouts, and cabbage contain glucosinolates, which can be converted into polyphenolic compounds with potential health benefits. Onions and garlic are known for their high content of flavonols, particularly quercetin. Additionally, tomatoes are a notable source of polyphenol lycopene, which is responsible for their red color. Cereals, although commonly associated with carbohydrates, also contain polyphenols. Whole grains, such as wheat, rice, barley, and oats, are rich in phenolic acids, particularly ferulic acid. These compounds are mainly found in the bran and outer layers of the grains. However, during the refining process, these outer layers are often removed, resulting in a significant loss of polyphenols. Therefore, consuming whole grain products is recommended to maximize polyphenol intake. Nuts and seeds are also notable sources of polyphenols. For example, almonds, walnuts, and peanuts contain various types of polyphenols, including flavonoids and phenolic acids [79]. Flaxseeds, chia seeds, and sesame seeds are rich in lignans, which are polyphenolic compounds with antioxidant and hormone-balancing properties. Additionally, cocoa beans, the main ingredient in chocolate, contain high levels of flavanols, particularly epicatechin.

Herbs and spices are valued for their culinary and medicinal properties, and many of them are rich in polyphenols. Some popular examples include cinnamon, turmeric, ginger, oregano, thyme, and parsley. These herbs and spices are known to contain various polyphenolic compounds, such as flavonoids, phenolic acids, and essential oils. They not only add flavor and aroma to dishes but also contribute to their antioxidant and anti-inflammatory properties. The content and composition of polyphenols in plant sources can be influenced by several factors. Environmental factors, such as climate, soil composition, sunlight exposure, and growing conditions, can affect the synthesis and accumulation of polyphenols in plants. For example, berries grown in colder climates tend to have higher levels of anthocyanins, which serve as a protective mechanism against UV radiation [80]. Additionally, the stage of plant growth and maturity can also impact the polyphenol content. For instance, young and tender leaves generally contain higher levels of polyphenols compared to older leaves. Furthermore, postharvest processing and storage conditions can affect the stability and composition of polyphenols. Exposure to heat, light, oxygen, and enzymatic activity can lead to the degradation of polyphenols. Therefore, proper handling, storage, and processing techniques should be employed to retain the polyphenol content and maximize its potential benefits.

7.4.2 Extraction Techniques: Solvent Extraction

Extraction techniques play a crucial role in obtaining polyphenols from plant sources for various applications. One of the commonly used methods is solvent extraction, which involves the use of organic solvents to dissolve and extract polyphenolic compounds from plant materials. This process allows for the efficient extraction of polyphenols while preserving their chemical integrity and bioactivity. In this discussion, we will explore the principles, factors affecting solvent extraction, and the advantages and limitations of this technique. Solvent extraction operates on the principle of solubility, where the polyphenols present in the plant matrix dissolve in an appropriate solvent to form a solution. The choice of solvent is a critical factor as it determines the selectivity and efficiency of extraction. Commonly used solvents include ethanol, methanol, acetone, and water, either individually or as mixtures. Ethanol and methanol are widely employed due to their ability to dissolve a wide range of polyphenols and their commercial availability. However, the selection of solvents depends on the specific polyphenols of interest, as some compounds may have a higher solubility in polar solvents like water or less polar solvents like ethyl acetate [81].

Several factors influence the efficiency of solvent extraction. One of the key factors is the solvent-to-solid ratio, which determines the concentration gradient and extraction kinetics. Increasing the solvent-to-solid ratio can enhance the extraction efficiency; however, there is an optimum ratio beyond which further increase may not significantly improve the extraction. The particle size of the plant material also affects extraction efficiency, as smaller particle sizes provide a larger surface area for solvent penetration and extraction. Extraction time and temperature are crucial parameters that influence the yield and composition of extracted polyphenols. Prolonged extraction times can lead to the degradation of heat-sensitive polyphenols, while excessively high temperatures may denature the compounds. Optimization of extraction conditions, such as time and temperature, is essential to maximize extraction efficiency while minimizing potential degradation. In addition to these factors, the presence of co-extracted compounds can affect the selectivity and purity of extracted polyphenols. Co-extracted compounds include sugars, organic acids, pigments, and other secondary metabolites. Proper purification techniques, such as liquid–liquid extraction, solid-phase extraction, or chromatographic methods, may be necessary to remove impurities and isolate the desired polyphenols.

There are several innovative approaches such as ultrasound-assisted extraction (UAE), the utilization of green solvents, and the combination of solvent extraction with other techniques like microwave-assisted extraction (MAE) and enzyme-assisted extraction (EAE) [82]. Ultrasound-assisted extraction (UAE) has emerged as a promising technique for polyphenol extraction. In this method, ultrasonic waves are applied during the extraction process, leading to several beneficial effects. The ultrasound waves create cavitation bubbles in the solvent, which subsequently collapse, generating localized high temperatures and pressures. These physical effects enhance the mass transfer between the solvent and plant materials, disrupting the cell walls and improving the release of polyphenols. As a result, UAE leads to higher extraction yields and reduced extraction time compared to conventional solvent extraction methods. Additionally, UAE has been shown to improve the extraction of specific polyphenols with low solubility, thereby expanding the range of extractable compounds.

The utilization of green solvents has gained considerable attention in recent years due to their safety, sustainability, and reduced environmental impact. Green solvents include water, supercritical carbon dioxide (SC–CO_2), and natural deep eutectic solvents (NADES). Water is the most abundant and environmentally friendly solvent, making it an attractive option for polyphenol extraction. However, the low solubility of some polyphenols in water can limit their efficacy. Supercritical carbon dioxide, on the other hand, offers unique properties such as low viscosity, high diffusivity, and tunable solvent power. SC–CO_2 can effectively extract polyphenols while being easily separable from the extract. NADES are a newer class of green solvents formed by mixing two or more components, typically composed of hydrogen bond donors and acceptors. NADES have shown promise in extracting polyphenols, offering high solubility and selectivity for specific compounds. The combination of solvent extraction with other techniques has also been explored to enhance extraction efficiency and obtain specific polyphenolic compounds. MAE involves the application of microwave energy, which generates heat within the solvent and plant matrix, facilitating the release of polyphenols [83]. MAE has demonstrated advantages such as shorter extraction times, improved yields, and increased selectivity for certain polyphenols. EAE utilizes enzymes to break down cell walls and facilitate the release of polyphenols. Enzymes such as cellulases, hemicellulases, and pectinases have been used in combination with solvent extraction to enhance extraction efficiency, especially for polyphenols trapped within plant cell walls.

Solvent extraction offers several advantages to polyphenol extraction. It is a versatile and widely applicable technique that can be easily scaled up for industrial production. Solvents used in extraction are generally safe and commercially available. Moreover, solvent extraction allows for the extraction of a wide range of polyphenols with different chemical properties, making it suitable for various applications. However, solvent extraction has some limitations. The use of organic solvents raises concerns regarding the environmental impact and potential toxicity. Therefore, solvent recycling and the development of green solvents are being explored to minimize the environmental footprint. Another limitation is the potential loss of thermally sensitive or volatile polyphenols during the extraction process. Extraction efficiency can also vary depending on the plant material and its matrix complexity, which may require the use of additional extraction techniques or modifications to optimize the extraction process.

7.4.3 Extraction Techniques: Solid-Phase Extraction

SPE is a widely used technique for the extraction of polyphenols from plant sources. It offers numerous advantages over other extraction methods, making it a preferred choice for researchers and industry professionals involved in polyphenol extraction and analysis. One of the primary advantages of solid-phase extraction is its ability to achieve high selectivity and purification of target compounds. This is made possible by the choice of sorbent material and its functional groups, which can be tailored to selectively interact with specific polyphenols. Different types of sorbents, such as reversed-phase, normal-phase, ion-exchange, and molecularly imprinted polymers (MIPs), can be employed depending on the desired selectivity and characteristics of the polyphenols [84].

The selectivity of the sorbent ensures that unwanted compounds are efficiently removed, resulting in a cleaner polyphenol extract. The versatility of solid-phase extraction is another advantage that contributes to its widespread use. It can accommodate a wide range of sample matrices, including liquid samples, solid samples, and complex mixtures. Solid-phase extraction can be adapted to handle different sample types by employing appropriate sample preparation techniques such as extraction, filtration, and evaporation. This flexibility allows for the extraction of polyphenols from diverse sources such as plant tissues, beverages, and food products without compromising the efficiency of the process.

Convenience and automation are key features of solid-phase extraction. The technique can be easily automated using automated liquid handling systems, ensuring reproducibility and reducing the labor-intensive nature of the extraction. By using prepacked cartridges or disks containing the sorbent material, the extraction process becomes standardized and less prone to human error. Moreover, solid-phase extraction allows for the simultaneous extraction of multiple samples, making it a time-efficient option for high-throughput analysis. Practical considerations, including scalability and cost-effectiveness, further contribute to the popularity of solid-phase extraction. The method can be easily scaled up or down to accommodate different sample volumes, making it suitable for both laboratory-scale and industrial-scale applications [85]. Additionally, the relatively low cost of the sorbent materials and the reusability of certain sorbents make solid-phase extraction an economically viable option for large-scale polyphenol extraction.

The solid-phase extraction process typically involves several steps. First, the sorbent material is conditioned to ensure its proper activation and equilibration, which is essential for achieving optimal extraction efficiency. Next, the sample is loaded onto the sorbent bed, where the polyphenols selectively adsorb onto the sorbent material. Washing is performed to remove any unwanted compounds that may have been retained during the loading step, further purifying the polyphenols of interest. Finally, elution is carried out to recover the target polyphenols, typically using an appropriate solvent or solvent mixture that allows for the desorption of the polyphenols from the sorbent material. Recent advancements in solid-phase extraction techniques have further improved its efficiency and applicability. The development of new sorbent materials with enhanced selectivity and capacity has expanded the range of polyphenols that can be effectively extracted. Molecularly imprinted polymers (MIPs) have gained significant attention as sorbents due to their specific binding sites that can mimic natural receptors, resulting in highly selective extraction of target polyphenols. These advancements contribute to the continuous refinement of solid-phase extraction methods and open up new possibilities for the extraction of polyphenols from various plant sources [86].

7.4.4 Extraction Techniques: Ultrasound-Assisted Extraction

UAE is a technique that has gained significant attention in recent years for the extraction of polyphenols from plant sources. This method involves the application of ultrasonic waves during the extraction process, which enhances mass transfer, disrupts cell walls, and improves the release of polyphenols from plant materials. UAE offers several advantages over conventional extraction methods, including increased extraction efficiency, reduced extraction time, and improved sustainability. One of the key advantages of UAE is its ability to enhance mass transfer. The application of ultrasonic waves causes cavitation, which generates localized high temperatures and pressures. This phenomenon leads to the formation and collapse of microbubbles, resulting in the disruption of cell walls and the release of intracellular polyphenols. The cavitation also promotes the penetration of solvents into the plant matrix, facilitating the extraction process and increasing the yield of polyphenols. The enhanced mass transfer in UAE is attributed to various mechanisms. Firstly, the rapid expansion and collapse of microbubbles create strong shock waves that disrupt the cell walls, releasing the polyphenols. Secondly, the microstreaming and jetting effect generated by cavitation improves the contact between the solvent and the plant material, facilitating the diffusion of polyphenols into the solvent. Lastly, the acoustic streaming effect induces convective flows that enhance the mixing and transport of solutes, leading to improved extraction efficiency [87].

Another advantage of the UAE is its ability to reduce extraction time. Compared to traditional extraction methods, UAE significantly shortens the extraction time required to obtain the desired polyphenols. The application of ultrasound waves increases the extraction rate by accelerating the release of polyphenols from the plant matrix. The shorter extraction time not only increases productivity but also minimizes the degradation and oxidation of sensitive polyphenols, resulting in higher-quality extracts. Furthermore, UAE offers sustainability benefits. It is a green extraction technique that reduces the use of organic solvents and energy consumption compared to conventional methods. The enhanced mass transfer and extraction efficiency of UAE enable the use of smaller solvent volumes and shorter extraction times, contributing to reduced solvent consumption. Additionally, the lower extraction temperatures associated with UAE decrease energy requirements and minimize the thermal degradation of polyphenols, preserving their bioactive properties. These sustainability aspects make UAE an attractive option for the extraction of polyphenols, aligning with the growing demand for environmentally friendly extraction processes.

The application of UAE is not limited to specific plant sources. It can be applied to various botanical species, including fruits, vegetables, herbs, spices, and other plant materials. This versatility allows for extracting polyphenols from multiple plant sources, each with its unique composition and polyphenolic profile. The choice of plant source can be tailored based on the specific polyphenols of interest and their intended applications. The extraction conditions in UAE can be optimized to maximize the extraction efficiency and yield of polyphenols. Parameters such as ultrasonic power, frequency, extraction temperature, extraction time, and solvent composition can be adjusted to achieve the desired extraction performance. The optimization process involves determining the optimal combination of these parameters that ensures efficient polyphenol extraction while minimizing the degradation of bioactive compounds. Researchers have recently explored various modifications and innovations in the UAE to improve its performance further. One such modification is using different extraction modes, including direct, indirect, and simultaneous ultrasound, which can be employed based on the nature of the plant material and the desired extraction efficiency. Additionally, the combination of UAE with other extraction techniques, such as MAE and EAE, has been investigated to enhance extraction efficiency and obtain specific polyphenolic compounds synergistically.

The combination of UAE with MAE involves applying both ultrasonic waves and microwave radiation simultaneously. This combination exploits the complementary effects of ultrasound and microwave to improve the extraction efficiency and shorten the extraction time. Microwave radiation generates internal heat, leading to a rapid increase in temperature, while ultrasound enhances mass transfer and disrupts the plant matrix. The synergistic effect of these techniques results in a more efficient extraction process. Similarly, the combination of UAE with EAE involves using enzymes to break down the plant cell walls and facilitate the release of polyphenols. The application of ultrasound waves enhances enzymatic activity by promoting better contact between the enzymes and the plant material. The combination of enzyme-assisted extraction with UAE can result in higher polyphenol yields and reduce the enzyme dosage required for extraction. In terms of analytical techniques, UAE extracts can be analyzed using various methods, including high-performance liquid chromatography (HPLC), gas chromatography (GC), mass spectrometry (MS), and nuclear magnetic resonance (NMR). These techniques allow for the identification and quantification of polyphenolic compounds in the extracts, providing valuable information about their composition and bioactivity [88].

7.4.5 Advantages and Limitations of Different Extraction Methods

Extraction methods are crucial in obtaining bioactive compounds, including polyphenols, from plant sources. Different extraction techniques offer advantages and limitations that need to be considered based on the specific requirements of the target compounds and the desired application. In this discussion, we will explore the advantages and limitations of three commonly used extraction methods: solvent extraction, SPE, and UAE. Solvent extraction is one of the most widely used methods for polyphenol extraction. It involves the use of organic solvents to dissolve and extract polyphenols from the plant material. One of the primary advantages of solvent extraction is its simplicity and versatility. It can be easily performed using common laboratory equipment and a wide range of solvents, allowing for the extraction of polyphenols from various plant sources. Solvent extraction also offers high extraction efficiency, enabling the recovery of a large amount of polyphenols. Moreover, solvent extraction is suitable for large-scale applications due to its scalability. However, solvent extraction has several limitations that need to be addressed. One limitation is the potential extraction of unwanted compounds, such as pigments, sugars, and proteins, which can affect the purity of the polyphenol extract.

Additional purification steps may be required to remove these interfering compounds. Solvent extraction also requires the use of organic solvents, which can be expensive and pose environmental concerns due to their toxicity and flammability. The selection of solvents should be carefully considered to ensure the safety and sustainability of the extraction process.

SPE is a technique that utilizes solid materials, known as sorbents, to selectively adsorb and retain target compounds from complex mixtures. SPE offers several advantages over solvent extraction. One of the key advantages is its high selectivity and purification capabilities. The choice of sorbent material and its functional groups can be tailored to selectively interact with specific polyphenols, enabling the removal of interfering compounds and purification of the target polyphenols. SPE also allows for the concentration of polyphenols, making it suitable for the analysis and further processing of extracts. Furthermore, SPE offers convenience and automation, as the technique can be easily automated using prepacked cartridges or disks containing sorbent material. However, SPE has certain limitations that need to be considered. The availability and cost of specific sorbent materials can be a limiting factor, especially for large-scale applications. The selectivity of the sorbent may vary depending on the polyphenolic compounds of interest, and different sorbents may be required for different polyphenols. Additionally, the sample preparation step before SPE, such as filtration and evaporation, may be time-consuming and labor-intensive. Despite these limitations, SPE remains a valuable technique for the selective extraction and purification of polyphenols. UAE is a technique that utilizes ultrasonic waves to enhance the extraction process. UAE offers several advantages over conventional extraction methods. One of the primary advantages is its ability to enhance mass transfer and extraction efficiency. The application of ultrasound waves disrupts the plant cell walls, facilitating the release of intracellular polyphenols and improving their solubility in the solvent. UAE also reduces the extraction time, allowing for faster extraction processes and minimizing the degradation of sensitive polyphenols. Moreover, UAE is considered a greener alternative to traditional extraction methods, as it reduces the use of organic solvents and energy consumption. Despite its advantages, UAE has certain limitations that need to be addressed. The equipment required for UAE can be costly and may only be readily available in some laboratories. The optimization of ultrasound parameters, such as power, frequency, and extraction time, is crucial to achieving optimal extraction efficiency, which can be a complex and time-consuming process. Applying ultrasound waves may generate heat, which can degrade heat-sensitive polyphenols. Therefore, temperature control in UAE is important to prevent excessive heating. Furthermore, the cavitation effect induced by ultrasound can cause erosion and damage to the plant material, potentially leading to extract contamination.

7.5 Dyeing Mechanisms of Polyphenols

7.5.1 Adsorption as a Dyeing Mechanism

Polyphenols, with their diverse chemical structures and functional groups, can interact with textile fibers through adsorption, making them suitable candidates for natural dyeing. Several factors, including the chemical nature of the polyphenol, the fiber composition, pH, temperature, and dyeing conditions, influence the adsorption of polyphenols onto the fiber surface. Understanding the mechanism of polyphenol adsorption is crucial for achieving effective and durable coloration [82, 84, 89]. In this discussion, we will explore the mechanism of polyphenol adsorption, the factors affecting adsorption, and the significance of polyphenol adsorption in dyeing processes. Polyphenols possess aromatic rings and hydroxyl groups, which contribute to their dyeing properties. The aromatic rings provide conjugated electron systems, allowing polyphenols to absorb light in the visible spectrum and exhibit vibrant colors. On the other hand, the hydroxyl groups contribute to hydrogen bonding and other interactions with the fiber surface, facilitating adsorption. The adsorption process can involve electrostatic interactions, van der Waals forces, hydrogen bonding, and covalent bonding, depending on the chemical properties of the polyphenol and the fiber.

One of the key factors influencing polyphenol adsorption is the compatibility between the polyphenol and the fiber. Different fibers have different chemical compositions and surface characteristics, affecting polyphenols'affinity and adsorption behavior. Natural fibers like cotton, wool, and silk have hydrophilic surfaces, making them more receptive to polyphenols through hydrogen bonding and van der Waals forces. With their hydrophobic surfaces, synthetic fibers like polyester and nylon may require pretreatment or mordants to improve polyphenol adsorption. The pH of the dyeing medium plays a significant role in polyphenol adsorption. It affects the ionization state of the polyphenol and the fiber surface, thereby influencing their electrostatic interactions. For instance, acidic conditions are generally favorable for the adsorption of polyphenols onto protein fibers, where the positively charged polyphenol molecules are attracted to the negatively charged fiber surface. Alkaline conditions, on the other hand, can promote the adsorption of polyphenols onto cellulose fibers. Temperature also influences polyphenol adsorption. Higher temperatures enhance the rate of polyphenol diffusion and increase the adsorption capacity. The increase in temperature promotes molecular motion, facilitating the penetration of polyphenol molecules into the fiber structure. However, excessively high temperatures can lead to thermal degradation of both the polyphenol and the fiber, compromising color fastness and fiber properties. Therefore, the selection of dyeing temperature should consider both polyphenol diffusion and thermal stability.

The duration of the dyeing process affects polyphenol adsorption as well. Longer dyeing times allow for more polyphenol molecules to come into contact with the fiber surface and establish stronger interactions [72]. However, extended dyeing times can lead to dye exhaustion and increased energy consumption. Optimizing the dyeing time is crucial to ensure sufficient polyphenol uptake and desired color intensity. The concentration of polyphenols in the dye bath also plays a role in adsorption. Higher polyphenol concentrations provide a greater number of dye molecules available for adsorption, leading to higher dye uptake and stronger coloration. However, there is a limit to the polyphenol concentration beyond which excessive dye aggregation or precipitation may occur, resulting in uneven dye distribution and reduced color fastness. Finding the optimal polyphenol concentration is important to achieve uniform coloration without compromising dyeing performance.

In addition to these factors, the presence of auxiliary chemicals, such as mordants and additives, can significantly influence polyphenol adsorption in dyeing. Mordants are substances that form complexes with polyphenols and enhance their affinity for fibers. Common mordants include metallic salts, such as alum, iron salts, and copper salts. These mordants form coordination complexes with polyphenols and fibers, facilitating adsorption and improving color fastness. Additives, such as surfactants and salts, can modify the surface tension and ionic strength of the dye bath, influencing the adsorption behavior of polyphenols. Polyphenol adsorption as a dyeing mechanism offers several advantages. Firstly, polyphenols are abundant natural compounds found in various plant sources, making them environmentally friendly alternatives to synthetic dyes. They offer opportunities for sustainable and eco-friendly dyeing processes. Secondly, the complex chemical structures of polyphenols contribute to their color stability and resistance to fading, resulting in long-lasting and durable coloration. Polyphenol-dyed textiles can exhibit excellent color fastness properties. Polyphenols also possess antioxidant and antimicrobial properties, which can be beneficial for dyeing applications. The presence of polyphenols on the fiber surface can provide antioxidant protection against oxidative damage and improve the overall durability of the textile. Additionally, the antimicrobial properties of certain polyphenols can contribute to the prevention of microbial growth and the development of odor in textile products.

7.5.2 Covalent Bonding in Dyeing with Polyphenols

Covalent bonding plays a significant role in the dyeing process with polyphenols, contributing to the durability and colorfastness of the dyed textiles. Unlike adsorption, which involves reversible interactions between the dye molecules and the fiber surface, covalent bonding involves the formation of strong chemical bonds between the dye molecules and the fiber molecules. In this discussion, we will explore the mechanism of covalent bonding in dyeing with polyphenols, its advantages, and its significance in achieving long-lasting coloration. Polyphenols contain functional groups such as hydroxyl (–OH), carboxyl (–COOH), and amino ($-NH_2$) groups, which have the potential to form covalent bonds with reactive sites on the fiber molecules [71, 74, 76, 90]. During the dyeing process, these functional groups undergo chemical reactions with the fiber, resulting in the formation of covalent bonds. The most common fiber reactive groups involved in covalent bonding with polyphenols are hydroxyl groups on cellulose fibers, amino groups on protein fibers, and carboxyl groups on synthetic fibers.

One of the key advantages of covalent bonding in dyeing is the improved wash fastness and colorfastness of the dyed textiles. Covalent bonds are strong and stable, making them resistant to washing, rubbing, and light exposure. The formation of covalent bonds between the dye molecules and the fiber molecules ensures that the color remains firmly attached to the fiber, even after repeated laundering and exposure to environmental factors. This leads to enhanced color durability and longevity of the dyed textiles. Covalent bonding also offers advantages in terms of color reproducibility and control. The formation of covalent bonds between polyphenols and fibers is a well-defined chemical process, allowing for precise control over the dyeing conditions and the resulting color. By adjusting factors such as pH, temperature, and dyeing time, the extent of covalent bonding can be optimized to achieve the desired color intensity and shade. This level of control is particularly important in industries where color consistency is crucial, such as textile manufacturing and dyeing. Furthermore, covalent bonding contributes to the improved colorfastness properties of the dyed textiles. The strong chemical bonds formed between the dye molecules and the fiber molecules prevent color bleeding, fading, and migration. Covalently bonded dyes remain securely attached to the fiber, even when exposed to harsh conditions such as heat, light, and friction. This ensures that the dyed textiles maintain their original color intensity and appearance over an extended period. The covalent bonding mechanism in dyeing with polyphenols offers compatibility with a wide range of fiber types. Polyphenols can form covalent bonds with natural fibers, such as cotton, wool, and silk, as well as synthetic fibers like polyester and nylon. This versatility makes polyphenols suitable for dyeing various textile materials, allowing for the production of dyed fabrics with different fiber compositions and properties.

However, it is important to note that the covalent bonding process requires specific conditions and chemical reactions to occur. The choice of dyeing conditions, including pH, temperature, and dyeing time, should be optimized to facilitate the formation of covalent bonds without causing damage to the fibers or the dyes. Care must be taken to avoid excessive heat or highly acidic or alkaline conditions, as they can lead to fiber degradation or undesirable side reactions.

7.5.3 Metal Complexation in Polyphenol Dyeing

Metal complexation is a fundamental process in polyphenol dyeing that involves the formation of coordination complexes between polyphenolic compounds and metal ions. This mechanism plays a crucial role in the coloration of textiles, offering unique dyeing properties and advantages. In this discussion, we will explore the process of metal complexation in polyphenol dyeing, its significance, and its applications in the textile industry. Metal complexation occurs due to metal-binding sites in polyphenolic compounds, which can form coordination bonds with metal ions. Polyphenols contain functional groups such as hydroxyl (–OH), carboxyl (–COOH), and catechol (–O-phenol-O–) groups, which have a strong affinity for metal ions. The metal ions involved in complexation can be transition metals such as iron (Fe), copper (Cu), chromium (Cr), or other metal ions with a high affinity for polyphenols [50, 52, 56, 57, 91]. The complexation process begins with coordinating metal ions with the polyphenolic ligands. The metal ions act as Lewis acids, accepting electron pairs from the Lewis base sites on the polyphenolic compounds. This coordination results in metal-polyphenol complexes, where the metal ion is coordinated with multiple polyphenolic ligands. The formation of these complexes leads to color development on the textile fibers, resulting in various shades and hues.

Metal complexation in polyphenol dyeing offers several advantages and unique dyeing properties. One of the key advantages is the broad color spectrum that can be achieved. Metal ions can interact with polyphenols in different ways, leading to the formation of complexes with distinct colors. For example, iron (Fe) complexes often produce shades of brown, while copper (Cu) complexes can generate green or blue colors. This versatility allows for a wide range of color options in textile dyeing. Another advantage of metal complexation is the enhanced colorfastness and durability of the dyed textiles. The coordination bonds formed between the metal ions and polyphenols are relatively strong, providing improved wash fastness and light fastness properties. The metal-polyphenol complexes are less prone to fading or bleeding compared to other dyeing mechanisms, resulting in long-lasting coloration on the fibers. Metal complexation also offers excellent light stability and resistance to environmental factors. The coordination bonds between the metal ions and polyphenols contribute to the stability of the dye molecules, preventing color fading or degradation upon exposure to sunlight or other sources of UV radiation. This property is particularly important in applications where textiles are subjected to outdoor or high-light conditions, such as outdoor clothing or upholstery. Furthermore, metal complexation can enhance the antimicrobial and antioxidant properties of dyed textiles. Certain metal-polyphenol complexes have been found to exhibit antimicrobial activity, inhibiting the growth of bacteria and fungi on the fabric surface. This antimicrobial effect can be beneficial in applications such as medical textiles, where the prevention of microbial contamination is crucial. Additionally, the antioxidant properties of polyphenols can be enhanced by metal complexation, providing additional benefits in terms of protection against oxidative damage.

The application of metal complexation in the textile industry is diverse, ranging from traditional dyeing techniques to advanced functional finishes. Metal-complex dyes can be applied to textiles through various dyeing methods, including exhaust dyeing, pad dyeing, and printing. In exhaust dyeing, the textile material is immersed in a dye bath containing metal-complex dyes, allowing the complexes to adsorb onto the fiber surface. Pad dyeing involves applying the metal-complex dyes directly onto the fabric using a padding machine, followed by fixation through heat or chemical treatment. Printing techniques can also be used to create intricate designs and patterns using metal-complex dyes. Additionally, metal complexation can be incorporated into functional finishes for textiles. For example, metal-complex-based finishes can be applied to textiles to impart additional properties such as flame retardancy, water repellence, or UV protection [92]. The metal-polyphenol complexes contribute to the desired functional properties while simultaneously providing coloration.

Despite the advantages and applications of metal complexation in polyphenol dyeing, there are certain considerations and limitations to be aware of. One limitation is the potential for metal ion migration or leaching from the dyed textiles. If not properly fixed or treated, the metal ions in the complexes may be released from the fibers, leading to color bleeding or potential environmental concerns. Proper fixation techniques, such as postdyeing treatments or the use of appropriate mordants, are necessary to minimize metal ion migration. Moreover, the use of metal-complex dyes may pose challenges in terms of cost and sustainability. Some metal ions used in dyeing processes, such as chromium (Cr), are considered environmentally hazardous and may have restrictions or regulations in place. Therefore, it is important to consider the environmental impact and sustainability aspects when using metal-complex dyes.

7.6 Applications of Polyphenols as Natural Dyes

7.6.1 Polyphenols in Fabric Dyeing

The use of polyphenols as natural dyes in fabric dyeing has gained significant attention due to their unique dyeing properties, environmental friendliness, and potential health benefits. In this section, we will review relevant studies and research articles to further explore the role of polyphenols in fabric dyeing. Polyphenols are widely distributed in various plant sources, including fruits, vegetables, tree bark, flowers, and leaves. These compounds are known for their diverse chemical structures, which contribute to a wide range of colors that can be obtained through fabric dyeing. Flavonoids, tannins, and anthocyanins are among the most commonly used polyphenolic compounds in dyeing processes. The sustainability advantages of using polyphenols as natural dyes have been emphasized in several studies. Synthetic dyes often require large amounts of water, energy, and chemicals during production, leading to environmental pollution and health hazards. In contrast, polyphenols are derived from renewable plant sources and can be extracted using environmentally friendly methods [93, 94]. For example, Li et al. (2018) investigated the extraction of polyphenols from persimmon peels using ultrasound-assisted extraction, which demonstrated the potential of using waste materials for dye production, thus promoting sustainability.

Polyphenols offer unique dyeing mechanisms that result in vibrant and durable colors on fabrics. The adsorption mechanism is the most common, where polyphenols bind to the fabric surface through weak interactions, such as hydrogen bonding and van der Waals forces. This mechanism allows for a wide range of colors and shades to be achieved, depending on the specific polyphenolic compounds used. Wang et al. (2019) studied the adsorption behavior of quercetin, a flavonoid polyphenol, on silk fabric and found that the dyeing process was influenced by factors such as pH, temperature, and dye concentration. The researchers concluded that quercetin exhibited good dyeability and colorfastness on silk fabric, making it a promising natural dye candidate [95]. Covalent bonding is another dyeing mechanism used with polyphenols, leading to excellent colorfastness and durability. Wang et al. (2015) investigated the covalent bonding of tannic acid, a tannin polyphenol, with cotton fabric using citric acid as a cross-linking agent. The study demonstrated that the covalent bonding between tannic acid and cotton fabric enhanced the colorfastness properties, such as washing and rubbing fastness. Metal complexation is a less common but interesting dyeing mechanism where polyphenols form coordination complexes with metal ions, resulting in unique coloration and enhanced properties [96]. Anthocyanins, a class of polyphenols responsible for red, purple, and blue colors in fruits and flowers, have been studied extensively for their metal complexation properties. Li et al. (2017) investigated the dyeing of silk fabric with cyanidin, an anthocyanin compound, complexed with aluminum ions. The results showed that the metal complexation enhanced the color strength and lightfastness of the dyed silk fabric.

The extraction of polyphenols from plant sources is a critical step in fabric dyeing. Several extraction techniques have been explored to obtain high-quality polyphenolic extracts for dyeing. Solvent extraction is a commonly employed method where polyphenols are extracted using organic solvents such as ethanol or water. Ultrasound-assisted extraction has also been investigated as a promising technique to improve the extraction efficiency and yield of polyphenols. For instance, Jena et al. (2018) compared different extraction methods for obtaining polyphenols from *Terminalia chebula*, a medicinal plant, and found that ultrasound-assisted extraction exhibited higher extraction efficiency and higher total phenolic content. The dyeing process using polyphenols can be carried out through various methods, including exhaust dyeing, pad dyeing, and printing. Exhaust dyeing involves immersing the fabric in a dye bath containing the polyphenolic extract, allowing the dye molecules to adsorb onto the fabric surface. Pad dyeing, on the other hand, entails applying the dye directly onto the fabric using a padding machine, followed by fixation through heat or chemical treatment. Printing techniques, such as screen printing or block printing, can also be used to create intricate patterns and designs using polyphenolic dyes.

Shin et al. (2017) investigated the dyeing of cotton fabric using tannic acid through a pad dyeing method and observed that the dyeing process was influenced by factors such as the concentration of tannic acid, curing temperature, and curing time. Polyphenol dyes find applications in various textile sectors, including fashion textiles and technical textiles. In the fashion industry, there is a growing demand for sustainable and eco-friendly clothing, and the use of polyphenol dyes aligns with these requirements. The vibrant colors and unique dyeing mechanisms of polyphenols provide opportunities for creative and innovative designs in textiles and fashion. Several studies have explored the application of polyphenol dyes in fashion textiles, demonstrating their potential in creating sustainable and aesthetically pleasing garments.

In addition to fashion textiles, polyphenol dyes have been investigated for applications in technical textiles. For example, Zhang et al. (2016) studied the dyeing of silk fabric with natural dyes, including polyphenols extracted from tea leaves. The dyed silk fabric exhibited excellent UV protection properties, making it suitable for the production of sun-protective textiles. The antimicrobial properties of certain polyphenols have also been explored for medical textiles, where they can inhibit the growth of bacteria and promote wound healing. While polyphenols offer numerous advantages in fabric dyeing, there are also some limitations and considerations to be aware of. One limitation is the colorfastness of polyphenol dyes, particularly when using adsorption as the dyeing mechanism [97]. Polyphenols are generally less resistant to fading and washing compared to synthetic dyes. However, this limitation can be addressed through proper fixation techniques, such as post-dyeing treatments or the use of mordants, which improve the colorfastness and durability of polyphenol dyes. Zhang et al. (2019) investigated the colorfastness properties of silk fabric dyed with natural polyphenols and found that the colorfastness could be significantly enhanced by using a combination of alum and tannic acid as mordants. Another consideration in using polyphenols for fabric dyeing is the variability in polyphenol content and composition among different plant sources. This can lead to variations in color intensity and shade, making batch-to-batch consistency a challenge in commercial dyeing processes. Standardization of polyphenol extracts and optimization of dyeing parameters are essential to achieve consistent and reproducible results. Fuentes-Alventosa et al. (2018) studied the extraction and dyeing properties of polyphenols from olive leaves and found that the extraction conditions significantly influenced the color intensity and shade of the dyed cotton fabric.

7.6.2 Polyphenols in Leather Tanning

Polyphenols are a diverse group of compounds found abundantly in plants and have gained significant attention for their various applications in different industries, including leather tanning. Leather tanning is a process that converts raw hides or skins into durable, flexible, and usable leather materials. Traditionally, leather tanning involves the use of synthetic chemicals, which can have negative environmental impacts. However, the use of polyphenols as alternative tanning agents offers a more sustainable and environmentally friendly approach. In this discussion, we will explore the role of polyphenols in leather tanning, including their sources, tanning mechanisms, advantages, and potential challenges. Polyphenols can be derived from various plant sources, making them a renewable and sustainable alternative to synthetic tanning agents. Tannins are a common group of polyphenols used in leather tanning. They are widely distributed in different plant parts, including barks, fruits, leaves, and stems. Tannins can be categorized into two main groups: hydrolyzable tannins (e.g. gallotannins and ellagitannins) and condensed tannins (e.g. catechins and proanthocyanidins). Each type of tannin exhibits different tanning properties and characteristics. Gallotannins, found in plant sources such as oak barks and sumac, are widely used in leather tanning due to their strong tanning ability and affinity for collagen fibers [4, 6, 98]. Ellagitannins, which are present in plants such as quebracho and chestnut, also offer effective tanning properties. Condensed tannins, commonly found in plants like quebracho, mimosa, and quebracho, possess good tanning potential and are often used in combination with other tannins to achieve desired results. Polyphenols interact with collagen fibers in the hides or skins through a complex process known as tanning. The tanning mechanisms of polyphenols involve the formation of stable complexes between the tannins and collagen, leading to the conversion of raw hides or skins into leather. The most common mechanism of tanning with polyphenols is the formation of insoluble complexes between the tannins and collagen fibers. The hydroxyl groups of tannins interact with the amino acid residues of collagen, resulting in the cross-linking of collagen chains and the stabilization of the collagen structure. This process imparts the desired properties of strength, flexibility, and resistance to degradation to the leather. The use of polyphenols in leather tanning offers several advantages over traditional synthetic tanning agents. One of the key advantages is the environmental friendliness of polyphenols. Synthetic tanning agents, such as chromium salts, can have harmful effects on the environment and human health. Polyphenols, on the other hand, are derived from renewable plant sources and have lower toxicity, making them a sustainable and safer alternative.

Polyphenols also contribute to the production of high-quality leather with desirable properties. The tanning process with polyphenols results in leather materials that are soft, supple, and durable. The natural tanning agents interact with collagen fibers in a way that maintains the natural structure of the hide or skin, enhancing the aesthetic and functional characteristics of the leather. Furthermore, polyphenols have antimicrobial properties, which can contribute to the preservation of leather. Several studies have demonstrated the antimicrobial effects of tannins in leather tanning. Tannins inhibit the growth of bacteria and fungi, preventing the degradation of leather caused by microbial activity. This antimicrobial property is particularly beneficial for leather used in applications such as footwear, bags, and upholstery, where durability and hygiene are important factors. Despite the advantages of using polyphenols in leather tanning, there are also challenges and considerations that need to be addressed. One challenge is the variation in polyphenol content and composition among different plant sources. The tanning properties of polyphenols can vary depending on factors such as plant species, geographical origin, and extraction methods. This variation can affect the tanning efficiency and the quality of the final leather product. Therefore, it is essential to standardize the extraction processes and optimize the tanning parameters to achieve consistent results. Another consideration is the colorfastness of polyphenol-tanned leather. Polyphenols can impart different colors to leather depending on the type and concentration of tannins used. However, some polyphenols, especially those with low molecular weights, may exhibit poor colorfastness, leading to color fading or bleeding during use. The colorfastness of polyphenol-tanned leather can be improved through proper post-tanning treatments, such as the application of color-fixing agents or the use of additional dyeing techniques.

The cost of polyphenol-based tanning agents is another consideration. Compared to synthetic tanning agents, polyphenols can be more expensive, especially if high-quality extracts or purified tannins are used. However, advancements in extraction technologies, such as the development of efficient extraction methods and the utilization of by-products from other industries, can help reduce the production costs of polyphenol-based tanning agents.

7.6.3 Polyphenols in Hair Colorants

Polyphenols can be obtained from various plant sources, making them a renewable and eco-friendly option for hair colorants. Different plants contain varying types and concentrations of polyphenols, providing a wide range of color options for hair coloring. Some common sources of polyphenols for hair colorants include plant extracts such as henna, indigo, chamomile, black tea, and coffee. These plant extracts contain specific polyphenolic compounds that can impart color to the hair. Henna, derived from the plant *Lawsonia inermis*, is one of the most well-known natural hair colorants. The dye molecule in henna, called lawsone, has an affinity for hair proteins and can produce shades ranging from red to brown. Indigo, obtained from the plant *Indigofera tinctoria*, is another plant extract used for hair coloring. It contains the compound indigotin, which imparts blue to black shades on hair [8, 9]. Other plant extracts, such as chamomile, black tea, and coffee, contain polyphenols that can create shades of blonde, brown, or darker hues on the hair. Polyphenols interact with the hair shaft through various mechanisms to provide color. The specific mechanisms can differ depending on the type of polyphenols and the desired hair color. One of the main mechanisms is the adsorption of polyphenols onto the hair surface. Polyphenols can bind to the protein structures in the hair, forming temporary color complexes. These complexes can enhance the natural color of the hair or provide a completely different shade, depending on the specific polyphenols used. The adsorption mechanism is reversible and can be gradually washed out over time.

Another mechanism involves the oxidation of polyphenols on the hair shaft. In this process, the polyphenols undergo chemical reactions with atmospheric oxygen, resulting in the formation of colored compounds. This mechanism is commonly observed in the use of henna as a hair colorant. Lawsone, the primary dye molecule in henna, undergoes oxidation reactions in the hair, leading to the development of various shades of red, brown, or orange. The use of polyphenols in hair colorants offers several advantages over synthetic dyes. One of the key advantages is their natural and sustainable nature. Polyphenols are derived from plant sources, which are renewable and biodegradable. This makes them a safer and more environmentally friendly option compared to synthetic dyes, which often contain potentially harmful chemicals. Polyphenol-based hair colorants also provide a gentler alternative for coloring hair. Synthetic dyes can be harsh on the hair, causing damage, dryness, and brittleness. In contrast, polyphenols, particularly those with conditioning properties, can help nourish and protect the hair during the coloring process. Certain polyphenols, such as those found in plant extracts like chamomile or *aloe vera*, can even provide moisturizing and soothing effects on the scalp.

Furthermore, polyphenol-based hair colorants offer versatility in terms of color options. Different plant extracts contain specific polyphenols that produce a wide range of shades, allowing individuals to choose from natural-looking colors or more vibrant and unconventional tones. The color intensity can be customized by adjusting the concentration of the polyphenol extract or by combining multiple plant extracts. While polyphenols offer several advantages in hair colorants, there are also challenges and considerations that need to be addressed. One challenge is the limited color stability of polyphenol-based hair colorants compared to synthetic dyes. Polyphenols generally provide semi-permanent coloration, which gradually fades with each shampoo. This can be a disadvantage for individuals who prefer longer-lasting colors or desire more dramatic changes. However, some postcoloring treatments, such as the use of color-fixing agents or protective coatings, can help prolong color retention. Another consideration is the variability in color results due to individual hair characteristics. The natural color, porosity, and condition of the hair can affect the final color achieved with polyphenol-based colorants. For example, individuals with darker or coarser hair may experience more subtle color changes compared to those with lighter or finer hair. Additionally, the presence of previous color treatments or chemical processes on the hair may interact differently with polyphenols, leading to unpredictable color results. Conducting strand tests and considering the hair's characteristics can help manage expectations and achieve the desired color outcome.

7.6.4 Polyphenols in Natural Food Coloring

Natural food coloring has gained significant popularity in recent years due to increasing consumer demand for clean-label and healthier food products. Polyphenols, a diverse group of compounds found abundantly in plants, offer an excellent source of natural food coloring. These compounds not only provide vibrant and attractive colors to food but also offer potential health benefits, making them an appealing alternative to synthetic food dyes. In this discussion, we will explore the role of polyphenols in natural food coloring, including their sources, coloring mechanisms, advantages, and applications [11, 12, 15]. Polyphenols can be obtained from various plant-based sources, making them a renewable and sustainable option for natural food coloring. Different plants contain varying types and concentrations of polyphenols, providing a wide range of color options for food applications. Some common sources of polyphenols for natural food coloring include fruits, vegetables, herbs, spices, and botanical extracts. Fruits and vegetables are rich sources of polyphenols and offer a wide spectrum of colors. For example, anthocyanins, a class of polyphenols, are responsible for the red, purple, and blue colors in fruits such as berries, grapes, and cherries. Carotenoids, another group of polyphenols, contribute to the vibrant orange, red, and yellow hues in fruits and vegetables such as carrots, tomatoes, and bell peppers. Additionally, herbs and spices, such as turmeric, saffron, and paprika, contain polyphenols that provide distinct and intense colors to foods. Polyphenols interact with food matrices through various mechanisms to provide color. The specific mechanisms can differ depending on the type of polyphenols and the food system.

One of the primary mechanisms is the adsorption of polyphenols onto food surfaces. Polyphenols can bind to proteins, fibers, or other components in the food matrix, resulting in coloration. This mechanism is particularly observed in the use of plant extracts or concentrated polyphenol solutions as natural food colorants. The adsorption of polyphenols can be influenced by factors such as pH, temperature, and the presence of other food components. Another mechanism involves the enzymatic or chemical reactions of polyphenols with other food components. For example, polyphenols can undergo enzymatic oxidation in the presence of polyphenol oxidase enzymes, leading to the development of brown or dark colors. This mechanism is observed in fruits and vegetables when they are cut or bruised, resulting in the enzymatic browning process. Polyphenols offer several advantages as natural food colorants, making them an attractive alternative to synthetic dyes. One of the key advantages is their natural and sustainable nature. Polyphenols are derived from plant sources, which are renewable and biodegradable. This makes them a safer and more environmentally friendly option compared to synthetic food dyes, which often contain potentially harmful chemicals.

Additionally, polyphenols offer potential health benefits when used as food colorants. Many polyphenols possess antioxidant and anti-inflammatory properties, which are associated with various health benefits, including the prevention of chronic diseases. By incorporating polyphenol-rich natural food colorants into their diet, consumers can enhance their intake of beneficial compounds. Polyphenol-based natural food colorants also provide versatility in terms of color options. Different polyphenols exhibit a wide range of colors, allowing for the creation of vibrant and visually appealing food products. Moreover, the use of multiple polyphenols or the combination of polyphenols with other natural colorants can lead to a broader spectrum of colors and shades. Polyphenols have numerous applications in natural food coloring across a wide range of food products. They can be used in beverages, confectionery, bakery products, dairy products, and even savory foods. In beverages, polyphenols can be used to impart colors to fruit juices, soft drinks, teas, and functional beverages. The natural colors obtained from polyphenols can enhance the visual appeal of these beverages and provide additional health benefits.

Confectionery products, such as candies, gummies, and chocolates, can benefit from the vibrant colors offered by polyphenols. Polyphenol-based natural food colorants can replace synthetic dyes, ensuring that the confectionery products meet consumer preferences for clean-label and natural ingredients. Bakery products, including cakes, cookies, and pastries, can be enhanced with natural colors derived from polyphenols. These colors not only provide visual appeal but also contribute to the perception of naturalness and quality. Dairy products, such as yogurts and ice creams, can also be colored with polyphenols. The use of natural colorants aligns with consumer demands for clean-label products, particularly in the dairy industry, where artificial dyes have been traditionally used. Savory foods, such as sauces, dressings, and snacks, can also benefit from polyphenol-based natural food coloring. These colorants can enhance the visual appeal and differentiate products in the market while meeting the demand for natural and healthier food options.

7.7 Challenges and Future Prospects

7.7.1 Challenges in Polyphenol Dye Applications

Polyphenols face several challenges in their application as dyes. Firstly, the extraction process of polyphenols from plant sources can be complex and time-consuming. The choice of extraction solvents, extraction parameters, and plant material quality can greatly affect the yield and quality of the extracted polyphenols. Additionally, standardization of polyphenol extracts is crucial for achieving consistent dyeing results, as variations in polyphenol content can lead to inconsistencies in color intensity and shade. Color stability is another challenge in polyphenol dye applications. Polyphenol dyes, especially those obtained through adsorption mechanisms, can be susceptible to color fading over time. Factors such as exposure to light, heat, and washing can cause the color to degrade and lose vibrancy. This challenge is particularly significant in applications where long-lasting color retention is essential, such as textiles and leather. The limited color range of polyphenols is also a challenge. While polyphenols offer a wide range of colors, achieving certain shades, especially bright and vivid colors, can be challenging with polyphenols alone. Efforts are being made to overcome this limitation through the combination of polyphenols with other natural colorants or the use of innovative dyeing techniques.

Colorfastness and wash fastness are important properties in dye applications. Polyphenol dyes, particularly those relying on weak interactions such as adsorption, may exhibit lower colorfastness compared to synthetic dyes. This can result in color bleeding or fading during washing, exposure to sunlight, or contact with other chemicals. Strategies such as the use of mordants, pre- and posttreatments, and optimization of dyeing processes are employed to improve colorfastness. Reproducibility and batch-to-batch consistency are crucial in industrial-scale dye applications. However, the inherent variability in polyphenol content among different plant sources and batches can pose challenges in achieving consistent results. Standardizing the polyphenol extracts, optimizing dyeing parameters, and implementing quality control measures are necessary to ensure consistent color outcomes. Cost and availability are additional challenges. Polyphenol dyes may be more expensive compared to synthetic dyes due to extraction costs, limited availability of certain plant sources, and lower dyeing efficiency. The availability of specific polyphenols in sufficient quantities can also be a challenge, especially when there is increased demand for natural dyes. Finally, compatibility with different substrates is a consideration. Polyphenol dyes may not be universally compatible with all types of substrates. Different materials, such as fabrics, leathers, or food matrices, may exhibit variations in their chemical and physical properties, which can influence the dyeing process and the final color outcome.

Addressing these challenges requires ongoing research and collaboration between academia, industry, and regulatory bodies. Developing new extraction techniques, optimizing dyeing processes, improving color stability, enhancing colorfastness, and expanding the color range of polyphenol dyes are key areas of focus. By overcoming these challenges, the full potential of polyphenols as sustainable and environmentally friendly colorants can be realized.

7.7.2 Future Research Directions and Opportunities

The field of polyphenol research offers numerous exciting opportunities for future exploration and innovation. As the demand for sustainable and natural alternatives continues to grow, there are several key research directions that can further advance our understanding and utilization of polyphenols in various applications. In this discussion, we will explore some of the future research directions and opportunities in the field of polyphenol research. Extraction and purification techniques: Developing efficient, cost-effective, and environmentally friendly extraction and purification techniques for polyphenols is an ongoing research area. Improving the extraction yield, selectivity, and scalability of polyphenols from different plant sources can contribute to their widespread utilization in various industries. Exploring novel extraction methods, such as green extraction technologies and emerging techniques like microwave-assisted extraction and ultrasound-assisted extraction, can enhance the efficiency and sustainability of polyphenol extraction.

- *Structure–function relationship*: Understanding the structure–function relationship of polyphenols is crucial for optimizing their properties and applications. Investigating the impact of different structural features, such as the number and arrangement of phenolic rings, hydroxylation pattern, glycosylation, and esterification, on their biological activities, dyeing properties, antioxidant capacity, and stability can provide valuable insights for tailoring polyphenols for specific applications. Advanced analytical techniques, such as mass spectrometry, nuclear magnetic resonance spectroscopy, and computational modeling, can aid in unraveling the structure–function relationship of polyphenols.
- *Dyeing mechanisms and color stability*: Further research is needed to elucidate the dyeing mechanisms of polyphenols, particularly in different substrates and application conditions. Investigating the interactions between polyphenols and various materials, such as fabrics, leathers, and food matrices, can shed light on the factors influencing dye uptake, color development, and color stability. Understanding the degradation pathways of polyphenol dyes and developing strategies to improve their colorfastness and resistance to environmental factors, such as light, heat, and washing, are important for their practical applications.

- *Application-specific optimization*: Tailoring polyphenols for specific applications can unlock new possibilities. Optimizing the dyeing parameters, such as temperature, pH, dye concentration, and dyeing time, for different substrates can enhance the dyeing efficiency and color performance. Additionally, exploring the combination of polyphenols with other natural colorants, additives, or materials can lead to synergistic effects and expanded color options. Collaborations between researchers, textile scientists, chemists, and industry professionals are essential for translating polyphenol research into practical applications.
- *Valorization of by-products and waste streams*: The valorization of polyphenol-rich by-products and waste streams from various industries is an emerging research area. Finding innovative ways to extract and utilize polyphenols from agricultural residues, food processing by-products, and other waste streams not only contributes to waste reduction but also provides a sustainable source of polyphenols for various applications. Developing efficient and cost-effective techniques to recover and purify polyphenols from these waste streams can unlock their potential as valuable bioactive compounds and colorants.
- *Health and functional benefits*: Polyphenols are known for their health-promoting properties, including antioxidant, anti-inflammatory, and antimicrobial activities. Further research is needed to explore the potential health benefits of polyphenol-dyed textiles, such as their impact on skin health, UV protection, and wound healing. Additionally, investigating the functional properties of polyphenol-dyed materials, such as moisture management, thermal regulation, and odor control, can open up new avenues for their applications in functional textiles.
- *Sustainability and life cycle assessment*: Assessing the environmental impact and sustainability of polyphenol-based products is crucial for their wider adoption. Conducting life cycle assessments (LCAs) to evaluate the environmental footprint of polyphenol extraction, dyeing processes, and end products can provide valuable insights into their environmental performance compared to synthetic dyes. Integration of sustainability principles, such as circular economy concepts, into the design and development of polyphenol-based products can further enhance their sustainability credentials.
- *Regulatory aspects and consumer acceptance*: Addressing regulatory aspects and ensuring consumer acceptance are vital for the successful commercialization of polyphenol-based products. Understanding the regulatory requirements for polyphenol use in different industries, such as textiles, food, and cosmetics, is essential. Assessing the safety, allergenicity, and potential interactions of polyphenols with other chemicals or materials can provide the necessary data for regulatory compliance. Additionally, understanding consumer perceptions, preferences, and acceptance of polyphenol-based products through consumer studies and market research can guide product development and marketing strategies.

7.7.3 Nanotechnology for Enhancing Dyeing Efficiency and Stability

Nanotechnology has emerged as a promising field for enhancing dyeing efficiency and stability in various industries. By utilizing nanomaterials and nanoscale techniques, significant advancements have been made to overcome the limitations of traditional dyeing processes. In this discussion, we will explore the applications of nanotechnology in dyeing, focusing on the enhancement of dyeing efficiency and stability. We will discuss various nanomaterials and their roles in improving dye adsorption, color fastness, and durability. Additionally, we will delve into the challenges and prospects of nanotechnology in the field of dyeing. Nanotechnology offers unique opportunities to enhance dyeing efficiency by improving the adsorption of dyes onto substrates. One approach is the use of nanoparticles as dye carriers or nanocarriers. These nanoparticles, such as metal oxide nanoparticles (e.g. titanium dioxide and zinc oxide), carbon nanotubes, and clay nanoparticles, can adsorb dyes through physical interactions or chemical bonding. The large surface area and high adsorption capacity of nanoparticles enable efficient dye loading onto the substrate, resulting in enhanced dye uptake and color intensity. Moreover, the controlled release of dyes from nanoparticles can provide sustained dye release, prolonging the color vibrancy and reducing dye consumption.

In addition to nanoparticles, nanofibers and nanocomposite materials have been explored to improve dyeing efficiency [99–107]. Electrospun nanofibers, with their high surface area and interconnected porous structure, offer improved dye diffusion and adsorption properties. Dyes can easily penetrate the nanofiber matrix, leading to rapid dye uptake and enhanced color development. Furthermore, nanocomposite materials, consisting of nanofillers dispersed within a polymer matrix, exhibit enhanced dye adsorption due to the synergistic effects between the nanofillers and the polymer. For example, incorporating nano clay or carbon nanotubes into polymer matrices can enhance the adsorption capacity and color fastness properties of the resulting nanocomposite materials. Here are some global examples showcasing the application of nanotechnology in enhancing dyeing efficiency and stability:

- *Self-cleaning textiles*: Researchers at the University of California, Davis, developed a self-cleaning cotton fabric using a nanocoating of titanium dioxide (TiO_2) nanoparticles. The nanocoating enables the fabric to break down organic stains and pollutants upon exposure to light, resulting in self-cleaning properties and improved color retention.
- *Nanofiber-based dyeing*: Nanofiber-based dyeing techniques have been explored in several countries. In Taiwan, researchers from the National Taiwan University developed a nanofiber membrane with enhanced dye adsorption properties. The electrospun nanofibers provided a large surface area for dye diffusion, resulting in rapid and efficient dye uptake.

- *Nanoencapsulation of dyes*: In Brazil, researchers from the Federal University of Santa Catarina developed a nanocapsule-based dyeing system for textiles. They encapsulated dyes within nanocapsules, which were then incorporated into a textile matrix. The nanocapsules protected the dyes from external factors, resulting in improved colorfastness and durability.
- *Nanocomposite dyeing materials*: Researchers in South Korea developed nanocomposite materials for improved dye adsorption and stability. They incorporated clay nanoparticles into a polymer matrix and used the resulting nanocomposite for dyeing textiles. The nanocomposite materials exhibited enhanced dye adsorption capacity and colorfastness properties.
- *Nanomaterial-based antimicrobial dyeing*: In India, researchers from the Indian Institute of Technology (IIT) developed antimicrobial textiles using silver nanoparticles. The silver nanoparticles were incorporated into the dye formulation, imparting antimicrobial properties to the dyed materials. This approach is particularly valuable in healthcare settings where antimicrobial properties are crucial.

Nanotechnology also plays a crucial role in enhancing the stability and durability of dyed materials. One challenge in dyeing is poor colorfastness, which leads to color fading or bleeding during washing, exposure to light, or environmental conditions. Nanotechnology offers strategies to address these challenges. One approach is the use of nanocoatings or surface modification techniques to protect the dye molecules from external factors. By coating the dyed materials with thin layers of nanomaterials, such as silica, titania, or polymers, the dyes are encapsulated, providing a protective barrier against environmental stressors. This enhances colorfastness and prevents color fading or leaching. Another strategy to improve dye stability is through the development of self-cleaning or self-repairing materials using nanotechnology. Self-cleaning fabrics utilize nanomaterials, such as photocatalytic nanoparticles, that can decompose organic stains or pollutants upon exposure to light. This not only helps in maintaining the color and cleanliness of the fabric but also reduces the need for frequent washing, thereby increasing the longevity of the dyed material. Self-repairing materials incorporate nanocapsules or microcapsules containing dyes and repairing agents. When the dyed material is damaged, these capsules rupture, releasing the repairing agents and dyes to restore the color and structural integrity of the material.

Moreover, nanotechnology enables the development of functionalized nanomaterials with tailored properties for specific dyeing applications. For instance, the incorporation of antimicrobial nanoparticles, such as silver nanoparticles or metal oxide nanoparticles, into dye formulations can provide antimicrobial properties to the dyed materials. This is particularly relevant in industries such as healthcare, textiles, and cosmetics, where microbial growth and odor control are important considerations. Furthermore, the integration of nanoparticles with other functional additives, such as UV absorbers, flame retardants, or water repellents, can enhance the multifunctionality of dyed materials, expanding their applications and value. Despite the significant progress in nanotechnology for enhancing dyeing efficiency and stability, several challenges and considerations need to be addressed. One of the challenges is the potential toxicity and environmental impact of nanomaterials. The release of nanoparticles during the dyeing process or the disposal of nanomaterial-treated dyed materials may raise concerns regarding their impact on human health and the environment. Therefore, it is crucial to conduct comprehensive toxicity assessments and lifecycle analyses to ensure the safe and sustainable use of nanotechnology in dyeing applications.

Another challenge is the scalability and cost-effectiveness of nanotechnology-based dyeing processes. While laboratory-scale studies have demonstrated the potential of nanomaterials for improved dyeing, the upscaling of these processes to industrial levels requires careful optimization and consideration of cost and feasibility. The availability and cost of nanomaterials, as well as the compatibility with existing dyeing infrastructure, are important factors to be addressed for successful implementation. Furthermore, the standardization of nanotechnology-based dyeing processes and the development of industry-wide guidelines are essential. Establishing standardized protocols for the synthesis and characterization of nanomaterials, as well as the application of nanomaterials in dyeing, can ensure consistency, reproducibility, and comparability of results across different laboratories and industries. Collaboration between academia, industry, and regulatory bodies is necessary to establish guidelines and regulations for the safe and responsible use of nanotechnology in dyeing applications.

7.8 Conclusions, Challenges, and Future Perspectives

In conclusion, polyphenols offer significant potential as natural dyes for various applications. Their abundance in plant sources, diverse chemical structures, and unique dyeing mechanisms make them attractive alternatives to synthetic dyes. The utilization of polyphenols in dyeing processes not only provides vibrant and sustainable color options but also offers additional benefits such as antioxidant and antimicrobial properties. Throughout this book chapter, we have explored the sources, extraction techniques, dyeing mechanisms, advantages, and applications of polyphenols in the field of dyeing. Polyphenols can be derived from a wide range of plant sources, including fruits, vegetables, barks, and flowers, allowing for a diverse palette of colors for dyeing applications. Extraction techniques such as solvent extraction, ultrasound-assisted extraction, and SPE have been developed to obtain high-quality polyphenolic extracts. The dyeing mechanisms of polyphenols involve adsorption, covalent bonding, and metal complexation, depending on the specific polyphenolic compounds and dyeing conditions.

The adsorption mechanism allows for a broad spectrum of colors and shades, while covalent bonding and metal complexation offer enhanced colorfastness and durability. The versatility of polyphenols in dyeing is exemplified by their applications in various industries, including textiles, leather tanning, hair colorants, food coloring, and more.

One of the key advantages of polyphenols in dyeing is their sustainability and environmental friendliness. Compared to synthetic dyes, polyphenols are derived from renewable plant sources, and their extraction processes often involve fewer harmful chemicals. This reduces the ecological footprint and potential negative impacts on human health associated with synthetic dyes. Furthermore, polyphenols offer unique properties such as antioxidant and antimicrobial activities, which can provide additional benefits to dyed materials. Polyphenols find applications in the textile industry, where the demand for natural and eco-friendly products is growing. They offer an opportunity to produce sustainable and socially responsible textiles that cater to consumers' preferences for environmentally friendly options. The vibrant colors and natural dyeing mechanisms of polyphenols allow for creative and innovative designs in fashion textiles, while their functional properties make them suitable for technical textiles such as medical textiles, sports textiles, and protective textiles.

However, there are also challenges and considerations to be addressed in using polyphenols for dyeing applications. One of the challenges is the colorfastness and durability of polyphenolic dyes, particularly when using adsorption as the dyeing mechanism. Polyphenols are generally less resistant to fading and washing compared to synthetic dyes. Therefore, proper fixation techniques, postdyeing treatments, or the use of mordants are necessary to improve the colorfastness and longevity of polyphenol dyes. Standardization of polyphenol extracts and optimization of dyeing parameters are also essential to achieve consistent and reproducible results. The variability in polyphenol content and composition among different plant sources can lead to variations in color intensity and shade, making batch-to-batch consistency a challenge in commercial dyeing processes. Therefore, efforts should be made to establish standardized procedures and quality control measures to ensure reliable and predictable dyeing outcomes.

Future research directions in polyphenol dyeing offer exciting opportunities for advancements. Exploring new plant sources and identifying novel polyphenolic compounds with desirable dyeing properties can expand the color palette and provide more options for dyeing applications. Furthermore, integrating nanotechnology and advanced materials can enhance polyphenol dyes' efficiency, stability, and functionality. Developing nanocoatings, nanofibers, and nanocomposites can improve dye adsorption, color retention, and durability of polyphenolic dyes, opening up new possibilities for their application in various industries.

References

1 Chen, X., He, Y., Fan, Y. et al. (2019). Nature-inspired polyphenol chemistry to fabricate halloysite nanotubes decorated PVDF membrane for the removal of wastewater. *Separation and Purification Technology* 212: 326–336.

2 De Matteis, V., Cascione, M., Rizzello, L. et al. (2021). Synergistic effect induced by gold nanoparticles with polyphenols shell during thermal therapy: macrophage inflammatory response and cancer cell death assessment. *Cancers (Basel)* 13 (14).

3 You, F., Xu, Y., Yang, X. et al. (2017). Bio-inspired Ni_2+−polyphenol hydrophilic network to achieve unconventional high-flux nanofiltration membranes for environmental remediation. *Chemical Communications* 53 (45): 6128–6131.

4 Punzo, A., Porru, E., Silla, A. et al. (2021). Grape pomace for topical application: green nades sustainable extraction, skin permeation studies, antioxidant and anti-inflammatory activities characterization in 3d human keratinocytes. *Biomolecules* 11 (8): 1–14.

5 Cheng, T.H., Liu, Z.J., Yang, J.Y. et al. (2019). Extraction of functional dyes from tea stem waste in alkaline medium and their application for simultaneous coloration and flame retardant and bioactive functionalization of silk. *ACS Sustainable Chemistry & Engineering* 7 (22): 18405–18413.

6 Husain, Q. (2006). Potential applications of the oxidoreductive enzymes in the decolorization and detoxification of textile and other synthetic dyes from polluted water: a review. *Critical Reviews in Biotechnology* 26 (4): 201–221.

7 Carvalho, S.S.F. and Carvalho, N.M.F. (2017). Dye degradation by green heterogeneous Fenton catalysts prepared in presence of *Camellia sinensis*. *Journal of Environmental Management* 187: 82–88.

8 Dritsa, V., Rigas, F., Natsis, K., and Marchant, R. (2007). Characterization of a fungal strain isolated from a polyphenol polluted site. *Bioresource Technology* 98 (9): 1741–1747.

9 SMYTH JD. (1954). A technique for the histochemical demonstration of polyphenol oxidase and its application to egg-shell formation in helminths and byssus formation in mytilus. *Journal of Cell Science* s3-95 (30): 139–152.

10 Enaru, B., Socaci, S., Farcas, A. et al. (2021). Novel delivery systems of polyphenols and their potential health benefits. *Pharmaceuticals* 14 (10): 1–28.

11 Spanidi, E., Karapetsas, A., Voulgaridou, G.P. et al. (2021). A new controlled release system for propolis polyphenols and its biochemical activity for skin applications. *Plants* 10 (2): 1–27.

12 Du, H., Yue, M., Huang, X. et al. (2022). Preparation, application and enhancement dyeing properties of ZnO nanoparticles in silk fabrics dyed with natural dyes. *Nanomaterials* 12 (22): 1–11.

13 Alharthi, F.A., Alghamdi, A.A., Alothman, A.A. et al. (2020). Green synthesis of ZnO nanostructures using salvadora persica leaf extract: applications for photocatalytic degradation of methylene blue dye. *Crystals* 10 (6).

14 Montané, X., Kowalczyk, O., Reig-Vano, B. et al. (2020). Current perspectives of the applications of polyphenols and flavonoids in cancer therapy. *Molecules* 25 (15).

15 Ouyang, Q., Kou, F., Tsang, P.E. et al. (2019). Green synthesis of Fe-based material using tea polyphenols and its application as a heterogeneous Fenton-like catalyst for the degradation of lincomycin. *Journal of Cleaner Production* 232: 1492–1498. https://doi.org/10.1016/j.jclepro.2019.06.043.

16 Mia, M.S., Yao, P., Zhu, X. et al. (2021). Degradation of textile dyes from aqueous solution using tea-polyphenol/Fe loaded waste silk fabrics as Fenton-like catalysts. *RSC Advances* 11 (14): 8290–8305.

17 Ding, Y., Morozova, K., Scampicchio, M., and Ferrentino, G. (2020). Non-Extractable polyphenols from food by-products: Current knowledge on recovery, characterisation, and potential applications. *Processes* 8 (8).

18 Sośnicka, M., Nowak, A., Czyżowska, A. et al. (2021). Polyphenolic herbal extract of cistus incanus as natural preservatives for sausages enriched with natural colors. *Processes* 9 (9).

19 Li, X.Q. and Tang, R.C. (2016). Crosslinked and dyed chitosan fiber presenting enhanced acid resistance and bioactivities. *Polymers (Basel)* 8 (4): 6–8.

20 Chhikara, N., Devi, H.R., Jaglan, S. et al. (2018). Bioactive compounds, food applications and health benefits of Parkia speciosa (stinky beans): a review. *Agriculture and Food Security* 7: 1–9. https://doi.org/10.1186/s40066-018-0197-x.

21 Wang, F., Gong, J., Zhang, X. et al. (2018). Preparation of biocolorant and eco-dyeing derived from polyphenols based on laccase-catalyzed oxidative polymerization. *Polymers (Basel)* 10 (2).

22 Paranhos de Araújo, V.A., da Luz, J.R.D., da Silva e Silva, N.S. et al. (2023). Optimization of polyphenol extraction with potential application as natural food preservatives from brazilian amazonian species *Dalbergia monetaria* and *Croton cajucara*. *Processes* 11 (3): 669.

23 Singh, K.K., Senapati, K.K., and Sarma, K.C. (2017). Synthesis of superparamagnetic Fe_3O_4 nanoparticles coated with green tea polyphenols and their use for removal of dye pollutant from aqueous solution. *Journal of Environmental Chemical Engineering* 5 (3): 2214–2221. https://doi.org/10.1016/j.jece.2017.04.022.

24 Lambrecht, L., Capablanca, L., Bou-Belda, E. et al. (2023). Enhancing polyphenols and tannins concentration on cotton dyed with red tea. *Sustainability* 15 (4): 1–13.

25 Chen, W., Zhang, Z., Shen, Y. et al. (2014). Effect of tea polyphenols on lipid peroxidation and antioxidant activity of litchi (*Litchi chinensis Sonn.*) fruit during cold storage. *Molecules* 19 (10): 16837–16850.

26 Li, N., Wang, Q., Zhou, J. et al. (2022). Insight into the progress on natural dyes: sources, structural features, health effects, challenges, and potential. *Molecules* 27 (10).

27 Gao, M., Xu, D., Gao, Y. et al. (2021). Mussel-inspired triple bionic adsorbent: facile preparation of layered double hydroxide@polydopamine@metal-polyphenol networks and their selective adsorption of dyes in single and binary systems. *Journal of Hazardous Materials* 420 (July): 126609. https://doi.org/10.1016/j.jhazmat.2021.126609.

28 Panak Balentić, J., Ačkar, Đ., Jokić, S. et al. (2018). Cocoa shell: a by-product with great potential for wide application. *Molecules* 23 (6): 1–14.

29 Kaya, S., Thakur, A., and Kumar, A. (2023). The role of in Silico/DFT investigations in analyzing dye molecules for enhanced solar cell efficiency and reduced toxicity. *Journal of Molecular Graphics & Modelling* 124 (June): 108536.

30 Thakur, A. and Kumar, A. (2023). Recent trends in nanostructured carbon-based electrochemical sensors for the detection and remediation of persistent toxic substances in real- time analysis. *Materials Research Express* 10: 034001.

31 Verma, C., Thakur, A., Ganjoo, R. et al. (2023, 2022). Coordination bonding and corrosion inhibition potential of nitrogen-rich heterocycles: azoles and triazines as specific examples. *Coordination Chemistry Reviews* 488: 215177. https://doi.org/10.1016/j.ccr.2023.215177.

32 Bashir, S., Thakur, A., Lgaz, H. et al. (2020). Corrosion inhibition efficiency of bronopol on aluminium in 0.5 M HCl solution: Insights from experimental and quantum chemical studies. *Surfaces and Interfaces* 20 (April): 100542. https://doi.org/10.1016/j.surfin.2020.100542.

33 Dhonchak, C. and Agnihotri, N. (2023). Computational Insights in the spectrophotometrically 4H-chromen-4-one complex using DFT method. *Biointerface Research in Applied Chemistry* 13 (4): 357.

34 Sharma, D. and Thakur, A. (2014). Synthesis, electrochemical, morphological, computational and corrosion inhibition studies of 3-(5-naphthalen-2-yl-[1,3,4]oxadiazol-2-yl)-pyridine against mild steel in 1 M HCl. *Asian Journal of Chemistry* 35 (5): 1079–1088.

35 Thakur, A., Kumar, A., Kaya, S. et al. (2022). Recent advancements in surface modification, characterization and functionalization for enhancing the biocompatibility and corrosion resistance of biomedical implants. *Coatings* 12: 1459.

36 Thakur, A. and Kumar, A. (2021). Sustainable inhibitors for corrosion mitigation in aggressive corrosive media: a comprehensive study. *Journal of Bio- and Tribo-Corrosion* 7 (2): 1–48. https://doi.org/10.1007/s40735-021-00501-y.

37 Bashir, S., Thakur, A., Lgaz, H. et al. (2019). Computational and experimental studies on phenylephrine as anti-corrosion substance of mild steel in acidic medium. *Journal of Molecular Liquids* 293: 111539.

38 Thakur, A. and Kumar, A. (2022). Recent advances on rapid detection and remediation of environmental pollutants utilizing nanomaterials-based (bio) sensors. *Science of the Total Environment* 834 (January): 155219. https://doi.org/10.1016/j.scitotenv.2022.155219.

39 Thakur, A., Kaya, S., Abousalem, A.S., and Kumar, A. (2022). Experimental DFT and MC simulation analysis of Vicia Sativa weed aerial extract as sustainable and eco-benign corrosion inhibitor for mild steel in acidic environment. *Sustainable Chemistry and Pharmacy* 29 (July): 100785. https://doi.org/10.1016/j.scp.2022.100785.

40 Thakur, A., Kumar, A., Sharma, S. et al. (2022). Materials today: proceedings computational and experimental studies on the efficiency of Sonchus arvensis as green corrosion inhibitor for mild steel in 0 . 5 M HCl solution. *Materials Today: Proceedings* 66: 609–621. https://doi.org/10.1016/j.matpr.2022.06.479.

41 Thakur, A., Sharma, S., Ganjoo, R. et al. (2022). Anti-corrosive potential of the sustainable corrosion inhibitors based on biomass waste: a review on preceding and perspective research. *Journal of Physics Conference Series* 2267 (1): 012079.

42 Thakur, A. and SAVAŞ K, Kumar A. (2023). Recent trends in the characterization and application progress of nano-modified coatings in corrosion mitigation of metals and alloys. *Applied Sciences* 13: 730.

43 Kinfu, H.H. and Rahman, M.M. (2023). Separation performance of membranes containing ultrathin surface coating of metal-polyphenol network. *Membranes (Basel)* 13 (5).

44 Mohamed Khalith, S.B., Ramalingam, R., Karuppannan, S.K. et al. (2022). Synthesis and characterization of polyphenols functionalized graphitic hematite nanocomposite adsorbent from an agro waste and its application for removal of Cs from aqueous solution. *Chemosphere* 286 (P1): 131493. https://doi.org/10.1016/j.chemosphere.2021.131493.

45 Fraga-Corral, M., García-Oliveira, P., Pereira, A.G. et al. (2020). Technological application of tannin-based extracts. *Molecules* 25 (3): 1–27.

46 D'Antuono, I., Carola, A., Sena, L.M. et al. (2018). Artichoke polyphenols produce skin anti-age effects by improving endothelial cell integrity and functionality. *Molecules* 23 (11).

47 Giordano, A., Morales-Tapia, P., Moncada-Basualto, M. et al. (2022). Polyphenolic composition and antioxidant activity (ORAC, EPR and cellular) of different extracts of argylia radiata vitroplants and natural roots. *Molecules* 27 (3).

48 Bujak, T., Zagórska-dziok, M., Ziemlewska, A. et al. (2021). Antioxidant and cytoprotective properties of plant extract from dry flowers as functional dyes for cosmetic products. *Molecules* 26 (9): 1–25.

49 Kumar, H., Bhardwaj, K., Cruz-Martins, N. et al. (2021). Applications of fruit polyphenols and their functionalized nanoparticles against foodborne bacteria: A mini review. *Molecules* 26 (11): 1–19.

50 Xiao, C., Li, H., Zhao, Y. et al. (2020). Green synthesis of iron nanoparticle by tea extract (polyphenols) and its selective removal of cationic dyes. *Journal of Environmental Management* 275 (April): 111262. https://doi.org/10.1016/j.jenvman.2020.111262.

51 Moutinho, J., Gouvinhas, I., Domínguez-Perles, R., and Barros, A. (2023). Optimization of the extraction methodology of grape pomace polyphenols for food applications. *Molecules* 28 (9).

52 Battistin, M., Dissette, V., Bonetto, A. et al. (2020). A new approach to UV protection by direct surface functionalization of TiO_2 with the antioxidant polyphenol dihydroxyphenyl benzimidazole carboxylic acid. *Nanomaterials* 10 (2): 1–19.

53 Zaini, N.S., Karim, R., Abdull Razis, A.F., and Zawawi, N. (2022). Utilizing nutritional and polyphenolic compounds in underutilized plant seeds for health application. *Molecules* 27 (20).

54 de Oliveira Xavier Machado, T., IBM, P., da Silva Padilha, C.V. et al. (2021). New trends in the use of enzymes for the recovery of polyphenols in grape byproducts. *Journal of Food Biochemistry* 45 (5): 1–13.

55 Barba, F.J., Rajha, H.N., Debs, E. et al. (2022). Optimization of polyphenols' recovery from purple corn cobs assisted by infrared technology and use of extracted anthocyanins as a natural colorant in pickled turnip. *Molecules* 27 (16).

56 Feng, C., Wei, T., Qing, S. et al. (2021). Application of tea polyphenols and their effects on ultrafiltration effluent disinfection and microbial risk. *Water (Switzerland)* 13 (18): 1–13.

57 Zhao, S., Song, P., Wang, Z., and Zhu, H. (2018). The PEGylation of plant polyphenols/polypeptide-mediated loose nanofiltration membrane for textile wastewater treatment and antibacterial application. *Journal of the Taiwan Institute of Chemical Engineers* 82: 42–55. https://doi.org/10.1016/j.jtice.2017.11.005.

58 Xu, D., Liang, H., Zhu, X. et al. (2020). Metal-polyphenol dual crosslinked graphene oxide membrane for desalination of textile wastewater. *Desalination* 487 (April): 114503. https://doi.org/10.1016/j.desal.2020.114503.

59 Xie, Y., Huang, J., Dong, H. et al. (2020, 2019). Insight into performance and mechanism of tea polyphenols and ferric ions on reductive decolorization of malachite green cationic dye under moderate conditions. *Journal of Environmental Management* 261: 110226. https://doi.org/10.1016/j.jenvman.2020.110226.

60 Rather, L.J., Shabbir, M., Li, Q., and Mohammad, F. (2019). Coloration, UV protective, and antioxidant finishing of wool fabric via natural dye extracts: cleaner production of bioactive textiles. *Environmental Progress & Sustainable Energy* 38 (5): 1–9.

61 Sharma, V., Ayothiraman, S., and Dhakshinamoorthy, V. (2019). Production of highly thermo-tolerant laccase from novel thermophilic bacterium *Bacillus* sp.

PC-3 and its application in functionalization of chitosan film. *Journal of Bioscience and Bioengineering* 127 (6): 672–678. https://doi.org/10.1016/j.jbiosc.2018.11.008.

62 Latif, M.S., Abbas, S., Kormin, F., and Mustafa, M.K. (2019). Green synthesis of plant-mediated metal nanoparticles: the role of polyphenols. *Asian Journal of Pharmaceutical and Clinical Research* 75–84.

63 Arabaci, G. and Usluoglu, A. (2014). The enzymatic decolorization of textile dyes by the immobilized polyphenol oxidase from quince leaves. *Scientific World Journal* 2014.

64 Pavlović, M.D., Buntić, A.V., Mihajlovski, K.R. et al. (2014). Rapid cationic dye adsorption on polyphenol-extracted coffee grounds-A response surface methodology approach. *Journal of the Taiwan Institute of Chemical Engineers* 45 (4): 1691–1699.

65 Zhang, N., Jiang, B., Zhang, L. et al. (2019). Low-pressure electroneutral loose nanofiltration membranes with polyphenol-inspired coatings for effective dye/divalent salt separation. *Chemical Engineering Journal* 359: 1442–1452. https://doi.org/10.1016/j.cej.2018.11.033.

66 Parsaeian, M.R., Dadfarnia, S., Shabani, A.M.H., and Moghaddam, R.H. (2022). Synthesize of iron-polyphenol nanohydrogel from melia azedarach fruit extract and its application as a flocculant for coagulation and removal of disperse yellow 211. *Analytical and Bioanalytical Chemistry Research* 9 (1): 101–111.

67 Findura, P., Kocira, S., Hara, P. et al. (2020). Extracts from artemisia vulgaris l. In potato cultivation – preliminary research on biostimulating effect. *Agriculture* 10 (8): 1–14.

68 Yang, F., Sadam, H., Zhang, Y. et al. (2020). A de novo sacrificial-MOF strategy to construct enhanced-flux nanofiltration membranes for efficient dye removal. *Chemical Engineering Science* 225.

69 Xu, L.Q., Neoh, K.G., and Kang, E.T. (2018). Natural polyphenols as versatile platforms for material engineering and surface functionalization. *Progress in Polymer Science* 87: 165–196.

70 Dean, L.L. (2020). Extracts of peanut skins as a source of bioactive compounds: methodology and applications. *Applied Sciences* 10 (23): 1–26.

71 Rombaut, N., Chave, T., Nikitenko, S.I. et al. (2021). Modification of olive leaves' surface by ultrasound cavitation. Correlation with polyphenol extraction enhancement. *Applied Sciences* 11 (1): 1–12.

72 Liu, S., Fang, X., Lou, M. et al. (2021). Construction of loose positively charged nf membrane by layer-by-layer grafting of polyphenol and polyethyleneimine on the pes/fe substrate for dye/salt separation. *Membranes (Basel)* 11 (9).

73 Małgorzata, D., Aleksandra, P., Monika, T., and Ireneusz, K. (2019). A new black elderberry dye enriched in antioxidants designed for healthy sweets production. *Antioxidants* 8 (8): 2–11.

74 Di Mauro, M.D., Fava, G., Spampinato, M. et al. (2019). Polyphenolic fraction from olive mill wastewater: scale-up and in vitro studies for ophthalmic nutraceutical applications. *Antioxidants* 8 (10).

75 Kuppusamy, S., Thavamani, P., Megharaj, M., and Naidu, R. (2015). Bioremediation potential of natural polyphenol rich green wastes: a review of current research and recommendations for future directions. *Environmental Technology & Innovation* 4: 17–28. https://doi.org/10.1016/j.eti.2015.04.001.

76 de Araújo, F.F., de Paulo, F.D., Neri-Numa, I.A., and Pastore, G.M. (2021, 2020). Polyphenols and their applications: an approach in food chemistry and innovation potential. *Food Chemistry* 338: 127535. https://doi.org/10.1016/j.foodchem.2020.127535.

77 Wang, Z., Fang, C., and Mallavarapu, M. (2015). Characterization of iron-polyphenol complex nanoparticles synthesized by Sage (Salvia officinalis) leaves. *Environmental Technology & Innovation* 4: 92–97. https://doi.org/10.1016/j.eti.2015.05.004.

78 Singh, V. and Mishra, A.K. (2016). White light emission from an aqueous vegetable cocktail: application towards pH sensing. *Dye Pigment* 125 (April 2018): 362–366. https://doi.org/10.1016/j.dyepig.2015.10.017.

79 Wang, Q., Liu, W., Tian, B. et al. (2020). Preparation and characterization of coating based on protein nanofibers and polyphenol and application for salted duck egg yolks. *Food* 9 (4): 1–16.

80 Shahid-ul-Islam, R.L.J., Shabbir, M., Sheikh, J. et al. (2019). Exploiting the potential of polyphenolic biomordants in environmentally friendly coloration of wool with natural dye from Butea monosperma flower extract. *Journal of Natural Fibers* 16 (4): 512–523.

81 Ebrahimi, I. and Parvinzadeh, G.M. (2016). Extraction of polyphenolic dyes from henna, pomegranate rind, and Pterocarya fraxinifolia for nylon 6 dyeing. *Coloration Technology* 132 (2): 162–176.

82 Maiti, P. and Dunbar, G.L. (2018). Use of curcumin, a natural polyphenol for targeting molecular pathways in treating age-related neurodegenerative diseases. *International Journal of Molecular Sciences* 19.

83 Franco, R.T., Silva, A.L., Licea, Y.E. et al. (2021). Green synthesis of iron oxides and phosphates via thermal treatment of iron polyphenols synthesized by a *Camellia sinensis* extract. *Inorganic Chemistry* 60 (8): 5734–5746.

84 Latos-Brozio, M. and Masek, A. (2020). The application of (+)-catechin and polydatin as functional additives for biodegradable polyesters. *International Journal of Molecular Sciences* 21 (2).

85 Uzan, E., Nousiainen, P., Balland, V. et al. (2010). High redox potential laccases from the ligninolytic fungi *Pycnoporus coccineus* and *Pycnoporus sanguineus* suitable for white biotechnology: from gene cloning to enzyme characterization and applications. *Journal of Applied Microbiology* 108 (6): 2199–2213.

86 Kammerer, D.R., Kammerer, J., Valet, R., and Carle, R. (2014). Recovery of polyphenols from the by-products of plant food processing and application as valuable food ingredients. *Food Research International* 65 (PA): 2–12. https://doi.org/10.1016/j.foodres.2014.06.012.

87 Brudzyńska, P., Sionkowska, A., and Grisel, M. (2021). Plant-derived colorants for food, cosmetic and textile industries: a review. *Materials (Basel)* 14 (13): 1–18.

88 Śliwa, P. and Śliwa, K. (2021). Nanomicellar extraction of polyphenols – methodology and applications review. *International Journal of Molecular Sciences* 22 (21).

89 Jeon, J.R., Kim, E.J., Kim, Y.M. et al. (2009). Use of grape seed and its natural polyphenol extracts as a natural organic coagulant for removal of cationic dyes. *Chemosphere* 77 (8): 1090–1098. https://doi.org/10.1016/j.chemosphere.2009.08.036.

90 Fang, X., Wei, S., Liu, S. et al. (2022). Metal-coordinated nanofiltration membranes constructed on metal ions blended support toward enhanced dye/salt separation and antifouling performances. *Membranes (Basel)* 12 (3).

91 Shafiq, F., Siddique, A., Pervez, M.N. et al. (2021). Extraction of natural dye from aerial parts of argy wormwood based on optimized taguchi approach and functional finishing of cotton fabric. *Materials (Basel)* 14 (19).

92 Jin, L., Yun, D., Zhang, W. et al. (2023). Polyphenols coordinated with Cu (II) in an aqueous system build ion-channel coatings on hair surfaces. *Materials (Basel)* 16 (4).

93 Kharouf, N., Haikel, Y., and Ball, V. (2020). Polyphenols in dental applications. *Bioengineering* 7 (3): 1–27.

94 Andersen, A., Chen, Y., and Birkedal H. (2019). Bioinspired Metal – Polyphenol Materials: Self-Healing and Beyond.

95 Ebrahimi, P. and Lante, A. (2022). Environmentally friendly techniques for the recovery of polyphenols from food by-products and their impact on polyphenol oxidase: a critical review. *Applied Sciences* 12 (4).

96 Lee, J.Y., Aguilar, L.E., Park, C.H., and Kim, C.S. (2019). UV light assisted coating method of polyphenol caffeic acid and mediated immobilization of metallic silver particles for antibacterial implant surface modification. *Polymers (Basel)* 11 (7).

97 Biao, L., Tan, S., Meng, Q. et al. (2018). Green synthesis, characterization and application of proanthocyanidins-functionalized gold nanoparticles. *Nanomaterials* 8 (1).

98 Saif, S., Tahir, A., and Chen, Y. (2016). Green synthesis of iron nanoparticles and their environmental applications and implications. *Nanomaterials* 6 (11): 1–26.

99 Thakur, A., Kaya, S., and Kumar, A. (2021). Recent innovations in nano container-based self-healing coatings in the construction industry. *Current Nanoscience* 18 (2): 203–216.

100 Thakur, A., Kumar, A., Kaya, S. et al. (2022). Suppressing inhibitory compounds by nanomaterials for highly efficient biofuel production: a review. *Fuel* 312 (September 2021): 122934. https://doi.org/10.1016/j.fuel.2021.122934.

101 Kumar, A. and Thakur, A. (2020). Encapsulated nanoparticles in organic polymers for corrosion inhibition. *Corrosion Protection at the Nanoscale* 345–362. https://doi.org/10.1016/B978-0-12-819359-4.00018-0.

102 Parveen, G., Bashir, S., Thakur, A. et al. (2020). Experimental and computational studies of imidazolium based ionic liquid 1-methyl- 3-propylimidazolium iodide on mild steel corrosion in acidic solution Experimental and computational studies of imidazolium based ionic liquid 1-methyl-3-propylimidazolium. *Materials Research Express* 7 (1): 016510.

103 Thakur, A., Kaya, S., Abousalem, A.S. et al. (2022). Computational and experimental studies on the corrosion inhibition performance of an aerial extract of *Cnicus benedictus* weed on the acidic corrosion of mild steel. *Process Safety and Environmental Protection* 161: 801–818. https://doi.org/10.1016/j.psep.2022.03.082.

104 Thakur, A., Kumar, A., and Zhang, R. (2023). Alcoholic beverage purification applications of activated carbon. In: *Activated Carbon: Progress and Applications* (ed. C. Verma and M.A. Quraishi). The Royal Society of Chemistry https://doi.org/10.1039/BK9781839169861-00152.

105 Kumar, A. and Thakur, A. (2023). Overview of the properties, applicability, and recent advancements of some natural products used as potential inhibitors in various corrosive systems. In: *Handbook of Research on Corrosion Sciences and Engineering*, 275–310. IGI Global.

106 Sandilya, S., Thakur, A., Suresh, S., and Kumar, A. (2021). Recent advances, synthesis and characterization of bio-based based polymers from natural sources. *Plant Cell Biotechnology and Molecular Biology* 22: 70–93.

107 Bashir, S., Thakur, A., Lgaz, H. et al. (2020). Corrosion inhibition performance of acarbose on mild steel corrosion in acidic medium: an experimental and computational study. *Arabian Journal for Science and Engineering* 45 (6): 4773–4783. https://doi.org/10.1007/s13369-020-04514-6.

8

Polyphenols for Packaging Application

Gianluca Viscusi and Giuliana Gorrasi

Department of Industrial Engineering, University of Salerno, Fisciano, SA, Italy

8.1 Introduction

Food packaging is essential for protecting and preserving processed foods, fruits, meats, vegetables, and dairy products. Moreover, they are insufficient to preserve the quality of food products, especially for foods requiring high gas permeability or perforations. One approach for reducing or mitigating food deterioration or foodborne outbreaks due to microbial pathogens concerns the use of natural and active substances [1]. So, the scientific and industrial community has prompted the design of a novel active packaging beyond its primary functions. One possibility could be placed in the use of additives. Among all materials, gas scavengers, moisture absorbents, absorbers of ethylene, antioxidants, and antimicrobial compounds have been applied to preserve the food quality and prevent spoilage due to air, moisture, and bacterial growth [2]. However, synthetic preservatives must be more suitable since they present drawbacks such as high toxicity and adverse health and environmental effects [3].

Concerns about food safety and related health risks are receiving considerable attention. For example, common chemical preservatives used in food are benzoic acids, sorbic acids, sulfur dioxide, nitrates and nitrites, and their derivative salts. Some of them appear to be potential allergens. Moreover, common preservatives could promote the formation of carcinogenic compounds of nitrosamines from nitrite and benzene from benzoic acid, causing adverse effects on the consumer's health [4]. To overcome these concerns, the research focuses on designing and producing novel packaging systems loaded with natural and functional active compounds that might inhibit oxidation and microbial spoilage [5] and prolong

Science and Engineering of Polyphenols: Fundamentals and Industrial Scale Applications,
First Edition. Edited by Chandrabhan Verma.
© 2024 John Wiley & Sons Inc. Published 2024 by John Wiley & Sons Inc.

the storage life of fresh food products. Consumers require packaging free of synthetic chemical additives while the demand for more natural products for food preservation is boosting. The typical natural compounds used in food packaging applications include some chemical substances derived from spices, oils, plants, animals, microflora, and herbs. For example, it has already been demonstrated that the growth of infectious organisms could be inhibited by the secondary metabolites found in spices and herbs [6].

These natural materials are often considered appealing ingredients since their interesting properties such as their bioactivities. Moreover, even if they are derived from natural matter, they must comply with food additive regulations (Regulation No. 450/2009) [7]. Among the naturally extracted compounds, phenolic compounds might be used since it is known that they could ensure antioxidant and antibacterial activities. Usually, phenolic compounds are derived from plant extracts [8] and used to fabricate edible coatings and active films. The main bioactive phenolic substances already used in food packaging include flavonoids, phenolic acids, tannins, lignans, and stilbenes. But other substances with minor effects are still used.

8.2 Active Packaging: Coating and Thin Films

Active food packagings prolongs packaged foods' shelf life, quality, and properties. Incorporating polyphenols can ensure antioxidant and antimicrobial activities and extension of the shelf life of food [9]. Food must be protected against oxygen, bacteria, moisture, and biochemical changes. Fabricating active packaging from biopolymers and naturally derived bioactive substances can be a potential strategy to address sustainability, food safety, and quality [10]. Edible films or coatings made of biopolymers have gained interest since their sustainability, friendliness, and food contact approval. Films and coatings produced from lipids, polysaccharides, and proteins are being studied as sustainable and environmentally friendly substitutes for conventional materials used for food packaging. Films are thin systems already applied to wrap food products. Research has been done to incorporate antimicrobial substances into food packaging to guarantee quality and delay food spoilage. This technique ensures the antimicrobial compounds are located at the food surface where microbes usually grow, speeding up the inhibition of spoilage microorganisms [11]. Besides, coatings are directly applied on the food surface. Usually, both are supposed to be directly consumed so edible polymers and food-grade additives must be used.

8.3 Phenolic Compounds

Phenolic compounds or polyphenols represent a class of phytochemicals derived from plants as secondary metabolites. These substances are widely present in vegetables, herbs, and spices. Most of them can be obtained from food by-products, so the spotlight on these compounds is motivated even by the potential economic value apart from their interesting properties [12]. Fruits such as lemons, pomegranates, grapes, cherries, berries, olive, broccoli, spinach, and cauliflower and low-value parts such as seeds, stones, peel, and pulp have been processed to obtain phenolic compounds [1, 13]. From a chemical point of view, polyphenols are aromatic compounds with at least one OH group joined to a benzene ring [14]. A simple classification of these compounds is reported in Figure 8.1. Table 8.1 reports the sources and properties of some polyphenols.

The antimicrobial and antioxidant properties of phenolic compounds have been well studied [15]. It is known that polyphenols can ensure the extension of the shelf-life and enhance food quality by migrating to food or absorbing antioxidative radicals. They can be recognized as natural additives able to delay the oxidation and the growth of microorganisms.

8.3.1 Antioxidant and Antimicrobial Activities

Since oxidation can occur, lipid rancidity, off-flavors, and a loss of color and flavor can affect food quality, so compounds with antioxidant properties could be added to retain organoleptic properties such as flavor and color while preventing metabolic alterations [16]. It is known that antioxidants can limit the occurrence of oxidative chain reactions inhibiting or delaying the oxidation in food. The most common ones are butylated hydroxytoluene, butylated hydroxyl anisole, propyl gallate, and *tert*-butyl hydroquinone [14]. But natural antioxidants can represent an alternative strategy investigated by the scientific community [17]. Polyphenols, flavonoids, phenolic acids, tannins, coumarins, quinones, terpenoids, and alkaloids can act as antioxidant and antimicrobial agents. They can be easily found in everyday foodstuffs such as grape, kiwi, orange, prunes, garlic, olives, berries, pepper, cherries, clove, citrus, vegetables, tea, herbs, chocolate, coffee, and wine. The use of these compounds is an interesting strategy to inhibit the growth of foodborne pathogens and the activity of microorganisms as a protection against fungal, bacterial infection, and yeast.

8.3.2 Polymeric Matrices for Polyphenols-Based Food Packaging

So far, different polysaccharides have been already studied to develop active packaging in the form of films/edible coatings able to play the role of carriers of phenolic compounds for food preservation. It is already known that polyphenols can bind with polysaccharides through H-bonds and hydrophobic interactions. Hydrophobic interactions may exist between the polysaccharide's methyl groups and the polyphenols' heterocyclic C-ring. Besides, hydrogen bonds between hydroxyl groups of polysaccharides and phenols can occur. These kinds of interactions could lead to a modification of features of films (i.e. moisture sensitivity,

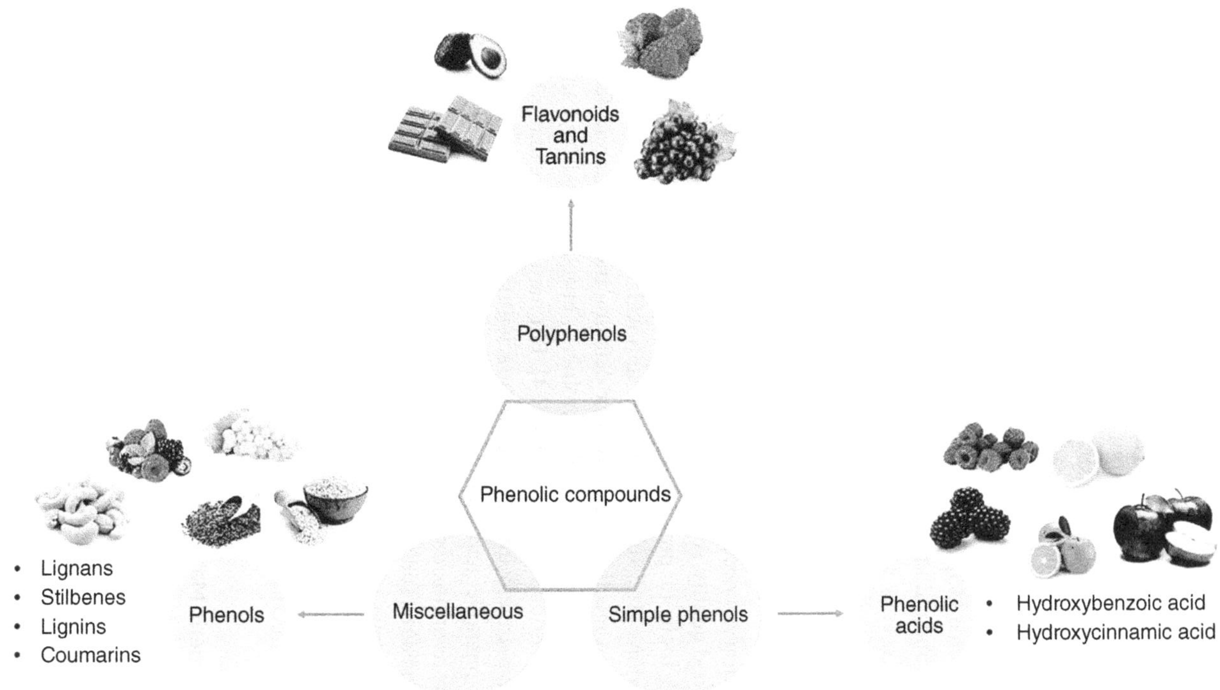

Figure 8.1 Naturally occurring phenolic compounds.

Table 8.1 Typical phenolic compounds, their sources, and their properties.

Phenolic compounds	Sources	Properties
Flavonoids	Berries, cacao, grapes, tea, spinach, olives, cherries, citrus fruits	Antimicrobial, antioxidant, antifungal
Tannins	Tea, coffee, chocolate, apple, wine	Antimicrobial, antifungal, antioxidant
Lignans	Flaxseed, sesame, berries, legumes	Antioxidant, antimicrobial
Stilbenes	Grapes, almonds, pine, peanuts	Antioxidant, anti-infective
Lignins	Spruce, cotton, hemp, pine, birch	Antioxidant, antimicrobial
Coumarins	Cinnamon, cloves, celery, apricots	Antioxidant, antimicrobial
Phenolic and hydroxycinnamic acid	Oranges, mushrooms, tea, propolis, persimmon, mustard	Antioxidant, antimicrobial, anti-infection

solubility, mechanical and UV-barrier properties) [18]. These phenol-polysaccharide interactions might enhance the performances of polysaccharide-based systems, but further investigation is required to highlight their nature better. No deep studies describe the interactions between polyphenols and polysaccharides [19]. Some examples of polymer-polyphenols systems with specific applications are reported in Table 8.2.

8.4 Classifications and Applications of Polyphenolic Compounds

Phenolics are chemical compounds showing an aromatic ring with at least one OH group. Their chemical structure can vary widely. They are classified into two groups: flavonoids [29] and non-flavonoids [30].

8.4.1 Flavonoids

Flavonoids are divided into different subgroups: anthocyanins, flavonols, flavanols, flavanones, flavonones, and isoflavones [30, 31]. Chemically, flavonoids show a phenol compound having a phenyl benzopyran chemical structure with a carbon skeleton of a C6–C3–C6 connected to a chroman ring [32]. Usually, three

Table 8.2 Example of polymeric packagings incorporating polyphenols with targeted applications.

Polymer	Source of phenolic compounds	Properties	Targeted food product	Reference
Nanocellulose	*Acacia mearnsii* bark	Air barrier High density Surface hydrophobicity	—	[20]
Cellulose	*Schinopsis lorenzii*	Anti-oxidative UV-adsorbing	—	[21]
Ethyl cellulose	Grape tannin	Antioxidant activity	—	[22]
Hydroxypropylcellulose polyethylene oxide	Gallic acid	Oxidation delay	Walnut	[23]
Chitosan	Kombucha tea	Antioxidant activity Retardation of lipid oxidation Microbial growth	Minced beef	[24]
Polyvinyl alcohol	Apple pomace	Antioxidant activity Delay of lipid oxidation	Soybean oil	[25]
Starch	Red cabbage	Antioxidant activity	Ground beef	[26]
Pectin-fish gelatin	Olives	Antioxidant activity Suppression of lipid oxidation	Beef meat	[27]
Methyl cellulose	Oregano (*Origanum vulgare*) and Clove Bud (*Syzygium aromaticum*)	Antimicrobial activity	Bread	[28]

or more hydroxyl groups are joined to their backbone structure [33]. They can be found in herbs, berries, tea, cacao, grapes, olives, citrus fruits, spinach, soybeans, and cherries. The main interest in flavonoids could be placed in their antimicrobial, antioxidant, and antifungal activities of improving the safety of packaged food or their quality by inhibiting microbial growth [34]. However, the main limit related to their use concerns the stability after the extraction process, which makes it quite challenging.

8.4.1.1 Flavanones

More than 160 flavanones from 36 plant families have been described [35]. Figure 8.2 reports the chemical structure of the main flavanones.

Flavanones are present in tomato, potato, spices, rosemary, lemon, peppermint, orange, raspberry, strawberry, and plum [36]. Another interesting flavanone is naringin (Figure 8.3), whose naringenin is its aglycone form.

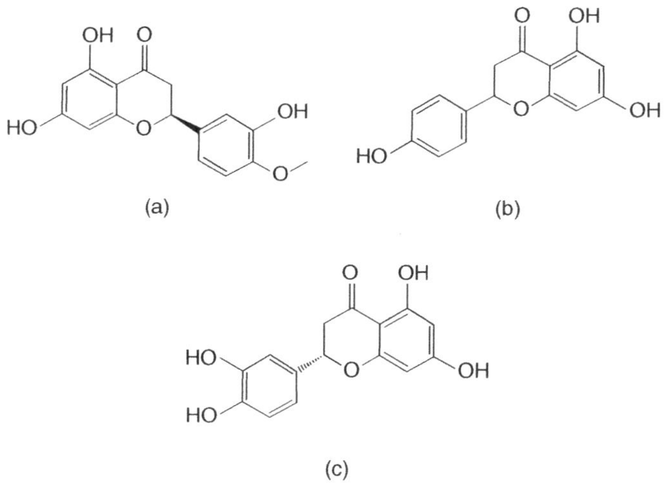

Figure 8.2 Chemical structures of hesperetin (a), naringenin (b), and eriodictyol (c).

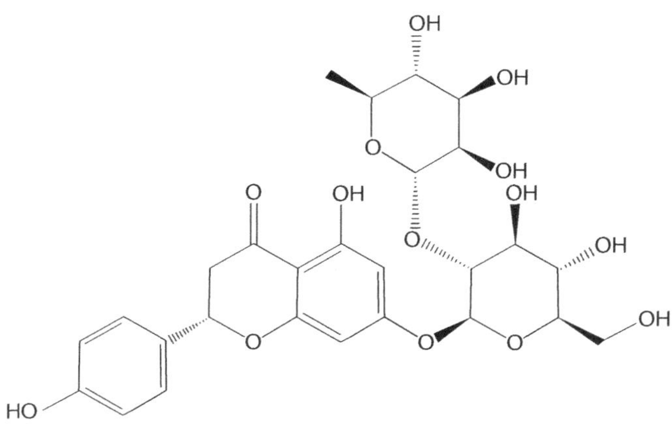

Figure 8.3 Chemical structure of naringin.

It is mainly found in mandarin, lemon, orange, and grapefruit [37]. Naringin possesses three rings (two are benzene rings) connected by a three-carbon chain and two rhamnose units attached at the C7 [38]. It shows some interesting properties, including antioxidant capacity, anti-inflammatory, light blocking activities. Moreover, it has also been found to have an antimicrobial effect against bacteria as well as different viruses and fungi [39–42]. It was proved from previous studies that naringin could inhibit the quorum sensing-regulated factors in *Pseudomonas* spp. [43]. Its use in food packaging is still under investigation. Li et al. prepared polylactic acid-naringin electrospun fiber and showed QS inhibition activity against Pseudomonas fluorescens with intelligent spoilage-sensitive properties. These features make the developed packaging suitable for salmon freshness keeping and monitoring [44]. Other interesting researches concern the use of naringin with cellulose [42] and chitosan [45].

8.4.1.2 Flavonols

Flavonols are substances found in apple, onions, saffron, persimmon, berries, tea, and red wine [46]. Quercetin and kaempferol are the main compounds. Quercetin (Figure 8.4) (3,3′,4′5,7-Pentahydroxyflavone) is known to possess antibacterial as well as antioxidant properties [47]. Quercetin contains two benzene rings that are linked with a pyrene ring. It includes five hydroxyl groups in the structure responsible for the biological properties [48]. It is mainly found in apples, onions, and berries.

Quercetin is a naturally occurring yellow-colored plant phenolic compound [49]. Quercetin exhibits very high antioxidant activity as a powerful radical scavenger of reactive oxygen species (ROS), such as oxygen, peroxynitrite, and nitric oxide radicals [50]. The strong antioxidant action of quercetin is due to the existence of a catechol group, a hydroxyl group, and a double bond bonded to oxygen in the pyrene ring [51]. Quercetin can repress the growth of both Gram-positive and negative bacterial strains such as *Shigella sonei*, *Pseudomonas aeruginosa*, *Streptococcus sanguis*, *Staphylococcus aureus*, and so on [52]. Adding quercetin to food packaging

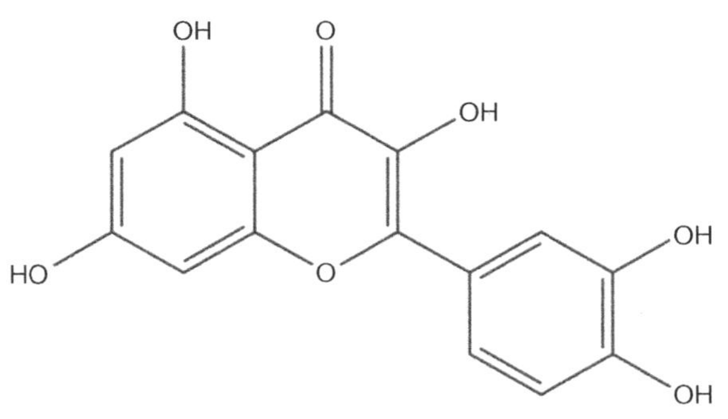

Figure 8.4 Structure of quercetin.

films or coating improved functional and physical performances [53]. Many research works have recently been published on quercetin, including various polymer-based films using solution casting or electrospinning [54, 55].

Yadav et al. prepared quercetin-starch-based chitosan-gelatin films (Figure 8.5), proving their antioxidant and antibacterial properties. Moreover, the films protected foods from UV radiation [47].

Luzi et al. produced gallic acid and quercetin-based poly(vinyl alcohol) by solvent casting process as intelligent and active food packaging systems. The films

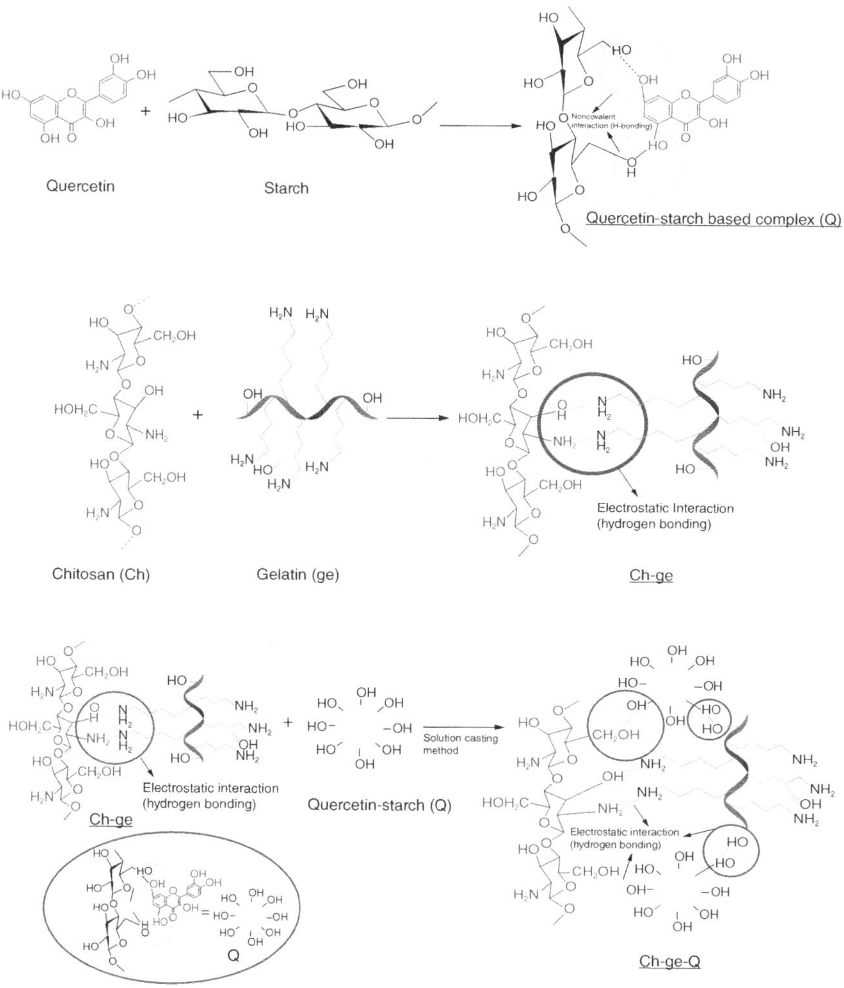

Figure 8.5 Schematization of films of chitosan and gelatin loaded with quercetin-starch-based complex. Reproduced from [47].

showed a radical scavenging activity of 91% for quercetin meanwhile a value of 95% was recorded for gallic acid [56]. Lopusiewicz et al. proposed poly(butylene succinate)-based films loaded with quercetin. The loaded films showed noticeable changes in color, and UV-blocking effect with free radicals (DPPH, O_2^-, ABTS) scavenging and some bactericidal activities [57]. Lou et al. investigated the effects of kafirin-quercetin films on the total microbiological count of cod fillets. According to their findings, all cod samples significantly grew in the total microbial count with storage time. However, the trend for samples wrapped in composite films increased less quickly than for those left unwrapped. The results showed that kafirin-quercetin films might boost the antibacterial activity of film and prevent microbiological development in fish products.

Other quercetin-loaded polymeric systems for food packaging concern the use of hordein/chitosan electrospun fibers [58], cassava starch-carboxymethyl cellulose as oxidation retardant of lard and delaying system of discoloration of pork [53], chitosan [59], polypropylene, ethylene vinyl acetate, ethylene vinyl alcohol, low-density polyethylene [60], carboxymethyl cellulose, gelatin, and poly(lactic acid) [61], chitosan/choline chloride and citric acid [54], kafirin [62], poly(butylene adipate-co-terephthalate)/thermoplastic starch-based [63] and carboxy methyl chitosan [64]. Kaempferol (3,5,7-trihydroxy-2-(4-hydroxyphenyl)-4*H*-chromen-4-one) (Figure 8.6) is a natural flavonol derived from beans, kale, leek, tea, broccoli, cabbage, strawberries, and grapes. It shows antioxidant, anti-microbial, anti-inflammatory, and anti-allergic activities. Usually, it is used for cosmetics and food-based applications. This compound has been incorporated in chitosan [65] and the system showed good oxygen and water barrier ability, mechanical properties, and high DPPH radical scavenging activity.

8.4.1.3 Flavanols

The chemical structure of flavanols presents a 3,3′,4,5,7-pentahydroxyflavone with two steric forms of (+)-catechin and its enantiomer [66]. The main catechins, considering the gallic acid conjugates, are reported in Table 8.3 [67].

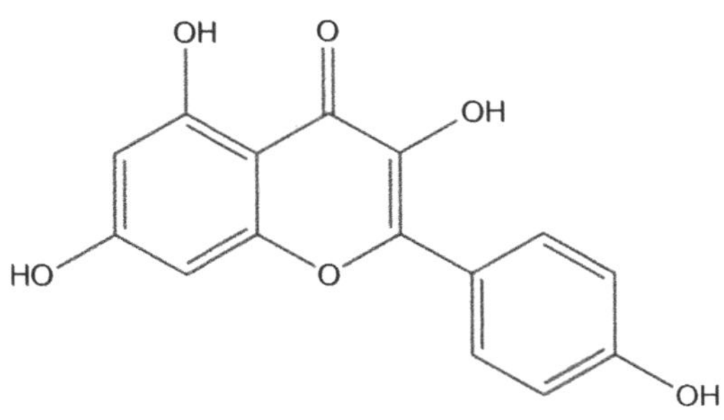

Figure 8.6 Structure of kaempferol.

Table 8.3 Chemical formulas and structures of catechins.

Compounds	Chemical formula	Chemical structure
(+)-Catechin	$C_{15}H_{14}O_6$	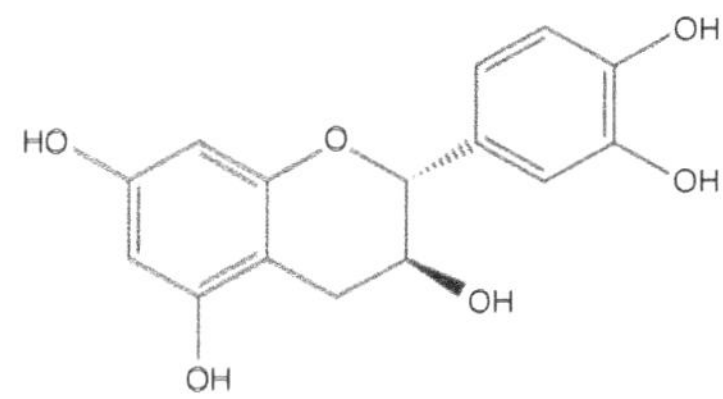
(−)-Epicatechin	$C_{15}H_{14}O_6$	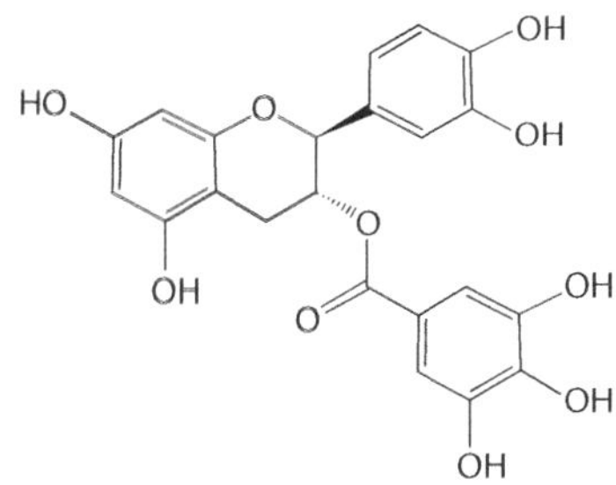
(−)-Catechin gallate	$C_{22}H_{18}O_{10}$	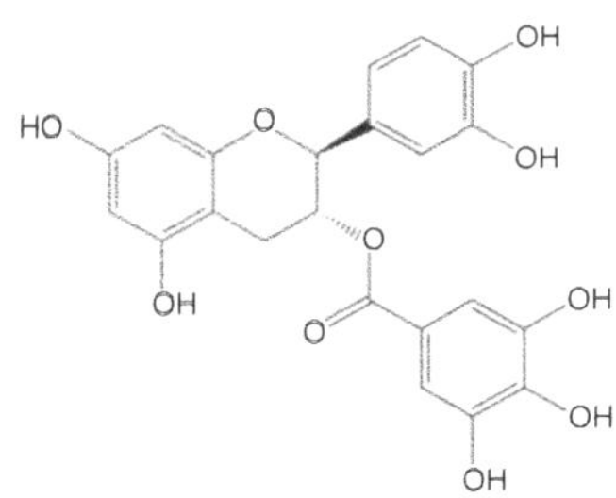
(−)-Epicatechin gallate	$C_{22}H_{18}O_{10}$	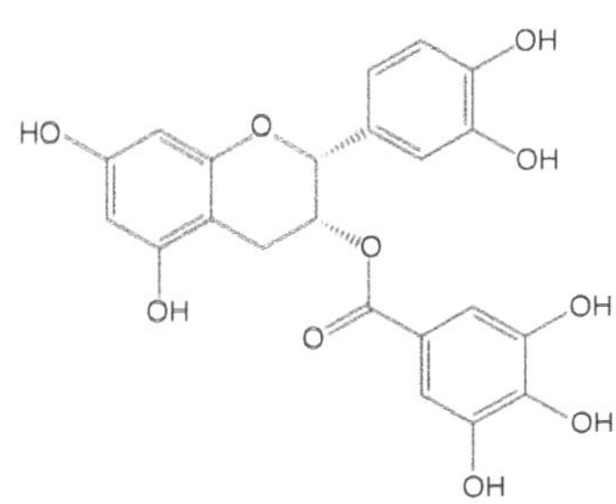

(Continued)

Table 8.3 (Continued)

Compounds	Chemical formula	Chemical structure
(−)-Gallocatechin	$C_{45}H_{38}O_{21}$	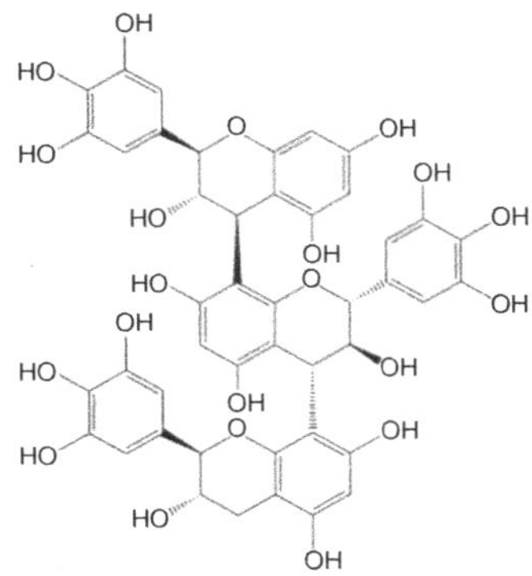
(−)-Epigallocatechin	$C_{15}H_{14}O_7$	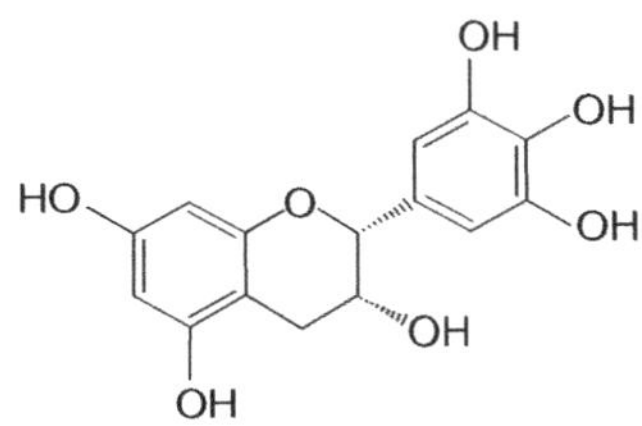
(−)-Gallocatechin gallate	$C_{22}H_{18}O_{11}$	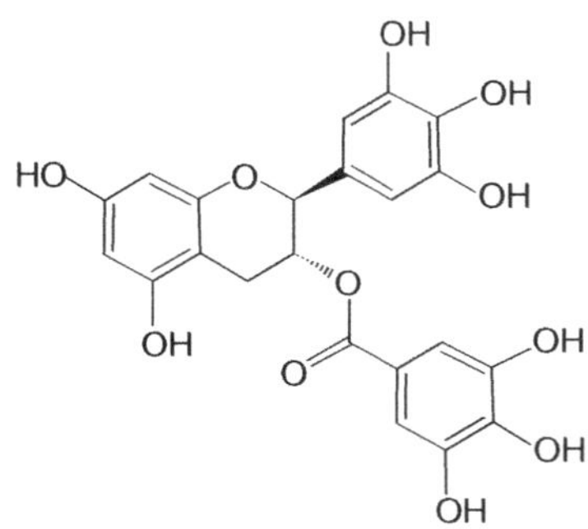
(−)-Epigallocatechin gallate	$C_{22}H_{18}O_{11}$	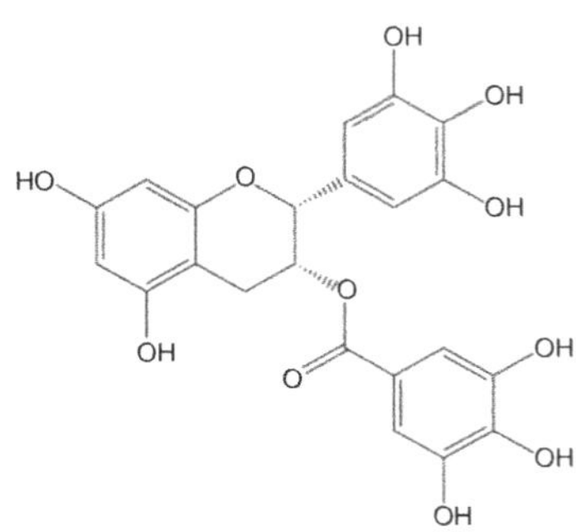

Catechins extracted from tea act as antioxidant and antimicrobial activities [68]. Moreover, they even showed health benefits [69]. Tea catechins might be used as a food additive or active component in food packaging applications for their antioxidant activity especially with fatty foods. Catechins have already been used to prolong the preservation of pork sausages [70], white shrimp [71], and brined sardines [72]. Table 8.4 summarizes some applications of catechins for food preservation.

8.4.1.4 Anthocyanins

Anthocyanins can be considered water-soluble flavonoids, which confer the typical colors of fruits and flowers. Anthocyanins are the glycoside form of anthocyanidins with a typical C6–C3–C6 basic skeleton, with two aromatic rings separated by a six-membered heterocyclic ring containing oxygen [86]. So far, more than

Table 8.4 Use of catechins in food packaging applications.

Source	Polymer	Properties	Reference
Black and green tea	Chitosan	Antioxidant activity	[73]
Green tea	Chitosan	Antioxidant activity	[74]
Green tea, grape, and ginger	Fish skin gelatin	UV light barrier Moisture resistance Antioxidant activity	[75]
Grape extract	Chitosan	Antioxidant activity Low water solubility	[76]
Tea	Chitosan	Moisture barrier property Antioxidant activity	[77]
Grape seed and green tea	Myofibrillar protein	Improvement of physical properties	[78]
Green tea	Gelatin	High tensile strength Light barrier Water vapor barrier	[79]
Camellia sinensis L.	Agar/fish gelatin	Radical scavenging ability	[80]
Catechins	Casein/sodium caseinate	Antioxidant activity	[81]
Catechins	Zein	Antioxidant activity	[82]
Green tea	EVOH	Antioxidant activity	[67]
Green tea	Whey protein	Colloidal stability	[83]
Tea	Chitosan/poly (γ-glutamic acid)	Antioxidant activity	[84]
Tea	Gelatin	Antioxidant activity	[85]

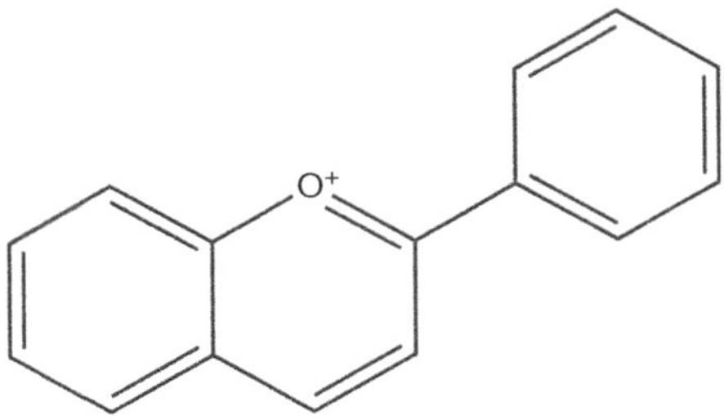

Figure 8.7 Structure of flavylium ion.

500 kinds of anthocyanins have been classified. The flavylium ion is the basic structural unit of anthocyanins (Figure 8.7).

Pelargonidin, delphinidin, cyanidin, peonidin, petunidin, and malvidin belong to the basic structures of anthocyanidins [87]. Some sources of anthocyanins include chokeberries, strawberry, elderberries, red raspberry, beans, pomegranates, orange, cabbage, and red onions [88]. Anthocyanins show high antioxidant power and health benefits. Moreover, they have been used as time-temperature or gas indicators [89] or to monitor the pH change due to food deterioration induced by microorganisms [90]. This is due to the behavior of anthocyanins in solution with different pHs, leading to a color change from red flavylium to purple/blue quinoidal base, colorless hemiketal, or yellow chalcone with the increase in pH value (Figure 8.8) [92, 93].

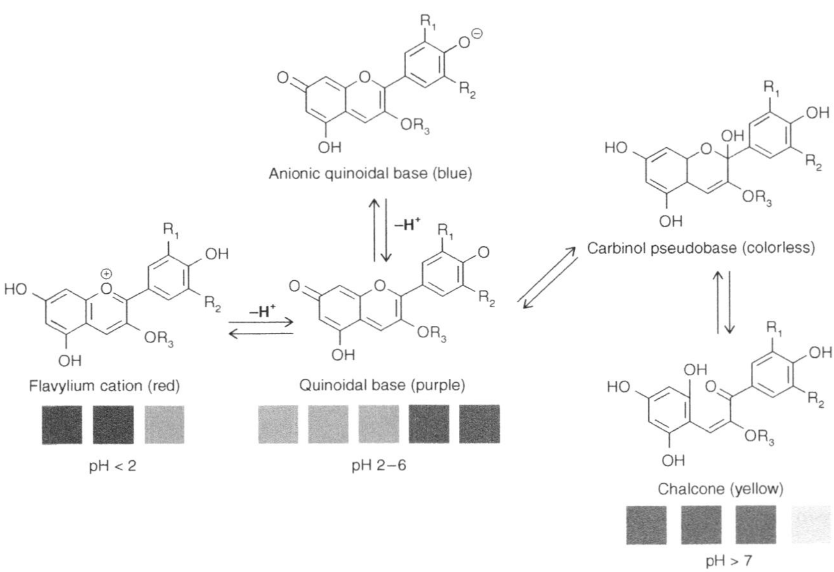

Figure 8.8 Structural transformation of anthocyanins as a function of pH. Source: Ref. [91]/MDPI/CCBY 4.0/Public domain.

Since this property, anthocyanins have been loaded into polymeric films for developing smart functional packaging. Table 8.5 summarizes the use of anthocyanins in food packaging applications.

Table 8.5 Use of anthocyanins in food packaging applications.

Anthocyanins source	Polymer	Properties and use	Reference
Red cabbage	Chitosan/PVA	Time–temperature indicators	[94]
Carrot	Cellulose/chitosan	Monitoring spoilage in pasteurized milk	[95]
Red cabbage	Bacterial cellulose	pH indicator label	[96]
Purple sweet potato	Gellan gum	Active and intelligent packaging	[97]
Rosemary extract	PVA/starch	Intelligent food packaging for milk spoilage	[98]
Red barberry	Methylcellulose/chitin	Monitoring fish fillet freshness	[99]
Black soybean seed coat extract	Chitosan	Antioxidant and visible pH-sensing	[100]
Roselle anthocyanins	Starch/polyvinyl alcohol	Monitoring of pork freshness	[101]
Purple sweet potato	Agar/potato starch	Monitoring spoilage point of pork samples	[102]
Amaranth extract	Hydroxypropyl methylcellulose	Storage of salmon oil	[103]
Bayberry extract	Cassava starch	Monitoring the freshness of pork	[104]
Blackberry pulp power	Arrowroot starch	Packaging with antioxidant activity	[105]
Black carrot	Bacterial cellulose	Monitoring the freshness of fish	[106]
Black rice bran extract	Gelatin	Monitoring the freshness of hairtail and shrimp	[107]
Blueberry extract and Elderberry extract	Furcellaran	Monitoring the freshness of fish	[108]
Blueberry and red grape skin pomace extract	Chitosan/ Carboxymethyl cellulose	Monitoring the freshness of chicken	[109]
Cranberry extract	Gelatin	Storage of olive oil	[110]
Grape extract	Cassava starch	Monitoring the freshness of pork	[111]

(Continued)

Table 8.5 (Continued)

Anthocyanins source	Polymer	Properties and use	Reference
Grape pomace extract	Guar gum	Storage of pomegranate arils	[112]
Red radish extract	Starch/gelatin	Monitoring the freshness of chicken and white prawn	[113]
Vitis amurensis husk extract	Tara gum	Monitoring the freshness of fish	[114]

8.4.1.5 Flavones

Flavones possess a basic chemical structure made of two benzene rings joined through a heterocyclic pyrone ring. The main components are luteolin (a) and apigenin (b) (Figure 8.9).

They can be found in cabbage, acerola, apricot, bean, cardon, dandelion, artichoke, apple, papaya, mango, and onion [115]. For example, luteolin encapsulated in oil-in-water nanoemulsions was loaded into the chitosan-based matrix. Luteolin seemed to improve barrier and mechanical properties. Its antioxidant activity is extended up to 10 days, demonstrating its potential application in the food industry [116].

8.4.2 Non-flavonoids

8.4.2.1 Tannins

Tannins are a class of water-soluble chemical compound extracted from tea, coffee, berries, chocolate, apples, and wine. Their molecular weight varies between 0.5 and 3 kDa and their polyphenolic structure consists of 12–16 phenolic groups and 5–7 aromatic rings per 1000 units of weight. These compounds attract interest since they are renewable, safe, and environmentally friendly, presenting interesting antioxidant and antimicrobial features. From a chemical point of view, they

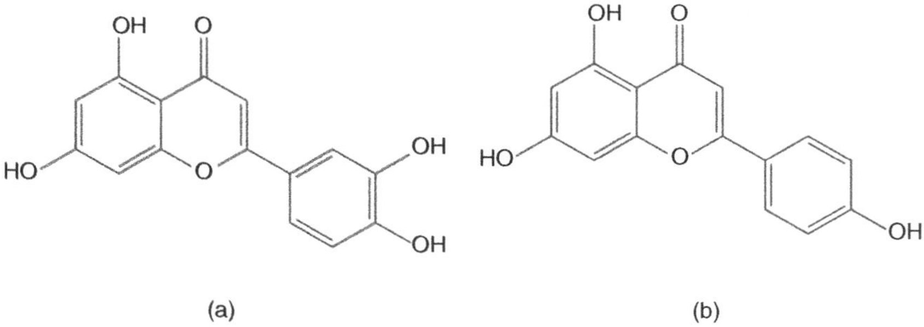

Figure 8.9 Chemical structures of (a) luteolin and (b) apigenin.

can be divided into hydrolysable and condensed tannins based on their resistance to hydrolysis [117]. They are considered safe (GRAS) additives showing interesting properties such as antifungal, UV absorption, antioxidant, and antimicrobial activities [1].

Huang et al. fabricated regenerated cellulose loaded with tannin from black wattle films with improved properties. The produced films appeared to possess antioxidant (80–90% of radical scavenging activity), UV-blocking, hydrophobicity (up to 75°), and water vapor barrier properties (400 g/m^2 day). The schematization of the incorporation of tannin onto regenerated cellulose film is reported in Figure 8.10[118].

Yu et al. produced methylcellulose films (Figure 8.11) loading tea catechins, demonstrating improved physical properties (i.e., water vapor permeability and mechanical resistance). The research proved the crosslinking effect via the formation of hydrophobic interactions and hydrogen bonds, which is responsible for the improved properties. Then, they demonstrated that tea catechins could either protect against β-carotene degradation induced by light adsorption or scavenge free radicals to reduce the degradation of β-carotene induced by free radical-mediated lipid peroxidation [119].

Cano et al. used different tannin extracts to fabricate active films. Based on the results, the authors demonstrated that films with tannins could be applied to avoid the occurrence of oxidative reactions or to control bacterial growth [120]. Konstantin et al. prepared novel systems based on tannin-induced precipitation of gelatin grafted onto surface-modified cellulose nanofibrils. The polymeric system showed high mechanical performances (tensile strength = 33 MPa, modulus of Young = 310 MPa, strain at failure = 22%), antioxidant, and UV-blocking properties [121]. Li et al. produced films based on tannin-loaded and cationic cellulose nanofibrils. Anti-oxidant properties, UV-shielding ability, higher thermal stability, and tensile strength (160 ± 9 MPa) are the main properties of the produced films [21]. Olejar et al. fabricated films based on ethyl cellulose and grape tannin. A low amount of grape tannins favored the increase in antioxidant activity without an alteration of physical properties. A higher amount of tannins (3.0%) further enhanced antioxidant activity [22]. Many papers have already reported the use of tannins against different fungi and bacteria (Table 8.6).

8.4.2.2 Phenolic Acids

Phenolic acids represent non-flavonoid polyphenolic substances in tea, red wine, berries, mango, nuts, and citrus fruits. Chemically, they present a carboxyl group joined to a benzene ring [36]. Based on carboxylic acids, phenolic acids could be divided into hydroxybenzoic (derived from benzoic acid) (C6–C1, gallic, *p*-hydroxybenzoic, syringic, vanillic, and ellagic acids) and hydroxycinnamic acids (derived from cinnamic acids) (C6–C3, caffeic, *p*-coumaric chlorogenic, and ferulic acids) [117]. Some of them are represented in Table 8.7.

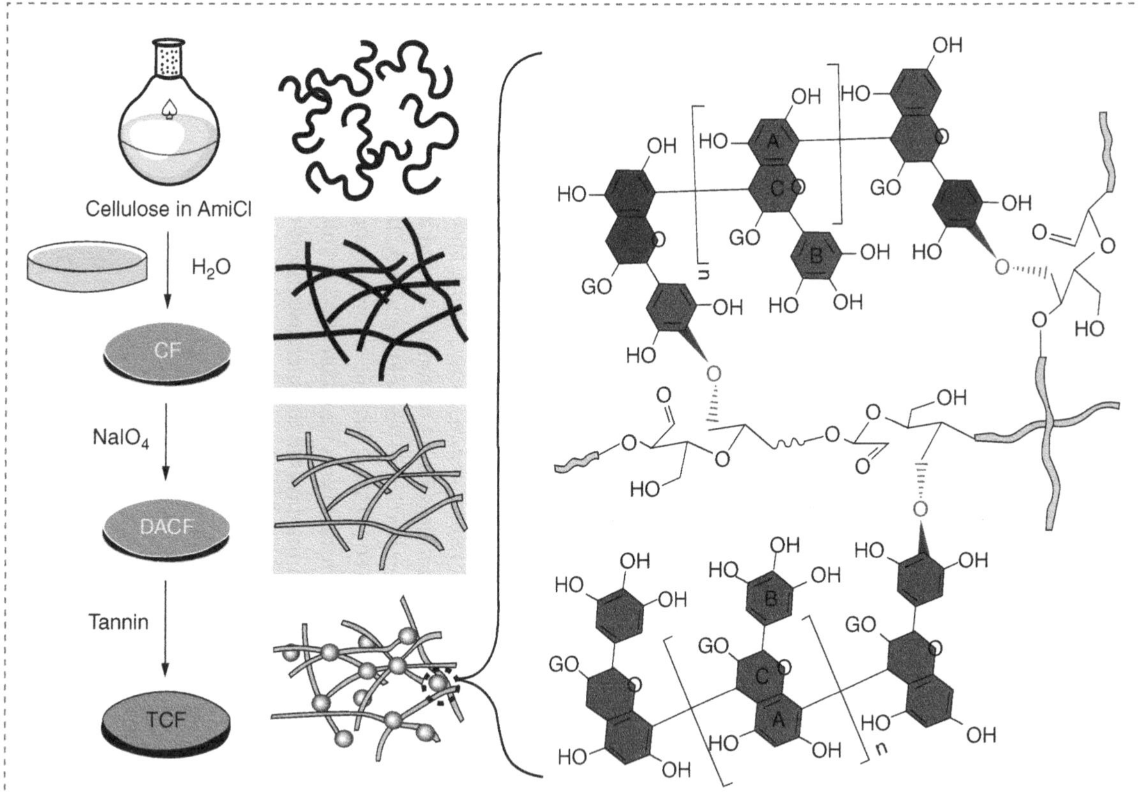

Figure 8.10 Schematization for incorporation of tannin onto regenerated cellulose. Reproduced from [118].

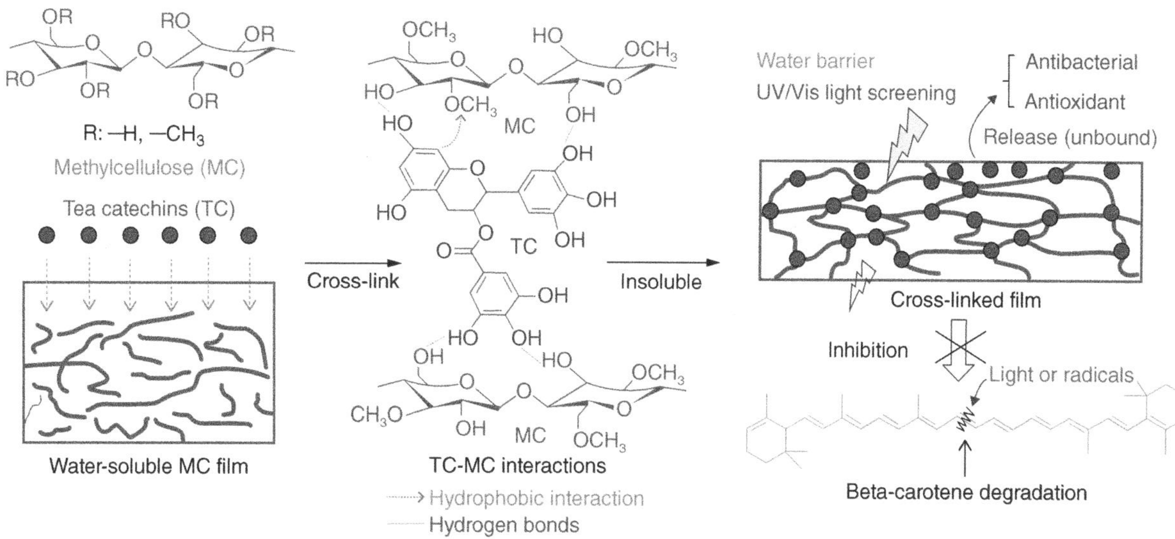

Figure 8.11 Fabrication of tea catechins-cross-linked methylcellulose films. Reproduced from [119].

Table 8.6 Effect of tannins and flavonoids against fungi and bacteria.

Source	Bacteria or fungi	Reference
Prunus cerasoides	*Klebsiella pneumoniae, Staphylococcus aureus, C. albicans P. aeruginosa*	[122]
Combretum album	*Bacillus licheniformis Pseudomonas aeruginosa, Bacillus subtilis, Pseudomonas fluorescens, Bacillus mycoides, Escherichia coli, Pseudomonas putida, B. licheniformis*	[123]
Tapinanthus bangwensis	*Staphylococcus aureus, Escherichia coli, Staphylococcus saprophyticus, Candida albicans*	[124]
Salacia oblonga	*Microsporum canis, Micrococcus luteus, Mycobacterium tuberculosis, Trichophyton interdigitale, Microsporum gypseum*	[125]
Piper beetle	*E. faecalis, L. fermentum, L. salivarius, S. sobrinus, S. mutans*, Gram-negative bacteria, *A. actinomycetemcomitans F. nucleatum*	[126]
Acacia mearnsii De Wild.	*E. coli, K. pneumoniae, P. vulgaris, S. marcescenes*	[127]
Gallnut	*E. coli, Staphylococcus aureus Aspergillus niger, Penicillium digitatum*	[128]

Hydroxycinnamic acids can be found in tea, fruits, vegetables, cocoa, wine, whole grains, and coffee [129]. In food packaging applications, phenolic acids are known to show antioxidant, anti-inflammatory, and antibacterial activities against *Bacillus cereus, E. coli, Staphylococcus aureus, Aspergillus flavus*, and *Aspergillus parasiticus* [130, 131]. Aydogdu et al. produced gallic acid/hydroxypropyl methylcellulose/polyethylene oxide electrospun nanofiber as active packaging materials for decreasing the oxidation of walnut during storage [23]. Cao et al. studied using a nisin/gallic acid/chitosan coating and high oxygen-modified atmosphere packaging to preserve the quality of fresh pork loin. The designed coating showed color protection, reduced protein, lipid oxidation, and bactericidal effect [132]. Vilela et al. produced chitosan and ellagic acid films (Figure 8.12). Solvent casting was used to fabricate films, while antimicrobial activity against *Staphylococcus aureus* and *Pseudomonas aeruginosa bacteria* was demonstrated [133].

Apart from hydroxycinnamic acids, phenolic compounds with C6–C3 carbon skeletons are known as phenylpropanoids, which can be classified as curcuminoids, coumarins, and chromones. They are derived from cereal grains, coffee, fruits, herbs, potatoes, and mushrooms [134].

Table 8.7 Chemical formulas and structures of phenolic acids.

Hydroxybenzoic acids		Hydroxycinnamic acids	
Chemical name and formula	**Chemical structure**	**Chemical name and formula**	**Chemical structure**
Gallic $C_7H_6O_5$	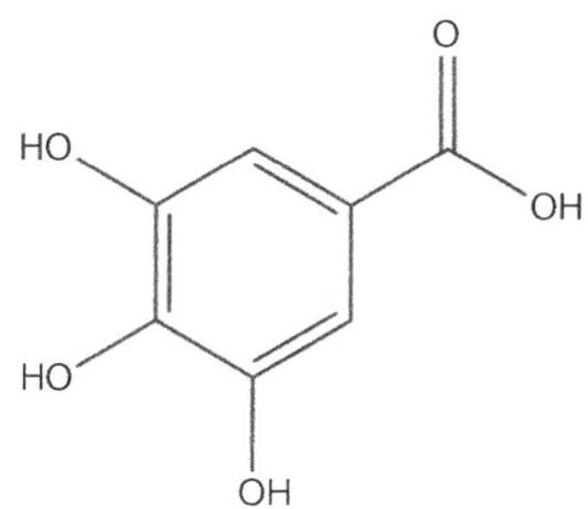	Caffeic $C_9H_8O_4$	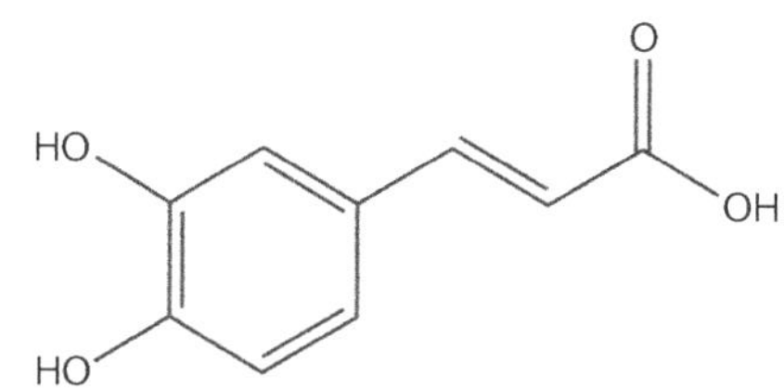
Sinapic $C_{11}H_{12}O_5$	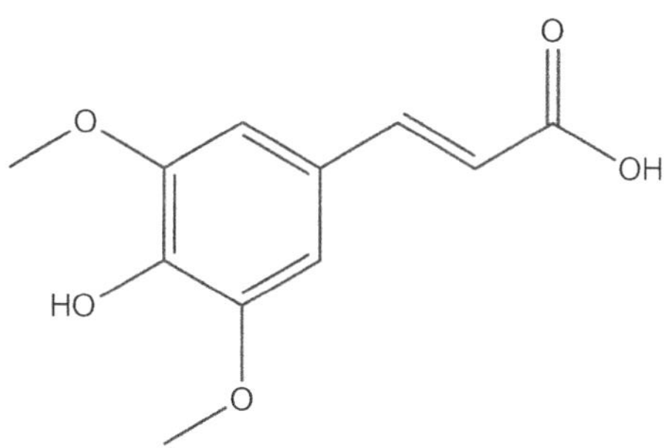	Chlorogenic $C_{16}H_{18}O_9$	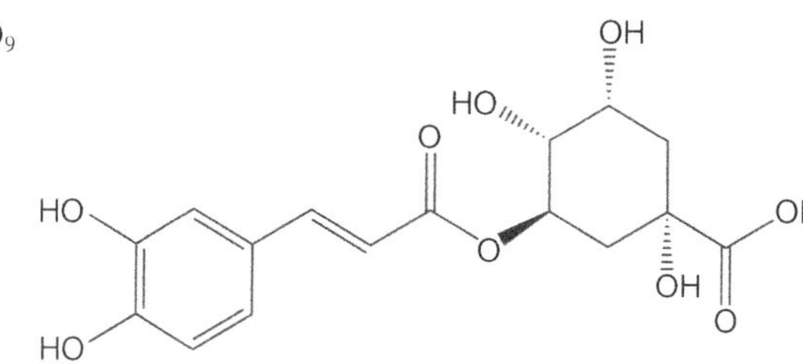

(*Continued*)

Table 8.7 (Continued)

Hydroxybenzoic acids		Hydroxycinnamic acids	
Chemical name and formula	**Chemical structure**	**Chemical name and formula**	**Chemical structure**
Vanillic $C_8H_8O_4$	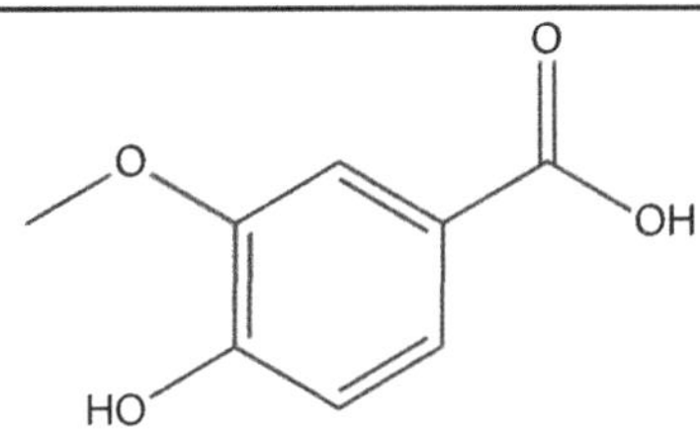	*p*-Coumaric $C_9H_8O_3$	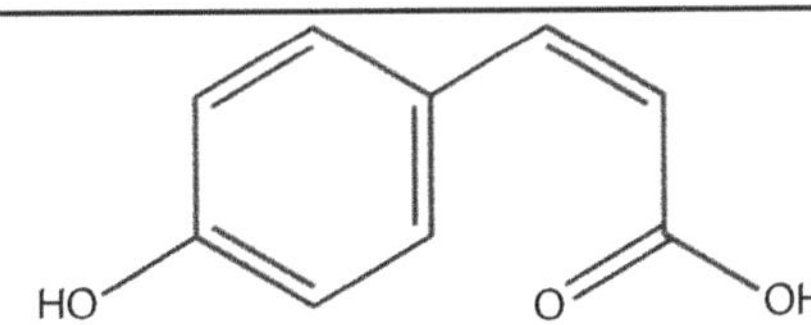
Ellagic $C_{14}H_6O_8$	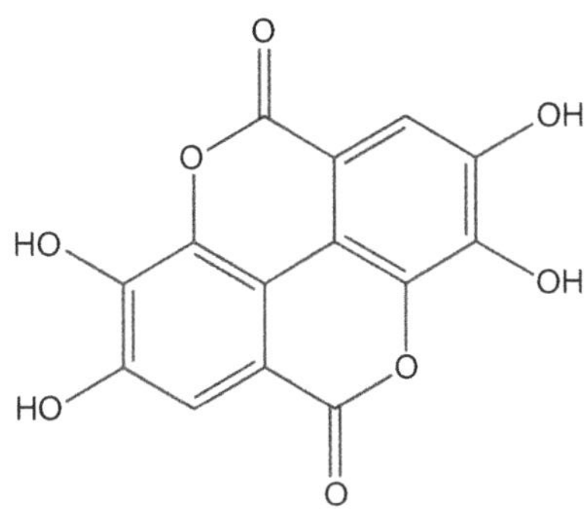	Ferulic $C_{10}H_{10}O_4$	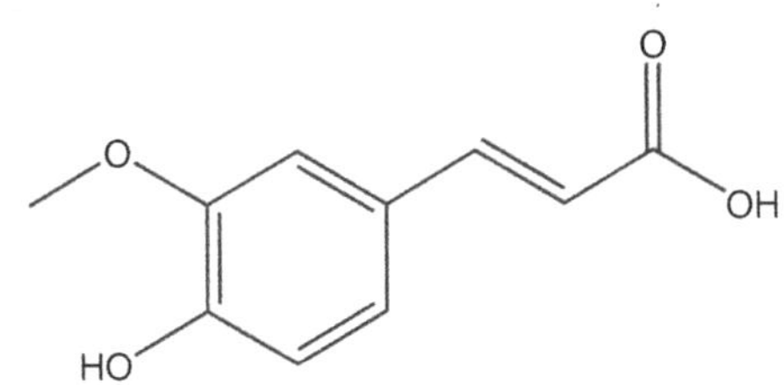

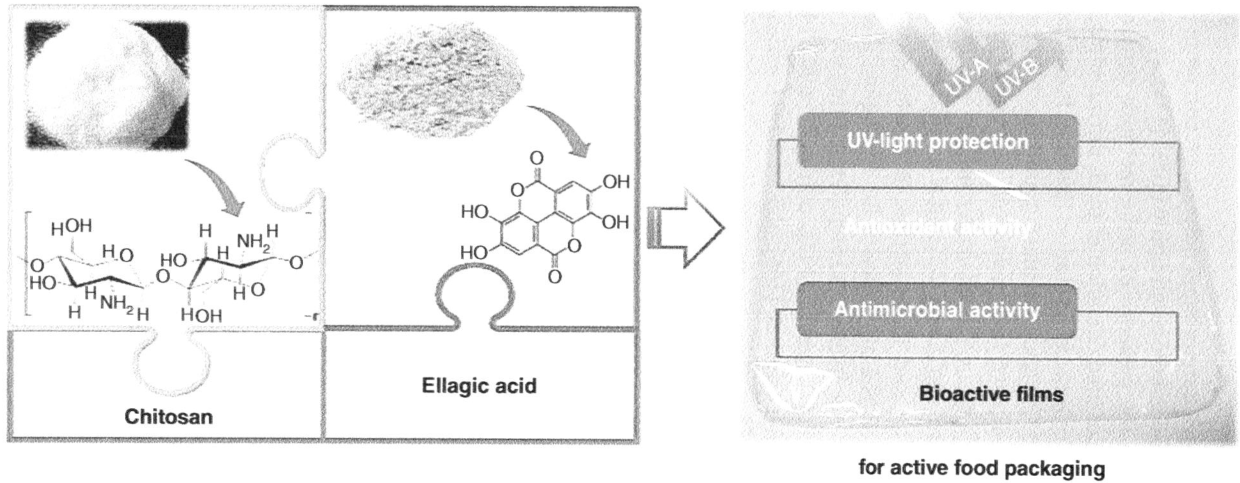

Figure 8.12 Bioactive films based on chitosan and ellagic acid. Source: Ref. [133]/Reproduced with permission from Elsevier.

8.4.2.3 Curcuminoids

Curcumin is the most studied curcuminoid isolated from *Curcuma longa* L. (Figure 8.13).

Turmeric contains 3–5% curcuminoid derivatives: curcumin (70%), demethoxy curcumin (20%), and bisdemethoxy curcumin (10%) [135]. Its antioxidant activity can be associated with the presence of ortho-methoxy curcumin groups [136], allowing for preventing of lipid oxidation of foods during storage. So far, curcumin's effect on different foods has been studied (biscuits, bread, pasta, cheese, milk, sausage, patties, etc.) [137]. It is known that curcumin is effective against *S. haemolyticus*, *Staphylococcus aureus*, *Escherichia coli*, *Proteus mirabilis*, *Streptococcus pyogenes*, *Enterococcus faecalis*, *Acinetobacter lwoffii*, and *Pseudomonas aeruginosa*. Moreover, curcumin can detect CO_2 and NH_3 production during food spoilage [136]. Since its color and its properties, it can be used as colorants and preservative [138] as well as a pH indicator for developing intelligent packaging systems. Curcumin has been loaded into biopolymer films such as cellulose [139], chitosan [140], starch [141], pectin [142], polylactic acid [143], gelatin [144], guar [145], carboxymethyl cellulose [146], alginate [147], and carrageenan [148].

8.4.2.4 Coumarins

Coumarin is a compound in cinnamon, tonka beans, celery, cloves, and apricots. Chemically, it contains an aromatic ring fused to a condensed lactone ring and shows interesting properties (i.e., antifungal, anticoagulant, antimicrobial, and antioxidant effects). Polymeric systems containing coumarin have already been studied and designed in various fields, such as biochemical, medicinal, and electrooptical materials [88]. Still, its use in active packaging has yet to be studied fully. Benbettaieb et al. produced chitosan and fish gelatin-based systems containing coumarin with antioxidant properties [89]. Some organic compounds containing coumarin moiety are known to exhibit biological activities (i.e., anticoagulant, anti-inflammatory, antibacterial, antifungal, antiviral, anticancer, and antioxidant properties) [92]. One of them, 7-hydroxy coumarins, a compound with vanilla flavor, has been loaded into PVA-starch film by solvent casting technique to fabricate an active packaging with good mechanical, barrier, and antioxidant properties [93].

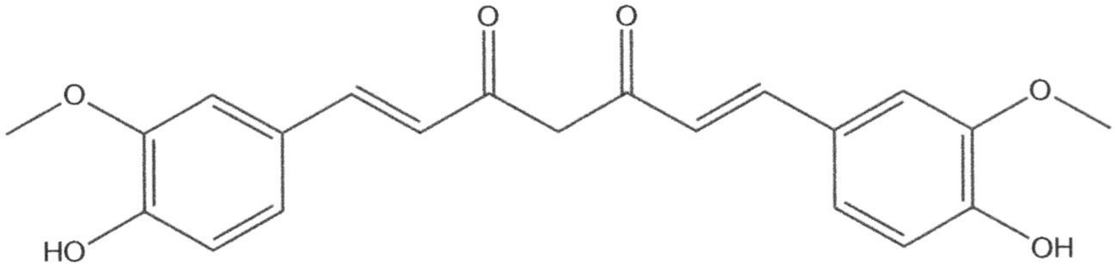

Figure 8.13 Structure of curcumin.

8.4.2.5 Lignans

Lignans constitute secondary plant metabolites with a complex phytoestrogen-related chemical composition. From a chemical point of view, they are diphenolic compounds obtained from the combination of two phenylpropanoid C6–C3 units at the β and β' carbon atoms. They can be linked to additional ether, lactone, or carbon bonds [18, 19]. They can be mainly derived from whole grains, oilseeds, legumes, and berries. For example, flaxseed lignans appear attractive for developing functional foods [149]. Their application in food science and nutrition could be related to their noticeable bioactivity. At the same time, their antioxidant capacities are due to their function as hydrogen donors and complex divalent transition metal cations [150].

8.4.2.6 Stilbenes

Stilbenes are complex chemical structures with biological activities. From a chemical point of view, the system is C6–C2–C6 with two benzene rings joined by a double bond. They can be found in almonds, grapes, pine, sorghum, peanuts, cranberries, mulberries, plum, and wine berries and are exploited for their bioactive components [36]. Among more than 400 natural stilbenes, trans-resveratrol (Figure 8.14) derived from grape skin is the most known and used [151]. Apart from being a natural phytoalexin, it showed many properties extensively studied.

Silva et al. produced an active film based on cellulose acetate and hydroxyethylcellulose while using resveratrol and its inclusion complex with hydroxypropyl-γ-cyclodextrin as active agents. The film showed the antimicrobial activity of these films against *Campylobacter jejuni*, *Campylobacter coli*, and *Arcobacter butzleri* [152]. Aminzare et al. produced active packaging based on carboxymethyl cellulose films adding resveratrol and eugenol as active compounds. The findings showed the antioxidant and antagonistic antibacterial effects of the produced films with a higher effect on Gram+ compared to Gram− [153].

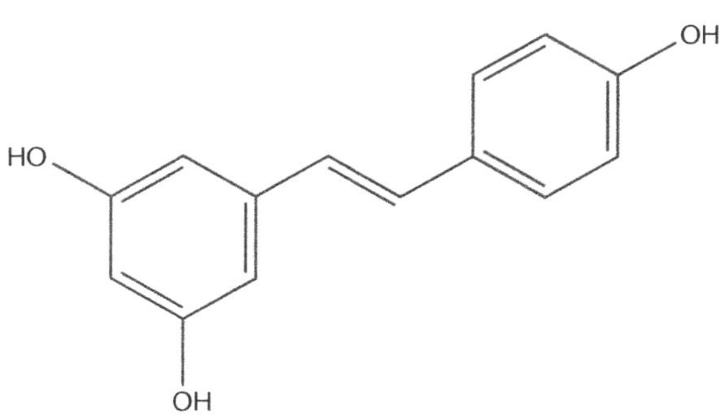

Figure 8.14 Chemical structure of trans-resveratrol.

8.4.2.7 Lignins

Lignin is a natural polyphenol with different properties (reinforcement, UV protection, antioxidation, antimicrobial) [154, 155]. Chemically, lignins show an aromatic, highly cross-linked structure with many functional groups. The antioxidant activity is ascribed to phenolic hydroxyl groups, aliphatic hydroxyl groups, and low molecular weight [156]. Yang et al. designed star-like lignin microparticles grafting onto PLA with a good UV barrier, low overall migration value, and antioxidation behavior [157]. Lignin NPs have been studied as filler for improving antioxidant and antibacterial properties of different polymeric films such as polyvinyl alcohol/chitosan [158, 159], glycidyl methacrylate-g-poly (lactic acid) [160], wheat gluten [161], and polylactic (acid) [162].

8.5 Action Mechanism

The action mechanism of polyphenols has already been investigated. It has been found that the mechanism of polyphenols' functions is based on single-electron transfer, free radical scavenging, hydrogen atom transfer, and metal chelation [163]. As far as oxidation concerns, it initiates with unsaturated fatty acid in the presence of a catalyst (copper, iron, light, heat). Oxidation starts with auto-oxidation and ends with termination through a chain reaction. The presence of free radicals can determine a nucleic acid breakdown, a double bond formation in unsaturated fatty acids as well as protein degradation. It follows that some compounds responsible for rancid aroma and flavors are formed, such as hydroperoxide, conjugated dienes, aldehyde, and ketone. So, active packaging with antioxidant properties is designed to inhibit lipid oxidation, preserving food shelf life and ensuring the quality of packaged food. The active polyphenolic compounds can diffuse towards the food surface, promoting different reactions inside the food. The pathway to lipid peroxidation is reported as Scheme 8.1 (*R* is a lipid alkyl group).

Initiation

$$RH \rightarrow R* + H*$$

Propagation

$$R* + O_2 \rightarrow ROO*$$

$$ROO* + RH \rightarrow ROOH + R*$$

Termination

$$ROO* + R* \rightarrow ROOR$$

$$R* + R* \rightarrow RR$$

Scheme 8.1 Pathway to lipid peroxidation.

$$ROO^* + AH \rightarrow ROOH + A^*$$

$$R^* + AH \rightarrow RH + A^*$$

$$RO^* + AH \rightarrow ROH + A^*$$

$$ROO^* + A^* \rightarrow ROOA$$

$$RO^* + A^* \rightarrow ROA$$

$$A^* + A^* \rightarrow A\text{–}A$$

Scheme 8.2 Antioxidation reactions with free radicals.

The presence of some factors (light, moisture, O_2, and metal ions) affects the initial radicalization. After that, autoxidation is verified with lipid radical oxidation or peroxide radicalization of high-lipid foods at the propagation stage. Finally, the termination stage occurs, and the lipid-free radicals may recombine with peroxide radicals or themselves. The presence of an antioxidant compound limits undesirable lipid oxidation reactions. The antioxidation reactions with free radicals (AH, antioxidant molecules) are reported as Scheme 8.2:

The antioxidant mechanism is based on scavenging reactive oxygen and nitrogen-free radical species [164]. Besides, polyphenolic compounds can exert even an antimicrobial action. Generally, the antimicrobial action is based on the lag phase extension and lowered live cell number of target bacteria. But the precise mechanism is yet to be precise. It is known that the phenolic group contains OH groups in their structures responsible for inhibitory activity since they can interact with the cell membrane of bacteria to disrupt its structure [165]. One mechanism of antimicrobial compounds regards the destabilization of the cytoplasmic membrane by —OH groups acting as proton exchangers and pH gradient lowering, determining cell death. Moreover, another explanation could refer to the size of the alkyl or alkenyl group since the alkyl groups associated C=C bond, with high electronegativity, interfere with electron transfer and react with protein and nucleic acid, determining an improvement of antimicrobial activity [166].

8.6 Limits and Future Predictions

Consumers are aware of the adverse effects of synthetic compounds on human health. If we want to discuss active packaging, the kind of sources of the active components are crucial since the consumers significantly consider important aspects such as safety, economic aspects, and viability. It is fundamental to the safety of the food products in contact with the active packaging. Despite the numerous advantages, efforts need to be done regarding the active food packaging applications. For example, the extraction of the active components is still greatly discussed. Although the efficacy of phenolic compounds has already been proved and appears promising, challenges concern the practical applications under different environmental conditions [167]. Regarding phenolic compounds' use, their stability, use at high temperatures, and interaction must still be addressed through adequate studies. So far, the main challenge involves preserving the physical properties of the packaging materials and the organoleptic properties of phenolic compounds able to affect the sensory qualities of packaged food. Phenolic compounds, which are abundant and play the role of natural additives, are becoming increasingly important. However, despite the benefits, the commercial use of edible films/coatings with polyphenolic compounds is quite limited.

8.7 Conclusions

Active packaging is attracting noticeable attention in the last decades as a promising solution for preserving foods and improving quality, safety, and stability by limiting chemical intervention in the food material. The main idea is to incorporate active compounds, which even leads to an assurance to the consumers. The natural active compounds contribute to the design of a sustainable approach to the food processing industry. So, the phenolic compounds, which are quite abundant and naturally available, can be considered alternatives to synthetic additives due to their antioxidant and antimicrobial characteristics. In conclusion, this chapter provides an overview of phenolic compounds' properties and applications by highlighting the properties of the different subgroups. A focus was placed on the potentiality of phenolic compounds to limit discoloration, oxidation, and microbiological degradation in food packaging. Source, features, properties, mechanisms of phenolic compounds have been discussed thoroughly.

References

1 Deshmukh, R.K. and Gaikwad, K.K. (2022). Natural antimicrobial and antioxidant compounds for active food packaging applications. *Biomass Conversion and Biorefinery* 2022 (1): 1–22. https://doi.org/10.1007/S13399-022-02623-W.

2 Ahmed, I., Lin, H., Zou, L. et al. (2017). A comprehensive review on the application of active packaging technologies to muscle foods. *Food Control* 82: 163–178. https://doi.org/10.1016/J.FOODCONT.2017.06.009.

3 Bhargava, N., Sharanagat, V.S., Mor, R.S., and Kumar, K. (2020). Active and intelligent biodegradable packaging films using food and food waste-derived bioactive compounds: a review. *Trends in Food Science and Technology 105*: 385–401. https://doi.org/10.1016/J.TIFS.2020.09.015.

4 Mustapha, F., Jai, J., Hamidon, F. et al. (2017). Antimicrobial agents from Malaysian plants and their potential use in food packaging material: review. *Chemical Engineering Research Bulletin 19*: 57–66. https://doi.org/10.3329/CERB.V19I0.33797.

5 Yong, H. and Liu, J. (2021). Active packaging films and edible coatings based on polyphenol-rich propolis extract: a review. *Comprehensive Reviews in Food Science and Food Safety* 20: 2106–2145. https://doi.org/10.1111/1541-4337.12697.

6 Negi, P.S. (2012). Plant extracts for the control of bacterial growth: efficacy, stability and safety issues for food application. *International Journal of Food Microbiology 156*: 7–17. https://doi.org/10.1016/J.IJFOODMICRO.2012.03.006.

7 Valdés, A., Mellinas, A.C., Ramos, M. et al. (2014). Natural additives and agricultural wastes in biopolymer formulations for food packaging. *Frontiers in Chemistry* 2: 6. https://doi.org/10.3389/FCHEM.2014.00006/BIBTEX.

8 Cano, A., Andres, M., Chiralt, A., and González-Martinez, C. (2020). Use of tannins to enhance the functional properties of protein based films. *Food Hydrocolloids* 100: 105443. https://doi.org/10.1016/J.FOODHYD.2019.105443.

9 Kuai, L., Liu, F., Chiou, B.S. et al. (2021). Controlled release of antioxidants from active food packaging: a review. *Food Hydrocolloids* 120: 106992. https://doi.org/10.1016/J.FOODHYD.2021.106992.

10 Umaraw, P., Munekata, P.E.S., Verma, A.K. et al. (2020). Edible films/coating with tailored properties for active packaging of meat, fish and derived products. *Trends in Food Science and Technology* 98: 10–24. https://doi.org/10.1016/J.TIFS.2020.01.032.

11 Holley, R.A. and Patel, D. (2005). Improvement in shelf-life and safety of perishable foods by plant essential oils and smoke antimicrobials. *Food Microbiology* 22: 273–292. https://doi.org/10.1016/J.FM.2004.08.006.

12 Lourenço, S.C., Moldão-Martins, M., and Alves, V.D. (2019). Antioxidants of natural plant origins: from sources to food industry applications. *Molecules* 24: https://doi.org/10.3390/MOLECULES24224132.

13 Gorrasi, G., Viscusi, G., Gerardi, C. et al. (2022). Physicochemical and antioxidant properties of white (Fiano Cv) and red (Negroamaro Cv) grape pomace skin based films. *Journal of Polymers and the Environment* 30: 3609–3621. https://doi.org/10.1007/S10924-022-02463-9.

14 Singh, A.K., Kim, J.Y., and Lee, Y.S. (2022). Phenolic compounds in active packaging and edible films/coatings: natural bioactive molecules and novel packaging ingredients. *Molecules* 27: 7513. https://doi.org/10.3390/MOLECULES27217513.

15 Salgado, P.R., Ortiz, C.M., Musso, Y.S. et al. (2015). Edible films and coatings containing bioactives. *Current Opinion in Food Science* 5: 86–92. https://doi.org/10.1016/J.COFS.2015.09.004.

16 Saltmarsh, M. (2020). Food additives and why they are used. *Saltmarsh's Essential Guide to Food Additives* 1–9. https://doi.org/10.1039/9781839161063-00001.

17 Albuquerque, B.R., Heleno, S.A., Oliveira, M.B.P.P. et al. (2021). Phenolic compounds: current industrial applications, limitations and future challenges. *Food & Function* 12: 14–29. https://doi.org/10.1039/D0FO02324H.

18 Renard, C.M.G.C., Watrelot, A.A., and Le Bourvellec, C. (2017). Interactions between polyphenols and polysaccharides: mechanisms and consequences in food processing and digestion. *Trends in Food Science and Technology* 60: 43–51. https://doi.org/10.1016/J.TIFS.2016.10.022.

19 Dobson, C.C., Mottawea, W., Rodrigue, A. et al. (2019). Impact of molecular interactions with phenolic compounds on food polysaccharides functionality. *Advances in Food and Nutrition Research* 90: 135–181. https://doi.org/10.1016/BS.AFNR.2019.02.010.

20 Missio, A.L., Mattos, B.D., Ferreira, D.d.F. et al. (2018). Nanocellulose-tannin films: from trees to sustainable active packaging. *Journal of Cleaner Production* 184: 143–151. https://doi.org/10.1016/J.JCLEPRO.2018.02.205.

21 Li, P., Sirviö, J.A., Haapala, A. et al. (2019). Anti-oxidative and UV-absorbing biohybrid film of cellulose nanofibrils and tannin extract. *Food Hydrocolloids* 92: 208–217. https://doi.org/10.1016/J.FOODHYD.2019.02.002.

22 Olejar, K.J., Ray, S., Ricci, A., and Kilmartin, P.A. (2014). Superior antioxidant polymer films created through the incorporation of grape tannins in ethyl cellulose. *Cellulose* 21: 4545–4556. https://doi.org/10.1007/S10570-014-0447-4/FIGURES/8.

23 Aydogdu, A., Sumnu, G., and Sahin, S. (2019). Fabrication of gallic acid loaded hydroxypropyl methylcellulose nanofibers by electrospinning technique as active packaging material. *Carbohydrate Polymers* 208: 241–250. https://doi.org/10.1016/J.CARBPOL.2018.12.065.

24 Ashrafi, A., Jokar, M., and Mohammadi Nafchi, A. (2018). Preparation and characterization of biocomposite film based on chitosan and kombucha tea as active food packaging. *International Journal of Biological Macromolecules* 108: 444–454. https://doi.org/10.1016/J.IJBIOMAC.2017.12.028.

25 Gaikwad, K.K., Lee, J.Y., and Lee, Y.S. (2016). Development of polyvinyl alcohol and apple pomace bio-composite film with antioxidant properties for active food packaging application. *Journal of Food Science and Technology* 53: 1608–1619. https://doi.org/10.1007/S13197-015-2104-9.

26 Ribeiro Sanches, M.A., Camelo-Silva, C., da Silva Carvalho, C. et al. (2021). Active packaging with starch, red cabbage extract and sweet whey: characterization and application in meat. *LWT - Food Science and Technology* 135: 110275. https://doi.org/10.1016/J.LWT.2020.110275.

27 Bermúdez-Oria, A., Rodríguez-Gutiérrez, G., Rubio-Senent, F. et al. (2019). Effect of edible pectin-fish gelatin films containing the olive antioxidants hydroxytyrosol and 3,4-dihydroxyphenylglycol on beef meat during refrigerated storage. *Meat Science 148*: 213–218. https://doi.org/10.1016/J.MEATSCI.2018.07.003.

28 Otoni, C.G., Avena-Bustillos, R.J., Azeredo, H.M.C. et al. (2017). Recent advances on edible films based on fruits and vegetables—a review. *Comprehensive Reviews in Food Science and Food Safety 16*: 1151–1169. https://doi.org/10.1111/1541-4337.12281.

29 Kumar, S. and Pandey, A.K. (2013). Chemistry and biological activities of flavonoids: an overview. *Scientific World Journal* 2013: https://doi.org/10.1155/2013/162750.

30 Kabera, J., Semana, E., Mussa, A., and He, X. (2014). Plant secondary metabolites: biosynthesis, classification, function and pharmacological properties. *Journal of Pharmacy and Pharmacology* 2 (7): 377–392.

31 Taherimehr, M., YousefniaPasha, H., Tabatabaeekoloor, R., and Pesaranhajiabbas, E. (2021). Trends and challenges of biopolymer-based nanocomposites in food packaging. *Comprehensive Reviews in Food Science and Food Safety 20*: 5321–5344. https://doi.org/10.1111/1541-4337.12832.

32 Manessis, G., Kalogianni, A.I., Lazou, T. et al. (2020). Plant-derived natural antioxidants in meat and meat products. *Antioxidants (Basel, Switzerland)* 9: 1–30. https://doi.org/10.3390/antiox9121215.

33 Pereira, D.M., Valentão, P., Pereira, J.A., and Andrade, P.B. (2009). Phenolics: from chemistry to biology. *Molecules* 14: 2202–2211. https://doi.org/10.3390/MOLECULES14062202.

34 Neri-Numa, I.A., Arruda, H.S., Geraldi, M.V. et al. (2020). Natural prebiotic carbohydrates, carotenoids and flavonoids as ingredients in food systems. *Current Opinion in Food Science 33*: 98–107. https://doi.org/10.1016/J.COFS.2020.03.004.

35 Brahmachari, G. (2008). Naturally occurring flavanones: an overview. *Natural Product Communications* 3: 1934578X0800300820.

36 Durazzo, A., Lucarini, M., Souto, E.B. et al. (2019). Polyphenols: a concise overview on the chemistry, occurrence, and human health. *Phytotherapy Research* 33: 2221–2243. https://doi.org/10.1002/PTR.6419.

37 Stabrauskiene, J., Kopustinskiene, D.M., Lazauskas, R., and Bernatoniene, J. (2022). Naringin and naringenin: their mechanisms of action and the potential anticancer activities. *Biomedicine* 10: https://doi.org/10.3390/biomedicines10071686.

38 Barreca, D., Gattuso, G., Bellocco, E. et al. (2017). Flavanones: citrus phytochemical with health-promoting properties. *BioFactors* 43: 495–506. https://doi.org/10.1002/BIOF.1363.

39 Liu, Z., Pan, Y., Li, X. et al. (2017). Chemical composition, antimicrobial and anti-quorum sensing activities of pummelo peel flavonoid extract. *Industrial Crops and Products* 109: 862–868. https://doi.org/10.1016/J.Indcrop.2017.09.054.

40 Huyut, Z., Beydemir, Ş., and Gülçin, I. (2017). Antioxidant and antiradical properties of selected flavonoids and phenolic compounds. *Biochemistry Research International* 2017: https://doi.org/10.1155/2017/7616791.

41 Alam, M.A., Subhan, N., Rahman, M.M. et al. (2014). Effect of citrus flavonoids, naringin and naringenin, on metabolic syndrome and their mechanisms of action. *Advances in Nutrition* 5: 404–417. https://doi.org/10.3945/AN.113.005603.

42 Guzman-Puyol, S., Hierrezuelo, J., Benítez, J.J. et al. (2022). Transparent, UV-blocking, and high barrier cellulose-based bioplastics with naringin as active food packaging materials. *International Journal of Biological Macromolecules* 209: 1985–1994. https://doi.org/10.1016/J.IJBIOMAC.2022.04.177.

43 Annapoorani, A., Umamageswaran, V., Parameswari, R. et al. (2012). Computational discovery of putative quorum sensing inhibitors against LasR and RhlR receptor proteins of *Pseudomonas aeruginosa*. *Journal of Computer-Aided Molecular Design 26*: 1067–1077. https://doi.org/10.1007/S10822-012-9599-1/FIGURES/6.

44 Li, T., Wang, D., Ren, L. et al. (2021). Preparation of PH-sensitive polylactic acid-naringin coaxial electrospun fiber membranes for maintaining and monitoring salmon freshness. *International Journal of Biological Macromolecules* 188: 708–718. https://doi.org/10.1016/J.IJBIOMAC.2021.08.087.

45 Iturriaga, L., Olabarrieta, I., Castellan, A. et al. (2014). Active naringin-chitosan films: impact of UV irradiation. *Carbohydrate Polymers* 110: 374–381. https://doi.org/10.1016/J.CARBPOL.2014.03.062.

46 Bataglion, G.A., Da Silva, F.M.A., Eberlin, M.N., and Koolen, H.H.F. (2015). Determination of the phenolic composition from Brazilian tropical fruits by UHPLC–MS/MS. *Food Chemistry* 180: 280–287. https://doi.org/10.1016/J.FOODCHEM.2015.02.059.

47 Yadav, S., Mehrotra, G.K., Bhartiya, P. et al. (2020). Preparation, physicochemical and biological evaluation of quercetin based chitosan-gelatin film for food packaging. *Carbohydrate Polymers* 227: 115348. https://doi.org/10.1016/J.CARBPOL.2019.115348.

48 Anand David, A.V., Arulmoli, R., and Parasuraman, S. (2016). Overviews of biological importance of quercetin: a bioactive flavonoid. *Pharmacognosy Reviews* 10: 84. https://doi.org/10.4103/0973-7847.194044.

49 Roy, S.; Ezati, P.; Khan, A.; Rhim, J.W. New opportunities and advances in quercetin-added functional packaging films for sustainable packaging applications: a mini-review. 2023, https://doi.org/10.1080/10408398.2023.2200553.

50 Lesjak, M., Beara, I., Simin, N. et al. (2018). Antioxidant and anti-inflammatory activities of quercetin and its derivatives. *Journal of Functional Foods* 40: 68–75. https://doi.org/10.1016/J.JFF.2017.10.047.

51 Wang, W., Sun, C., Mao, L. et al. (2016). The biological activities, chemical stability, metabolism and delivery systems of quercetin: a review. *Trends in Food Science and Technology* 56: 21–38. https://doi.org/10.1016/J.TIFS.2016.07.004.

52 Nguyen, T.L.A. and Bhattacharya, D. (2022). Antimicrobial activity of quercetin: an approach to its mechanistic principle. *Molecules* 27. https://doi.org/10.3390/molecules27082494.

53 Tongdeesoontorn, W., Mauer, L.J., Wongruong, S. et al. (2020). Physical and antioxidant properties of Cassava starch–carboxymethyl cellulose incorporated with quercetin and TBHQ as active food packaging. *Polymers* 12: 366. https://doi.org/10.3390/POLYM12020366.

54 Jakubowska, E., Gierszewska, M., Szydłowska-Czerniak, A. et al. (2023). Development and characterization of active packaging films based on chitosan, plasticizer, and quercetin for repassed oil storage. *Food Chemistry* 399: 133934. https://doi.org/10.1016/J.FOODCHEM.2022.133934.

55 Viscusi, G., Paolella, G., Lamberti, E. et al. (2023). Quercetin-loaded polycaprolactone-polyvinylpyrrolidone electrospun membranes for health application: design, characterization, modeling and cytotoxicity studies. *Membranes* 13: 242. https://doi.org/10.3390/membranes13020242.

56 Luzi, F., Pannucci, E., Santi, L. et al. (2019). Gallic acid and quercetin as intelligent and active ingredients in poly(vinyl alcohol) films for food packaging. *Polymers* 11: 1999. https://doi.org/10.3390/polym11121999.

57 Łopusiewicz, Ł., Zdanowicz, M., Macieja, S. et al. (2021, 13). Development and characterization of bioactive poly(butylene-succinate) films modified with quercetin for food packaging applications. *Polymers* 1798. https://doi.org/10.3390/polym13111798.

58 Li, S., Yan, Y., Guan, X., and Huang, K. (2020). Preparation of a hordein-quercetin-chitosan antioxidant electrospun nanofibre film for food packaging and improvement of the film hydrophobic properties by heat treatment. *Food Packaging and Shelf Life* 23: 100466. https://doi.org/10.1016/J.FPSL.2020.100466.

59 Souza, M.P., Vaz, A.F.M., Silva, H.D. et al. (2015). Development and characterization of an active chitosan-based film containing quercetin. *Food and Bioprocess Technology* 8: 2183–2191. https://doi.org/10.1007/S11947-015-1580-2/TABLES/4.

60 Chen, X., Lee, D.S., Zhu, X., and Yam, K.L. (2012). Release kinetics of tocopherol and quercetin from binary antioxidant controlled-release packaging films. *Journal of Agricultural and Food Chemistry* 60: 3492–3497. https://doi.org/10.1021/jf2045813/asset/images/large/jf-2011-045813_0004.jpeg.

61 Ezati, P. and Rhim, J.W. (2021). Fabrication of quercetin-loaded biopolymer films as functional packaging materials. *ACS Applied Polymer Materials* 3: 2131–2137. https://doi.org/10.1021/acsapm.1c00177/suppl_file/ap1c00177_si_001.pdf.

62 Lou, D., Tong, L., Kang, X. et al. (2021). Preparation and characterization of kafirin-quercetin film for packaging cod during cold storage. *Journal of Texture Studies* 52: 71–80. https://doi.org/10.1111/JTXS.12560.

63 Yang, F., Chen, G., Li, J. et al. (2022). Effects of quercetin and organically modified montmorillonite on the properties of poly(butylene adipate-co-terephthalate)/thermoplastic starch active packaging films. *ACS Omega* 8: 663–672. https://doi.org/10.1021/acsomega.2c05836/asset/images/large/ao2c05836_0008.jpeg.

64 Bai, R., Zhang, X., Yong, H. et al. (2019). Development and characterization of antioxidant active packaging and intelligent Al^{3+}-sensing films based on carboxymethyl chitosan and quercetin. *International Journal of Biological Macromolecules* 126: 1074–1084. https://doi.org/10.1016/J.IJBIOMAC.2018.12.264.

65 Zhang, N., Bi, F., Xu, F. et al. (2020). Structure and functional properties of active packaging films prepared by incorporating different flavonols into chitosan based matrix. *International Journal of Biological Macromolecules* 165: 625–634. https://doi.org/10.1016/j.ijbiomac.2020.09.209.

66 Bernatoniene, J. and Kopustinskiene, D.M. (2018). The role of catechins in cellular responses to oxidative stress. *Molecules* 23: 965. https://doi.org/10.3390/molecules23040965.

67 Opez De Dicastillo, C.L., Ner, C., Alfaro, P. et al. (2011). Development of new antioxidant active packaging films based on ethylene vinyl alcohol copolymer (EVOH) and green tea extract. *Journal of Agricultural and Food Chemistry* 59: 7832–7840. https://doi.org/10.1021/jf201246g.

68 Turkmen, N., Velioglu, Y.S., Sari, F., and Polat, G. (2007). Effect of extraction conditions on measured total polyphenol contents and antioxidant and antibacterial activities of black tea. *Molecules* 12: 484–496. https://doi.org/10.3390/12030484.

69 Khan, N. and Mukhtar, H. (2007). Tea polyphenols for health promotion. *Life Sciences* 81: 519–533. https://doi.org/10.1016/J.LFS.2007.06.011.

70 Siripatrawan, U. and Noipha, S. (2012). Active film from chitosan incorporating green tea extract for shelf life extension of pork sausages. *Food Hydrocolloids* 27: 102–108. https://doi.org/10.1016/J.FOODHYD.2011.08.011.

71 Nirmal, N.P. and Benjakul, S. (2011). Retardation of quality changes of pacific white shrimp by green tea extract treatment and modified atmosphere packaging during refrigerated storage. *International Journal of Food Microbiology* 149: 247–253. https://doi.org/10.1016/j.ijfoodmicro.2011.07.002.

72 López-De-Dicastillo, C., Gómez-Estaca, J., Catalá, R. et al. (2012). Active antioxidant packaging films: development and effect on lipid stability of brined sardines. *Food Chemistry* 131: 1376–1384. https://doi.org/10.1016/j.foodchem.2011.10.002.

73 Peng, Y., Wu, Y., and Li, Y. (2013). Development of tea extracts and chitosan composite films for active packaging materials. *International Journal of Biological Macromolecules* 59: 282–289. https://doi.org/10.1016/J.IJBIOMAC.2013.04.019.

74 Siripatrawan, U. and Harte, B.R. (2010). Physical properties and antioxidant activity of an active film from chitosan incorporated with green tea extract. *Food Hydrocolloids* 24: 770–775. https://doi.org/10.1016/j.foodhyd.2010.04.003.

75 Li, J.H., Miao, J., Wu, J.L. et al. (2014). Preparation and characterization of active gelatin-based films incorporated with natural antioxidants. *Food Hydrocolloids* 37: 166–173. https://doi.org/10.1016/j.foodhyd.2013.10.015.

76 Ferreira, A.S., Nunes, C., Castro, A. et al. (2014). Influence of grape pomace extract incorporation on chitosan films properties. *Carbohydrate Polymers 113*: 490–499. https://doi.org/10.1016/j.carbpol.2014.07.032.

77 Wang, L., Dong, Y., Men, H. et al. (2013). Preparation and characterization of active films based on chitosan incorporated tea polyphenols. *Food Hydrocolloids* 32: 35–41. https://doi.org/10.1016/j.foodhyd.2012.11.034.

78 Nie, X., Gong, Y., Wang, N., and Meng, X. (2015). Preparation and characterization of edible myofibrillar protein-based film incorporated with grape seed procyanidins and green tea polyphenol. *LWT-Food Science and Technology* 64: 1042–1046. https://doi.org/10.1016/J.LWT.2015.07.006.

79 Wu, J., Chen, S., Ge, S. et al. (2013). Preparation, properties and antioxidant activity of an active film from silver carp (hypophthalmichthys molitrix) skin gelatin incorporated with green tea extract. *Food Hydrocolloids* 32: 42–51. https://doi.org/10.1016/j.foodhyd.2012.11.029.

80 López de Lacey, A.M., Giménez, B., Pérez-Santín, E. et al. (2012). Bioaccessibility of green tea polyphenols incorporated into an edible agar film during simulated human digestion. *Food Research International* 48: 462–469. https://doi.org/10.1016/J.FOODRES.2012.04.024.

81 Helal, A., Tagliazucchi, D., Conte, A., and Desobry, S. (2012). Antioxidant properties of polyphenols incorporated in casein/sodium caseinate films. *International Dairy Journal* 25: 10–15. https://doi.org/10.1016/j.idairyj.2011.12.002.

82 Arcan, I. and Yemenicioğlu, A. (2011). Incorporating phenolic compounds opens a new perspective to use zein films as flexible bioactive packaging materials. *Food Research International* 44: 550–556. https://doi.org/10.1016/j.foodres.2010.11.034.

83 von Staszewski, M., Jagus, R.J., and Pilosof, A.M.R. (2011). Influence of green tea polyphenols on the colloidal stability and gelation of WPC. *Food Hydrocolloids* 25: 1077–1084. https://doi.org/10.1016/J.foodhyd.2010.10.004.

84 Tang, D.W., Yu, S.H., Ho, Y.C. et al. (2013). Characterization of tea catechins-loaded nanoparticles prepared from chitosan and an edible polypeptide. *Food Hydrocolloids* 30: 33–41. https://doi.org/10.1016/j.foodhyd.2012.04.014.

85 Chen, Y., Tsai, G., Tang, D., and Peng, Y. (2010). Novel technology for the preparation of self-assembled catechin/gelatin nanoparticles and their characterization. *Journal of Agricultural and Food Chemistry* 58: 6728–6734. https://doi.org/10.1021/jf1005116.

86 Castañeda-Ovando, A., Pacheco-Hernández, M.d.L., Páez-Hernández, M.E. et al. (2009). Chemical studies of anthocyanins: a review. *Food Chemistry* 113: 859–871. https://doi.org/10.1016/j.foodchem.2008.09.001.

87 Yong, H. and Liu, J. (2020). Recent advances in the preparation, physical and functional properties, and applications of anthocyanins-based active and intelligent packaging films. *Food Packaging and Shelf Life* 26: 100550. https://doi.org/10.1016/j.fpsl.2020.100550.

88 Rajauria, G. and Tiwari, B. (2017). *Fruit Juices: Extraction, Composition, Quality and Analysis*. USA: Academic Press.

89 Saliu, F. and Della Pergola, R. (2018). Carbon dioxide colorimetric indicators for food packaging application: applicability of anthocyanin and poly-lysine mixtures. *Sensors and Actuators B: Chemical* 258: 1117–1124. https://doi.org/10.1016/j.snb.2017.12.007.

90 Neves, D., Andrade, P.B., Videira, R.A. et al. (2022). Berry anthocyanin-based films in smart food packaging: a mini-review. *Food Hydrocolloids* 133: 107885. https://doi.org/10.1016/J.foodhyd.2022.107885.

91 Abedi-Firoozjah, R., Yousefi, S., Heydari, M. et al. (2022). Application of red cabbage anthocyanins as PH-sensitive pigments in smart food packaging and sensors. *Polymers* 14: 1629. https://doi.org/10.3390/polym14081629.

92 Sousa, D., Basílio, N., Oliveira, J. et al. (2022). A new insight into the degradation of anthocyanins: reversible versus the irreversible chemical processes. *Journal of Agricultural and Food Chemistry* 70: 656–668. https://doi.org/10.1021/acs.jafc.1c06521/asset/images/large/jf1c06521_0014.jpeg.

93 Viscusi, G., Lamberti, E., Gerardi, C. et al. (2022). Encapsulation of grape (*Vitis vinifera* L.) Pomace polyphenols in soybean extract-based hydrogel beads as carriers of polyphenols and PH-monitoring devices. *Gels* 8: 734. https://doi.org/10.3390/GELS8110734.

94 Pereira, V.A., de Arruda, I.N.Q., and Stefani, R. (2015). Active chitosan/PVA films with anthocyanins from *Brassica oleracea* (red cabbage) as time–temperature indicators for application in intelligent food packaging. *Food Hydrocolloids 43*: 180–188. https://doi.org/10.1016/j.foodhyd.2014.05.014.

95 Ebrahimi Tirtashi, F., Moradi, M., Tajik, H. et al. (2019). Cellulose/chitosan pH-responsive indicator incorporated with carrot anthocyanins for intelligent food packaging. *International Journal of Biological Macromolecules* 136: 920–926. https://doi.org/10.1016/J.ijbiomac.2019.06.148.

96 Pourjavaher, S., Almasi, H., Meshkini, S. et al. (2017). Development of a colorimetric pH indicator based on bacterial cellulose nanofibers and red cabbage (Brassica oleracea) extract. *Carbohydrate Polymers* 156: 193–201. https://doi.org/10.1016/j.carbpol.2016.09.027.

97 Wei, Y.C., Cheng, C.H., Ho, Y.C. et al. (2017). Active gellan gum/purple sweet potato composite films capable of monitoring pH variations. *Food Hydrocolloids* 69: 491–502. https://doi.org/10.1016/j.foodhyd.2017.03.010.

98 Mustafa, P., Niazi, M.B.K., Jahan, Z. et al. (2020). PVA/starch/propolis/anthocyanins rosemary extract composite films as active and intelligent food packaging materials. *Journal of Food Safety* 40: e12725. https://doi.org/10.1111/JFS.12725.

99 Sani, M.A., Tavassoli, M., Hamishehkar, H., and McClements, D.J. (2021). Carbohydrate-based films containing pH-sensitive red barberry anthocyanins: application as biodegradable smart food packaging materials. *Carbohydrate Polymers* 255: 117488. https://doi.org/10.1016/j.carbpol.2020.117488.

100 Wang, X., Yong, H., Gao, L. et al. (2019). Preparation and characterization of antioxidant and pH-sensitive films based on chitosan and black soybean seed coat extract. *Food Hydrocolloids* 89: 56–66. https://doi.org/10.1016/j.foodhyd.2018.10.019.

101 Zhang, J., Zou, X., Zhai, X. et al. (2019). Preparation of an intelligent pH film based on biodegradable polymers and Roselle anthocyanins for monitoring pork freshness. *Food Chemistry* 272: 306–312. https://doi.org/10.1016/j.foodchem.2018.08.041.

102 Choi, I., Lee, J.Y., Lacroix, M., and Han, J. (2017). Intelligent pH indicator film composed of agar/potato starch and anthocyanin extracts from purple sweet potato. *Food Chemistry* 218: 122–128. https://doi.org/10.1016/j.foodchem.2016.09.050.

103 Akhtar, M.J., Jacquot, M., and Desobry, S. (2014). Effect of HPMC-anthocyanin packaging color and oxygen permeability on salmon oil preservation. *Food and Bioprocess Technology* 7: 93–104. https://doi.org/10.1007/s11947-013-1075-y/figures/7.

104 Yun, D., Cai, H., Liu, Y. et al. (2019). Development of active and intelligent films based on Cassava starch and Chinese bayberry (*Myrica rubra* Sieb. et Zucc.) anthocyanins. *RSC Advances* 9: 30905–30916. https://doi.org/10.1039/C9RA06628D.

105 Nogueira, G.F., Fakhouri, F.M., Velasco, J.I., and de Oliveira, R.A. (2019). Active edible films based on arrowroot starch with microparticles of blackberry pulp obtained by freeze-drying for food packaging. *Polymers* 11: 1382. https://doi.org/10.3390/POLYM11091382.

106 Moradi, M., Tajik, H., Almasi, H. et al. (2019). A novel pH-sensing indicator based on bacterial cellulose nanofibers and black carrot anthocyanins for monitoring fish freshness. *Carbohydrate Polymers* 222: 115030. https://doi.org/10.1016/J.CARBPOL.2019.115030.

107 Ge, Y., Li, Y., Bai, Y. et al. (2020). Intelligent gelatin/oxidized chitin nanocrystals nanocomposite films containing black rice bran anthocyanins for fish freshness monitorings. *International Journal of Biological Macromolecules* 155: 1296–1306. https://doi.org/10.1016/J.IJBIOMAC.2019.11.101.

108 Jamróz, E., Kulawik, P., Guzik, P., and Duda, I. (2019). The verification of intelligent properties of Furcellaran films with plant extracts on the stored fresh Atlantic Mackerel during storage at 2 °C. *Food Hydrocolloids* 97: 105211. https://doi.org/10.1016/j.foodhyd.2019.105211.

109 Kurek, M., Hlupić, L., Ščetar, M. et al. (2019). Comparison of two pH responsive color changing bio-based films containing wasted fruit pomace as a source of colorants. *Journal of Food Science 84*: 2490–2498. https://doi.org/10.1111/1750-3841.14716.

110 Wang, S., Xia, P., Wang, S. et al. (2019). Packaging films formulated with gelatin and anthocyanins nanocomplexes: physical properties, antioxidant activity and its application for olive oil protection. *Food Hydrocolloids* 96: 617–624. https://doi.org/10.1016/J.FOODHYD.2019.06.004.

111 Golasz, L.B., da Silva, J., and da Silva, S.B. (2013). Film with anthocyanins as an indicator of chilled pork deterioration. *Food Science and Technology* 33: 155–162. https://doi.org/10.1590/S0101-20612013000500023.

112 Saurabh, C.K., Gupta, S., and Variyar, P.S. (2018). Development of guar gum based active packaging films using grape pomace. *Journal of Food Science and Technology* 556: 55. 1982–1992, https://doi.org/10.1007/S13197-018-3112-3.

113 Chayavanich, K., Thiraphibundet, P., and Imyim, A. (2020). Biocompatible film sensors containing red radish extract for meat spoilage observation. *Spectrochimica Acta Part A Molecular and Biomolecular Spectroscopy 226*: 117601. https://doi.org/10.1016/J.SAA.2019.117601.

114 Ma, Q., Ren, Y., Gu, Z., and Wang, L. (2017). Developing an intelligent film containing *Vitis amurensis* Husk extracts: the effects of pH value of the film-forming solution. *Journal of Cleaner Production* 166: 851–859. https://doi.org/10.1016/J.JCLEPRO.2017.08.099.

115 Siriamornpun, S. and Kaewseejan, N. (2017). Quality, bioactive compounds and antioxidant capacity of selected climacteric fruits with relation to their maturity. *Science Horticulture (Amsterdam) 221*: 33–42. https://doi.org/10.1016/J.SCIENTA.2017.04.020.

116 Bi, F., Qin, Y., Chen, D. et al. (2021). Development of active packaging films based on chitosan and nano-encapsulated luteolin. *International Journal of Biological Macromolecules 182*: 545–553. https://doi.org/10.1016/J.IJBIOMAC.2021.04.063.

117 Mark, R., Lyu, X., Lee, J.J.L. et al. (2019). Sustainable production of natural phenolics for functional food applications. *Journal of Functional Foods 57*: 233–254. https://doi.org/10.1016/J.JFF.2019.04.008.

118 Huang, X., Ji, Y., Guo, L. et al. (2022). Incorporating tannin onto regenerated cellulose film towards sustainable active packaging. *Industrial Crops and Products* 180: 114710. https://doi.org/10.1016/J.INDCROP.2022.114710.

119 Yu, S.H., Tsai, M.L., Lin, B.X. et al. (2015). Tea catechins-cross-linked methylcellulose active films for inhibition of light irradiation and lipid peroxidation induced β-carotene degradation. *Food Hydrocolloids 44*: 491–505. https://doi.org/10.1016/J.FOODHYD.2014.10.022.

120 Cano, A., Contreras, C., Chiralt, A., and González-Martínez, C. (2021). Using tannins as active compounds to develop antioxidant and antimicrobial chitosan and cellulose based films. *Carbohydrate Polymer Technologies and Applications* 2: 100156. https://doi.org/10.1016/J.CARPTA.2021.100156.

121 Kriechbaum, K. and Bergström, L. (2020). Antioxidant and UV-blocking leather-inspired nanocellulose-based films with high wet strength. *Biomacromolecules* 21: 1720–1728. https://doi.org/10.1021/acs.biomac.9b01655/asset/images/large/bm9b01655_0006.jpeg.

122 Arora, D.S. and Mahajan, H. (2019). Major phytoconstituents of *Prunus cerasoides* responsible for antimicrobial and antibiofilm potential against some reference strains of pathogenic bacteria and clinical isolates of MRSA. *Applied Biochemistry and Biotechnology* 188: 1185–1204. https://doi.org/10.1007/S12010-019-02985-4/FIGURES/5.

123 Burman, S., Bhattacharya, K., Mukherjee, D., and Chandra, G. (2018). Antibacterial efficacy of leaf extracts of Combretum album Pers. against some pathogenic bacteria. *BMC Complementary and Alternative Medicine* 18: 1–8. https://doi.org/10.1186/S12906-018-2271-0/FIGURES/1.

124 Sakyiamah, M.M., Gordon, P.K., Bolah, P. et al. (2023). Assessment of the phytochemical composition and antimicrobial properties of *Tapinanthus bangwensis* leaves hosted by the branches of *Persea americana*. *BMC Complementary Medicine and Therapies* 23: 34. https://doi.org/10.1186/s12906-023-03860-w/figures/3.

125 Surekha, C., Srikanth, R., Thupurani, M.K. et al. (2022). Antimicrobial activities of Salacia oblonga wall leaf and root extracts against different bacterial strains and fungal isolates. *Current Microbiology* 79: 1–10. https://doi.org/10.1007/S00284-022-02888-4/TABLES/5.

126 Teanpaisan, R., Kawsud, P., Pahumunto, N., and Puripattanavong, J. (2017). Screening for antibacterial and antibiofilm activity in thai medicinal plant extracts against oral microorganisms. *Journal of Traditional and Complementary Medicine* 7: 172–177. https://doi.org/10.1016/J.JTCME.2016.06.007.

127 Ogawa, S. and Yazaki, Y. (2018). Tannins from Acacia mearnsii De Wild. bark: tannin determination and biological activities. *Molecules* 23: 837. https://doi.org/10.3390/molecules23040837.

128 Aloui, H., Deshmukh, A.R., Khomlaem, C., and Kim, B.S. (2021). Novel composite films based on sodium alginate and gallnut extract with enhanced antioxidant, antimicrobial barrier and mechanical properties. *Food Hydrocolloids* 113: 106508. https://doi.org/10.1016/j.foodhyd.2020.106508.

129 Santana-Gálvez, J., Cisneros-Zevallos, L., and Jacobo-Velázquez, D.A. (2017). Chlorogenic acid: recent advances on its dual role as a food additive and a nutraceutical against metabolic syndrome. *Molecules* 22: 358. https://doi.org/10.3390/molecules22030358.

130 Stojković, D., Petrović, J., Soković, M. et al. (2013). In situ antioxidant and antimicrobial activities of naturally occurring caffeic acid, p-coumaric acid and rutin, using food systems. *Journal of the Science of Food and Agriculture 93*: 3205–3208. https://doi.org/10.1002/JSFA.6156.

131 Shahidi, F. and Ambigaipalan, P. (2015). Phenolics and polyphenolics in foods, beverages and spices: antioxidant activity and health effects—a review. *Journal of Functional Foods* 18: 820–897. https://doi.org/10.1016/J.JFF.2015.06.018.

132 Cao, Y., Warner, R.D., and Fang, Z. (2019). Effect of chitosan/nisin/gallic acid coating on preservation of pork loin in high oxygen modified atmosphere packaging. *Food Control* 101: 9–16. https://doi.org/10.1016/J.FOODCONT.2019.02.013.

133 Vilela, C., Pinto, R.J.B., Coelho, J. et al. (2017). Bioactive chitosan/ellagic acid films with UV-light protection for active food packaging. *Food Hydrocolloids* 73: 120–128. https://doi.org/10.1016/J.FOODHYD.2017.06.037.

134 Neelam, Khatkar, A., and Sharma, K.K. (2019). Phenylpropanoids and its derivatives: biological activities and its role in food, pharmaceutical and cosmetic industries. 60: 2655–2675. https://doi.org/10.1080/10408398.2019.1653822.

135 Sabet, S., Rashidinejad, A., Melton, L.D., and McGillivray, D.J. (2021). Recent advances to improve curcumin oral bioavailability. *Trends in Food Science and Technology* 110: 253–266. https://doi.org/10.1016/J.TIFS.2021.02.006.

136 Aliabbasi, N., Fathi, M., and Emam-Djomeh, Z. (2021). Curcumin: a promising bioactive agent for application in food packaging systems. *Journal of Environmental Chemical Engineering* 9: 105520. https://doi.org/10.1016/J.JECE.2021.105520.

137 Munekata, P.E.S., Pateiro, M., Zhang, W. et al. (2021). Health benefits, extraction and development of functional foods with curcuminoids. *Journal of Functional Foods 79*: 104392. https://doi.org/10.1016/J.JFF.2021.104392.

138 Ma, Q., Du, L., and Wang, L. (2017). Tara gum/polyvinyl alcohol-based colorimetric NH_3 indicator films incorporating curcumin for intelligent packaging. *Sensors and Actuators B: Chemical* 244: 759–766. https://doi.org/10.1016/J.SNB.2017.01.035.

139 Luo, N., Varaprasad, K., Reddy, G.V.S. et al. (2012). Preparation and characterization of cellulose/ curcumin composite films. *RSC Advances* 2: 8483–8488. https://doi.org/10.1039/C2RA21465B.

140 Wu, C., Zhu, Y., Wu, T. et al. (2019). Enhanced functional properties of biopolymer film incorporated with curcurmin-loaded mesoporous silica nanoparticles for food packaging. *Food Chemistry 288*: 139–145. https://doi.org/10.1016/J.FOODCHEM.2019.03.010.

141 Liu, D., Dang, S., Zhang, L. et al. (2021). Corn starch/polyvinyl alcohol based films incorporated with curcumin-loaded pickering emulsion for application in intelligent packaging. *International Journal of Biological Macromolecules* 188: 974–982. https://doi.org/10.1016/J.IJBIOMAC.2021.08.080.

142 Ezati, P. and Rhim, J.W. (2020). pH-responsive pectin-based multifunctional films incorporated with curcumin and sulfur nanoparticles. *Carbohydrate Polymers* 230: 115638. https://doi.org/10.1016/J.CARBPOL.2019.115638.

143 Roy, S. and Rhim, J.W. (2020). Preparation of bioactive functional poly(lactic acid)/curcumin composite film for food packaging application. *International Journal of Biological Macromolecules* 162: 1780–1789. https://doi.org/10.1016/J.IJBIOMAC.2020.08.094.

144 Etxabide, A., Uranga, J., Guerrero, P., and de la Caba, K. (2017). Development of active gelatin films by means of valorisation of food processing waste: a review. *Food Hydrocolloids 68*: 192–198. https://doi.org/10.1016/J.FOODHYD.2016.08.021.

145 Aydogdu, A., Radke, C.J., Bezci, S., and Kirtil, E. (2020). Characterization of curcumin incorporated guar gum/orange oil antimicrobial emulsion films. *International Journal of Biological Macromolecules 148*: 110–120. https://doi.org/10.1016/J.IJBIOMAC.2019.12.255.

146 Roy, S. and Rhim, J.W. (2020). Carboxymethyl cellulose-based antioxidant and antimicrobial active packaging film incorporated with curcumin and zinc oxide. *International Journal of Biological Macromolecules 148*: 666–676. https://doi.org/10.1016/J.IJBIOMAC.2020.01.204.

147 Bojorges, H., Ríos-Corripio, M.A., Hernández-Cázares, A.S. et al. (2020). Effect of the application of an edible film with turmeric (*Curcuma longa* L.) on the oxidative stability of meat. *Food Science & Nutrition 8*: 4308–4319. https://doi.org/10.1002/FSN3.1728.

148 Lai, D., Zhou, F., Zhou, A. et al. (2022). Comprehensive properties of photodynamic antibacterial film based on κ-carrageenan and curcumin-β-cyclodextrin complex. *Carbohydrate Polymers 282*: 119112. https://doi.org/10.1016/J.CARBPOL.2022.119112.

149 Ho, C.H.L., Cacace, J.E., and Mazza, G. (2007). Extraction of lignans, proteins and carbohydrates from flaxseed meal with pressurized low polarity water. *LWT- Food Science and Technology 40*: 1637–1647. https://doi.org/10.1016/J.LWT.2006.12.003.

150 Touré, A. and Xueming, X. (2010). Flaxseed lignans: source, biosynthesis, metabolism, antioxidant activity, bio-active components, and health benefits. *Comprehensive Reviews in Food Science and Food Safety 9*: 261–269. https://doi.org/10.1111/J.1541-4337.2009.00105.X.

151 Fabbrini, M., D'amico, F., Barone, M. et al. (2022). Polyphenol and tannin nutraceuticals and their metabolites: how the human gut microbiota influences their properties. *Biomolecules 12*: https://doi.org/10.3390/BIOM12070875.

152 Silva, Â., Duarte, A., Sousa, S. et al. (2016). Characterization and antimicrobial activity of cellulose derivatives films incorporated with a resveratrol inclusion complex. *LWT 73*: 481–489. https://doi.org/10.1016/J.LWT.2016.06.043.

153 Aminzare, M., Moniri, R., Hassanzad Azar, H., and Mehrasbi, M.R. (2022). Evaluation of antioxidant and antibacterial interactions between resveratrol and eugenol in carboxymethyl cellulose biodegradable film. *Food Science & Nutrition 10*: 155–168. https://doi.org/10.1002/FSN3.2656.

154 Yang, W., Fortunati, E., Gao, D. et al. (2018). Valorization of acid isolated high yield lignin nanoparticles as innovative antioxidant/antimicrobial organic materials. *ACS Sustainable Chemistry & Engineering 6*: 3502–3514. https://doi.org/10.1021/ACSSUSCHEMENG.7B03782/ASSET/IMAGES/LARGE/SC-2017-037827_0007.JPEG.

155 Zadeh, E.M., O'Keefe, S.F., and Kim, Y.T. (2018). Utilization of lignin in biopolymeric packaging films. *ACS Omega 3*: 7388. https://doi.org/10.1021/ACSOMEGA.7B01341.

156 Vinardell, M.P., Ugartondo, V., and Mitjans, M. (2008). Potential applications of antioxidant lignins from different sources. *Industrial Crops and Products 27*: 220–223. https://doi.org/10.1016/J.INDCROP.2007.07.011.

157 Yang, W., Weng, Y., Puglia, D. et al. (2020). Poly(lactic acid)/lignin films with enhanced toughness and anti-oxidation performance for active food packaging. *International Journal of Biological Macromolecules* 144: 102–110. https://doi.org/10.1016/J.IJBIOMAC.2019.12.085.

158 Yang, W., Owczarek, J.S., Fortunati, E. et al. (2016). Antioxidant and antibacterial lignin nanoparticles in polyvinyl alcohol/chitosan films for active packaging. *Industrial Crops and Products 94*: 800–811. https://doi.org/10.1016/J.INDCROP.2016.09.061.

159 Yang, W., Fortunati, E., Bertoglio, F. et al. (2018). Polyvinyl alcohol/chitosan hydrogels with enhanced antioxidant and antibacterial properties induced by lignin nanoparticles. *Carbohydrate Polymers 181*: 275–284. https://doi.org/10.1016/J.CARBPOL.2017.10.084.

160 Yang, W., Dominici, F., Fortunati, E. et al. (2015). Effect of lignin nanoparticles and masterbatch procedures on the final properties of glycidyl methacrylate-g-poly (lactic acid) films before and after accelerated UV weathering. *Industrial Crops and Products 77*: 833–844. https://doi.org/10.1016/J.INDCROP.2015.09.057.

161 Yang, W., Kenny, J.M., and Puglia, D. (2015). Structure and properties of biodegradable wheat gluten bionanocomposites containing lignin nanoparticles. *Industrial Crops and Products* 74: 348–356. https://doi.org/10.1016/J.INDCROP.2015.05.032.

162 Yang, W., Fortunati, E., Dominici, F. et al. (2015). Effect of processing conditions and lignin content on thermal, mechanical and degradative behavior of lignin nanoparticles/polylactic (acid) bionanocomposites prepared by melt extrusion and solvent casting. *European Polymer Journal* 71: 126–139. https://doi.org/10.1016/J.EURPOLYMJ.2015.07.051.

163 Barbosa, C.H., Andrade, M.A., Vilarinho, F. et al. (2021). Active edible packaging. *Encyclopedia* 1: 360–370. https://doi.org/10.3390/encyclopedia1020030.

164 Zhou, L. and Elias, R.J. (2013). Understanding antioxidant and prooxidant mechanisms of phenolics in food lipids. *Lipid Oxidation: Challenges in Food Systems* 297–321. https://doi.org/10.1016/B978-0-9830791-6-3.50012-6.

165 Xue, J., Davidson, P.M., and Zhong, Q. (2013). Thymol nanoemulsified by whey protein-maltodextrin conjugates: the enhanced emulsifying capacity and antilisterial properties in milk by propylene glycol. *Journal of Agricultural and Food Chemistry 61*: 12720–12726. https://doi.org/10.1021/JF4043437/ASSET/IMAGES/LARGE/JF-2013-043437_0006.JPEG.

166 Moleyar, V. and Narasimham, P. (1986). Antifungal activity of some essential oil components. *Food Microbiology 3*: 331–336. https://doi.org/10.1016/0740-0020(86)90017-1.

167 Malhotra, B., Keshwani, A., and Kharkwal, H. (2015). Antimicrobial food packaging: potential and pitfalls. *Frontiers in Microbiology* 6: 611. https://doi.org/10.3389/FMICB.2015.00611/BIBTEX.

9

Polyphenols for Textile-Based Application

Mohd Yusuf[1], *Vikrant Jayant*[1], *and Wasim Khan*[2]

[1] *Department of Natural and Applied Sciences, School of Science and Technology, Glocal University, Mirzapur Pole, Saharanpur, Uttar Pradesh, India*
[2] *Department of Petroleum Engineering, School of Science and Technology, Glocal University, Mirzapur Pole, Saharanpur, Uttar Pradesh, India*

9.1 Introduction

Polyphenols are naturally occurring compounds found in nature-based flora and fauna that can be classified into several classes based on their chemical structure. They are long-lasting bioresource products of multiple hydroxyl groups linked to an aromatic ring. Polyphenols have several industrial uses due to their distinct biological and chemical properties. The largest class of polyphenols is flavonoids, which include compounds such as quercetin and catechins. Phenolic acids, such as caffeic and ferulic acids, are common polyphenols found in fruits, vegetables, and grains. Stilbenes, including resveratrol and pterostilbene, are found in grapes, berries, and other fruits [1, 2]. Lignans, such as secoisolariciresinol and matairesinol, are abundant in flaxseed, sesame seeds, and whole grains. Tannins, such as ellagitannins and proanthocyanidins, are found in fruits, nuts, and tea. Each class of polyphenols has unique chemical and biological properties, contributing to their health benefits and potential applications in various industries. Polyphenols have a wide range of industrial applications due to their unique biological and chemical characteristics. They are potent antioxidants that protect materials from oxidative damage and have anti-inflammatory and antibacterial characteristics that may be used to treat illnesses and prevent infections [3]. They may also be used as natural dyes for textiles, offering a more sustainable and ecologically friendly alternative to synthetic dyes. Polyphenols are also employed as nutraceuticals, giving cardiovascular health, cognitive function, and cancer prevention

Science and Engineering of Polyphenols: Fundamentals and Industrial Scale Applications,
First Edition. Edited by Chandrabhan Verma.
© 2024 John Wiley & Sons Inc. Published 2024 by John Wiley & Sons Inc.

advantages [4, 5]. Because of their diverse uses, they are a focus of research and development in a variety of sectors, including food preservation, cosmetics, packaging materials, and nutraceuticals. The different chemical structure of polyphenols is shown in Figures 9.1 and 9.2.

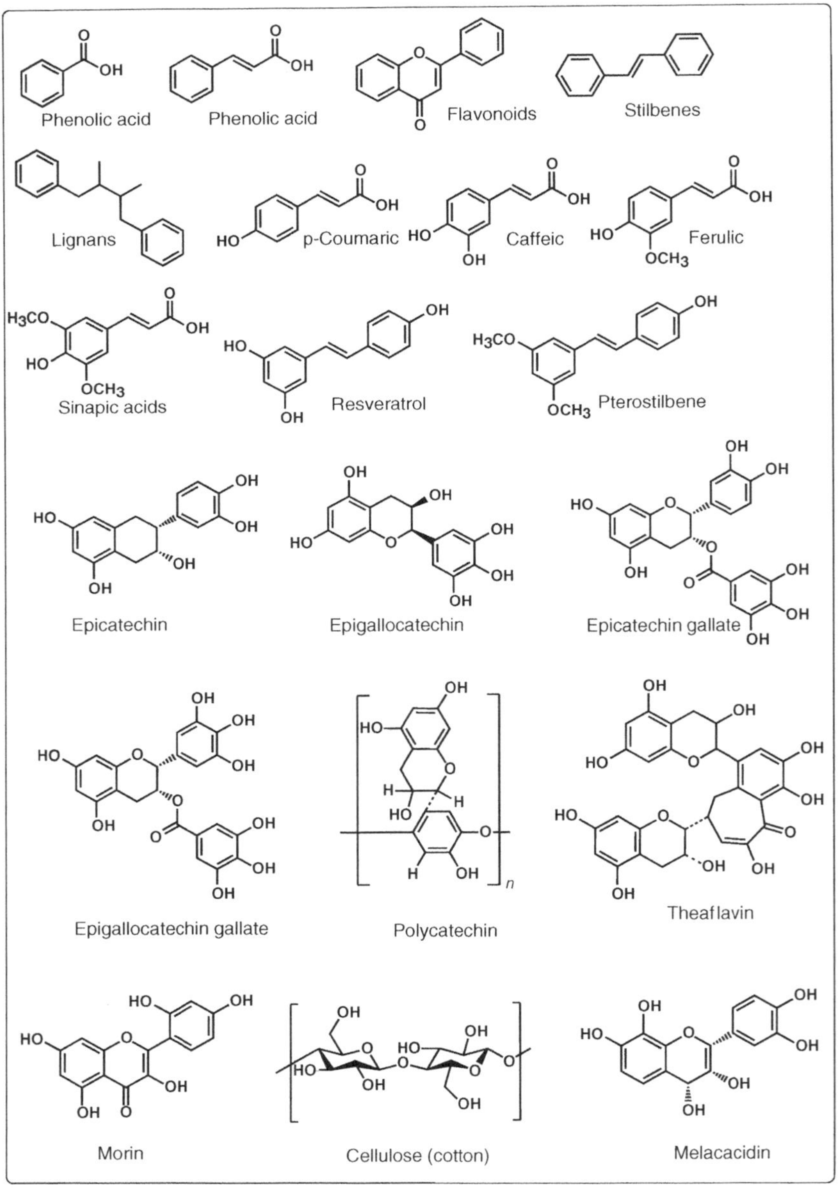

Figure 9.1 Some chemical structures of polyphenols are under different classes.

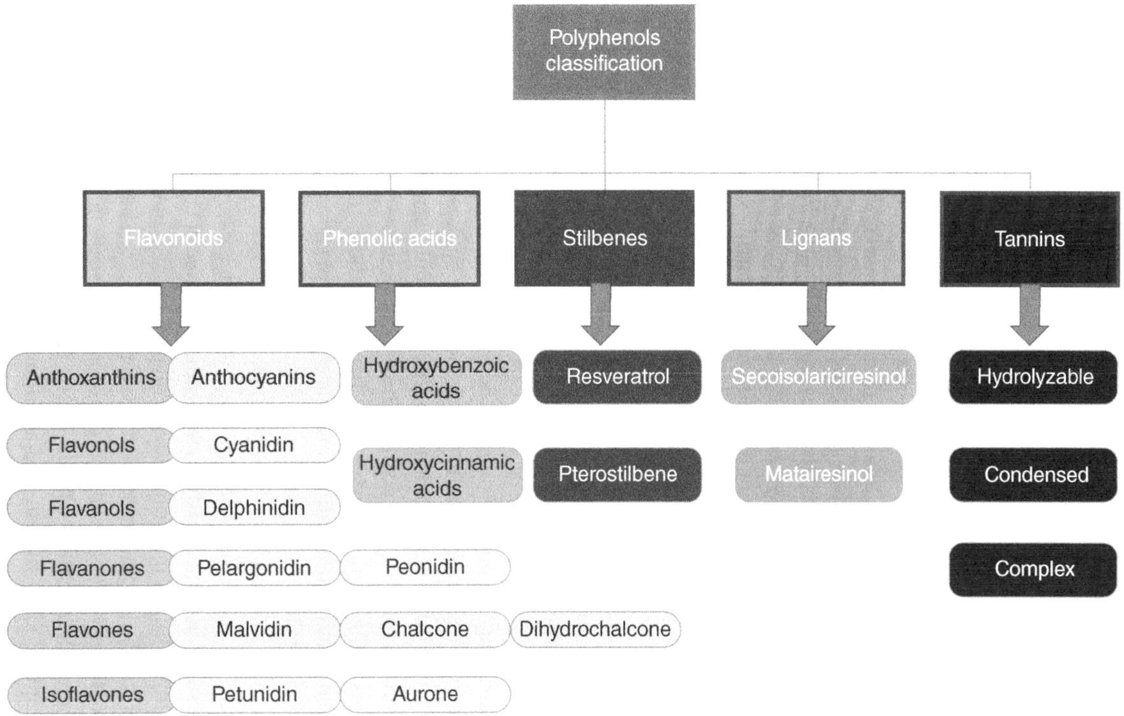

Figure 9.2 Representation of classification of polyphenols.

Polyphenols are a diverse category of plant bioactive compounds that, due to their potent antioxidant and antiradical properties, can protect cells from oxidative damage. Amino acids (tyrosine), essential oils, phenolic acids (e.g. vanilla, gallic, coumaric, and ferulic acids), flavonoids (e.g. catechins, quercetin, rutin, anthocyanidins), and tannin are examples of PF. Polyphenols have been investigated as natural dyes for textile applications due to their ability to produce vibrant colors and sober hues. They also inherited their environmentally compatible and potential health benefits [6–9]. Polyphenols can be extracted from various plant sources, such as tea, coffee, and fruits, and used to dye natural fibers such as cotton, wool, and silk [10].

Polyphenol-based dyes have several advantages over synthetic dyes. They are biodegradable and non-toxic, making them more environmentally friendly [11]. They also have potential health benefits, such as antioxidant and anti-inflammatory properties. Also, polyphenol-based dyes can create unique and vibrant colors that are difficult to achieve with synthetic dyes. The dyeing process typically involves extracting the polyphenols from the plant material using solvents such as water or ethanol and then treating the fabric with the dye solution. The dyeing process can be optimized by adjusting the treatment's pH, temperature, and duration. The resulting colors can range from yellow to red, depending on the specific polyphenol used and the conditions of the dyeing process. Furthermore, some challenges are associated with using polyphenols as textile dyes [7, 12, 13]. They can be less stable than synthetic dyes, meaning that they may fade more quickly or change color over time. They can also be more expensive than synthetic dyes, and extracting and purifying the polyphenols can be time-consuming.

The use of polyphenols as natural dyes for textile applications is an area of active research, and further developments in this area could lead to a more sustainable and environmentally friendly textile industry [8]. However, they also have potential applications in the textile industry. Polyphenols can be used as natural dyes, antimicrobial agents, and antioxidants for textile-based applications. As natural dyes, polyphenols can provide an alternative to synthetic dyes, which often have negative environmental impacts [14]. Polyphenols can be extracted from various plant sources, such as tea, coffee, and fruits, and used to dye natural fibers such as cotton, wool, and silk. These natural dyes can produce a range of colors, from yellow to red, and are known for their vibrant hues. Polyphenols also have antimicrobial properties, which can be helpful for textile-based applications. They can inhibit the growth of bacteria, fungi, and other microorganisms, making them ideal for use in clothing, bedding, and other textiles, where hygiene is essential. Polyphenols can create fabrics resistant to odor, mildew, and additional microbial growth. In addition, polyphenols are powerful antioxidants that protect textiles from damage caused by ultraviolet (UV) radiation, oxidation, and other environmental factors [15, 16].

They can be used as a natural alternative to synthetic antioxidants, which can harm the environment [17, 18]. However, polyphenols have great potential for textile-based applications, offering natural and sustainable alternatives to synthetic compounds. Further research and development could lead to a more environmentally friendly and sustainable textile industry.

9.2 Polyphenols in Textile-Based Applications

9.2.1 Flavonoids

Flavonoids are naturally occurring polyphenolic substances in plants such as fruits, vegetables, and herbs. They have been extensively researched for possible health advantages such as antioxidant and anti-inflammatory qualities. Flavonoids have also demonstrated interest in textile-based applications. Flavonoids are one of the most widespread dye families of polyphenolic chemicals produced in nature [7, 19]. Flavonoids are yellow-coloring chemicals that have a wide range of applications. In other words, flavonoids have been employed as a coloring agent in food, textiles, art, leather, cosmetics, and other applications. Many flavonoids are recognized antioxidant compounds, and flavonoid or flavonoids are found in many plants and fruits. Flavonoid molecules are found in nearly 90% of yellow natural pigments. Since ancient times, flavonoids have been employed for the textile dyeing of art artifacts (carpet, kilim, textiles, etc.). In this sense, flavonoids are the most often utilized natural dye components in natural dyeing.

Polyphenolic compounds are a diverse group of naturally occurring phytochemicals in numerous plant species. These compounds are characterized by their chemical structure, which includes multiple phenol rings. Researchers have identified and illustrated over 8000 different polyphenolic compounds from various plants. Flavonoids are the largest and most extensively studied subclass of polyphenolic compounds. They include compounds like quercetin, kaempferol, and catechins commonly found in fruits, vegetables, tea, and wine. Flavonoids are known for their antioxidant, anti-inflammatory, and anticancer properties, among other health-promoting effects [11, 18]. Identifying and characterizing over 8000 polyphenolic compounds highlight plant-based diets' incredible diversity and potential health benefits [20, 21]. These compounds continue to be an active area of research as scientists uncover their specific mechanisms of action and explore their applications in medicine, nutrition, and functional foods.

Flavonoids, a polyphenolic chemical, have received much attention and recognition for their numerous uses in the nutraceutical, pharmaceutical, medical, and cosmetic sectors. Because of their different chemical structures and biological functions,

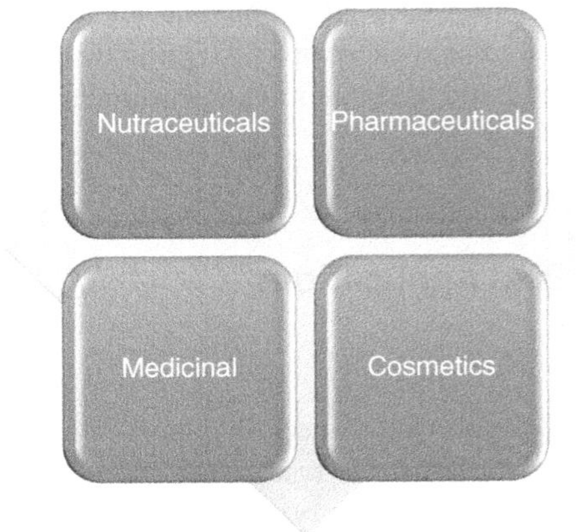

Figure 9.3 Some additional aesthetic functionalities towards the applications of flavonoids.

they are valuable components in various goods. Here are some additional aesthetic functionalities concerning the applications of flavonoids (Figure 9.3):

Nutraceuticals: Because of their potential health advantages, flavonoids are widely used in the nutraceutical business. They are frequently found in nutritional supplements and functional meals. Flavonoids have antioxidant capabilities that help protect cells from free radical oxidative damage. Their antioxidant action helps to promote general health, reduce the risk of chronic illnesses, and improve immunological function.

Pharmaceuticals: Flavonoids exhibit a range of pharmacological activities, making them valuable in pharmaceutical formulations. They have been studied for their anti-inflammatory, antimicrobial, anticancer, and neuroprotective properties. Certain flavonoids, such as quercetin and epicatechin, have demonstrated anti-inflammatory effects by inhibiting inflammatory enzymes and reducing the production of inflammatory molecules. These compounds have also shown potential in preventing and treating cancer by inhibiting tumor growth and inducing cell death in cancer cells.

Medicinal: Flavonoids have been utilized in traditional medicine systems for centuries. They are known for their therapeutic effects and have been used to treat various ailments. For example, flavonoids like rutin and hesperidin strengthen blood vessels and improve blood circulation. Quercetin has been investigated for its anti-allergic properties, while kaempferol has shown potential in managing diabetes by reducing blood sugar levels. Moreover, flavonoids have been studied for their cardiovascular benefits, including their ability to lower blood pressure, reduce cholesterol levels, and improve heart health.

Cosmetics: Flavonoids are increasingly used in cosmetic and skincare products due to their antioxidant and anti-aging properties. These compounds help protect the skin from damage caused by environmental factors, such as UV radiation and pollution, by neutralizing free radicals. Flavonoids can also support collagen production, enhance skin elasticity, and reduce the appearance of wrinkles and age spots. Additionally, some flavonoids possess anti-inflammatory properties, which can soothe and calm the skin. Flavonoids have become a subject of extensive research as scientists uncover their mechanisms of action and explore their potential applications. Their broad range of biological activities and their natural origin make them attractive candidates for various industries. Continued investigation into flavonoids holds promise for developing novel therapies, functional foods, and cosmetic formulations that harness their beneficial properties.

The flavones are responsible for the golden colors of textiles in all museums. Flavonoids were employed in ancient times to color paintings, murals, icons, and other objects. As a result, flavonoid molecules and flavonoid-based colors may be found everywhere, from ancient times to the present. After the pre-operation of the dyeing method on textile materials, dyes containing flavonoids are most likely employed for the dyeing of mordanted silk and wool, etc. The flavonoid coloring spectrum on cloth can range from yellow to brown. Flavonoids are now employed in environmentally safe and sustainable textile dyeing. They are also employed in the pharmaceutical and cosmetic sectors [22].

Flavonoids in textiles as finish enhancers for improving the fabric's performance have been utilized since prehistoric times. Flavonoids, such as quercetin and kaempferol, have been shown to impart various colors to textiles, including yellow, orange, and red. Incorporating flavonoids into textiles could provide a natural and sustainable alternative to synthetic dyes. Flavonoids also have UV-absorbing properties, making them suitable for textile-based applications requiring protection against UV radiation [23]. For example, fabrics treated with flavonoid extracts have been shown to provide UV protection to the wearer. This property could be useful in developing textiles used in outdoor activities or for garments that require protection against UV radiation. Dyed wool/cotton fabrics with these dyes have been shown to improve their tensile strength and tear resistance and provide antimicrobial and antifungal properties to fabrics, making them a potential alternative to synthetic antimicrobial agents.

Additionally, their use may require additional processing steps, such as fabric pre-treatment, to ensure optimal flavonoid absorption. Despite their potential benefits, there are challenges associated with using flavonoids in textile-based applications. These compounds can be sensitive to pH, temperature, and light, affecting their stability and quality. Flavonoids offer a promising alternative to synthetic chemicals in various textile-based applications. Their potential advantages include environmental sustainability, health benefits, and potential for cost savings. Continued research and development in this field will be crucial to address the challenges associated with their use and to explore new opportunities for their application in the textile industry [23, 24].

When used to create functional textiles for health care, polyamide fiber must have antioxidant and antibacterial qualities. Using an adsorption technique, three natural flavonoids (baicalin, quercetin, and rutin) were employed to impart antioxidant and antibacterial properties to polyamide fiber concurrently. The chemical structures of flavonoids were addressed concerning their adsorption capabilities, adsorption processes, and antioxidant and antibacterial properties. Using the adsorption technique, three natural flavonoids were used to impart antioxidant and antibacterial activity to polyamide fiber concurrently. The pH of baicalin solutions significantly impacts their adsorption on polyamide fiber. The three flavonoids' adsorption followed the pseudo-second-order kinetic model. Three flavonoid's adsorption isotherms suit the Langmuir-Nernst model. Quercetin had the maximum affinity and adsorption amount to polyamide fiber, followed by baicalin, but rutin had limited adsorption capabilities due to a glycosidic moiety. Compared to baicalin and rutin, quercetin gave powerful antioxidant and antibacterial activities to polyamide fiber, and these functionalities were resistant to washing. It shows that treatments based on natural flavonoids can achieve simultaneous antioxidant and antibacterial functionalization of polyamide fiber [25].

9.2.2 Phenolic Acids

Phenolic acids are polyphenolic chemicals in various plant sources, including fruits, vegetables, and grains. They have antioxidant and antibacterial characteristics, making them attractive candidates for textile-based applications. Natural dyes are one possible application for phenolic acids in textiles. When employed as natural dyes, several phenolic acids, such as ferulic acid and caffeic acid, have been proven to generate yellow to brown hues on fabrics. These dyes have various benefits over synthetic dyes, including biodegradability, low toxicity, and cost-saving potential [26–28].

Phenolic acids are also antibacterial, making them appropriate for textile-based applications requiring antimicrobial protection. Ferulic acid, for example, has been demonstrated to suppress bacterial development in cotton and wool materials. This feature might be beneficial in creating textiles used in healthcare settings or clothing that must be washed often. Despite their potential benefits, there are several drawbacks to use phenolic acids in textile-based applications [29]. These chemicals are potentially unstable and may deteriorate over time, compromising the color and quality of the fabric. Furthermore, their utilization may necessitate extra processing procedures, such as fabric pre-treatment, to guarantee adequate phenolic acid absorption. Furthermore, phenolic acids have a significant potential application in textiles as finishes for increasing fabric performance. For example, coating cotton fabric with ferulic acid has increased its tensile strength and abrasion resistance [30]. Phenolic acid-based treatments might also protect textiles from UV rays, making them a viable alternative to synthetic UV protectants. As a result, phenolic acids represent a possible alternative to synthetic chemicals in various textile-based applications. Their potential benefits include sustainability for the environment, health benefits, and the chance to save money. Continued research and development in this subject will be critical in addressing the issues connected with their usage and exploring new avenues for their application in the textile sector.

9.2.3 Stilbenes

Stilbenes are a polyphenolic chemical found in plants such as grapes, berries, and peanuts. Stilbenes are classified structurally into two types based on the presence of a 1,2-diphenylethylene nucleus: monomeric stilbenes and oligomeric stilbenes [31, 32]. They are well known for their antioxidant and UV-protective qualities, making them viable options for textile-based applications. UV protectants are another possible application for stilbenes in textiles. Stilbenes like resveratrol have been demonstrated to absorb UV radiation, shielding the skin from damaging rays. Incorporating stilbenes into textiles might give a natural and environmentally friendly alternative to synthetic UV protectants [33, 34]. Resveratrol, for example, has been found to suppress the growth of germs and fungus on cotton materials. This feature might be necessary to create textiles used in healthcare settings or clothing that requires regular washing. Furthermore, the hidden application of stilbenes in textiles is as finishes to improve fabric performance. For example, coating cotton fabric with resveratrol has increased tensile strength and abrasion resistance. Stilbene-based treatments may also be used to color fabrics, giving them a viable alternative to synthetic dyes [35].

Despite its potential benefits, using stilbenes in textile-based applications needs to be improved. These substances might be unstable and deteriorate over time, reducing the quality of the cloth [26]. Furthermore, their use may necessitate different processing processes, such as fabric pre-treatment, to enable efficient stilbene absorption. As a result, Stilbenes provide a viable alternative to synthetic chemicals in various textile-based applications. Their potential benefits include environmental sustainability, health benefits, and the chance to save money. Continued research and development in this subject will be critical in addressing the issues connected with their usage and exploring new avenues for their application in the textile industry [33].

Stilbenes occur in many plant species, including grapes, berries, peanuts, and certain medicinal herbs, and the most well-known stilbenes are resveratrol, which has gained significant attention for its potential health benefits. Stilbenes exhibit a range of properties, including antioxidant, anti-inflammatory, and anticarcinogenic effects, described below:

Antioxidant properties: Stilbenes, including resveratrol, are potent antioxidants that help to neutralize harmful free radicals in the body, which are unstable molecules that can cause cellular damage and contribute to aging, chronic diseases, and cancer. By scavenging free radicals, stilbenes protect cells from oxidative stress and maintain their normal functioning [36].

Anti-inflammatory properties: Stilbenes possess anti-inflammatory properties, which can help reduce inflammation in the body [37]. Chronic inflammation is associated with various diseases, including cardiovascular diseases, diabetes, and cancer. Stilbenes can inhibit the activity of inflammatory enzymes and modulate inflammatory signaling pathways, leading to a decrease in inflammation and its detrimental effects.

Anticarcinogenic properties: Stilbenes have been studied for their potential anticarcinogenic effects. They exhibit several mechanisms that may help prevent or inhibit the growth of cancer cells. Stilbenes can interfere with carcinogenesis's initiation, promotion, and progression stages. They may inhibit the proliferation of cancer cells, induce apoptosis (programmed cell death), and inhibit the formation of blood vessels that supply tumors (angiogenesis) [38]. Moreover, stilbenes have been shown to have anti-mutagenic properties, which means they can help prevent DNA damage that can lead to cancer development.

It is important to note that while stilbenes, such as resveratrol, have demonstrated promising effects in preclinical studies and some early clinical trials, further research is needed to understand their potential benefits and optimize their use fully. Additionally, the bioavailability of stilbenes can be challenging, as they may be rapidly metabolized and eliminated from the body. Therefore, the effective delivery and absorption of stilbenes for therapeutic purposes are areas of ongoing investigation [39].

9.2.4 Lignans

Lignans are a class of polyphenolic compounds found in various plant-based sources, including flaxseed, sesame seeds, and berries. They have gained attention for their potential health benefits and antioxidant properties, making them a potential candidate for textile-based applications. Lignans are distinguished by the presence of two linked phenylpropanoid units, and their concentration in the diet varies. They are very complex bioactive polyphenolic phytochemicals produced by the union of two coniferyl alcohol residues [40]. Lignans also exhibited significant antimicrobial properties, making them suitable for textile-based applications requiring protection against microbial growth. For example, lignan-rich extracts from flaxseed have been shown to inhibit the growth of bacteria and fungi on cotton fabrics [41, 42]. This property could be useful in developing textiles used in healthcare settings or for garments that require frequent washing. And so, they have latent application in textiles to impart functional finishing for improving the fabric's performance. For example, coating cotton fabric with SDG has improved its tensile strength and abrasion resistance. Lignan-based finishes could also provide UV protection to fabrics, making them a potential alternative to synthetic UV protectants. Considering their potential benefits, there are certain drawbacks to using lignans in textile-based applications [43, 44]. These chemicals may be pH, temperature, and light sensitive, affecting their stability and quality. Furthermore, their use may necessitate extra processing processes, such as fabric pre-treatment, to enable effective lignan absorption. Furthermore, lignans provide a viable alternative to synthetic chemicals in various textile-based applications. Their potential benefits include environmental sustainability, health benefits, and the chance to save money. Continued research and development in this subject will be critical in addressing the issues connected with their usage and exploring new avenues for their application in the textile industry [45].

9.2.5 Tannins

Tannins are a polyphenolic substance found in plants such as fruits, vegetables, and trees. They have been used for ages in the tanning and coloring of leather. In recent years, there has been a surge of interest in using tannins in textile-based applications, such as natural dyes, finishes, and functional coatings. Tannins can impart various colors to textiles, including brown, black, and grey [46, 47]. Using tannins as natural dyes provides a sustainable alternative to synthetic dyes, which can be environmentally harmful and adversely affect human health. Tannins can also be used as finishes for textiles to improve their performance. For example, cotton fabrics treated with tannin extracts have been shown to have improved tensile strength, tear resistance, and abrasion resistance. Tannin-based finishes can also provide antimicrobial and antifungal properties to textiles, making them a potential alternative to synthetic antimicrobial agents [7, 9, 48, 49].

In addition to their functional properties, tannins have unique aesthetic properties that are functionally important in textile-based applications. Tannins can create a range of natural textures, including wrinkling and stiffening, which can be used for decorative purposes in textiles. Tannins offer a promising alternative to synthetic chemicals in various textile-based applications. Their potential advantages include environmental sustainability, health benefits, and unique aesthetic and functional properties [50]. Tannic acid (TA) and Fe(III) coordination assembly were used to create highly hydrophobic cotton textiles, which were then treated with 1-octadecylamine. SEM research revealed that the unique Fe(III)/TA metalorganic system coated the cotton fabrics and changed the surface roughness, making the textiles hydrophobic [51].

9.3 Specific Functional Finishing Using Polyphenols

9.3.1 UV Protection

UV protection fabrics using polyphenols involve the incorporation of these natural compounds into textiles during the manufacturing process. When applied to textiles, polyphenols, including flavonoids and tannins, can act as UV-protective agents [52–56]. They may be integrated into fabrics during production and have inherent photoprotective qualities. The following three references explore the use of polyphenols in textiles for UV protection:

Selection: Various polyphenols can be utilized for UV protection, including flavonoids (e.g. quercetin, kaempferol, rutin) and tannins (e.g. ellagitannins, catechins). These polyphenols are commonly found in plants and have demonstrated UV-absorbing properties.

Fabric treatment: The incorporation of polyphenols into fabrics can be achieved through different methods, such as dyeing, finishing, or coating. The technique employed depends on factors like the type of polyphenol, fabric composition, and desired UV protection level. Polyphenols are sometimes combined with other compounds, such as binders or encapsulating agents, to enhance their stability and adhesion to the fabric.

UV absorption mechanism: Polyphenols act as UV filters by absorbing UV radiation and dissipating it as heat, thereby reducing the amount of UV radiation reaching the skin. This absorption occurs due to conjugated double bonds within the polyphenol structure, allowing them to absorb specific wavelengths of UV light.

UV protection evaluation: The effectiveness of polyphenol fabrics can be evaluated through standardized testing methods. One commonly used method is the Ultraviolet Protection Factor (UPF), which measures the amount of UV radiation transmitted through the fabric. The higher the UPF value, the greater the fabric's UV protection.

Durability and maintenance: UV protection fabrics using polyphenols can vary depending on fabric type, polyphenol concentration, and laundering practices. It is important to consider the stability of polyphenols and their ability to withstand repeated washing and exposure to environmental conditions. Some polyphenols may exhibit fading or loss of UV-absorbing properties over time, necessitating reapplication or retreatment of the fabric [57].

It is important to note that while polyphenols have demonstrated promise for UV protection, their precise efficiency and durability may differ depending on the fabric's makeup, the application method, and the type of polyphenol used. The field is still undergoing development and more study. Further research and development in UV protection fabrics using polyphenols is ongoing to enhance their efficacy, durability, and commercial viability.

9.3.2 Remediation of Organic Synthetic Dyes

Despite being widely used across many sectors, organic synthetic dyes can harm the environment and human health. The following are some possible negative consequences linked to organic synthetic dyes:

Environmental pollution: Improper disposal of dye-containing wastewater can lead to environmental pollution. Synthetic dyes can persist in the environment and accumulate in water bodies, soil, and sediments. They can disrupt the balance of ecosystems, inhibit the growth of aquatic organisms, and degrade water quality.

Toxicity to organisms: Many organic synthetic dyes have toxic properties and can harm living organisms. They may exhibit acute or chronic toxicity to aquatic life, such as fish, algae, and invertebrates, impacting their growth, reproduction, and survival. Some dyes and their degradation products may also be toxic to terrestrial organisms and plants.

Carcinogenicity: Certain organic synthetic dyes, particularly azo dyes, have been associated with potential carcinogenic properties. Some azo dyes can break down into aromatic amines, which are classified as carcinogens and may pose a risk to human health if exposed to high concentrations or for prolonged periods.

Skin sensitization and allergic reactions: Some individuals may experience skin sensitization or allergic reactions upon contact with certain synthetic dyes. This can manifest as skin irritation, redness, itching, or dermatitis. Sensitization can concern individuals working in industries that handle or come into contact with dyes, such as textile manufacturing.

Occupational health risks: Workers producing, handling, or applying synthetic dyes may face occupational health risks. Exposure to dye dust, vapors, or contaminated workplace environments can lead to respiratory problems, eye irritation, and skin disorders. Long-term exposure to certain dyes may increase the risk of respiratory diseases and other chronic health conditions.

It is crucial to note that not all organic synthetic dyes display the same degree of harmful effects due to the wide range of their chemical makeup and characteristics. Manufacturers, regulatory agencies, and businesses have been putting steps in place to lessen the impact of synthetic dyes on the environment and human health, such as wastewater treatment, replacing dyes with safer alternatives, and adhering to legal requirements. To reduce the negative consequences of manufacturing, using, and disposing of organic synthetic dyes, carrying on with research, monitoring, and adopting sustainable practices [58].

The ability of polyphenols to decolorize and eliminate organic synthetic dyes from different aqueous environments has been investigated. These substances efficiently treat wastewater polluted with dyes because they have significant reducing and adsorption properties. Ahmad et al. [59] explored the decolorization of organic synthetic dyes using polyphenols and found that the decolorization process was significantly enhanced. This work concluded the adsorption kinetics, equilibrium studies, and the active role of polyphenols in the decolorization process. Consequently, Stavrinou et al. [31] described the adsorption and decolorization potential of activated carbon prepared from banana peel for a cationic dye. Furthermore, a study was performed to investigate the decolorization efficiency, factors affecting the process, and the role of tannin acid in the decolorization mechanism. As a result, positive outcomes were noticed using TA residues [60] and tea residues in treating dye-contaminated wastewater [61].

9.3.3 Interfacial Assembly of Metal–Phenolic Networks for Hair Dyeing

Raman and UV–vis absorption spectroscopy are used to explore the mechanism of self-assembled metal phenolic networks (MPNs) for hair coloring. In vitro cell investigations and in vivo skin toxicity testing show that MPNs outperform p-phenylenediamine regarding biocompatibility and hair regrowth. The proposed hair coloring approach eliminates the use of harmful compounds included in typical hair colors while causing minor damage to hair structures and tensile strength. Rapid and simple coating methods are critical in surface engineering and functionalizing materials for various applications. We provide a safe, fast, and environmentally friendly hair coloring procedure for natural grey hair based on optimized MPN production. MPNs made of gallic acid operate well, and the coloring may be adjusted by changing the metal ion types. MPN-based hair coloring is resistant to repeated washing with a detergent solution (at least 50 times) without color fading and can be discolored in an acidic solution (pH = 2). Furthermore, GA–Fe (II) complexes exhibited no substantial cell cytotoxicity or skin toxicity. MPN dyes may be used as a replacement for current hair dyeing ingredients [53].

9.3.4 Anti-Odor Finishing

For defense against diseases in varied settings, creating antimicrobial fabrics and textiles that can consistently inhibit a wide range of bacteria is essential. The use of hazardous or harsh chemicals, harsh synthesis conditions, complex and/or time-consuming processing, problems with discoloration and reduced breathability, and other factors make it challenging to develop antimicrobial fabrics [62]. Polyphenols have been researched for their antibacterial and deodorizing abilities in textiles as an anti-odor finishing treatment method [63–66]. While polyphenols have shown promise as anti-odor agents, their efficiency may vary depending on the exact polyphenol employed, the concentration, and the target odor-causing bacteria. Polyphenols can act against odor-causing bacteria in several ways. They may disrupt the bacterial cell membrane, inhibit enzymatic activities, or interfere

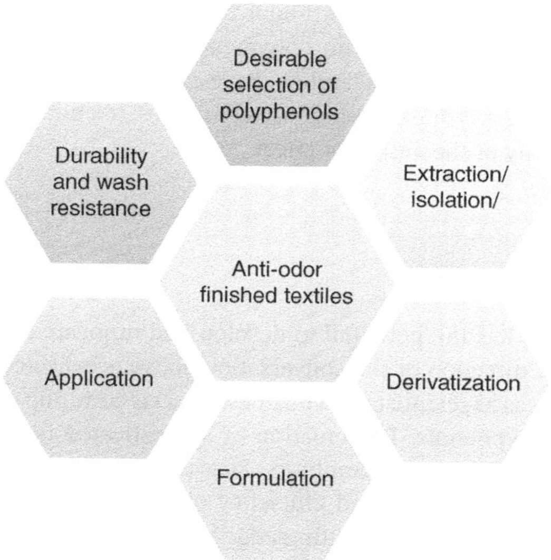

Figure 9.4 Summary of how polyphenols can be utilized for anti-odor finishing.

with bacterial metabolism. These actions help control the growth of odor-causing bacteria and reduce the formation of unpleasant odors. More research and development are being conducted to improve polyphenol-based textile anti-odor treatments. A broad summary of how polyphenols can be utilized for anti-odor finishing is provided in Figure 9.4:

Desirable selection of polyphenols: Different polyphenols have varying antimicrobial properties, so selecting appropriate polyphenols is crucial. Some commonly studied polyphenols for this purpose include catechins (found in green tea), tannins (found in various plant sources), and flavonoids (found in fruits and vegetables).

Extraction/isolation: Polyphenols can be extracted from natural sources or synthesized in the laboratory, depending on the specific polyphenol used. Extraction methods often involve using solvents like water, ethanol, or both.

Derivatization: Extracted and isolated products can be modified via additive reagents of several oate (containing ester moiety).

Formulation: The extracted or synthesized polyphenols are typically formulated into a solution that can be easily applied to the textile material. The formulation may include additional ingredients to enhance stability and adherence to the fabric.

Application: The polyphenol solution is applied to the fabric using various techniques such as padding, spraying, or exhaustion. The fabric is typically dried or cured afterward to ensure proper fixation of the polyphenols onto the textile fibers.

Durability and wash resistance: One challenge in anti-odor finishing is ensuring the durability of the polyphenol treatment. The polyphenols should ideally remain active even after multiple washes. Researchers have explored different methods, such as cross-linking agents or encapsulation techniques, to enhance the wash resistance and longevity of the anti-odor effect.

9.4 Future Outlook

The use of polyphenols in textiles has the potential to develop and innovate the industry significantly. The future direction of the conversation above is expected to be influenced by several key areas as scientists and business experts continue to investigate the possibilities of polyphenols. The creation of sophisticated polyphenol extraction methods will be one area of emphasis. Researchers will study cutting-edge techniques to increase the yield and efficiency of extraction from diverse natural sources. This might entail using cutting-edge solvents, enzymatic reactions, or environmentally friendly extraction techniques, producing polyphenols with improved characteristics for textile applications. Future evaluations will continue to place a high priority on sustainability and environmental friendliness. The industry's increasing focus on environmentally friendly practices will encourage the development of greener dyeing and finishing methods and procedures that use less water, energy, and waste. Future studies will assess the environmental effects of fabrics treated with polyphenols and look for ways to increase their sustainability. Future assessments will also have a big emphasis on novel textile applications. Although UV protection is well known, researchers will look into new or unexplored uses for polyphenols. Antimicrobial fabrics, antioxidant qualities, or self-healing characteristics may fall under this category. The synergistic interactions of polyphenols with other functional additives or coatings will also be studied to open up new possibilities. A crucial area of study will be how to make polyphenol-treated fabrics more resilient and long-lasting. Future studies will examine ways to increase the stability of polyphenols on textiles, evaluate performance after several washings and exposure to the environment, and look at tactics to keep them working overtime. By doing this, it will be made sure that textiles treated with polyphenol remain functional for the duration of their useful lives.

Future usage of polyphenols in textiles will also consider biodegradability and the circular economy. As textiles near the end of their useful lives, reviews will examine how polyphenols act in the environment and look at methods to hasten their decomposition or reuse in new products. This will aid in the development of sustainable and circular textile systems. Additionally, the future uses of textiles treated with polyphenols for health and well-being will be investigated. The effects of these fabrics on skin health, wound healing, and perhaps even relaxation or sleep will be covered in later reviews. Unlocking the full potential of polyphenols in enhancing well-being via textiles will need a better understanding of the processes behind these potential benefits and assessing their effectiveness.

9.5 Conclusion

This high-tech era is founded on "*beauty with artificial intelligence*" and attractive items keeping eco-friendliness. Many studies are being done to create a beautiful, helpful textile because of its appearance and artificial intelligence to send signals in response to changes in the surrounding environment. Natural dyes from polyphenols are used to create valuable and intelligent fabrics along with usefulness, innovative and functional dyes synthesis and utilizing the associated sustainability and environmental point of view. Polyphenols have antioxidant, antibacterial, and UV protection qualities and may be included in textile finishing, dyeing, and printing. Polyphenols have the potential to minimize dependency on synthetic chemicals while also improving the sustainability of the textile industry. However, further study is needed to investigate polyphenols' potential and optimize their use in textile-based products. The analysis emphasizes the need to integrate natural chemicals such as polyphenols into textile applications and transition to a more environmentally friendly, sustainable textile business and circular economy globally.

Acknowledgement

Prof. (Dr.) S. K. Sharma, Pro-Vice Chancellor, Glocal University, is acknowledged for his critical suggestions.

References

1 Bijak, M., Ziewiecki, R., Saluk, J. et al. (2014). Thrombin inhibitory activity of some polyphenolic compounds. *Medicinal Chemistry Research* 23: 2324–2337.
2 Patra, J.K., Das, G., Lee, S. et al. (2018). Selected commercial plants: a review of extraction and isolation of bioactive compounds and their pharmacological market value. *Trends in Food Science & Technology* 82: 89–109.
3 Khan, M.I., Ahmad, A., Khan, S.A. et al. (2011). Assessment of antimicrobial activity of catechu and its dyed substrate. *Journal of Cleaner Production* 19 (12): 1385–1394.
4 Kumari, A. and Bharadvaja, N. (2023). A comprehensive review on algal nutraceuticals as prospective therapeutic agent for different diseases. *3 Biotech* 13 (2): 44. https://doi.org/10.1007/s13205-022-03454-2.

5 Yusuf, M., Shahid, M., Khan, M.I. et al. (2015). Dyeing studies with henna and madder: a research on effect of tin (II) chloride mordant. *Journal of Saudi Chemical Society* 19 (1): 64–72.

6 Yusuf, M., Ahmad, A., Shahid, M. et al. (2012). Assessment of colorimetric, antibacterial and antifungal properties of woollen yarn dyed with the extract of the leaves of henna (*Lawsonia inermis*). *Journal of Cleaner Production 27*: 42–50.

7 Yusuf, M., Shabbir, M., and Mohammad, F. (2017). Natural colorants: historical, processing and sustainable prospects. *Natural Products and Bioprospecting 7*: 123–145.

8 Yusuf, M., Rani, S., Chawla, U. et al. (2022). Modern perspectives on adiponectin: targeting obesity, diabetes, and cancer together using herbal products. *Biointerface Research in Applied Chemistry* 13 (2): 1–16.

9 Shahid, M., Ahmad, A., Yusuf, M. et al. (2012). Dyeing, fastness and antimicrobial properties of woolen yarns dyed with gallnut (*Quercus infectoria* Oliv.) extract. *Dyes and Pigments* 95 (1): 53–61.

10 Yusuf, M. (ed.) (2018). *Handbook of Renewable Materials for Coloration and Finishing*. Hoboken, UK: Scrivener Publishing.

11 Yusuf, M., Khan, M.A., and Mohammad, F. (2016). Investigations of the colourimetric and fastness properties of wool dyed with colorants extracted from Indian madder using reflectance spectroscopy. *Optik: International Journal for Light and Electron Optics* 127 (15): 6087–6093.

12 Khan, S.A., Ahmad, A., Khan, M.I. et al. (2012). Antimicrobial activity of wool yarn dyed with *Rheum emodi* L.(Indian Rhubarb). *Dyes and Pigments* 95 (2): 206–214.

13 Yusuf, M., Mohammad, F., and Shabbir, M. (2017). Eco-friendly and effective dyeing of wool with anthraquinone colorants extracted from *Rubia cordifolia* roots: optimization, colorimetric and fastness assay. *Journal of King Saud University, Science* 29 (2): 137–144.

14 Yusuf, M. (2019). Synthetic dyes: a threat to the environment and water ecosystem. In: *Textiles and Clothing: Environmental Concerns and Solutions* (ed. M. Shabbir), 11–26. Hoboken, UK: Scrivener Publishing.

15 Sahiner, M., Yilmaz, A.S., Gungor, B. et al. (2022). Therapeutic and nutraceutical effects of polyphenolics from natural sources. *Molecules* 27 (19): 6225.

16 Rana, A., Samtiya, M., Dhewa, T. et al. (2022). Health benefits of polyphenols: a concise review. *Journal of Food Biochemistry* 46 (10): e14264.

17 Yusuf, M., Aijaz, M., Keserwani, N. et al. (2022). Ethnomedicinal, pharmacological and commercial perspectives of *Laccifer lacca* body exudate (LBE). *Letters in Applied NanoBioScience* 12: 1–10.

18 Aijaz, M., Keserwani, N., Yusuf, M. et al. (2022). Chemical, biological, and pharmacological prospects of caffeic acid. *Biointerface Research in Applied Chemistry* 13: 324.

19 Maity, S.K. and Halder, M. (2019). Application of natural flavonoids in textile finishing. *Advances in Colloid and Interface Science* 265: 26–41.

20 Das, S. and Padhy, S.K. (2018). Flavonoids as natural colorants in textiles: a review. *International Journal of Fashion Design, Technology and Education* 11 (2): 253–261.

21 Zhou, Y. and Tang, R.C. (2017). Natural flavonoid-functionalized silk fiber presenting antibacterial, antioxidant, and UV protection performance. *ACS Sustainable Chemistry & Engineering* 5 (11): 10518–10526.

22 Deveoglu, O. (2022). A Review on onion skin, a natural dye source. *Journal of Textiles, Coloration and Polymer Science* 19 (2): 307–319.

23 Nambela, L., Haule, L.V., and Mgani, Q. (2020). A review on source, chemistry, green synthesis and application of textile colorants. *Journal of Cleaner Production* 246: 119036.

24 Grifoni, D., Bacci, L., Di Lonardo, S. et al. (2014). UV protective properties of cotton and flax fabrics dyed with multifunctional plant extracts. *Dyes and Pigments* 105: 89–96.

25 Li, H., Gao, Y., Zhang, Q. et al. (2019). A review on the application of tannin in textile. *Journal of Cleaner Production* 227: 221–233.

26 Mohanapriya, M. and Rajendran, V. (2019). Antimicrobial finishing of textiles using natural phenolic acids: a review. *Journal of Natural Fibers* 16 (1): 56–72.

27 Sadowska-Rociek, A., Szymczyk, K., and Kyzioł-Komosińska, J. (2021). Application of phenolic acids in the finishing of textiles–A review. *Fibres & Textiles in Eastern Europe* 29 (1): 107–117.

28 Hassan, M.M., Saad, M., and Abdel-Mohsen, A.M. (2019). Natural dyeing of cotton with ferulic acid and its derivatives. *Journal of Natural Fibers* 16 (7): 1023–1033.

29 Kothari, V.K., Tyagi, R., and Pathak, P. (2014). Polyphenols as natural colorants for textile dyeing and printing. *Coloration Technology* 130 (2): 83–99.

30 Shanmugam, S., Jayasubramanian, V., and Rajendran, V. (2020). Application of ferulic acid on cotton fabrics for improved mechanical properties and color characteristics. *Journal of Natural Fibers* 17 (7): 978–991.

31 Stavrinou, A., Aggelopoulos, C.A., and Tsakiroglou, C.D. (2018). Exploring the adsorption mechanisms of cationic and anionic dyes onto agricultural waste peels of banana, cucumber and potato: adsorption kinetics and equilibrium isotherms as a tool. *Journal of Environmental Chemical Engineering* 6 (6): 6958–6970.

32 Ruparelia, K.C., Patel, R.B., and Patel, A.R. (2021). Applications of stilbenes in textile industry–a review. *Journal of Cleaner Production* 280: 124482.

33 Balakrishnan, K., Uthirappan, M., and Arumugam, V. (2019). Natural UV protection in textiles using resveratrol. *Natural Product Research* 33 (8): 1208–1211.

34 Chiappetta, G. and Grieco, F. (2019). Potential use of plant polyphenols in the textile industry. *Molecules* 24 (4): 763.

35 Gupta, R., Shukla, P., and Agarwal, A. (2015). Polyphenolic extract of *Cicer arietinum* L.(chickpea) exhibits antimicrobil activity against *Klebsiella pneumoniae*. *Journal of Textiles and Polymers* 3 (2): 20–25.

36 Osamudiamen, P.M., Oluremi, B.B., Oderinlo, O.O. et al. (2020). Trans-resveratrol, piceatannol and gallic acid: potent polyphenols isolated from *Mezoneuron benthamianum* effective as anticaries, antioxidant and cytotoxic agents. *Scientific African* 7: e00244.

37 Wang, J., Song, Y., Chen, Z., and Leng, S.X. (2018). Connection between systemic inflammation and neuroinflammation underlies neuroprotective mechanism of several phytochemicals in neurodegenerative diseases. *Oxidative Medicine and Cellular Longevity* 1972714: 1–16.

38 Yi, J., He, Q., Peng, G., and Fan, Y. (2022). Improved water solubility, chemical stability, antioxidant and anticancer activity of resveratrol via nanoencapsulation with pea protein nanofibrils. *Food Chemistry* 377: 131942.

39 Berlanga, P., Muñoz, L., Piqueras, M. et al. (2016). miR-200c and phospho-AKT as prognostic factors and mediators of osteosarcoma progression and lung metastasis. *Molecular Oncology* 10 (7): 1043–1053.

40 Rauf, A., Patel, S., Imran, M. et al. (2018). Honokiol: an anticancer lignan. *Biomedicine & Pharmacotherapy* 107: 555–562.

41 Bhatti, I.A., Aslam, S., Hussain, T., and Hussain, M.A. (2019). Lignans as sustainable natural colorants for textiles. *Journal of Cleaner Production* 239: 118028.

42 Cano, A., Hernández, J.A., and García-Carmona, F. (2000). Synthetic lignin and its use as a textile dye. *Dyes and Pigments* 47 (3): 213–223.

43 Niqui-Arroyo, J.L., Navarro, M., and Cabedo, L. (2018). Lignin-based nanomaterials for sustainable applications in the textile industry: an overview. *Journal of Cleaner Production* 201: 595–606.

44 Salame, M., Pascual, J., Larrayoz, M.A. et al. (2016). Antibacterial and antioxidant cotton textiles coated with Lignocellulosic extracts from flax shives. *Cellulose* 23 (2): 1051–1065.

45 Wani, I.A., Gani, A., Sogi, D.S. et al. (2017). Recent advances in use of natural colorants in textile dyeing. *Journal of Cleaner Production* 142: 2898–2905.

46 Eichhorn, B. and Winter, T. (2016). Tannins in textiles: from traditional dye to biotechnological application. *Applied Microbiology and Biotechnology* 100 (23): 9863–9879.

47 Kaur, J. and Kaur, J. (2021). Tannins in textiles: a review. *Journal of Textile Association* 82 (2): 70–85.

48 Li, Y.D., Guan, J.P., Tang, R.C. et al. (2019). Application of natural flavonoids to impart antioxidant and antibacterial activities to polyamide fiber for health care applications. *Antioxidants* 8 (8): 301.

49 Liu, Y. and Wang, C. (2021). Application of natural tannins in textile dyeing and finishing. *Journal of Textile Research* 42 (6): 86–93.

50 Martins, I.M., Barros, A.S., and Delgado, R. (2017). Tannins as potential natural dyeing agents: review. *Reviews in Chemical Engineering* 33 (4): 365–387.

51 Gu, S., Yang, L., Huang, W. et al. (2017). Fabrication of hydrophobic cotton fabrics inspired by polyphenol chemistry. *Cellulose* 24: 2635–2646.

52 Rehan, M., Ibrahim, G.E., Mashaly, H.M. et al. (2022). Simultaneous dyeing and multifunctional finishing of natural fabrics with *Hibiscus* flowers extract. *Journal of Cleaner Production* 374: 133992.

53 Geng, H., Zhuang, L., Li, M. et al. (2020). Interfacial assembly of metal–phenolic networks for hair dyeing. *ACS Applied Materials & Interfaces* 12 (26): 29826–29834.

54 Wang, T., Zhao, J., Yang, Z. et al. (2022). Polyphenolic sunscreens for photoprotection. *Green Chemistry* 24 (9): 3605–3622.

55 Cheng, T.H., Liu, Z.J., Yang, J.Y. et al. (2019). Extraction of functional dyes from tea stem waste in alkaline medium and their application for simultaneous coloration and flame retardant and bioactive functionalization of silk. *ACS Sustainable Chemistry & Engineering* 7 (22): 18405–18413.

56 Kesbiç, F.I. and Gültepe, N. (2023). Bioactive components, sun protective properties, and total phenolic contents of halobacterial extracts. *Biochemical Systematics and Ecology* 108: 104647.

57 Sothornvit, R. and Krochta, V. (2016). UV protection of bio-based films made of polyphenol extracts from rice bran. *Food Hydrocolloids* 57 (16): 23–29.

58 Jóźwiak, T., Filipkowska, U., Struk-Sokołowska, J. et al. (2021). The use of spent coffee grounds and spent green tea leaves for the removal of cationic dyes from aqueous solutions. *Scientific Reports* 11 (1): 9584.

59 Ahmad, M.A., Puad, N.A.A., and Bello, O.S. (2014). Kinetic, equilibrium and thermodynamic studies of synthetic dye removal using pomegranate peel activated carbon prepared by microwave-induced KOH activation. *Water Resources and Industry* 6: 18–35.

60 Lopes, E.C., Santos, S.C., Pintor, A.M. et al. (2019). Evaluation of a tannin-based coagulant on the decolorization of synthetic effluents. *Journal of Environmental Chemical Engineering* 7 (3): 103125.

61 Zheng, X., Xie, X., Yu, C. et al. (2019). Unveiling the activating mechanism of tea residue for boosting the biological decolorization performance of refractory dye. *Chemosphere* 233: 110–119.

62 Richardson, J.J., Liao, W., Li, J. et al. (2022). Rapid assembly of colorless antimicrobial and anti-odor coatings from polypfhenols and silver. *Scientific Reports* 12 (1): 2071.

63 Gulati, R., Sharma, S., and Sharma, R.K. (2022). Antimicrobial textile: recent developments and functional perspective. *Polymer Bulletin* 79 (8): 5747–5771.

64 Yusuf, M. and Shahid, M. (ed.) (2022). *Emerging Technologies for Textile Coloration*. Boca Raton: CRC Press.

65 Rathinamoorthy, R. and Thilagavathi, G. (2022). Herbal antibacterial agents as odour control finish in textiles. In: *Odour in Textiles: Generation and Control* (ed. G. Thilagavathi and R. Rathinamoorthy), 117–142. Boca Raton: CRC Press.

66 Yusuf, M. (2017). Natural dyes from indigoid-rich plants: an overview. In: *Plant-Based Natural Products: Derivatives and Applications* (ed. Shahid-ul-Islam), 27–46. Hoboken, UK: Scrivener Publishing.

10

Polyphenols for Sensing/Biosensing Application

Elia Grueso[1] *and Rosa M. Giráldez-Pérez*[2]

[1] *University of Seville, Department of Physical Chemistry, Seville, Spain*
[2] *University of Córdoba, Departments of Cellular Biology, Physiology and Immunology, Córdoba, Spain*

10.1 Introduction

Polyphenols are polyhydroxyphenols with multiple phenolic structural units in their skeleton widely distributed in nature and from which can be derived strong antioxidant properties. For this reason, most of them are composed of natural compounds and plants, and one of the most studied is the phytochemicals class. Polyphenols from plants are secondary plant metabolites with various chemical structures. These polyphenols generally contain carbohydrate units and acylated sugars at different positions in their system. However, those in a smaller proportion corresponding to synthetic or semi-synthetic polyphenols should also be considered [1].

The most common polyphenol classification method is based on considering the number of phenolic rings and the structural elements or functional groups attached to the coils [2]. Other classifications are based on their sources of origin or primary biological functions [3]. Attending to the type based on the number of phenolic rings, the main classes are flavonoids and no flavonoid compounds that include phenolic acids, lignans, and stilbenes (Figure 10.1) [4-5]. Different subclasses can be considered in the case of flavonoids, including anthocyanidins, anthocyanins, flavanols, flavanones, flavones, flavonols, and isoflavones. The structural diversity of flavonoid compounds is given by differences in their hydroxylation pattern and the central pyrin ring that can present distinct oxidation states [6]. The phenolic acids subclass also includes a wide range of compounds, predominantly benzoic acid and cinnamic acid derivatives, constituting the main form of no-flavonoid compounds [7]. Importantly, this kind of

Science and Engineering of Polyphenols: Fundamentals and Industrial Scale Applications,
First Edition. Edited by Chandrabhan Verma.
© 2024 John Wiley & Sons Inc. Published 2024 by John Wiley & Sons Inc.

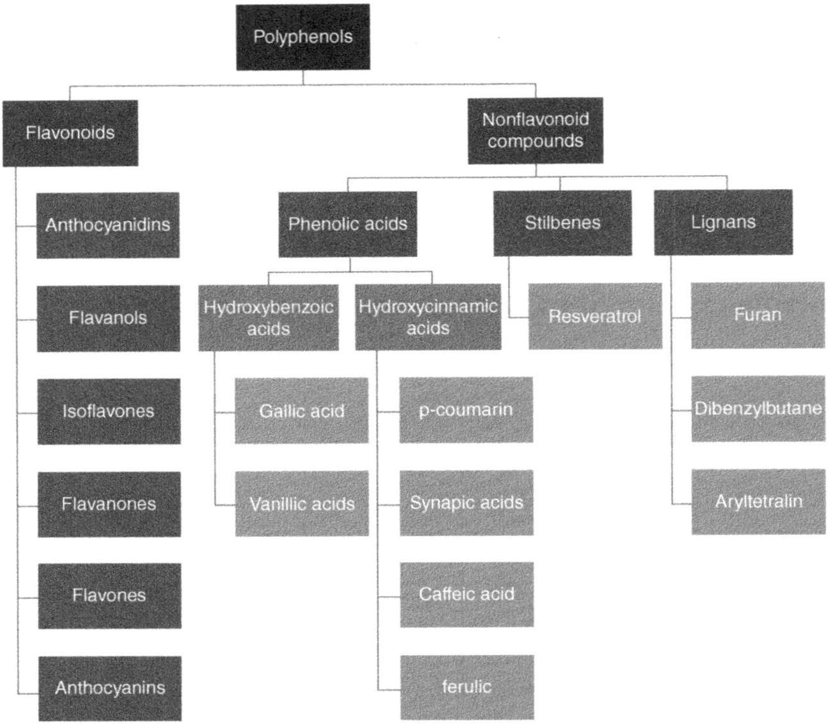

Figure 10.1 Classification of principal polyphenolic compounds.

polyphenols derivatives are generally found conjugated with other structures such as glucose, quinic acid, other polyphenols, or other components of the plant origin. They can be subdivided into hydroxybenzoic acids, including gallic and vanillic acids, and hydroxycinnamic acids, including synaptic acids, ferulic, *p*-coumaric, and caffeic acids. The structure of stilbenes is characterized by the presence of two phenyl moieties linked by methylene groups [6]. Among stilbenes, resveratrol structure is worthy of mention due to its excellent antioxidant properties [8] and its role in strengthening muscle cells [9]. Besides, resveratrol is a dietary polyphenol found in natural fruits or food products, and it can be distributed in two isomeric forms, trans and cis. Its medical applications highlight its use in treating diabetes [10] and cancer [11]. Finally, lignans are characterized by two propyl benzene units linked together between the β-position in C8 of the propane side chains with different substituents in the C9 and C9′positions. They are mainly accessible and widely distributed in angiosperms, gymnosperm, and many kinds of food such as legumes, seeds, and vegetable oils. Among lignans, highlight furan, dibenzylbutane, and aryltetralin subclasses [12]. Another detailed classification based on the Phenol-Explorer [13–15] databases is presented in Figure 10.2.

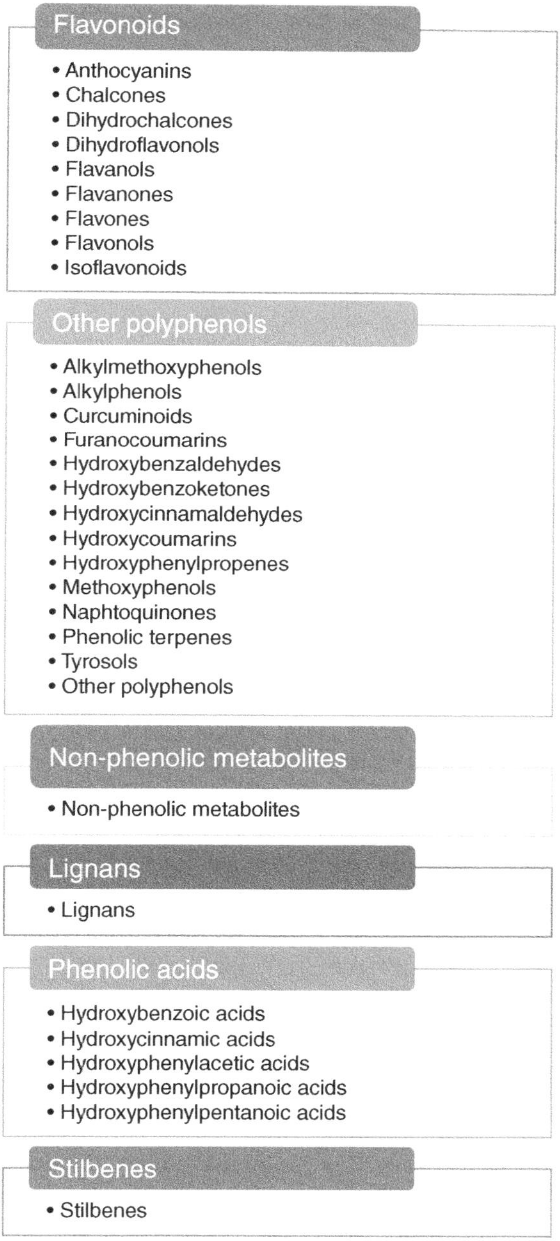

Figure 10.2 Classification of the types of Polyphenols. Based on the Phenol-Explorer database [13–15].

The Phenol-Explorer database lists the content of approximately 500 polyphenols in some 400 foods. The database shows pharmacokinetic data of approximately 400 metabolites in biofluids after eating foods rich in polyphenols. It also provides data on the losses of these elements by cooking some 155 foods, the retention factors being the changes in the polyphenol content after food transformation.

10.2 Electrochemical Biosensors of Phenolic Compounds Based on Polyphenol Oxidases

In conjunction with proteins, polyphenols are especially relevant for electrochemical biosensor fabrication. These structures' combined use can serve for sensing phenolic compounds in different fluids and matrices [16]. For example, some electrochemical biosensors have been constructed using polyphenol oxidases (PPO). As is known, a biosensor is a device used to recognize or detect interactions between some biological components like enzymes, cells, bacteria, viruses, or microorganisms as targets. The obtained measurable signal originated in such an interaction depends on the target concentration and the media used. Precisely, based on this variation, it is possible to quantify the target concentration in a specific sample [17]. Two components constitute the biosensor (Figure 10.3): (i) the bioreceptor, which is it is own biochemical dispositive for analyte recognition, and (ii) the transducer, whose primary role is to convert the chemical or physical signal provided by the bioreceptor in the electrochemical response [18].

In this concern, PPO constitute an attractive option in biosensing applications due to their capacity to mediate electron-transfer reactions even without the participation of other cofactors. Thus, different electrochemical biosensors have been constructed to detect phenolic compounds based on PPO [16, 19]. Detection of phenolic compounds is important for many applications in human health, the food industry, and the environment. In human health, they have been fundamentally

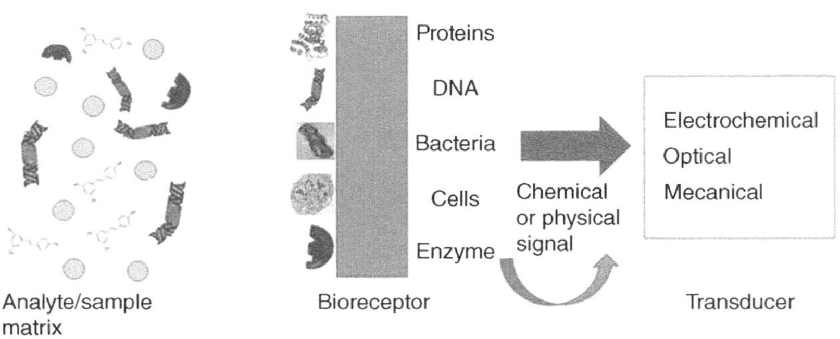

Figure 10.3 The basic scheme of the main components of a biosensor.

used as endocrine disruptors, antioxidants, or anticancer drugs [20]. In contrast, from the environmental perspective, their rapid detection is helpful because some are chemical pollutants. Phenolic compounds are present in industrial residues and effluent sources like dyes, plastics, or pharmaceutics drugs, being also toxic to human health and the reproductive process [21]. Another concern regarding phenolic compound determination comes from a food perspective, as they are analytical tools for food quality control. Despite several techniques frequently used to detect polyphenols, such as UV-visible and fluorescence spectroscopy, chemiluminescence, electrophoresis, or chromatography, they could be more practical [22]. In this sense, they involve complex and extensive sample preparation protocols, personnel with high qualifications to acquire and analyze the data, and expensive reactants. Thus, electrochemical biosensors based on PPO and proteins constitute a beautiful alternative to overcome these problems. In such a way, PPO and proteins form part of the active layer for which some enzymes like tyrosinase and laccase are the most used. The obtained devices are portable and permit the in situ analysis of the polyphenols, thus providing clear advantages.

Different strategies can be used to immobilize PPO in the matrix that conforms to the biosensor. Among them highlight the encapsulation method, adsorption, covalent binding, cross-linking, and entrapment [23]. Moreover, the enzyme used for preparing the biosensor should satisfy some special criteria: (i) the enzyme should catalyze the analyte reaction selectively, at least one reagent or product should be electroactive in the case of amperometric or voltammetry-based biosensors; (ii) the working potential should be the lowest possible by selecting the appropriate enzyme to improve biosensor selectivity, and (iii) Both the matrix and the enzyme should be compatible to avoid problems derived from stability, sensitivity, or the presence of possible denaturation processes in the enzyme molecule.

In this regard, the type of matrix used to stabilize the biosensor and immobilize the active layer plays a vital role in optimizing biosensor response. Different materials can be used for this purpose such as polyvinylpyrrolidine (PVP), chitosan biopolymer, starch macromolecule, bovine serum albumin (BSA) protein, β-cyclodextrin or cellulose-based biopolymers [24]. Once the general aspects of biosensors have been numbered, we briefly described some examples of biosensors for polyphenol determination based on PPO, focusing on the sensors constructed based on the oxidases tyrosinase (Tyr) and laccase (Lac). Tyr and Lac are the enzymes most used to monitor antioxidants like phenolic compounds.

10.2.1 Electrochemical Biosensors of Phenolic Compounds Based on Tyrosinase (Tyr)

Tyrosinase is a copper-containing oxidase enzyme that catalyzes the fabrication of phenolic pigments like melanin. This enzyme is one of the most commonly used for sensing polyphenols in samples of different nature. Some examples of tyrosinase as a biosensor for phenolic compounds are given in Table 10.1.

Table 10.1 Comparison of Tyrosinase biosensors with different architectures for Polyphenols determination.

Biosensor type	Technique	Sample and solvent conditions	Linear range	LCR	Polyphenol	Ref.
Cross-linking SPE/Tyr/HAS/GA	Amperometric	Natural infusions 0.1 M acetate buffer, pH 4.5	[1.0–109.0] μM [5.82–13.1] μM	26 μM 0.71 μM	Catechol	[25]
Tyr/PO_4-Ppy/Pt	Amperometric	Environmental waters 0.1 M PBS buffer, pH 7.0	[10–120] μM	0.84 μM 1.71 μM 8.54 μM 5.61 μM 5.24 μM 7.86 μM	Catechol Phenol 2-bromophenol 2-chlorophenol 2-iodophenol 2-fluorophenol	[26]
Tyr/GO/GA/GCE	Amperometric	Real samples of wine. Deoxygenated 0.1 M PBS buffer, pH 7.0	[0.05–50] μM	0.03 μM	Catechol	[27]
GPH-MnPc-Tyr/SPE	Voltammetric Square-wave voltammetric	Real samples of green coffee beans. 0.1 M PBS buffer, pH 7.0	[0.1–678.5] μM [0.1–678.5] μM	1.4 μM 0.23 μM	Chlorogenic acid	[28]
PDT/PPO	Voltammetric	0.1 M sodium phosphate buffer solution (pH 6.5)	[10–100] μM [10–200] μM	5 μM 5 μM 50 μM 500 μM 100 μM 500 μM 500 μM	p-cresol 4-chlorophenol Phenol 3-chlorophenol Dopamine Epinephrine L-Dopa	[33]
CoP-Tyr	Voltammetric	Commercially dopamine medication. 0.1 M NAPB buffer solution (pH 8.0)	[2–30] μM	0.43 μM	Dopamine	[38]

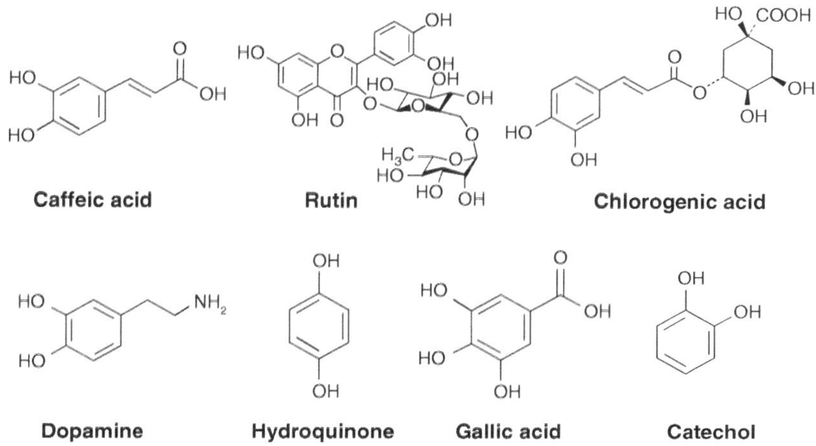

Figure 10.4 Structures of the phenolic compounds detected by using an electrochemical technique based on PPO.

Moreover, in Figure 10.4 are some examples of phenolic compound structures detected using PPO. A tyrosinase-based amperometry biosensor was developed for detecting catechol using different immobilization methods. For this purpose, the entrapment with water-soluble polyvinyl alcohol, cross-linking using glutaraldehyde (GA), or using both GA and human serum albumin (HSA) were tested. Among them, the second cross-linking method was the most effective, giving a better linear calibration range (LCR) and sensitivity. This method employed bifunctional reactants contributing to intermolecular bond formation enzyme-enzyme for sensing catechol. The phenolic compound catechol is an essential molecule from a biological point of view found in lignins and resins. It has two hydroxyl groups attached to a benzene ring and can be used as dyeing, photographic developer, or antiseptic. The screen-printed carbon electrode (SPE)/Tyr/HAS/GA electrochemical biosensor presented a wide linear range of catechol detection responses from 1 to 109 μM and a detection limit of 26 μM [25]. The matrix comprised natural infusions of *Salvia microphylla*, *Lippia dulcis*, and *Lippia alba*. In another study, catechol and other polyphenols were determined using a tyrosinase amperometric-based biosensor employing an electropolymerized phosphate-doped polypyrrole film designated Tyr/PO_4–Ppy/Pt [26]. The polypyrrole film was electrochemically synthesized in this configuration on a Pt disk electrode from a 0.1 M pyrrole and 0.1 M phosphate buffer solution (PBS) (pH 7) water solution. The tyrosinase-based sensor yields good bioactivity and high sensitivity for phenolic compound determination in environmental waters. In this condition, catechol was determined with the lowest detection limit of 0.84 μM, a value lower than the one obtained by Rodríguez-Sevilla et al. [25].

The response current of the biosensor with catechol concentration in PBS was linear from 10 to 120 μM. The other studied phenolic compounds, phenol, 2-bromophenol, 2-chlorophenol, 2-iodophenol, and 2-fluorophenol, were also determined with limits of detection of 1.71, 8.54, 5.61, 5.24, and 7.86 μM, respectively. The Tyr/PO_4–Ppy/Pt biosensor showed high sensitivity, versatility, and stability for sensing phenolic compounds with promising applications in the fields of environment and industry. In a more recent study, a graphene oxide-modified tyrosinase electrode (Tyr/GO/GA/GCE) was designed for phenol derivatives quantification like catechol based on the use of the covalent bonding between graphene oxide and Tyr via the –OH group of glutaraldehyde [27]. The sensor response was proved in different real samples of wine, showing its great versatility for polyphenol determination. In the specific case of catechol, a wide sensing linear range from 0.05 to 50 μM was obtained, with a LCR of 0.03 μM in a deoxygenated 0.1 M PBS (pH 7.0) as solvent media. Compared with other previous biosensors based on Tyr, the TYR/GA/GO/GCE biosensors highlight several advantages such as easy construction protocol, high response rate and response time, extended linear range, high stability, and long lifetime.

Another frequently determined phenolic compound based on the use of Tyr is chlorogenic acid (CGA). CGA, 5-*O*-caffeoylquinic acid, is a polyphenol commonly used as a dietary product and in cats as a curative drug against oxidative stress. This compound plays an active key role in treating several diseases like cancer. In recent work, a novel tyrosinase biosensor based on the use of SPE, also modified with graphene and manganese phthalocyanine (GPH-MnPc-Tyr/SPE), has been developed to detect CGA [28]. Tyr was immobilized using a cross-linking technique with glutaraldehyde technique, and the response to CGA content was measured by voltammetry (CV) and square-wave voltammetry (SWV). The obtained LCR measured in a 0.1 M (pH 7.0) PBS varies with the technique used, giving a limit of detection of 1.4 and 0.23 μM with CV and SWV techniques, respectively. The response with CGA content was linear at 0.1 to 678.51 μM. The method was applied to real samples of green coffee beans and complex matrices containing CGA, giving better biosensitivity than other sensors and biosensors for CGA determination previously published [29–32]. Thus, the authors concluded that the inmovilization of Tyr produced a synergic effect in CGA biodetection.

The method reported by Védrine et al. is worth mentioning due to its versatility for detecting a large number of polyphenolic compounds, where dopamine sensing is accomplished among them. In this sense, several phenolic compounds and different herbicides were determined based on an amperometric Tyr-based biosensor that was constructed employing an electrogenerated polythiophene film (PEDT) in a glassy carbon electrode as entrapment support [33]. Dopamine (DA) is a relevant neurotransmitter implicated in many medical diseases like Parkinson's, Alzheimer's, attention deficit hyperactivity disorder, or schizophrenia [34]. Monophenolic and di-phenolic compounds were detected by this method with different sensitivity. In the case of mono-phenolic compounds, the sensitivity decreased in the following order: p-cresol>4-chlorophenol>phenol>3-chlorophenol, and this order was: catechol>dopamine>epinephrine>L-Dopa for ortho-diphenol derivatives. The sensor response was linear in the range of 10–100 μM and from 10–200 μM for mono- and ortho-derivate, respectively. The limit of detection was 5 μM for p-cresol and 4-chlorophenol, 50 μM for phenol, 100 μM for dopamine, and 500 μM for epinephrine, L-Dopa, and 3-chlorophenol. The authors conclude that pH and temperature optimization can help gain sensitivity compared with other immobilization tyrosinase-based biosensors reported in the literature [35–37]. DA detection was also accomplished with a novel Tyr-based electrochemical biosensor based on the use of gold modified with cobalt(II)-porphyrin (CoP) film as an electrochemical mediator and enzyme support for Tyr cross-linking [38]. The sensor has a detection limit of 0.43 μM and a linear concentration response in the 2–30 μM range. The current response was optimized at pH 8.0 and 0.1 M NAPB solution. The sensor response was investigated in the presence of commercially available dopamine drugs, giving good recovery and suitable reproducibility parameters. Taking into account the results given in Table 10.1, it can be concluded that the Tyr/GO/GA/GCE complex system was the most convenient Tyr biosensor to detect catechol with a LCR of 0.03 μM and in the case of dopamine a low LCR was obtained for the CoP-Tyr biosensor in which the porphyrin macrocycle compound was used improving tyrosinase biocompatibility and reproducibility.

10.2.2 Electrochemical Biosensors of Phenolic Compounds Based on Laccase (Lac)

The enzyme laccase belongs to the blue multi-copper-oxidase family; these enzymes highlight their ability to oxidize various organic substrates and reduce molecular oxygen to water [39]. This ability to catalyze phenolic and nonphenolic compounds oxidation in conjunction with the possibility of producing the reduction O_2- H_2O makes this enzyme fundamental in detecting phenolic compounds and derived commercial applications. Different methods to detect phenolic compounds based on the Lac enzyme were summarized in Table 10.2. As shown in the table, the rutin flavonoid was determined based on a voltammetric Lac biosensor in acetate buffer at pH 4.0 employing as an immobilization method the encapsulation in microspheres of chitosan cross-linked with tripolyphosphate [40]. The analytic curve shows two linear ranges from 0.59 to 3.92 μM and 5.82 to 13.1 μM, with detection limits of 0.062 and 0.71 μM, respectively. Another work determined rutin based on a voltammetric biosensor containing laccase and imidazolium bis(trifluoromethylsulfonyl)imide ionic liquid [41]. In these case, the linearity ranges in response to rutin concentration was observed by using different configurations (from 4.77 to 46.2 μM, from 5.84 to 53.6 μM and from 5.84 to 53.6 μM by

Table 10.2 Comparison of Laccase biosensors with different architectures for Polyphenols determination.

Biosensor type	Technique	Sample and solvent conditions	Linear range	LCR	Analyte	Ref.
Microspheres of chitosan cross-linked with tripolyphosphate	Voltammetric	Acetate buffer, pH 4.0	[0.59–3.92] μM [5.82–13.1] μM	0.062 μM 0.71 μM	Rutin	[40]
Lac/imidazolium bis(trifluoromethylsulfonyl) imide ionic liquid	Voltammetric	Commercial and pharmaceutical samples. 0.1 M acetate buffer, pH 5.0	[4.77–46.2] μM (BMI × Tf(2)N-Lac) [5.84–53.6] μM DMI × Tf(2)N-Lac) [5.84–53.6] μM (DMI × Tf(2)N-Lac)	0.45 μM 0.69 μM 0.78 μM	Rutin	[41]
Lac/Si/MWCNTs/SPE	Voltammetric	0.1 M PBS buffer, pH 6.0	[1.3–85.5] μM	0.42 μM	Dopamine	[42]
SiO_2-PA nanocomposites	Amperometric	Pharmaceuticals and rabbit blood serum samples. 0.1 M PBS, pH 6.0	[0.99–103.1] μM	0.26 μM	Dopamine	[43]
Lac-halloysite nanotube and imidazolium zwitterionic surfactant	Voltammetric	Pharmaceuticals samples. 0.1 M Phosphate buffer, pH 7.0	[0.99–67.8] μM	0.252 μM	Dopamine	[44]
Nafion/Lac-A-CZUF/GCE	Voltammetric	0.1 M acetate buffer, pH 5.0	[0.166–7.0] μM	0.166 μM	Catechol	[45]
Lac–AuNPs cross-linked with fibers (CZUF)	Amperometric	0.1 M sodium phosphate buffer (pH 5.0)	[0.20–209.7] μM	0.085 μM	Catechol	[46]
Lac - polyether sulphone membrane	Amperometric	12% ethanol acetate buffer, pH [3.5–4.5]	[10–80] μM [10–60] μM	10 μM 10 μM	Caffeic acid Catechin	[47]
CGE-laccase	Amperometric	White, rose, and red wines. Citrate buffer, pH 5.0	[0.0056–0.56] μM [0.0176–4.7] μM	0.004 μM 0.0153 μM	Caffeic acid Gallic acid	[48]

using BMI·Tf2N-laccase, DMI·Tf2N-laccase and TDMI·Tf2N-laccase, respectively). The sensor response was measured in commercial and pharmaceutical samples employing a 0.1 M acetate buffer solution at pH 5.0, giving detection limits that vary from 0.45 to 0.78 μM, depending on the dispositive used. Hydrophobic ionic liquids provide high sensitivity, stability, and reproducibility in the sensor response.

Regarding dopamine detection (DA), a laccase/Si/MWCNTs/SPE biosensor was constructed based on the Lac enzyme, enabling the in situ determination of DA with good sensitivity, selectivity, and reproducibility [42]. For this configuration, silica spheres on the MWCNTs-doped screen-printed electrode were employed for immobilizing Lac. A linear response was obtained measuring in PBS 0.1 M at pH 5.0 that goes from 1.3 to 85.5 μM, with a LCR of 0.42 μM. DA was detected using a laccase-based amperometric biosensor that employed SiO_2-PA nanocomposites, which confers a biocompatible microenvironment for Lac, providing good electroconduction and maintaining enzymatic activity [43]. The sensor has a wide linear range response from 0.99 to 103.10 μM with a detection limit of 0.26 μM in 0.1 M PBS (pH 6.0). The suitability of the biosensor was tested in pharmaceutical and blood serum samples, giving excellent results. A recent study successfully determined DA using a laccase-based voltammetric biosensor based on a laccase-halloysite nanotube and imidazolium zwitterionic surfactant [44]. Using these types of surfactants in the dispositive enhances both the catalytic activity and stability of the biosensor. The calibration plot showed a linear range from 0.99 to 67.8 μM employing the SWV technique, with a detection limit of 0.252 μM n PBS (0.1 M, pH 7.0). The biosensor was proven in real pharmaceutical samples, showing its excellent efficacy in biological recognition processes.

Regarding catechol, this polyphenol was determined using a biosensor based on the preparation of laccase–gold nanoparticles cross-linked with zein ultrafine fibers (CZUF) in actual samples [45]. The voltammetric biosensor provides a wide linear response range from 0.166 to 7 μM with a limit detection limit of 0.166 μM in 0.1 M acetate buffer solution (pH 5.0). On the other hand, catechol content was determined based on a laccase amperometric biosensor immobilizing the enzyme on graphene-cellulose microfibers (GR-CMF) and using screen-printed modified carbon electrodes (SPCE) [46]. In this new configuration, the sensor was a linear response in the range from 0.2 to 209.7 μM, with a limit of detection of 0.085 μM measured in 0.1 M sodium phosphate buffer at pH 5.0. The higher sensitivity selectivity toward catechol in the presence of active biomolecules and phenolic compounds could be due to the complex strategy used for Lac immobilization. In a more recent study, catechin, caffeic acids, and other polyphenols (gallic acid, malvidin, quercetin, rutin, and trans-resveratrol) were determined based on laccase-based amperometric biosensor where Lac was immobilized on a polyether sulphone membrane, at pH 4.5. The samples were measured in aqueous and 12% ethanol acetate buffer [47]. Despite sensitivities of 0.02443 $mA\,M^{-1}$ for catechin and 0.02401 mA M^{-1} for caffeic acid, this parameter cannot be measured for the other polyphenols tested. This negative result for other polyphenols can be related to the existence of a reaction between the substrate solution and the own membrane that interferes with the measurement. Caffeic acid and gallic acid were also quantified based on another laccase biosensor immobilized via cross-linking with glutaraldehyde onto a glassy carbon electrode [48]. The analytes were measured in real samples of white, rose, and red wines with a linear range of response from 0.0056 to 0.56 μM and LCR of 0.004 μM for caffeic acid and a range of 0.0176 to 4.7 μM and LCR of 0.0153 for gallic acid, measured in citrate buffer at pH 5.0. The biosensor showed good analytical performance with possible applications in the wine industry.

10.2.3 Other Biosensors for Phenolic Compounds Quantification

Other examples of biosensors of interest in the bibliography that used Lac OR Tyr in their structure are collected in Table 10.3. Recently, a novel fiber-optic sensor for the detection of phenolic compounds in water media at pH 7.0 has been developed based on the immobilization of Tyr in polyacrylamide gel (LPFG coated with enzyme-entrapped gel layer) [49]. Polyphenols like catechol, 4-chlorophenol, m-cresol, and phenol were determined with high sensitivity. The sensor response was followed by registering the changes in the resonance wavelength with polyphenol concentrations due to the enzymatic reaction between the target and Tyr. The sensor gives a linear response from 1 to 800 μM, with a limit of detection of 6.8 μM for catechol. Caffeic acid and L-DOPA were recently quantified using an optical biosensor that employs a novel platform based on a laccase-hydrophobic chimera protein [50]. Both polyphenols were tested in different solvents with a wide range of linearity range of 5–1000 μM and 10–1000 μM for L-DOPA in buffer and plasma media, respectively, and that goes from 50–4000 μM for caffeic acid in the buffer. The limit of detection was in the order of μM concentration.

Using nanoparticles and nanostructures as material incorporated in the matrices has recently been especially relevant. This is not only for its essential role in their interaction with biomolecules used in the active layer but also to enhance the analytical device response improving specific parameters such as conductivity and surface area, among others. These aspects contribute to the increase in the electrochemical signal and avoid problems related to overpotentials. In this sense, a novel nanohybrid structure based on the assembly of an amperometric Lac biosensor and graphene oxide was developed by Boujakhrout et al. [51]. A two-step modification approach was needed to obtain the Lac/GC–rGO structure: (i) first, the nanostructure was formed via functionalization by covalent interaction of the graphene oxide with reactive epoxy groups of (3-glycidyloxypropyl) trimethoxysilane followed by cross-linking with glycol chitosan molecule, and (ii) second, the Lac enzyme was covalently immobilized through cross-linking

Table 10.3 Other biosensors based on Lac and Tyr for Polyphenols determination.

Biosensor type	Technique	Sample and solvent conditions	Linear range	LCR	Analyte	Ref.
Tyr-LPFG	Transmission spectroscopy	Air-saturated 0.1 M Phosphate buffer, pH 7.0	[1–800] μM	6.8 μM	Catechol	[49]
				62.7 μM	4-chlorophenol	
				52.8 μM	m-cresol	
				89.1 μM	phenol	
Lac-chimera POXA1b-Vmh2	UV-visible spectrophotometry (A_{420nm})	0.1 M Na-citrate buffer, pH 3.0 and plasma	[5–1000] μM	3 μM	L-Dopa (buffer)	[50]
			[10–1000] μM	2 μM	L-Dopa (plasma)	
			[50–4000] μM	1 μM	Caffeic acid (buffer)	
Lac/GC–rGO	Amperometric	Herbal tea samples. 0.1 M sodium phosphate buffer, pH 6.0	[0.2–15] μM	76 nM	Catechol	[51]
			[5.0–30.0] μM	2.5 μM	Hydroquinone	
			[9.9–30.0] μM	5.0 μM	3,4-xylenol	
			[5.0–30.0] μM	4.9 μM	Phenol	
			[5.0–19.0] μM	4.3 μM	4-chloro-1-naphthol	
Lac/AA/$LuPc_2$	Voltammetric	NaCl 0.1 M, phosphate buffer 0.01 M, pH 7.0 water solution.	[4–150] μM	5.10 μM	Vanillic acid	[52]
				0.459 μM	Catechol	
				0.589 μM	Caffeic acid	
				0.518 μM	Hydroquinone	
				0.0418 μM	Gallic acid	
				1.87 μM	Pyrogallol	
PtNPs, BOT and Lac (1 : 6 : 1)	Voltammetric	Brewed coffees. 0.1 M Phosphate buffer, pH 4.0	[0.56–7.3] μM	0.18 μM	Chlorogenic acid	[53]
[(CHI)-(AuNP)-(CHI)-$(CuPcS)]_2$-Tyr	Amperometric	0.01 M phosphate buffer, pH 7	[2.4–20.0] μM	0.855 nM	Catechol	[54]

with glutaraldehyde in the nanostructure surface formed in the first step. Different phenolic compounds were tested using this nanohybrid, giving distinct responses in commercial herbal tea samples. For instance, in the case of catechol and hydroquinone, the linear range was from 0.2 to 15 μM and 5.0 to 30 μM, respectively, with a LCR of 76 nM and 2500 nM. A similar response to hydroquinone was registered for the other polyphenols studied, such as 3,4-Xylenol, Phenol, and 4-Cloro-1-naphtol. The amperometric behavior of the dispositive for sensing catechol is remarkable regarding the low LCR obtained and its stability and sensitivity. In early 2014, Rodríguez-Mendez and coworkers designed a multiple-enzyme multisensory system for sensing monophenols, orto-phenols, and triphenols phenolic compounds [52]. The bioelectronics tongue was composed of Tyr and Lac phenol oxidases incorporated into a biomimetic platform constituted by an arachidic acid Langmuir-Blodgett film in which lutetium bisphthalocyanine ($LuPc_2$) acted as an electron mediator. As a result, an array of three electrodes was obtained and designed as ascorbic acid (AA)/$LuPc_2$, Tyr/AA/$LuPc_2$, and Lac/AA/$LuPc_2$. The detection limits differed depending on the electrode-type, where Lac/AA/$LuPc_2$ offered better sensitivity for all the phenols explored. In such a way, the sensor showed a linear response from 4 to 150 μM, with detection limits varying from 0.0418 and 5.10 μM registered for gallic acid and vanillic acid, respectively. This voltammetric multisensory system is noted for its ability to discriminate among distinct phenolic compounds due to their chemical nature with high selectivity. More recently, another biosensor based on nanoparticles and Lac was developed by Salamanca-Neto and coworkers. In this case, graphite oxide, platinum nanoparticles, and laccase as enzymes were mixed to form a novel material for sensing CGA [53]. The sensor showed a linear response from 0.56 to 7.3 μM and a detection limit of 0.18 μM. The sensor was proven in natural coffee beverages and exhibited an excellent anti-interference response. The samples were also measured using the HPLC technique as a reference, giving results in good agreement with voltammetric determination. In another approximation, a combination between chitosan/AuNPs/phthalocyanine composite (LbL films) and Tyr or Lac phenol oxidases were designed to detect catechol, where LbL films were formed alternating layers of the chitosan cationic linker with the anionic electrocatalytic materials [54]. Subsequently, the nanostructure was formed via the deposition of the enzyme onto the LbL films. Among the different combinations of materials, the platform type $[(CHI)\text{-}(AuNPs)\text{-}(CHI)\text{-}(CuPcS)]_2$-Tyr is worthy of note, showing a LCR of 0.855 nM for sensing catechol and a linear range of detection from 2.4 to 20 μM measured by the amperometric technique in 0.01 M phosphate buffer pH 7. The authors concluded that the effective combination of materials and the use of the LbL technique is crucial to improve electrochemical detection via the efficient transmission of charges, higher roughness, porosity, and the increase in the number of sensing sites. The synthesized platform improves the LCRs obtained concerning previous similar platforms reported in the literature [52, 55–61].

In recent years, attention has also been focused on the sensors based on electropolymerization of natural phenolic compounds [62]. Combining electropolymerized natural phenolic antioxidants and carbon nanomaterials offers optimum electrochemical sensors for analytical applications. For instance, in early 1999, Cizewski and coworkers synthesized a modified-platinum electrode with eugenol (4-allyl-2-methoxyphenol) electropolymerized films [63]. The Polyeugenol/Pt system was then used to quantify dopamine (DA) over AA using differential pulse voltammetry (DPV) technique, employing a PBS at pH 7.4 as solvent media. As a result, DA was determined with a detection limit of 0.1 μM in a linear dynamic range from 0.1 to 50 μM. In another approach, a polyrutin/PIGE sensor was used to detect specific polyphenols using the DPV technique [64]. The poly rutin (Ru) modified paraffin-impregnated graphite electrode (WGE) designed using the electrochemical methods offers the possibility to detect epinephrine (EP), serotonin (5-HT), and AA in A 0.1 M PBS solution at pH 4.0 with good stability, sensitivity, and selectivity. As a result, a linear response range of [3–90] μM, [0.3–9.0] μM and [2.0–60] μM and a limit of detection of 0.8, 0.1 and 1.0 μM was obtained for EP, 5-HT, and AA, respectively. In 2014, Devadas and coworkers designed a Polycurcumin/GCE sensor for the determination of epinephrine (EP) and p-acetoaminophenol (AP) using a voltammetric technique [65]. As a result, a linear response of [4.97–230.76] μM and [0.99–230.76] μM and a low limit of detection of 0.05 and 0.1 μM were registered for EP and AP, respectively, in phosphate buffer solvent at pH 8. The sensor was also proved for the simultaneous determination of EP and AP in adrenaline injection ($1\,mg\,ml^{-1}$) solutions and paracetamol tablets to evaluate the feasibility of the dispositive. The designed poly CM (curcumin) film-modified GCE (glassy carbon electrode) electrode showed excellent stability, reproducibility, and rapid response toward EP and AP polyphenols simultaneous determination with promising applications in sensing actual samples. In 2015, Wei and coworkers designed a catechin-modified carbon paste electrode (polycatechin/CPE) for sensing dopamine and serotonin by the amperometric technique [66]. Linear concentration ranges of [10–780] nM and [30–2340] nM were obtained for DA and 5-HT, respectively, where the LCR was 0.5 and 3 nM for DA and 5-HT, respectively, measured in 0.1 M PBS buffer at pH 7.0. The resulting biosensor and more investigation into real-time interactions between catechin and neurotransmitters could help understand catechin's role in nervous excitement.

On the other hand, some sensor platforms based on carbon nanomaterials and electropolymerized natural phenolic antioxidants are worth mentioning because of their promising applications in quantifying neuromediators. For instance, GCE/MWCNTs electrode combined with poly(vanillic acid) as a sensitive layer resulted in a valuable platform for detecting DA, AA, and uric acid [67]. The amperometric sensor showed a linear response range of [5–120] μM and a detection limit of 4.5 nM for DA in urine samples. In another example, quercetin flavonoid was successfully detected in medicinal herbs extracts using a combination of

GCE/MWCNTs electrode and poly(gallic acid) in phosphate buffer at pH 6.0. [68]. As a result, quercetin was determined in a LCR of [0.075–25] μM with the lowest detection limit of 0.054 μM. In another approach, an MWNT (multiwalled carbon nanotube) electrode modified with multi-walled carbon nanotubes and poly-quecetin was designed to detect gallic acid, catechin, and epigallocatechin gallate (EGCG) phenolic compounds by using the voltammetric technique [69]. The samples were evaluated in phosphate buffer, pH 7.0 solution, giving a linear range of [0.50–10], [0.10–10], and [0.050–10] μM for gallic acid, catechin, and EGCG, respectively, with the detection limits of 0.10, 0.024 and 0.014 μM. The results showed that the new platform could be used as an alternative measure of tea's antioxidant properties. In conclusion, the high number and diversity of phenolic antioxidants found in nature and their capacity to electro-oxidate and polymerize on the surface of an electrode make this type of sensor very useful in biomedicine, food control, or environment control. Thus, in light of these promising characteristics, novel electrochemical sensors are expected to be designed in the future. However, this kind of sensor's selectivity and possible interferences in the measurement must be further evaluated.

10.3 Polyphenols for Metal Ions Sensing Applications

As it is known, polyphenols have been extensively used for metal ions complexation and subsequent precipitation. There are a high number of protocol in the industry that uses tannins or tannin-modified materials to produce metal ions complexation [70–71]. For instance, several centuries ago, the blue-blanch complexes obtained by Fe(III)/tannin complexation were employed for dyeing leather, clothes, and hair. Moreover, tannins and iron have been employed to fabricate writing inks [72–73]. In the early, McDonald and coworkers used tannins, low tannins and low molecular weight phenols for the precipitation of copper(II) and zinc(II) in wood preservation treatments [70]. Another application uses tannin and phenolic compounds to recover metal ions from wastewater [74]. Moreover, their role in biology is worthy of note, specifically in the processes related to iron bioavility or aluminum and copper mineral absorption, which are fundamental in plant defense and human nutrition [75]. Thus, a better understanding of the mechanism involved in plant iron deprivation requires a thoughtful study of polyphenols/iron(III) interactions. More recently, taking as a starting point these previous studies related to polyphenols/metal ions complexation and precipitations, numerous chemosensors based on polyphenols have been developed to detect metal ions in different media. In this section, some of the results of these studies are reviewed, paying particular attention to some specifications such as the type of matrix and solvent used for measurement, the limit of detection, and possible interferences.

In early 2000, Wu and coworkers synthesized an array-based polyphenol for sensing Fe^{3+}, Cu^{2+}, Co^{2+}, and Ni^{2+} metal ions. The polymer array was composed of different phenolic homopolymers and copolymers, taking mixtures of five different phenolic monomers (p-cresol, p-phenyl phenol, p-methoxyphenol, p-hydroxyphenylacetic acid, or p-hydroxybenzoic acid) as starting point, and then generating the array via peroxidase-catalyzed oxidative reactions [76]. The sensing procedure was based on the fluorescence changes registered in the intrinsic polyphenol fluorescence upon excitation at 322 nm and the addition of different ion metal content that can be added separately or mixed. The fluorescence was measured in a 50 Mm Bis-Tris propane buffer, pH 7.0. The maximum emission wavelength was variable (from 360.8 to 444.8 nm) depending on the polyphenol used as a monomer in the sensor-array fabrication. This approach permits distinguishing among metal ions at concentrations as low as 10–50 μM. Despite the method could help detect metal ions in the environment, in this early study, important parameters like the LCR, or possible interferences were not investigated or controlled. Another similar study was carried out by Kim and coworkers in which phenolic polymers containing fluorescein or calcein as fluorescent probes were prepared for sensing distinct metal ions such as Cu^{2+}, Fe^{2+}, Zn^{2+}, Mn^{2+}, and Fe^{3+} [77]. The polymer-based system was generated using soybean hull peroxidase (SBP) as a biocatalyst, where fluorescent probes were directly incorporated into the polymeric structures. In this case, the method permits distinguishing metal ions concentrations as low as 10 μM. The samples were measured in 50 mM Bis-Tris buffer (pH 7.0) at 496 nm wavelength. The possible interference of highly concentrated solutions of NaCl was tested as it could be an important parameter to control in detecting metal ions in seawater. The results showed that even at high NaCl concentrations of 0.5 M, the fluorescent behavior of the system does not change, supporting the method's validity.

Coumarin derivatives have long been used to detect cationic and anionic metal ions due to their intense and stable fluorescence emission, excellent optical properties, structural flexibility, and great biocompatibility. For this reason, many fluorescent chemosensors for sensing ions include these polyphenols molecules [78]. As an example of this, in 2009, a coumarin derivative-based sensor molecule (4-Chloro-3-nitro coumarin) was synthesized by Suresh and Das for sensing calcium and magnesium ions, among others [79]. The binding Ca^{2+}/coumarin and Mg^{2+}/coumarin strongly enhanced the fluorescence emission of the polyphenol, while a weak quenching was observed for other metal ions like Hg^{2+}, Ni^{2+}, Fe^{3+}, and Co^{2+}. A maximum in the fluorescent emission was monitored at 370 nm upon excitation at 315 nm in acetonitrile media. A detection limit of 50 ppb and a linearity range from 1 to 50 μM were found to detect both metal ions. In more recent work, a base Schiff of coumarin was synthesized for evaluating selective detection of Al^{3+} with possible applications in live cell imaging and diagnosis [80]. In addition to the harmful effects of Al^{3+} in the aquatic environment, exposure to high concentrations of this metal is highly toxic to human life. Its high toxicity converts this metal ion into one of the principal causes of distinct neurological diseases like Alzheimer's or Parkinson's [81]. The sensing method is based on following the increase in the fluorescence emission of the system at 461 nm as a function of Al^{3+} in different solvent conditions. It was checked that several ions like Fe^{3+} and Cr^{3+} could slightly interfere with the measurement procedure. The method generated a linear response in an aqueous buffer (0.01 M HEPES buffer: methanol 1:9; v/v; pH 7.4) from 1 to 60 μM concentration interval, with a detection limit for Al^{3+} of 1.34 μM. Moreover, the developed chemosensor can also detect Al^{3+} ions in human cervical cancer cells.

In another example, Liu and coworkers designed a coumarin derivative-based chemosensor to detect Ag^{+} and Ce^{3+} metal ions selectively in an aqueous solution by UV-visible and fluorescence spectroscopy [82]. Different metal ions were analyzed in a 0.1 M HEPES buffer EtOH/H_2O (9:1, v, v) solution, pH 7.16 to evaluate them as possible interferences. Thus, when Ag^{+} was added to the solution containing the coumarin dye, an absorption band with a maximum of over 580 nm appeared, and the color of the solution changed from brown to pale yellow. However, the other metal ions (Co^{2+}, Li^{+}, Zn^{2+}, Fe^{3+}, or Al^{3+}, among others) analyzed in the same condition do not produce changes in the spectra, proving the sensor's selectivity for Ag^{+} ions. In the case of Ce^{3+} ions, this was detected also over other metal ions including Fe^{3+}, Al^{3+}, Cr^{3+}, or La^{3+} following the changes in the fluorescence emission band at 350 nm. As a result, the LCR of Ce^{3+} ions was 0.207 μM. The linear range response and the LCR of Ag^{+} were not calculated.

Garcia-Beltrán and coworkers designed an affordable sensor based on a coumarin derivative to detect Hg^{2+} selectively by UV-visible and fluorescent techniques [83]. The detection limit was 0.22 and 4.5 μM calculated using the colorimetric and the fluorescent sensor, respectively, and a solution of 20 mM PBS buffer, pH 6.0 as solvent media. The sensor response was proved in the presence of the essential biothiols, glutathione, and cysteine, evaluating its possible applications in imaging living cells. Subsequently, Mondal and coworkers designed a triazole-bridged coumarin-benzimidazole ESIPT-based chemosensor to detect Cd^{2+} ions via fluorescence spectroscopy in C6 and Hep G2 rat glial and human liver cells, respectively [84]. A LCR of 14 nM was calculated in a pH range from 5 to 9 following the formation of the complexes between the coumarin derivative and Cd^{2+} ions. Other ions, such as Ni^{2+}, Cu^{2+}, Ag^{+}, and Cr^{3+}, were tested in ACN (acetonitrile) solvent to discard them as possible interferences. These metal ions did not produce changes in the fluorescence emission of the system at 485 nm, proving the selectivity of the chemosensor for detecting Cd^{2+} ions. In another study, Cs^{+} ion was selectively detected using a combination of calixarene bis (crown-6-ether) and coumarin, resulting in two water-soluble fluorescent chemosensors [85]. As increasing concentrations of Cs^{+} were added to the solution, an increase in the fluorescent emission together with a red shift in the UV-visible spectra of both coumarin-based sensors was registered, giving a limit of detection of 1.6 and 0.77 μM for each coumarin structure.

In recent work, Zhang and coworkers evaluated the interaction of procyanidins [PC, epicatechin 16 (4→8) catechin] and the catechin monomer with distinct metal ions by fluorescence spectroscopy [86]. The complexation efficiency was calculated by measuring the Stern-Volmer constant (K_{SV}) at 320 nm of each PC-ion complex in acetate buffer solution, pH 6.0. It was proved that adding the BSA to the samples enhanced the complexation efficiency by increasing the value of K_{SV}. Cu^{2+}, Zn^{2+}, Sn^{3+}, Al^{3+}, and Fe^{3+} metal ions were tested at concentrations as low as 1.0, 1.0, 0.2, 0.2, and 0.8 μM, respectively. However, the limit of detection was not calculated.

Recently, the more complex matrix from natural extracts containing polyphenols has been evaluated as a possible sensor for metal ions. In particular, flavonoids in phytochemical plants have been proven as they exhibit intense fluorescence spectra when excited with UV light. For instance, leaves of sage (*Salvia officinalis*), a plant rich in flavonoids and other polyphenolic compounds, were extracted and used to detect a high variety of metal ions [87]. Ge^{4+}, Mg^{2+}, K^{+}, and Na^{+} were strongly quenched, while Cr^{3+}, Pb^{2+}, and Co^{2+} were highlighted by their excellent intense fluorescence when excited at 365 nm in ultrapure water media. The fluorescent behavior of Co^{2+} ions was studied as a model system for proving the effectivity of the sensor in detecting metal ions. The lowest concentration registered for this cation was 0.2 mg l^{-1}. However, this work did not examine the LCR and the possible interferences for the measurement. Thus, further studies are needed to validate its possible practical use as a fluorescent sensor of metal ions. In another recent work, Silva-De Hoyos and coworkers employed biogenic gold nanoparticles to detect aqueous–divalent metal ions [88]. The nanosystem was synthesized via green synthesis of Camellia Sinensis (green tea) as a reducing and capping agent for the metal gold core. This plant extract is known for being rich in polyphenols like catechol and epicatechin, among others, and its use in green synthesis produces stable nanoparticles with a surface plasmon band (SPR) centered between 529 and 536 nm that confers to the solution a pink-reddish color. The obtained colloidal nanoparticles can be used to detect Ca^{2+}, Sr^{2+}, Cu^{2+}, and Zn^{2+} by using the UV-visible technique due to the formation of stable complexes with these metal ions. As a consequence, the position and the intensity of the SPR band are modified upon increasing ion concentrations. The samples were measured from 1 to 1000 μM concentrations, and it was established that a 1000 μM concentration was optimal for the determination. However, further studies are needed again to set the detection limit and evaluate the possible interferences in the measurement process.

10.4 Polyphenols and Human Health

Oxidative stress damages macromolecules such as proteins, carbohydrates, lipids, and nucleic acids. It also alters the functionality of cell membranes, as well as the production of enzymes and gene induction, among other essential processes for organisms. Stay away from the disease [89]. Naturally, there are situations where these imbalances occur between the generation of molecules or parts of them, which usually have one or more unpaired electrons in their outermost orbital and are, therefore, very reactive. These processes are, among others, intense exercise, ischemia, hyperoxia situations, external agents such as ultraviolet light, ionizing radiation, and other environmental pollutants [90, 91]. To maintain balance and oxidative stress is not maintained, the body must have antioxidants. These antioxidants must counteract the action of reactive species; that is, they are compounds that oppose the oxidizing action of oxygen and free radicals [92–94]. These compounds include catalase, superoxide dismutase, glutathione peroxidase, and other molecules with the ability to directly destroy free radicals such as AA, glutathione, and alpha-tocopherol, among others, being the proteases present in cells, which will be responsible for eliminating proteins that free radicals have damaged. There is a category of tertiary antioxidants that help repair damaged biomolecules [91, 95–100]. As mentioned before, many polyphenols can be found in food that gives a great variety of metabolites after eating foods that are not rich in polyphenols. However, they have been used in the industry for quite some time for dyes, and adhesives, among other uses [101]. Lately, interest in polyphenols, especially flavonoids, has increased considerably, mainly due to the discovery of their antioxidant properties and side effects. It is vital in preventing various chronic diseases [102–106]. These compounds add powerful antioxidant actions to bioelements such as antioxidant enzymes and vitamins, complementing and adding to the functions of antioxidant vitamins and enzymes, minimizing oxidative stress caused by the increase in reactive oxygen species [107, 108]. Therefore, polyphenols could be attributed to the ability to contribute to the protection and prevention of cardiovascular, metabolic, and neurodegenerative diseases and some types of cancer [94, 109–115] (Figure 10.5).

The investigations to verify the effects of polyphenols focus on studying the consumption of these compounds and demonstrating their impact on health and disease; sometimes, the relative contribution of a polyphenol from food is compared to others [116, 117]. To be metabolized, microorganisms from the digestive system, specifically from the colon, are needed, producing low molecular weight polyphenolic structures, which allow them to be absorbed so that they can be used.

Therefore, these microorganisms may be responsible for the health effects of consuming foods rich in polyphenols, after being processed, instead of the original compounds found in food [118]. The structure of polyphenols affects the degree and rate of intestinal absorption, affecting the metabolites that circulate

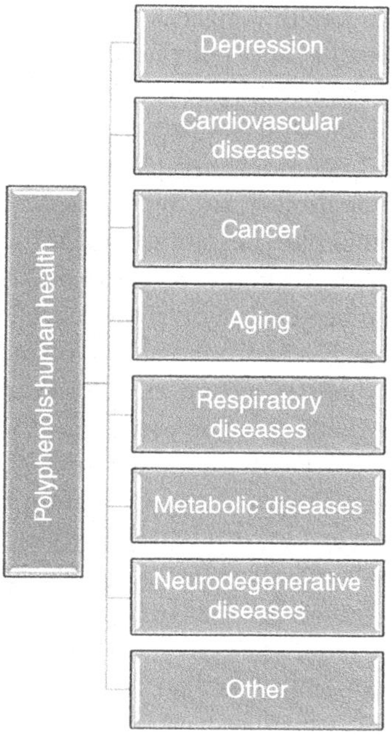

Figure 10.5 Protective and preventive effects of polyphenols on health.

in the plasma [119]. Although the microbiota can be subdivided into different enterotypes of different bacterial genera, they all seem to share a high functional uniformity [120], regardless of factors such as age, sex, and body mass index, among others [121]. Among other factors, the diet's composition significantly impacts the design and function of the intestinal microbiota, from breastfeeding to solid food intake throughout life [122, 123]. Several studies show that consuming foods with polyphenols can protect against depression [94, 124]. Some investigations focus their work on the incidence of diet and nutrition in mental health. Some studies relate diet and food to the pathways of the microbiome, the immune system, and inflammation with the pathophysiology of depression [125–127]. The most effective include isoflavones from soy, tea, and cocoa flavanols, curcumin and hydroxycinnamic acid from coffee, and walnut flavonols, among others [128–145]. Polyphenols influence the modulation of transmission at the level of the synapses, which is why they help to improve cognitive function due to their antioxidant action, acting on signaling pathways in which they interact with molecules whose action is carried out between mechanisms of synaptic plasticity and cellular metabolism. An excellent strategy to avoid oxidative damage would be to choose the right foods, such as diets rich in polyphenols, to obtain the benefits they offer the body, including neurological protection [94].

Among the new therapies to improve health, capsules have been developed for structures composed of a set of proteins and polyphenols, conjugates and nanoparticles with self-assembly, as a vehicle material in the administration of drugs, as therapy in the treatment of different diseases such as infectious diseases, cancer, colon diseases, heart disease and neurodegenerative diseases [146–150]. These nanocomplexes have high biocompatibility and biodegradability of their components; due to the protein-polyphenol particles, they had a response to biological enzymes with low toxicity [146]. An example of these new tools is the encapsulation of curcumin, which could be equivalent to the use of other polyphenols. Curcumin, or diferuloylmethane, is a hydrophobic polyphenol extracted from the turmeric rhizome. It is regularly used as a condiment for the preparation of different dishes and as a natural medicine in some countries [151, 152] and classified as another polyphenol, sub-class curcuminoids, molecular weight 368.3799 and chemical formula $C_{21}H_{20}O_6$ (Figure 10.6). Curcumin has various anticancer functions. It could inhibit, at some level, a wide variety of tumor cells by interfering, among other mechanisms, with apoptosis, cell cycle development, cancer cell inflammation, and their proliferation [153–162]. Being hydrophobic, administration to humans is a handicap since it has low aqueous solubility and presents physicochemical instability [163–169]. It is also very susceptible to pH changes. To avoid these problems, possible strategies have been investigated, such as the encapsulation of curcumin in nanoparticles, thus protecting the drug from its degradation and allowing easy absorption. In various investigations, to administer the active principles of drugs, they have been used from polymeric materials to macromolecules [170–172]. Polyesters are biodegradable polymers and, in addition, biocompatible, therefore, applicable to human health. Among them is PLGA or poly(lactic-*co*-glycolic acid) which is the most widely used polymer in

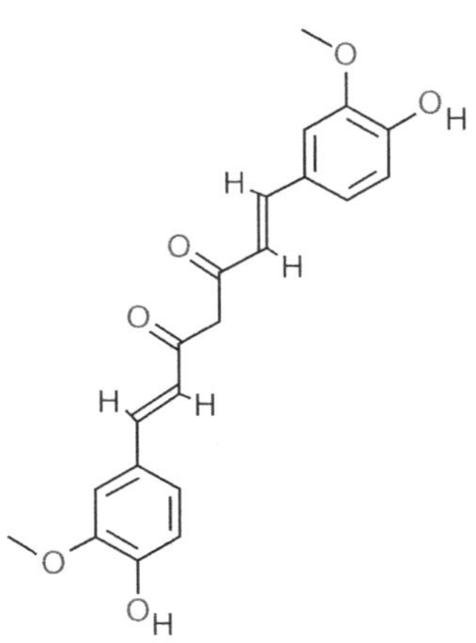

Figure 10.6 Chemical structure of Curcumin [181].

nanomedicine, being a product certified as a safe substance for application in medicine [173–175]. By breaking down into two elements, natural metabolites whit lactic and glycolic acid, they are easily excreted by the body, which entails low toxicity [176]. We illustrate one of the techniques for the encapsulation of curcumin in PLGA, one of the most used currently, and that could be extended to other polyphenols. These complexes consist of an emulsification/solvent evaporation process. To do this, the curcumin, medicine, and polymer are subjected to a solution using an organic solvent or a mixture of solvents and an emulsion in a medium containing a surfactant. Secondly, there is an evaporation of the solvent, obtaining a precipitation of solid and hard particles due to the reduction of the solvent concentration in the dispersed organic phase. To eliminate the remains, both free drug and possible surfactant, the preparation is subjected to a series of centrifugation cycles followed by washings or dialysis. Freeze drying minimizes solvent residue after the evaporation step. Additionally, it can be combined with the mini emulsion technique to reduce droplet size to the submicron range [177–179]. To enhance the production of curcumin and PLGA microparticles as a format to apply the nanocomplex intraperitoneal, multiplexed liquid-actuated co-flow (LDCF) has been used. For the LDCF, a set of three pumps is used, which confer a continuous flow during the internal phase of the curcumin in 1,2-dihydroxypropane, also consisting centrally of a device, a coaxial needle, a manifold, and a pressure chamber [180]. This allows the collection of the liquid very slowly, which helps the solvent evaporation [180].

Curcumin encapsulated in nanoparticles composed of PLGA, has been studied in multiple works related to breast cancer [182–188], ovarian cancer [189, 190], cervical cancer [191, 192], prostate cancer [189], colorectal cancer [193], brain cancer [194], liver cancer [195], and leukemia [195], among others. Encapsulation allows interaction with many ligands that are specific to cancer, such as antibodies, proteins, antigens, receptors, peptides, and other polymeric coatings, such as poly (ethylene glycol), poly (ε-caprolactone), chitosan, dextran, among others. However, in vitro studies gave better results than in vivo compared to the free drug [196].

On the other hand, among the new strategies in the fight against cancer, specifically for the therapy and multimodal imaging of tumors [197], are the metal-phenolic networks (MPN). These techniques consist of self-assembly from metal ions and polyphenols, with good biocompatibility and bio-adhesiveness [197–199]. These structures could modify their surface so that they could be attached to different substrates without the influence of composition or size [200]. Therefore, they could be used as photothermal support in multimodal engineering theranostics, such as ions such as iron and tannic acid (TA). Tannic acid is a natural polyphenol that contains numerous aromatic ring structures of catechol and pyrogallol, which give it a series of chemical properties such as water solubility suitable for chemical modification in an ecological way; it is both hydrolyzable and biodegradable, aided by the breaking of the ester bond between two pyrogallol, in addition to the adhesiveness underwater [201]. The chelating properties of metals, which can act in biological processes by functional chelation of metals, are also used as sensors [202, 203]. In addition, with plenty of odihydroxy and trihydroxy aromatic rings, this naturally extracted compound can reduce noble metal ions from metal nanoparticles, such as those made of gold or silver [201]. Nanovesicle-assisted FeIII/TA (PNV@FeIII/TA) is used for tumor-specific photoactivated uses, including NIR (near infrared) photothermal therapy with complete tumor removal, photothermal imaging, photoacoustic imaging, and T1-MRI (longitudinal relaxation-magnetic resonance) imaging. PNV@FeIII/TA can be used for T2-MRI (transverse relaxation-magnetic resonance) imaging by further doping Mn(II) and NIR (near-infrared) fluorescence imaging by encapsulating a hydrophilic NIR fluoroprobe [204]. In addition, other metals, including Al, V, Cr, Zn, Zr, Rh, Mn, Co, Ni, and Cu, can interact with TA to form functional MPN networks on particle substrates. For example, the RhIII-TA complex with a catalytic function for the hydrogenation of quinolones, or the one created by MnII-TA, is very effective for generating magnetic resonance images in vivo [205].

For all of the above, colloidal particles, made from polyphenols and polymeric or biomolecular compounds, with both noncovalent and covalent interactions, allow for a possible design with specific functional properties and a capacity to exert multiple responses in the elimination of oxidizing free radicals, in addition to allowing these designs. Incorporate drug elements that make image formation possible, among others. It constitutes a proper tool for its advantages in diagnosing and treating diseases such as cancer.

10.5 Conclusions and Future Perspectives

This work aimed to provide a critical state of art on the development of distinct biosensors for polyphenols quantification on different media and using them as metal ionic sensors. Regarding the sensing of polyphenols, the evaluation of polyphenol concentrations continues to be a hot area of research due to the direct implications of these compounds in degenerative and chronic diseases and the field of food stability and quality control of foods and beverages. However, polyphenols are implicated in a wide variety of reactions due to their significant reactivity conferred by the existence of multiple functional groups available in their structure and their natural endogenous. In this context, the electrochemical methods can provide a certain degree of selectivity, sensitivity, and long-term stability. In particular, the design of bioelectrochemical and optical sensors based on tyrosinase and laccase polyphenols was highlighted, among others, due to their excellent biocompatibility and selectivity. In recent years, combining nanomaterials and electrochemical sensors improved the biosensors general properties despite the complexity. However, most of the procedures reported in the bibliography are validated in water or buffer media at a specific pH of work and temperature conditions. These methods should be validated in actual samples. Moreover, the evaluation of possible interferences for polyphenol quantifications is relatively scarce. For instance, in the specific case of Lacasse-based sensors, these systems suffer from interference because laccase is not a phenolic-specific enzyme. Instead, tyrosinase is an enzyme specific for phenolic compounds, but it has the inconvenience of presenting lower long-term stability than the analogous laccase-based sensor [206]. Thus, these issues should be considered in developing new biosensors adapted to the real world. In particular, the matrix effect should be better analyzed about the possible use of polyphenols in medicine and clinical trials due to enzymatic interactions, reactions with foods, and genetic characteristics that could interfere with health claims [207].

Concerning the use of polyphenols as sensors or as constituents of sensing platforms for detecting metal ions, very similar problems can be found. The methods are frequently evaluated in nonreal media, and the possible interferences have yet to be assessed. In this area of research, studies are usually found in which the linear range of the sensor response needs to be evaluated or the detection limit needs to be calculated. Moreover, when the sensing method is based on a natural extract containing polyphenols, the possibility of varying the relative concentration of polyphenols in the extract among different preparations is not analyzed. These aspects constitute a possible source of error and variability in the optimization of the detection method in which the intensity of the signal or even the position of the maximum band analyzed could change depending on the preparation used.

References

1 Tijjani, H., Zangoma, M.H., Mohammed, Z.S. et al. (2020). Polyphenols: classifications, biosynthesis and bioactivities. In: *Functional Foods and Nutraceuticals* (ed. C. Egbuna and G. Dable Tupas), 389–414. Cham: Springer.

2 Butterfield, D.A., Castegna, A., Pocernich, C.B. et al. (2002). Nutritional approaches to combat oxidative stress in Alzheimer's disease. *The Journal of Nutritional Biochemistry* 13 (8): 444–461.

3 Tsao, R. (2010). Chemistry and biochemistry of dietary polyphenols. *Nutrients* 2 (12): 1231–1246.

4 Manach, C., Scalbert, A., Morand, C. et al. (2004). Polyphenols: food sources and bioavailability. *American Journal of Clinical Nutrition* 79 (5): 727–747.

5 Shahidi, F. and Naczk, M. (2003). *Phenolics in Food and Nutraceuticals*, 2e, 1–576. CRC Press.

6 Singla, R.K., Dubey, A.K., Garg, A. et al. (2019). Natural polyphenols: chemical classification, definition of classes, subcategories, and structures. *Journal of AOAC International* 102 (5): 1397–1400.

7 Chang, J., Reiner, J., and Xie, J. (2005). Progress on the chemistry of dibenzocyclooctadiene lignans. *Chemical Reviews* 105 (12): 4581–4609.

8 Olas, B., Wachowicz, B., Saluk-Juszczak, J. et al. (2001). Antioxidant activity of resveratrol in endotoxin-stimulated blood platelets. *Cell Biology and Toxicology* 17 (2): 117–125.
9 Dolinsky, V.W., Jones, K.E., Sidhu, R.S. et al. (2012). Improvements in skeletal muscle strength and cardiac function induced by resveratrol during exercise training contribute to enhanced exercise performance in rats. *The Journal of Physiology* 590 (11): 2783–2799.
10 Su, M., Zhao, W., Xu, S., and Weng, J. (2022). Resveratrol in treating diabetes and its cardiovascular complications: a review of its mechanisms of action. *Antioxidants* 11 (6), Article 1085.
11 Ko, J.H., Sethi, G., Um, J.Y. et al. (2017). The role of resveratrol in cancer therapy. *International Journal of Molecular Sciences* 18 (12), Article 2589.
12 MacRae, W.D. and Towers, G.H.N. (1984). Biological activities of lignans. *Phytochemistry* 23 (6): 1207–1220.
13 Neveu, V., Perez-Jiménez, J., Vos, F. et al. (2010). Phenol-Explorer: an online comprehensive database on polyphenol contents in foods. *Database (Oxford)* 2010, Article bap024.
14 Rothwell, J.A., Urpi-Sarda, M., Boto-Ordoñez, M. et al. (2012). Phenol-Explorer 2.0: a major update of the Phenol-Explorer database integrating data on polyphenol metabolism and pharmacokinetics in humans and experimental animals. *Database (Oxford)* 2012, Article bas031.
15 Rothwell, J.A., Perez-Jimenez, J., Neveu, V. et al. (2013). Phenol-Explorer 3.0: a major update of the Phenol-Explorer database to incorporate data on the effects of food processimng on polyphenol content. *Database* 2013, Article bat070.
16 Raymundo-Pereira, P.A., Silva, T.A., Caetano, F.R. et al. (2020). Polyphenol oxidase-based electrochemical biosensors: a review. *Analytica Chimica Acta* 1139: 198–221.
17 Dzyadevych, S.V., Arkhypova, V.N., Soldatkin, A.P. et al. (2008). Amperometric enzyme biosensors: past present and future. *IRBM* 29 (2–3): 171–180.
18 Baranwal, J., Barse, B., Gatto, G. et al. (2022). Electrochemical sensors and their applications: a review. *Chemosensors* 10 (9), Article 363.
19 Munteanu, I.G. and Apetrei, C. (2022). A review on electrochemical sensors and biosensors used in assessing antioxidant activity. *Antioxidants* 11 (3), Article 584.
20 Higgins, L.M. and Llanos, E. (2015). A healthy indulgence? Wine consumers and the health benefits of wine. *Wine Economics and Policy* 4 (1): 3–11.
21 Muazu, N.D., Zubair, J.M., and Alagha, O. (2017). Removal of phenolic compounds from water using sewage sludge-based activated carbon adsorption: a review. *International Journal of Environmental Research and Public Health* 14 (10): 3–34.
22 Sýs, M., Metelka, R., Frangu, A. et al. (2017). Electrochemical study of Trametes versicolor laccase compatibility to different polyphenolic substrates. *Chemosensors* 5 (1), Article 9.

23 Nguyen, H.H., Lee, S.H., Lee, U.J. et al. (2019). Immobilized enzymes in biosensor applications. *Materials* 12 (1), Article 121.

24 Long, J.J., Benoudjit, A.M., Arris, F.A. et al. (2018). Polymers in biosensors. In: *Multifaceted Protocol in Biotechnology* (ed. A. Amid, S. Sulaiman, D. Jimat, and N. Azmin), 151–165. Springer.

25 Rodríguez-Sevilla, E., Ramírez-Silva, M.-T., Romero-Romo, M. et al. (2014). Electrochemical quantification of the antioxidant capacity of medicinal plants using biosensors. *Sensors* 14 (8): 14423–14439.

26 Apetrei, C., Rodríguez-Méndez, M.L., and De Saja, J.A. (2011). Amperometric tyrosinase based biosensor using an electropolymerized phosphate-doped polypyrrole film as an immobilization support. Application for detection of phenolic compounds. *Electrochimica Acta* 56 (24): 8919–8925.

27 Wang, Y., Zhai, F., Hasebe, Y. et al. (2018). A highly sensitive electrochemical biosensor for phenol derivatives using a graphene oxide-modified tyrosinase electrode. *Bioelectrochemistry* 122: 174–182.

28 Munteanu, I.G. and Apetrei, C. (2022). Tyrosinase-based biosensor-a new tool for chlorogenic acid detection in nutraceutical formulations. *Materials* 15 (9), Article 3221.

29 Ma, X., Yang, H., Xiong, H. et al. (2016). Electrochemical behavior and determination of chlorogenic acid based on multi-walled carbon nanotubes modified screen-printed electrode. *Sensors* 16 (11), Article 1797.

30 Fernandes, S.C., Moccelini, S.K., Scheeren, C.W. et al. (2009). Biosensor for chlorogenic acid based on an ionic liquid containing iridium nanoparticles and polyphenol oxidase. *Talanta* 79 (2): 222–228.

31 Moccelini, S.K., Spinelli, A., and Vieira, I.C. (2008). Biosensors based on bean sprout homogenate immobilized in chitosan microspheres and silica for determination of chlorogenic acid. *Enzyme and Microbial Technology* 43 (4–5): 381–387.

32 Munteanu, I.G. and Apetrei, C. (2021). Electrochemical determination of chlorogenic acid in nutraceuticals using voltammetric sensors based on screen-printed carbon electrode modified with graphene and gold nanoparticles. *International Journal of Molecular Sciences* 22 (16), Article 8897.

33 Védrine, C., Fabiano, S., and Tran-Minh, C. (2003). Amperometric tyrosinase based biosensor using an electrogenerated polythiophene film as an entrapment support. *Talanta* 59 (3): 535–544.

34 Paris, I., Cardenas, S., Lozano, J. et al. (2007). Aminochrome as a preclinical experimental model to study degeneration of dopaminergic neurons in Parkinson's disease. *Neurotoxicity Research* 12: 125–134.

35 Cosnier, S. and Innocent, C. (1992). A novel biosensor elaboration by electropolymerization of an adsorbed amphiphilic pyrrole-tyrosinase enzyme layer. *Journal of Electroanalytical Chemistry* 328 (1–2): 361–366.

36 Hedenmo, M., Narváez, A., Domínguez, E., and Katakis, I. (1997). Improved mediated tyrosinase amperometric enzyme electrodes. *Journal of Electroanalytical Chemistry* 425 (1–2): 1–11.
37 Deng, Q., Guo, Y., and Dong, S. (1996). Cyro-hydrogel for the construction of a tyrosinase-based biosensor. *Analytica Chimica Acta* 319 (1–2): 71–77.
38 Florescu, M. and David, M. (2017). Tyrosinase-based biosensors for selective dopamine detection. *Sensors* 17 (6), Article 1314.
39 Solomon, E.I., Sundaram, U.M., and Machonkin, T.E. (1996). Multicopper oxidases and oxygenases. *Chemical Reviews* 96 (7): 2563–2606.
40 Fernandes, S.C., de Oliveira, I.R.W.Z., Fatibello-Filho, O. et al. (2008). Biosensor based on laccase immobilized on microspheres of chitosan crosslinked with tripolyphosphate. *Sensors and Actuators B: Chemical* 133 (1): 202–207.
41 Franzoi, A.C., Migowski, P., Dupont, J., and Vieira, I.C. (2009). Development of biosensors containing laccase and imidazolium bis(trifluoromethylsulfonyl)imide ionic liquid for the determination of rutin. *Analytica Chimica Acta* 639 (1–2): 90–95.
42 Li, Y., Zhang, L., Li, M. et al. (2012). A disposable biosensor based on immobilization of laccase with silica spheres on the MWCNTs-doped screen-printed electrode. *Chemistry Central Journal* 6, Article 103.
43 Wang, K., Liu, P., Ye, Y. et al. (2014). Fabrication of a novel laccase biosensor based on silica nanoparticles modified with phytic acid for sensitive detection of dopamine. *Sensors and Actuators B: Chemical* 197: 292–299.
44 Oliveira Decarli, N., Zapp, E., de Souza, B.S. et al. (2022). Biosensor based on laccase-halloysite nanotube and imidazolium zwitterionic surfactant for dopamine determination. *Biochemical Engineering Journal* 186: 1–9.
45 Chen, X., Li, D., Li, G. et al. (2015). Facile fabrication of gold nanoparticle on zein ultrafine fibers and their application for catechol biosensor. *Applied Surface Science* 328: 444–452.
46 Palanisamy, S., Ramaraj, S.K., Chen, S. et al. (2017). A novel laccase biosensor based on laccase immobilized graphene-cellulose microfiber composite modified screen-printed carbon electrode for sensitive determination of catechol. *Scientific Reports* 7, Article number: 41214.
47 Gomes, S.A.S.S. and Rebelo, M.J.F. (2003). A new laccase biosensor for polyphenols determination. *Sensors* 3 (6): 166–175.
48 Gamella, M., Campuzano, S., Reviejo, A.J., and Pingarrón, J.M. (2006). Electrochemical estimation of the polyphenol index in wines using a laccase biosensor. *Journal of Agricultural and Food Chemistry* 54 (21): 7960–7967.
49 Mishra, S.K. and Chiang, K.S. (2020). Phenolic-compounds sensor based on immobilization of tyrosinase in polyacrylamide gel on long-period fiber grating. *Optics and Laser Technology* 131, Article 106464.
50 Sorrentino, I., Giardina, P., and Piscitelli, A. (2019). Development of a biosensing platform based on a laccase-hydrophobin chimera. *Applied Microbiology and Biotechnology* 103 (7): 3061–3071.

51 Boujakhrout, A., Jimenez-Falcao, S., Martínez-Ruiz, P. et al. (2016). Novel reduced graphene oxide–glycol chitosan nanohybrid for the assembly of an amperometric enzyme biosensor for phenols. *Analyst* 141: 4162–4169.

52 Medina-Plaza, C., De Saja, J.A., and Rodriguez-Mendez, M.L. (2014). Bioelectronic tongue based on lipidic nanostructured layers containing phenol oxidases and lutetium bisphthalocyanine for the analysis of grapes. *Biosensors & Bioelectronics* 57: 276–283.

53 Salamanca-Neto, C.A., Marcheafave, G.G., Scremin, J. et al. (2020). Chemometric-assisted construction of a biosensing device to measure chlorogenic acid content in brewed coffee beverages to discriminate quality. *Food Chemistry* 315, Article 126306.

54 Salvo-Comino, C., González-Gil, A., Rodriguez-Valentin, J. et al. (2020). Biosensors platform based on chitosan/AuNPs/phthalocyanine composite films for the electrochemical detection of catechol. The role of the surface structure. *Sensors* 20 (7), Article 2152. https://doi.org/10.3390/s20072152.

55 Albuquerque, Y.D.T. and Ferreira, L.F. (2007). Amperometric biosensing of carbamate and organophosphate pesticidesutilizing screen-printed tyrosinase-modified electrodes. *Analytica Chimica Acta* 596 (2): 210–221.

56 Cerrato-Alvarez, M., Bernalte, E., Bernalte-García, M.J., and Pinilla-Gila, E. (2019). Fast and direct amperometric analsis of polyphenols in beers using tyrosinase-modified screen-printed gold nanoparticles biosensors. *Talanta* 193: 93–99.

57 Albayati, S.A.R., Kashanian, S., Nazari, M., and Rezaei, S. (2019). Novel fabrication of a laccase biosensor to detect phenolic compounds using a carboxylated multiwalled carbon nanotubeon the electropolymerized support. *Bulletin of Materials Science* 42, Article 187.

58 Apetrei, C., Alessio, P., Constantino, C.J.L. et al. (2011). Biomimetic biosensor based on lipidic layers containing tyrosinase and lutetiumbisphthalocyanine for the detection of antioxidants. *Biosensors & Bioelectronics* 26: 2513–2519.

59 Ozsoz, M., Erdern, A., Kilinc, E., and Gokgunnec, L. (1996). Mushroom-based cobalt phthalocyanine dispersed amperometric biosensor for the determination of phenolic compounds. *Electroanalysis* 8 (2): 147–150.

60 Polatoglu, I. (2019). Electrochemical sensing platform based on tyrosinase immobilized magnetite chitosan nanobiocomposite film and its application as catechol biosensor. *Journal of the Electrochemical Society* 166: B1620–B1629.

61 Bracamonte, M.V., Linárez-Pérez, O.E., López-Teijelo, M. et al. (2014). Quaternized chitosan mediated assembly of gold nanoparticles multilayers. *Electrochimica Acta* 146: 178–185.

62 Ziyatdinova, G., Guss, E., and Yakupova, E. (2021). Electrochemical sensors based on the electropolymerized natural phenolic antioxidants and their analytical application. *Sensors* 21 (24), Article 8385.

63 Ciszewski, A. and Milczarek, G. (1999). Polyeugenol-modified platinum electrode for selective detection of dopamine in the presence of ascorbic acid. *Analytical Chemistry* 71 (5): 1055–1061.

64 Jin, G.-P., Chen, Q.-Z., Ding, Y.-F., and He, J.-B. (2007). Electrochemistry behavior of adrenalin, serotonin and ascorbic acid at novel poly rutin modified paraffin-impregnated graphite electrode. *Electrochimica Acta* 52 (7): 2535–2541.

65 Devadas, B., Rajkumar, M., and Chen, S.M. (2014). Electropolymerization of curcumin on glassy carbon electrode and its electrocatalytic application for the voltammetric determination of epinephrine and p-acetoaminophenol. *Colloids and Surfaces, B: Biointerfaces* 116: 674–680.

66 Wei, J., He, J., Chen, C., and Wang, X. (2015). A catechin-modified carbon paste electrode for electrocatalytic determination of neurotransmitters. *Analytical Methods* 7: 5641–5648.

67 Silva, L.V.D., Silva, F.A.S., Kubota, L.T. et al. (2016). Amperometric sensor based on carbon nanotubes and electropolymerized vanillic acid for simultaneous determination of ascorbic acid, dopamine, and uric acid. *Journal of Solid State Electrochemistry* 20 (9): 2389–2393.

68 Ziyatdinova, G., Guss, E., and Budnikov, H. (2018). Poly(gallic acid)/MWNT-modified electrode for the selective and sensitive voltammetric determination of quercetin in medicinal herbs. *Journal of Electroanalytical Chemistry* 821: 73–81.

69 Ziyatdinova, G., Kozlova, E., and Budnikov, H. Polyquercetin/MWNT-modified electrode for the determination of natural phenolic antioxidants. *Electroanalysis* 29 (11): 2610–2619.

70 McDonald, M., Mila, I., and Scalbert, A. (1996). Precipitation of metal ions by plant polyphenols: optimal conditions and origin of precipitation. *Journal of Agricultural and Food Chemistry* 44 (2): 599–606.

71 Seavell, A.J. (1978). Anticorrosive properties of mimosa (wattle) tannin. *Journal of the Oil & Colour Chemists' Association* 61: 439–462.

72 Grimshaw, J. (1976). Phenolic aralkylamines, monohydric alcohols, monocarbaldehydes, monoketones and monocarboxylic acids. In: *Rodd's Chemistry*, 2e (ed. S. Coffey), 141–202. Elsevier: Amsterdam.

73 Meunier, L. and Vaney, C. (1903). *La Tannerie*. Paris: Gauthier-Villars.

74 Randall, J.M., Garrett, V.H., Bermann, R., and Waiss, A.C. (1974). Use of bark to remove heavy metal ions from waste solutions. *Forest Products Journal* 24 (9): 80–84.

75 Scalbert, A. et al. (1999). Polyphenols, metal ion complexation and biological consequences. In: *Plant Polyphenols 2. Basic Life Sciences*, vol. 66 (ed. G.G. Gross, R.W. Hemingway, T. Yoshida, and S.J. Branham). Boston, MA: Springer.

76 Wu, X., Kim, J., and Dordick, J.S. (2000). Enzymatically and combinatorially generated array-based polyphenol metal ion sensor. *Biotechnology Progress* 16 (3): 513–516. https://doi.org/10.1021/bp0000104.

77 Wu, X., Kim, J., and Dordick, J.S. (2000). Enzymatically and combinatorially generated array-based polyphenol metal ion sensor. *Biotechnology Progress* 16 (3): 13–16.

78 Pooja, Pandey, H., Aggarwal, S. et al. (2022). Coumarin-based chemosensors for metal ions detection. *Asian The Journal of Organic Chemistry* 11, Article e202200455.

79 Suresh, M. and Das, A. (2009). New coumarin-based sensor molecule for magnesium and calcium ions. *Tetrahedron Letters* 50 (42): 5808–5812.

80 Sen, B., Sheet, S.K., Thounaojam, R. et al. (2017). A coumarin based Schiff base probe for selective fluorescence detection of Al^{3+} and its application in live cell imaging. *Spectrochimica Acta. Part A, Molecular and Biomolecular Spectroscopy* 173: 537–543.

81 Fasman, G.D. (1996). Aluminum and alzheimer's disease: model studies. *Coordination Chemistry Reviews* 149: 125–165.

82 Liu, M., Xu, Z., Song, Y. et al. (2018). A novel coumarin-based chemosensor for colorimetric detection of Ag(I) ion and fluorogenic sensing of Ce(III) ion. *Journal of Luminescence* 198: 337–341.

83 García-Beltrán, O., Rodríguez, A., Trujillo, A. et al. (2015). Synthesis and characterization of a novel fluorescent and colorimetric probe for the detection of mercury(II) even in the presence of relevant biothiols. *Tetrahedron Letters* 56: 5761–5766.

84 Mondal, S. and Dey, S. (2021). Strategy to design a flexible and macromolecular sensor to bind Cd^{2+} ions: a complete photophysical analysis and bio-imaging study. *ACS Omega* 6 (42): 27936–27945.

85 Pham, X.-Q., Jonusauskaite, L., Depauw, A. et al. (2019). Sensitive and selective detection of uranyl ions based on aggregate-breaking mechanism. *Journal of Photochemistry and Photobiology A: Chemistry* 364: 355–362.

86 Zhang, L., Guan, Q., and Zhang, H. (2022). Complexation of metal ions by propyl gallate and procyanidins: a spectroscopic and quantum chemical calculation method. *Research Square* 2022: 1–14.

87 Sargin, İ. (2019). On fluorescent sensing of metal ions using water extracts of Salvia officinalis. *Cumhuriyet Science Journal* 40 (4): 830–837.

88 Silva-De Hoyos, L.E., Sánchez-Mendieta, V., Vilchis-Nestor, A.R. et al. (2019). Plasmonic sensing of aqueous-divalent metal ions by biogenic gold nanoparticles. *Journal of Nanomaterials* 2019, Article 9846729.

89 Roche, E. and Romero-Alvira, D. (1997). Introducción a la bioquímica y citotoxicidad del desequilibrio oxidativo. (II) Alteraciones oxidativas en las macromoléculas biológicas. In: *Cardiología, estrés oxidativo, nutrición y biología molecular. Bases y aplicaciones sobre el estrés oxidativo, aspectos nutricionales y de la biología molecular en cardiología* (ed. D. Romero-Alvira and E. Roche), 91–104. ENE Ediciones.

90 Roche, E. and Romero-Alvira, D. (1997). Introducción a la bioquímica y citotoxicidad del desequilibrio oxidativo. (I) Especies activas del oxígeno. In: *Cardiología, estrés oxidativo, nutrición y biología molecular. Bases y aplicaciones sobre el estrés oxidativo, aspectos nutricionales y de la biología molecular en cardiología* (ed. D. Romero-Alvira and E. Roche), 55–88. ENE Ediciones.

91 Veiga Álvarez, E., Aguilar Doreste, J.Á., Clavo Varas, B., and Llanes Menéndez, L. (1997). Radicales libres, formación y daño celular: el sistema antioxidante como protector frente a los radicales libres. *Análisis clínicos: órgano de expresión científica y profesional de la Asociación Española de Farmacéuticos Analistas* 22 (89): 201–216.

92 Pelleymounter, M.A., Cullen, M.J., and Wellman, C.L. (1995). Characteristics of BDNF-induced weight loss. *Experimental Neurology* 131 (2): 229–238.

93 Tsuchida, A., Nakagawa, T., Itakura, Y. et al. (2001). The effects of brain-derived neurotrophic factor on insulin signal transduction in the liver of diabetic mice. *Diabetologia* 44 (5): 555–566.

94 Gomez-Pinilla, F. and Nguyen, T.T.J. (2012). Natural mood foods: the actions of polyphenols against psychiatric and cognitive disorders. *Nutritional Neuroscience* 15 (3): 127–133.

95 Wolff, S.P., Garner, A., and Dean, R.T. (1986). Free radicals, lipids and protein degradation. *Trends in Biochemical Sciences* 11 (1): 27–31.

96 Romero Alvira, D., Villalba Martín, M.P., Mur Villacampa, M. et al. (1990). The importance of anti-oxidants in human nutrition | Importancia de los antioxidantes en la alimentación humana. *Medicina Clínica (Barcelona)* 94 (2): 69–75.

97 Scandalios, J.G. (1992). *Molecular Biology of Free Radical Scavenging Systems*. Cold Spring Harbor Press.

98 Ballester, M., Riera, J., Castañer, J., and Casulleras, M. (1978). The reduction of inert free radicals with ascorbic acid. The reaction with perchlorotriphenylmethyl radical. *Tetrahedron Letters* 19 (7): 643–644.

99 Davies, K.J.A. (1986). Intracellular proteolytic systems may function as secondary antioxidant defenses: an hypothesis. *Journal of Free Radicals in Biology & Medicine* 2 (3): 155–173.

100 Roche, E. and Romero-Alvira, D. (1997). Introducción a la bioquímica y citotoxicidad del desequilibrio oxidativo. (III) Sistemas de defensa y reparación. In: *Cardiología, estrés oxidativo, nutrición y biología molecular. Bases y aplicaciones sobre el estrés oxidativo, aspectos nutricionales y de la biología molecular en cardiología* (ed. D. Romero-Alvira and E. Roche), 105–123. ENE Ediciones.

101 Quideau, S., Deffieux, D., Douat-Casassus, C., and Pouységu, L. (2011). Plant polyphenols: Chemical properties, biological activities, and synthesis. *Angewandte Chemie, International Edition* 50 (3): 586–621.

102 Sileika, T.S., Barrett, D.G., Zhang, R. et al. (2013). Colorless multifunctional coatings inspired by polyphenols found in tea, chocolate, and wine. *Angewandte Chemie, International Edition* 52 (41): 10766–10770.

103 Krishnan, G. (1951). Phenolic tanning and pigmentation of the cuticle in carcinus maenas. *Journal of Cell Science* 3 (19): 333–342.

104 Huber, B., Eberl, L., Feucht, W., and Polster, J. (2003). Influence of polyphenols on bacterial biofilm formation and quorum-sensing. *Zeitschrift fur Naturforschung - Section C Journal of Biosciences* 58 (11–12): 879–884.

105 Arts, I.C. and Hollman, P.C. (2005). Polyphenols and disease risk in epidemiologic studies. *The American Journal of Clinical Nutrition* 81 (1 Suppl): 317S–325S.

106 Scalbert, A., Manach, C., Morand, C. et al. (2005). Dietary polyphenols and the prevention of diseases. *Critical Reviews in Food Science and Nutrition* 45 (4): 287–306.

107 Williams, R.J., Spencer, J.P.E., and Rice-Evans, C. (2004). Flavonoids: antioxidants or signalling molecules? *Free Radical Biology & Medicine* 36 (7): 838–849.

108 Dong, Z., Surh, Y.-J., Cadenas, E., and Packer Packer, L. (2009). *Dietary Modulation of Cell Signaling Pathways.* CRC Press.

109 Cardona, F., Andrés-Lacueva, C., Tulipani, S. et al. (2013). Benefits of polyphenols on gut microbiota and implications in human health. *Journal of Nutritional Biochemistry* 24 (8): 1415–1422.

110 West, N.R. and Powrie, F. (2015). Immunotherapy not working? check your microbiota. *Cancer Cell* 28 (6): 687–689.

111 Vétizou, M., Pitt, J.M., Daillère, R. et al. (2015). Anticancer immunotherapy by CTLA-4 blockade relies on the gut microbiota. *Science (1979)* 350 (6264): 1079–1084.

112 Von Schwartzenberg, R.J. and Turnbaugh, P.J. (2015). Siri, what should i eat? *Cell* 163 (5): 1051–1052.

113 Siddharth, J., Holway, N., and Parkinson, S.J. (2013). A western diet ecological module identified from the "humanized" mouse microbiota predicts diet in adults and formula feeding in children. *PLoS One* 8 (12), Article e83689.

114 Pinzone, M.R., Celesia, B.M., Di Rosa, M. et al. (2012). Microbial translocation in chronic liver diseases. *International Journal of Microbiology*, Article 694629.

115 Schwiertz, A., Taras, D., Schäfer, K. et al. (2010). Microbiota and SCFA in lean and overweight healthy subjects. *Obesity* 18 (1): 190–195.

116 Manach, C., Williamson, G., Morand, C. et al. (2005). Bioavailability and bioefficacy of polyphenols in humans. I. Review of 97 bioavailability studies. *The American Journal of Clinical Nutrition* 81 (1 Suppl): 230S–242S.

117 Williamson, G. and Manach, C. (2005). Bioavailability and bioefficacy of polyphenols in humans. II. Review of 93 intervention studies. *The American Journal of Clinical Nutrition* 81 (1 Suppl): 243S–255S.

118 Appeldoorn, M.M., Vincken, J.-P., Gruppen, H., and Hollman, P.C.H. (2009). Procyanidin dimers A1, A2, and B2 are absorbed without conjugation or methylation from the small intestine of rats. *Journal of Nutrition* 139 (8): 1469–1473.

119 Scalbert, A. and Williamson, G. (2000). Dietary intake and bioavailability of polyphenols. *Journal of Nutrition* 130 (8 Suppl): 2073S–2085S.

120 Arumugam, M., Raes, J., Pelletier, E. et al. (2011). Enterotypes of the human gut microbiome. *Nature* 473 (7346): 174–180.

121 Kurokawa, K., Itoh, T., Kuwahara, T. et al. (2007). Comparative metagenomics revealed commonly enriched gene sets in human gut microbiomes. *DNA Research* 14 (4): 169–181.

122 David, L.A., Maurice, C.F., Carmody, R.N. et al. (2014). Diet rapidly and reproducibly alters the human gut microbiome. *Nature* 505 (7484): 559–563.

123 Daïen, C.I., Pinget, G.V., Tan, J.K., and Macia, L. (2017). Detrimental impact of microbiota-accessible carbohydrate-deprived diet on gut and immune homeostasis: An overview. *Frontiers in Immunology* 8: (MAY), Article 548.

124 Bayes, J., Schloss, J., and Sibbritt, D. (2020). Effects of polyphenols in a mediterranean diet on symptoms of depression: a systematic literature review. *Advances in Nutrition* 11 (3): 602–615.

125 Jacka, F.N. (2017). Nutritional psychiatry: where to next? *eBioMedicine* 17: 24–29.

126 Foster, J.A. and McVey Neufeld, K.-A. (2013). Gut–brain axis: how the microbiome influences anxiety and depression. *Trends in Neurosciences* 36 (5): 305–312.

127 Angelino, D., Godos, J., Ghelfi, F. et al. (2019). Fruit and vegetable consumption and health outcomes: an umbrella review of observational studies. *International Journal of Food Sciences and Nutrition* 70 (6): 652–667.

128 Sathyapalan, T., Beckett, S., Rigby, A.S. et al. (2010). High cocoa polyphenol rich chocolate may reduce the burden of the symptoms in chronic fatigue syndrome. *Nutrition Journal* 9 (1), Article 55.

129 Nina Estrella, R.E., Landa, A.I., Lafuente, J.V., and Gargiulo, P.A. (2014). Effects of antidepressants and soybean association in depressive menopausal women. *Acta Poloniae Pharmaceutica. Drug Research* 71 (2): 323–327.

130 Atteritano, M., Mazzaferro, S., Bitto, A. et al. (2014). Genistein effects on quality of life and depression symptoms in osteopenic postmenopausal women: A 2-year randomized, double-blind, controlled study. *Osteoporosis International* 25 (3): 1123–1129.

131 Lopresti, A.L., Maes, M., Maker, G.L. et al. (2014). Curcumin for the treatment of major depression: a randomised, double-blind, placebo controlled study. *Journal of Affective Disorders* 167: 368–375.

132 Panahi, Y., Badeli, R., Karami, G.-R., and Sahebkar, A. (2015). Investigation of the efficacy of adjunctive therapy with bioavailability-boosted curcuminoids in major depressive disorder. *Phytotherapy Research* 29 (1): 17–21.

133 Yu, J.-J., Pei, L.-B., Zhang, Y. et al. (2015). Chronic supplementation of curcumin enhances the efficacy of antidepressants in major depressive disorder: a randomized, double-blind, Placebo-Controlled Pilot Study. *Journal of Clinical Psychopharmacology* 35 (4): 406–410.

134 Ibero-Baraibar, I., Perez-Cornago, A., Ramirez, M.J. et al. (2016). An increase in plasma homovanillic acid with cocoa extract consumption is associated with the alleviation of depressive symptoms in overweight or obese adults on an energy restricted diet in a randomized controlled trial. *Journal of Nutrition* 146 (4): 897S–904S.

135 Pribis, P. (2016). Effects of walnut consumption on mood in young adults—A randomized controlled trial. *Nutrients* 8 (11), Article 668.

136 Hirose, A., Terauchi, M., Akiyoshi, M. et al. (2016). Low-dose isoflavone aglycone alleviates psychological symptoms of menopause in Japanese women: a randomized, double-blind, placebo-controlled study. *Archives of Gynecology and Obstetrics* 293 (3): 609–615.

137 Ruusunen, A., Lehto, S.M., Tolmunen, T. et al. (2010). Coffee, tea and caffeine intake and the risk of severe depression in middle-aged Finnish men: the Kuopio ischaemic heart disease risk factor study. *Public Health Nutrition* 13 (8): 1215–1220.

138 Lucas, M., Mirzaei, F., Pan, A. et al. (2011). Coffee, caffeine, and risk of depression among women. *Archives of Internal Medicine* 171 (17): 1571–1578.

139 Omagari, K., Sakaki, M., Tsujimoto, Y. et al. (2014). Coffee consumption is inversely associated with depressive status in Japanese patients with type 2 diabetes. *Journal of Clinical Biochemistry and Nutrition* 55 (2): 135–142.

140 Pham, N.M., Nanri, A., Kurotani, K. et al. (2014). Green tea and coffee consumption is inversely associated with depressive symptoms in a Japanese working population. *Public Health Nutrition* 17 (3): 625–633.

141 Yu, S., Guo, X., Yang, H. et al. (2015). Soybeans or soybean products consumption and depressive symptoms in older residents in rural Northeast China: a cross-sectional study. *The Journal of Nutrition, Health & Aging* 19 (9): 884–893.

142 Su, Q., Yu, B., He, H. et al. (2016). Nut consumption is associated with depressive symptoms among Chinese adults. *Depression and Anxiety* 33 (11): 1065–1072.

143 Navarro, A.M., Abasheva, D., Martínez-González, M.Á. et al. (2018). Coffee consumption and the risk of depression in a middle-aged cohort: the SUN project. *Nutrients* 10 (9), Article 1333.

144 Miyake, Y., Tanaka, K., Okubo, H. et al. (2018). Soy isoflavone intake and prevalence of depressive symptoms during pregnancy in Japan: baseline data from the Kyushu Okinawa Maternal and Child Health Study. *European Journal of Nutrition* 57 (2): 441–450.

145 Yu, B., Yu, F., Su, Q. et al. (2018). A J-shaped association between soy food intake and depressive symptoms in Chinese adults. *Clinical Nutrition* 37 (3): 1013–1018.

146 Shin, M., Lee, H.-A., Lee, M. et al. (2018). Targeting protein and peptide therapeutics to the heart via tannic acid modification. *Nature Biomedical Engineering* 2 (5): 304–317.

147 Ouberai, M., Dumy, P., Chierici, S., and Garcia, J. (2009). Synthesis and biological evaluation of clicked curcumin and clicked KLVFFA conjugates as inhibitors of β-amyloid fibril formation. *Bioconjugate Chemistry* 20 (11): 2123–2132.

148 Lomova, M.V., Brichkina, A.I., Kiryukhin, M.V. et al. (2015). Multilayer capsules of bovine serum albumin and tannic acid for controlled release by enzymatic degradation. *ACS Applied Materials & Interfaces* 7 (22): 11732–11740.

149 Kilic, E., Novoselova, M.V., Lim, S.H. et al. (2017). Formulation for oral delivery of lactoferrin based on bovine serum albumin and tannic acid multilayer microcapsules. *Scientific Reports* 7, Article 44159.

150 Lau, H.H., Murney, R., Yakovlev, N.L. et al. (2017). Protein-tannic acid multilayer films: a multifunctional material for microencapsulation of food-derived bioactives. *Journal of Colloid and Interface Science* 505: 332–340.

151 Brouk, B. (1975). *Plants Consumed by Man*. Academic Press.

152 Nahar, L. and Sarker, S.D. (2007). Phytochemistry of the genus Curcuma. In: *Turmeric: The Genus Curcuma*, 71–106. CRC Press.

153 Gupta, S.C., Patchva, S., and Aggarwal, B.B. (2013). Therapeutic roles of curcumin: lessons learned from clinical trials. *AAPS Journal* 15 (1): 195–218.

154 Ravindran, J., Prasad, S., and Aggarwal, B.B. (2009). Curcumin and cancer cells: how many ways can curry kill tumor cells selectively? *The AAPS Journal* 11 (3): 495–510.

155 Kammath, A.J., Nair, B., Sreelekshmi, P., and Nath, L.R. (2021). Curry versus cancer: potential of some selected culinary spices against cancer with in vitro, in vivo, and human trials evidences. *Journal of Food Biochemistry* 45 (3), Article e13285.

156 Anand, P., Sundaram, C., Jhurani, S. et al. (2008). Curcumin and cancer: An "old-age" disease with an "age-old" solution. *Cancer Letters* 267 (1): 133–164.

157 Kunnumakkara, A.B., Anand, P., and Aggarwal, B.B. (2008). Curcumin inhibits proliferation, invasion, angiogenesis and metastasis of different cancers through interaction with multiple cell signaling proteins. *Cancer Letters* 269 (2): 199–225.

158 Gupta, S.C., Kim, J.H., Kannappan, R. et al. (2011). Role of nuclear factor-kB-mediated inflammatory pathways in cancer-related symptoms and their regulation by nutritional agents. *Experimental Biology and Medicine* 236 (6): 658–671.

159 Vallianou, N.G., Evangelopoulos, A., Schizas, N., and Kazazis, C. (2015). Potential anticancer properties and mechanisms of action of curcumin. *Anticancer Research* 35 (2): 645–651.

160 Zheng, J., Zhou, Y., Li, Y. et al. (2016). Spices for prevention and treatment of cancers. *Nutrients* 8 (8), Article 495.

161 Giordano, A. and Tommonaro, G. (2019). Curcumin and cancer. *Nutrients* 11 (10), Article 2376.

162 Mansouri, K., Rasoulpoor, S., Daneshkhah, A. et al. (2020). Clinical effects of curcumin in enhancing cancer therapy: a systematic review. *BMC Cancer* 20 (1), Article 791.

163 Priyadarsini, K.I. (2014). The chemistry of curcumin: from extraction to therapeutic agent. *Molecules* 19 (12): 20091–20112.

164 Wang, Y.-J., Pan, M.-H., Cheng, A.-L. et al. (1997). Stability of curcumin in buffer solutions and characterization of its degradation products. *Journal of Pharmaceutical and Biomedical Analysis* 15 (12): 1867–1876.

165 Khurana, A. (1988). High performance liquid chromatographic analysis of curcuminoids and their photo-oxidative decomposition compounds in curcuma longal. *Journal of Liquid Chromatography* 11 (11): 2295–2304.

166 Wahlström, B. and Blennow, G. (1978). A study on the fate of curcumin in the rat. *Acta Pharmacol Toxicol (Copenh)* 43 (2): 86–92.

167 Sharma, R.A., Gescher, A.J., and Steward, W.P. (2005). Curcumin: the story so far. *European Journal of Cancer* 41 (13): 1955–1968.

168 Couvreur, P., Fattal, E., Alphandary, H. et al. (1992). Intracellular targeting of antibiotics by means of biodegradable nanoparticles. *Journal of Controlled Release* 19 (1–3): 259–267.

169 Mahran, R.I., Hagras, M.M., Sun, D., and Brenner, D.E. (2017). Bringing curcumin to the clinic in cancer prevention: a review of strategies to enhance bioavailability and efficacy. *AAPS Journal* 19 (1): 54–81.

170 Ghalandarlaki, N., Alizadeh, A.M., and Ashkani-Esfahani, S. (2014). Nanotechnology-applied curcumin for different diseases therapy. *BioMed Research International* 2014, Article 394264e.

171 Coelho, J.F., Ferreira, P.C., Alves, P. et al. (2010). Drug delivery systems: advanced technologies potentially applicable in personalized treatments. *The EPMA Journal* 1 (1): 164–209.

172 D'Angelo, N.A., Noronha, M.A., Kurnik, I.S. et al. (2021). Curcumin encapsulation in nanostructures for cancer therapy: a 10-year overview. *International Journal of Pharmaceutics* 604, Article 120534.

173 Astete, C.E. and Sabliov, C.M. (2006). Synthesis and characterization of PLGA nanoparticles. *Journal of Biomaterials Science. Polymer Edition* 17 (3): 247–289.

174 Swider, E., Koshkina, O., Tel, J. et al. (2018). Customizing poly(lactic-*co*-glycolic acid) particles for biomedical applications. *Acta Biomaterialia* 73: 38–51.

175 Danhier, F., Ansorena, E., Silva, J.M. et al. (2012). PLGA-based nanoparticles: an overview of biomedical applications. *Journal of Controlled Release* 161 (2): 505–522.

176 Avgoustakis, K. (2008). Polylactic-co-glycolic acid (PLGA). In: *Encyclopedia of Biomaterials and Biomedical Engineering*. CRC Press.

177 Musyanovych, A., Schmitz-Wienke, J., Mailänder, V. et al. (2008). Preparation of biodegradable polymer nanoparticles by miniemulsion technique and their cell interactions. *Macromolecular Bioscience* 8 (2): 127–139.

178 Staff, R.H., Lieberwirth, I., Landfester, K., and Crespy, D. (2012). Preparation and characterization of anisotropic submicron particles from semicrystalline polymers. *Macromolecular Chemistry and Physics* 213 (3): 351–358.

179 Nava-Arzaluz, M.G., Pinon-Segundo, E., Ganem-Rondero, A., and Lechuga-Ballesteros, D. (2012). Single emulsion-solvent evaporation technique and modifications for the preparation of pharmaceutical polymeric nanoparticles. *Recent Patents on Drug Delivery & Formulation* 6 (3): 209–223.

180 Dwivedi, P., Yuan, S., Han, S. et al. (2018). Core–shell microencapsulation of curcumin in PLGA microparticles: programmed for application in ovarian cancer therapy. *Artificial Cells, Nanomedicine, and Biotechnology* 46 (sup3): S481–S491.

181 National Center for Biotechnology Information (2023). *Compound Summary for CID 969516*. Curcumin.

182 Verderio, P., Bonetti, P., Colombo, M. et al. (2013). Intracellular drug release from curcumin-loaded PLGA nanoparticles induces G2/M block in breast cancer cells. *Biomacromolecules* 14 (3): 672–682.

183 Meena, R., Kumar, S., Kumar, R. et al. (2017). PLGA-CTAB curcumin nanoparticles: fabrication, characterization and molecular basis of anticancer activity in triple negative breast cancer cell lines (MDA-MB-231 cells). *Biomedicine & Pharmacotherapy* 94: 944–954.

184 Mukerjee, A., Ranjan, A.P., and Vishwanatha, J.K. (2016). Targeted nanocurcumin therapy using annexin A2 antibody improves tumor accumulation and therapeutic efficacy against highly metastatic breast cancer. *Journal of Biomedical Nanotechnology* 12 (7): 1374–1392.

185 Gerke, V. and Moss, S.E. (2002). Annexins: from structure to function. *Physiological Reviews* 82 (2): 331–371.

186 Sinha, P., Hütter, G., Köttgen, E. et al. (1998). Increased expression of annexin I and thioredoxin detected by two-dimensional gel electrophoresis of drug resistant human stomach cancer cells. *Journal of Biochemical and Biophysical Methods* 37 (3): 105–116.

187 Wang, Y., Serfass, L., Roy, M.-O. et al. (2004). Annexin-I expression modulates drug resistance in tumor cells. *Biochemical and Biophysical Research Communications* 314 (2): 565–570.

188 Sharma, M.R., Rothman, V., Tuszynski, G.P., and Sharma, M.C. (2006). Antibody-directed targeting of angiostatin's receptor annexin II inhibits Lewis Lung Carcinoma tumor growth via blocking of plasminogen activation: possible biochemical mechanism of angiostatin's action. *Experimental and Molecular Pathology* 81 (2): 136–145.

189 Yallapu, M.M., Gupta, B.K., Jaggi, M., and Chauhan, S.C. (2010). Fabrication of curcumin encapsulated PLGA nanoparticles for improved therapeutic effects in metastatic cancer cells. *Journal of Colloid and Interface Science* 351 (1): 19–29.

190 Yallapu, M.M., Maher, D.M., Sundram, V. et al. (2010). Curcumin induces chemo/radio-sensitization in ovarian cancer cells and curcumin nanoparticles inhibit ovarian cancer cell growth. *Journal of Ovarian Research* 3 (1), Article 11.

191 Nair, K.L., Thulasidasan, A.K.T., Deepa, G. et al. (2012). Purely aqueous PLGA nanoparticulate formulations of curcumin exhibit enhanced anticancer activity with dependence on the combination of the carrier. *International Journal of Pharmaceutics* 425 (1–2): 44–52.

192 Punfa, W., Yodkeeree, S., Pitchakarn, P. et al. (2012). Enhancement of cellular uptake and cytotoxicity of curcumin-loaded PLGA nanoparticles by conjugation with anti-P-glycoprotein in drug resistance cancer cells. *Acta Pharmacologica Sinica* 33 (6): 823–831.

193 Li, L., Xiang, D., Shigdar, S. et al. (2014). Epithelial cell adhesion molecule aptamer functionalized PLGA-lecithin-curcumin-PEG nanoparticles for targeted drug delivery to human colorectal adenocarcinoma cells. *International Journal of Nanomedicine* 9 (1): 1083–1096.

194 Zhang, X., Li, X., Hua, H. et al. (2017). Cyclic hexapeptide-conjugated nanoparticles enhance curcumin delivery to glioma tumor cells and tissue. *International Journal of Nanomedicine* 12: 5717–5732.

195 Chen, C., Yang, W., Wang, D.-T. et al. (2014). A modified spontaneous emulsification solvent diffusion method for the preparation of curcumin-loaded PLGA nanoparticles with enhanced in vitro anti-tumor activity. *Frontiers of Materials Science* 8 (4): 332–342.

196 Feltrin, F.D.S., Agner, T., Sayer, C., and Lona, L.M.F. (2022). Curcumin encapsulation in functional PLGA nanoparticles: a promising strategy for cancer therapies. *Advances in Colloid and Interface Science* 300, Article 102582.

197 Liu, P., Shi, X., Zhong, S. et al. (2021). Metal-phenolic networks for cancer theranostics. *Biomaterials Science* 9 (8): 2825–2849.

198 Zhang, Z., Xie, L., Ju, Y., and Dai, Y. (2021). Recent advances in metal-phenolic networks for cancer theranostics. *Small* 17 (43), Article e2100314.

199 Xie, L., Li, J., Wang, L., and Dai, Y. (2022). Engineering metal-phenolic networks for enhancing cancer therapy by tumor microenvironment modulation. *Wiley Interdiscip Rev Nanomed Nanobiotechnol*, Article e1864.

200 Ejima, H., Richardson, J.J., and Caruso, F. (2017). Metal-phenolic networks as a versatile platform to engineer nanomaterials and biointerfaces. *Nano Today* 12: 136–148.

201 Chen, C., Yang, H., Yang, X., and Ma, Q. (2022). Tannic acid: a crosslinker leading to versatile functional polymeric networks: a review. *RSC Advances* 12 (13): 7689–7711.

202 Yoshida, T., Lai, T.C., Kwon, G.S., and Sako, K. (2013). PH-and ion-sensitive polymers for drug delivery. *Expert Opinion on Drug Delivery* 10 (11): 1497–1513.

203 Jehle, F., Fratzl, P., and Harrington, M.J. (2018). Metal-tunable self-assembly of hierarchical structure in mussel-inspired peptide films. *ACS Nano* 12 (3): 2160–2168.

204 Liu, T., Zhang, M., Liu, W. et al. (2018). Metal ion/tannic acid assembly as a versatile photothermal platform in engineering multimodal nanotheranostics for advanced applications. *ACS Nano* 12 (4): 3917–3927.

205 Gossuin, Y., Gillis, P., Hocq, A. et al. (2009). Magnetic resonance relaxation properties of superparamagnetic particles. *Wiley Interdisciplinary Reviews. Nanomedicine and Nanobiotechnology* 1 (3): 299–310.

206 Tarasov, A., Stozhko, N., Bukharinova, M., and Khamzina, E. (2023). Biosensors based on phenol oxidases (laccase, tyrosinase, and their mixture) for estimating the total phenolic index in food-related samples. *Life* 13 (2), Article 291.

207 Bertelli, A., Biagi, M., Corsini, M. et al. (2021). Polyphenols: from theory to practice. *Food* 10 (11), Article 2595.

11

Polyphenol-based Hydrogels and Nanocomplexes: Fundamental, Properties, and Industrial Applications

Salima El Yakhlifi[1] *and Vincent Ball*[1,2]

[1] *Institut National de la Santé et de la Recherche Médicale, Unité Mixte de Recherche, Strasbourg, Cedex, France*
[2] *Université de Strasbourg, Faculté de Chirurgie Dentaire, Strasbourg, France*

11.1 Introduction

Polyphenols, essentially classified as condensed or hydrolyzable polyphenols [1], are widespread in plants and fruits where they are secondary metabolites with other functions to protect the plant against prey. They are mainly concentrated in tea leaves and red wine grapes. Such molecules have been used for a long time in human technology, particularly in tanning leather. More and more polyphenols are used for their biological applications related to their interactions with proteins and polysaccharides. Among the biological effects of interest are the antioxidant, antibacterial, and anticancer activities. However, the direct administration of polyphenols is problematic because of their high reactivity with oxidants and nucleophiles to which most proteins belong. An easily perceptible effect of polyphenols is mouth astringency when consuming some fruits, juices, or wines [2]. Many polyphenols interact strongly with proteins from the saliva, notably proline-rich proteins [3] and essential salivary proline-rich proteins [4], to produce small aggregates or particulates encapsulating some water from the saliva with an immediate mouthfeel effect. In addition, a high concentration of polyphenols may kill bacteria, have a positive impact, and induce worse cytotoxicity to cells from the host. It is hence of tremendous importance to develop formulations from which the polyphenols may be slowly released and become available for targeted

Science and Engineering of Polyphenols: Fundamentals and Industrial Scale Applications,
First Edition. Edited by Chandrabhan Verma.
© 2024 John Wiley & Sons Inc. Published 2024 by John Wiley & Sons Inc.

biological application without kilpling healthy cells. In this chapter, we will consider three ways to load catechols/polyphenols in hydrogels and related highly hydrated soft materials:

(i) The covalent grafting of polyphenols on polymers can undergo gelation.
(ii) The addition of polyphenols to hydrogel formulations and their subsequent chemical reaction with reactive groups on the polymer under oxidative conditions.
(iii) The use of polyphenols to precipitate polymers and to form bioactive coacervates to produce homogeneous films at solid–liquid interfaces.

The different crosslinking modalities and mechanisms of adhesion of those materials have already been reviewed [5]. Still, we will primarily focus on strategies (ii) and (iii) for which much less literature is available.

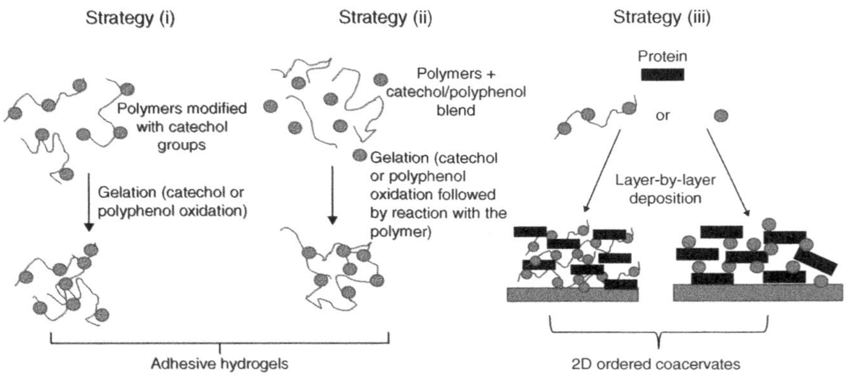

Scheme 11.1 The three strategies described in this chapter to incorporate catechols or polyphenols in hydrogels and two-dimensional-ordered nanocomplexes or coacervates. Disks represent catechols/polyphenols.

These three ways to incorporate polyphenols in materials are directly related to the large repertoire of chemical modalities of such molecules: they have the properties of weak acids/bases, rich redox chemistry, they can behave as nucleophiles/electrophiles and chelate metallic cations in a *pH-dependent* manner [6]. The three strategies to incorporate catechols or polyphenols in gels or nanocomplexes/coacervates are described in Scheme 11.1.

11.2 Polyphenols Grafted on Molecules Able to Undergo Gelation

There are natural polymers, namely proteins, containing hydroxylated L-tyrosine residues affording strong adhesion between the mussel plaque and solid substrates in wet conditions as well as strong cohesion between such proteins (*mefps* for mussel foot proteins) [7]. The different ways to graft polyphenols on natural or synthetic polymer backbones have been widely reviewed and will not be addressed here [8]. These chemical strategies allowed to control of the grafting density of catechol or other polyphenol groups on the selected polymer. Dopamine, for instance, was grafted (at a grafting density between 10% and 15%) on the carboxylic acid groups of alginic acid using the EDC-NHS chemistry [9]. Early investigators in this field used catechol-modified peptides [10] and polystyrene modified with catechol groups [11]. All these catechol-modified polymers can be used as adhesives under wet conditions, as reviewed elsewhere [12]. Some representative examples of adhesive catechol-modified polymers and their adhesive properties towards different solid materials or biological tissues are listed in Table 11.1.

Table 11.1 Adhesion strength of gels made from polymers with covalently bound catechol groups.

Adhesive material	Substrates	Adhesion strength	Ref.
Poly(lysine.HBr_4-$DOPA_1$), $Mn = 1.13\times 10^5$ g.mol^{-1}	Al_2O_3, H_2O_2 as the oxidant	4.29 ± 0.81 MPa	[10]
3,4-dihydroxystyrene:styrene copolymer with 3.4/96.6% composition	Aluminum, 1 h curing at room temperature and 22 h at 55 °C	IO_4^- as the oxidant: 0.9 ± 0.1 Mpa $Cr_20_7^{2-}$ as the oxidant: 1.2 ± 0.5 MPa	[11]
Dopamine modified silk	Aluminum	135 ± 15 kPa	[13]
A mixture of proteins from Mytilus edulis	Porcine skin	Dry conditions (25 °C, 40% relative humidity) after 24 h: 0.33 ± 0.17 MPa Wet conditions (37 °C, 80% relative humidity) after 24 h: 0.93 ± 0.32 MPa	[14]
Complex coacervates are made from mixtures of copolymers, the anionic one containing phosphate and dopamine groups and the cationic one containing primary amino groups.	Cortical bone	120 kPa	[15]
Poly[(3,4-dihydroxymandelic acid)$_x$-co-(lactic acid)$_{1-x}$]	Aluminum, dry state	2.6 ± 0.4 Mpa	[16]
	Aluminum, wet state	1.0 ± 0.3 MPa	
	Sanded steel, dry state	1.7 ± 0.5 MPa	
	Teflon, dry state	0.32 ± 0.05 MPa	

(Continued)

Table 11.1 (Continued)

Adhesive material	Substrates	Adhesion strength	Ref.
4-arm PEG ($M_w = 10^4$ g mol^{-1}) modified with catechol + collagen (0.1% w/v) + hydroxyapatite (2.5% w/v). Crosslinking with 120 mM $NaIO_4$	Skin tissue covered with blood	40 kPa	[17]
4-arm PEG (150 mg mL^{-1} + Laponite (0–2 wt%), $NaIO_4$/dopamine ratio of 0.5	Bovine pericardium	Laponite free gel: 3.5 ± 1.2 kPa Laponite at 2 wt%: 7.9 ± 1.8 kPa	[18]
8-arm PEG-modified with dopamine blended with gelatin microgel particles (at various amounts), oxidized with $NaIO_4$	Bovine pericardium (lap shear tests)	2.0 ± 0.3kPa without gelatin 3.8 ± 0.5 kPa with 7.5% (w/w) gelatin microgel	[19]
4-arm PEG-modified with dopamine or nitrodopamine	Bovine pericardium, in wet conditions	Nitrodopamine: 4.0 ± 1.0 kPa at pH = 6.7 and 7.4 Dopamine: 1.8 ± 0.2 kPa at pH 6.7 8.0 ± 1.5 kPa at pH = 7.4	[20]
4-arm PEG modified with dopamine, with $NaIO_4$-loaded liposomes. Gelation was thermally triggered	Bovine skin	35 ± 12.5 kPa	[21]
Chitin modified with pyrogallol, crosslinked with $NaIO_4$	Bovine skin	215 kPa	[22]
Gelatin modified with 2,3,4-trihydroxybenzaldehyde	PMMA	136.7 ± 1.4 kPa	[23]
	Iron	147.3 ± 6.3 kPa	
	Glass	92.9 ± 7.4 kPa	
	Pig skin	56.5 ± 4.4 kPa	
The polymer containing dopamine and ureido-4-pirimidinoneside chains plus 2-ethyhexyl side chains (to induce phase microphase separation)	Bovine pericardium	122.4 ± 42.4 kPa	[24]
Alginate was modified with 15% of the monomers modified with catechols	Titanium and gingiva	30 ± 5 kPa	[25]
Block copolymers with hydrolyzable lactide monomers, two acrylate end groups (for UV-triggered crosslinking), and one end group ending with DOPA (from 1 to 5)	TiO_2	Polymer with one DOPA end group: 212 ±139 mJ m^{-2} Polymer with 5 DOPA end groups: 413±123 mJ m^{-2}	[26]

Very interesting for industrial applications are catechol-containing polymers [26], which undergo crosslinking through photochemical activation using end-grafted acrylates and where the other end on the polymer contains DOPA groups (from 1 to 5) affording adhesion towards TiO_2-coated glass slides. Unfortunately, it is pretty difficult to compare these data (Table 11.1) because of the different used substrates, the different curing times, and the different retraction rates between the two materials with the adhesive in between. An excellent example of this peeling rate effect is provided with poly(HEMA-co-DMA), where the amount of modified hydroxyethylmethacrylate (HEMA) with dopamine methacrylate (DMA) varied from 0 to 10%, playing the role of adhesive between different materials (Au, Pt, Al_2O_3, Parylene-C, glass, and PTFE) (Figure 11.1) [27]. Nevertheless, Feng et al. [17] provided a fair

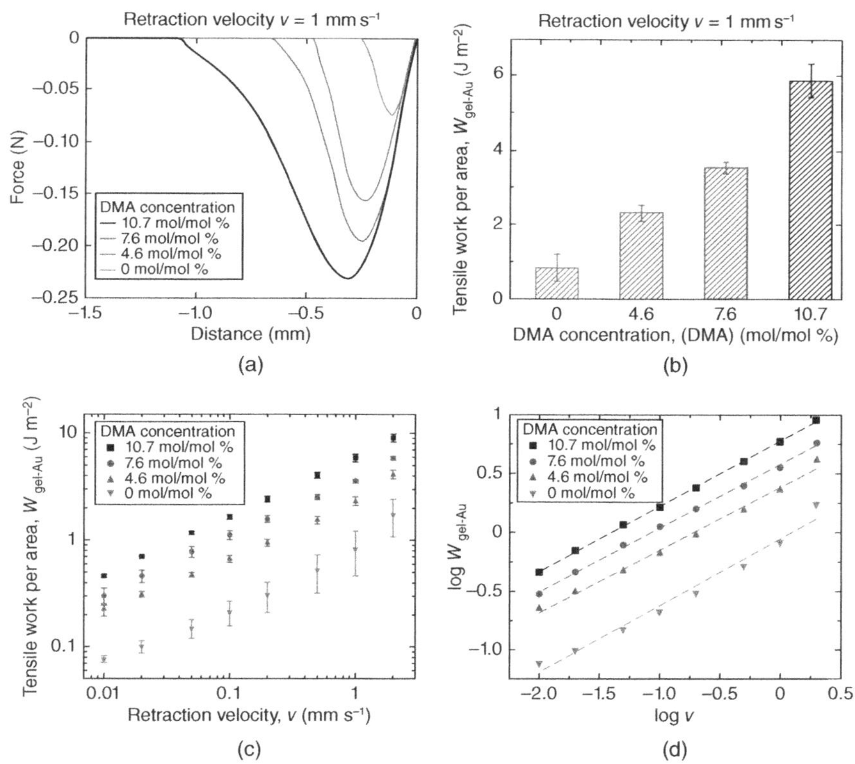

Figure 11.1 (a) Force–displacement curves of poly(HEMA-*co*-DMA) hydrogels as a function of the molar percentage of dopamine-modified monomer units, the adhesion substrate being Au, with a constant retraction speed of 1 mm/s. (b) Tensile work per unit area of the indentor between Au and the poly(HEMA-*co*-DMA) hydrogels as a function of the molar percentage of dopamine-modified monomer units. (c) Values of the tensile work as a function of the retraction velocity *v* on Au substrates for different molar percentages of dopamine-modified monomer units as indicated in the inset. (d) The same data are represented in a double logarithmic representation. The dashed lines correspond to linear regressions to the data. Source: Reproduced from reference [27] with authorization from Wiley-VCH.

comparison between PEG-dopamine, fibrin and cyanoacrylates for their adhesive strength on porcine skin: their PEG-dopamine polymer (blended with collagen and hydroxyapatite) (100 ± 20 kPa) performed much better than fibrin glue (10 ± 5 kPa), but slightly less than cyanoacrylate (200 ± 40 kPa) [17].

Such catechol-modified polymers can also spontaneously form gels at the water/air interface (Figure 11.2) without stirring. Indeed, the catechol moieties undergo faster oxidation, followed by crosslinking, at the water/air interface than in the bulk of the solution [28]. The obtained membranes are anisotropic and robust: adhering firmly to the vial in which they are formed and able to sustain the weight of the underlying solution (Figure 11.2f).

The chemical modification of polymers with catechol groups offers the advantage of controlling the adhesive properties. Still, it requires some synthesis and, often, some purification steps to remove the unbound catechol groups. In addition, those steps have to be performed in deoxygenated solvents to preclude the

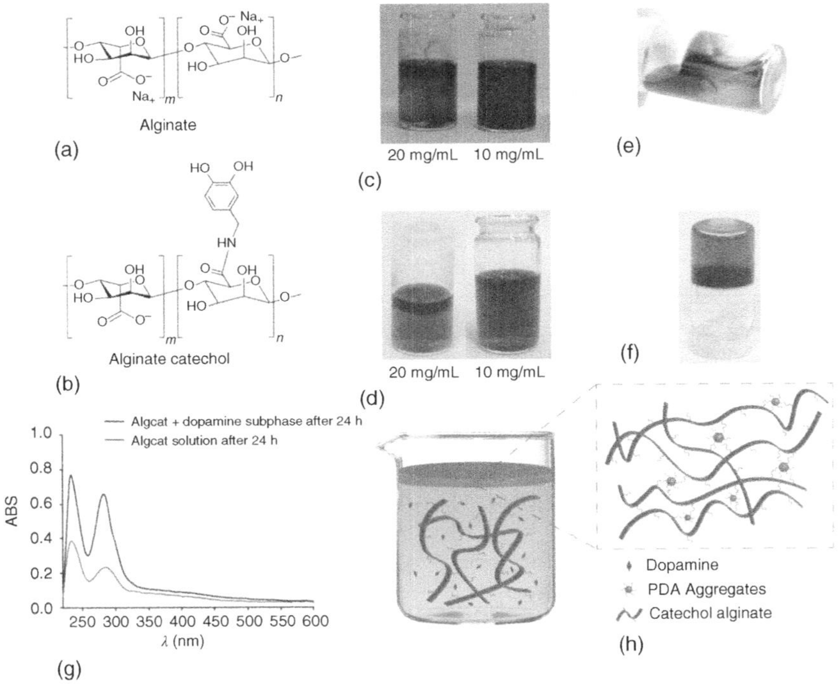

Figure 11.2 Chemical structures of (a) sodium alginate and (b) Alg-Cat (sodium alginate with grafted catechol). Pictures of (c) sodium alginate and (d) Alg-Cat solutions after 24 h of oxidation. Pictures of an alginate (e) and an Alg-Cat (f) membrane sticky to the vial. (g) UV-Vis spectra of the solution containing both Alg-Cat and dopamine in the subphase and Alg-Cat solution alone after 24 hours of reaction. (h) Schematic representation of the covalent cross-links formed between Alg-Cat and PDA aggregates. Source: Florian Ponzio et al. [28]/Reproduced with permission from American Chemical Society.

oxidation of the used catechols, catecholamines, or polyphenols. Therefore, the assumption to blend catechols, catecholamines, or polyphenols with polymers containing nucleophilic groups emerged. Indeed, even before the description of the Alg-Cat-based membranes [28], it was found that a dopamine solution containing branched poly(ethyleneimine) in the presence of oxygen and the absence of stirring yields a robust free standing and Janus-like membrane at the water/air interface [29]. The following paragraph will describe the adhesive properties of those polymer-based blends.

11.2.1 Polyphenols Blended with Molecules Able to Undergo Gelation

The blending of tannic acid (Figure 11.3a) with 4-arm poly(ethylene glycol) (PEG) yields strongly adhesive hydrogels whose adhesion strength depends mainly on the nature of the PEG end groups. The highest adhesion strength is obtained with $-NH_2$ modified PEG (Figure 11.3d), as expected based on the higher nucleophilicity of this group compared with hydroxyl or thiol groups.

Since this first study, some investigations have investigated polymer-polyphenol blends' adhesion strength. The obtained results are summarized in Table 11.2. Concerning their adhesive properties, most of these blends, prepared in a one-pot manner, are competitive with the gels made from catechol-modified polymers (data in Table 11.1).

Interestingly, bovine gelatin gels (10% (w/v)) remain stable up to 50 °C when blended with 10 mM catechol (or dopamine) after oxidation with 10 mM $NaIO_4$, whereas the unmodified gelatin gels undergo a gel-sol transition above 30–31 °C [32]. Such a catechol/dopamine afforded thermal stability opens the route for the biological evaluation of these gelatin/catechol blends. In addition, the storage modulus (determined by shear rheometry), the adhesion strength (between stainless steel plates), and the cohesive–adhesive energy of gelatin hydrogels blended with catechol or dopamine display a maximum for catechol (or dopamine) concentrations close to 20 mM, when the blends are oxidized with $NaIO_4$ at a constant concentration of 10 mM (Figure 11.4).

Note that polyphenols extracted from plants have also been used to crosslink gelatin-based hydrogels for applications as food additives [34]. The hydrogels blending biopolymers and polyphenols should display many analogies in structure and properties with polyphenol-based coacervates.

11.2.2 Controlled Protein–Polyphenol Coacervates Based on LBL Assembly

Protein–polyphenol mixtures usually form coacervates, as mentioned in the introduction, particularly for proline-rich proteins. Usually, an increase in the polyphenol's molecular mass allows larger aggregates to be created with given proteins [35].

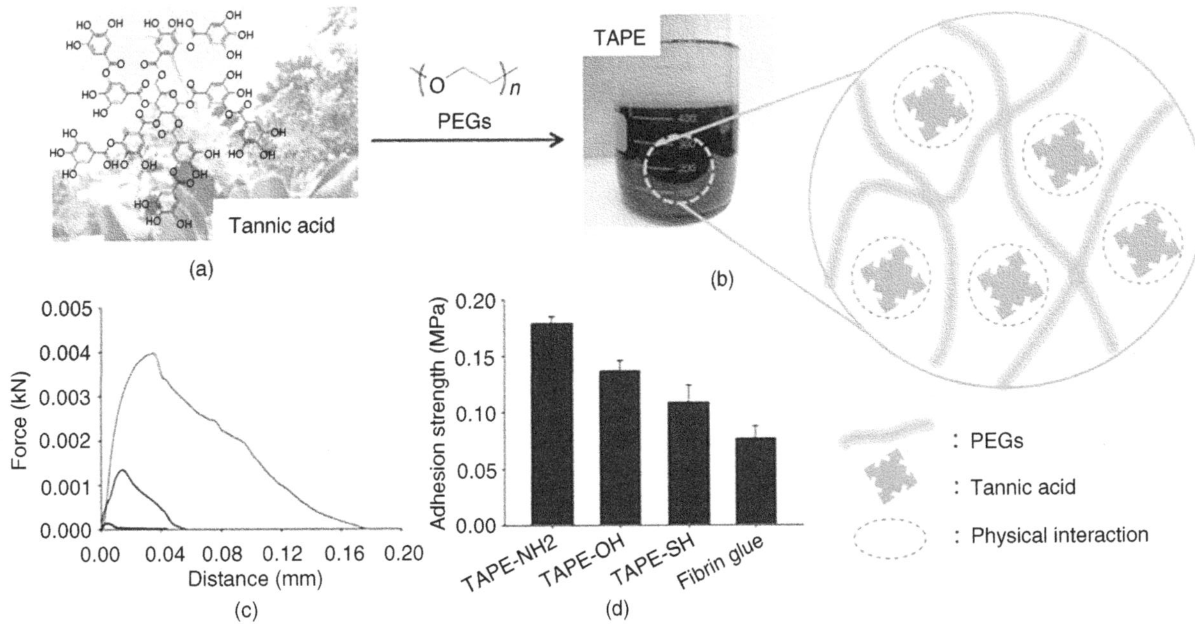

Figure 11.3 Principle of the formation of an adhesive by intermixing tannic acid (TA) and polyethyleneoxide (PEG). (a) Structure of TA. (b) Picture representing the phase separation occurring upon mixing TA and PEG. (c) Force distance curves (acquired with a universal testing machine) on porcine skin with TA + PEG-NH_2 (upper curve: the end groups of PEG being primary amino groups) compared with TA alone (middle curve) and PEG-NH_2 alone (lower curve). Adhesion strength of the different tested adhesives on porcine skin. Source: Keumyeon Kim et al. [30]/Reproduced from John Wiley & Sons, Inc.

Table 11.2 Adhesion strength of gels made from polymers blended with catechol or polyphenols.

4-arm PEG-modified with $-NH_2$ or OH groups *mixed* with Tannic acid (TA)	Epidermic side of porcine skin	TA+PEG-NH_2: 0.17 ± 0.01 MPa TA+PEG-OH: 0.13 ± 0.01MPa TA+PEG-SH: 0.11 ± 0.01MPa	[30]
Poly(N-vinylpyrrolidone) *mixed* with Tannic acid	Glass	3.71 MPa	[31]
Gelatin *blended* with 20 mM pyrocatechol or 20 mM dopamine + 10 mM $NaIO_4$, 3h of gelation	Stainless steel	Dopamine: 89 ± 5 kPa Catechol: 99 ± 14 kPa	[32]
Gelatin *blended* with 10 mM pyrocatechol + 30 mM $NaIO_4$, 3h of gelation	Stainless steel	80 ± 10 kPa	[33]

The interactions between proteins and polyphenols are dominated mainly by hydrogen bonding and hydrophobic interactions [36, 37]. In contrast, electrostatic interactions are often of minor importance except at pH values higher than 8 where the hydroxyl groups carried on phenolic compounds become deprotonated. However, the balance between hydrophobic interactions and hydrogen bonding is difficult to appreciate. The best way to estimate the importance of hydrophobic interactions is either to investigate them at various temperatures (an increase in the "binding constant" being typical of entropy-driven processes) or by changing the polarity of the solvent (the hydrophobic interaction being minored in apolar solvents and disappears in the vacuum).

The conformational mobility of the polyphenols is of significant importance in their interactions with proteins, with a stronger affinity for a given protein when the flexibility is the highest (for a polyphenol with a given amount of hydroxyl groups and phenolic rings). This research field is highly complex because of the difficulty to characterize the structure of the obtained complexes or coacervates and requires most often the correlation of many experimental techniques [37]. In addition, those coacervates may be very polydisperse and only useful as emulsions or pastes. The protein–polyphenol coacervates are also intensively described in the literature [37]. Hence, we will focus on an easy and controllable self-assembly method to get controlled protein–polyphenol gel-like structures at solid–liquid interfaces. To that aim, the widely described layer-by-layer deposition method (LBL) allows depositing films of controlled thickness from mutually interacting species: proteins, polyelectrolytes, nanoparticles, and molecules able to establish coordination bonds [38, 39, 40].

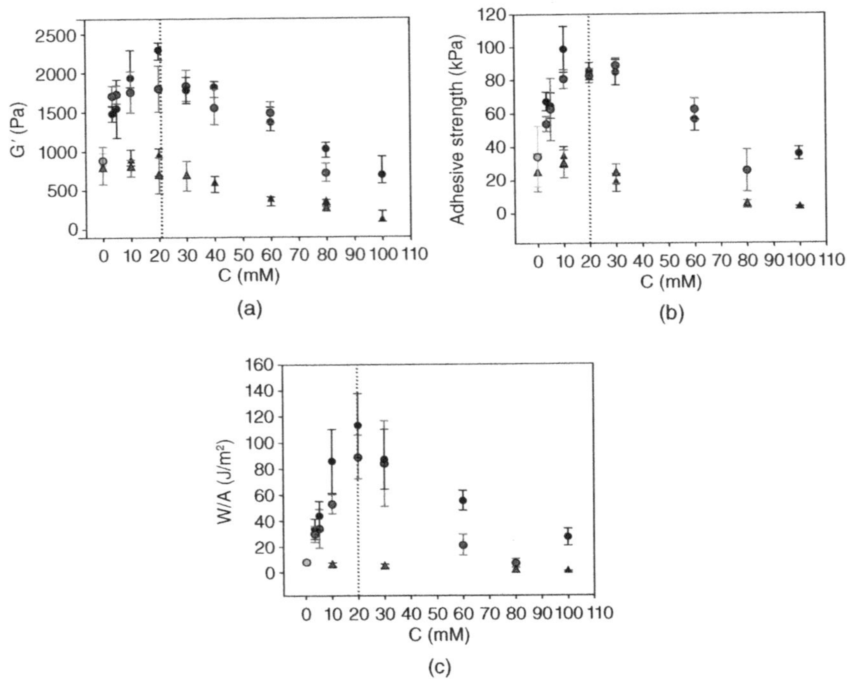

Figure 11.4 G10C(D)yN10 denote hydrogels made from a 10% (w/v) gelatin sol with y mM of catechol C (or dopamine (D)) and the addition of sodium periodate (N) to reach 10 mM. (a): Final storage modulus (G′) after 3 h of gelation kinetics for G10CyN10 (circles) or G10DyN10 (circles) as a function of their concentration and the same experiments without $NaIO_4$ triggered oxidation (triangles). The circles corresponding to C=0 are the data for pure gelatin gels without $NaIO_4$ and without catechol/dopamine. (b): Adhesion strength, (c): and debonding energy as function of the catechol or dopamine concentration for gelatin 10% w/v with 10 mM $NaIO_4$ (circles) or without $NaIO_4$ (triangles). The triangles corresponding to C=0 are the data for pure gelatin gels at 10% (w/v) and the circles corresponding to C=0 are the data for the 10% (w/v) gelatin gels but with 10 mM added $NaIO_4$. In all the graphs, the error bars correspond to ±1 standard deviation over three independent experiments. The dotted lines in all panels are aimed to identify, qualitatively, the optimal catechol/dopamine concentration in terms of gel elasticity and adhesiveness. Source: Reproduced from Back et al. [32] with authorization from MDPI.

The LBL films can grow linear or exponential [40] with respect to the number of deposition cycles, where a deposition cycle corresponds to the successive deposition of the two reacting partners. Usually, the linearly growing films display an ordered (yet non-crystalline) arrangement of the building blocks. In contrast, the exponentially growing film is highly hydrated and displays many analogies with physical gels, namely, a high mobility of the integrated polymers [40]. From this point of view, it was interesting to ask if such behaviors could occur in the case of LBL film deposition, where one of the deposited molecules is a polyphenol. Indeed, the LBL deposition method has been applied to polyphenols and a vast repertoire of molecules able to interact with them, mostly as in polyphenol–protein-based coacervates through hydrogen bonds and hydrophobic interactions [41]. The thickness of LBL films made from gelatin and tannic acid displayed a peculiar ionic strength dependence [42] with a linear growth behavior as a function of the number of deposited layer pairs.

Enzymes like alkaline phosphatase could be incorporated in LBL film with alginate covalently modified with catechol groups [9] (Figure 11.5). The enzymatic activity was found to be a linear function of the number of deposited layer pairs

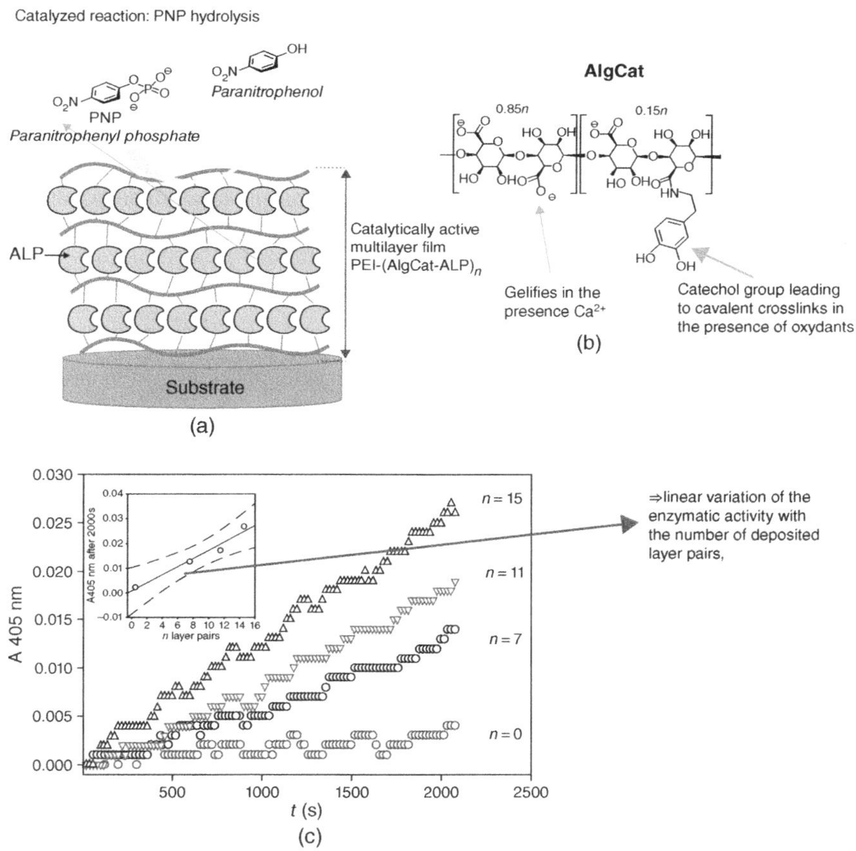

Figure 11.5 (a) Schematic representation of the layer-by-layer deposition of alginate modified with catechol groups (Alg-Cat) and alkaline phosphatase (Alp) to build-up (Alg-Cat–ALP)$_n$ multilayer films. (b) Schematic representation of the two crosslinking modes allowed by Alg-Cat. (c) Evolution with time of the enzymatic activity of (Alg-Cat–Alp)$_n$ films with the number of deposited layer pairs. The inset represents the evolution of the slope of the curves with the number of deposited layer pairs. Source: Reproduced with modification from Longo et al. [9].

(Figure 11.5c), suggesting that the enzymatic activity is not affected by the Alg-Cat polymer and that the substrate (paranitrophenol phosphate) diffuses through the whole thickness of the LBL film.

However, the alternate deposition of the same enzyme, alkaline phosphatase (Alp) and tannic acid (TA), to yield PEI-(Alp-TA)*n* LBL films delivers an exponentially growing film in which only the enzymes in direct contact with the solution remain active [43]. In addition, after contact with sodium periodate to oxidize TA and to create covalent bonds between the formed quinones and amines on the enzyme, all the enzymatic activity is lost [43]. The first considered case, namely the (Alg-Cat–Alp)*n* films, is akin to the class (i) of materials (Scheme 11.1), those where the catechol or polyphenol is already bound to the polymer, whereas the PEI-(Alp-TA)*n* films belong to the class (ii) of materials, where the enzyme and the polyphenol are simply mixed. The finding is that total enzymatic activity is conserved in the first case. In contrast, only the enzyme not fully embedded with polyphenols remains active, suggesting that a dilution strategy of the binding moiety, the catechol or polyphenol, is mandatory to get functional materials. However, this assumption has to be checked for other combinations of enzymes and catechols/polyphenols.

11.3 Conclusions and Perspectives

This chapter describes three immobilization strategies of catechols/polyphenols in hydrogels or 2D-ordered assemblies (Scheme 11.1). The use of polymers able to undergo gelation and grafted with catechols (strategy (i)) allows to produce biocompatible adhesives, useful for biomedical applications whose adhesion strength is competitive with cyanoacrylates and much better than fibrin glues. However, the preparation of such conjugated polymers requires chemical synthesis (sometimes in several steps) in the absence of dissolved oxygen (which would oxidize the catechol groups) and purification. From this point of view, the single mixing of catechols/polyphenols and suited polymers (strategy (ii)), able to undergo gelation and react with oxidized catechols/polyphenols, offers the alternative of a one-pot preparation and, according to the investigated systems, adhesion strength comparable to that of catechol modified polymers. However, much more research effort is required in this field, and the obtained hydrogels' cytotoxicity must be systematically investigated. Using polyphenol or catechol-modified polymers (according to strategy (i)) allows obtaining 2D coacervates with proteins at solid–liquid interfaces through the LBL deposition method. Such coatings (strategy (iii)) can be used as platforms for biosensors, as layers affording antioxidant properties, or for protecting metals against corrosion. To reach such a goal, the stability of the obtained coatings should be improved both from a mechanical point of view and from the point of view of enzymatic stability over a prolonged duration.

References

1 Quideau, S., Deffieux, D., Douat-Casassus, C., and Pouységu, L. (2011). Plant polyphenols: chemical properties, biological activities, and synthesis. *Angewandte Chemie, International Edition* **50**: 586–621.
2 Bate-Smith, E.C. (1973). Haem analysis of tannins:_the concept of relative astringency. *Phytochemistry* **12**: 907–912.
3 Bennick, A. (1982). Salivary proline-rich proteins. *Molecular and Cellular Biochemistry* **45**: 83–99.
4 Hagerman, A.E. and Butler, L.G. (1981). The specificity of the proanthocyanidin-protein interaction. *The Journal of Biological Chemistry* **256**: 4494–4497.
5 Zhang, W., Wang, R., Sun, Z.M. et al. (2020). Catechol-funtionalized hydrogels: biomimetic design, adhesion mechanism, and biomedical application. *Chemical Society Reviews* **49**: 433–464.
6 Yang, J., Cohen Stuart, M.A., and Kamperman, M. (2014). Jack of all trades: versatile catechol crosslinking mechanisms. *Chemical Society Reviews* **43**: 8271–8298.
7 Lee, B.P., Messersmith, P.B., Israelachvili, J.N., and Waite, J.H. (2011). *Annual Review of Materials Research* **41**: 99–132.
8 Faure, E., Falentin-Daudré, C., Jérôme, C. et al. (2013). Catechols as versatile platforms in polymer chemistry. *Progress in Polymer Science* **38**: 236–270.
9 Longo, J., Garnier, T., Mateescu, M. et al. (2015). Stable bioactive enzyme containing multilayer films based on covalent crosslinking from mussel inspired adhesives. *Langmuir* **31**: 12447–12454.
10 Yu, M. and Deming, T.J. (1998). Synthetic polypeptide mimics of marine adhesives. *Macromolecules* **31**: 4739–4745.
11 Westwood, G., Horton, T.N., and Wilker, J.J. (2007). Simplified polymer mimics of cross-linking adhesive proteins. *Macromolecules* **40**: 3960–3964.
12 Moulay, S. (2014). Dopa/catechol-tethered polymers: bioadhesives and biomimetic adhesive materials. *Polymer Reviews* **54**: 436–513.
13 Burke, K.A., Roberts, D.C., and Kaplan, D.L. (2016). Silk fibroin aqueous-based adhesives inspired by mussel adhesive proteins. *Biomacromolecules* **17**: 237–245.
14 Ninan, L. (2003). Adhesive strength of marine mussel extracts on porcine skin. *Biomaterials* **24**: 4091–4099.
15 Shao, H., Bachus, K.N., and Stewart, R.J.A. (2009). Water-borne adhesive modeled after the sandcastle glue of *P. Californica*: a water-borne adhesive modeled after the sandcastle glue of *P. Californica*. *Macromolecular Bioscience* **9**: 464–471.
16 Jenkins, C.L., Siebert, H.M., and Wilker, J.J. (2017). Integrating mussel chemistry into a bio-based polymer to create degradable adhesives. *Macromolecules* **50**: 561–568.

17 Feng, J., Ton, X.-A., Zhao, S. et al. (2017). Mechanically reinforced catechol-containing hydrogels with improved tissue gluing performance. *Biomimetics* **2**: art 23.

18 Liu, Y., Meng, H., Konst, S. et al. (2014). Injectable dopamine-modified poly(ethylene glycol) nanocomposite hydrogel with enhanced adhesive property and bioactivity. *ACS Applied Materials & Interfaces* **6**: 16982–16992.

19 Li, Y., Meng, H., Liu, Y. et al. (2016). Gelatin microgel incorporated poly(ethylene glycol)-based bioadhesive with enhanced adhesive property and bioactivity. *ACS Applied Materials & Interfaces* **8**: 11980–11989.

20 Cencer, M., Murley, M., Liu, Y., and Lee, B.P. (2015). Effect of nitro-functionalization on the cross-linking and bioadhesion of biomimetic adhesive moiety. *Biomacromolecules* **16**: 404–410.

21 Burke, S.A., Ritter-Jones, M., Lee, B.P., and Messersmith, P.B. (2007). Thermal gelation and tissue adhesion of biomimetic hydrogels. *Biomedical Materials* **2**: 203–210.

22 Oh, D.X., Kim, S., Lee, D., and Hwang, D.S. Tunicate-mimetic nanofibrous hydrogel adhesive with improved wet adhesion. *Acta Biomaterialia* **20**, 104–112.

23 Han, N., Xu, Z., Cui, C. et al. (2020). Fe^{3+}-crosslinked pyrogallol-tethered gelatin adhesive hydrogelwith antibacterial activity for wound healing. *Biomaterials Science* **8**: 3164–3172.

24 Balkenende, D.W.R., Winkler, S.M., Li, Y., and Messersmith, P.B. (2020). Supramolecular cross-links in mussel-inspired tissue adhesives. *ACS Macro Letters* **9**: 1439–1445.

25 Mateescu, M., Baixe, S., Garnier, T. et al. (2015). Antibacterial peptide-based gel for prevention of medical implanted-device infection. *PLoS One* **10**: e0145143.

26 Lee, B.P., Chao, C.-Y., Nunalee, F.N. et al. (2006). Rapid gel formation and adhesion in photocurable and biodegradable block copolymers with high DOPA content. *Macromolecules* **39**: 1740–1748.

27 Wu, H., Sariola, V., Zhao, J. et al. (2016). Composition-dependent underwater adhesion of catechol bearing hydrogels. *Polymer International* **65**: 1355–1359.

28 Ponzio, F., Le Houerou, V., Zafeiratos, S. et al. (2017). Robust alginate-catechol @polydopamine free standing membranes obtained from the water/air interface. *Langmuir* **33**: 2420–2426.

29 Hong, S., Schaber, C.F., Dening, K. et al. (2014). Air/water interfacial formation of freestanding, stimuli responsive, self-healing catecholamine Janus-faced microfilms. *Advanced Materials* **26**: 7581–7587.

30 Kim, K., Shin, M., Koh, M.-Y. et al. (2015). TAPE: a medical adhesive inspired by a ubiquitous compound in plants. *Advanced Functional Materials* **25**: 2402–2410.

31 Nam, H.G., Nam, M.G., Yoo, P.J., and Kim, J.-H. (2019). Hydrogen bonding-based strongly adhesive coacervate hydrogels synthesized using poly(*N*-vinylpyrrolidone) and tannic acid. *Soft Matter* **15**: 785–791.

32 Back, F., Mathieu, E., Betscha, C. et al. (2022). Optimization of the elasticity and adhesion of catechol-or dopamine-loaded gelatin gels under oxidative conditions. *Gels* **8**: art.210.

33 Back, F., Ball, V., and Arntz, Y. (2021). Influence of the $NaIO_4$ concentration on the gelation and the adhesive strength of pyrocatechol/pyrogallol containing gelatin hydrogels. *Frontiers in Materials* **8**: 671451.

34 Strauss, G. and Gibson, S.M. (2004). Plant phenolics as cross-linkers of gelatin gels and gelatin-based coacervates for use as food ingredients. *Food Hydrocolloids* **18**: 81–89.

35 Haslam, E. (1974). Polyphenol-protein interactions. *Biochemical Journal* **139**: 285–288.

36 Oh, H.I., Hoff, J.E., Armstrong, G.S., and Haff, L.A. (1980). Hydrophobic interaction in tannin-protein complexes. *Journal of Agricultural and Food Chemistry* **28**: 394–398.

37 Le Bourvellec, C. and Renard, C.M.G.C. (2012). Interactions between polyphenols and macromolecules: quantification methods and mechanisms. *Critical Reviews in Food Science and Nutrition* **52**: 213–248.

38 Decher, G. (1997). Fuzzy nanoassemblies: toward layered polymeric nanocomposites. *Science* **277**: 1232–1237.

39 Borges, J. and Mano, J.F. (2014). Molecular interactions driving the layer-by-layer assembly of multilayers. *Chemical Reviews* **114**: 8883–8942.

40 Lavalle, P., Voegel, J.C., Vautier, D. et al. (2011). Dynamic aspects of films prepared by a sequential deposition of species: perspectives for smart and responsive materials. *Advanced Materials* **23**: 1191–1221.

41 Reitzer, F., Allais, M., Ball, V., and Meyer, F. (2018). Polyphenols at interfaces. *Advances in Colloid and Interface Science* **257**: 31–41.

42 Ringwald, C. and Ball, V. (2016). Step-by-step deposition of gelatin B and tannic acid displays a peculiar ionic strength dependance at pH 5. *RSC Advances* **6**: 4730–4738.

43 Allais, M., Meyer, F., and Ball, V. (2018). Multilayered films made from tannic acid and alkaline phosphatase with enzymatic activity and electrochemical behavior. *Journal of Colloid and Interface Science* **512**: 722–729.

12

Polyphenol-based Materials as Sensors: Recent Trends in Chemo- and Bio-sensing

Jing Wei[1], *Youyou Feng*[1], *and Gen Wang*[2]

[1]*The Key Laboratory of Biomedical Information Engineering of Ministry of Education, School of Life Science and Technology, Xi'an Jiaotong University, Xi'an, P. R. China*
[2]*School of Environmental and Municipal Engineering, Xi'an University of Architecture and Technology, Xi'an P. R. China*

12.1 Introduction

A sensor is a device that can measure a physical quantity and convert it into a readable signal. Sensors are implemented in various aspects of daily life. There are over ten sensors inside just one mobile phone. There may be hundreds of sensors on a car. Planes, ships, and high-speed trains even have thousands of sensors. Various laboratory tests in hospitals require sensing, safety production in factories requires sensing, food safety requires sensing, and environmental monitoring is also the case. It can be said that sensors exist in every device in daily life, maintaining the regular operation of human life and ensuring human health and safety [1–5]. The core component of the sensor is the sensing material. It is the key component to realize signal acquisition and conversion. The innovation of sensing materials is of great significance for improving the performance of sensors and expanding their application fields. By adjusting the chemical composition and structure of sensing materials, the performance of sensors can be regulated. Sensing technology based on nanomaterials has been studied for decades. Nanomaterials have minor volume effects, surface effects, quantum size effects, quantum tunneling effects, and dielectric confinement. Therefore, nanomaterials have unique properties that conventional materials do not possess, which are essential for outstanding sensing performance. Moreover, nanomaterials as

Science and Engineering of Polyphenols: Fundamentals and Industrial Scale Applications, First Edition. Edited by Chandrabhan Verma.
© 2024 John Wiley & Sons Inc. Published 2024 by John Wiley & Sons Inc.

promising sensing materials have focused increasing attention due to their tunable composition and structure, ease of modification, and ability to bind various biological molecules.

Chemical and biological sensing, as a major branch of sensors, have significant research value in industry, military, food, environment, and especially medicine. In industrial production, chemical sensors are mainly used for monitoring product quality and environmental safety [1, 6–10]. In the military, chemical sensors are used to detect toxic chemical weapons to ensure the safety of soldiers [11]. In the food industry, it is used to detect illegal food additives. It can also monitor the freshness of food [12–14]. In terms of the environment, chemical sensors can monitor air or water pollution [15–17]. In biomedicine, chemical and biological sensors can detect biochemical indicators from blood, urine, sweat, feces, and respiratory gas. Diagnosis and early screening of diseases can be achieved by detecting disease markers [18–19]. According to differences in signals, chemical or biological sensing is generally classified into electrochemical sensing [20–21], optical sensing [22], thermal sensing, and magnetic sensing [23]. Depending on the type of object to be tested, sensors can be divided into ion sensing, small molecule sensing, and biomolecular sensing. Currently, nanomaterials have been widely studied in chemical and biological sensing. The excellent feature of combining biomolecules has given them rich and diverse sensing properties. Among the various sensing materials, polyphenol-based nanomaterials are one emerging sensing material. Due to their excellent physical and chemical properties, polyphenols have been widely used as building blocks of functional nanomaterials.

This chapter focuses on preparing polyphenol-based nanomaterials for chemical and biological sensing applications. It is specifically divided into the following parts. Firstly, the physical and chemical properties of plant polyphenols are introduced. The advantages of polyphenols in the preparation of functional nanomaterials are elaborated. Secondly, different polyphenol-based nanomaterials' properties and preparation methods are presented, including metal-polyphenol composites and their derived carbon and metal oxide materials. Thirdly, the application of polyphenol-based nanomaterials in chemical and biological sensing to detect different analytes is covered. Finally, the prospects and challenges of polyphenol-based nanomaterials in sensing applications are summarized.

12.2 Polyphenol

Polyphenol is a general definition for a class of compounds with multiple phenolic hydroxyl groups. Polyphenols are extremely common in nature and are mainly secondary metabolites of plants [24–25]. They are widely present in leaves, fruits and their products (wine, tea, coffee, chocolate, etc.). The molecular weight of plant polyphenols ranges from 500 to 4000 Da. On average, there are 5–7 aromatic rings and 12–16 phenolic hydroxyl groups per 1000 Da [25]. Based on the molecular structure of polyphenols, they can be divided into proanthocyanins, gallo- and ellagitannins and phlorotannins [25, 26]. Abundant phenolic hydroxyl groups endow polyphenols with unique physicochemical properties. First, polyphenols have antioxidant characteristics as phenolic hydroxyls can readily oxidize to quinone structures by oxygen and free radicals [26–29]. Therefore, polyphenols are often used in health care and skin care products to combat free radicals produced by aging in the human body. Because of such antioxidant properties, polyphenol-rich tea and wine are hugely popular [28]. Besides, the phenolic hydroxyl groups of polyphenols can bind to biological macromolecules, such as proteins, polysaccharides, nucleic acids, glycoproteins and cell membranes [30–33], through hydrophobic interaction and hydrogen bonding [34]. The unique feature renders polyphenol antibacterial properties through binding with bacterial enzymes or combining with bacterial membranes [27]. Additionally, polyphenols can chelate with metal ions via forming coordination bonds between metal ions and phenolic hydroxyl groups [35]. The strength of coordination bonding varies with different pH values [36–37]. The higher valence state metal ions will oxidize the polyphenol to a quinone during the coordination. Finally, polyphenols also have common chemical properties of phenols. They can covalently crosslink with other molecules, such as addition reaction, esterification reaction, amidation reaction, etc. [38]. The chemical structures of some common polyphenols are given in Figure 12.1.

Due to the ability to interact with various molecules and ions, polyphenols have been widely used as building blocks for functional nanomaterials [38–42]. On the one hand, polyphenols can be converted to polyphenol polymers of different morphologies (one-dimensional, two-dimensional, or three-dimensional nanomaterials) and structures (vesicles, colloidal spheres, crystals, etc.). It can also synthesize metal-polyphenol coordination polymers. On the other hand, carbon materials and metal oxides can be prepared using polyphenol polymers as a precursor. Different derivative nanomaterials are prepared by calcinating polyphenol polymers in the air or inert atmosphere. In addition, polyphenol polymer materials can serve as carriers to load different small molecules, biological macromolecules, or nanoparticles (e.g. noble metal nanoparticles). Polyphenol polymers and their derived materials exhibited unique physical and chemical properties, allowing wide application in sensing, catalysis, biomedical, environmental, energy and other fields. In the following parts, we introduce polyphenol polymers' synthesis strategy, properties, and their derived materials. Then, their applications in sensing will be presented and discussed.

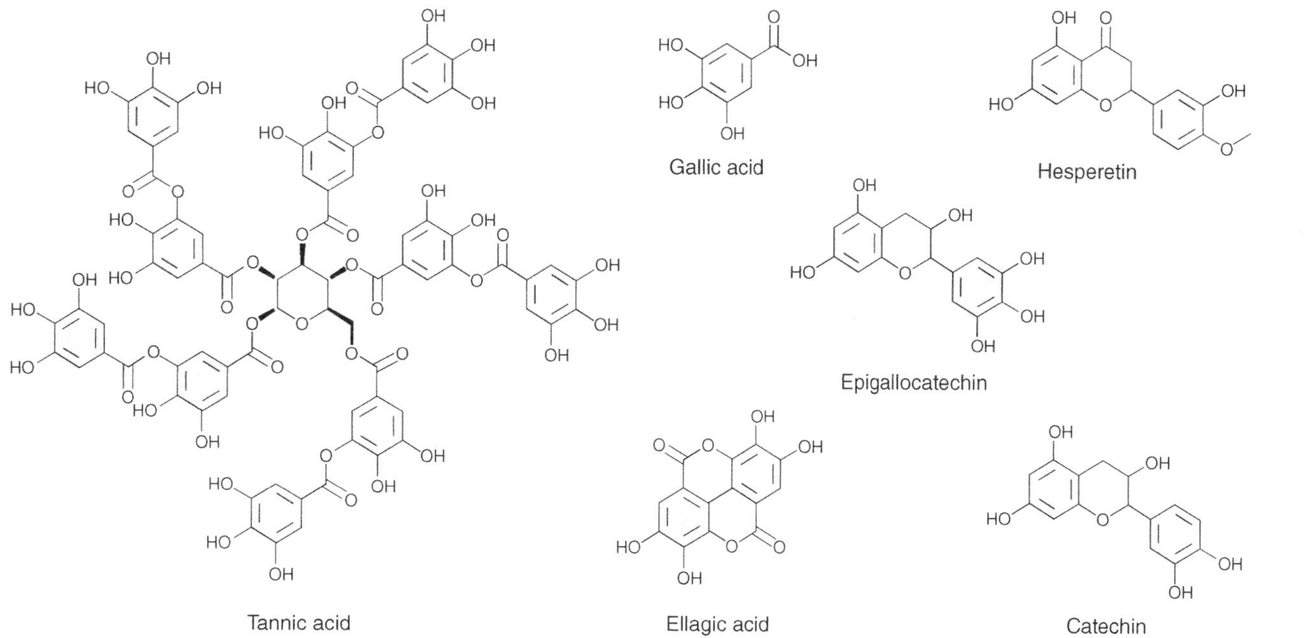

Figure 12.1 Chemical structures of some common polyphenols.

12.3 Polyphenol-based Nanomaterials and Their Derivatives

Polyphenols can combine with other components through covalent (e.g., esterification reaction, addition reaction, amidation reaction, *etc.*) and non-covalent ways (e.g. hydrophobic interaction, electrostatic interaction, hydrogen bond, coordination bond, etc.) [43–44]. Due to their unique physical and chemical properties, polyphenols are excellent building blocks for the synthesis of functional nanomaterials [39, 42–45]. Metal-polyphenol composites have been widely studied with different morphologies and components. Meanwhile, metal-polyphenol composites are also excellent precursors for further conversion into carbon and metal oxide materials. The general synthesis methods of these materials are discussed below.

12.3.1 Metal-polyphenol Coordination Polymers

There are lots of catechol and pyrogallol groups in polyphenols. They can coordinate with different metal ions and form specific nanostructures through self-assembly. Inspired by mussel biomimicry, polyphenols have good adhesion properties [32, 46]. Moreover, polyphenols have good adhesion capability on diverse inorganic or organic substrates (polystyrene, calcium carbonate, plastic, glass, Au, PDMS, etc.) [37]. Further removal of templates will result in metal-polyphenol coordination polymers with thin film or capsule structures. Ejima et al. [37] prepared metal-polyphenol complex membranes and capsules for the first time based on the coordination reaction between tannic acid and metal ions. Using polystyrene as a substrate, tannic acid and Fe^{3+} were first coordinated on the surface of polystyrene, forming a uniform thin film through self-assembly. Metal-polyphenols can be evenly formed on planar, spherical, and ellipsoidal structures of polystyrene. After removing the template, metal-polyphenol films and capsules were obtained. Later, Ejima et al. [47] further expanded the category of metal-polyphenol coordination polymer (Figure 12.2). They used 18 metal species (i.e., Al, V, Cr, Mn, Fe, Co, Ni, Cu, Zn, Zr, Mo, Ru, Rh, Cd, Ce, Eu, Gd, and Tb) to coordinate with polyphenols to prepare metal-polyphenol coordination capsules. The film thicknesses and fluorescence behaviors can be tuned by regulating the assembly processes. The prepared capsules showed potential applications in PET, MRI, drug delivery, and catalysis. Metal-polyphenol coordination polymer films can also be synthesized through mechanochemical methods. Kang et al. [35] developed a method for preparing metal-polyphenol coordination polymer films on metallic planes. Tannic acid was used as a polyphenol ligand and coated on a metallic planar substrate by ball-on-disk mode friction. It has active phenolic hydroxyl groups that can coordinate with metal substrates. Metal ions diffuse within the film. The content

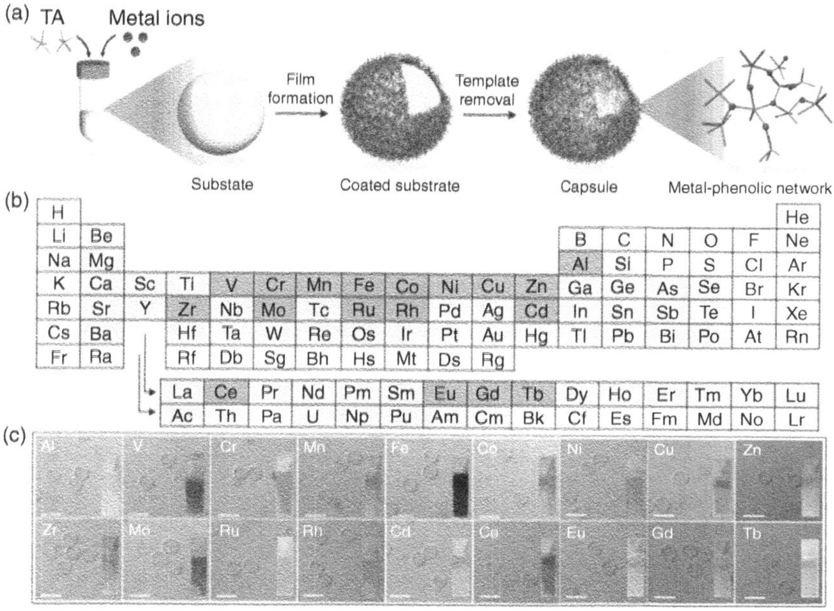

Figure 12.2 Scheme of metal-polyphenol networks prepared by 18 kinds of metal ions and their DIC images. Source: Reproduced with permission from Ref. [47]. Copyright © 2014, John Wiley & Sons, Inc.

of metal ions gradually decreases as the thickness of the film increases. Thicker film structures were formed through simple polymerization and π–π stacking between aromatic rings of tannic acid.

In addition, metal-polyphenol complexes can form crystals. Similar to metal-organic frameworks, metal-polyphenol crystals can be obtained by regulating the coordination bond between polyphenols and metal ions. Yang et al. [48] prepared metal-phenolic mesocrystals. Ellagic acid was used as a polyphenol ligand, and Zn^{2+} as a crosslinking agent (Figure 12.3). "Peanut"-like metal-phenolic mesocrystals were obtained. It showed crystal structure according to the X-ray diffraction analysis. Wei et al. [49] also reported the synthesis of Co-tannic acid mesocrystals. Firstly, tannic acid and Co^{2+} coordinated to form a metal-polyphenol coordination polymer with an amorphous structure. Then, the Co-tannic acid polymer was hydrothermally treated to form a crystal structure. During the hydrothermal process, organic ligands underwent oxidative polymerization. The polymer was further cross-linked to form crystals. Huang et al. [50] also prepared a spiny metal-polyphenol crystal using Cu or Zn as coordination atoms and tannic acid as ligands. At lower ammonia concentrations, the material exhibited needle-like morphology. As the concentration of ammonia increases, the material is further assembled to form a sea urchin-like structure.

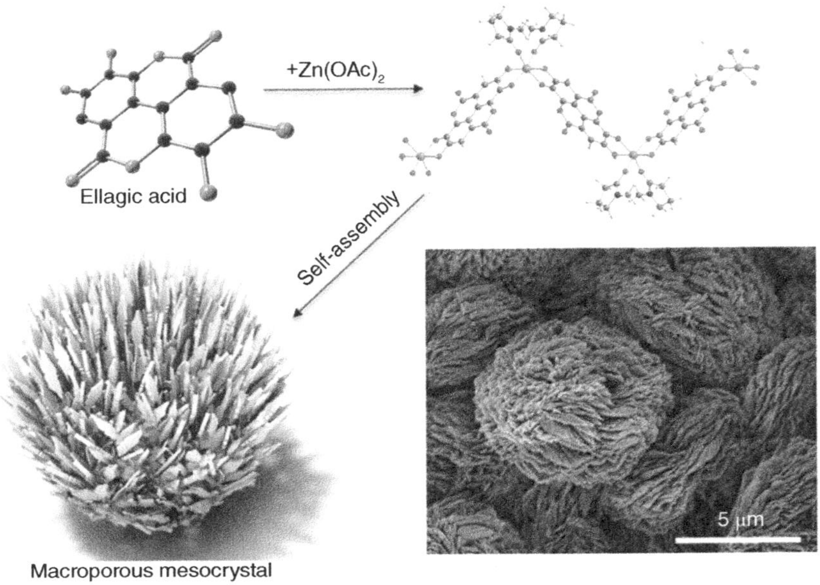

Figure 12.3 Scheme of ellagic acid derived microporous mesocrystal synthesis and its SEM image. Source: Reproduced with permission from Ref. [48]. Copyright © 2015, American Chemical Society.

By the synthesis of phenolic resins, polyphenols can also form polymers through covalent crosslinking with various molecules. Wei et al. [51] prepared metal-polyphenol coordination polymer colloidal spheres using the sol–gel method (Figure 12.4). Tannic acid was applied as a polyphenol ligand. Different metal ions (Ce^{3+}, Fe^{3+}, Al^{3+}, Fe^{2+}, Ni^{2+}, Co^{2+}, Cu^{2+} and Zn^{2+}) were served as the crosslinking agents. Firstly, tannic acid and formaldehyde cross-linked in an alkaline ethanol solution to form oligomers. The subsequent addition of metal ions resulted in the coordination of the oligomers. Metal-polyphenol colloidal spheres were obtained after hydrothermal treatment.

The abundant phenolic hydroxyl groups in tannic acid can coordinate with many metal ions. In the absence of formaldehyde, tannic acid coordinates with metal ions to form a crystal structure. The addition of formaldehyde can crosslink with tannic acid to suppress the coordination between phenolic hydroxyl groups and metal ions, forming amorphous spherical morphology. The metal-polyphenol colloidal spheres prepared by this sol–gel method have a uniform diameter (~300 nm) and show promising application potential in sensing.

More metal-polyphenol composites have been developed based on the self-assembly between polyphenols and metal ions. Lin et al. [52] prepared ordered mesoporous metal-polyphenol network particles using PS-*b*-PEO as a template. Polyphenols and metal ions were first assembled into the mesopores of polymer

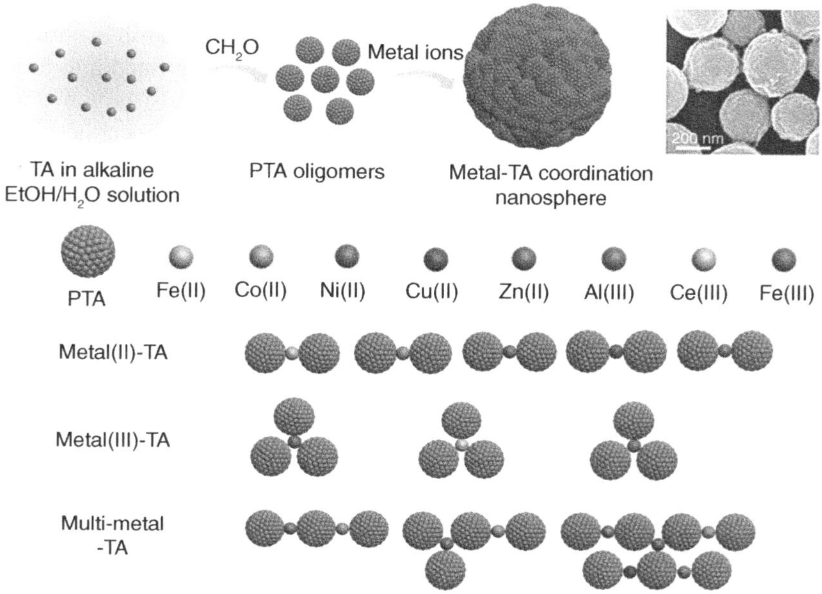

Figure 12.4 Scheme of the synthesis of metal-tannic coordination nanosphere and its SEM image. Source: Reproduced with permission from Ref. [51]. Copyright © 2018, John Wiley & Sons, Inc.

cubes. Then, metal-polyphenol network particles with ordered mesoporous structures were obtained by removing the mesoporous polymer template with tetrahydrofuran. The types of polyphenols were adjustable ((−)-epigallocatechin gallate and gallic acid were both possible). Different metal ions (Zr^{4+}, Cu^{2+} and Fe^{3+}) were also used to synthesize mesoporous metal-polyphenol network particles. The average diameter of the metal-polyphenol network particle is $2.4 \pm 0.6\,\mu m$. The abundant mesopores of the network particle can be used for loading various proteins.

12.3.2 Metal-polyphenol Coordination Polymers Derived Carbon Nanomaterials

Polyphenols are natural reagents extracted from plants. Therefore, they are excellent precursors for the synthesis of carbon materials that are more environmentally friendly. There have been many reports on the preparation of various functional carbon nanomaterials using polyphenols. Pan et al. [53] recently successfully transformed metal-phenolic networks into metal-carbon networks. Tannic acid was used as a ligand. The metal ions were used as crosslinkers to coordinate TA on the surface of $CaCO_3$ templates. Subsequently, the organic fraction was converted to carbon in an inert atmosphere (e.g. argon), while the metal ions were reduced to metallic species. By calcination in an inert atmosphere, many polyphenol-based nanomaterials have been transformed into functional carbon materials. Metal-polyphenol complex crystals can also be converted into carbon materials. Yang et al. [48] converted "peanut"-like phenolic metal composites to carbon materials via calcination, which have hierarchical pore architecture and high BET-specific surface area (842 m^2/g). Wei et al. [49] developed a metal/N-doped carbon composite using Co-tannic acid crystals and dicyandiamide as the precursor. Co^{2+}, tannic acid and dicyandiamide were used as metal, C, and N sources, respectively. After calcination in N_2, Co/N-doped carbon composite with excellent ORR and OER activity were obtained. Wang et al. [54] prepared nanoporous carbon spheres using Zn-tannic acid as a precursor. Zinc ions acted as a pore-forming agent, which firstly reduced to metallic Zn and then vaporized from carbon at a high calcination temperature in N_2. The obtained carbon spheres have a high specific surface area (527–2221 m^2/g).

Zhang et al. [55] developed ordered mesoporous carbon from tannins via a mechanical balling strategy. Traditional synthetic mesoporous carbon materials generally use highly toxic phenol and formaldehyde. Zhang utilized mechanical synthesis methods with tannins as a carbon source. By ball milling, tannins, block copolymers, and metal ions were mechanically assembled to form metal-polyphenol precursors. After calcination under nitrogen gas, ordered mesoporous carbon materials were obtained. Block copolymer micelles acted as pore-forming agents. When Zn^{2+} was selected as the crosslinking agent, mesoporous carbon without metal can be obtained (Figure 12.5). The strategy also prepared Ni-loaded ordered mesoporous carbon materials when Ni^{2+} was used.

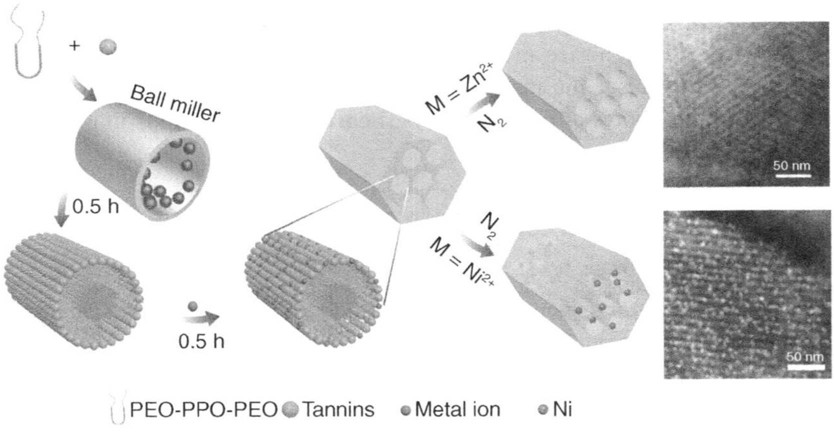

Figure 12.5 Scheme of tannin-derived ordered mesoporous carbon by the mechanochemical assembly mediated coordination crosslinking method and its relative TEM images. Source: Reproduced from Ref. [55]. Copyright © 2017, Springer Nature/CC BY 4.0.

12.3.3 Metal-polyphenol Coordination Polymers Derived Metal Oxide Nanomaterials

Compared to metal-organic framework materials, most metal-polyphenol coordination polymers show amorphous structures with tunable compositions and morphologies. Therefore, tailoring the compositions and properties of the derived materials is more reliable. The metal ions are oxidized and crystallized during calcination to form metal oxide nanomaterials with porous structures. Using this advantage, Wang et al. [56] prepared a mesoporous metal oxide spheres family. Firstly, they prepared metal-polyphenol coordination spheres. Due to polyphenols' metal ion chelating properties, the metal ion content in nanospheres was very high. Subsequently, metal-polyphenol coordination spheres were calcined in air to prepare mesoporous metal oxide spheres with different compositions. Different mesoporous metal oxide spheres (ZnO, Al_2O_3, Co_3O_4, Fe_2O_3, and CuO) were obtained by adjusting the types of metal ions in the precursor. The structure of the obtained metal oxide spheres varied according to the type of metal ions. When the crystallization temperature (T_c) of metal oxide was lower than the decomposition temperature (T_d) of polyphenol polymers, metal-polyphenol coordination spheres transformed into mesoporous metal oxide spheres (such as ZnO). When T_c was higher than T_d, for example, the crystallization temperature of Al_2O_3 was above 600 °C, and the organic part of the spheres completely decomposed before crystallization. Finally, a solid mesoporous metal oxide with a crystal structure was formed. When T_c and T_d were close, the crystallization process and the decomposition of the organic part co-occurred (Figure 12.6). Metal-polyphenol complex nanospheres tended to form hollow structures, such as Co_3O_4, CuO, and Fe_2O_3. Using polyphenols as precursors is a universal method for synthesizing metal oxide nanospheres with broad application prospects.

Wang et al. [57] prepared mesoporous metal oxide nanospheres with tunable components and structures based on the above studies. Binary (Fe-Co, Ni-Zn, and Ni-Co), ternary (Ni-Co-Mn and Ni-Co-Zn), and even quinary (Ni-Co-Fe-Cu-Zn) metal ions can be introduced into metal-polyphenol coordination polymers. After calcination treatment, hollow multi-layered metal oxide spheres were prepared. Wang et al. [58] also prepared high-entropy metal oxide nanospheres using a similar method. They adjusted the type of metal ions (Ni-Co-Cr-Fe-Mn) in the metal-polyphenol coordination polymer. After calcination, spherical Ni-Co-Cr-Fe-Mn high-entropy oxide with single crystal phase and spinel structure was obtained.

12.4 Sensing Applications

Polyphenols play an essential role in the preparation of functional materials. Polyphenol polymers and their derivatives have been widely studied in sensing

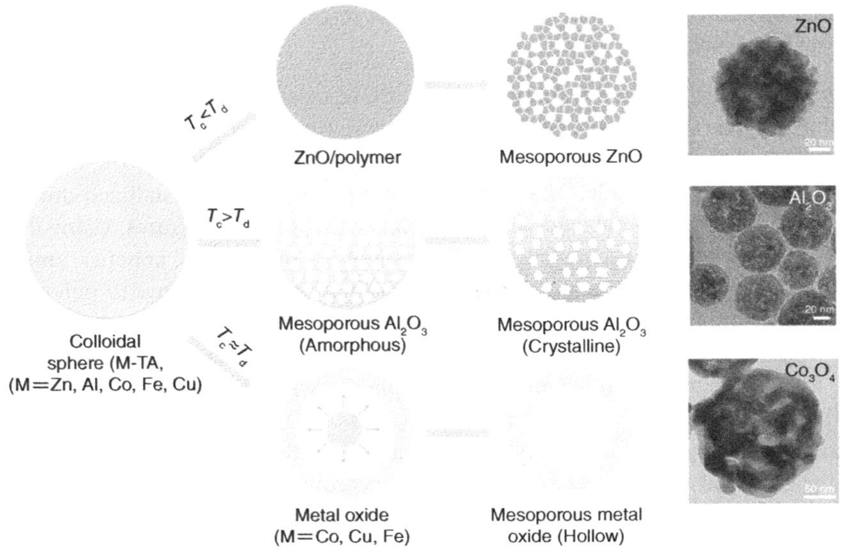

Figure 12.6 The formation process of different morphologies of metal-tannic acid derived mesoporous metal oxide spheres and the relative TEM images. Source: Reproduced with permission from Ref. [56]. Copyright © 2018, John Wiley & Sons, Inc.

applications. Polyphenol-based nanomaterials have been applied in different sensing methods, such as electrochemical, colorimetric, fluorescence, surface-enhanced Raman scattering (SERS), etc. The detection objects are also diverse, including cell detection, biomacromolecule detection, small molecule compound detection, ion detection, etc. It is significant in safety production, environment monitoring and medical diagnosis. The sensing applications of polyphenol-based nanomaterials for different detection objects are discussed below.

12.4.1 Ion Sensing

12.4.1.1 pH Sensing

pH measurement is one of the most general types of ion detection. Currently, the methods for pH detection are well established. In addition to conventional pH detection scenarios, the detection of pH in sweat has been implemented in many wearable devices [59]. Adhesive materials can improve the sensitivity of sweat detection in wearable devices. Lin et al. [60] prepared self-adhesive tannin/reduced graphene oxide (rGO) supramolecular composites. The abundant phenolic hydroxyl groups of polyphenols conferred adhesion to the material. Phenolic hydroxyl groups are pH-sensitive. The pH detection was achieved based on the proton-coupled electron transfer effect. Reduced graphite oxide combined with polyphenol through π–π stacking, thus enhancing the electron transfer efficiency. Due to the excellent coating performance of polyphenol/rGO, it can be directly deposited on flexible substrates to obtain wearable pH sensors with good stability and selectivity. Even if common ions and molecules are present in sweat, it can detect pH without interference. The pH response also remained stable when the electrode was bent. This wearable pH sensor can monitor the pH value of sweat during exercise (Figure 12.7). This work utilizes the adhesion of polyphenols to enhance the contact between wearable devices and skin, providing ideas for the research of other wearable devices. Zhang et al. [61] synthesized Au nanoparticles for SERS analysis of sweat composition based on a metal-phenolic network. Fe-tannic acid can reduce $AuCl_4^-$ to obtain Au/MPN nanomaterials with a film-like structure. The SERS sensing platform can detect the sweat pH of volunteers after exercise, taking Raman fingerprints of other metabolites simultaneously.

12.4.1.2 Metal Ions Sensing

Heavy metal ions are seriously affecting the ecological environment and public health. Detecting metal ions in water, food, and the human body is significant. Cr ions are common heavy metal ions. Compared with Cr^{3+}, Cr^{4+} is more bio-toxic. It can gradually accumulate in the human body. Fang et al. [62] prepared Fe_3O_4 nanoparticles for electrochemical detection of Cr^{4+}, modified with tannins, heparin, and polyaniline. Tannin plays an essential role in the adsorption of Cr^{4+} due to its excellent coordination property. Differential pulse voltammetry curves performed the Cr^{4+} detection after coating the material on a glassy carbon electrode (Figure 12.8). The detection limit and sensitivity are 0.0358 ppm and 18.56 μA ppm^{-1} cm^{-2}, respectively.

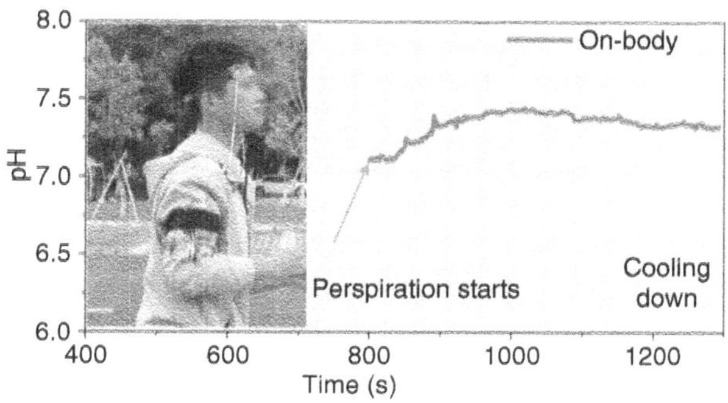

Figure 12.7 The flexible pH sensor's real-time sweat test curve and the photograph show that a runner used the sensor. Source: Reproduced with permission from Ref. [60]. Copyright © 2022, Elsevier.

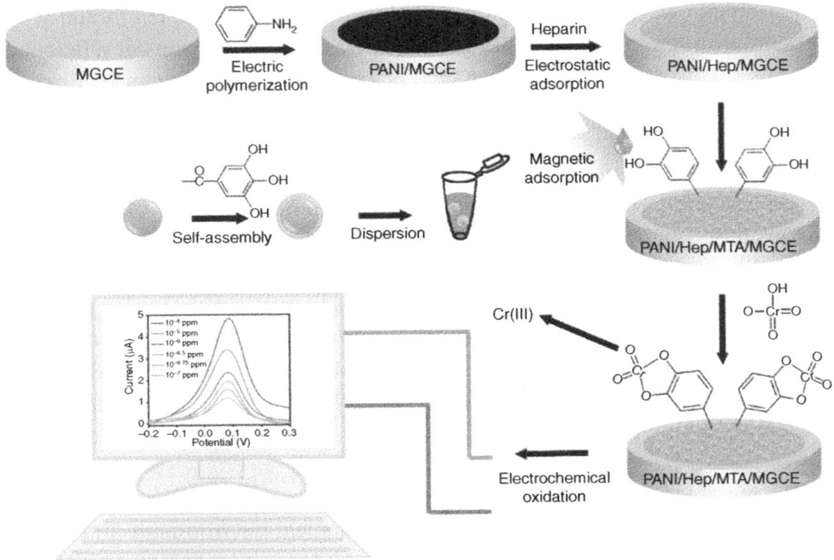

Figure 12.8 Scheme of the synthesis of polyaniline/heparin/tannin modified Fe_3O_4 magnetic glassy carbon electrode and its detection progress of Cr^{4+}. Source: Reproduced with permission from Ref. [62]. Copyright © 2020, Elsevier.

Jiang et al. [63] prepared gold nanoplates to detect lead ions. They used gallic acid as a capping and reducing agent. Gallic acid underwent directional adsorption during the growth of gold nanoplates, assisting in forming plate-like structures. Due to the edge effect of gold nanoplates, they significantly improved the electrochemical sensing performance for lead ions. Compared to bare GCE electrodes, the gold nanoplate-modified electrode reduced the impedance of the electrode and amplified the current signal. The electrode also maintained good selectivity in the presence of interfering cadmium ions. This gold nanoplate, prepared with the assistance of gallic acid, is expected to be used to detect lead ions in polluted wastewater.

Pavadai et al. [64] developed copper phenolic coordination nanosheets for detecting Cd^{2+} and S^{2-} in the environment. Copper oxide nanosheets and tannic acid were added to the alkaline ethanol aqueous solution and used to prepare copper phenolic coordination nanosheets (Cu-PCNS) via hydrothermal progress. FAM fluorescent groups tagged guanine-thymine-rich single-strand aptamers could be adsorbed on the surface of Cu-PCNS via π–π stacking, hydrogen bonding and electrostatic interactions. Based on the fluorescence response energy transfer effect, the probe's fluorescence was quenched. The aptamer probe formed a hairpin structure when Cd^{2+} or S^{2-} ions were present. The FAM fluorescent groups detached from Cu-PCNS, leading to fluorescence recovery. This fluorescence "ON-OFF" strategy was successfully achieved to detect Cd^{2+} and S^{2-}. This method can effectively distinguish Cd^{2+} from other metal ions with good selectivity. The Cd^{2+} and S^{2-} detection limits are 0.3 and 3.3 nM, respectively.

Shen et al. [65] prepared carbon quantum dots (CQDs) to detect Cr^{6+} in wastewater. Using hydrothermal methods, natural wattle-bark-tannin was directly converted into CQDs. The obtained CQDs were naturally non-toxic and fluorescent. Then, the CQDs-PVA hydrogel material was formed by compounding the carbon quantum dots with polyvinyl alcohol (PVA). PVA hydrogel can prevent the aggregation of CQDs, avoiding fluorescence quenching and improving performance. CQDs can specifically detect Cr^{6+} and Co^{2+} with a detection limit of 1.37 and 0.66 μM, respectively. The CQDs were proved to detect Cr^{6+} in actual environment water.

Taleb et al. [66] developed a paper-based sensor to detect Fe ions. The main component of filter paper was cellulose. After modification with hydrophilic tannic acid, the filter paper can coordinate with Fe ions. The coordination complex appeared purple-black. As the increment of Fe^{3+} concentration in the solution increased, the color of the filter paper gradually deepened. The prepared paper-based sensor was susceptible to Fe^{3+} with a linear relationship ranging from 10 to 190 ppm.

12.4.2 Molecule Sensing

12.4.2.1 Formaldehyde Sensing

Formaldehyde is a familiar carcinogen that is present in adhesives. Borah et al. [67] prepared the GA-AuNP@Tollens complex to detect formaldehyde and benzaldehyde in food. Firstly, chloroauric acid was reduced to Au nanoparticles to form GA-AuNP. Then, Tollens' agent was added to form GA-AuNP@Tollens complex. The complex was very sensitive to formaldehyde. In the presence of formaldehyde or benzaldehyde, the Ag^0 obtained by reduction enabled color difference of the system. The sensing of formaldehyde or benzaldehyde can be achieved through colorimetric detection. The detection limit is 20.08 nM for formaldehyde and 0.12 μM for benzaldehyde, which is lower than those reported for other studies. The linear range is 10–150 nM and 0.15–0.75 μM for formaldehyde and benzaldehyde, respectively. In addition to detecting formaldehyde in food, there is also significant research value in detecting formaldehyde in the air, especially in detecting furniture quality.

Feng et al. [68] prepared an Ag_2O/SnO_2 sphere to detect formaldehyde at low operating temperatures (Figure 12.9). By utilizing the coordination ability and reducibility of polyphenols, Sn^{2+} and Ag^+ were integrated into metal-polyphenol nanospheres, which were then calcined in air to obtain Ag_2O/SnO_2 spheres

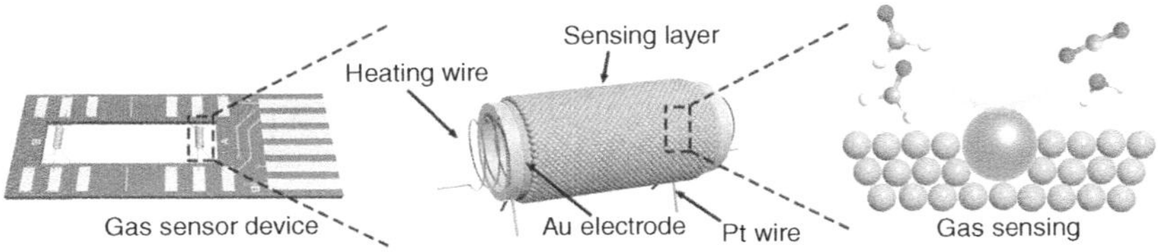

Figure 12.9 Schematic illustration of formaldehyde detection progress by Ag_2O/SnO_2 gas sensor. Source: Reproduced with permission from Ref. [68]. Copyright © 2022, American Chemical Society.

derived from Sn/Ag-TA coordination nanospheres. Oxygen anions (e.g. O^-, O^{2-}, O_2^-) were adsorbed on the material surface. Formaldehyde molecules reacted with these oxygen anions by oxidation-reduction reactions, leading to sensor resistance changes. The detection limit is relatively low (23.6 ppb). The operating temperature is 75 °C, which is lower than most reported metal oxide semiconductor-based formaldehyde sensors. Compared with SnO_2 without Ag_2O loading, the Ag_2O/SnO_2 spheres showed better selectivity for formaldehyde due to the catalytic effect of Ag_2O.

12.4.2.2 Hydrogen Peroxide Sensing

Hydrogen peroxide (H_2O_2) is an oxidant and bleaching agent. The rapid detection of hydrogen peroxide is crucial as it is associated with various physiological and pathological processes in organisms. Cheng et al. [69] prepared a paper-based sensor for detecting H_2O_2. The sensor used hollow mesoporous copper oxide as sensitive material. Firstly, Cu^{2+} coordinated with tannic acid to form a Cu-tannic acid coordination polymer. Air oxidized it to create mesoporous copper oxide nanozyme with hollow structures. CuO has peroxidase-like properties and can catalyze H_2O_2 to produce hydroxyl radical, which can oxidize 3,3′,5,5′-tetramethylbenzidine (TMB) to form oxidized TMB (oxTMB) with blue color. The color change of the system can detect the presence of H_2O_2. The signal of a paper-based sensor can be directly observed by human eyes or read by a camera on a smartphone for fast and convenient detection (Figure 12.10). Electrochemical methods have also been used to detect H_2O_2. Tian et al. [70] prepared Cu nanoparticles via the hydrothermal method using gallic acid as the reducing agent. This synthesis of Cu nanoparticles is green and environmentally friendly. The Cu nanoparticles modified electrode exhibited good H_2O_2 detection performance, stability, and linear relationship. The linear detection range is 1 to 20 mM with a sensitivity of $5.74\,\mu A/mM/cm^2$.

12.4.2.3 Ethanol Sensing

The detection of ethanol is also of great significance in daily life. Typically, the detection of ethanol gas can be used for checking drunk driving. The mesoporous

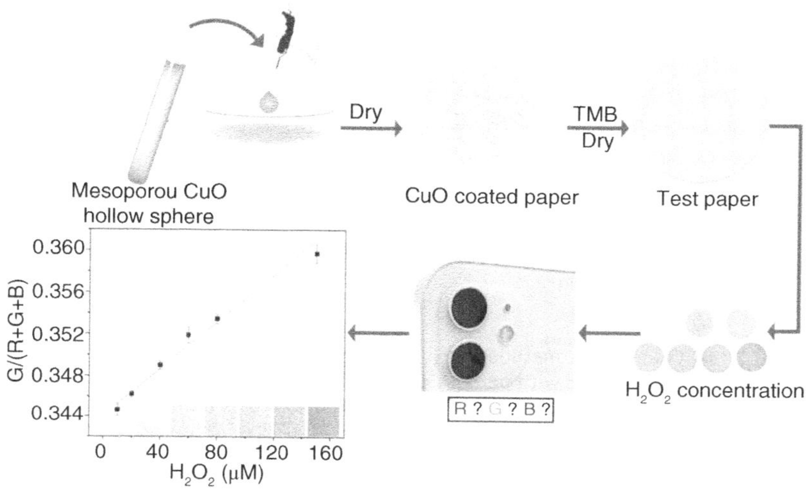

Figure 12.10 Scheme of CuO hollow sphere nanozyme-based paper colorimetric sensor for H_2O_2 detection. Source: Reproduced from Ref. [69]. Copyright © 2021, MDPI/ CC BY 4.0.

ZnO material prepared by Wang et al. [56] exhibited excellent ethanol gas sensing performance. ZnO nanospheres were derived from Zn-tannic acid coordination polymer and coated on the electrode surface. It exhibited a good selectivity to ethanol at a low working temperature (250 °C) with a linear detection range from 10–500 ppm. The unique mesoporous structure of ZnO nanospheres enabled outstanding sensing performance. SnO_2 also showed exceptional performance in ethanol sensing. Feng et al. [71] prepared SnO_2 nanospheres gas sensing materials using Sn-tannic acid as a precursor. By changing the amount of block copolymer F127 used in the synthesis process, the diameter of SnO_2 nanospheres could be controlled. Due to the mesoporous structure (~5.8 nm) and ultra-high specific surface area (80.9–185.6 m^2/g) of the nanospheres, the prepared SnO_2 showed high exposure of active sites and thus exhibited good selectivity toward ethanol.

12.4.2.4 Melamine Sensing

Polyphenol-based materials are also widely used in food safety sensing. Melamine is an illegal food additive often used in milk to pass it off as high-protein milk. Melamine can cause kidney failure and even death. It is necessary to detect melamine in milk. Liu et al. [72] developed monodispersed Ag nanoparticles to detect melamine (Figure 12.11a). They used gallic acid to reduce Ag^+ to nanoparticles. In the presence of melamine, the phenolic hydroxyl group of gallic acid on the surface of Ag nanoparticles can form hydrogen bonds with the amino group of

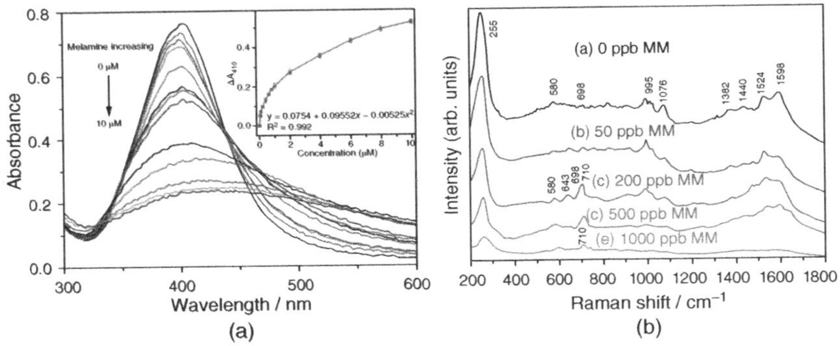

Figure 12.11 (a) Colorimetric method and (b) Raman "turn-off" method for detecting melamine. Source: Reproduced with permission from Ref. [72]. Copyright © 2021, the authors. Published by American Chemical Society. Reproduced with permission from Ref. [73]. Copyright © 2022, Elsevier.

melamine, leading to the aggregation of Ag nanoparticles and a change in the color of the reaction solution. The sensing of melamine can be achieved through colorimetric measurement with a detection limit of 0.099 μM. Das et al. [73] developed gold nanoparticles to detect melamine in milk. The detection was based on colorimetric and Raman "turn-off" methods (Figure 12.11b). Gallic acid was used as a capping and reducing agent to prepare Au nanoparticles with tunable size and morphology. In the presence of melamine, Au nanoparticles aggregated, and the solution color changed from red-purple to blue-gray, enabling colorimetric detection. At the same time, the aggregation of materials led to increases in particle size, gradually reducing the Raman signal for SERS detection with a limit of 1.15 ppb. The potential application of melamine detection was proved by analyzing melamine in milk samples.

12.4.2.5 Monofluoroacetic Acid Sensing

The residues of pesticides or rodenticides in food seriously threaten public health. Monofluoroacetic acid is a typical rodenticide. Sun et al. [74] developed a strategy for SERS detection of monofluoroacetic acid. Firstly, the metal-phenolic supramolecular networks were coated on the anodic aluminum oxide films. Subsequently, the polyphenols deposited Ag nanostructures reductively on the electrode surface. Thiosallic acid was used as a Raman probe that reacts with monofluoroacetic acid. As the concentration of monofluoroacetic acid increases, the SERS signal decreases. Finally, the detection of monofluoroacetic acid was achieved. The limit of detection is 50 nM. The sensing performance of this method was verified by detecting monofluoroacetic acid in food and biological samples.

12.4.2.6 Clenbuterol Sensing

Illegal food additives (such as ractopamine and clenbuterol) are also dangers that cannot be ignored in food safety sensing. Liu et al. [75] developed a paper-based sensor for detecting ractopamine and clenbuterol using the lateral flow immunoassay method (Figure 12.12). They synthesized Fe-tannic acid nanozymes as signal tags. The peroxidase-like activity of Fe-tannic acid nanozymes can catalyze hydrogen peroxide to generate ROS that can further oxidize TMB and produce colorimetric signals. A point-of-care detection method was designed with ractopamine and clenbuterol as targets. The detection limit for ractopamine and clenbuterol is 0.015 and 0.156 ng/ml, respectively. The detection of clenbuterol in natural samples (e.g. pork and pork liver) was explored and achieved.

12.4.2.7 Trimethylamine and Triethylamine Sensing

During the seafood's death and decay, bacteria degrade oxidized trimethylamine, forming fishy-smelling trimethylamine. Therefore, to timely detect the freshness of seafood, the detection of trimethylamine is critical. Chen et al. [76] prepared Au-ZnO sensing material via a self-template method for detecting trimethylamine. Tannic acid was first cross-linked with formaldehyde in an alkaline solution to form a polyphenol polymer. Then, Zn^{2+} and $AuCl_4^-$ ions were added to form metal-polyphenol coordination complexes. The complex was calcinated in the air to remove organic parts. Zn^{2+} underwent oxidization and crystallization and converted into ZnO. Finally, mesoporous Au-ZnO materials were obtained. The material had a high specific surface area (46.8 m^2/g) and a large pore size

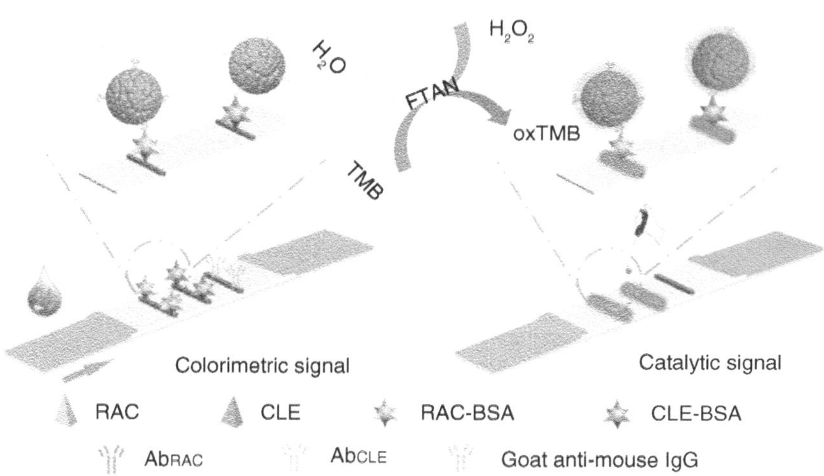

Figure 12.12 Scheme of dual-readout multiplex lateral flow immunoassay for detecting ractopamine and clenbuterol. Source: Reproduced with permission from Ref. [75]. Copyright © 2022, Elsevier.

(12.9 nm). Compared with ZnO without Au modification, the sensing response was significantly improved with a low detection limit (0.4 ppm) and fast response (12 s). Feng et al. [77] also synthesized Au/In_2O_3 nanospheres to detect triethylamine. Au specifications are highly dispersed within the material, and their sensing performance is 9.6 times higher than that of In_2O_3. The operating temperature of Au/In_2O_3 was also reduced compared to In_2O_3.

12.4.2.8 Glutathione Sensing

Polyphenol-based nanomaterials have wide applications in the biomedical field, which is beneficial for the diagnosis and early prevention of diseases. Glutathione is an essential antioxidant in organisms and participates in many physiological processes. It is a significant biomarker of many diseases, such as cancer, Alzheimer's disease, Parkinson's disease, and so on. Therefore, detecting glutathione is of great significance for early diagnosis and prevention. Li et al. [78] prepared silver nanoparticles (Ag NPs)/copper polyphenol colloid spheres for the detection of glutathione (Figure 12.13). Firstly, copper-polyphenols coordination polymers were prepared in alkaline ethanol-water with the assistance of formaldehyde. Subsequently, silver nanoparticles were modified onto colloidal spheres through the reducibility of polyphenols. By changing the content of Ag^+, the coverage of Ag nanoparticles on the surface of colloidal spheres could be effectively regulated. Silver nanoparticles/copper polyphenol colloidal spheres can catalyze TMB to oxidized TMB (oxTMB), which response highly to SERS. In the presence of glutathione, oxTMB was reduced, and the signal output was decreased. With the assistance of TMB, sensitive SERS detection of glutathione can be achieved with a linear range from 1 to 100 μM and a detection limit of 0.1 μM.

12.4.2.9 Acetaminophen Sensing

Acetaminophen is a common drug for the treatment of fever and pain conditions. Excessive use of acetaminophen would lead to liver failure. Yan et al. [79] prepared

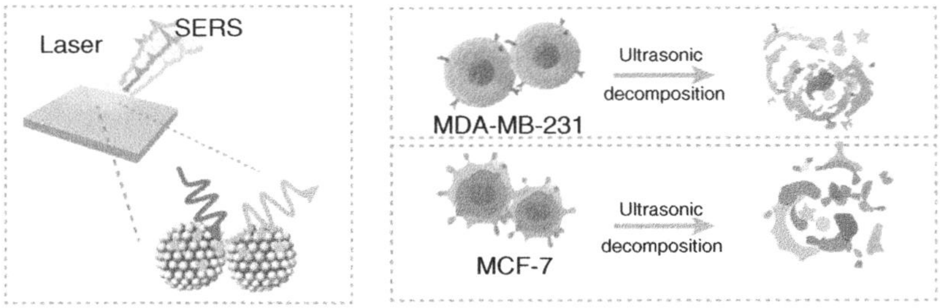

Figure 12.13 Scheme of silver nanoparticles/copper polyphenol colloid spheres for the SERS detection of intracellular glutathione. Source: Reproduced with permission from Ref. [78]. Copyright © 2023, Elsevier.

3D-hierarchical mesoporous carbon nanosheets (hMCNS) microspheres for the electrochemical detection of acetaminophen. They first trained the Zn-gallic acid coordination precursor and calcined them under a nitrogen atmosphere to obtain hMCNS. The excellent conductivity and adjustable porosity of hMCNS facilitate electron transfer. The detection limit of the hMCNS electrode for acetaminophen is 3.5 nM, and the linear detection range is 0.005–5.0 μM. The feasibility of detecting acetaminophen by hMCNS was verified using rat plasma samples, revealing a promising application prospect in therapeutic drug monitoring.

12.4.2.10 Nucleic Acid Sensing

DNA carries the genetic information of organisms. The detection of DNA is of great significance in disease diagnosis. Baykal et al. [80] prepared gallic acid (GA)-titanium dioxide nanocomposites to detect guanine using an electrochemical method. GA was modified on TiO_2 to synthesize the composites and then cross-linked with chitosan (CHIT) to obtain CHIT-[$GATiO_2$]. Cyclic voltammetry (CV) and electrochemical impedance spectroscopy (EIS) were used to detect. The detection limit is 0.95 μg/mL, and the linear range is 2–10 μg/mL. Metal-polyphenol coordination materials have the ability of fluorescence quenching and can be directly utilized as a nucleic acid detection platform. Wei et al. [51] developed metal-polyphenol coordination nanospheres for fluorescence detection of miR-21. The abundant sp^2 hybrid rings of polyphenols were beneficial for the adsorption of DNA probes. Meanwhile, metal-polyphenol coordination spheres can quench the fluorescence of DNA probes. The target miR-21 can bind to the probe and detach from the material, resulting in fluorescence recovery. Similar to amorphous metal-polyphenol complex colloidal spheres, metal-polyphenol crystals can also achieve nucleic acid detection. Metal-polyphenol crystal has similar property. Huang et al. [50] synthesized spiny metal-polyphenol crystals for miR-21 detection with a 5–60 nM linear range. It can effectively distinguish triple-base, double-base, or even single-base DNA mismatches. Metal oxides, in addition to polymers and carbon materials, can also be used for DNA detection. Wang et al. [54] implemented nanoporous carbon spheres derived from metal-polyphenol complexes for nuclear acid detection. Due to their ultra-high specific surface area and fluorescence quenching ability, nanoporous carbon spheres became excellent fluorescence sensing platforms. FAM-labeled single-stranded DNA can be adsorbed on the surface of carbon spheres via π–π stacking. The carbon spheres can form double bonds with single-stranded DNA probes in the presence of target DNA. The interaction between bases was more substantial than that between single-stranded DNA and carbon spheres. After the DNA probe detached from the carbon sphere, fluorescence was restored. The fluorescence intensity increased with the increase in the concentration of target DNA. There was an excellent linear relationship from 0.5–10 nM. The detection limit is 0.36 nM. This fluorescence sensing platform can effectively distinguish target DNA with base mismatches.

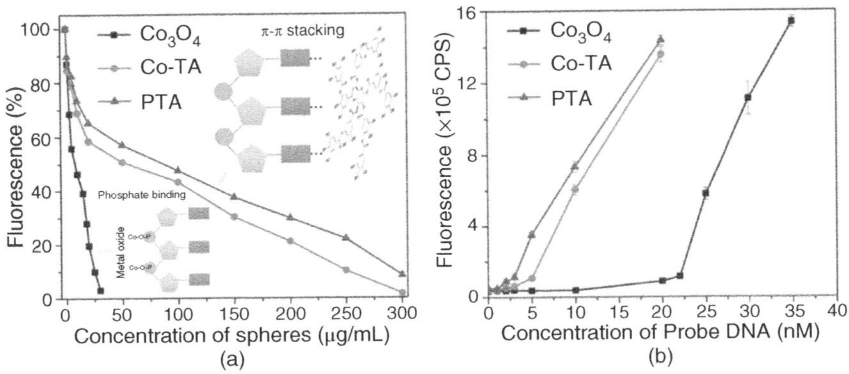

Figure 12.14 (a) The fluorescence intensities of DNA probes and (b) DNA loading capacity detected by different concentrations of PTA, Co-TA and Co_3O_4. Source: Reproduced with permission from Ref. [56]. Copyright © 2018, WILEY-VCH Verlag GmbH & Co. KGaA, Weinheim.

Wang et al. [56] prepared Co_3O_4 nanospheres derived from Co-tannic acid complexes. They used FAM-labeled fluorescent probes to detect target DNA (Figure 12.14). The Co_3O_4 nanospheres showed excellent fluorescence quenching ability with a detection limit of 0.19 nM. The linear range is 0–10 nM. The high-entropy metal oxide nanospheres derived from metal polyphenols have also been applied in the fluorescence detection of DNA [58].

12.4.2.11 Protein Sensing

Fang et al. [81] prepared metal-phenolic nanoparticles modified potassium ion (K^+)-mediated J-aggregate K-PTC MOF (PTCA, perylene-3,4,9,10-tetracarboxylic) for detecting neuron-specific enolase based on electrochemiluminescence (ECL) method. PEG, BSA, Fe^{3+} and TA were self-assembled to form metal-phenolic bioactive nanoparticles (MPNPs) to achieve accurate detection. MPNPs regulated ECL signals and quenched the sensor signal. The exposed phenolic hydroxyl groups of polyphenols exerted antioxidant properties to reduce the free radicals ($SO_4^{\cdot -}$ and $OH^{\cdot -}$) involved in luminescence, leading to quenching fluorescence (Figure 12.15). When MPNPs bound to K-PTC through immune reaction, the ECL signals were quenched due to the spatial isolation effect. Under the regulation of MPNPs, the detection of neuron-specific enolase has been realized by the ECL method. The linear range is 10 pg/mL–50 ng/mL. The detection exhibited a low limit of 7.4 pg/mL, which is expected to be used for preventing and detecting small-cell lung cancer.

The detection of disease markers is also essential for the diagnosis and early screening of diseases. Huang et al. [82] developed an electrochemical method to detect nasopharyngeal carcinoma biomarkers using metal-polyphenol complex

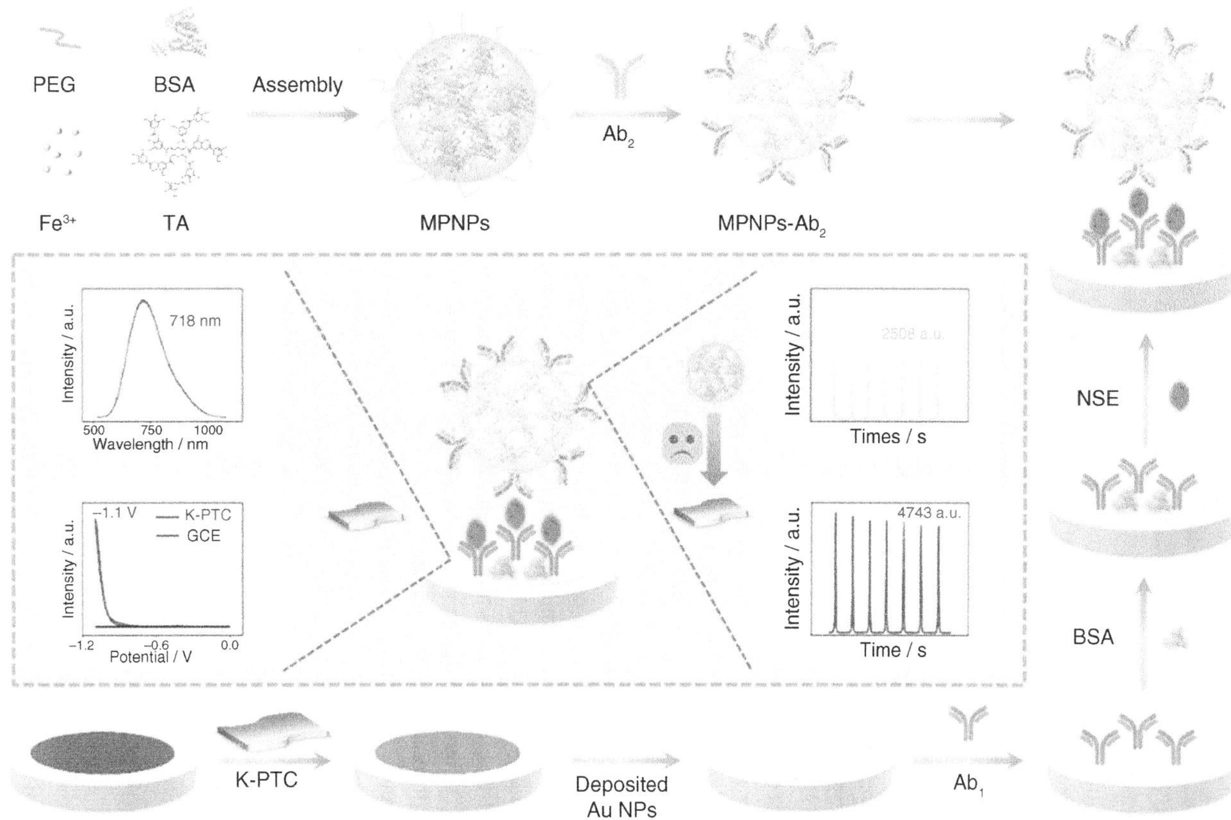

Figure 12.15 Scheme of J-Aggregate K-PTC MOF regulated by metal-phenolic nanoparticles for neuron-specific enolase (a marker of small-cell lung cancer) analysis. Source: Reproduced with permission from Ref. [81]. Copyright © 2022, American Chemical Society.

materials. Soft metal-polyphenol capsules for detecting Epstein-Barr virus were prepared using a template method. Firstly, tannic acid was adsorbed on the surface of $CaCO_3$ and coordinated with metal ions to form soft metal-polyphenol capsules. Then, anti-human IgA was modified on the surface to include a metal probe. The surface of the gold electrode was coated with Epstein-Barr virus capsid antigen. BSA was used as a blocking agent. The gold electrode was incubated with the analyte solution and bound to the metal probe. Finally, HNO_3 was used to destroy the metal probe and electrochemical detection was realized. The square wave voltammetry method realized the Epstein-Barr virus capsid antigen-antibody detection. The linear range was 1 fM to 1 nM. The limit of detection was 0.46 fM.

12.5 Summary and Outlook

With the rapid development of sensing technology, the demand for novel sensing materials is ever-increasing. Due to their unique physicochemical properties, polyphenols have been demonstrated to be excellent building blocks for preparing sensing materials. This chapter summarizes the synthesis methods of different types of polyphenol-based nanomaterials. Specifically, the synthesis methods and properties of metal-polyphenol coordination capsules, crystals, colloidal spheres, and their derived carbon and metal oxide materials are introduced. These polyphenol-based nanomaterials have been widely used in chem- and bio-sensing. The sensing application can be divided into ion sensing (e.g. pH detection and metal ion detection), small molecule compound sensing and biological macromolecule sensing (e.g. DNA and proteins). Polyphenol-based nanomaterials are crucial in energy, environment, military, biomedicine and food safety applications. Designing suitable polyphenol-based nanomaterials is essential for developing different kinds of sensors.

Currently, sensors are being developed toward miniaturization, portability, and intelligence [83–84]. The mainstream combines sensing materials with microelectronics, artificial intelligence, and other techniques to design more intelligent, sensitive, and stable sensors. It poses a higher challenge for the design and synthesis of different kinds of novel sensing materials. As an integral part of polyphenol-based nano-sensing materials, they need to be further investigated in the future.

Acknowledgments

This work was financially supported by the National Key Research and Development Program (2022YFE0100400), the National Natural Science Foundation of China (No. 21701130), and the Key Research and Development Program of Shaanxi (Program No. 2021GY-225).

References

1 Dincer, C., Bruch, R., Costa-Rama, E. et al. (2019). Disposable sensors in diagnostics, food, and environmental monitoring. *Advanced Materials* 31 (30): e1806739.

2 Gu, Y., Zhang, T., Chen, H. et al. (2019). Mini review on flexible and wearable electronics for monitoring human health information. *Nanoscale Research Letters* 14 (1): 263.

3 Guntner, A.T., Abegg, S., Konigstein, K. et al. (2019). Breath sensors for health monitoring. *ACS Sensors* 4 (2): 268–280.

4 Joe, H.-E., Yun, H., Jo, S.-H. et al. (2018). A review on optical fiber sensors for environmental monitoring. *International Journal of Precision Engineering and Manufacturing - Green Technology* 5 (1): 173–191.

5 Kaushik, A., Kumar, R., Arya, S.K. et al. (2015). Organic-inorganic hybrid nanocomposite-based gas sensors for environmental monitoring. *Chemical Reviews* 115 (11): 4571–4606.

6 Lin, P. and Yan, F. (2012). Organic thin-film transistors for chemical and biological sensing. *Advanced Materials* 24 (1): 34–51.

7 Liu, Y., Dong, X., and Chen, P. (2012). Biological and chemical sensors based on graphene materials. *Chemical Society Reviews* 41 (6): 2283–2307.

8 Mao, S., Chang, J., Pu, H. et al. (2017). Two-dimensional nanomaterial-based field-effect transistors for chemical and biological sensing. *Chemical Society Reviews* 46 (22): 6872–6904.

9 Zhao, Y., Li, X.G., Zhou, X., and Zhang, Y.N. (2016). Review on the graphene based optical fiber chemical and biological sensors. *Sensors and Actuators, B: Chemical* 231: 324–340.

10 Yang, Q., Li, J., Wang, X. et al. (2018). Strategies of molecular imprinting-based fluorescence sensors for chemical and biological analysis. *Biosensors & Bioelectronics* 112: 54–71.

11 Cruz-Martinez, H., Rojas-Chavez, H., Montejo-Alvaro, F. et al. (2021). Recent developments in graphene-based toxic gas sensors: a theoretical overview. *Sensors* 21 (6): 1992.

12 Shen, Y., Wei, Y., Zhu, C. et al. (2022). Ratiometric fluorescent signals-driven smartphone-based portable sensors for onsite visual detection of food contaminants. *Coordination Chemistry Reviews* 458: 214442.

13 Wang, P.L., Xie, L.H., Joseph, E.A. et al. (2019). Metal-organic frameworks for food safety. *Chemical Reviews* 119 (18): 10638–10690.

14 Yousefi, H., Su, H.M., Imani, S.M. et al. (2019). Intelligent food packaging: a review of smart sensing technologies for monitoring food quality. *ACS Sensors* 4 (4): 808–821.

15 Das, R., Vecitis, C.D., Schulze, A. et al. (2017). Recent advances in nanomaterials for water protection and monitoring. *Chemical Society Reviews* 46 (22): 6946–7020.

16 Rebelo, P., Costa-Rama, E., Seguro, I. et al. (2021). Molecularly imprinted polymer-based electrochemical sensors for environmental analysis. *Biosensors & Bioelectronics* 172: 112719.

17 Kim, H.N., Ren, W.X., Kim, J.S., and Yoon, J. (2012). Fluorescent and colorimetric sensors for detection of lead, cadmium, and mercury ions. *Chemical Society Reviews* 41 (8): 3210–3244.

18 Jayanthi, V., Das, A.B., and Saxena, U. (2017). Recent advances in biosensor development for the detection of cancer biomarkers. *Biosensors & Bioelectronics* 91: 15–23.

19 Broza, Y.Y., Zhou, X., Yuan, M. et al. (2019). Disease detection with molecular biomarkers: from chemistry of body fluids to nature-inspired chemical sensors. *Chemical Reviews* 119 (22): 11761–11817.

20 Teymourian, H., Parrilla, M., Sempionatto, J.R. et al. (2020). Wearable electrochemical sensors for the monitoring and screening of drugs. *ACS Sensors* 5 (9): 2679–2700.

21 Rasheed, T. and Rizwan, K. (2022). Metal-organic frameworks based hybrid nanocomposites as state-of-the-art analytical tools for electrochemical sensing applications. *Biosensors & Bioelectronics* 199: 113867.

22 Schaferling, M. (2012). The art of fluorescence imaging with chemical sensors. *Angewandte Chemie, International Edition* 51 (15): 3532–3254.

23 Fenzl, C., Hirsch, T., and Wolfbeis, O.S. (2014). Photonic crystals for chemical sensing and biosensing. *Angewandte Chemie, International Edition* 53 (13): 3318–3335.

24 Tsao, R. (2010). Chemistry and biochemistry of dietary polyphenols. *Nutrients* 2 (12): 1231–1246.

25 Quideau, S., Deffieux, D., Douat-Casassus, C., and Pouysegu, L. (2011). Plant polyphenols: chemical properties, biological activities, and synthesis. *Angewandte Chemie, International Edition* 50 (3): 586–621.

26 Hussain, T., Tan, B., Yin, Y. et al. (2016). Oxidative stress and inflammation: what polyphenols can do for us? *Oxidative Medicine and Cellular Longevity* 2016: 7432797.

27 Heleno, S.A., Martins, A., Queiroz, M.J., and Ferreira, I.C. (2015). Bioactivity of phenolic acids: metabolites versus parent compounds: a review. *Food Chemistry* 173: 501–513.

28 Du, G.J., Zhang, Z., Wen, X.D. et al. (2012). Epigallocatechin gallate (EGCG) is the most effective cancer chemopreventive polyphenol in green tea. *Nutrients* 4 (11): 1679–1691.

29 Singh, B.N., Shankar, S., and Srivastava, R.K. (2011). Green tea catechin, epigallocatechin-3-gallate (EGCG): mechanisms, perspectives and clinical applications. *Biochemical Pharmacology* 82 (12): 1807–1821.

30 Tardy, B.L., Richardson, J.J., Nithipipat, V. et al. (2019). Protein adsorption and coordination-based end-tethering of functional polymers on metal-phenolic network films. *Biomacromolecules* 20 (3): 1421–1428.

31 Kim, B.J., Han, S., Lee, K.B., and Choi, I.S. (2017). Biphasic supramolecular self-assembly of ferric ions and tannic acid across interfaces for nanofilm formation. *Advanced Materials* 29 (28): 1700784.

32 Li, W., Bing, W., Huang, S. et al. (2015). Mussel byssus-like reversible metal-chelated supramolecular complex used for dynamic cellular surface engineering and imaging. *Advanced Functional Materials* 25 (24): 3775–3784.

33 Yun, G., Richardson, J.J., Capelli, M. et al. (2019). The biomolecular corona in 2D and reverse: Patterning metal–phenolic networks on proteins, lipids, nucleic acids, polysaccharides, and fingerprints. *Advanced Functional Materials* 30 (1): 1905805.

34 Rahim, M.A., Kristufek, S.L., Pan, S. et al. (2019). Phenolic building blocks for the assembly of functional materials. *Angewandte Chemie, International Edition* 58 (7): 1904–1927.

35 Kang, J., Bai, G., Ma, S. et al. (2019). On-site surface coordination complexation via mechanochemistry for versatile metal-phenolic networks films. *Advanced Functional Materials* 6 (5): 1801789.

36 Xu, W., Pan, S., Noble, B.B. et al. (2022). Site-selective coordination assembly of dynamic metal-phenolic networks. *Angewandte Chemie, International Edition* 61 (34): e202208037.

37 Ejima, H., Richardson, J.J., Liang, K. et al. (2013). One-step assembly of coordination complexes for versatile film and particle engineering. *Science* 341 (6142): 154–157.

38 Guo, Y., Sun, Q., Wu, F.G. et al. (2021). Polyphenol-containing nanoparticles: synthesis, properties, and therapeutic delivery. *Advanced Materials* 33 (22): e2007356.

39 Ejima, H., Richardson, J.J., and Caruso, F. (2017). Metal-phenolic networks as a versatile platform to engineer nanomaterials and biointerfaces. *Nano Today* 12: 136–148.

40 Faure, E., Falentin-Daudré, C., Jérôme, C. et al. (2013). Catechols as versatile platforms in polymer chemistry. *Progress in Polymer Science* 38 (1): 236–270.

41 Kharissova, O.V., Dias, H.V., Kharisov, B.I. et al. (2013). The greener synthesis of nanoparticles. *Trends in Biotechnology* 31 (4): 240–248.

42 Zhou, J., Lin, Z., Ju, Y. et al. (2020). Polyphenol-mediated assembly for particle engineering. *Accounts of Chemical Research* 53 (7): 1269–1278.

43 Wu, D., Zhou, J., Creyer, M.N. et al. (2021). Phenolic-enabled nanotechnology: versatile particle engineering for biomedicine. *Chemical Society Reviews* 50 (7): 4432–4483.

44 Feng, Y., Li, P., and Wei, J. (2022). Engineering functional mesoporous materials from plant polyphenol based coordination polymers. *Coordination Chemistry Reviews* 468: 214649.

45 Xu, L.Q., Neoh, K.-G., and Kang, E.-T. (2018). Natural polyphenols as versatile platforms for material engineering and surface functionalization. *Progress in Polymer Science* 87: 165–196.

46 Lee, B.P., Messersmith, P.B., Israelachvili, J.N., and Waite, J.H. (2011). Mussel-inspired adhesives and coatings. *Annual Review of Materials Research* 41: 99–132.

47 Guo, J., Ping, Y., Ejima, H. et al. (2014). Engineering multifunctional capsules through the assembly of metal-phenolic networks. *Angewandte Chemie, International Edition* 53 (22): 5546–5551.

48 Yang, S.J., Antonietti, M., and Fechler, N. (2015). Self-assembly of metal phenolic mesocrystals and morphosynthetic transformation toward hierarchically porous carbons. *Journal of the American Chemical Society* 137 (25): 8269–8273.

49 Wei, J., Liang, Y., Hu, Y. et al. (2016). Hydrothermal synthesis of metal-polyphenol coordination crystals and their derived metal/N-doped carbon composites for oxygen electrocatalysis. *Angewandte Chemie, International Edition* 55 (40): 12470–12474.

50 Huang, H., Qin, J., Wang, G. et al. (2018). Synthesis of spiny metal-phenolic coordination crystals as a sensing platform for sequence-specific detection of nucleic acids. *CrystEngComm* 20 (47): 7626–7630.

51 Wei, J., Wang, G., Chen, F. et al. (2018). Sol–gel synthesis of metal-phenolic coordination spheres and their derived carbon composites. *Angewandte Chemie, International Edition* 57 (31): 9838–9843.

52 Lin, Z., Zhou, J., Cortez-Jugo, C. et al. (2020). Ordered mesoporous metal-phenolic network particles. *Journal of the American Chemical Society* 142 (1): 335–341.

53 Pan, S., Goudeli, E., Chen, J. et al. (2021). Exploiting supramolecular dynamics in metal-phenolic networks to generate metal-oxide and metal-carbon networks. *Angewandte Chemie, International Edition* 60 (26): 14586–14594.

54 Wang, G., Qin, J., Zhao, Y., and Wei, J. (2019). Nanoporous carbon spheres derived from metal-phenolic coordination polymers for supercapacitor and biosensor. *Journal of Colloid and Interface Science* 544: 241–248.

55 Zhang, P., Wang, L., Yang, S. et al. (2017). Solid-state synthesis of ordered mesoporous carbon catalysts via a mechanochemical assembly through coordination cross-linking. *Nature Communications* 8: 15020.

56 Wang, G., Qin, J., Zhou, X. et al. (2018). Self-template synthesis of mesoporous metal oxide spheres with metal-mediated inner architectures and superior sensing performance. *Advanced Functional Materials* 28 (51): 1806144.

57 Wang, G., Zhou, X., Qin, J. et al. (2019). General synthesis of mixed semiconducting metal oxide hollow spheres with tunable compositions for low-temperature chemiresistive sensing. *ACS Applied Materials & Interfaces* 11 (38): 35060–35067.

58 Wang, G., Qin, J., Feng, Y. et al. (2020). Sol-gel synthesis of spherical mesoporous high-entropy oxides. *ACS Applied Materials & Interfaces* 12 (40): 45155–45164.

59 Bariya, M., Nyein, H.Y.Y., and Javey, A. (2018). Wearable sweat sensors. *Nature Electronics* 1 (3): 160–171.

60 Lin, K., Xie, J., Bao, Y. et al. (2022). Self-adhesive and printable tannin-graphene supramolecular aggregates for wearable potentiometric pH sensing. *Electrochemistry Communications* 137: 107261.

61 Zhang, X., Wang, X., Ning, M. et al. (2022). Fast synthesis of Au nanoparticles on metal-phenolic network for sweat SERS analysis. *Nanomaterials* 12 (17): 2977.

62 Fang, D., Xu, T., Fang, L. et al. (2021). A blood compatible, high-efficient sensor for detection of Cr(VI) in whole blood. *Sensors and Actuators, B: Chemical* 329: 129219.

63 Jiang, X., Ma, J., Jiang, G. et al. (2021). Preparation of gold nanoplates using ortho carbonyl compounds as capping agents for electrochemical sensing of lead ions. *Nanoscale Research Letters* 16 (1): 57.

64 Pavadai, R. and Perumal, P. (2022). Versatile sensing platform of an innovative copper oxide-assisted Cu-phenolic coordination nanosheet-mediated fluorophore-tagged GT-rich SSA-based fluorescence ON-OFF biosensor for the subsequent detection of Cd^{2+} and $S^{2}-$ ions. *New Journal of Chemistry* 46 (7): 3431–3447.

65 Shen, J., Gu, H., He, Z., and Lin, W. (2023). Wattle-bark-tannin-derived carbon quantum dots as multi-functional nanomaterials for intelligent detection of Cr^{6+} ions, bio-imaging, and fluorescent ink applications. *Industrial and Engineering Chemistry Research* 62 (8): 3622–3634.

66 Taleb, M.A., Kalam, A., Alghamdi, N.A., and Salem, W.M. (2022). Green and sustainable tannin-immobilized cellulose dipstick for metallochromic sensing of ferric cations. *Journal of Molecular Liquids* 358: 119180.

67 Borah, N., Gogoi, D., Ghosh, N.N., and Tamuly, C. (2023). GA-AuNP@Tollens' complex as a highly sensitive plasmonic nanosensor for detection of formaldehyde and benzaldehyde in preserved food products. *Food Chemistry* 399: 133975.

68 Feng, B., Feng, Y., Li, Y. et al. (2022). Synthesis of mesoporous Ag_2O/SnO_2 nanospheres for selective sensing of formaldehyde at a low working temperature. *ACS Sensors* 7 (12): 3963–3972.

69 Cheng, D., Qin, J., Feng, Y., and Wei, J. (2021). Synthesis of mesoporous CuO hollow sphere nanozyme for paper-based hydrogen peroxide sensor. *Biosensors* 11 (8): 258.

70 Tian, W., Ding, X., Jiang, F. et al. (2022). Green preparation of cu nanoparticles via gallic acid applied to H_2O_2 detection. *Journal of Electronic Materials* 51 (4): 1752–1758.

71 Feng, B., Feng, Y., Qin, J. et al. (2021). Self-template synthesis of spherical mesoporous tin dioxide from tin-polyphenol-formaldehyde polymers for conductometric ethanol gas sensing. *Sensors and Actuators, B: Chemical* 341: 192265.

72 Liu, X., Wang, J., Wang, Y. et al. (2021). In situ functionalization of silver nanoparticles by gallic acid as a colorimetric sensor for simple sensitive determination of melamine in milk. *ACS Omega* 6 (36): 23630–23635.

73 Das, A., Chadha, R., Chalke, B., and Maiti, N. (2022). Gold nanoparticle based colorimetric and Raman "turn-off" sensing of melamine in milk. *Colloids and Surfaces A: Physicochemical and Engineering Aspects* 651: 129717.

74 Sun, J., Xue, D., Shan, W. et al. (2021). In situ growth large area silver nanostructure on metal phenolic network coated NAAO film and its SERS sensing application for monofluoroacetic acid. *ACS Sensors* 6 (6): 2129–2135.

75 Liu, S., Shu, R., Ma, J. et al. (2022). Mussel-inspired Fe-based tannic acid nanozyme: a renewable bioresource-derived high-affinity signal tag for dual-readout multiplex lateral flow immunoassay. *Chemical Engineering Journal* 446: 137382.

76 Chen, Y., Li, Y., Feng, B. et al. (2022). Self-templated synthesis of mesoporous Au-ZnO nanospheres for seafood freshness detection. *Sensors and Actuators, B: Chemical* 360: 131662.

77 Feng, B., Wu, Y., Chen, Y. et al. (2022). Polyphenol-mediated synthesis of mesoporous Au-In_2O_3 nanospheres for room-temperature detection of triethylamine. *ACS Applied Nano Materials* 5 (7): 9688–9697.

78 Li, Y., Li, P., Chen, Y. et al. (2023). Interfacial deposition of Ag nanozyme on metal-polyphenol nanosphere for SERS detection of cellular glutathione. *Biosensors & Bioelectronics* 228: 115200.

79 Yan, Y., Liu, Z., Xie, P. et al. (2023). Sensitive electrochemical assay of acetaminophen based on 3D-hierarchical mesoporous carbon nanosheets. *Journal of Colloid and Interface Science* 634: 509–520.

80 Baykal, B., Kadikoylu, G., Senturk, H. et al. (2021). Preparation and characterization gallic acid-titanium dioxide nanocomposites for biosensing application on voltammetric detection of DNA. *Journal of Electroanalytical Chemistry* 892: 115262.

81 Fang, J., Dai, L., Feng, R. et al. (2023). High-performance electrochemiluminescence of a coordination-driven J-aggregate K-PTC MOF regulated by metal-phenolic nanoparticles for biomarker analysis. *Analytical Chemistry* 95 (2): 1287–1293.

82 Huang, J., Meng, J., Chen, S. et al. (2020). A soft metal-polyphenol capsule-based ultrasensitive immunoassay for electrochemical detection of Epstein-Barr (EB) virus infection. *Biosensors & Bioelectronics* 164: 112310.

83 Bandodkar, A.J., Jeerapan, I., and Wang, J. (2016). Wearable chemical sensors: present challenges and future prospects. *ACS Sensors* 1 (5): 464–482.

84 Han, S.T., Peng, H., Sun, Q. et al. (2017). An overview of the development of flexible sensors. *Advanced Materials* 29 (33): 1700375.

Part III

Polyphenols for Life Science Applications

13

Polyphenols in Control of COVID-19

Elyor Berdimurodov[1,2,3,], Khasan Berdimuradov[4], Ashish Kumar[5], Omar Dagdag[6], Mohamed Rbaa[7,8], and Bhawana Jain[9]*

[1] Chemical & Materials Engineering, New Uzbekistan University, Tashkent, Uzbekistan
[2] Medical School, Central Asian University, Tashkent, Uzbekistan
[3] Faculty of Chemistry, National University of Uzbekistan, Tashkent, Uzbekistan
[4] Faculty of Industrial Viticulture and Food Production Technology, Shahrisabz branch of Tashkent Institute of Chemical Technology, Shahrisabz, Uzbekistan
[5] Nalanda College of Engineering, Bihar Engineering University, Department of Science, Technology and Technical Education, Government of Bihar, India
[6] Department of Mechanical Engineering, Gachon University, Seongnam, Republic of Korea
[7] The Higher Institute of Nursing Professions and Health Techniques of Casablanca, Morocco
[8] Laboratory of Materials, Nanotechnology, and Environment, Faculty of Sciences, Mohammed V University in Rabat, Agdal-Rabat, Morocco
[9] SIDDHACHALAM LABORATORY, Raipur, India

13.1 Introduction

13.1.1 Overview of COVID-19 and Polyphenols

COVID-19, caused by the novel coronavirus SARS-CoV-2, first emerged in Wuhan, China in December 2019 and quickly escalated into a global pandemic. The World Health Organization (WHO) declared COVID-19 a Public Health Emergency of International Concern on January 30, 2020, and later characterized it as a pandemic on March 11, 2020. Since then, the virus has spread worldwide, affecting millions and causing significant morbidity and mortality [1, 2]. The primary symptoms of COVID-19 include fever, cough, shortness of breath, and loss of taste or smell, although the disease can manifest with a wide range of symptoms and severity. Some individuals may experience mild to moderate symptoms, while

* Corresponding author

Science and Engineering of Polyphenols: Fundamentals and Industrial Scale Applications, First Edition. Edited by Chandrabhan Verma.
© 2024 John Wiley & Sons Inc. Published 2024 by John Wiley & Sons Inc.

others develop severe or critical illness, often leading to respiratory distress, organ failure, and death. Specific populations, such as the elderly and those with underlying health conditions, are at higher risk of severe outcomes [3, 4].

Transmission of the virus primarily occurs through respiratory droplets produced when an infected person coughs, sneezes, talks, or breathes. The virus can also be spread by touching contaminated surfaces and the face. To mitigate the spread of the virus, various prevention measures have been implemented worldwide, including social distancing, wearing masks, frequent handwashing, and widespread vaccination campaigns. The impact of the COVID-19 pandemic on global health has been pro-found. Healthcare systems have been overwhelmed, and there have been significant social, economic, and mental health repercussions. The pandemic has further emphasized the importance of scientific research, international collaboration, and the development of effective prevention and treatment strategies to manage and eventually overcome the crisis. As COVID-19 continues to impact the world, our understanding of the virus and its effects has evolved. Here, we'll delve into more details about the virus, the immune response, variants, and current treatments [5, 6].

SARS-CoV-2 is an enveloped, single-stranded, positive-sense RNA virus belonging to the family Coronaviridae. The virus has four main structural proteins: spike (S), envelope (E), membrane (M), and nucleocapsid (N). The spike protein plays a crucial role in viral entry into host cells by binding to the angiotensin-converting enzyme 2 (ACE2) receptor. The life cycle of SARS-CoV-2 begins with the attachment of the viral spike protein to the host cell's ACE2 receptor, followed by the fusion of the viral and host membranes. The viral RNA is then released into the host cell's cytoplasm, translated into viral proteins, and replicated. Newly synthesized viral RNA and proteins assemble into new virions, removed from the host cell to infect other cells [7, 8].

The immune response to COVID-19 is complex and involves both the innate and adaptive immune systems. Initially, the innate immune system detects the virus. It mounts a rapid response, characterized by the release of cytokines and the activation of immune cells such as macrophages and natural killer (NK) cells. This early response aims to limit viral replication and spread. The adaptive immune response, which includes the activation of T and B cells, follows the innate response. T cells help eliminate infected cells, while B cells produce neutralizing antibodies that block the virus from entering host cells. Effective immune responses can lead to the clearance of the virus and the development of lasting immunity [9, 10]. As the virus continues to circulate, it undergoes mutations, leading to the emergence of new variants. Some of these variants have been classified as Variants of Concern (VOCs) due to their potential impact on transmissibility, disease severity, and vaccine effectiveness. Notable VOCs include Alpha (B.1.1.7), Beta (B.1.351), Gamma (P.1), and Delta (B.1.617.2). Health authorities closely monitor these variants and their implications for public health measures and vaccine development [11, 12].

The management of COVID-19 has evolved, with various treatments being explored to reduce the severity of the illness and improve patient outcomes. Some of the current treatments include antiviral medications such as remdesivir, immunomodulatory drugs like dexamethasone, and monoclonal antibody therapies. Vaccines have been developed unprecedentedly, with multiple vaccines authorized for emergency use worldwide. These vaccines use different platforms, such as mRNA (Pfizer-BioNTech and Moderna), viral vectors (Oxford-AstraZeneca and Johnson & Johnson), and inactivated viruses (Sinopharm and Sinovac). Large-scale vaccination campaigns are ongoing to achieve widespread immunity and eventually bring the pandemic under control. Despite these advances, the COVID-19 pandemic is not yet over, and continued research, surveillance, and public health measures remain crucial in our efforts to overcome this global crisis [13, 14].

13.1.2 Polyphenols: Definition and General Properties

Polyphenols are a diverse group of naturally occurring compounds found primarily in plants. They are characterized by multiple phenol units in their chemical structure. Polyphenols play essential roles in plants, including defense against pathogens and contributing to various plant-based foods' color, taste, and aroma. Polyphenols can be classified based on their chemical structure and the number of phenol rings. The main classes of polyphenols are [15–17]:

1) *Flavonoids*: This is the largest and most extensively studied group of polyphenols. Flavonoids can be divided into subclasses: flavonols, flavones, isoflavones, flavanones, flavanols (catechins), and anthocyanidins. They are found in various fruits, vegetables, and beverages, such as tea, red wine, and cocoa.
2) *Phenolic acids*: Phenolic acids are divided into two main groups: hydroxybenzoic acids and hydroxycinnamic acids. They are commonly found in various fruits, vegetables, whole grains, and spices.
3) *Lignans*: These polyphenols are found in high concentrations in seeds, particularly flaxseeds, as well as in whole grains, nuts, and some fruits and vegetables. Lignans exhibit phytoestrogenic properties and have been studied for their potential health benefits.
4) *Stilbenes*: The most well-known stilbene is resveratrol, found in red wine, grapes, and some berries. Stilbenes have gained attention for their potential health-promoting effects and antioxidant properties.
5) *Tannins*: Tannins are a heterogeneous group of polyphenols that can be further categorized into hydrolysable tannins and condensed tannins (also known as proanthocyanidins). They are found in tea, wine, and various fruits, contributing to their astringency and mouthfeel.

These different classes of polyphenols have unique chemical structures and properties, which influence their bioavailability, metabolism, and potential health effects.

13.1.3 Natural sources and extraction

Polyphenols are abundant in a wide range of plant-based foods and beverages. Some of the most common sources of polyphenols include [18–20]:

1) *Fruits*: Berries (e.g. blueberries, strawberries, raspberries), grapes, apples, cherries, citrus fruits (e.g. oranges, lemons), and pomegranates are rich in polyphenols such as anthocyanins, flavonols, and flavanones.
2) *Vegetables*: Leafy greens (e.g. spinach, kale), onions, broccoli, artichokes, and red cabbage contain significant amounts of flavonoids and phenolic acids.
3) *Nuts and seeds*: Walnuts, almonds, hazelnuts, and flaxseeds are good sources of lignans, flavonoids, and phenolic acids.
4) *Whole grains*: Whole grains like wheat, barley, oats, and rye are rich in phenolic acids, particularly ferulic acid.
5) *Legumes*: Soybeans, lentils, and chickpeas contain isoflavones and other flavonoids.
6) Spices and herbs: Cloves, peppermint, oregano, thyme, and rosemary are highly concentrated sources of various polyphenols.
7) *Beverages*: Tea (especially green tea), red wine, coffee, and cocoa are rich in various polyphenols like catechins, proanthocyanidins, and phenolic acids.

Polyphenols can be extracted from plant sources using various methods, depending on the type of polyphenol, the plant matrix, and the desired application. Some common extraction techniques include:

1) *Solvent extraction*: This method involves using water, ethanol, methanol, or their combinations to dissolve and extract polyphenols from plant materials. The choice of solvent depends on the polarity of the targeted polyphenols and the plant matrix.
2) *Solid-phase extraction*: This technique uses a solid sorbent material to selectively adsorb and desorb polyphenols from a liquid sample, allowing for their concentration and purification.
3) *Ultrasound-assisted extraction*: This method involves using ultrasound energy to facilitate the release of polyphenols from plant tissues, increasing extraction efficiency and reducing extraction time.
4) *Supercritical fluid extraction*: This technique uses supercritical fluids, such as supercritical carbon dioxide, as solvents to extract polyphenols. This method can provide high extraction efficiency and selectivity while avoiding potentially harmful organic solvents.
5) *Enzyme-assisted extraction*: This approach uses enzymes to break down cell walls and improve the release of polyphenols from plant tissues, enhancing extraction efficiency.

The choice of extraction method depends on factors such as the type and concentration of polyphenols, the complexity of the plant matrix, and the intended application of the extracted polyphenols (e.g. food, supplements, or pharmaceuticals).

13.1.4 General Health Benefits

Polyphenols have been widely studied for their potential health benefits, which are believed to be primarily attributed to their antioxidant, anti-inflammatory, and anti-microbial properties. Some of the general health benefits associated with polyphenol consumption include [21–23]:

1) *Antioxidant activity*: Polyphenols can neutralize harmful free radicals and reactive oxygen species in the body, protecting cells from oxidative damage that can contribute to chronic diseases, aging, and inflammation.
2) *Anti-inflammatory effects*: Polyphenols can modulate inflammatory pathways by inhibiting pro-inflammatory cytokine production and promoting anti-inflammatory mediators' production. This may help reduce the risk of chronic inflammatory diseases, such as cardiovascular disease, diabetes, and certain types of cancer.
3) *Cardiovascular health*: Consuming polyphenol-rich foods has been associated with a reduced risk of cardiovascular diseases. Polyphenols may help improve endothelial function, lower blood pressure, reduce inflammation, and inhibit platelet aggregation, contributing to better cardiovascular health.
4) *Cancer prevention*: Some polyphenols have been shown to exhibit anti-cancer properties through various mechanisms, such as inhibiting cancer cell growth, inducing apoptosis (programmed cell death), and interfering with tumor angiogenesis (the formation of new blood vessels that support tumor growth).
5) *Neuroprotective effects*: Polyphenols have been studied for their potential to protect the brain from oxidative stress and inflammation, which are believed to contribute to neurodegenerative diseases like Alzheimer's and Parkinson's. Some polyphenols may help improve cognitive function, prevent neuronal damage, and reduce the risk of age-related cognitive decline.
6) *Gut health*: Polyphenols can modulate the gut microbiota, promoting beneficial bacteria's growth and inhibiting harmful bacteria's growth. This may help maintain a healthy gut environment and support overall digestive health.
7) *Blood sugar control*: Some polyphenols have been shown to improve insulin sensitivity and glucose metabolism, which may help regulate blood sugar levels and reduce the risk of type 2 diabetes.
8) *Weight management*: Some studies suggest that polyphenols may play a role in weight management by modulating fat metabolism, reducing fat absorption, and enhancing energy expenditure.

It is important to note that the health benefits associated with polyphenol consumption can vary depending on the type, amount, and bioavailability of the polyphenols and individual factors such as genetics, age, and overall health status. To maximize the health benefits of polyphenols, consuming a diverse diet rich in fruits, vegetables, whole grains, legumes, nuts, and other polyphenol-containing foods is recommended.

13.2 Antiviral Properties of Polyphenols

13.2.1 Mechanisms of Action

13.2.1.1 Inhibition of Viral Entry

Polyphenols have been shown to exhibit antiviral properties against various viruses, including influenza, herpes simplex, HIV, and hepatitis C. One of the primary mechanisms by which polyphenols exert their antiviral effects is inhibiting viral entry into host cells. This can be achieved through several strategies [24–26]:

1) *Blocking viral attachment*: Polyphenols can interfere with the initial interaction between the virus and host cell receptors, preventing the virus from attaching to the cell surface. This can be achieved by binding to viral surface proteins or host cell receptors, thereby obstructing the virus's and host cells' interfaces.
2) *Inhibition of viral fusion and internalization*: Some polyphenols can disrupt the fusion of the viral envelope with the host cell membrane, a crucial step for entering enveloped viruses such as influenza and HIV. This can be accomplished by interacting with viral surface proteins, host cell receptors, or the membrane lipid bilayer, altering their conformation or function and blocking viral fusion and internalization.
3) *Interference with viral endocytosis*: For viruses that enter host cells through endocytosis, such as hepatitis C and some strains of influenza, polyphenols may inhibit the endocytic process by interacting with host cell proteins involved in endocytosis or by altering the intracellular environment, such as pH or ion concentrations, that are essential for endocytosis and viral uncoating.
4) *Inhibition of viral replication*: Some polyphenols can inhibit the replication of viral RNA or DNA once the virus has entered the host cell. This can be achieved by directly interacting with viral enzymes, such as reverse transcriptase, protease, or polymerase, or by modulating host cell signaling pathways critical for viral replication.

Polyphenols can effectively prevent the infection and spread of viruses by targeting one or multiple steps in the viral entry process. The specific mechanisms of action and efficacy may vary depending on the type of polyphenol, the virus, and the host cell. Further research is needed to better understand the antiviral properties of different polyphenols and their potential applications in preventing and treating viral infections. For example, epigallocatechin-3-gallate and green tea extract (EGCG/GTE) (Figure 13.1) supplementation could provide a multitarget approach to help manage COVID-19 by regulating the activity of key transcription factors and their target genes, potentially reducing inflammation and preventing lung fibrosis. However, further research is needed to confirm these effects and determine the most effective and safe use of these supplements in COVID-19 [17].

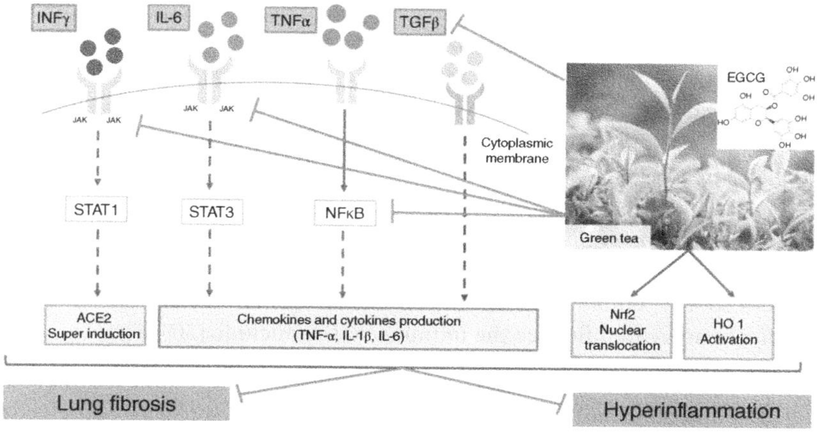

Figure 13.1 Exploring how green tea extract and epigallocatechin-3-gallate (EGCG) may confer beneficial effects in patients with coronavirus disease 2019 (COVID-19). Source: Menegazzi et.al. [17]/from MDPI / CC BY 4.0.

13.2.1.2 Interference with Viral Replication

Polyphenols can exhibit antiviral properties by interfering with different stages of the viral replication process. Once the virus has entered a host cell, it must replicate its genetic material and produce viral proteins to assemble new virus particles. Polyphenols can target various aspects of this process, including [27–29]:

1) *Inhibition of viral enzymes*: Many viruses rely on specific enzymes to replicate their genetic material, such as reverse transcriptase for retroviruses like HIV or RNA-dependent RNA polymerase (RdRp) for RNA viruses like influenza and hepatitis C. Some polyphenols can directly bind to and inhibit these essential viral enzymes' activity, preventing viral genetic material replication.
2) *Interference with viral protein synthesis*: Viral proteins are crucial for the assembly of new virus particles, and some polyphenols can inhibit the production or function of these proteins. This can be achieved by interfering with the translation of viral mRNA, disrupting the folding or stability of viral proteins, or modulating host cell signaling pathways that regulate protein synthesis.
3) *Inhibition of viral assembly and release*: The assembly of new virus particles and their subsequent release from infected cells are critical for viral spread. Polyphenols can interfere with these processes by disrupting the interaction between viral proteins, inhibiting the maturation of virus particles, or modulating host cell machinery involved in the assembly and release of virions.
4) *Modulation of host cell signaling pathways*: Some polyphenols can exert antiviral effects by modulating host cell signaling pathways essential for viral replication. For example, polyphenols may inhibit the activation of certain transcription factors, kinases, or other signaling molecules required to express viral genes or regulate host cell processes that support viral replication.

By targeting one or more steps in the viral replication process, polyphenols can effectively inhibit the production and spread of various viruses. The specific mechanisms of action and potency may depend on the type of polyphenol, the virus, and the host cell. Further research is needed to better understand the antiviral properties of different polyphenols and their potential applications in preventing and treating viral infections.

13.2.1.3 Modulation of Host Immune Response

In addition to their direct antiviral effects, polyphenols can modulate the host immune response, enhancing the body's defense against viral infections. Some ways polyphenols can influence the immune system include [30, 31]:

1) *Immunomodulation*: Polyphenols can modulate the immune system by regulating the production and activity of various immune cells, such as macrophages, T cells, and B cells. This may involve enhancing the activation and proliferation of immune cells, promoting the production of specific cytokines, or regulating the balance between pro-inflammatory and anti-inflammatory responses.
2) *Enhancing innate immunity*: The innate immune system provides the first line of defense against viral infections, and polyphenols can enhance the antiviral activity of innate immune cells. For example, polyphenols may promote the production of type I interferons, crucial for the antiviral response, or stimulate the activity of NK cells, which can directly target and eliminate virus-infected cells.
3) *Supporting adaptive immunity*: Polyphenols can also support the adaptive immune response, which generates a specific and long-lasting defense against viral pathogens. Some polyphenols may enhance the activation and proliferation of T cells, which are critical for recognizing and eliminating virus-infected cells, or promote the production of virus-specific antibodies by B cells.
4) *Reducing inflammation*: While inflammation is an important component of the immune response to viral infections, excessive inflammation can damage tissue and contribute to disease progression. Polyphenols can help regulate the inflammatory response by inhibiting the production of pro-inflammatory cytokines and promoting the release of anti-inflammatory mediators, thereby reducing inflammation and tissue damage.
5) *Antioxidant effects*: Oxidative stress can impair the immune response and contribute to viral infections. With their antioxidant properties, polyphenols can neutralize harmful free radicals and reactive oxygen species, protecting immune cells from oxidative damage and enhancing their function.

By modulating the host immune response, polyphenols may help prevent and control viral infections, reduce disease severity, and promote recovery. The specific mechanisms of action and efficacy of polyphenols in modulating the immune response may depend on the type of polyphenol, the virus, and the host's immune status. Further research is needed to understand better the immunomodulatory effects of different polyphenols and their potential applications in preventing and treating viral infections.

13.3 Polyphenols Active Against COVID-19

13.3.1 Examples of Polyphenols with Potential Antiviral Effects

Polyphenols are a diverse group of compounds that have shown antiviral properties against various viruses, including coronaviruses such as SARS-CoV-2, responsible for COVID-19. Several polyphenols have been studied for their potential antiviral effects against COVID-19, and while more research is needed to confirm their efficacy and safety, some examples include [32, 33]:

1) *Resveratrol*: Resveratrol (Figure 13.2) is a stilbenoid found in grapes, red wine, peanuts, and other plant sources. It has been reported to inhibit the replication of SARS-CoV-2 in vitro by targeting the viral main protease (Mpro), an enzyme essential for viral replication. Resveratrol has anti-inflammatory and antioxidant effects, possibly contributing to its potential antiviral activity against COVID-19 [15].

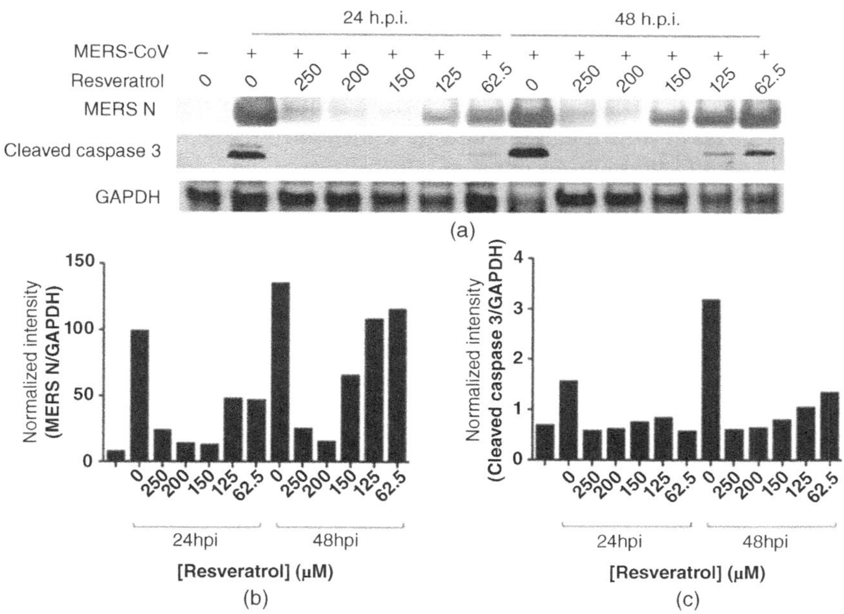

Figure 13.2 Resveratrol decreased MERS-induced cell apoptosis. Vero E6 cells were infected with MERS-CoV at an M.O.I. of 0.1 and treated with resveratrol for 24 and 48 hours before protein lysates were collected. (a) Protein expression levels of MERS nucleocapsid and cleaved caspase 3 were assessed through western blotting, with glyceraldehyde 3-phosphate dehydrogenase (GAPDH) as a loading control. (b) and (c) Protein expression levels were quantified and normalized to GAPDH expression. Both MERS nucleocapsid and cleaved caspase 3 expressions were diminished by resveratrol in a dose-dependent manner. Source: Lin et al. [15]/Springer Nature / CCBY 4.0 / Public domain.

2) *Quercetin*: Quercetin (Figure 13.3) is a flavonoid in many fruits, vegetables, and grains. It has been shown to inhibit the entry of SARS-CoV-2 into host cells by binding to the viral spike protein and blocking its interaction with the human ACE2 receptor. Quercetin also exhibits anti-inflammatory and antioxidant properties, which may help modulate the immune response and reduce the severity of COVID-19.
3) *EGCG*: EGCG is a catechin found in green tea and has been studied for its antiviral properties against various viruses, including coronaviruses. It has been suggested that EGCG may inhibit SARS-CoV-2 by binding to the viral spike protein, interfering with viral entry and replication. Additionally, its anti-inflammatory and antioxidant properties may help support the immune response against COVID-19.
4) *Curcumin*: Curcumin is a polyphenol found in turmeric and has been investigated for its potential antiviral effects against SARS-CoV-2. It has been proposed that curcumin may inhibit SARS-CoV-2 replication by targeting viral enzymes, such as the Mpro and the RdRp. Furthermore, curcumin's anti-inflammatory and antioxidant properties may help modulate the immune response to the virus.
5) *Baicalin*: Baicalin is a flavonoid found in the roots of *Scutellaria baicalensis* (Chinese skullcap) and has been shown to exhibit antiviral activity against SARS-CoV-2 in vitro. It may inhibit viral replication by targeting the viral Mpro and the RdRp. Baicalin also has anti-inflammatory and antioxidant effects, which may contribute to its potential activity against COVID-19.

These examples represent only a small subset of the numerous polyphenols that may have potential antiviral effects against COVID-19. It is important to note that while in vitro and computational studies provide valuable insights, more research, including in vivo studies and clinical trials, is needed to validate the efficacy and safety of these polyphenols for the prevention and treatment of COVID-19.

13.3.2 Studies and Evidence Supporting Their Activity Against SARS-CoV-2

While growing evidence suggests that certain polyphenols may have potential antiviral activity against SARS-CoV-2, it is important to note that many of these studies are preliminary. More research is needed to validate their efficacy and safety. Here are some examples of studies and evidence supporting the activity of various polyphenols against SARS-CoV-2 [1, 3–5]:

1) *Quercetin*: In silico (computational) studies have shown that quercetin may bind to the SARS-CoV-2 spike protein, potentially blocking its interaction with the human ACE2 receptor and inhibiting viral entry. Quercetin has also been shown to inhibit the viral Mpro in in vitro studies.

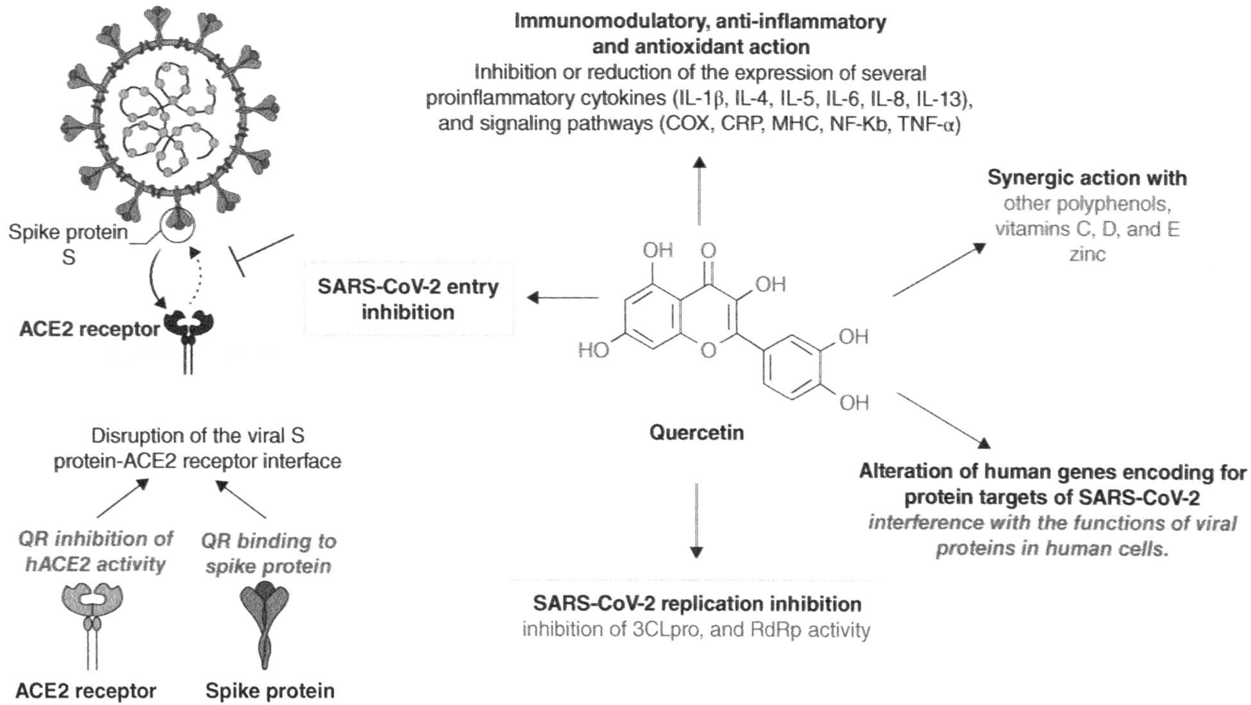

Figure 13.3 Quercetin's potential as a preventative, mitigative, and therapeutic agent for COVID-19 infection is due to its multifactorial beneficial action. Source: Gasmi et al. [9]/MDPI/CCBY 4.0.

2) *Resveratrol*: In vitro studies have demonstrated that resveratrol can inhibit the replication of SARS-CoV-2 by targeting the viral Mpro. Additionally, resveratrol has been shown to reduce the production of pro-inflammatory cytokines in SARS-CoV-2-infected cells, suggesting potential anti-inflammatory effects.
3) *EGCG:* In silico studies have suggested that EGCG may inhibit SARS-CoV-2 by binding to the viral spike protein, potentially interfering with viral entry and replication. Further research is needed to confirm these findings in vitro and in vivo.
4) *Curcumin*: Computational studies have proposed that curcumin may inhibit SARS-CoV-2 replication by targeting viral enzymes, such as the Mpro and the RdRp. In vitro studies are needed to validate these findings.
5) *Baicalin*: In vitro studies have shown that baicalin can inhibit SARS-CoV-2 replication by targeting the viral Mpro and the RdRp. Further research is needed to confirm these findings in vivo and explore the potential therapeutic applications of baicalin in COVID-19 treatment.

These studies highlight the potential antiviral activity of various polyphenols against SARS-CoV-2. However, it is crucial to note that many of these findings are based on in vitro experiments or computational predictions. More research, including in vivo studies and clinical trials, is needed to confirm the antiviral efficacy of these polyphenols in humans and to determine their safety, optimal dosages, and potential applications in the prevention and treatment of COVID-19.

13.4 Potential Applications of Polyphenols in COVID-19 Prevention and Treatment

13.4.1 Dietary Sources and Supplements

13.4.1.1 Foods Rich in Polyphenols

Polyphenols are naturally occurring compounds found in various plant-based foods and beverages. Incorporating these foods into a balanced and diverse diet may help support the immune system and contribute to overall health, potentially aiding in preventing and managing viral infections, including COVID-19. Some foods and beverages rich in polyphenols include [6, 10, 11]:

1) *Fruits*: Many fruits contain polyphenols, particularly flavonoids and phenolic acids. Examples include berries (e.g. blueberries, strawberries, raspberries, blackberries), apples, cherries, grapes, citrus fruits (e.g. oranges, lemons, grapefruits), plums, and pomegranates.
2) *Vegetables*: Certain vegetables are also good sources of polyphenols. Examples include spinach, kale, onions, artichokes, broccoli, asparagus, and red cabbage.

3) *Nuts and seeds*: Nuts and seeds are rich in various polyphenols, including flavonoids, phenolic acids, and lignans. Examples include almonds, walnuts, hazelnuts, pecans, flaxseeds, and sunflower seeds.
4) *Legumes*: Some legumes contain significant polyphenols, such as isoflavones, flavonoids, and phenolic acids. Examples include soybeans, lentils, chickpeas, and various types of beans.
5) *Whole grains*: Whole grains are a source of polyphenols, particularly phenolic acids and lignans. Examples include whole wheat, brown rice, oats, barley, and quinoa.
6) *Spices and herbs*: Many spices and herbs are rich in polyphenols, including curcumin, flavonoids, and phenolic acids. Examples include turmeric, cloves, cinnamon, oregano, rosemary, thyme, and sage.
7) *Tea*: Both green and black teas are rich in catechins, a type of flavonoid. Green tea, in particular, is a good source of EGCG, one of the most studied polyphenols for its potential antiviral effects.
8) *Cocoa and dark chocolate*: Cocoa and dark chocolate (with a high cocoa content) are rich in flavanols, a type of flavonoid. These compounds have various health benefits, including antioxidant and anti-inflammatory effects.
9) *Red wine*: Red wine, in moderation, can be a source of resveratrol, a stilbenoid with potential antiviral properties. It is important to note that excessive alcohol consumption can have negative health effects and should be avoided.
10) *Extra virgin olive oil*: Extra virgin olive oil is a good source of oleuropein, hydroxytyrosol, and other phenolic compounds, associated with various health benefits, including antioxidant and anti-inflammatory effects.

In this study [33] (Figure 13.4), the researchers discovered that CC/Fr.2 (a specific fraction of Cinnamomi Cortex) contains inhibitors of SARS virus infection. The fraction was also found to inhibit the internalization of the transferrin receptor, suggesting interference with clathrin-dependent endocytosis, a process some viruses utilize to enter host cells.

In addition to consuming polyphenol-rich foods, some people may take polyphenol supplements for their potential health benefits. It is important to note that the efficacy and safety of polyphenol supplements can vary, and it is always recommended to consult with a healthcare professional before starting any new supplement regimen.

13.4.1.2 Polyphenol Supplements and Their Potential Benefits

Polyphenol supplements are concentrated sources of these bioactive compounds extracted from various plant sources. They are available in various forms, including

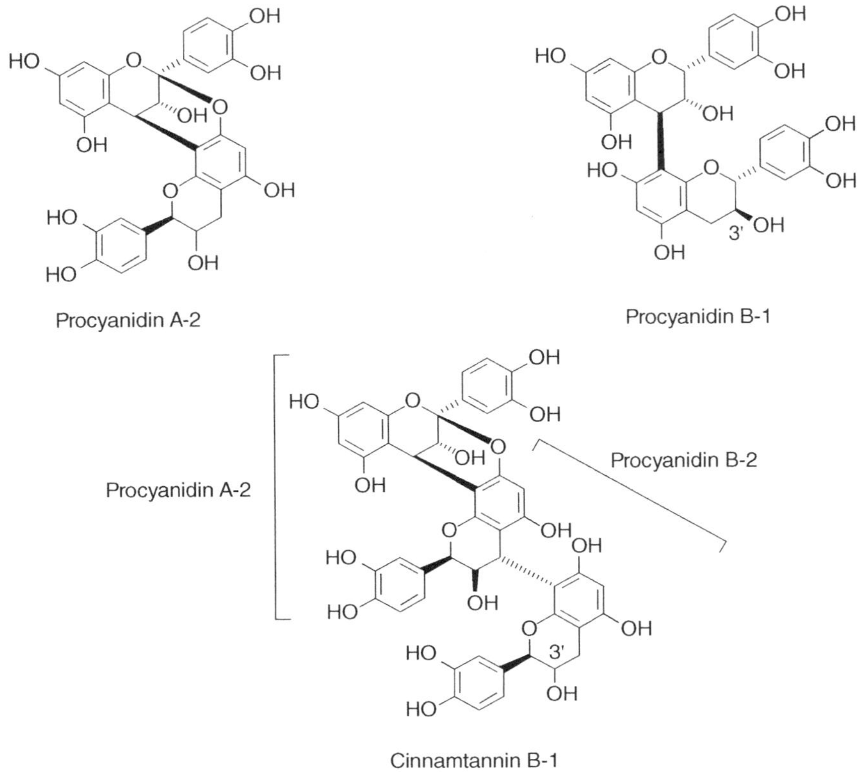

Figure 13.4 Various polyphenols from plant extracts: Procyanidin A2, procyanidin B1, and cinnamtannin B1. Source: Zhuang et al. [33]/Elsevier.

capsules, tablets, and powders. Some potential benefits of polyphenol supplements may include [16, 24]:

1) *Antioxidant effects*: Polyphenols are well-known for their antioxidant properties, which help neutralize free radicals that can cause cellular damage and contribute to aging and various diseases. Consuming polyphenol supplements may help boost the body's antioxidant defenses.
2) *Anti-inflammatory effects*: Many polyphenols have been shown to have anti-inflammatory properties, which may help reduce inflammation in the body and support overall health. This could be particularly beneficial for individuals with chronic inflammatory conditions or those looking to maintain a healthy immune system.
3) *Cardiovascular health*: Some polyphenols, such as resveratrol and flavonoids, have been linked to improved cardiovascular health. They may help reduce blood pressure, improve blood vessel function, and decrease the risk of heart disease.

4) *Cancer prevention*: Some studies have shown that certain polyphenols, such as EGCG and curcumin, may have anticancer properties, potentially inhibiting the growth and spread of cancer cells. However, more research is needed to determine the effectiveness of polyphenol supplements in cancer prevention and treatment.
5) *Brain health*: Polyphenols, especially flavonoids, have been associated with improved cognitive function and a reduced risk of neurodegenerative diseases, such as Alzheimer's and Parkinson's. They may help protect brain cells from oxidative stress and inflammation, which can contribute to cognitive decline.
6) *Gut health*: Polyphenols may have prebiotic effects, promoting the growth of beneficial gut bacteria and contributing to a healthy gut microbiome. This may help support digestion, immune function, and overall health.
7) *Potential antiviral effects*: Several polyphenols have shown potential antiviral activity against various viruses, including SARS-CoV-2. While more research is needed to determine their efficacy in preventing and treating COVID-19, polyphenol supplements may offer some immune support against viral infections.

It is important to note that the potential benefits of polyphenol supplements can vary depending on the specific compound, dosage, and individual factors. Moreover, the efficacy and safety of these supplements are only sometimes well-established, and whole foods may provide a broader range of nutrients and health benefits. Always consult a healthcare professional before starting any new supplement regimen, and prioritize a balanced diet rich in whole, plant-based foods to obtain a diverse array of polyphenols.

13.4.2 Drug Development

13.4.2.1 Polyphenols as Lead Compounds for Drug Development

Polyphenols possess many biological activities, including antioxidant, anti-inflammatory, and antiviral effects, making them attractive candidates for drug development. Their structural diversity and ability to modulate various molecular targets in the human body have led researchers to investigate their potential as lead compounds for developing new therapeutics. Several polyphenols have shown promising results in preclinical studies; some have even advanced to clinical trials. Examples include [24]:

1) *EGCG*: The major catechin in green tea has shown potential anticancer, antiviral, and neuroprotective effects in preclinical studies. Some clinical trials have investigated its potential for cancer prevention and treatment, as well as neurodegenerative diseases like Alzheimer's.
2) *Curcumin*: The active component of turmeric has demonstrated anti-inflammatory, antioxidant, and anticancer properties in preclinical studies. Some clinical trials have explored its potential as an adjunct therapy for cancer, arthritis, and other inflammatory conditions.

3) *Resveratrol*: This stilbenoid, found in red wine and grapes, has garnered attention for its potential cardiovascular, anticancer, and neuroprotective effects. Clinical trials have investigated its potential role in preventing and treating various diseases, including heart disease, cancer, and Alzheimer's.
4) *Quercetin*: This flavonoid, found in various fruits and vegetables, has shown potential antioxidant, anti-inflammatory, and antiviral properties. Clinical trials have explored its potential benefits in treating chronic fatigue syndrome, allergy symptoms, and viral infections, including COVID-19.

13.4.2.2 Challenges and Progress in the Development of Polyphenol-based Therapeutics

Despite their potential, there are several challenges in developing polyphenol-based therapeutics:

1) *Bioavailability*: The bioavailability of many polyphenols is often low, meaning that only a small fraction of the ingested compound is absorbed and reaches the target tissue. This can limit their therapeutic potential. Researchers are developing novel formulations and delivery systems to enhance the bioavailability of polyphenols, such as nanoformulations, liposomes, and solid lipid nanoparticles.
2) *Pharmacokinetics*: Polyphenols are often metabolized and excreted rapidly from the body, which can reduce their effectiveness. Developing modified polyphenol derivatives with improved pharmacokinetic properties is a potential strategy to overcome this challenge.
3) *Selectivity and specificity*: The broad range of biological activities and molecular targets of polyphenols can be both a strength and a limitation. While this may contribute to potential synergistic effects, it raises concerns about off-target effects and potential toxicity. Further research is needed to optimize the selectivity and specificity of polyphenol-based therapies (Figure 13.5) [29].
4) *Clinical evidence*: While many polyphenols have shown promising results in preclinical studies, clinical evidence for their efficacy and safety still needs to be improved. More well-designed clinical trials are required to validate polyphenols' therapeutic potential and identify optimal dosages and treatment regimens.

Despite these challenges, the development of polyphenol-based therapeutics continues to progress, driven by advances in drug delivery, formulation, and chemical modification techniques. As our understanding of the molecular mechanisms underlying their biological activities deepens, the potential of polyphenols as lead compounds for drug development is expected to grow.

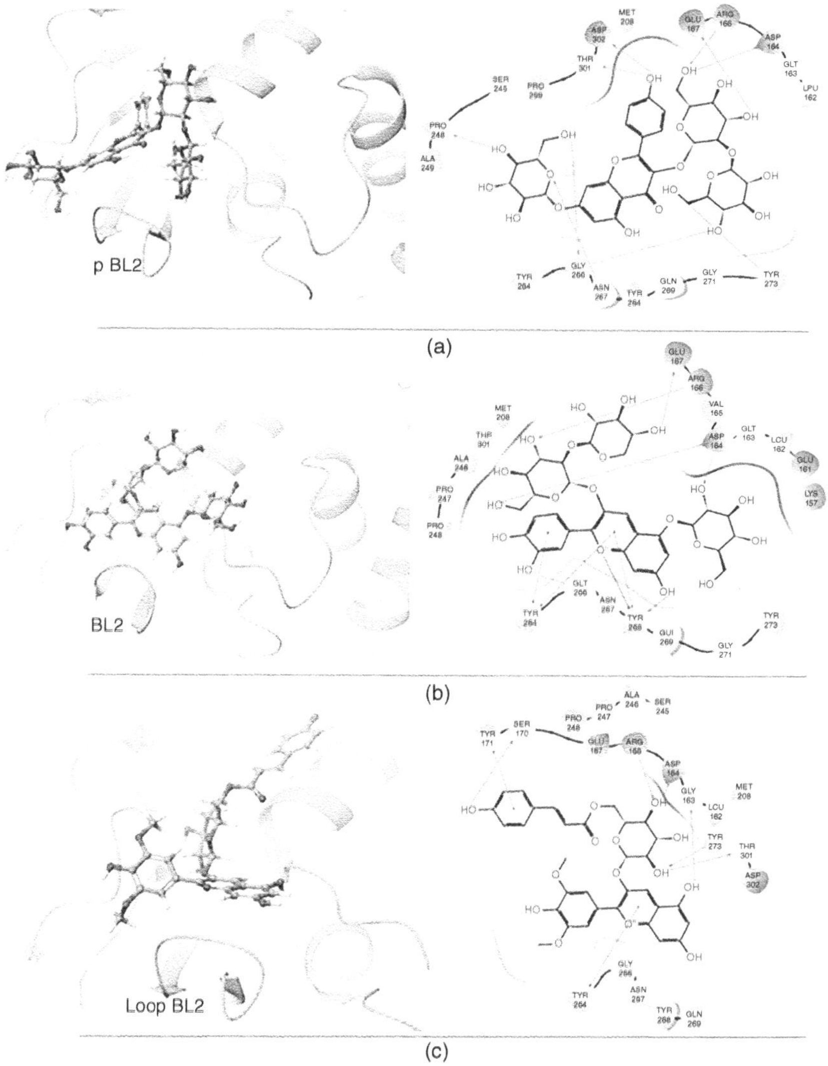

Figure 13.5 The docking positions and 2D ligand–protein interaction diagrams of 6W9C and the top three ligands are A, Kaempferol 3-O-sophoroside 7-O-glucoside; B, Cyanidin 3-O-sambubioside 5-O-glucoside; C, Malvidin 3-O-(6″-p-coumaroyl-glucoside). A pink arrow indicates the hydrogen bond, a green line represents the π–π stacking interaction, and the π–cation interaction is shown by a red line. Source: Wu et al. [29]/Elsevier.

13.5 Clinical Trials and Research on Polyphenols and COVID-19

13.5.1 Summary of Completed Trials

There have been limited completed clinical trials investigating the efficacy of polyphenols against COVID-19. However, some ongoing studies and preliminary results have been reported. It is essential to remember that the situation is rapidly evolving, and new trial outcomes may have emerged since then.

1) *Quercetin*: One of the most studied polyphenols in COVID-19 is quercetin, a flavonoid with reported antiviral and anti-inflammatory properties. A few clinical trials have investigated its potential as a preventative measure or a treatment for COVID-19, often combined with other agents, such as vitamin C or zinc. Preliminary results from some trials have suggested potential benefits, but more research is needed to confirm these findings and determine optimal dosages and treatment regimens.
2) *EGCG*: The major catechin in green tea has shown potential antiviral activity against SARS-CoV-2 in preclinical studies. While no completed clinical trials have investigated EGCG for COVID-19 treatment, some ongoing trials aim to evaluate its potential benefits in this context.
3) *Resveratrol*: Resveratrol, a stilbenoid found in red wine and grapes, has been reported to have potential antiviral and anti-inflammatory effects. Preclinical studies have suggested that it may inhibit SARS-CoV-2 replication. A few clinical trials were underway as of
4) *Curcumin*: Curcumin, the active component of turmeric, has demonstrated potential anti-inflammatory and antiviral properties in preclinical studies. Some clinical trials were registered to evaluate the role of curcumin in COVID-19 prevention.

13.5.2 Outcomes and Implications for Clinical Practice

Due to the limited number of completed clinical trials and the rapidly evolving nature of COVID-19 research, it is challenging to draw definitive conclusions about the potential benefits of polyphenols in preventing or treating COVID-19. While some preliminary evidence suggests that certain polyphenols, such as quercetin, may have potential benefits, more well-designed clinical trials are needed to confirm these findings and establish their clinical implications. Until more conclusive evidence is available, following established public health guidelines and recommendations for COVID-19 prevention and treatment, including vaccination, social distancing, masking, and proper hygiene practices, is necessary. Incorporating polyphenol-rich foods into a balanced diet may support overall health and immune function, but they should not be considered a primary strategy for COVID-19 prevention or treatment.

13.5.3 Potential Future Directions in Research

Future research on polyphenols and COVID-19 may focus on [24, 29]:

Elucidating mechanisms of action: Further studies are needed to understand the molecular mechanisms through which polyphenols exert their potential antiviral and anti-inflammatory effects. This knowledge could help optimize the design of polyphenol-based therapeutics and inform the development of targeted interventions. Optimizing formulations and delivery methods: To overcome the challenges of low bioavailability and rapid metabolism, researchers may continue to explore novel formulations and drug delivery systems, such as nanotechnology-based approaches, to enhance the therapeutic potential of polyphenols.

Evaluating combination therapies: Polyphenols may be most effective when combined with other agents, such as vitamins, minerals, or antiviral drugs. Future research could focus on identifying synergistic combinations and determining the optimal dosages and treatment regimens for COVID-19 prevention or treatment. Conducting well-designed clinical trials: To provide more conclusive evidence on the efficacy and safety of polyphenols in the context of COVID-19, well-designed, randomized, and controlled clinical trials are needed. These trials should include diverse populations and focus on clinically relevant outcomes, such as symptom severity, hospitalization rates, and mortality. Assessing long-term effects: As the COVID-19 pandemic evolves and long-term consequences, such as long COVID, become more apparent, research may focus on the potential role of polyphenols in preventing or alleviating these complications.

13.6 Safety and Limitations of Polyphenols in COVID-19 Management

13.6.1 Safety Concerns and Side Effects

Polyphenols are generally considered safe and offer various health benefits when consumed through a balanced diet rich in fruits, vegetables, and other plant-based foods. However, when taken as supplements or at high concentrations, some potential safety concerns and side effects may arise [28, 32]:

1) *Potential side effects and toxicity*: Most polyphenols are well-tolerated, but high doses, especially in supplement form, can cause side effects in some individuals. These side effects may include gastrointestinal discomfort, nausea, diarrhea, and headaches. Some polyphenols, like EGCG, have been associated with liver toxicity in very high doses.

2) *Interactions with medications and other supplements*: Polyphenols can interact with certain medications and other supplements, potentially altering their effectiveness or causing adverse effects. For example, green tea extract (EGCG) may inhibit the absorption of beta-blockers, a class of medications used to treat high blood pressure. Quercetin may interact with blood thinners like warfarin, increasing the risk of bleeding. It is essential to consult a healthcare professional before using polyphenol supplements, particularly if you are taking medications or have underlying health conditions.
3) *Individual variability*: The safety and effectiveness of polyphenols can vary between individuals due to genetics, age, sex, and overall health status. People with specific health conditions, such as kidney or liver disease, may be more susceptible to adverse effects from polyphenol supplements.
4) *Purity and quality of supplements*: The quality and purity of polyphenol supplements can vary widely, potentially leading to inaccurate dosing or contamination with harmful substances. Choosing reputable brands and following the recommended dosages is essential to minimize potential risks.

13.6.2 Limitations in COVID-19 Management

While polyphenols have shown potential antiviral and anti-inflammatory properties in preclinical studies, their limitations in COVID-19 management include:

1) *Limited clinical evidence*: Clinical evidence supporting using polyphenols for COVID-19 prevention or treatment remains limited. In this context, more well-designed clinical trials are needed to establish their efficacy and safety.
2) *Bioavailability and pharmacokinetics*: The bioavailability of many polyphenols is often low, and they are rapidly metabolized and excreted from the body. These factors can limit their therapeutic potential. Further research is needed to develop formulations and delivery systems that enhance the bioavailability and pharmacokinetics of polyphenols.
3) *Lack of specificity*: Polyphenols can interact with multiple molecular targets, which may contribute to their broad range of biological activities. However, this lack of specificity can also lead to off-target effects and potential toxicity. More research is required to optimize the selectivity and specificity of polyphenol-based therapies for COVID-19.

In summary, while polyphenols have shown promising antiviral and anti-inflammatory properties in preclinical studies, their safety and efficacy in COVID-19 management remain uncertain. Following established public health guidelines for COVID-19 prevention and treatment is essential, including vaccination, social distancing, and hygiene practices. Polyphenol-rich foods can be incorporated into a balanced diet to support overall health, but they should not be considered a primary strategy for COVID-19 prevention or treatment.

13.7 Conclusion

In preclinical studies, polyphenols, a group of naturally occurring compounds in various plant-based foods, have shown potential antiviral and anti-inflammatory properties. These properties have sparked interest in their potential role in preventing and treating COVID-19. Some ongoing clinical trials investigate the potential benefits of various polyphenols, such as quercetin, EGCG, resveratrol, and curcumin, in patients with COVID-19. The outcomes of these trials will be crucial in establishing the efficacy and safety of these compounds in this context. Despite their potential therapeutic benefits, polyphenols also have limitations, including low bioavailability, potential side effects or toxicity at high doses, and interactions with certain medications. Additional research is needed to address these limitations and optimize the therapeutic potential of polyphenols in COVID-19 management. In conclusion, while polyphenols may offer some potential benefits in COVID-19, more well-designed clinical trials are needed to confirm their efficacy and safety. Until more conclusive evidence is available, following established public health guidelines for COVID-19 prevention and treatment is essential. Incorporating polyphenol-rich foods into a balanced diet can support overall health and immune function, but they should not be considered a primary strategy for COVID-19 prevention or treatment.

References

1 Adem, Ş., Eyupoglu, V., Ibrahim, I.M. et al. (2022). Multidimensional in silico strategy for identification of natural polyphenols-based SARS-CoV-2 main protease (Mpro) inhibitors to unveil a hope against COVID-19. *Computers in Biology and Medicine* 145: 105452.

2 Adem, S., Eyupoglu, V., Sarfraz, I., Rasul, A., and Ali, M. (2020). Identification of potent COVID-19 main protease (Mpro) inhibitors from natural polyphenols: an in silico strategy unveils a hope against CORONA.

3 Alzahrani, Q.E., Gillis, R., Harding, S.E. et al. (2022). Potential of the triad of fatty acids, polyphenols, and prebiotics from Cucurbita against COVID-19 in diabetic patients: a review. *Journal of Reports in Pharmaceutical Sciences* 11 (1): 28–40.

4 Bernini, R. and Velotti, F. (2021). Natural polyphenols as immunomodulators to rescue immune response homeostasis: quercetin as a research model against severe COVID-19. *Molecules* 26 (19): 5803.

5 Bettuzzi, S., Gabba, L., and Cataldo, S. (2021). Efficacy of a polyphenolic, standardized green tea extract for the treatment of COVID-19 syndrome: a proof-of-principle study. *Covid* 1 (1): 2–12.

6 de Oliveira, G.V., Volino-Souza, M., dos Santos, K.S. et al. (2021). A perspective on the use of polyphenols nano-formulation as a nutritional strategy to manage

the symptoms of the infected patient with COVID-19. *Research, Society and Development* 10 (13): e400101321471-e.
7 Di Pierro, F., Derosa, G., Maffioli, P. et al. (2021). Possible therapeutic effects of adjuvant quercetin supplementation against early-stage COVID-19 infection: a prospective, randomized, controlled, and open-label study. *International Journal of General Medicine* 2359–2366.
8 El-Missiry, M.A., Fekri, A., Kesar, L.A., and Othman, A.I. (2021). Polyphenols are potential nutritional adjuvants for targeting COVID-19. *Phytotherapy Research* 35 (6): 2879–2889.
9 Gasmi, A., Mujawdiya, P.K., Lysiuk, R. et al. (2022). Quercetin in the prevention and treatment of coronavirus infections: a focus on SARS-CoV-2. *Pharmaceuticals* 15 (9).
10 Ghosh, R., Chakraborty, A., Biswas, A., and Chowdhuri, S. (2021). Evaluation of green tea polyphenols as novel corona virus (SARS CoV-2) main protease (Mpro) inhibitors–an in silico docking and molecular dynamics simulation study. *Journal of Biomolecular Structure and Dynamics* 39 (12): 4362–4374.
11 Giovinazzo, G., Gerardi, C., Uberti-Foppa, C., and Lopalco, L. (2020). Can natural polyphenols help in reducing cytokine storm in COVID-19 patients? *Molecules* 25 (24): 5888.
12 Jankovics, I., Borsos, M., Mirani, S., and Dénes, B. (2021). Early use of polyphenol-rich Cistus creticus extract containing nasopharyngeal spray is associated with significantly shorter duration of symptoms in mild COVID-19 patients: A retrospective case–control study. *Journal of Community Medicine and Public Health Reports* 2 (9).
13 Levy, E., Delvin, E., Marcil, V., and Spahis, S. (2020). Can phytotherapy with polyphenols serve as a powerful approach for the prevention and therapy tool of novel coronavirus disease 2019 (COVID-19)? *American Journal of Physiology. Endocrinology and Metabolism* 319 (4): E689–E708.
14 Limanaqi, F., Busceti, C.L., Biagioni, F. et al. (2020). Cell clearing systems as targets of polyphenols in viral infections: potential implications for COVID-19 pathogenesis. *Antioxidants* 9 (11): 1105.
15 Lin, S.-C., Ho, C.-T., Chuo, W.-H. et al. (2017). Effective inhibition of MERS-CoV infection by resveratrol. *BMC Infectious Diseases* 17 (1): 144.
16 Mehany, T., Khalifa, I., Barakat, H. et al. (2021). Polyphenols as promising biologically active substances for preventing SARS-CoV-2 : A review with research evidence and underlying mechanisms. *Food Bioscience* 40: 100891.
17 Menegazzi, M., Campagnari, R., Bertoldi, M. et al. (2020). Protective effect of epigallocatechin-3-gallate (EGCG) in diseases with uncontrolled immune activation: could such a scenario be helpful to counteract COVID-19? *International Journal of Molecular Sciences* 21 (14).

18 Mhatre, S., Srivastava, T., Naik, S., and Patravale, V. (2021). Antiviral activity of green tea and black tea polyphenols in prophylaxis and treatment of COVID-19: A review. *Phytomedicine* 85: 153286.

19 Milton-Laskibar, I., Trepiana, J., Macarulla, M.T. et al. (2022). Potential usefulness of Mediterranean diet polyphenols against COVID-19-induced inflammation: a review of the current knowledge. *Journal of Physiology and Biochemistry* 79 (2): 1–12.

20 Mishra, C.B., Pandey, P., Sharma, R.D. et al. (2021). Identifying the natural polyphenol catechin as a multi-targeted agent against SARS-CoV-2 for the plausible therapy of COVID-19: an integrated computational approach. *Briefings in Bioinformatics* 22 (2): 1346–1360.

21 Mulu, A., Gajaa, M., and Woldekidan, H.B. (2021). The impact of curcumin derived polyphenols on the structure and flexibility COVID-19 main protease binding pocket: a molecular dynamics simulation study. *PeerJ* 9: e11590.

22 Musazadeh, V., Karimi, A., Jafarzadeh, J. et al. (2022). The favorable impacts of silibinin polyphenols as adjunctive therapy in reducing the complications of COVID-19: a review of research evidence and underlying mechanisms. *Biomedicine & Pharmacotherapy* 154: 113593.

23 Noor, N., Gani, A., Gani, A. et al. (2021). Exploitation of polyphenols and proteins using nanoencapsulation for anti-viral and brain boosting properties–Evoking a synergistic strategy to combat COVID-19 pandemic. *International Journal of Biological Macromolecules* 180: 375–384.

24 Paraiso, I.L., Revel, J.S., and Stevens, J.F. (2020). Potential use of polyphenols in the battle against COVID-19. *Current Opinion in Food Science* 32: 149–155.

25 Rakshit, M., Muduli, S., Srivastav, P.P., and Mishra, S. (2022). Pomegranate peel polyphenols prophylaxis against SARS-CoV-2 main protease by in-silico docking and molecular dynamics study. *Journal of Biomolecular Structure and Dynamics* 40 (23): 12917–12931.

26 Santos, J.C., Ribeiro, M.L., and Gambero, A. (2021). The impact of polyphenols-based diet on the inflammatory profile in COVID-19 elderly and obese patients. *Frontiers in Physiology* 11: 612268.

27 Singh, S., Sk, M.F., Sonawane, A. et al. (2021). Plant-derived natural polyphenols as potential antiviral drugs against SARS-CoV-2 via RNA-dependent RNA polymerase (RdRp) inhibition: an in-silico analysis. *Journal of Biomolecular Structure and Dynamics* 39 (16): 6249–6264.

28 Tallei, T.E., Niode, N.J., Idroes, R. et al. (2021). A comprehensive review of the potential use of green tea polyphenols in the management of COVID-19. *Evidence-based Complementary and Alternative Medicine* 2021.

29 Wu, Y., Pegan, S.D., Crich, D. et al. (2021). Polyphenols as alternative treatments of COVID-19. *Computational and Structural Biotechnology Journal* 19: 5371–5380.

30 Xu, L., Ho, C.-T., Liu, Y. et al. (2022). Potential application of tea polyphenols to the prevention of COVID-19 infection: Based on the gut-lung axis. *Frontiers in Nutrition* 9: 899842.

31 Yang, C.-Y., Chen, Y.-H., Liu, P.-J. et al. (2022). The emerging role of miRNAs in the pathogenesis of COVID-19: protective effects of nutraceutical polyphenolic compounds against SARS-CoV-2 infection. *International Journal of Medical Sciences* 19 (8): 1340.

32 Zhang, Z., Zhang, X., Bi, K. et al. (2021). Potential protective mechanisms of green tea polyphenol EGCG against COVID-19. *Trends in Food Science and Technology* 114: 11–24.

33 Zhuang, M., Jiang, H., Suzuki, Y. et al. (2009). Procyanidins and butanol extract of Cinnamomi Cortex inhibit SARS-CoV infection. *Antiviral Research* 82 (1): 73–81.

14

Polyphenols for Dietary Polyphenols Application

Elyor Berdimurodov[1,2,3,], Khasan Berdimuradov[4], Ilyos Eliboev[3], Azimjon Choriev[5], Anvar Khamidov[3], Mukhriddin Yusufov[3], and Chinmurot Yodgorov[3]*

[1] *Chemical & Materials Engineering, New Uzbekistan University, Tashkent, Uzbekistan*
[2] *Medical School, Central Asian University, Tashkent, Uzbekistan*
[3] *Faculty of Chemistry, National University of Uzbekistan, Tashkent, Uzbekistan*
[4] *Faculty of Industrial Viticulture and Food Production Technology, Shahrisabz Branch of Tashkent Institute of Chemical Technology, Shahrisabz, Uzbekistan*
[5] *Faculty of Chemistry-Biology, Karshi State University, Karshi, Uzbekistan*

14.1 Introduction

Products containing anthracene from polyphenols and drugs derived from them are utilized as excipients in medicine. They affect the large intestine and increase its peristalsis. The impact of anthraglycosides is felt approximately 8 hours after taking the medication. Aglycones have a weaker effect than their glycosides. Alizarin and its products have been discovered to have a diuretic impact and to dissolve kidney, urinary tract, and bladder stones. Consequently, drugs containing these compounds are employed to dissolve body stones composed of oxalate, phosphate, and ureate compounds and to expel urine. Among polyphenols, flavonoids primarily exhibit the effects of vitamin P, reducing the permeability and fragility of blood vessels [1–3]. The sum of the flavonoids in some plants also possesses laxative and diuretic properties. Pure flavonoids and their combined preparations, as well as medicinal products made from plants and products containing flavonoids, are used to treat diseases caused by vitamin P deficiency and vascular permeability disorders. They are also used for treating blood pressure reduction, sedative, cardiac cardiotonic, and certain

* Corresponding author

Science and Engineering of Polyphenols: Fundamentals and Industrial Scale Applications, First Edition. Edited by Chandrabhan Verma.
© 2024 John Wiley & Sons Inc. Published 2024 by John Wiley & Sons Inc.

cancer conditions, acting as a cure for diseases, as well as a laxative and diuretic. Polyphenols, coumarins, furocoumarins, and preparations obtained from plants containing compounds of this group exhibit anticoagulant, spasmolytic, cardiovascular dilation, anti-malignant, and other effects. Therefore, these drugs treat thrombosis, spasms, and certain inoperable cancer types. Peicedanin was isolated from the plant for cancer treatment [4–6].

Over the past 10 years, heightened curiosity has arisen regarding the potential health advantages of plant polyphenols from dietary sources. A continually growing number of research have outlined their antioxidative characteristics and connected this to their suggested function in preventing human illnesses. Plant polyphenols are prevalent in the human diet, particularly in fruits, vegetables, and legumes, and have been consistently linked to a reduced risk of cancer, cardiovascular disease, and other chronic illnesses. Polyphenols are a category of phytochemicals that likely serve as active ingredients in fruits, vegetables, and legumes. However, quantifying the benefits provided by polyphenols has proven to be highly challenging due to several factors, such as: (i) insufficient knowledge about the absorption levels and metabolic fate of individual polyphenols in specific foods; (ii) difficulties in characterizing and measuring regular food consumption; (iii) limited information on the polyphenol content of specific foods in widely-used food composition databases; and (iv) an imbalanced understanding of the metabolic fate of individual polyphenols [7, 8].

Polyphenols provide healing advantages due to their antioxidative characteristics and regulatory function in cellular communication. Compounds like curcumin exhibit neuroprotective properties in experimental frameworks of central nervous system (CNS) disorders. Nutritional enhancement with curcumin grants neuroprotection in the genetically modified mouse representation of Alzheimer's disease (AD). Curcumin can neutralize reactive elements, inhibit protein clustering, and stimulate the growth of new neurons. Curcumin has demonstrated neuroprotective capabilities in Parkinson's disease (PD). Analogous to curcumin, resveratrol and tea polyphenols also safeguard against neurodegenerative processes in living organisms [9–11].

Our food contains various hydroxycinnamic acids, anthocyanins, phloridzin, chlorogenic acid, flavanols, and quercetin glycosides, among other complex combinations of polyphenols. Similarly, wine, like other beverages, comprises stilbenes, flavanols, anthocyanins, proanthocyanidins, hydroxycinnamates, and phenolic acids. While some polyphenols, like quercetin, are present in many plant sources, including tea, vegetables, fruits, grains, and legumes. Bacterial and microfloral enzymes may subject these polyphenols to demethylation, dihydroxylation, and deglycosylation [12–14].

Natural polyphenolic compounds rank among the most potent bioactive substances in the nutraceutical realm. These compounds, found in vegetables, fruits, and other naturally occurring items, are derived from plants and feature numerous hydroxyl groups on aromatic rings. There are five distinct types of polyphenols: tannins, phenolic acids, flavonoids, and stilbenes (Figure 14.1). Among these, tea polyphenols (catechins), resveratrol (stilbenes), and curcumin (diferuloylmethanes) have been the focus of extensive research [15].

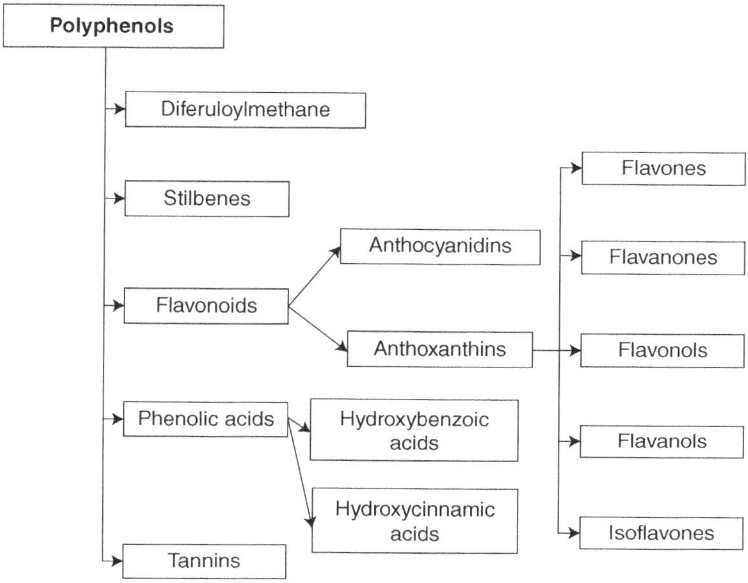

Figure 14.1 Polyphenols: categorization and role. Source: Pandareesh et al. [15]/ Elsevier.

14.2 Main Part

14.2.1 Dietary Polyphenols in the Therapy of the Central Nervous System (CNS)

Natural compounds rich in dietary polyphenols may benefit the CNS, affecting memory, cognition, and other brain processes in healthy people and those with neuropsychiatric diseases. After receiving Concord grape juice (which contains polyphenols like anthocyanins and flavanols) for 16 weeks, participants with mild cognitive impairment had their memory and brain activity analyzed. The study found that people who drank grape juice had better cognitive performance and more active brains than those who drank a placebo. To ascertain long-term effects, researchers examined the connection between midlife polyphenol consumption across all classes and cognitive ability assessed 13 years later. The consumption of hydroxybenzoic acids, flavonols, theaflavins, and catechins was favorably related to language and verbal memory, particularly episodic memory, according to the cognitive performance assessment of 2574 middle-aged people. According to this material, taking high amounts of particular polyphenols, such as flavonoids and phenolic acids, may help maintain verbal memory in the long run. Spatial memory appeared to benefit from the injection of Provinols™, polyphenolic substances derived from red wine, in a comparable clinical experiment looking at the effects of Provinols on advanced human brain processes [16, 17].

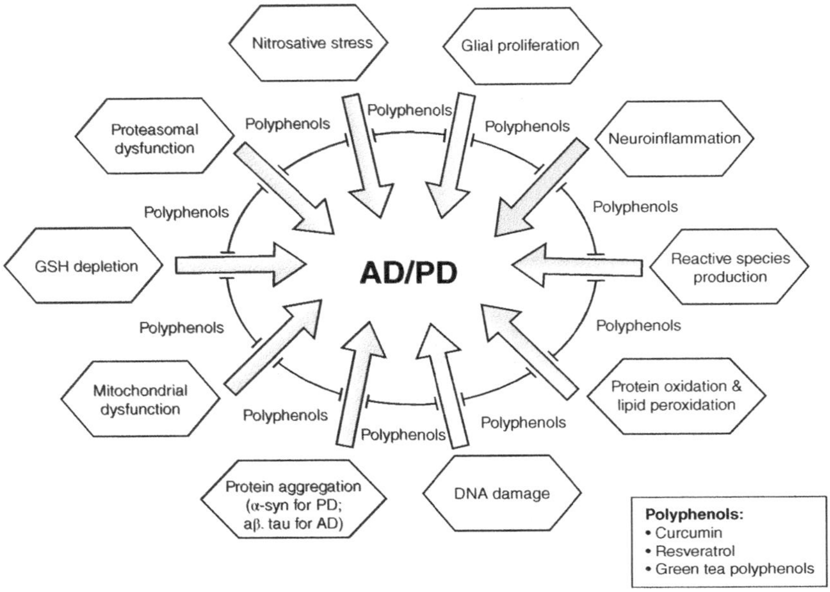

Figure 14.2 Shared degenerative processes contribute to developing Parkinson's disease (PD) and Alzheimer's disease (AD), with polyphenols providing neuroprotection. Examples of polyphenols include curcumin, resveratrol, and tea polyphenols, with the majority possessing the potential to target all pathways. Source: Pandareesh et al. [15]/ Elsevier.

In experimental models for CNS disorders, particularly neurodegenerative conditions such as PD and AD, curcumin, resveratrol, and tea-derived polyphenols are among the most extensively studied (Figure 14.2). Curcumin is permeable to the blood–brain barrier (BBB), offering the potential for treating PD when administered at higher dosages in individuals. In the 6-hydroxydopamine (6-OHDA) PD model, curcumin elevates striatal dopamine levels while protecting dopaminergic neurons. Additionally, curcumin reduces dopamine neurotoxicity and increases the number of dopaminergic neurons in the substantia nigra (SN), suggesting a potential role in neurogenesis [18, 19].

Curcumin offers neuroprotection via anti-apoptotic pathways and preservation of mitochondria. It stops c-jun phosphorylation, hinders JNK activation, and turns off caspase-3. Curcumin inhibits pro-apoptotic proteins, neutralizes cytochrome c discharge, and activates genes that prevent apoptosis. It exhibits antioxidant properties and reduces inducible nitric oxide synthase. Gamma-glutamyl cysteine ligase, glutathione synthesis enzyme expression, and indirectly activating the Nuclear Factor (erythroid-derived 2)-like 2/antioxidant response element pathway, which results in phase II antioxidant enzyme induction, are all ways that curcumin prevents lipid peroxidation and improves antioxidant function. In contrast, curcumin blocks α-synuclein aggregation in vitro. It improves α-synuclein solubility and disassembles preformed α-synuclein fibrils, making them harmless. Furthermore, curcumin stimulates iron regulatory proteins and reduces ferritin and hepcidin levels, leading to iron chelation [20, 21].

Resveratrol shares the capability with curcumin to target various PD pathways. The 1-methyl-4-phenyl-1,2,3,6-tetrahydropyridine model protects against motor dysfunction, oxidative damage, and neuronal death and safeguards SN dopaminergic neurons and striatal dopamine in vivo. Resveratrol protects against oxidative stress and 1-methyl-4-phenylpyridinium toxicity, which causes dopaminergic neurodegeneration and neuroinflammation. It also reduces neurotoxicity and inflammation caused by 6-OHDA in SN neurons. Furthermore, resveratrol defends against neurotoxicity induced by lipopolysaccharide. Resveratrol provides neuroprotection in two distinct models of dopamine neurotoxicity: (i) the α-synuclein model of PD, which enhances locomotor activity and extends lifespan and (ii) a cellular model of PD [22, 23].

These results have been supported by research using various polyphenols. For instance, resveratrol can control the generation of nitric oxide, which encourages vasodilation, improves blood flow, and impacts brain function. Resveratrol administration caused dose-dependent increases in cerebral blood flow (CBF) during task performance in a clinical trial assessing its effect on CBF and cognitive performance. However, cognitive function remained primarily unaffected. An analogous epigallocatechin-3-O-gallate (EGCG) study in healthy adult humans showed that a single dosage of EGCG in healthy people could change CBF parameters without impacting mood or cognitive performance. The diminished effect on human brain function could be due to insufficient bioavailability. In contrast to placebo and resveratrol taken alone, co-supplementing with the pepper alkaloid piperine improved the effects of resveratrol on CBF during the completion of memory tests. Considering its impact on CBF, resveratrol's bioefficacy is correlated with its bioavailability, and this may also hold for other polyphenols, according to this information [24–26].

14.2.2 Bioavailability, Excretion, Metabolic Transformation, and Absorption of Polyphenols

Individual differences in dietary polyphenol absorption are significant, and it is likely quite constrained as a result of glycosylation patterns and levels of polymerization. For instance, before being absorbed as aglycones, dietary flavonoids, among the least absorbable glycosides, must first be deglycosylated by mammalian glucosidases in the small intestine. Due to phase I and II metabolism of oxidized and conjugated flavonoids, aglycone levels in the plasma are typically small and fluctuate widely. This considerable change results in forms different from those in the diet that reach the plasma and tissues. The main goal of a bioavailability study is to pinpoint the active metabolites created during dietary polyphenol's absorption. Moreover, it is essential to determine if metabolism affects active metabolites' biological functions and establish their presence. For example, enterolactone, enterodiol, and equol are phytoestrogen metabolites generated by intestinal bacteria. The effects of enterolactone, enterodiol, and equol on estrogens can be either agonistic or antagonistic, but equol enhances the original isoflavone's phytoestrogenic capabilities. The cause of this variation is the heterogeneity in the intestinal microbiota among individuals [27, 28].

Direct interactions with other food ingredients such as alcohol, fat, fiber, carbs, and proteins influence dietary polyphenol'absorption ability. Flavonoids are better absorbed when there is dietary fat present. The capacity of polyphenol metabolic products to enter cells is hampered by the binding of plasma proteins (like albumin). Flavonoids and their metabolites interact at the membrane or are incorporated into the cytoplasm, determining how they affect cells. The binding of quercetin derivatives to serum albumin decreases the body's ability to excrete it. A similar strong affinity for blood proteins is shown by EGCG as well. By interacting with albumin and other blood proteins, EGCG may lengthen the half-life in the bloodstream [29, 30].

The in vivo absorption capacity of green tea polyphenols depends on the method of delivery. In humans, oral catechins are absorbed, metabolized, and primarily eliminated within 24 hours. Catechins were found as free, sulfated, and glucuronidated molecules in rat plasma and bile. However, after administering tea polyphenols, only 0.1% of EGCG was available [31].

On the other hand, resveratrol has unfavorable pharmacokinetic characteristics since it is light-sensitive and easily oxidized. After being metabolized in the liver and gut, resveratrol and its derivatives have a greater than 1% absorption capacity. However, repeatedly increasing or giving the dosage has no discernible impact on bioavailability. Within 15 minutes of entering the bloodstream, resveratrol transforms into glucuronides and sulfates, and its metabolites circulate in the serum for up to 9 hours. With the maximum plasma concentrations of resveratrol and its metabolites at 491 90 ng/mL (about 2 M) and a plasma half-life of 9.2 0.6 hours, resveratrol was at least 70% absorbed in humans after oral ingestion (of 25 mg). Although the majority of the oral dosage was found in the urine, only trace levels of unmodified resveratrol (5 < ng/mL) were found in the plasma [32, 33].

After consuming either pure substances or a meal containing a known amount of polyphenols, the levels of their metabolites in plasma and urine can be examined to determine the ability of polyphenolic compounds to be absorbed. It can also be inferred from the rise in plasma's antioxidant capacity after eating foods high in polyphenols. However, studies on polyphenol bioavailability using a single-dose design—an in vivo technique where the blood concentration of the examined polyphenol briefly increases—mainly reflect the body's capacity to absorb the polyphenol from dietary sources. However, due to the cumulative effects of the polyphenol combination present in the diet, a complete assessment of polyphenols' absorption and biological activity cannot be done with a single dose. Furthermore, only after repeated treatments do, polyphenols and flavonoids show continued accumulation in the blood and tissues. For instance, in humans, the peak level of plasma flavonoids occurs between 1 and 6 hours after consumption, with a half-life ($t1/2$) of 1 and 28 hours for their elimination. Similarly, those who eat a low-flavonoid diet have very low plasma quercetin levels, but people who are fasting frequently have levels between 50 and 80 nmol/L. However, supplementing with 500 mg of rutin daily for 6 weeks raised plasma quercetin levels (165 nmol/L). Therefore, it is hypothesized that dietary polyphenols are absorbed when plasma and cell concentrations of polyphenols considerably increase after swallowing them in tiny amounts (Figure 14.3) [34].

The effectiveness of different dietary polyphenol'absorption and consumption varies. For instance, curcumin was found in mice's plasma 15 minutes after eating, peaked at 1 hour, and declined after 6 hours. During this procedure, curcumin first changed into dihydrocurcumin, then into tetrahydrocurcumin, and lastly into conjugates of monoglucuronides. In a different investigation, curcumin and its metabolites were found in the liver and blood in very small concentrations after oral dosing of healthy subjects [35].

14.2.3 Dietary Polyphenols in Cancer Treatments

The second-most lethal and third-most frequently diagnosed cancer in the world is colorectal cancer (CRC). The majority of CRC cases are mostly impacted by dietary factors, even though it has a variety of causes. Numerous studies in recent years have linked a preventive role for polyphenols and foods high in these chemicals (such as fruits and vegetables) in the battle against CRC. Polyphenols have been shown to operate as chemopreventive agents by obstructing cancer development, spread, and evolution. The fundamental chemopreventive properties of different polyphenols, including xanthohumol, resveratrol, catechin, epicatechin, epigallocatechin-3-gallate, chrysin, myricetin, rutin, quercetin, and against CRC, as seen in cellular culture models, are outlined in this review. According to the results discussed in this article, these compounds exhibit anti-inflammatory and antioxidant properties, inhibit proliferation, angiogenesis, and metastasis, and slow cellular growth by inducing cell cycle arrest and apoptosis. As a result, these impacts cover several molecular and biochemical action pathways yet to be fully understood. As a result, it is essential to use caution when attempting to extrapolate the results of CRC cell line research to human situations [36–38].

It is currently unknown which molecular pathways, particularly at physiological levels, underlie the cancer-preventing effects of polyphenols, despite the data shown by the cellular studies emphasized in this article. To get around this problem, research should concentrate on (i) more human clinical trials to confirm and measure the bioavailability, safety, potency, and anti-cancer properties of polyphenols and (ii) examination of polyphenol bioavailability in animal subjects to emphasize the anti-cancer properties of their gut metabolites (in particular) [36, 39].

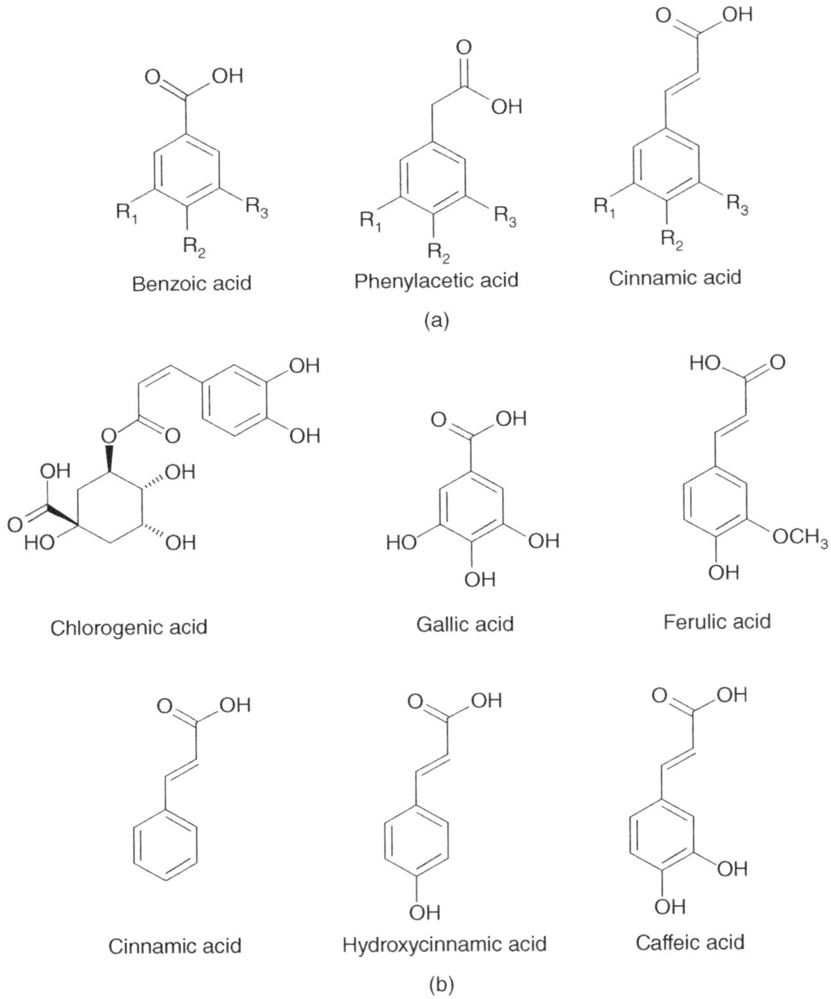

Figure 14.3 Molecular structure of phenolic acids. Source: Rzepecka-Stojko et al. [34]/ MDPI/CC BY 4.0.

Since they restore average cell growth by controlling proliferation, programmed cell death, blood vessel development, metastasis, and inflammation, as well as focusing on various molecular and biochemical pathways implicated in tumor progression, polyphenols have the potential to be used as cancer-prevention agents for CRC therapy. Nevertheless, as these experimental components have yet to be confirmed in humans, care must be taken to translate findings from CRC cell line studies to human subjects. Deriving conclusions from cell culture research findings requires taking into account a number several including the specific CRC cell line used, the polyphenol concentrations examined (which are typically much higher than those achievable in the human body), the stability and oxidation of polyphenols, being redox-sensitive substances, and the duration of cell exposure to the polyphenol, among others [36, 40].

Numerous studies have shown how important polyphenols with high antioxidant content are in preventing the spread and development of experimental pancreatic cancer models. Although information on the safety and benefits of these polyphenols during cancer treatment is mainly missing, preliminary cell lines and animal studies suggest potential benefits. Notably, the ability of these polyphenols to enhance the efficiency of conventional chemotherapeutic medications and other natural agents against experimental pancreatic cancer makes it conceivable to test them in clinical settings. Further preclinical and clinical studies are required to confirm the use of such polyphenols as adjunctive agents, to ascertain their ideal dosages and timetables when integrated with cancer treatment regimens, and to ascertain their excellent timings because a sizable number of cancer patients undergoing active therapies utilize antioxidant-dense substances from various natural sources, not all of which may be advantageous. Additional preclinical research describing these polyphenols' pharmacokinetic and pharmacodynamic properties is needed to assess the bioavailability of related compounds and metabolites. So that their anti-tumor efficiency may be examined both alone and in combination with medicinal medications, these evaluations would help to improve the dose recommendations for these natural polyphenols. A bioinformatics study is also required to examine the targets of these polyphenols' pharmacological inhibitors and corroborate the mechanisms in cellular and preclinical experiments (Figure 14.4) [41].

A practical way of living, which integrates tolerable stress levels, well-rounded nourishment, controlled food consumption, and regular exercise, is vital in thwarting cancer. A tactic comparable to the proactive chemoprevention of heart-related disorders might apply to cancer prevention. Measurable risk elements, such as elevated cholesterol and increased blood pressure, have been identified in cardiovascular conditions. Notably, notable advancements have been achieved in alleviating them through medications that reduce cholesterol levels and decrease blood pressure (Figure 14.5).

As a result, the obstacle involves identifying comparable quantifiable risk aspects for breast cancer and eradicating or reducing them. Adopting Asian and Mediterranean eating patterns and reducing or eliminating exposure to risk factors like dietary alcohol or lipids are practical examples that do not entail considerable costs. Regular consumption of fruits and vegetables high in polyphenols may also be a substitute for cancer prevention. Since cancer is generally considered a chronic illness, prolonged administration of cancer-preventing agents is crucial [42, 43].

Nutritional polyphenols boast an extensive and widespread history of utilization worldwide. Although informal epidemiological data have conventionally

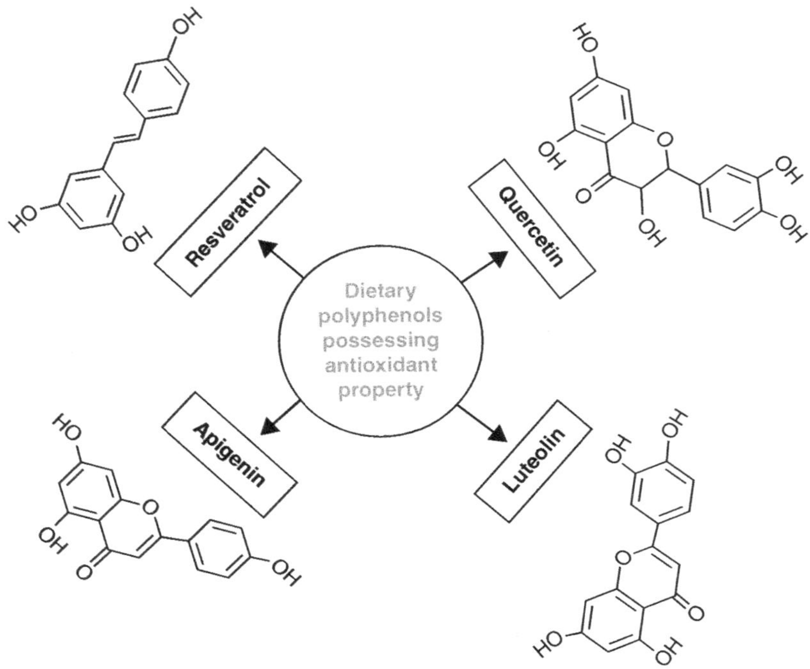

Figure 14.4 Used dietary polyphenols in antioxidant applications. Source: Thyagarajan et al. [41]/MDPI/CC BY 4.0.

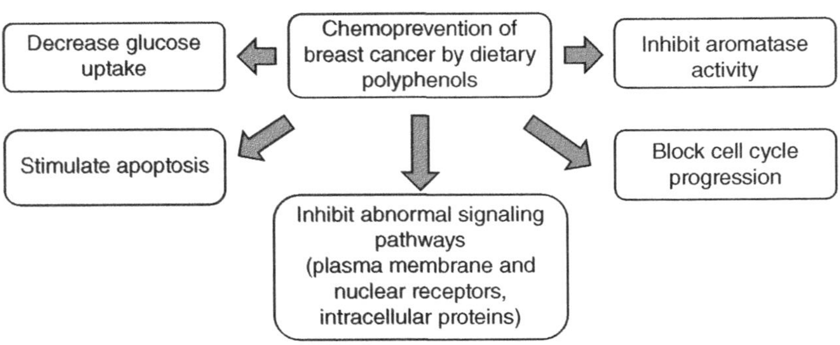

Figure 14.5 Dietary polyphenols in chemoprevention of breast cancer. Source: Mocanu et al. [42]/MDPI/CC BY 4.0.

supported the concept of varied diets correlating with enhanced well-being, modern experimental results from diverse preclinical and clinical studies firmly endorse the hypothesis that nutritional polyphenols could possess potentially beneficial outcomes on a multitude of health concerns, including cancer. This review article presents a novel perspective on the possible chemoprevention offered by dietary practices and nutritional agents. The all-encompassing information gathered here suggests that the positive effects of various nutritional polyphenols may be partially ascribed to their epigenetic features, such as modifications in DNA methylation patterns, management of histone changes, and shifts in the expression of specific miRNAs. While the health advantages of nutritional polyphenols in humans are generally considered promising, there are unique obstacles and restrictions in the available data for a more profound understanding of the molecular mechanisms responsible for these effects and potential interactions among different polyphenols and other nutritional elements. In vitro models have significantly aided in comprehending the polyphenols' mediated control of the epigenetic framework; however, more in vivo data are needed for most of these nutritional compounds. Therefore, until sufficient preclinical and clinical data are gathered regarding the epigenetic modifications prompted by particular healthy polyphenols, prudence must be exercised when interpreting and extrapolating the significance of existing in vitro findings. Once such evidence is substantiated, the next crucial phase would involve identifying the optimal dosages of these "nutritional nutraceuticals" to attain various positive outcomes in human subjects. Further clinical exploration is necessary to evaluate the safety profile of different amounts of nutritional polyphenols, and more foundational scientific inquiries are needed to expand our knowledge of the molecular mechanisms that underlie the chemopreventive properties of various nutritional polyphenols. It is genuinely exciting to witness that we have finally begun to examine the molecular foundations supporting the "merits" of specific diets and diet-related factors that have persisted for centuries. The fact that hundreds of nutritional polyphenols are being analyzed from an "epigenomic"standpoint showcases our eagerness and faith in secure and natural agents for cancer chemoprevention. Undoubtedly, the current evidence is limited, and a lengthy and treacherous path lies before us; however, given the promise and potential of these polyphenols, it is logical to contemplate the prospect that some of these substances may become indispensable for cancer chemoprevention in the future [44–46].

14.2.4 Dietary Polyphenols in Inflammation Application

A considerable collection of research indicates that tumor development is associated with inflammation. Nuclear factor-kappa B (NF-κ B), a key regulator of infection and inflammation, has been identified as an essential facilitator in the progression where inflammation can transform into cancer. Nutritional polyphenols have exhibited anti-inflammatory and anti-cancer characteristics, partially owing to their capacity to suppress NF-κ B activation. Dietary polyphenols exhibit potential in combating and preventing cancer, partly due to their anti-inflammatory characteristics. Inflammation has been associated with cancer development and progression, and polyphenols have demonstrated their ability to suppress the activation of critical inflammatory signaling pathways like NF-κ B. Although more research is necessary to comprehensively decipher the molecular mechanisms responsible for the anti-cancer actions of dietary polyphenols, their natural and safe nature positions them as a promising direction for future cancer prevention approaches [47–49].

Focusing on numerous inflammation-related cellular communication systems, specifically MAPKs, NF-κB, PI3K/Akt, and Wnt/β-catenin, utilizing dietary polyphenols has gained recognition as an innovative strategy for preventing and addressing chronic inflammatory illnesses. Organic polyphenols that can hinder molecular and cellular processes associated with inflammation show potential for their use in preventing and managing ongoing inflammatory issues. We have investigated the effects of several commonly occurring polyphenols (gallic acid, resveratrol, isoliquiritigenin, genistein, curcumin, quercetin, kaempferol, and apigenin) on NF-B, MAPK, PI3K/Akt, and Wnt cell signaling systems, which are linked to chronic inflammatory diseases. Polyphenols' control over critical signaling elements in various pathways may successfully prevent the start and spread of inflammatory conditions. These pathways interact to exchange signaling components, which generally change the composition of cytokines and other inflammatory mediators. Integrating signals from many sources within a responsive inflammatory network will influence a collective inflammation-related response. To design future functional foods aimed at preventing and treating inflammatory disorders, it will be essential to understand the mechanisms involved in altering these signaling networks through dietary polyphenols [50, 51].

Inflammatory bowel disorder (IBD) is a mysterious, ongoing, and recurrent gastrointestinal tract inflammation resulting from the dysregulation of the mucosal immune system. The most prevalent antioxidants in the human diet are polyphenols, secondary plant components in all fruits and vegetables. Dietary polyphenols have been shown to have positive impacts on health. This study uses cellular activities and pharmacological considerations to assess the potential of commonly used polyphenols, particularly in IBD. These include bilberry anthocyanin, curcumin, EGCG, and green tea polyphenols; naringenin; olive oil polyphenols; pomegranate polyphenols; ellagic acid; quercetin; and resveratrol. Dietary polyphenols have been proven to help control IBD by lowering inflammatory cytokines and enzymes, enhancing antioxidant defenses, and decreasing inflammatory pathways and the cellular signaling processes linked to them. Additional preclinical and clinical research is needed to thoroughly understand dietary polyphenols' safety, bioavailability, and bioefficacy in IBD patients [52, 53].

14.3 Conclusion and Outlook

Polyphenols provide healing advantages due to their antioxidative characteristics and regulatory function in cellular communication. Compounds like curcumin exhibit neuroprotective properties in experimental frameworks of CNS disorders. Nutritional enhancement with curcumin grants neuroprotection in the genetically modified mouse representation of AD. Curcumin can neutralize reactive elements, inhibit protein clustering, and stimulate the growth of new neurons. Curcumin has demonstrated neuroprotective capabilities in PD. Analogous to curcumin, resveratrol, and tea polyphenols also safeguard against neurodegenerative processes in living organisms. For their potential to treat CNS disorders utilizing experimental models, particularly neurodegenerative illnesses like PD and AD, curcumin, resveratrol, and tea polyphenols are among those that have received the most attention. When used in high dosages, curcumin exhibits low toxicity in humans and can potentially treat PD. It can penetrate the BBB. In the 6-OHDA model of PD, curcumin raises striatal dopamine levels and safeguards dopaminergic neurons.

Curcumin also prevents dopamine neurotoxicity and increases the number of SN dopaminergic neurons, suggesting a potential function in neurogenesis. Because of glycosylation patterns and polymerization levels, dietary polyphenols' ability to be absorbed varies significantly between people and is probably relatively constrained. Globally, CRC is the second-deadliest and third-most common disease to be diagnosed. Although various contributing factors exist, dietary factors are responsible for most CRC cases. In recent years, polyphenols and foods high in them (such as fruits and vegetables) have been linked to a protective function in battling CRC. Polyphenols work as chemopreventive agents by preventing cancer initiation, progression, and development.

References

1 Zhu, H., Song, D., and Zhao, X. (2021). Potential applications and primary mechanism of action of dietary polyphenols against hyperuricemia: a review. *Food Bioscience* 43: 101297.

2 Rudrapal, M., Mishra, A.K., Rani, L. et al. (2022). Nanodelivery of dietary polyphenols for therapeutic applications. *Molecules* 27 (24): 8706.

3 Peron, G., Hidalgo-Liberona, N., González-Domínguez, R. et al. (2019). Exploring the molecular pathways behind the effects of nutrients and dietary polyphenols on gut microbiota and intestinal permeability: a perspective on the potential of metabolomics and future clinical applications. *Journal of Agricultural and Food Chemistry* 68 (7): 1780–1789.

4 Squillaro, T., Cimini, A., Peluso, G. et al. (2018). Nano-delivery systems for encapsulation of dietary polyphenols: an experimental approach for neurodegenerative diseases and brain tumors. *Biochemical Pharmacology* 154: 303–317.

5 Ishii, T., Ishikawa, M., Miyoshi, N. et al. (2009). Catechol type polyphenol is a potential modifier of protein sulfhydryls: development and application of a new probe for understanding the dietary polyphenol actions. *Chemical Research in Toxicology* 22 (10): 1689–1698.

6 Wan, M.L.Y., Co, V.A., and El-Nezami, H. (2021). Dietary polyphenol impact on gut health and microbiota. *Critical Reviews in Food Science and Nutrition* 61 (4): 690–711.

7 Spencer, J.P.E., Abd El Mohsen, M.M., Minihane, A.-M., and Mathers, J.C. (2008). Biomarkers of the intake of dietary polyphenols: strengths, limitations and application in nutrition research. *The British Journal of Nutrition* 99 (1): 12–22.

8 Patra, S., Pradhan, B., Nayak, R. et al. (2021). Dietary polyphenols in chemoprevention and synergistic effect in cancer: clinical evidences and molecular mechanisms of action. *Phytomedicine* 90: 153554.

9 Magrone, T. and Jirillo, E. (2011). Potential application of dietary polyphenols from red wine to attaining healthy ageing. *Current Topics in Medicinal Chemistry* 11 (14): 1780–1796.

10 Yang, C.S., Landau, J.M., Huang, M.-T., and Newmark, H.L. (2001). Inhibition of carcinogenesis by dietary polyphenolic compounds. *Annual Review of Nutrition* 21 (1): 381–406.

11 Rastogi, H. and Jana, S. (2016). Evaluation of physicochemical properties and intestinal permeability of six dietary polyphenols in human intestinal colon adenocarcinoma Caco-2 cells. *European Journal of Drug Metabolism and Pharmacokinetics* 41: 33–43.

12 Osawa, T. (1999). Protective role of dietary polyphenols in oxidative stress. *Mechanisms of Ageing and Development* 111 (2–3): 133–139.

13 Ohara, M. and Ohyama, Y. (2014). Delivery and application of dietary polyphenols to target organs, tissues and intracellular organelles. *Current Drug Metabolism* 15 (1): 37–47.

14 Bracci, L., Fabbri, A., Del Cornò, M., and Conti, L. (2021). Dietary polyphenols: promising adjuvants for colorectal cancer therapies. *Cancers* 13 (18): 4499.

15 Pandareesh, M.D., Mythri, R.B., and Srinivas Bharath, M.M. (2015). Bioavailability of dietary polyphenols: factors contributing to their clinical application in CNS diseases. *Neurochemistry International* 89: 198–208.

16 Magrone, T., Russo, M.A., and Jirillo, E. (2016). Role of immune cells in the course of central nervous system injury: modulation with natural products. *Current Pharmaceutical Design* 22 (6): 701–708.

17 Ashaari, Z., Hadjzadeh, M.-A.-R., Hassanzadeh, G. et al. (2018). The flavone luteolin improves central nervous system disorders by different mechanisms: a review. *Journal of Molecular Neuroscience* 65: 491–506.

18 Granda, H. and de Pascual-Teresa, S. (2018). Interaction of polyphenols with other food components as a means for their neurological health benefits. *Journal of Agricultural and Food Chemistry* 66 (31): 8224–8230.

19 Jaskiw, G.E., Xu, D., Obrenovich, M.E., and Donskey, C.J. (2022). Small phenolic and indolic gut-dependent molecules in the primate central nervous system: levels vs. bioactivity. *Metabolomics* 18(1): 8.

20 Arcusa, R., Carillo, J.Á., Cerdá, B. et al. (2023). Ability of a polyphenol-rich nutraceutical to reduce central nervous system lipid peroxidation by analysis of oxylipins in Urine: a randomized, double-blind, placebo-controlled clinical trial. *Antioxidants* 12 (3): 721.

21 Bouayed, J. (2010). Polyphenols: a potential new strategy for the prevention and treatment of anxiety and depression. *Current Nutrition & Food Science* 6 (1): 13–18.

22 Perrone, L., Sampaolo, S., and Melone, M.A.B. (2020). Bioactive phenolic compounds in the modulation of central and peripheral nervous system cancers: facts and misdeeds. *Cancers* 12 (2): 454.

23 Furlan, V. and Bren, U. (2021). Insight into inhibitory mechanism of PDE4D by dietary polyphenols using molecular dynamics simulations and free energy calculations. *Biomolecules* 11 (3): 479.

24 Gomez-Pinilla, F. and Gomez, A.G. (2011). The influence of dietary factors in central nervous system plasticity and injury recovery. *PM & R : The Journal of Injury, Function, and Rehabilitation* 3 (6): S111–S116.

25 Liu, S., Cheng, L., Liu, Y. et al. (2023). Relationship between dietary polyphenols and gut microbiota: new clues to improve cognitive disorders, mood disorders and circadian rhythms. *Food* 12 (6): 1309.

26 Panickar, K.S. (2013). Effects of dietary polyphenols on neuroregulatory factors and pathways that mediate food intake and energy regulation in obesity. *Molecular Nutrition & Food Research* 57 (1): 34–47.

27 Croft, K.D. (2016). Dietary polyphenols: antioxidants or not? *Archives of Biochemistry and Biophysics* 595: 120–124.

28 Luca, S.V., Macovei, I., Bujor, A. et al. (2020). Bioactivity of dietary polyphenols: the role of metabolites. *Critical Reviews in Food Science and Nutrition* 60 (4): 626–659.

29 Ahmad, N., Qamar, M., Yuan, Y. et al. (2022). Dietary polyphenols: extraction, identification, bioavailability, and role for prevention and treatment of colorectal and prostate cancers. *Molecules* 27 (9): 2831.

30 Tomás-Barberán, F.A., Selma, M.V., and Espín, J.C. (2016). Interactions of gut microbiota with dietary polyphenols and consequences to human health. *Current Opinion in Clinical Nutrition and Metabolic Care* 19 (6): 471–476.

31 Shivashankara, K.S. and Acharya, S.N. (2010). Bioavailability of dietary polyphenols and the cardiovascular diseases. *The Open Nutraceuticals Journal* 3 (1).

32 Santhakumar, A.B., Battino, M., and Alvarez-Suarez, J.M. (2018). Dietary polyphenols: structures, bioavailability and protective effects against atherosclerosis. *Food and Chemical Toxicology* 113: 49–65.

33 Marín, L., Miguélez, E.M., Villar, C.J., and Lombó, F. (2015). Bioavailability of dietary polyphenols and gut microbiota metabolism: antimicrobial properties. *BioMed Research International* 2015.

34 Rzepecka-Stojko, A., Stojko, J., Kurek-Górecka, A. et al. (2015). Polyphenols from bee pollen: structure, absorption, metabolism and biological activity. *Molecules* 20 (12): 21732–21749.

35 Kawabata, K., Yoshioka, Y., and Terao, J. (2019). Role of intestinal microbiota in the bioavailability and physiological functions of dietary polyphenols. *Molecules* 24 (2): 370.

36 Araújo, J.R., Gonçalves, P., and Martel, F. (2011). Chemopreventive effect of dietary polyphenols in colorectal cancer cell lines. *Nutrition Research* 31 (2): 77–87.

37 Gu, H.-F., Mao, X.-Y., and Du, M. (2020). Prevention of breast cancer by dietary polyphenols – role of cancer stem cells. *Critical Reviews in Food Science and Nutrition* 60 (5): 810–825.

38 Chatterjee, K., Mukherjee, S., Vanmanen, J. et al. (2019). Dietary polyphenols, resveratrol and pterostilbene exhibit antitumor activity on an HPV E6-positive cervical cancer model: an in vitro and in vivo analysis. *Frontiers in Oncology* 9: 352.

39 Sheikh, I., Sharma, V., Tuli, H.S. et al. (2020). Cancer chemoprevention by flavonoids, dietary polyphenols and terpenoids. *Biointerface Research in Applied Chemistry* 11 (1): 8502–8537.

40 Khan, H., Ullah, H., Castilho, P.C.M.F. et al. (2020). Targeting NF-κB signaling pathway in cancer by dietary polyphenols. *Critical Reviews in Food Science and Nutrition* 60 (16): 2790–2800.

41 Thyagarajan, A., Forino, A.S., Konger, R.L., and Sahu, R.P. (2020). Dietary polyphenols in cancer chemoprevention: implications in pancreatic cancer. *Antioxidants* 9 (8).

42 Mocanu, M.-M., Nagy, P., and Szöllősi, J. (2015, 2012). Chemoprevention of breast cancer by dietary polyphenols. *Molecules* 22578–22620.

43 Musial, C., Siedlecka-Kroplewska, K., Kmiec, Z., and Gorska-Ponikowska, M. (2021). Modulation of autophagy in cancer cells by dietary polyphenols. *Antioxidants* 10 (1): 123.

44 Link, A., Balaguer, F., and Goel, A. (2010). Cancer chemoprevention by dietary polyphenols: promising role for epigenetics. *Biochemical Pharmacology* 80 (12): 1771–1792.

45 De Silva, S.F. and Alcorn, J. (2019). Flaxseed lignans as important dietary polyphenols for cancer prevention and treatment: chemistry, pharmacokinetics, and molecular targets. *Pharmaceuticals* 12 (2): 68.

46 Lall, R.K., Syed, D.N., Adhami, V.M. et al. (2015). Dietary polyphenols in prevention and treatment of prostate cancer. *International Journal of Molecular Sciences* 16 (2): 3350–3376.

47 Guo, W., Kong, E., and Meydani, M. (2009). Dietary polyphenols, inflammation, and cancer. *Nutrition and Cancer* 61 (6): 807–810.

48 Saracila, M., Panaite, T.D., Papuc, C.P., and Criste, R.D. (2021). Heat stress in broiler chickens and the effect of dietary polyphenols, with special reference to willow (Salix spp.) bark supplements – a review. *Antioxidants* 10 (5): 686.

49 Pawlowska, E., Szczepanska, J., Koskela, A. et al. (2019). Dietary polyphenols in age-related macular degeneration: protection against oxidative stress and beyond. *Oxidative Medicine and Cellular Longevity* 2019.

50 Jantan, I., Haque, M.A., Arshad, L. et al. (2021). Dietary polyphenols suppress chronic inflammation by modulation of multiple inflammation-associated cell signaling pathways. *The Journal of Nutritional Biochemistry* 93: 108634.

51 Aravind, S.M., Wichienchot, S., Tsao, R. et al. (2021). Role of dietary polyphenols on gut microbiota, their metabolites and health benefits. *Food Research International* 142: 110189.

52 Farzaei, H. and M, Rahimi R, Abdollahi M. (2015). The role of dietary polyphenols in the management of inflammatory bowel disease. *Current Pharmaceutical Biotechnology* 16 (3): 196–210.

53 Shakoor, H., Feehan, J., Apostolopoulos, V. et al. (2021). Immunomodulatory effects of dietary polyphenols. *Nutrients* 13 (3): 728.

15

Polyphenols for Health Management and Disease Control Applications

Shveta Sharma[1,5], *Manu Sharma*[2], *Richika Ganjoo*[3], *and Ashish Kumar*[4]

[1] *Department of Chemistry, Government College Una, Affiliated to Himachal Pradesh University, Una, India*
[2] *Department of Orthopaedics, Maharishi Markandeshwar Medical College and Hospital, Solan, HP, India*
[3] *Department of Chemistry, School of Chemical Engineering and Physical Sciences, Lovely Professional University, Punjab, India*
[4] *Nalanda College of Engineering, Bihar Engineering University, Department of Science, Technology and Technical Education, Government of Bihar, India*
[5] *Department of Chemistry, School of Chemical Engineering and Physical Sciences, Lovely Professional University, Punjab, India*

15.1 Introduction

Polyphenols are chemical substances that exist naturally and are made up of several units of phenol. Polyphenols are the secondary intermediates of cellular metabolism that are generated by plants. These intermediates of cellular metabolism have played an important part in both industrial and medicinal uses [1, 2]. Stilbenes, phenolic acids, flavonoids, and lignans are the four categories into which thousands of phenolic compounds, which are plant secondary intermediates of cellular metabolism, have been categorized as a result of previous research [3]. A significant portion of them has been recognized in human nutrition; in fact, they are the plant-derived phytochemicals that are present in the highest abundance in our food supply [4]. While certain polyphenols (PP), like quercetin, may be found in almost all plant sources (fruits and vegetables), others are far less common. For instance, citrus fruits have a high concentration of flavanones, while apples have an exceptionally high phloridzin content [5, 6]. Polyphenols may be found in a wide variety of foods, including fruits (apples, cherry, citrus fruits etc.), vegetables (carrot, broccoli, chicory, beans), grains

Science and Engineering of Polyphenols: Fundamentals and Industrial Scale Applications,
First Edition. Edited by Chandrabhan Verma.
© 2024 John Wiley & Sons Inc. Published 2024 by John Wiley & Sons Inc.

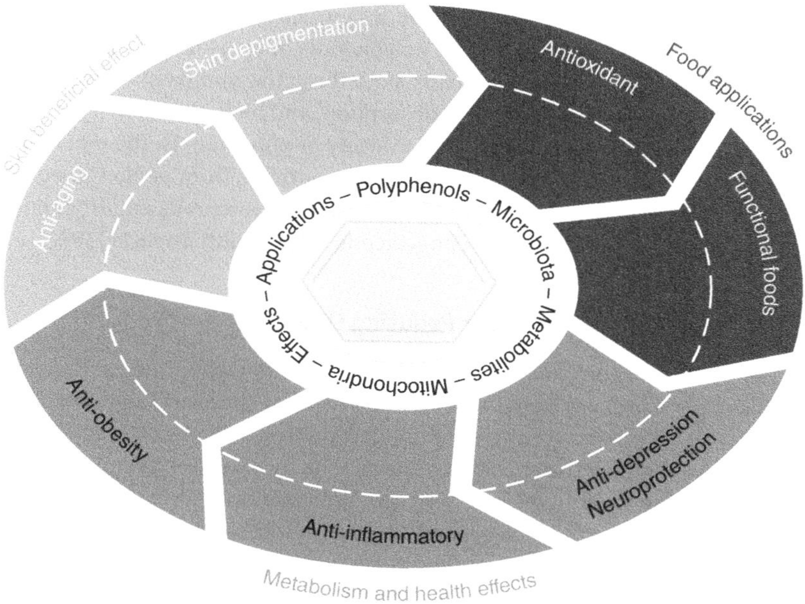

Figure 15.1 Benefits of using polyphenols (PP). Source: Rajha et al. [6]/CCBY 4.0/ Public domain.

(rye, wheat, oats), legumes (sprouts), dry fruits (such as almonds, flax seeds, chestnuts, and walnuts), herbs and spices (such as clove, cinnamon) and beverages (wine, tea etc.) [7, 8]. PP is well-known for its wide range of biological properties, including those used to relieve pain, reduce inflammation, and bring down a high temperature. They are effective in treating malignant and killing bacteria or stopping them from growing and causing disease [9]. PP can be utilized in various ways like in medicinal industry, food, and cosmetic industry, and Figure 15.1 demonstrates the usage of PP in various fields.

In this chapter the usage of PP in various sectors including health, cosmetic are elaborated.

15.2 Health Benefits of Polyphenols

Scientists have theorized that phenolic compounds possess remarkable antibacterial as well as antimicrobial characteristics. Polyphenols have been linked to a wide variety of health benefits according to their properties. Several studies have shown a connection between polyphenols and a reduced risk of developing chronic illnesses. It is well known that the Mediterranean diet and its primary components, including fruits and vegetables, wine, extra virgin olive oil, almonds, and even green tea and chocolate, contain a significant quantity of polyphenols. Nevertheless, the protective benefits of polyphenols in vivo are contingent on the amount of PPs to be extracted from foods, both the active and passive transport of polyphenols, energy generated by food absorption, and biological activity with the intended site. When a person consumes a variety of polyphenols, the metabolites that are found in their plasma and urine will be variable from person to person. This may be due to differences in the makeup of the gut microbiota and/or genetic polymorphisms, and it may also be connected to clear health impacts [10].

15.2.1 Usage of Polyphenols for Reducing Obesity

Since 1980, when just one-sixth of the world's population was afflicted, the prevalence of obesity has skyrocketed to about one-third of the world's population in 2016 [11]. This growth has occurred at an alarmingly quick rate. Obesity is characterized by an energy imbalance that is brought on by a consistent and prolonged deficit between the amounts of calories consumed and burned, and it is linked to a wide range of degenerative disorders. As a result, obesity has developed into a problem that affects both society and the public's health and calls for both prevention and therapeutic measures. Recent research has brought to light the possibility of a link between alterations in the makeup of the gut microbiota and the development of obesity [12, 13]. There is evidence that polyphenol extract from plants, which is taken orally, modifies the microbial makeup of the digestive tract. This, in turn, helps reduce the risk of obesity by regulating the metabolism of lipids and carbohydrates [14–16]. Innovative anti-obesity techniques that lead to a large decrease in food intake and/or an increase in energy expenditure while maintaining high levels of efficiency, security, and selectivity are desperately needed in today's society. Within the realm of therapeutic approaches that are aimed at preventing and treating obesity, the discovery of leptin in 1994 ushered in a new era. However, the administration of leptin is ineffective in reducing the body weight of obese animals that are not leptin-deficient but rather have high levels of circulating leptin coupled with a loss of responsiveness to this hormone. This is the case because high levels of circulating leptin are associated with a loss of responsiveness to leptin [17]. In addition, the broad range of preventive actions that polyphenols have been shown to have on obesity models, including their ability to inhibit numerous of the complex diseases that are linked to obesity, enabled us to formulate the hypothesis that polyphenols or fruits high in polyphenols may have the ability to regulate brain and peripheral leptin signaling and, as a result, restore leptin sensitivity. Zhang et al. [18] investigated *Camellia nitidissima* Chi flower extracts rich in PP and found that the extract had the property to work against obesity. *C. nitidissima* Chi is a multifunctional edible plant with many polyphenol compounds such as procyanidins, apigenins, and quercetins [19–21]. In vitro, the anti-obesity effects of the extract have been proved [21] and in this study in vivo has been investigated. Sample was a rat to which a high-fat diet (HFD) was given, and various factors such as food-taking capacity, after taking HFD, the total weight gained by rat, etc., were the various parameters that were taken into consideration. Various tests like oral glucose tolerance test (OGTT), along with biochemical measurements and liver tissue histological analysis were performed. Results revealed that extract successfully reduced the weight of rats. The main cause was that it decreased the appetite, and interaction was shown with gut microbiota.

Ardid-Ruiz et al. [22] studied resveratrol for its anti-obesity properties, and the purpose of this research was twofold: first, to determine whether or not resveratrol may decrease fat deposition by raising the levels of their sensitivity to leptin, and second, to determine which resveratrol-derived circulating substance after breakage are possibly implicated in these metabolic effects. The study was conducted on male Wistar rats divided into equal groups with six rats in each group, and two types of diets were given to the groups: one with a high-fat percentage and the other with a low-fat percentage. Up to the very end of the investigation, weekly measurements of body weight were taken. In addition, the day before the animals were put to death, their entire body composition was determined using nuclear magnetic resonance (NMR). When compared to the group that had a conventional meal, those who consumed an HFD from the cafeteria for a period of 12 weeks consistently developed obesity. This was shown (Figure 15.2) by a substantially larger body weight gain (50.1% higher) and a significantly higher total body fat mass (124.3% higher, as determined by NMR scanning). When compared to the CAF group, the animals who were given RSV at a dose of 200 mg/kg per day had a body weight growth that was 17% lower. There was a reduction in the total body fat mass, and at this amount of RSV, the hyperleptinemia caused by an HFD was reversed.

Further, the glucose, serum, and insulin levels were also controlled at normal levels by this much amount of RSV. In order to conduct a more in-depth investigation into the impact of RSV from a multidimensional perspective, principal component analysis (PCA) was carried out. As a consequence of this, the PCA score plots for both diet groups in which all the animals were categorized on the basis of their diet (Figure 15.2g) and at a specific dose of RSV could only had the capacity to bring back the metabolic changes caused by the diet (Figure 15.2h). RSV reduced lipid content present in liver, muscle, and adipose tissues, which was because of the diet taken. Also, RSV maintains the leptin transcription in adipose tissues. In a rat model of diet-induced obesity, it was shown that RSV is capable of reversing the disturbance of metabolic parameters as well as the lipid profile. However, further research is necessary to gain a better understanding of the molecular pathways via which RSV affects leptin signaling in obesity. In conclusion, the metabolites that are produced by the microbiota in the gut might be a partial explanation for the contribution that the greatest dosage of RSV makes to the reduction of the metabolic change that is induced by obesity. Therefore,

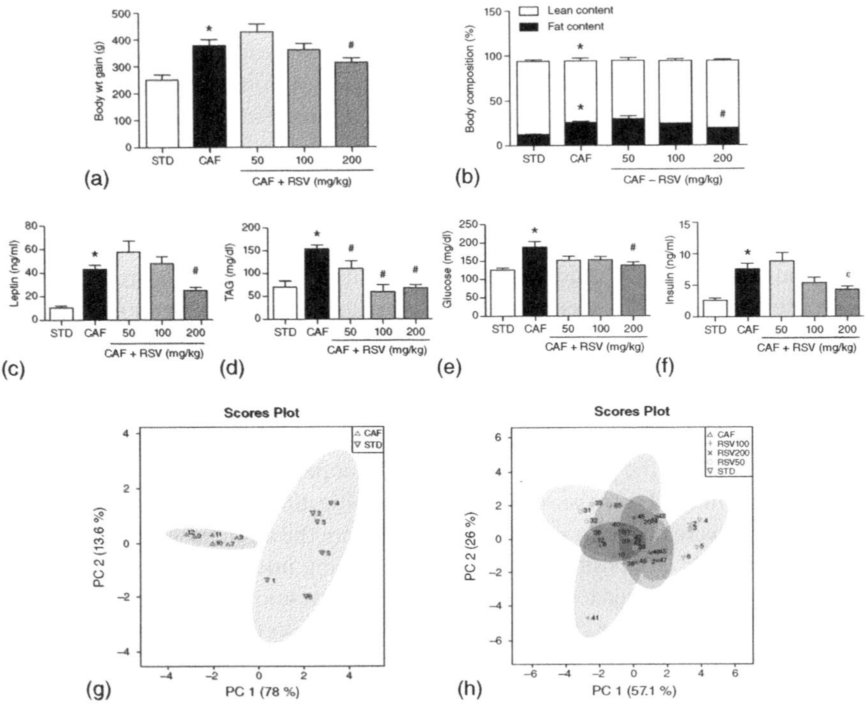

Figure 15.2 Various parameters showing different metabolic activities. (a) weight gain and (b-f) % composition calculated (g-h) clusters in accordance with studied biometric parameters. Source: Ardid-Ruiz et al. [22]/MDPI/CCBY 4.0.

polyphenols are a highly interesting category of naturally occurring compounds that have the potential to be used in an effort to combat obesity. Because of this, it is practical to emphasize their potential applicability for use in functional food formulations geared toward weight loss.

15.2.2 Usage of Polyphenol for Prebiotics Ingredients

Polyphenols were just included in the class of substances known as prebiotics owing to their capacity to interact with gut microorganisms and the host and/or affect the set of microbial metabolites that are accessible to the host [23, 24]. Steed et al. [25] got to know about deaminotyrosine (DAT), a metabolite that ultimately helped in increasing the survival of mice with influenza mortality. Jiao et al. [16] reported that treatment of blueberry polyphenol extract (PPE) over a period of 12 weeks prevented mice on an HFD from gaining weight and normalized their degradation of fatty acids (fundamental biological molecules) and the building blocks of more structurally complex lipids. In addition, it was found that the improved lipid metabolism of the animals that were given the phenolic extract was connected with the regulation of the makeup of the intestinal microbiota, which is useful in the digestion of many host nutrients, drugs, etc. This observation was made after the animals were given the phenolic extract. In a study performed by Chen et al. [26] mice were fed a diet containing PP, which is contained in green tea, for a period of seven weeks, which resulted in a changed composition of the microbiota in the cecon and colon. This alteration was associated with the decrease of blood glucose in the animals that were given the treatment. In a similar fashion a USA-based company used cranberry powder for the preparation of shakes, yogurts, etc. [27]. In addition to the significant effect that PP has on mitochondria and in the prevention of neurodegenerative diseases, it has also been established that PP has a significant effect on the gut microbiota via the contact that they have with that [28]. PP have been shown to interact with the metabolism, physiology, and health of the host, which plays an important part in the protection against infectious diseases are caused by pathogens, which can be bacteria, fungi, protozoa, worms, viruses, or even infectious proteins called prions. In the regulation of a disturbance in the balance between the production of reactive oxygen species and antioxidant defenses, infectious diseases can be triggered. Use of probiotics is one of the strategies to manage microbiota. It was studied that the microbiota produced substances made or used when the body breaks down food, drugs or chemicals, or its own tissue such as hydrogen sulfide (H_2S), that can affect the mitochondrial respiratory chain, which produces 90% of the ATP required for cell sustenance [6]. Diet and/or metabolites may be used to re-organize and regulate both the quality of the human microbiota and the dynamics of mitochondria in order to bring the system back into balance and homeostasis. It makes a difference to the quality of the metabolites, whether they come from "good" or "bad" gut bacteria (as given in Figure 15.3).

15.2.3 Usage of Polyphenol for the Treatment of Hyperpigmentation

The improper activity of melanocytes may lead to the overproduction of melanin, which can then contribute to the development of a number of other pigmentary skin conditions. Inhibition of tyrosinase, a type of enzyme that catalyzes the essential step in the production of melanin, is one of the most prevalent methods used for the management of skin pigmentation. Panzella and Napolitano [29] emphasize the tyrosinase-inhibiting characteristics of a variety of natural and synthetic phenolic compounds that have been introduced into the scientific literature during the last five years. The search for skin lighteners and agents that are able to treat hyperpigmentation diseases is expanding at a consistent rate, notably in Asian nations. The majority of the approaches that have been investigated thus far center on the regulation of the activity of tyrosinase, which is the primary and rate-limiting enzyme involved in the manufacture of melanins, which are the pigments

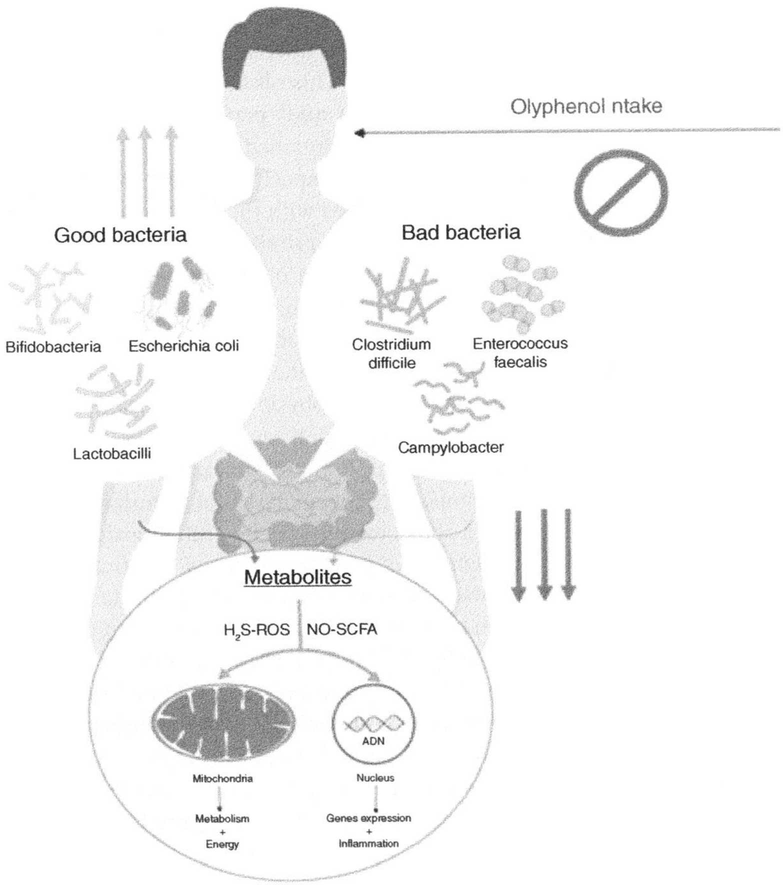

Figure 15.3 Phenol-rich diet affecting the host. Source: Rajha et al. [6]/CCBY 4.0/ Public domain.

that give mammals their characteristic dark skin pigmentation [30–33]. When developing a new tyrosinase inhibitor, in addition to product performance, the degree to which a substance can cause damage to a cell, dissolution, the transport of chemicals from the outer surface of the skin both into the skin and into circulation, and stability, other critical issues that should be carefully considered include adverse effects derived from the action of the enzyme itself on the product. These effects have been largely documented in the recent case of tyrosine kinase inhibitors for the treatment of cancer. With reference to rhododendrol, a depigmenting chemical currently available that has the potential to induce leukoderma. Tyrosinase for mammals is very difficult to collect, so studies have been carried out on fungal (mushroom) tyrosinase. Results revealed that tyrosinase, which is the core enzyme system of melanogenesis and is responsible for catalyzing the rate-limiting phase of the process, has become one of the most significant targets for the treatment of hyperpigmentary diseases that have either clinical or aesthetic significance. As a consequence of this, during the last several years, a remarkable number of inhibitors have been found that operate at the level of tyrosinase to exert their effect. Even though there is a wide variety of natural inhibitors, it is noteworthy to note that the majority of tyrosinase inhibitors found in nature are of the phenolic chemical class. Numerous researchers have produced innovative synthetic inhibitors while also designing appropriate frameworks that have been inspired by the structure of natural substances. Even though a large variety of tyrosinase inhibitors derived from natural and synthetic sources have been identified, there are only a select handful of them that are recognized to be safe substances in addition to having an established record of efficacy [34–37].

From the past studies taken and summarized for the purpose of this review, it seems that there have been no substantial breakthroughs made regarding the toxicity concerns associated with tyrosinase inhibitors. The relevant studies only reported the findings of in vitro cytotoxicity tests. Because of this, it is very necessary to investigate the efficiency and security of inhibitors by, for example, determining whether or not the proposed inhibitor is a substrate of tyrosinase that is changed after coming into contact with the enzyme. Before the substance might be considered for clinical trials in the perspectives of its use in medicinal and aesthetic applications, in vivo investigations utilizing reconstructed three-dimensional human epidermis should be carried out. These tests have largely replaced testing on animal models. In vivo experiments should be carried out before clinical trials. Once this first level of safety assessment has been passed, in vivo experiments using reconstructed three-dimensional human epidermis should be carried out.

Shiji et al. [38] performed a clinical trial, the effects of continuous apple polyphenol (AP) administration on facial skin problems and pigmentation generated by ultraviolet (UV) irradiation in healthy women volunteers were evaluated. The skin is subject to damage after exposure to ionizing radiation and a variety of chemical agents as a result of several stressors, one of which is the everyday exposure to ultraviolet radiation from the sun. Particularly, ultraviolet radiation from the sun may cause erythema, inflammation, hyperpigmentation, immunosuppression, and even skin cancer in exposed areas of the body's skin. Damaged skin structures cause an increase in trans-epidermal water loss, which in turn leads to a reduction in the amount of water and moisture that is contained inside the skin. UV irradiation at moderate levels has the effect of speeding up the process of melanogenesis, which is the production of melanin by the skin's melanocytes. On the other hand, UV irradiation at high doses and repetitive exposure to UV light causes skin cancer and photoaging. Free radicals and reactive oxygen species (the superoxide anion radical and the hydrogen peroxide radical) are both produced by UV light [39]. During a human clinical experiment, researchers found that orally administering AP was able to suppress UV irradiation-induced skin pigmentation. The purpose of this study was to evaluate whether or whether the administration of AP ameliorated the pigmentation produced by UV irradiation. For this purpose, 65 healthy women were randomly assigned to either the placebo or AP group. However, three individuals withdrew out of the research due to personal reasons, and three people failed to attend study visits for the purpose of measurement. As a result, these participants were not included in the analysis of the study's effectiveness. As a result, the information presented here comes from 59 different individuals (systematic diagram of the trial in Figure 15.4).

Same ingestion was found in placebo and AP group, according to these findings, the oral administration of AP was able to prevent the production of erythema that was brought on by UV irradiation. Similarly, the effect of AP on the melanin value was also evaluated. No difference was found between both groups. AP group was protected by pigmentation caused by UV, also the reduction in skin lightening was also controlled for AP group. There was no discernible impact made by the AP intervention on any of the other face skin condition criteria that were evaluated.

15.2.4 Usage of Polyphenol for the Treatment of Allergies

Inflammation is the starting point for a number of different biological processes, and in recent years, it has come to people's attention that there is a complicated relationship between inflammatory events and allergic responses. Brown seaweeds collected from various sources contain polyphenols that have been identified as having high value as bioactive chemicals and as being powerful modulators of various biochemical pathways connected to the disruption of homeostasis in major chronic illnesses. Corona et al. [40] studied phlorotannins are a kind of polyphenol that may be found in high concentrations in brown seaweeds. There is currently a lack of knowledge on the bioavailability of seaweed polyphenols, and there is very minimal data regarding the bioactivity of seaweed polyphenols in vivo. Therefore, the anti-inflammatory and anti-allergic properties of these samples in vitro were investigated. Results show that PP metabolites may treat a wide variety of disorders, including obesity, depression, inflammation, and allergic reactions, by interacting with mitochondria and the microbiota in the gut.

15.2.4.1 Usage of Polyphenol for the Preparation of Cosmetics

The notion of employing eumelanins, which are a darker kind of skin pigment, is widely known to help organisms cope with molecular damage caused by sunlight. It was discovered that these pigments have a significant property to scavenge free radicals, which was attributed to the said part of the pigment that was produced from the natural products through enzymatic reactions precursor known as 5,6-dihydroxyindole-2-carboxylic acid (DHICA). However, due to the unfavorable characteristics of DHICA melanin, its usage in skincare solutions that use

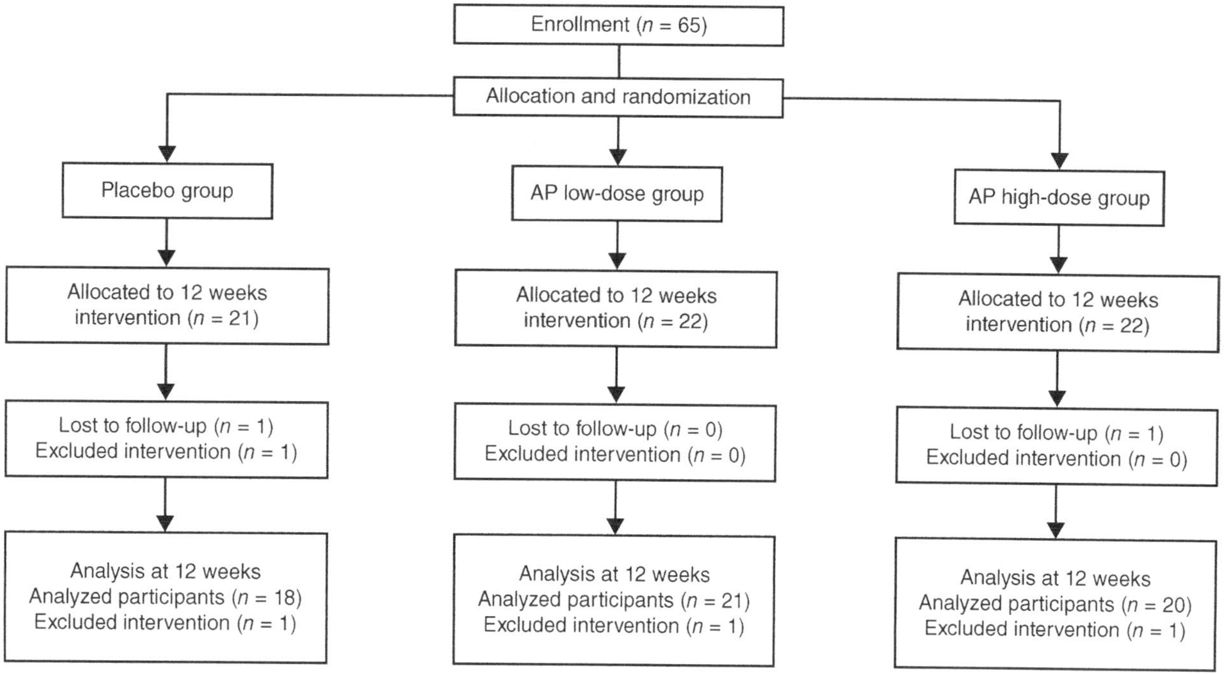

Figure 15.4 Diagram showing the trial. Source: Shoji et al. [38]/CCBY 4.0/Public domain.

sophisticated, active ingredients to directly support or care for the symptoms of various skin conditions is limited or avoided. In order to circumvent this constraint, a polymeric polyphenolic unit with a similar structure was manufactured by oxidizing methyl ester of DHICA, also known as methyl ester of 5,6-dihydroxyindole-2-carboxylic acid (MeDHICA), under biomimetic circumstances. This was done in order to produce MeDHICA. This pigment had a more favorable solubility profile, and In antioxidant experiments, the MeDHICA pigment performed very well, and it was shown to be able to inhibit the buildup of reactive oxygen species (ROS) in keratinocytes that had been exposed to UVA radiation [41, 42].

15.3 Conclusion and Future Outlook

PP (derived from plants) is needed by the human body in very small quantity. It is stated that they are full of advantages for health and are high in antioxidant content. Various studies are there to show the usage of PP in reducing weight, treating cancer and diabetes, blood pressure control, stress management, and treatment of heart diseases. It is important to point out that clinical investigations need to be conducted in order to determine the margins of acceptable concentrations for the intake of these compounds. This is necessary due to the fact that these substances may have harmful effects when they are ingested in large amounts. However, the precise mechanism and the manner in which they are shown to be so beneficial to the human body are not yet fully understood, and studies are still being carried out. It has been proven in a number of different studies that people may improve their general health by increasing their consumption of vegetables that are high in polyphenol content on a regular basis. The existing information and comprehension of how polyphenols function is quite limited, and additional investigations and research are necessary to completely grasp the functioning mechanism of polyphenols in order to acquire knowledge about their biological role in a manner that is more effective. It will need a great deal of research, both at the molecular and cellular levels, in order to understand the role that polyphenols play in the prevention of different diseases.

References

1 Rasouli, H., Farzaei, M.H., and Khodarahmi, R. (2017). Polyphenols and their benefits: a review. *International Journal of Food Properties 20* (2): 1700–1741.

2 Singla, R.K., Dubey, A.K., Garg, A. et al. (2019). Natural polyphenols: chemical classification, definition of classes, subcategories, and structures. *Journal of AOAC International 102* (5): 1397–1400.

3 Guo, W., Kong, E., and Meydani, M. (2009). Dietary polyphenols, inflammation, and cancer. *Nutrition and Cancer 61* (6): 807–810.

4 Wan, M.L.Y., Co, V.A., and El-Nezami, H. (2021). Dietary polyphenol impact on gut health and microbiota. *Critical Reviews in Food Science and Nutrition 61* (4): 690–711.

5 Manach, C., Scalbert, A., Morand, C. et al. (2004). Polyphenols: food sources and bioavailability. *The American Journal of Clinical Nutrition 79* (5): 727–747.

6 Rajha, H.N., Paule, A., Aragonès, G. et al. (2022). Recent advances in research on polyphenols: effects on microbiota, metabolism, and health. *Molecular Nutrition & Food Research 66* (1): 2100670.

7 Chauha N, D., Kumar, K., Kumar, S., & Kumar, H. (2018). Effect of incorporation of oat flour on nutritional and organoleptic characteristics of bread and noodles. *Current Research in Nutrition and Food Science Journal*, *6*(1), 148-156.

8 Cheynier, V. (2012). Phenolic compounds: from plants to foods. *Phytochemistry Reviews 11* (2-3): 153–177.

9 Malathy, R., Prabakaran, M., Kalaiselvi, K. et al. (2020). Comparative polyphenol composition, antioxidant and anticorrosion properties in various parts of Panax ginseng extracted in different solvents. *Applied Sciences 11* (1): 93.

10 Tresserra-Rimbau, A., Arranz, S., and Vallverdu-Queralt, A. (2017). New insights into the benefits of polyphenols in chronic diseases. *Oxidative Medicine and Cellular Longevity 2017*.

11 Chooi, Y.C., Ding, C., and Magkos, F. (2019). The epidemiology of obesity. *Metabolism 92*: 6–10.

12 Dao, M.C. and Clément, K. (2018). Gut microbiota and obesity: concepts relevant to clinical care. *European Journal of Internal Medicine 48*: 18–24.

13 Dalby, M.J., Ross, A.W., Walker, A.W., and Morgan, P.J. (2017). Dietary uncoupling of gut microbiota and energy harvesting from obesity and glucose tolerance in mice. *Cell Reports 21* (6): 1521–1533.

14 Palau-Rodriguez, M., Tulipani, S., Isabel Queipo-Ortuno, M. et al. (2015). Metabolomic insights into the intricate gut microbial–host interaction in the development of obesity and type 2 diabetes. *Frontiers in Microbiology 6*: 1151.

15 Zhang, Z., Peng, X., Li, S. et al. (2014). Isolation and identification of quercetin degrading bacteria from human fecal microbes. *PLoS One 9* (3): e90531.

16 Jiao, X., Wang, Y., Lin, Y. et al. (2019). Blueberry polyphenols extract as a potential prebiotic with anti-obesity effects on C57BL/6 J mice by modulating the gut microbiota. *The Journal of Nutritional Biochemistry 64*: 88–100.

17 Friedman, J.M. (2019). Leptin and the endocrine control of energy balance. *Nature Metabolism 1* (8): 754–764.

18 Zhang, H.L., Wu, Q.X., and Qin, X.M. (2020). *Camellia nitidissima* Chi flower extract alleviates obesity and related complications and modulates gut microbiota composition in rats with high-fat-diet-induced obesity. *Journal of the Science of Food and Agriculture 100* (12): 4378–4389.

19 Yang, R., Guan, Y., Wang, W. et al. (2018). Antioxidant capacity of phenolics in *Camellia nitidissima* Chi flowers and their identification by HPLC Triple TOF MS/MS. *PLoS One 13* (4): e0195508.

20 Lin, J.N., Lin, H.Y., Yang, N.S. et al. (2013). Chemical constituents and anticancer activity of yellow camellias against MDA-MB-231 human breast cancer cells. *Journal of Agricultural and Food Chemistry 61* (40): 9638–9644.

21 Zhang, H.L., Wu, Q.X., and Qin, X.M. (2019). *Camellia nitidissima* Chi flower extracts inhibit α-amylase and α-glucosidase: in vitro by analysis of optimization of addition methods, inhibitory kinetics and mechanisms. *Process Biochemistry 86*: 177–185.

22 Ardid-Ruiz, A., Ibars, M., Mena, P. et al. (2018). Potential involvement of peripheral leptin/STAT3 signaling in the effects of resveratrol and its metabolites on reducing body fat accumulation. *Nutrients 10* (11): 1757.

23 de Paulo Farias, D., de Araújo, F.F., Neri-Numa, I.A., and Pastore, G.M. (2019). Prebiotics: trends in food, health and technological applications. *Trends in Food Science & Technology 93*: 23–35.

24 Gibson, G.R., Hutkins, R., Sanders, M.E. et al. (2017). Expert consensus document: The International Scientific Association for Probiotics and Prebiotics (ISAPP) consensus statement on the definition and scope of prebiotics. *Nature Reviews Gastroenterology & Hepatology 14* (8): 491–502.

25 Steed, A.L., Christophi, G.P., Kaiko, G.E. et al. (2017). The microbial metabolite desaminotyrosine protects from influenza through type I interferon. *Science 357* (6350): 498–502.

26 Chen, T., Liu, A.B., Sun, S. et al. (2019). Green tea polyphenols modify the gut microbiome in db/db mice as co-abundance groups correlating with the blood glucose lowering effect. *Molecular Nutrition & Food Research 63* (8): 1801064.

27 de Araújo, F.F., de Paulo Farias, D., Neri-Numa, I.A., and Pastore, G.M. (2021). Polyphenols and their applications: an approach in food chemistry and innovation potential. *Food Chemistry 338*: 127535.

28 Zhao, Y. and Jiang, Q. (2021). Roles of the polyphenol–gut microbiota interaction in alleviating colitis and preventing colitis-associated colorectal cancer. *Advances in Nutrition 12* (2): 546–565.

29 Panzella, L. and Napolitano, A. (2019). Natural and bioinspired phenolic compounds as tyrosinase inhibitors for the treatment of skin hyperpigmentation: recent advances. *Cosmetics 6* (4): 57.

30 Zolghadri, S., Bahrami, A., Hassan Khan, M.T. et al. (2019). A comprehensive review on tyrosinase inhibitors. *Journal of Enzyme Inhibition and Medicinal Chemistry 34* (1): 279–309.

31 Kumari, S., Thng, S.T.G., Verma, N.K., and Gautam, H.K. (2018). Melanogenesis inhibitors. *Acta Dermato-Venereologica 98* (10): 924–931.

32 Pillaiyar, T., Namasivayam, V., Manickam, M., and Jung, S.H. (2018). Inhibitors of melanogenesis: an updated review. *Journal of Medicinal Chemistry 61* (17): 7395–7418.

33 Lee, S.Y., Baek, N., and Nam, T.G. (2016). Natural, semisynthetic and synthetic tyrosinase inhibitors. *Journal of Enzyme Inhibition and Medicinal Chemistry 31* (1): 1–13.

34 Ito, S. and Wakamatsu, K. (2018). Biochemical mechanism of rhododendrol-induced leukoderma. *International Journal of Molecular Sciences 19* (2): 552.

35 Gabe, Y., Miyaji, A., Kohno, M. et al. (2018). Substantial evidence for the rhododendrol-induced generation of hydroxyl radicals that causes melanocyte cytotoxicity and induces chemical leukoderma. *Journal of Dermatological Science 91* (3): 311–316.

36 Abe, Y., Okamura, K., Kawaguchi, M. et al. (2016). Rhododenol-induced leukoderma in a mouse model mimicking Japanese skin. *Journal of Dermatological Science 81* (1): 35–43.

37 Mann, T., Gerwat, W., Batzer, J. et al. (2018). Inhibition of human tyrosinase requires molecular motifs distinctively different from mushroom tyrosinase. *Journal of Investigative Dermatology 138* (7): 1601–1608.

38 Shoji, T., Masumoto, S., Moriichi, N. et al. (2020). Administration of apple polyphenol supplements for skin conditions in healthy women: a randomized, double-blind, placebo-controlled clinical trial. *Nutrients 12* (4): 1071.

39 Bickers, D.R. and Athar, M. (2006). Oxidative stress in the pathogenesis of skin disease. *Journal of Investigative Dermatology 126* (12): 2565–2575.

40 Corona, G., Ji, Y., Anegboonlap, P. et al. (2016). Gastrointestinal modifications and bioavailability of brown seaweed phlorotannins and effects on inflammatory markers. *British Journal of Nutrition 115* (7): 1240–1253.

41 Panzella, L., Ebato, A., Napolitano, A., and Koike, K. (2018). The late stages of melanogenesis: Exploring the chemical facets and the application opportunities. *International Journal of Molecular Sciences 19* (6): 1753.

42 Liberti, D., Alfieri, M.L., Monti, D.M. et al. (2020). A melanin-related phenolic polymer with potent photoprotective and antioxidant activities for dermo-cosmetic applications. *Antioxidants 9* (4): 270.

16

Polyphenols as Antimicrobial (Disinfectant and Pest control) Applications

Vrinda[1], *Garima Narang*[1], *Garima Pandey*[2], *Prashant Singh*[1], *and Kamlesh Kumari*[3]

[1] *Department of Chemistry, Atma Ram Sanatan Dharma College, University of Delhi, New Delhi, India*
[2] *Department of Chemistry, SRM Institute of Science & Technology, Modinagar, India*
[3] *Department of Zoology, University of Delhi, New Delhi, India*

16.1 Introduction

Microorganisms are a type of minute organism that cannot be seen with the naked eye but may be seen using a light or electron microscope. Microorganisms can be categorized according to a wide range of criteria. Bacteria, viruses, fungi, and certain algae make up the majority of microorganisms [1]. Many harmful infections are brought on by microbes (Figure 16.1). Many microorganisms infect crop plants with the disease, which reduces crop yield and hurts crop output. Plant fungal infections pose a significant threat to food safety. Additionally, consuming fungus-contaminated nuts, grains, and their derivatives may lead to intoxication in humans [2]. Polyphenolic compounds are recognized for their potential efficacy against various human viruses and their anti-inflammatory and antimicrobial activities. Polyphenols, a group of secondary metabolites, play a role in scavenging hydrogen peroxide in plant cells [3]. These metabolites have significant roles in plants pigmentation, development and reproduction, defense against ultraviolet (UV) radiation, and tolerance to diseases and herbivores, and many other nutrient metabolites have been identified as chemicals that may be responsible for the positive impacts on plants' health in addition to their positive effects on people [4]. According to the structure of their basic skeleton, they have been divided into many groups or classes, ranging from very simple molecules like

Science and Engineering of Polyphenols: Fundamentals and Industrial Scale Applications, First Edition. Edited by Chandrabhan Verma.
© 2024 John Wiley & Sons Inc. Published 2024 by John Wiley & Sons Inc.

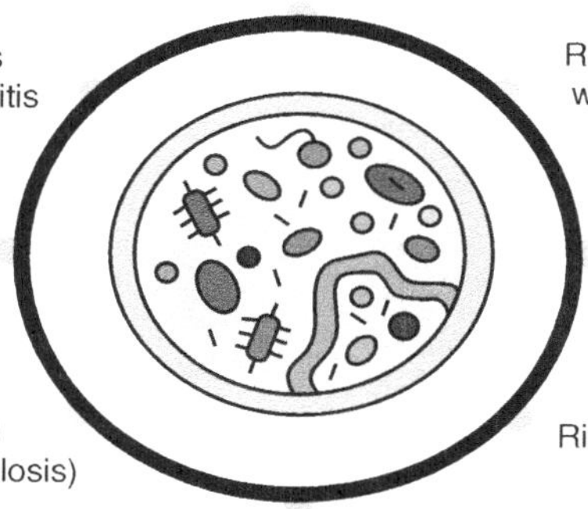

Figure 16.1 Various diseases caused by microorganisms in plants, animals, and humans. Source: https://www.canva.com/design/DAFcx92_YxE/-pftyjnFs8UaNvU6XVTO1g/edit?utm_content=DAFcx92_YxE&utm_campaign=designshare&utm_medium=link2&utm_source=sharebutton.

phenolic acids to polymerized molecules with relatively high molecular masses like hydrolyzable and condensed tannins [4]. Following are a few examples of polyphenolic compounds with applications in pest management, such as disinfectants, and therapeutic effects for cardiovascular disease, neurodegenerative disorders, cancer, and obesity. Table 16.1 explains various polyphenolic compounds, their chemical formula formulas, subclass, and the source from which they can be extracted.

In this chapter, we will examine the utilization of polyphenolic compounds as antimicrobial agents and the use of these substances to control pests and serve as disinfectants.

16.2 Pest Management

Insects have evolved mechanisms to combat plant toxins, including avoidance, degradation, sequestration, and excretion. These adaptations have fueled a co-evolutionary arms race and co-diversification between plants and insects. Pests are currently one of the biggest problems in agriculture. There are several ways to lessen or get rid of pests, but each comes with its own set of drawbacks. Halogenated hydrocarbons and organophosphates are synthetic pesticides readily available on the market. These pesticides are regarded to have more toxicological properties than most natural compounds due to their longer half-lives in the environment. There is a rising need for environmentally friendly, toxicologically safe,

Table 16.1 Various polyphenolic compounds with their formula, structure, sub-class, and source of extraction.

Name	Chemical formula	Structure	Sub class	Source	Ref.
Catechin	$C_{15}H_{14}O_6$	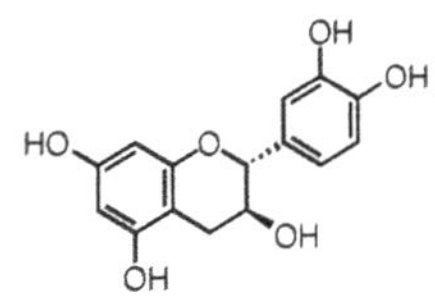	Flavan-3-ols	Grapes	[5]
procyanidin B2	$C_{30}H_{26}O_{12}$	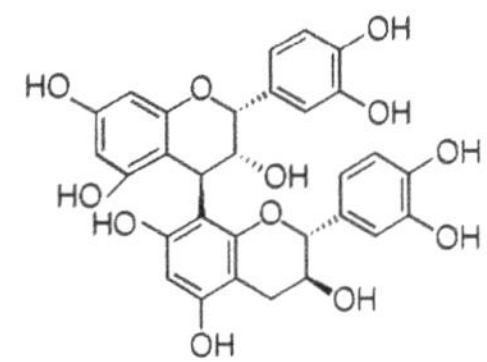	Flavan-3-ols	Grapes	[5]
Apigenin	$C_{15}H_{10}O_5$	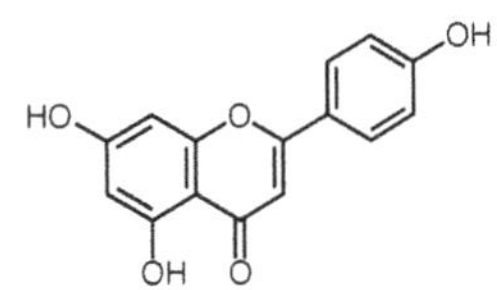	Flavones	Vegetables and fruits such as parsley, onions, oranges and herbs (snakeberry, thyme)	[6]
Cyanidin	$C_{15}H_{11}O_6^+$	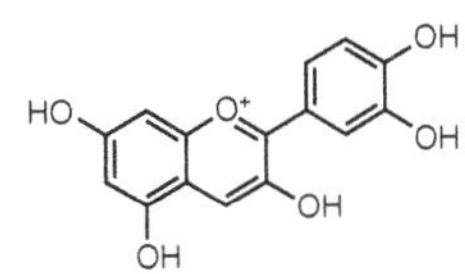	Anthocyanidins	Berries, apples, oranges	[7]

Table 16.1 (Continued)

Name	Chemical formula	Structure	Sub class	Source	Ref.
p-Coumaric acid	$HOC_6H_4CH{=}CHCO_2H$	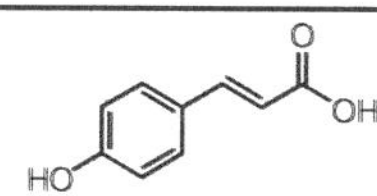	Cinnamic acid	Peanut or groundnut (*Arachis hypogea* L.)	[8]
Resveratrol	$C_{14}H_{12}O_3$	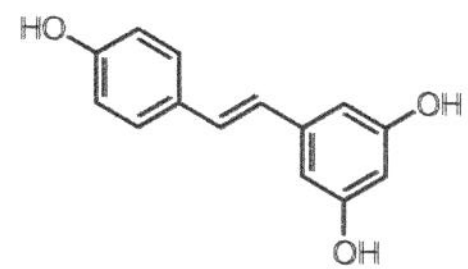	Stilbenes	Peanuts, grapes, blueberries	[9]
Scoparone	$C_{11}H_{10}O_4$	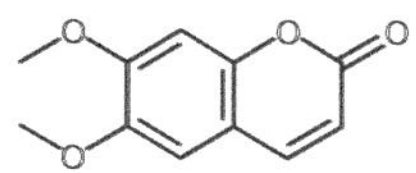	Coumarins	Chinese medicinal Artemisia capillaris Thunb	[10]

more targeted, and effective pesticides due to the aforementioned factors and other factors, and in this regard, botanicals play a significant role [11]. Pesticides are a class of chemicals that are widely employed in agriculture and many other sectors of the global economy. High toxicity toward pests and perhaps low toxicity toward other creatures, particularly humans and aquatic organisms, should be the hallmarks of chemical compounds used as pesticides [12, 13]. The structures of toxic pesticides Lindane, HCH, DDD and DDE are given in (Figure 16.2).

On socioeconomic and ecological functions, pest management strategies might have consequential effects [14]. Many secondary metabolites (polyphenolic compounds) from plants exhibit various behaviors in a wide range of insects, and these natural substances are now highly regarded for creating pesticides [15]. Polyphenolic substances play crucial roles in herbivore resistance. Chlorogenic acid, tannic acid, and methyl jasmonate are plant phenolic compounds that exhibit toxicity against insect pests and adversely affect important physiological processes [16]. Table 16.2 has some polyphenolic compounds that can be used for pest management along with their structure, sub-class, source, and insecticidal activity against the pest.

Some examples of polyphenolic compounds which can be used for pest management are given below, along with details on their synthesis and mode of action.

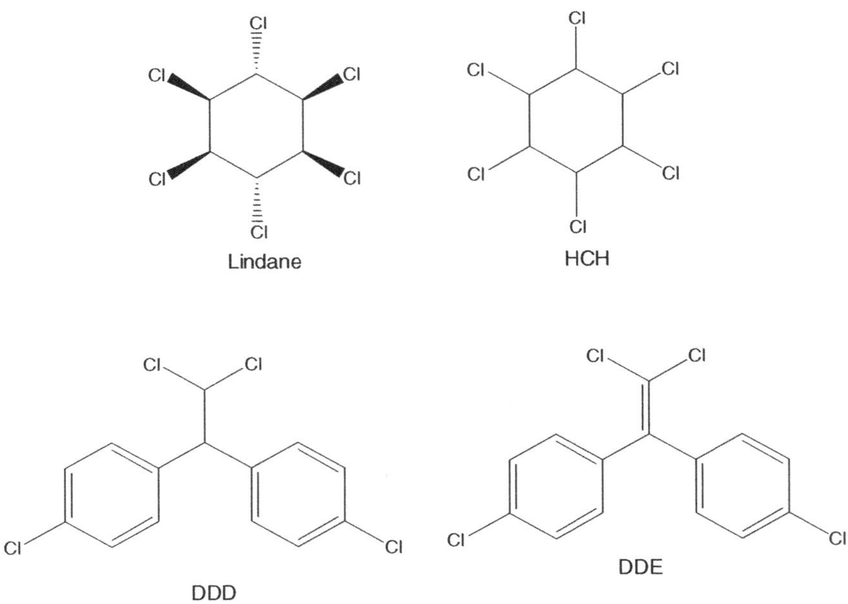

Figure 16.2 Chemical structure of various toxic pesticides.

Table 16.2 Various polyphenolic Compounds for pest management.

Polyphenolic compound	Structure	Sub-class	Source	Insecticidal activity against	Ref.
p-Coumaric acid	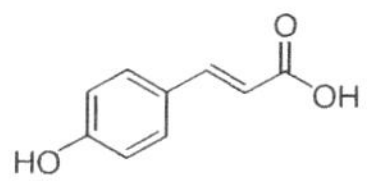	Cinnamic acid	*T. diversifolia*	*T. castaneum* and *S. oryzae*	[17]
Quercetin	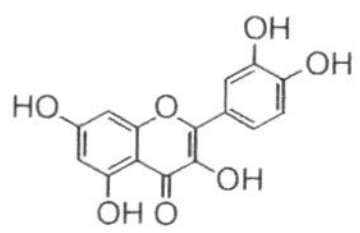	Flavonols	*Platanus orientalis* L., *Ginkgo biloba* L.	*Hyphantria cunea* Drury	[18]
kaempferol	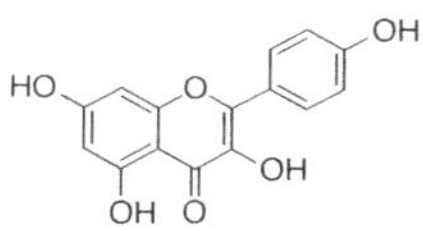	Flavonols	Dragea volubilis	*Helicoverpa armigera*	[19, 20]
Caffeic acid	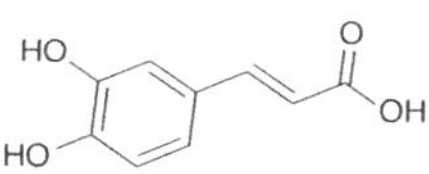	Cinnamic acid	Fruits and vegetables, such as apples, blueberries, and peanut	tobacco cutworm, *Spodoptera litura* (Fabricius)	[21, 22]
Gallo catechin	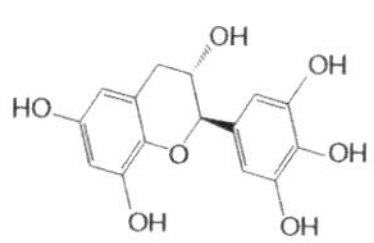	Flavonols	Banana (*Musa spp.*)	Tea aphid, *Toxoptera aurantii* (Boyer de Fonscolombe) polyphagous pest	[23, 24]

16.2.1 *p*-Coumaric Acid

p-Coumaric acid (p-CA) is a secondary metabolite in plants and mushrooms derived from the amino acid tyrosine or phenylalanine. It exhibits several biological properties, including antioxidant, anti-inflammatory, and anticancer effects, and is a crucial building block for phenolic acids, flavonoids, and stilbenes [25]. The production of p-CA can occur through two pathways: direct non-oxidative deamination of l-tyrosine by tyrosine ammonia lyase (TAL) or deamination of l-phenylalanine through phenylalanine ammonia-lyase (PAL) followed by hydroxylation catalyzed by cinnamate-4-hydroxylse (C4H) [26]. C4H, a cytochrome P450 hydroxylase, requires a functional electron transfer system to function properly, and its production in heterologous prokaryotic systems is often inadequate. Therefore, the most efficient method for producing p-CA is through the deamination of l-tyrosine by TAL. The enzymatic synthesis of *p*-coumaric acid using TAL and PAL is provided below (Figure 16.3).

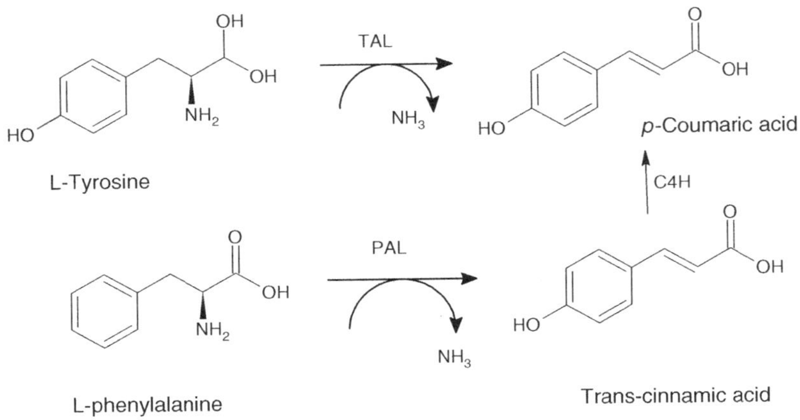

Figure 16.3 Enzymatic synthesis by TAL and PAL of *p*-coumaric acid/trans-cinnamic acid [26]. TAL: tyrosine ammonia lyase; PAL: phenylalanine ammonia-lyase; C4H: cinnamate-4-hydroxylse.

Suppressing α-glucosidase is described as the mode of action of *p*-coumaric acid [27]. Pesticides pose a serious threat to people and animals because of their widespread usage and bioaccumulation in the environment. A naturally occurring dietary polyphenolic compound with anti-inflammatory and antioxidant properties is *p*-coumaric acid [28]. Fipronil is a lipophilic agent, and most pesticides are highly lipid soluble, which can result in their accumulation in the phospholipids of the plasma membrane and affect erythrocytes, leading to decreased RBC and hemoglobin [28]. Pesticides are well-known for their ability to increase the production of reactive oxygen species (ROS), causing changes in the cellular antioxidant defense system and impairment of critical biomolecules such as proteins and lipids. As a result, tissues are more vulnerable to oxidative damage due to lipid peroxidation or impaired bio-membrane function. Recent research has shown that *p*-coumaric acid, a phenolic acid, has significant antioxidant properties that can help reduce ROS and lipid peroxidation while maintaining cell membrane integrity. *p*-Coumaric acid's antioxidant activity of *p*-coumaric acid is most likely due to its capacity to chelate metal ions, scavenge reactive oxygen and nitrogen species (ROS and RON), boost antioxidant enzymes, and repair oxidative damage to biomolecules. Furthermore, it has been demonstrated that *p*-coumaric acid activates nuclear factor erythroid-derived 2-like 2 (Nrf2), which regulates the expression of numerous antioxidant and detoxifying genes. This activation, in turn, can reduce malondialdehyde (MDA) formation, lowering lipid peroxidation. *p*-Coumaric acid's capacity to decrease oxidative stress and lipid peroxidation shows its promise as a therapeutic agent in treating oxidative stress-related disorders. These findings also imply that *p*-coumaric acid might be used as a dietary supplement or a functional food component to increase the antioxidant status of those prone to oxidative stress [28].

16.2.2 Quercetin

Quercetin (QE) is a natural flavonoid in substantial concentrations in the human diet. It is usually considered safe for eating, with no known adverse effects. Quercetin is often present in the human diet in the form of glycosides, the most numerous of which are quercetin-3-*O*-D-glucopyranoside (isoquercitrin) and quercetin-3-O-rutinoside. Direct extraction of quercetin is not practical due to the complex matrix and variety of glycosylated forms found in natural sources. Rutin, a commercial chemical derived from various plant materials, is frequently used as a starting material in the production of quercetin. Rutin is typically obtained from the Brazilian tree Fava d'anta, Dimorphandra mollis [29]. Figure 16.4 shows the bioconversion of rutin to quercetin using the catalyst rutinosidase.

Blocking a-amylase activity was described as the mode of action for quercetin. Quercetin's anticholinesterase activities were improved by coordination complexes with metal ions, with the quercetin-Fe complex exhibiting the strongest inhibitory impact [30]. Stronger – glucosidase inhibition was seen in fisetin and quercetin metal complexes (iron and zinc) than in their components. The quercetin-iron complex demonstrated the highest inhibition [30]. Due to its ability to scavenge free radicals and bind metal ions, QE exhibits a variety of pharmacological actions under pathological conditions, including anti-oxidative and anti-inflammatory

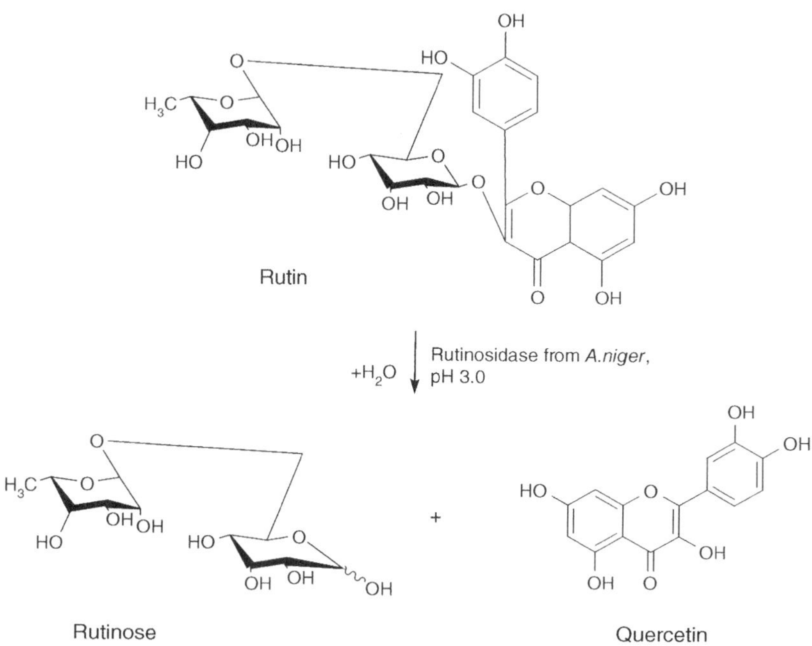

Figure 16.4 Utilizing rutinosidase bioconversion of rutin to quercetin.

characteristics. Moreover, QE might significantly suppress fibroblast activation, in part because of its capacity to inhibit the NF-kB signal transduction pathway, which is essential for controlling the inflammatory response [31]. Strong α-amylase inhibitory and antioxidant properties may be attributed to the quercetin glucoside in grape and peach peels [32].

16.2.3 Gallo Catechin

Catechins are a polyphenolic substance present in green tea extract obtained from young *Camellia sinensis* L. leaves that have been shown to have several health advantages, including antioxidant and cardioprotective qualities [33]. Gallo catechin is particularly significant among these catechins because of its distinct Michael acceptor moiety. Thanks to this moiety, Gallo catechin may create covalent adducts with cysteine residues, which are crucial in protein activity. By interacting with cysteine residues, gallo catechin may serve as an efficient and long-lasting modulator of enzyme performance. This shows that gallo catechin has a promising role in controlling protein activity, which might lead to additional therapeutic uses in various illness states [34].

Gallo catechin can be synthesized from Naringenin. The various steps in converting Naringenin to Gallo catechin are illustrated in Figure 16.5. Naringenin is first converted into penta-hydroxy flavanone with the help of Catalyst F3′5′H (flavonoid 3′,5′-hydroxylase). Penta-hydroxy flavanone thus formed is converted into dihydromyricetin via F3’H (flavanone 3-hydroxylase), which further gets converted to Leucodelphinidin by catalyst; DFR (dihydroflavonol-4-reducatse). Leucodelphinidin formed is finally converted to Gallo Catechin with the enzyme LAR (leucoanthocyanidin reductase).

16.3 Disinfectants

Antimicrobial products, also known as disinfectants, contain active ingredients such as triclosan, chlorhexidine, chlorine, iodine, alcohols, hydrogen peroxide, silver, and quaternary ammonium compounds. These active substances offer a broad range of antimicrobial action and are critical in various applications, such as the food sector, water treatment, and human and animal healthcare. In drinking water treatment systems, disinfection is the primary method of controlling pathogen transmission [36, 37]. Table 16.3 has some polyphenolic compounds that can be used for pest management along with their structure, sub-class, source and insecticidal activity against the pest.

Some examples of polyphenolic compounds are shown below, along with details on their synthesis and mode of action.

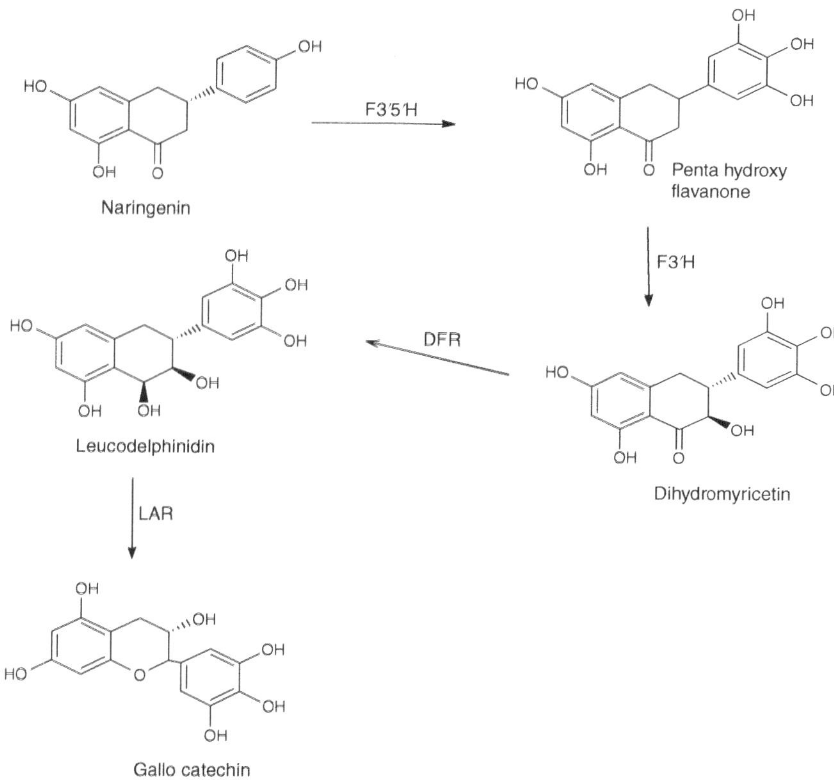

Figure 16.5 Synthesis of Gallo catechin from Naringenin [35].
F3′5′H = flavonoid 3′,5′-hydroxylase; F3′H = flavanone 3-hydroxylase; DFR = dihydroflavonol-4-reducatse; LAR = leucoanthocyanidin reductase.

16.3.1 Gallic Acid

Gallic acid (GA) is a polyphenolic substance with a predominant redox property and effective free radical deactivation. By using pathways for electron transfer, this scavenging capacity lowers oxidative damage. The use of GA and many of its derivatives as antibacterial, antioxidant, and anticancer agents is already widespread [46]. Several studies have found GA, a polyphenolic compound, in various natural sources, including Rosa chinensis Jacq. flowers have significant antibacterial activity against pathogenic Vibrios species. These pathogenic Vibrios are responsible for multiple severe diseases in humans and animals. GA has been found to inhibit the growth of Vibrio bacteria by disrupting the bacterial cell wall and membrane, preventing the bacteria from multiplying and causing infection.

Table 16.3 Various polyphenolic compounds are used as disinfectants.

Polyphenolic Compound	Structure	Sub-class	Source	Active against given Pathogens	Ref.
Gallic acid	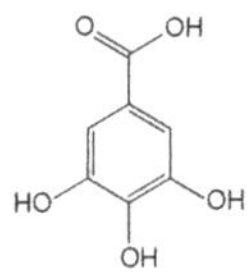	Benzoic acid	*Caesalpinia mimosoides* Lamk	Human Pathogens – *Escherichia coli*, *Staphylococcus aureus*, *Salmonella*	[38, 39]
Pinoresinol	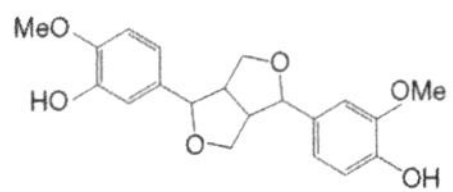	Lignans	*Eucommia ulmoides*, *Forsythia suspensa* and *Styrax* sp.	*E. coli*, *L. monocytogenes*, *Salmonella enteriditis*	[40, 41]
Morin	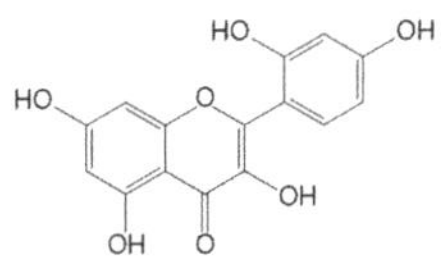	Flavone	Mulberry figs and oriental medicinal herbs	*Staphylococcus aureus* and *Bacillus cereus*	[42, 43]
(-) Epicatechin	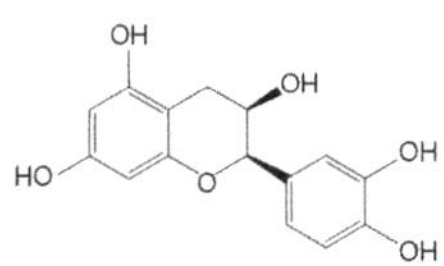	Flavonoids	Cocoa, green tea, black tea, peanuts, grapes, berries, apples, and dark chocolate	Actinobacteria and Firmicutes	[37, 44]
(+) catechin	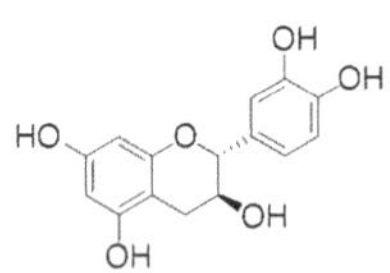	flavonoids	tamarind seeds	Bacteroidetes, Pseudomonas (Proteobacteria)	[37, 45]

GA has also been discovered to have remarkable properties that promote elasticity and plasticity. As a result, it can be used as a plasticizer, a substance added to materials to soften and make them more flexible. Brittleness and rigidity were common issues encountered while processing materials in traditional manufacturing processes. The addition of GA, on the other hand, has been shown to alleviate these issues, resulting in more pliable materials. GA has this property, which can be helpful in various industries, including food and pharmaceuticals [39]. GA is produced from 3-dehydroshikimate (DHS) by the action of shikimate dehydrogenase (AroE) [47]. Phosphoenolpyruvate, when it reacts with Erythrose-4-phosphate with the help of catalyst DAHP synthase (3-deoxy-D-arabino-heptulosonate 7-phosphate synthetase), forms D-Arabino heptulosonic acid 7-phosphate (DAHP). DAHP thus began converted into 3-Dehhydroquinic acid (DHQ) via enzyme DHQ synthetase (3-Dehydroquinate synthase), which further gets converted to 3-DHS (3-Dehydroshikimic acid) with DHQ dehydratase (3-Dehydroquinate Dehydratase). GA is formed from (3-DHS) via shikimate dehydrogenase (Figure 16.6) [48].

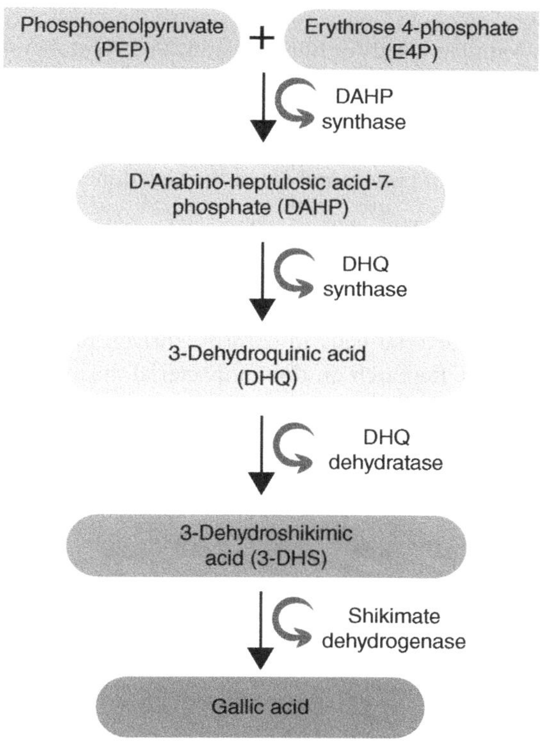

Figure 16.6 Synthesis of Gallic Acid [48]. https://www.canva.com/design/DAFf1b_6NXs/mm210G8nLM4E01ua0y5u4Q/edit?utm_content=DAFf1b_6NXs&utm_campaign=designshare&utm_medium=link2&utm_source=sharebutton.

Low molecular weight tri phenolic chemical GA (3,4,5-trihydroxybenzoic acid) can be found in monomeric and polymeric forms. Although not completely understood, its route of antimicrobial action may be related to the deactivation of numerous microbial proteins, including adhesions, enzymes, and cell envelope transport proteins. GA was discovered to have synergistic effects with several chemotherapeutic medications and as an anchor for creating novel pharmacological compounds with various therapeutic properties [49]. The color and fragrance of water are the first things that GA affects. Secondary reactions between GA and chlorine, a common disinfectant for water, can result in the formation of substances like chloroform and halo acetic acids. These substances are poisonous to people and can decrease the oxygen dissolved in water courses, which is necessary for aquatic organisms to survive. They are also carcinogenic, teratogenic, and mutagenic [50].

16.3.2 Pinoresinol

Pinoresinol is a biologically active lignan with a wide range of properties [51]. Pinoresinol is a dimer of coniferyl alcohol (Figure 16.6), is one of the physically simplest lignans and is typically found in woody or fibrous plants. Pinoresinol production by plants varies greatly. Young leaves, seeds, and various plant reproductive organs have all been discovered to contain high levels of pinoresinol. Because of the substance's antihelminthic, anti-inflammatory, and antifungal properties make it a defense agent [52]. Formation of pinoresinol dimer through dimerization of coniferyl alcohol (Figure 16.7) [53].

Pinoresinol has some antibacterial activity against Gram-positive and Gram-negative bacteria, including *S. aureus* and *B. subtilis*. The cell wall may be significantly damaged by pinoresinol. In Gram-positive bacteria, it breaks down the cell wall. Pinoresinol treatment causes bacterial cells to deform, shrivel, pit, stick together, and break in various places. Research on the antibacterial mechanism demonstrates that pinoresinol's main form of antibacterial action should involve direct cell wall destruction and an increase in the permeability of bacterial plasma membrane [52].

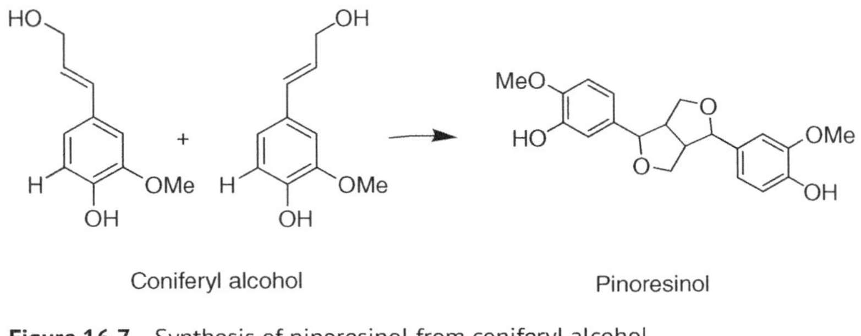

Figure 16.7 Synthesis of pinoresinol from coniferyl alcohol.

16.3.3 (+) Catechin

The health benefits of green tea are primarily attributed to its catechins, a subclass of flavonoids known as flavan-3-ols. Catechins are plants' secondary metabolites that induce antioxidant enzymes, block pro-oxidant enzymes, scavenge ROS, and produce phase II detoxifying enzymes [54]. The conversion of flavanones to leucoanthocyanidin is catalyzed by the enzymes dihydroflavonol 4-reductase (DFR), flavanone 3-hydroxylase (F3H), flavonoid 3'-hydroxylase (F3''H), flavonoid 3'5'-hydroxylase (F3'5'H), and flavanone 3-hydroxylase (F3H). Subsequently, leucoanthocyanidin 4-reductase (LAR) transforms leucoanthocyanidins into catechins [55]. Naringenin is first converted into dihydrokaempferol with the help of catalyst F3H (flavonoid 3 hydroxylase) [56] (Figure 16.8). Further, dihydrokaempferol formed is converted into dihydroquircetin via F3'H (flavanone 3-hydroxylase), which further gets converted to Leucocyanidin by catalyst; DFR (dihydroflavonol-4-reducatse). Leucodelphinidin formed is finally converted to (+) Catechin with the enzyme LAR (leucoanthocyanidin reductase). Large amounts of (+)-catechin can be found in a variety of fruits, red wine, green tea, and chocolate. Because of its polyphenolic structure, which encourages electron delocalization and neutralizes ROS molecules, it has an exceptional antioxidant effect [57].

16.4 Conclusion and Future Outlook

Polyphenolic compounds (anthocyanins, phenolic acids, flavonoids, lignans and others) are naturally occurring chemicals with complicated structures. They have several beneficial qualities, including antioxidant, anti-inflammatory, antibacterial, and anticancer effects. These molecules have found uses in a variety of sectors, including food science, where they play an important role in food stability, flavor profiles, and textures. They have also been discovered to offer therapeutic promise in the treatment of a variety of ailments such as cancer, diabetes, cardiovascular disease, arthritis, and neurodegenerative disorders. Due to their diverse biological activity, polyphenolic compounds have piqued the interest of the pharmaceutical, nutraceutical, and cosmetic sectors. Further, it is important to mention that pests or insects harm plants, agriculture, and food. Therefore, there is a need to address this problem. So, potential biological compounds like polyphenolic compounds could be promising candidates in pest management.

More than eight thousand polyphenolic compounds are reported for their different importance. The polyphenols market is increasing significantly daily as they can enhance the nutritional values of eating materials and others. They have a wide range of biological activities like antioxidant, antimicrobial, anti-cancerous, and many others. We all know that watering materials play an essential role in fighting diseases (non-communicable, neurodegenerative diseases, and obesity), so the polyphenolic compounds can be beneficial. Although, some of the polyphenolic compounds showed toxicity, if taken excessively. Therefore, there is a need

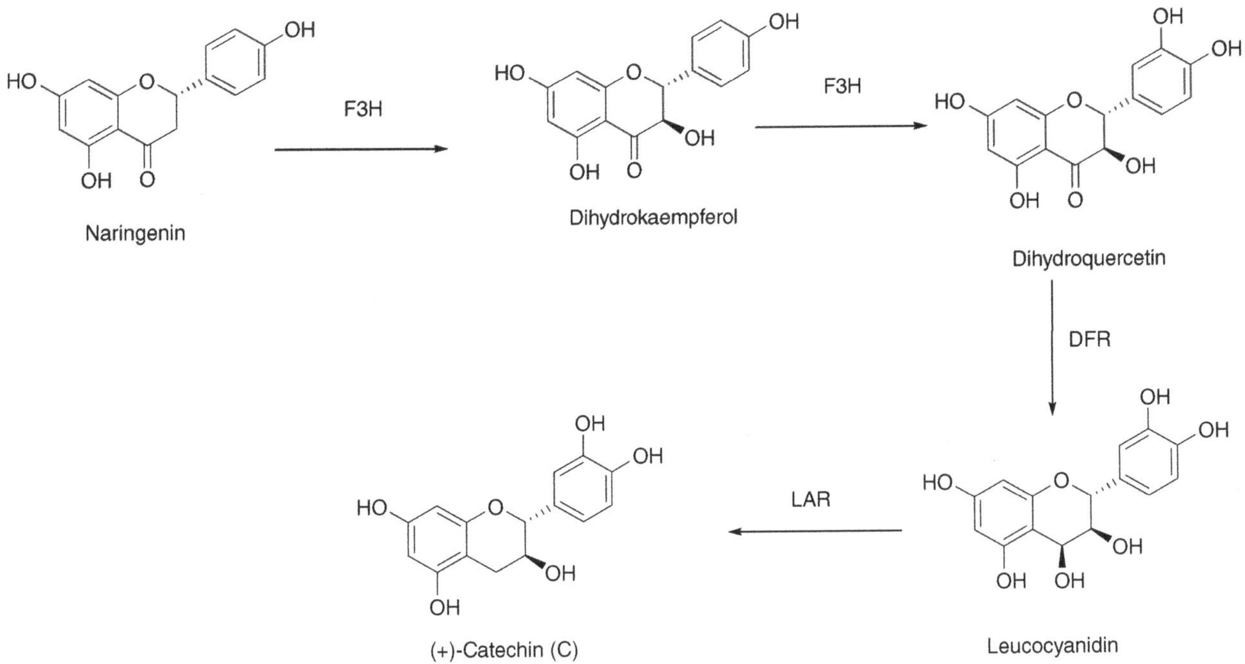

Figure 16.8 Synthesis of (+)-Catechin from Naringenin.

to take or consume in safe amounts. Further, there is a need to explore polyphenolic compounds in pest management thoroughly.

References

1 Zhang, J., Li, C., Rahaman, M. et al. (2022). A comprehensive review of image analysis methods for microorganism counting: from classical image processing to deep learning approaches. *Artificial Intelligence Review* 55 (4). Springer Netherlands: 2875–2944. https://doi.org/10.1007/s10462-021-10082-4.

2 Khan, M., Khan, A.U., Rafatullah, M. et al. (2022). Search for effective approaches to fight microorganisms causing high losses in agriculture: application of *P. lilacinum* metabolites and mycosynthesised silver nanoparticles. *Biomolecules* 12 (2): 174. https://doi.org/10.3390/biom12020174.

3 Sun, W. and Shahrajabian, M.H. (2023). Therapeutic potential of phenolic compounds in medicinal plants-natural health products for human health. *Molecules* 28 (4): 1845. https://doi.org/10.3390/molecules28041845.

4 González-Sarrías, A., Tomás-Barberán, F.A., and García-Villalba, R. (2020). Structural diversity of polyphenols and distribution in foods. *Diet Polyphenols* 1–29. https://doi.org/10.1002/9781119563754.ch1.

5 Andabaka, Ž., Šikuten, I., Tomaz, I. et al. (2022). Diversity of seed flavan-3-Ols in croatian native grapevine cultivars (*Vitis vinifera* L.) grown in coastal region. *Diversity* 14 (8): 1–11. https://doi.org/10.3390/d14080667.

6 Li, Z., Zhou, J., Ji, L. et al. (2022). Recent advances in the pharmacological actions of apigenin, its complexes, and its derivatives. *Food Review International* 39 (9): 6568–6601. https://doi.org/10.1080/87559129.2022.2122989.

7 Safdar, M.A., Aslam, R.M.N., Shakeel, A. et al. (2023). Cyanidin as potential anticancer agent targeting various proliferative pathways. *Chemical Biology & Drug Design* 101 (2): 438–452. https://doi.org/10.1111/cbdd.14173.

8 Massarioli, A.P., de Oliveira Sartori, A.G., Juliano, F.F. et al. (2023). Simulated gastrointestinal digestion/Caco-2 cell model to predict bioaccessibility and intestinal permeability of *p*-coumaric acid and p-coumaroyl derivatives in peanut. *Food Chemistry* 400: 134033. https://doi.org/10.1016/j.foodchem.2022.134033.

9 Duke, S.O. (2022). Benefits of resveratrol and pterostilbene to crops and their potential nutraceutical value to mammals. *Agriculture* 12 (3): 368. https://doi.org/10.3390/agriculture12030368.

10 Song, Q., Liu, H., Zhang, Y. et al. (2022). Lipidomics revealed alteration of the sphingolipid metabolism in the liver of nonalcoholic steatohepatitis mice treated with scoparone. *ACS Omega* 7 (16): 14121–14127. https://doi.org/10.1021/acsomega.2c00693.

11 Barak, D. (2022). Role of plant metabolites in plant protection and their potential in integrated pest management. *Pharmaceutical Innovation* 11 (5): 699–704. https://doi.org/10.22271/tpi.2022.v11.i5i.12469.

12 Trypuć, A.J. and Wiater, J. (2022). Protective effect of plant compounds in pesticides toxicity. *Journal of Environmental Health Science and Engineering* 20 (2): 1035–1045. https://doi.org/10.1007/s40201-022-00823-0.

13 Hassaan, M.A. and El Nemr, A. (2020). Pesticides pollution: classifications, human health impact, extraction and treatment techniques. *Egyptian Journal of Aquatic Research* 46 (3): 207–220. https://doi.org/10.1016/j.ejar.2020.08.007.

14 Zheng, R., Ma, Y., Liu, L. et al. (2022). Synergistic improvement of production, economic return and sustainability in the tea industry through ecological pest management. *Horticulturae* 8 (12): 1155. https://doi.org/10.3390/horticulturae8121155.

15 Cui, G., Yuan, H., He, W. et al. (2022). Synergistic effects of botanical curcumin-induced programmed cell death on the management of *Spodoptera litura* Fabricius with avermectin. *Ecotoxicology and Environmental Safety* 229: 113097. https://doi.org/10.1016/j.ecoenv.2021.113097.

16 Wang, R., Wang, R., Zhang, Q. et al. (2023). Toxicity, baseline of susceptibility, detoxifying mechanism and sublethal effects of chlorogenic acid, a potential botanical insecticide, on *Bemisia tabaci*. *Frontiers in Plant Science* 14: 1–9. https://doi.org/10.3389/fpls.2023.1150853.

17 Devi, T.B., Jena, S., Patra, B. et al. (2022). Acute and sub-acute toxicity evaluation of dihydro-*p*-coumaric acid isolated from leaves of *Tithonia diversifolia* Hemsl. A. Gray in BALB/c mice. *Frontiers in Pharmacology* 13 (November): 1–11. https://doi.org/10.3389/fphar.2022.1055765.

18 Gao, Y.-L., Pan, Z.-Y., Meng, X. et al. (2022). The effect of quercetin on the growth, development, nutrition utilization, and detoxification enzymes in *Hyphantria cunea* drury (lepidoptera: arctiidae). *Forests* 13 (11): 1945. https://doi.org/10.3390/f13111945.

19 Tyagi, S., Rathinam, M., Shashank, P.R. et al. (2022). Deciphering of pod borer [*Helicoverpa armigera* (Hübner)] resistance in *Cajanus platycarpus* (Benth.) offers novel insights on the reprogramming and role of flavonoid biosynthesis pathway. *Toxins (Basel)* 14 (7): 455. https://doi.org/10.3390/toxins14070455.

20 Sharma, A., Gill, N., and Kumar, R. (2022). Development and validation of thin layer chromatography-densitometric method for quantification of kaempferol and chlorogenic acid in methanolic extract of dragea volubilis. *Asian Pacific Journal of Health Sciences* 9: 112–116. https://doi.org/10.21276/apjhs.2022.9.4.23.

21 Zhao, X., Liu, Z., Liu, H. et al. (2022). Hybrid molecules based on caffeic acid as potential therapeutics: a focused review. *European Journal of Medicinal Chemistry* 243 (January): 114745. https://doi.org/10.1016/j.ejmech.2022.114745.

22 Punia, A., Singh, V., Thakur, A., and Singh, N. (2023). Heliyon Impact of caffeic acid on growth, development and biochemical physiology of insect pest, *Spodoptera litura* (Fabricius). *Heliyon* 9 (3): e14593. https://doi.org/10.1016/j.heliyon.2023.e14593.

23 Pokharel, S.S., Zhong, Y., Changning, L. et al. (2022). Influence of reduced N-fertilizer application on foliar chemicals and functional qualities of tea plants under *Toxoptera aurantii* infestation. *BMC Plant Biology* 22 (1): 1–19. https://doi.org/10.1186/s12870-022-03533-9.

24 Tongkaew, P., Tohraman, A., Bungaramphai, R. et al. (2022). Kluai Hin (*Musa sapientum* Linn.) peel as a source of functional polyphenols identified by HPLC-ESI-QTOF-MS and its potential antidiabetic function. *Scientific Reports* 12 (1): 1–8. https://doi.org/10.1038/s41598-022-08008-3.

25 Cai, M., Liu, J., Song, X. et al. (2022). De novo biosynthesis of *p*-coumaric acid and caffeic acid from carboxymethyl-cellulose by microbial co-culture strategy. *Microbial Cell Factories* 21 (1): 1–12. https://doi.org/10.1186/s12934-022-01805-5.

26 Cui, P., Zhong, W., Qin, Y. et al. (2020). Characterization of two new aromatic amino acid lyases from actinomycetes for highly efficient production of *p*-coumaric acid. *Bioprocess and Biosystems Engineering* 43 (7): 1287–1298. https://doi.org/10.1007/s00449-020-02325-5.

27 Golovinskaia, O. and Wang, C.K. (2023). The hypoglycemic potential of phenolics from functional foods and their mechanisms. *Food Science and Human Wellness* 12 (4): 986–1007. https://doi.org/10.1016/j.fshw.2022.10.020.

28 Bal, S.S., Leishangthem, G.D., Sethi, R.S., and Singh, A. (2022). *p*-Coumaric acid ameliorates fipronil induced liver injury in mice through attenuation of structural changes, oxidative stress and inflammation. *Pesticide Biochemistry and Physiology* 180 (October 2021): 104997. https://doi.org/10.1016/j.pestbp.2021.104997.

29 Kapešová, J., Petrásková, L., Markošová, K. et al. (2019). Bioproduction of quercetin and rutinose catalyzed by rutinosidase: novel concept of 'solid state biocatalysis'. *International Journal of Molecular Sciences* 20 (5): 1112. https://doi.org/10.3390/ijms20051112.

30 Peng, X., Liu, K., Hu, X. et al. (2023). Hesperetin-Cu(II) complex as potential α-amylase and α-glucosidase inhibitor: Inhibition mechanism and molecular docking. *Spectrochimica Acta, Part A: Molecular and Biomolecular Spectroscopy* 290 (September 2022): 122301. https://doi.org/10.1016/j.saa.2022.122301.

31 Hafez, D.A., Abdelmonsif, D.A., Aly, R.G. et al. (2022). Role of fennel oil/ quercetin dual nano-phytopharmaceuticals in hampering liver fibrosis: comprehensive optimization and in vivo assessment. *Journal of Drug Delivery Science and Technology* 69 (February): 103177. https://doi.org/10.1016/j.jddst.2022.103177.

32 Song, J., Jeong, J., Kim, E.H., and Hong, Y.S. (2023). A strategy for healthy eating habits of daily fruits revisited: a metabolomics study. *Current Research in Food Science* (6, October 2022): 100440. https://doi.org/10.1016/j.crfs.2023.100440.

33 Park, D.H., Park, J.Y., Kang, K.S., and Hwang, G.S. (2021). Neuroprotective effect of gallocatechin gallate on glutamate-induced oxidative stress in hippocampal HT22 cells. *Molecules* 26 (5): https://doi.org/10.3390/molecules26051387.

34 Ge, G., Zhang, Y., Zhu, G.H. et al. (2023). Discovery and characterization of the covalent SARS-CoV-2 3clpro Inhibitors from ginkgo biloba extract via integrating chemoproteomic and biochemical approaches. *SSRN Electronic Journal* 154796. https://doi.org/10.2139/ssrn.4332105.

35 Rippin, A., Sharma, K., and Beniwal, V. (2022). Biosynthesis and medicinal applications of proanthocyanidins: a recent update. *Biocatalysis and Agricultural Biotechnology* 45 (September 2021): 102500. https://doi.org/10.1016/j.bcab.2022.102500.

36 van Dijk, H.F.G., Verbrugh, H.A., and Ad hoc advisory committee on disinfectants of the Health Council of the Netherlands (2022). Resisting disinfectants. *Communication & Medicine* 2 (1): 1–5. https://doi.org/10.1038/s43856-021-00070-8.

37 Li, J., Feng, C., and Xin, B. (2022). Study on the effect of a tea polyphenol disinfectant on antibiotic resistance genes in drinking water. *Environmental Science: Water Research & Technology* 8 (12): 3101–3112. https://doi.org/10.1039/d2ew00118g.

38 Pi, J., Jin, P., Tang, X. et al. (2022). Microwave-assisted polyethylene glycol-based aqueous two-phase extraction of gallic acid and ellagic acid from *Euonymus alatus*: process optimization, quantification analysis and antioxidant activity. *Chemical Engineering and Processing Process Intensification* 172 (December 2021): 108772. https://doi.org/10.1016/j.cep.2021.108772.

39 Sun, X., Wang, Z., Kadouh, H., and Zhou, K. (2014). The antimicrobial, mechanical, physical and structural properties of chitosan-gallic acid films. *LWT* 57 (1): 83–89. https://doi.org/10.1016/j.lwt.2013.11.037.

40 Xie, Y., Fan, D., Wang, F., and Zhang, L. (2022). Pinoresinol diglucoside alleviates pressure overload-induced cardiac injury via the AMPK/SIRT3/RIG-1 pathway. Research Square, 1–19. https://doi.org/10.21203/rs.3.rs-2245921/v1.

41 Fancello, F., Multineddu, C., Santona, M. et al. (2022). Antimicrobial activities of virgin olive oils in vitro and on lettuce from pathogen-inoculated commercial quick salad bags. *Food Control* 133: 108657. https://doi.org/10.1016/j.foodcont.2021.108657.

42 Asati, V., Deepa, P.R., and Sharma, P.K. (2022). Desert legume *Prosopis cineraria* as a novel source of antioxidant flavonoids/isoflavonoids: biochemical characterization of edible pods for potential functional food development. *Biochemistry and Biophysics Reports* 29: 101210. https://doi.org/10.1016/j.bbrep.2022.101210.

43 Yaneva, Z., Beev, G., Rusenova, N. et al. (2022). Antimicrobial potential of conjugated lignin/morin/chitosan combinations as a function of system complexity. *Antibiotics* 11 (5): https://doi.org/10.3390/antibiotics11050650.

44 Dash, J.R., Pattnaik, G., Ghosh, G. et al. (2023). Protective effect of epicatechin in diabetic-induced peripheral neuropathy: a review. *Journal of Applied Pharmaceutical Science* 13 (1): 056–063. https://doi.org/10.7324/JAPS.2023.130105-1.

45 Wandee, R., Sutthanut, K., Songsri, J. et al. (2022). Tamarind seed coat: a catechin-rich source with anti-oxidation, anti-melanogenesis, anti-adipogenesis and anti-microbial activities. *Molecules* 27 (16): 5319. https://doi.org/10.3390/molecules27165319.

46 Mehmood, S., Kausar Janjua, N., Tabassum, S. et al. (2022). Cost effective synthesis approach for green food packaging coating by gallic acid conjugated gold nanoparticles from *Caesalpinia pulcherrima* extract. *Results in Chemistry* 4 (April): 100437. https://doi.org/10.1016/j.rechem.2022.100437.

47 Liu, L., Ma, X., Bilal, M. et al. (2022). Toxicity and inhibition mechanism of gallic acid on physiology and fermentation performance of *Escherichia coli*. *Bioresources and Bioprocessing* 9 (1): 1–16. https://doi.org/10.1186/s40643-022-00564-w.

48 Osman, N.I., Jaafar Sidik, N., and Awal, A. (2018). Efficient enhancement of gallic acid accumulation in cell suspension cultures of *Barringtonia racemosa* L. by elicitation. *Plant Cell, Tissue and Organ Culture* 135 (2): 203–212. https://doi.org/10.1007/s11240-018-1456-z.

49 Goc, A., Rath, M., and Niedzwiecki, A. (2022). Inhibition of borrelia burgdorferi-induced TLR2-NFκB canonical signaling by gallic acid through targeting the CD^{14+} adaptor protein and p65 molecule. *International Journal of Molecular Sciences* 23 (19): https://doi.org/10.3390/ijms231910987.

50 Pirozzi, D., Pansini, M., Marocco, A. et al. (2023). Adsorption of gallic acid by tailor-made magnetic metal-ceramic nanocomposites. *Journal of Molecular Liquids* 371: 121083. https://doi.org/10.1016/j.molliq.2022.121083.

51 Zhou, H., Huang, R., Su, T. et al. (2022). A c-MWCNTs/AuNPs-based electrochemical cytosensor to evaluate the anticancer activity of pinoresinol from *Cinnamomum camphora* against HeLa cells. *Bioelectrochemistry* 146 (April): https://doi.org/10.1016/j.bioelechem.2022.108133.

52 Zhou, H., Ren, J., and Li, Z. (2017: https://doi.org/10.1016/j.foodcont.2017.03.041). Highlights We found the pinoresinol attached to lignans for the first time in *Cinnamomum*. *Food Control* 79: 192–199.

53 Hemmati, S., Schmidt, T.J., and Fuss, E. (2007). (+) -Pinoresinol /(À) -lariciresinol reductase from *Linum perenne* Himmelszelt involved in the biosynthesis of justicidin B OH. *FEBS Letters* 581: 603–610. https://doi.org/10.1016/j.febslet.2007.01.018.

54 Abdel-Aal, E.S.M., Rabalski, I., Mats, L., and Rai, I. (2022). Identification and quantification of anthocyanin and catechin compounds in purple tea leaves and flakes. *Molecules* 27 (19): 1–15. https://doi.org/10.3390/molecules27196676.

55 Zhang, X., Jiang, X., He, Y. et al. (2019). AtHB2, a class II HD-ZIP protein, negatively regulates the expression of CsANS, which encodes a key enzyme in Camellia sinensis catechin biosynthesis. *Physiologia Plantarum* 166 (4): 936–945. https://doi.org/10.1111/ppl.12851.

56 Xiong, L., Li, J., Li, Y. et al. (2013). Plant physiology and biochemistry dynamic changes in catechin levels and catechin biosynthesis-related gene expression in albino tea plants (*Camellia sinensis* L.). *Plant Physiology and Biochemistry* 71: 132–143. https://doi.org/10.1016/j.plaphy.2013.06.019.

57 Cadoná, F.C., Dantas, R.F., de Mello, G.H., and Silva-Jr, F.P. (2022). Natural products targeting into cancer hallmarks: an update on caffeine, theobromine, and (+)-catechin. *Critical Reviews in Food Science and Nutrition* 62 (26): 7222–7241. https://doi.org/10.1080/10408398.2021.1913091.

17

Polyphenols for Antioxidant Application: Biochemistry of Polyphenols

Akbar Esmaeili

Department of Chemical Engineering, North Tehran Branch, Islamic Azad University, Tehran, Iran

17.1 Introduction

Free radicals are harmful compounds that cause changes in the body's cells, damage the cell's nucleus, and, finally, their death. These compounds are produced naturally in the body and increase due to environmental toxins [1]. Antioxidants are compounds that remove these harmful compounds from the body. Polyphenols are various compounds of plant origin. These compounds are the most common food antioxidants, and their daily consumption is about 1 g, more than other compounds known as antioxidants [2]. During the last decade, research on the effect of polyphenols on human health has increased, and their role in preventing chronic diseases such as heart disease, osteoporosis, diabetes, cancer, and neurodegenerative diseases such as Alzheimer's and Parkinson's is supported by researchers [3]. The effects of polyphenols are applied in the following ways: (i) Increasing antioxidant activity in plasma. (ii) Decrease in the concentration of (low-density lipoprotein or LDL) and increase in the concentration of (high-density lipoprotein or HDL) in plasma. (iii) Dilation of blood vessels and decrease in blood pressure. (iv) Increase the concentration of alkaline phosphatase and osteocalcin in the serum. (v) Decrease in the concentration of thromboxane B2. (vi) Increasing the plasma concentration of ascorbate and vitamin B6 [3]. The phenol group in polyphenol compounds receives an electron, forms stable phenoxyl radicals, and disrupts the reaction of oxidation chains in the cell [4]. Porphyromonas is considered a primary pathogen in chronic periodontitis. The colonization of this bacterium in the subgingival plaque is an initial stage.

Science and Engineering of Polyphenols: Fundamentals and Industrial Scale Applications,
First Edition. Edited by Chandrabhan Verma.
© 2024 John Wiley & Sons Inc. Published 2024 by John Wiley & Sons Inc.

Bacterial products can stimulate the production of cytokines and prostaglandins from the host cells, eventually destroying the tooth's protective structure.

Recently, the pathology of periodontal diseases has mentioned the role of products resulting from oxidative reactions and consequences resulting from lipid peroxidation. Paterniti and colleagues [5] stated that the antioxidant activity of the saliva of healthy people is 40–50% more than that of people with periodontitis; as a result of increasing production of free radicals and decreasing antioxidant activity, the balance between oxidant and non-oxidant factors is lost. The activation of NF-KB is under the control of the balance between oxidative and non-oxidative processes in the body. On the other hand, this factor plays a vital role in the expression of genes responsible for producing inflammatory mediators [6]. Polyphenols will probably reduce the expression of inflammatory mediators by suppressing their activation. These compounds prevent the infiltration of polymorphonuclear cells to the site of inflammation by reducing the incidence of binding molecules. Through this mechanism, mountain tea, with the scientific name *Hypericum perforatum*, exerts an anti-inflammatory effect on periodontal diseases in rats. This plant's anti-inflammatory and anti-depressant properties have been proven, and antibacterial and antiviral effects have also been attributed to it [5]. It has even been considered a cost-effective treatment for AIDS [7]. Two medicinal products are available in the Iranian market: Haipiran drops and Perforan tablets, both used in treating depression.

17.1.1 Types of Polyphenols

So far, more than eight thousand different types of polyphenols have been discovered, which can be classified into four main groups: (i) Flavonoids: Flavonoids make up about 60% of all polyphenols and are found in foods such as apples, onions, dark chocolate, and red cabbage. Famous flavonoids include quercetin, kaempferol, catechins, and anthocyanins. (ii) Phenolic acids: this polyphenol group makes up about 30% of all polyphenols. Among the famous phenolic acids, we can mention stilbenes and lignans, primarily found in fruits and vegetables, whole grains or whole grains, and seeds. (iii) Polyphenolic amides: This group of polyphenols includes capsaicinoids, which are found in chili peppers, and avenanthramides, which are found in oats. (iv) Other polyphenols: Some other polyphenols include ellagic acid in berries, curcumin in turmeric, and lingonin in flaxseeds, sesame seeds, and whole grains [8–10].

17.1.2 Plants and Fruits Derived Polyphenols

The amount of polyphenols in any food varies depending on where the plant is grown. It was developed and transported, how ripe. The most important source of polyphenols is tea, especially green tea. Some animal studies have shown that tea's polyphenols prevent cancer cell growth. However, scientists are still investigating how tea polyphenols affect human health. Usually, green tea is known as a source of antioxidants, but all teas contain polyphenols. Also, coffee has some amount of polyphenols among drinks. Moderate consumption of coffee has been shown to reduce the risk of type 2 diabetes, some cancers, heart problems, stroke, Alzheimer's, and Parkinson's. One of the reasons that coffee can reduce the risk of diseases is that its beans contain a lot of polyphenols. The amount of caffeine that each of us can tolerate is different. But generally, four cups of coffee daily is enough for an adult. In the case of coffee, you should also pay attention to sugar and cream because these two increase calories. These fruits also contain polyphenols: oranges, apples, grapes, peaches, grapefruit juice, cherries, blueberries, pomegranate juice, raspberries, cranberries, black currants, plums, Persian raspberries, strawberries, apricots, cherries, and pears. Among dried fruits, legumes, and grains with polyphenols, we can mention roasted soybeans, black beans, white beans, chestnuts, hazelnuts, walnuts, and hazelnuts almonds, walnuts, flax seeds, oats, whole wheat, and rye flour. In general, nuts are not high in carbohydrates, and even though they contain fat, their consumption cannot cause weight gain [11].

Research has shown that nuts can improve blood lipids and reduce inflammation, and legumes are a rich source of fiber, protein, vitamins, minerals, and polyphenols. Although beans are a source of carbohydrates, the high fiber in them keeps their glycemic load low. Research has shown that legumes have antioxidant properties. Legumes are generally known for their antioxidant benefits, but some, such as peas, red beans, cowpeas, etc., also have health benefits. Consumption of legumes plays a significant role in reducing the risk of cardiovascular diseases. Celery, onion, parsley, cabbage, and broccoli have powerful polyphenols in vegetables. Leafy vegetables are known as a valuable source of nutrients.

They are deficient in calories and packed with nutrients. These vegetables are rich in fiber, iron, vitamins A, C, and magnesium. They are also a rich source of polyphenols. Research in 2017 showed that the polyphenol in leafy vegetables could affect their iron. The more polyphenols in vegetables, the easier it is to absorb iron [12]. Polyphenols help the growth of intestinal bacteria, fight harmful bacteria in the digestive tract, and improve digestion. For example, evidence shows that tea extract is rich in polyphenols and increases the growth of bids and bacteria. Green tea polyphenols also fight against harmful bacteria such as *Clostridium difficile, Escherichia coli*, and salmonella and reduce the symptoms of gastrointestinal ulcers and improve inflammatory bowel disease. Polyphenols also help to increase the life of probiotics. However, more research is needed in this field [13].

The amount of polyphenols in any food varies depending on where the plant is grown. It was developed and transported. It is cooked or prepared. The most important source of polyphenols is tea, especially green tea. Some animal studies have shown that tea's polyphenols prevent cancer cell growth. However, scientists are still investigating how tea polyphenols affect human health. In the research conducted on humans than on animals. Usually, green tea is known as a source of antioxidants, but all teas contain polyphenols [14]. Also, coffee has some amount of polyphenols among drinks. Moderate consumption of coffee has been shown to reduce the risk of type 2 diabetes, some cancers, heart problems, stroke, Alzheimer's, and Parkinson's. One of the reasons that coffee can reduce the risk of diseases is that its beans contain a lot of polyphenols. The amount of caffeine that each of us can tolerate is different. But generally, four cups of coffee daily is enough for an adult. In the case of coffee, you should also pay attention to sugar and cream because these two increase calories.

These fruits also contain polyphenols: oranges, apples, grapes, peaches, grapefruit juice, cherries, blueberries, pomegranate juice, raspberries, cranberries, black currants, plums, Persian raspberries, strawberries, apricots, cherries, and pears [15]. Among dried fruits, legumes, and grains with polyphenols, we can mention roasted soybeans, black beans, white beans, chestnuts, hazelnuts, walnuts, and hazelnuts almonds, walnuts, flax seeds, oats, whole wheat, and rye flour. In general, nuts are not high in carbohydrates; even though they contain fat, their consumption cannot cause weight gain. Research has shown that nuts can improve blood lipids and reduce inflammation, and legumes are a rich source of fiber, protein, vitamins, minerals, and polyphenols. Although beans are a source of carbohydrates, the high fiber in them keeps their glycemic load low. Research has shown that legumes have antioxidant properties. Legumes are generally known for their antioxidant benefits, but some, such as peas, red beans, cowpeas, etc., also have health benefits. Consumption of legumes plays a significant role in reducing the risk of cardiovascular diseases [16].

Celery, onion, parsley, cabbage, and broccoli have powerful polyphenols in vegetables. Leafy vegetables are known as a valuable source of nutrients. They are deficient in calories and packed with nutrients. These vegetables are rich in fiber, iron, vitamins A, C, and magnesium. They are also a rich source of polyphenols. Research in 2017 showed that the polyphenol in leafy vegetables could affect their iron. The subject means that the more polyphenols in vegetables, the easier it is to absorb iron. More extensive research is needed to determine which vegetables have more polyphenols and iron. So, you try to use different types of them. Fats like dark chocolate, virgin olive oil, and sesame oil have polyphenols. Polyphenols in cocoa have been shown to contain antioxidant, anti-inflammatory, and antiarthritic properties. For this reason, cocoa is one of the foods that help prevent cardiovascular diseases. Cocoa products with the most cocoa, such as dark chocolate, can provide many polyphenols. Unfortunately, sweet chocolate drinks and chocolates contain added sugar, and you should keep a balance in their consumption. By consuming dark chocolate and introducing many polyphenols into your body, you have also blocked the way to harmful carbohydrates and calories. Unprocessed cocoa is a rich source of flavonoids. A study showed that cocoa consumption significantly reduces insulin resistance [17, 18].

17.1.3 Olive Oil-Based Polyphenols

Due to its high nutritional value, medicinal properties, and usability in cosmetics and health industries, Olive oil is one of the products whose demand is increasing daily in the world market. In the last 15 years, nearly 30% growth has been reported in the global production of olive oil. The countries of the administrative area alone account for 97% of the world's olive oil production. The best olive oil-producing countries are Spain, Italy, Greece, Turkey, Tunisia, Portugal, Morocco, and Algeria. In addition, olives are cultivated in the Middle East, the United States of America, Argentina, and Australia. Olive oil production is a significant economic activity, especially in Spain, Italy, and Greece [19, 20]. Three systems, including the traditional two-phase non-continuous application of centrifugation, are used to extract olive oil on an industrial scale. Three-phase centrifugation is often used today. The aqueous phase of the centrifuge systems is called the effluent of oil. The degrees produced in the three-phase system include olive oil. Solid cake is a combination of olive oil, effluent, and solid cake of pulp and olive kernels, which is transferred to other departments to extract the remaining oil and kernel oil. The output of the two-phase system is olive oil and semi-solid (watery) waste. Non-continuous process, two-phase waste is less but more concentrated (for every 1000 kg of olives, 0.5–1 m^2) compared to centrifuge processes (it produces 1–1.5 m^2/1000 kg of olives). The annual production of wastewater from olive oil factories in the world is estimated to be 10–30 million m^2 due to the high water consumption in the preparation of olive paste in the three-phase process; the output liquid of this system, compared to Biphasic, is the majority.

Depending on many factors, including the type, maturity, and origin of olives, weather conditions, cultivation or processing methods, and olive oil extraction methods, the characteristics and percentage of waste compounds of olive oil factories vary. This wastewater contains 83–94% water, 18–4% organic materials (including carbohydrates, lipids, phenolic compounds, proteins, etc.), and 0.5–2.4% inorganic materials. Biological oxygen demand (BOD) was amount eight. Solid materials better quality suspension is in 7. Chemical oxygen demand (COD) ranges are 35–110, 45–170, and 1–9 g/L. Olive oil effluent has a high organic content and carbon–nitrogen ratio. A high percentage of organic compounds in this wastewater are phenolic compounds [20].

Phenolic compounds are responsible for structural and protective functions in plants and contribute to the color, hardness, and bitterness of fruits or vegetables. In the olive oil extraction process, these compounds are released due to the destruction of the olive's physical structure. Also, some phenols are produced during extraction processes and enter the wastewater of these units. The concentration of phenolic compounds in olive oil effluent varies from 0.5 to 24 g per liter. Among the phenolic compounds in the sewage above, we can mention phenolic alcohols (such as tyrosol and hydroxytyrosol) and polyphenols such as phenolic acids, lignins, flavonoids, and tannins [20].

Antioxidant properties of phenols in olive oil wastewater have drawn attention to their reuse in pharmaceutical, food, and cosmetic industries. One of the antioxidant activities of these compounds is their ability to scavenge free radicals. It is related to the hardening of the arteries and also limits the oxidative damage of DNA. Hydroxytyrosol is one of the few foods approved by the European Food Safety Organization due to its high antioxidant capacity [21, 22]. More about this source texture text is required for additional translation information. These characteristics have made researchers try to recover valuable phenolic compounds from olive oil wastewater. However, phenolic compounds are on the list of priority hazardous toxic substances of the Environmental Protection Agency due to their harmful effects on plants and microorganisms. The high concentration of phenolic compounds in the effluents of olive oil extraction units causes: (i) changes in the color of streams and rivers, (ii) inhibits seed germination, (iii) inhibits the growth of plants, and (iv) prevents the growth of microorganisms. So, it is necessary to remove phenols from these effluents. Physiological, biological, and combined methods have been used to recover or remove phenolic compounds from this wastewater [23].

17.1.4 Extraction of Polyphenols

Polyphenols are substances found in plants and natural compounds. Polyphenols are antioxidants. They contain biological compounds that are known to be beneficial for health. Foods containing polyphenols reduce the risk of chronic disease. Cherries have a high polyphenol content. It contains polyphenols. Products made from these fruits also contain polyphenols. There are polyphenol plant extracts and food and medicinal supplements. Polyphenols are located in the cellular part of plants, producing polyphenol-rich foods released from plant cells. Ultrasonic extraction is a method for the cellular structures of plant compounds. Ultrasonic extraction from the thermal process prevents the deterioration of the forms of polyphenols. Ultrasound is a suitable method that improves extraction efficiency [15].

17.1.5 Tea-Related Polyphenols

Because of tea's many properties, it is recommended to drink it daily in many countries, not as a drink but as a herbal medicine. The bitter taste of tea is attributed to the presence of polyphenols. Polyphenols are a group of compounds found in abundance in tea, especially green tea, and have anti-cancer properties and prevent the occurrence of heart diseases. In recent decades, the popularity of complementary medicine has been increasing daily. Attempts to confirm the use of plants and plant products in preventing, controlling, and treating diseases have advanced research in laboratories and, in some cases, in the body environment. One of these medicinal plants is the tea plant. Due to humans' ancient knowledge of tea and its massive consumption, thousands of extensive research have been conducted on it, showing this valuable plant's importance. If fresh tea leaves are dried quickly, green tea is obtained. In this situation, tea is not allowed to ferment. The most important properties of green tea include preventing tooth decay due to high fluoride, preventing stomach and skin cancer, and strengthening the immune system. Chinese people have long used green tea as a medicinal drink. But if the fresh tea leaves are subjected to fermentation, changes in color, smell, taste, ingredients, and effects will occur compared to green tea. After fermentation, it is dried and packaged. Therefore, green and black tea are the same and differ in their actions. Catechins belong to the category of polyphenols. Green tea contains more catechins than black tea. Polyphenols are the most common food antioxidants essential in preventing chronic diseases, including cancer, by inhibiting free radicals. By attacking the cell's cytoplasmic membrane and destroying it, free radicals gain access to the DNA inside the cell and cause cancer by changing the DNA. In addition to eliminating free radicals, polyphenols also protect healthy cells by accelerating the death of cancer cells. According to the general chemical structure of catechins, the higher the number of hydroxyls (OH) groups, the higher the antioxidant capacity of polyphenols. Antioxidants protect the cell from potent oxidizing agents such as metal ions by forming a complex. The antioxidant property of green tea is five times higher due to the higher concentration of hydroxyl-rich polyphenols [3].

17.2 Antioxidant Activity of Polyphenols

The effect of polyphenols on bacterial biofilm formation plays a vital role in causing bacterial infections in humans. Biofilm allows bacteria to attack the host system and resist mechanical and chemical treatments [24, 25]. Previous studies have shown that polyphenols can prevent the attachment of pathogens such as *Escherichia coli* to the urinary tract mucosa and prevent infection. Also, they can inhibit *Helicobacter pylori*'s binding to the gastrointestinal mucosa's surface and prevent gastrointestinal ulcers [24, 25]. According to the study conducted by La and colleagues [26], polyphenols at a concentration of 50 μg/ml and more prevent the attachment of *Porphyromonas gingivalis* to oral epithelial cells and surfaces covered by extracellular matrix protein. The mechanism of this action is through these compounds' interference in the reaction between ligands and receptors. Therefore, they do not affect the already-formed biofilm. Because the mechanism of action of polyphenols is by inhibiting the attachment of bacteria to dental surfaces, bacterial resistance to these compounds will not occur [25]. These compounds prevent the formation of biofilm of *Fusobacterium nucleatum* and coaggregation and prevent *P. gingivalis* periodontopathogens [24].

Steinberg and colleagues [27] stated that polyphenols have an anti-caries effect by preventing the formation of biofilms of bacteria such as *S. mutants* and *Sciurus sabrinus*, similar to the development of polyphenols on periodontal pathogens. Excessive production of reactive oxygen species (ROS) may cause tissue damage and thus lead to the inflammatory process. The antioxidant activity of polyphenols depends on the structure of their functional groups. The number of hydroxyl groups significantly affects several mechanisms of antioxidant action, such as radical scavenging and metal ion adsorption ability [28].

17.3 Influence of Polyphenols on Antibacterial Biofilm

Biofilm formation plays a vital role in causing bacterial infections in humans. Biofilm allows bacteria to attack the host system and resist mechanical and chemical treatments [24, 25]. Previous studies have shown that polyphenols can prevent the attachment of pathogens such as *E. coli* to the urinary tract mucosa and prevent infection. Also, they can inhibit *H. pylori*'s binding to the gastrointestinal mucosa's surface and prevent gastrointestinal ulcers [24, 25]. According to the study conducted by La and colleagues [26], polyphenols at a concentration of 50 μg/ml and more prevent the attachment of *P. gingivalis* to oral epithelial cells and surfaces covered by extracellular matrix protein. The mechanism of this action is through these compounds' interference in the reaction between ligands and receptors. Therefore, they do not affect the already-formed biofilm. Because the mechanism of action of polyphenols is by inhibiting the attachment of bacteria to dental surfaces, bacterial resistance to these compounds will not occur [24, 25]. These compounds prevent the formation of biofilm of *F. nucleatum* and coaggregation and prevent *P. gingivalis* periodontopathogens [24, 25].

Steinberg and colleagues [27] stated that polyphenols have an anti-caries effect by preventing the formation of biofilms of bacteria such as *S. mutants* and *S. sabrinus*, similar to the development of polyphenols on periodontal pathogens. The stages of biofilm formation in bacteria follow a general model; In the first stage, the connection and establishment of the microorganism with the living or non-living surface happen. The primary relationship is with a non-specific character and is done by hydrophobic, van der Waals, and electrostatic bonds. In this case, the state of the microorganism changes from the planktonic form to the stationary form [29]. After the initial connection, the surface proteins of the microbe binding to the matrix specifically strengthen the attachment of the bacteria to the surface, and the multiplication of the bacteria forms a bacterial microcolony. Then the three-dimensional structure that is the mature biofilm is formed. This mature biofilm disintegrates after some time, but the released bacterial cells can cause infection and a new biofilm focus in other places [29].

In response to environmental changes, enzymes are secreted from the bacteria enclosed in the biofilm, affecting the extracellular matrix's composition. The matrix compounds create this structure's dried, completely resistant, and robust state. This feature helps bacterial cells to stick together. Bacterial cells must be together to establish intercellular communication, such as DNA transfer, often in the biofilm phase. Also, this extracellular matrix protects bacteria from the sting of antibacterial chemicals, oxidizing compounds, and the harmful effects of radiation. Although the microorganisms that make up the bio-layer also have planktonic life, the biochemistry of these microorganisms is different. Past investigations have shown that this change. The difference is due to the switching off of one group and the dynamic of another group of genes. For example, *Pseudomonas aeruginosa* needs alginate synthesis in the biolayer life mode; Therefore, alginate synthesizing genes become dynamic. Investigations show that the transcription of the algal gene, whose product is alginate, occurs four times more in the life of the biofilm than in the planktonic state [30].

Bacteria continuously secrete several sensory signaling molecules, such as acyl-homoserine lactone. At the same time, as the number of these bacteria increases, the concentration of symptom-giving substances also increases. When the amount of these substances reaches a threshold concentration, the bacteria respond and change their behavior by changing the activity of genes. This state is different in different species of bacteria. *P. aeruginosa* produces alginate. Genes that affect metabolic pathways and the production of pathogenic agents may be activated [31].

The research of the CBC Center in 2000 showed that the microorganisms in the bio-layer are more resistant to antimicrobial compounds than their free species (microorganisms with planktonic life). These studies show that more than 40% of the proteins in the cell wall of bacteria and microbes are different between planktonic and biolayer states. This protein change changes the antibiotic target sites on these microorganisms, so their response to antibiotics and death mechanisms differ. The investigations of this center in 2001 identified a sigma factor. It shows a kind of bacterial signal to change the biochemistry of biofilm life. This center has announced the inverse sigma factor; biofilms can be transformed into separate and planktonic cells [32]. For microorganisms, living in a biofilm has certain advantages. Microbial communities are usually much more resistant to stress. Potential stressors include water scarcity, increased or decreased environmental pH, or the presence of substances toxic to microbes, such as antibiotics, antimicrobials, or heavy metals.

There are many justifications for the high hardness and resistance of biofilms. For example, the EPS layer that covers them is a physical factor against the mentioned factors. This layer can help with water shortage or create resistance against ultraviolet rays that threaten microorganisms. In addition, harmful substances such as metals and antibiotics are either neutralized or inactivated by binding to EPS.

In addition, these substances lose their effective concentration in contact with surface cells before reaching the deeper cells of the biofilm. However, some antibiotics can penetrate EPS and invade different biofilm layers. Another defense mechanism is used here: the presence of physiologically dormant bacteria. All antibiotics need the cellular activities of microbes to be effective. If the bacteria are inactive, there is not much left for the antibiotic. Another method of protection against antibiotics is the presence of special bacterial cells called "persisters." These cells do not divide and are resistant to many antibiotics. A paper published in 2010 states that persisters resist antibiotics by producing substances that target their target structures. In general, microorganisms living with each other under the name of biofilm benefit from the presence of different cells in their community [32].

17.4 Significance of Polyphenols

Oxidation of lipids during food storage and processing causes the loss of nutritional quality of food and produces free radicals that, in addition to changing the taste, cause cancer. Therefore, various antioxidant compounds are used for the stability of fats. The activity of antioxidants in food preservation was known about one hundred years ago [33]. Antioxidants prevent the destruction of genetic material and, as a result, the occurrence of cancer. On the other hand, accelerating the elimination of toxins and cellular waste prevents diseases such as dementia, Alzheimer's, and advanced heart, liver, and heart problems due to cell destruction [34]. All body cells contain an atom with a positively charged nucleus and negatively charged electrons around it. When the body is exposed to air pollution, cigarette smoke, chemicals, pesticides, alcohol, drugs, toxic fumes, fried foods, unsaturated fats, etc., it can throw electrons out of orbit. In this case, the atom has lost an electron. It becomes unstable and tries to take electrons from the particles close to it to replace the missing electron, resulting in chain reactions. For this purpose, foods are divided into two categories according to the amount of fat:

- High-fat foods: such as cereals and foods made from meat.
- Low-fat foods: such as rice and some cereals.

The first step of isolating antioxidants from high-fat foods is extraction with petroleum ether-carbon tetrachloride or similar solvents. One point during extraction is that the solvent used must be pure because the presence of peroxide reduces the number of antioxidants. Most solvents used for extraction are ethanol, methanol, hexane, petroleum ether, and dimethyl sulfoxide. In some articles, vacuum steam and distillation have separated antioxidants from oil. When fat is low, steam distillation directly extracts the antioxidant. Proteins with high molecular weight, such as phospholipids, etc., are not removed by these solvents and remain: the formation of a complex between phospholipids and antioxidants in high-fat foods prevents the extraction of antioxidants. When the food is high in sugar and protein, we face such a problem, so in these cases, acidification or alkalization of the sample is used [35].

17.5 Biochemistry of Polyphenols

The largest class of phytochemicals is called polyphenols, and many of these substances can be found in foods made from plants. Diets high in polyphenols have been associated with several health advantages. This work aims to examine the chemistry and biochemistry of polyphenols through classification, extraction, separation, and analytical methods, their occurrence and biosynthesis in plants, and their biological activities and implications for human health. Due to their beneficial effects on human health, dietary polyphenols have drawn much interest from nutritionists, food scientists, and consumers. The function of polyphenols in protecting degenerative diseases, including malignancies, cardiovascular diseases, and neurological disorders, has received substantial support in recent research. As a line of defense against oxidative stress brought on by excess (ROS), polyphenols are potent antioxidants that support and enhance the roles of antioxidant vitamins and enzymes.

Antioxidant-increasing compounds are the plants' most important defense mechanisms against stress. The antioxidant defense system includes nonenzymatic antioxidants such as phenolic compounds, (i) beta-carotene, (ii) ascorbic acid, (iii) alpha-tocopherol, (iv) glutathione, (v) enzymes including superoxide dismutase, (vi) guaiacol peroxidase, (vii) ascorbate peroxidase, (viii) catalase, (ix) polyphenol oxidase, and (x) glutathione reductase (the cooperation of these components) [36]. The production of active oxygen species prevents or reduces the destructive effects of these compounds by collecting them (the entry of new technologies into the agricultural industry is the most critical factor in the creation of modern agriculture. Meanwhile, nanotechnology is an appropriate field for producing food and agricultural products. Enzymes include peroxidase, catalase, ascorbate peroxidase, and superoxide dismutase. Reducing reactive oxygen species and malondialdehyde lessens the adverse effects of stress [37]. Malondialdehyde is a product of the oxidation of fatty acids and phospholipids. The level of lipid peroxidation is used as an index to evaluate the severity of oxidative stress and damage to biological membranes [38].

The most important research done to reduce the speed of the aging process is related to nutrition studies and effective plants in maintaining and promoting health. Considering the many medicinal effects that these plants have shown, significant research is underway to clarify which active compounds are responsible for these positive effects. Phenolic compounds are among these compounds [39]. Phenols include over 8000 chemical compounds with diverse structures and properties [40]. Phenol substances generally have one or more aromatic rings with hydroxyl groups (OH). Three main groups are classified; simple phenols, which include phenolic acids, flavonoids, and tannins, and a different group, including compounds such as coumarins, stilbenes, and lignans. Phenolic acids, flavonoids, stilbenes, and lignins are the most phenolic compounds in plants [41].

Phenolic compounds are usually bound with other substances, such as glycosylated forms or esters, and of course, in tiny amounts, they are free. They are uniformly distributed in some plant tissue. Cell wall compounds such as polysaccharides and proteins. Connecting these compounds' stability under temperature conditions and oxidative damage is variable [42]. Recently, the interest in studying foods of plant origin as a source of natural antioxidants rich in polyphenol compounds has increased. It has been shown that the reviving properties and these compounds' ability to remove active oxygen species can significantly reduce the risk of diseases becoming chronic, including cancer, cardiovascular diseases, and Alzheimer's [43]. Its effective polyphenolic compounds are curcuminoids (Curcuminoids). The analysis of polyphenolic compounds depends on various factors, including their chemical nature, the size of the tested sample, storage time and storage conditions, the extraction method and the quality of the technique used, and the presence of interfering factors [44, 45].

A group of chemicals exist naturally in plants and are called plant chemicals. There are more than 500 unique polyphenols in nature. These chemicals are collectively known as phytochemicals. Polyphenols can have a natural structure like that found in plants and can have a semi-synthetic or synthetic form. Our discussion in this article is about polyphenols with natural design, polyphenols found in plant foods. One of the main benefits of polyphenols is the property to fight the effects of cell damage due to the presence of antioxidants. The depth and extent of the development of polyphenols are such that they can affect some genes and bacteria in the human body. There are about 500 types of polyphenols that you can get from foods. Some polyphenols can be classified into flavonoids, lignans, stilbenes, and phenolic acids. Since there are many polyphenols in each group, it is recommended to use a variety of plant foods to get different types of polyphenols. Due to their antioxidant properties, polyphenols fight the activity of free radicals and prevent cell damage. Since they have anti-inflammatory properties, they fight inflammation, one of the leading causes of many chronic diseases.

17.6 Conclusions and Future Perspectives

Polyphenols are one of the simple organic compounds. They consist of the connection of a hydroxyl group and a benzene ring. High phenol concentrations in living organisms, including humans, can cause skin burns – compounds with more than one hydroxyl group attached to the benzene ring. These compounds are extensively used as raw materials in many industries to produce valuable materials. In case of contact with the skin, it will enter the bloodless. Short-term exposure to phenol can cause eye irritation and headaches. People with skin contact with a lot of phenol, skin irritation, liver lesions, dark urine, and irregular heartbeat are observed. Natural polyphenols exhibit a pleiotropic action profile and can exert anti-endometriosis effects through comprehensive interactions with multiple endometriosis-related molecular targets. It includes cell proliferation, apoptosis, inflammation, oxidative stress, angiogenesis, and invasiveness. According to research, frozen fruits are rich in vitamins and antioxidants necessary for the body. Frozen fruits and vegetables can prevent all types of cancers, skin and hair problems, and even gastrointestinal disorders. Frozen fruits and vegetables have more nutrients than fresh ones. Vegetables and fruits such as broccoli, carrots, and berries have occurred. Examining the contents of frozen fruits, doctors have found a variety of vitamins, including vitamins, polyphenols, anthocyanins, and lutein in their structure.

References

1 Babitha, N., Swamy, D., and Chakrapani, S. (2009). Role of green tea as an antioxidant in periodontal disease. *Journal of Orofacial Sciences* 1 (2): 39.
2 Scalbert, A., Johnson, I.T., and Saltmarsh, M. (2005). Polyphenols: antioxidants and beyond. *The American Journal of Clinical Nutrition* 81 (1): 215S–217S.
3 Mudgal, V., Madaan, N., Mudgal, A., and Mishra, S. (2010). Dietary polyphenols and human health. *Asian Journal of Biochemistry* 5 (3): 154–162.
4 Pandey, K.B. and Rizvi, S.I. (2009). Plant polyphenols as dietary antioxidants in human health and disease. *Oxidative Medicine and Cellular Longevity* 2: 270–278.
5 Paterniti, I., Briguglio, E., Mazzon, E. et al. (2010). Effects of Hypericum perforatum, in a rodent model of periodontitis. *BMC Complementary and Alternative Medicine* 10 (1): 1–10.
6 Battino, M., Bullon, P., Wilson, M., and Newman, H. (1999). Oxidative injury and inflammatory periodontal diseases: the challenge of anti-oxidants to free radicals and reactive oxygen species. *Critical Reviews in Oral Biology and Medicine* 10 (4): 458–476.

7 Maury, W., Price, J.P., Brindley, M.A. et al. (2009). Identification of light-independent inhibition of human immunodeficiency virus-1 infection through bioguided fractionation of Hypericum perforatum. *Virology Journal* 6: 1–12.

8 Esmaeili, A. and Beni, A.A. (2014). A novel fixed-bed reactor design incorporating an electrospun PVA/chitosan nanofiber membrane. *Journal of Hazardous Materials* 280: 788–796.

9 Esmaeili, A. and Beni, A.A. (2015). Novel membrane reactor design for heavy-metal removal by alginate nanoparticles. *Journal of Industrial and Engineering Chemistry* 26: 122–128.

10 Esmaeili, A. and Ebrahimzadeh, F.M. (2016). Optimization and preparation of methylcellulose edible film combined with of ferulago angulata essential oil (FEO) nanocapsules for food packaging applications. *Flavour and Fragrance Journal* 31 (5): 341–349.

11 Esmaeili, A., Saremnia, B., and Kalantari, M. (2015). Removal of mercury (II) from aqueous solutions by biosorption on the biomass of *Sargassum glaucescens* and *Gracilaria corticata*. *Arabian Journal of Chemistry* 8 (4): 506–511.

12 Hormozi, N. and Esmaeili, A. (2019). Synthesis and correction of albumin magnetic nanoparticles with organic compounds for absorbing and releasing doxorubicin hydrochloride. *Colloids and Surfaces. B, Biointerfaces* 182: 110368.

13 Masoumi, S. and Esmaeili, A. (2020). New method of creating hybrid of buprenorphine loaded rifampin/polyethylene glycol/alginate nanoparticles. *International Journal of Biological Macromolecules* 159: 204–212.

14 Esmaeili, A. and Saremnia, B. (2016). Synthesis and characterization of NaA zeolite nanoparticles from Hordeum vulgare L. husk for the separation of total petroleum hydrocarbon by an adsorption process. *Journal of the Taiwan Institute of Chemical Engineers* 61: 276–286.

15 Esmaeili, A. and Singh, S. (2017). Optimization of in vitro release of an anticonvulsant using nanocapsule-based thermogels. *European Journal of Pharmaceutical Sciences* 99: 9–16.

16 Esmaeili, A., Sharafian, S., Safaiyan, S. et al. (2009). Biotransformation of one monoterpene by sporulated surface cultures of Aspergillus niger and Penicillium sp. *Natural Product Research* 23 (11): 1058–1061.

17 Esmaeili, A. and Khakpoor, M. (2012). Biological activities and chemical composition of solvent extracts of *Stoechospermum marginatum* (C. Agardh). *Acta Biochimica Polonica* 59 (4).

18 Soltanabadi, Z., Esmaeili, A., and Bambai, B. (2022). Fabrication of morphine detector based on quartz@ Au-layer biosensor. *Microchemical Journal* 175: 107127.

19 Morillo, J., Antizar-Ladislao, B., Monteoliva-Sánchez, M. et al. (2009). Bioremediation and biovalorisation of olive-mill wastes. *Applied Microbiology and Biotechnology* 82: 25–39.

20 Paraskeva, P. and Diamadopoulos, E. (2006). Technologies for olive mill wastewater (OMW) treatment: a review. *Journal of Chemical Technology & Biotechnology: International Research in Process, Environmental & Clean Technology* 81 (9): 1475–1485.

21 Rahmanian, N., Jafari, S.M., and Galanakis, C.M. (2014). Recovery and removal of phenolic compounds from olive mill wastewater. *Journal of the American Oil Chemists' Society* 91: 1–18.

22 Casalino, E., Calzaretti, G., Sblano, C. et al. (2002). Antioxidant effect of hydroxytyrosol (DPE) and Mn2+ in liver of cadmium-intoxicated rats. *Comparative Biochemistry and Physiology Part C: Toxicology & Pharmacology* 133 (4): 625–632.

23 Said, K.A.M., Ismail, A.F., Karim, Z.A. et al. (2021). A review of technologies for the phenolic compounds recovery and phenol removal from wastewater. *Process Safety and Environment Protection* 151: 257–289.

24 Bonifait, L. and Grenier, D. (2010). Cranberry polyphenols: potential benefits for dental caries and periodontal disease. *Journal of the Canadian Dental Association* 76: a130-a.

25 Labrecque, J., Bodet, C., Chandad, F., and Grenier, D. (2006). Effects of a high-molecular-weight cranberry fraction on growth, biofilm formation and adherence of *Porphyromonas gingivalis*. *The Journal of Antimicrobial Chemotherapy* 58 (2): 439–443.

26 La, V.D., Howell, A.B., and Grenier, D. (2010). Anti-porphyromonas gingivalis and anti-inflammatory activities of A-type cranberry proanthocyanidins. *Antimicrobial Agents and Chemotherapy* 54 (5): 1778–1784.

27 Steinberg, D., Feldman, M., Ofek, I., and Weiss, E.I. (2004). Effect of a high-molecular-weight component of cranberry on constituents of dental biofilm. *The Journal of Antimicrobial Chemotherapy* 54 (1): 86–89.

28 Esmaeili, A. and Sadeghi, E. (2014). The efficiency of Penicillium commune for bioremoval of industrial oil. *International Journal of Environmental Science and Technology* 11 (5): 1271–1276.

29 Fakhari, E. and Behfarnia, P. (2011). Effect of polyphenols on periodontal diseases. pp. 445–52.

30 Allison, D.G. (2000). *Community Structure and Co-operation in Biofilms*. Cambridge University Press.

31 An, D. and Parsek, M.R. (2007). The promise and peril of transcriptional profiling in biofilm communities. *Current Opinion in Microbiology* 10 (3): 292–296.

32 Hodgson, A., Nelson, S., Brown, M., and Gilbert, P. (1995). A simple in vitro model for growth control of bacterial biofilms. *The Journal of Applied Bacteriology* 79 (1): 87–93.

33 Esmaeili, A. and Ebrahimzadeh, M. (2015). Preparation of polyamide nanocapsules of Aloe vera L. delivery with in vivo studies. *AAPS PharmSciTech* 16: 242–249.

34 Abbas, A. and Shameel, M. (2008). Anatomical studies on *Stoechospermum marginatum* (Phaeophyta) from the coast of Pakistan. *Pakistan Journal of Botany* 40 (6): 2567–2572.

35 Abd El Mageid, M., Salama, N., Saleh, M. et al. (2009). Antioxidant and antimicrobial characteristics of red and brown algae extracts. In: *4th Conference on Recent Technologies in Agriculture.*

36 Aref, I., El Atta, H., El Obeid, M. et al. (2013). Effect of water stress on relative water and chlorophyll contents of Juniperus procera Hochst. ex Endlicher in Saudi Arabia. *Life Science Journal* 10 (4): 681–685.

37 Lei, Z., Mingyu, S., Chao, L. et al. (2007). Effects of nanoanatase TiO_2 on photosynthesis of spinach chloroplasts under different light illumination. *Biological Trace Element Research* 119: 68–76.

38 Israr, M. and Sahi, S.V. (2006). Antioxidative responses to mercury in the cell cultures of Sesbania drummondii. *Plant Physiology and Biochemistry* 44 (10): 590–595.

39 Shimatsu, A., Kakeya, H., Imaizumi, A. et al. (2012). Clinical application of "curcumin", a multi-functional substance. *Anti-Aging Medicine* 9 (2): 75–83.

40 Robbins, R.J. (2003). Phenolic acids in foods: an overview of analytical methodology. *Journal of Agricultural and Food Chemistry* 51 (10): 2866–2887.

41 Vermerris, W. and Nicholson, R. (2007). *Phenolic Compound Biochemistry.* Springer Science & Business Media.

42 Naczk, M. and Shahidi, F. (2004). Extraction and analysis of phenolics in food. *Journal of Chromatography. A* 1054 (1–2): 95–111.

43 Kurosawa, T., Itoh, F., Nozaki, A. et al. (2005). Suppressive effects of cacao liquor polyphenols (CLP) on LDL oxidation and the development of atherosclerosis in Kurosawa and Kusanagi-hypercholesterolemic rabbits. *Atherosclerosis* 179 (2): 237–246.

44 Shahidi, F. and Naczk, M. (2003). *Phenolics in Food and Nutraceuticals.* CRC Press.

45 Bahrami, M., Afshari, Z., Ahmadi, F. et al. (2013). Evaluation of phenolic content of turmeric hydroalcoholic extract in iran by singleton method. *SSU Journals* 21 (3): 281–290.

18

Phenolic Compounds from Marine Organisms and Their Biofunctional Properties

Jayasri Ravichandran[1], *Liya Babu*[2,3], *Kirubakaran Baskaran Jayalakshmi*[2], *Mangaiyarkarasi Manivel*[1], *Kamini Numbi Ramudu*[2,3], *Ganesan Ponesakki*[2,3], *and Nimish Mol Stephen*[1]

[1] *Department of Fish Processing Technology, Dr. M.G.R. Fisheries College and Research Institute, Tamil Nadu Dr. J. Jayalalithaa Fisheries University, Ponneri, Tamil Nadu, India*
[2] *Department of Biochemistry and Biotechnology, CSIR-Central Leather Research Institute (CLRI), Adyar, Chennai, Tamil Nadu, India*
[3] *Academy of Scientific and Innovative Research (AcSIR), Ghaziabad, India*

18.1 Introduction

In developed nations, food and lifestyle-related illnesses such as obesity, type 2 diabetes, cancer, and neurodegenerative disorders are increasing. Numerous scientific organizations have recently shown an increased interest in studying aquatic organisms for their different biofunctional properties, such as antioxidant, antimicrobial, and neuroprotective effects. Global epidemiological research has revealed that nations that regularly consume marine foods have long life span with lower rates of lifestyle disorders.

Phenolic compounds (PCs) are among the several secondary metabolites from marine organisms with diverse biological characteristics that have garnered the most attention. PCs such as phlorotannins (PTs) and bromophenols are predominantly found in marine sources. Marine species deserve attention as prospective, omnipresent, and widely distributed sources of PCs. Due to their health benefits and biological activities, these compounds can be used in the food and pharmaceutical industries, so it helps to benefit human health and disease prevention.

These polyphenolic compounds are widely distributed in marine sources such as red, brown, and green macroalgae (seaweeds), sponges, seaweeds, corals,

Science and Engineering of Polyphenols: Fundamentals and Industrial Scale Applications,
First Edition. Edited by Chandrabhan Verma.
© 2024 John Wiley & Sons Inc. Published 2024 by John Wiley & Sons Inc.

microalgae, and microbes. The main goal of this book chapter is to direct attention to the sources and classification of marine polyphenols (PP), and the biofunctional activities of the significant marine polyphenolic compounds.

18.2 Classification of Marine Phenols

Phenolic molecules are a varied class of compounds with an aromatic ring with one or more hydroxyl (–OH) groups connected to their structure (Figure 18.1). The term "phenol" specifically refers to a substructure that contains one phenolic hydroxyl group (–OH) attached to an aromatic ring. Catechol and resorcinol are examples of benzenediols, meaning they have two hydroxyl groups attached to an aromatic ring. Pyrogallol and phloroglucinol are benzenetriols, with three hydroxyl groups attached to an aromatic ring. PCs can vary widely in molecular size that ranges from simple molecules like hydroxycinnamic acids (HCAs), and flavonoids to more complex polymers. The molecular weight of PCs ranges from 125 to 650 kilodaltons (kDa). This wide range of molecular sizes contributes to PCs' diverse properties and functions in biological systems.

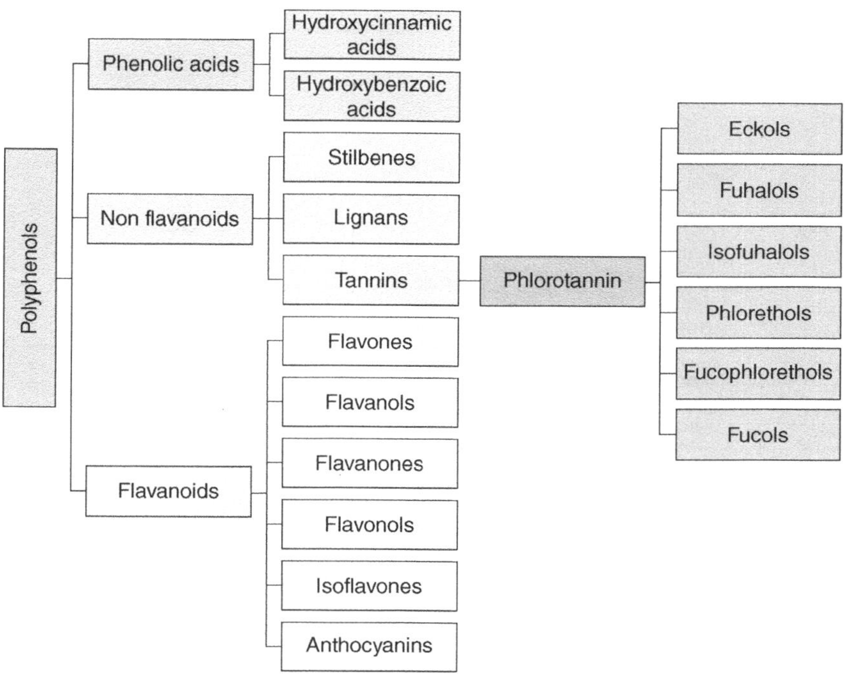

Figure 18.1 Classification of marine polyphenols.

18.2.1 Phenolic Acids

Phenolic acids (PAs) are mostly linked to other molecules, including organic acids, simple or complex carbohydrates, and other bioactive compounds like terpenoids and flavonoids. The chemical structures of some of the common phenolic acids present in marine organisms are presented in Figure 18.2. The classification of PAs is mostly based on the number of carbon atoms in the chain connected to the phenolic ring. These PAs comprise a single phenol ring with at least one functional carboxylic acid group. Based on the carbon chain, these phenolic acids are divided into three groups: C6–C1 for hydroxybenzoic acid (HBA) with one carbon

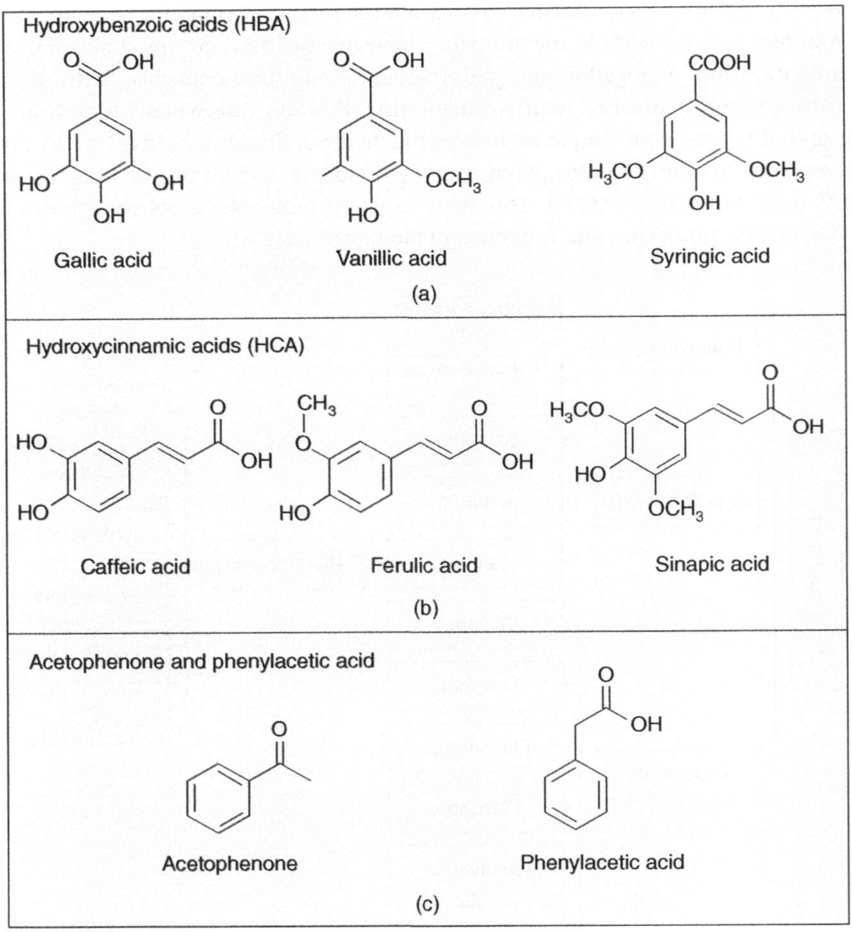

Figure 18.2 Chemical structures of some common phenolic acids present in marine organisms.

chain attached to the phenolic ring, C6–C2 for acetophenone and phenylacetic acids with two carbon chains, and C6–C3 for HCAs with three carbon chains attached to the phenol ring [1].

Some examples of HBAs are gallic acid, p-hydroxybenzoic, vanillic, syringic, and protocatechins. These HBAs differ regarding the fundamental structure, including the aromatic ring's methoxylation and hydroxylation. These compounds primarily exist as conjugates [2]. The HCA derivatives are p-coumaric (4-hydroxy) acids, ferulic (3-methoxy-4-hydroxy), caffeic (3,4-dihydroxycinnamic), and sinapic (3,5-dimethoxy-4-hydroxy) acids [3]. Coumarins in green seaweeds, including *Dasycladus vermicularis*, and vanillic acid derivatives in *Cladophora socialis* [4]. *Ascophyllum nodosum, Bifurcariabifurcata, and Fucus vesiculosus* are brown seaweeds that have been studied for their HBAs, rosmarinic acid, and quinic acid derivatives [5]. PAs such as salicylic acid, syringic, vanillic, protocatechuic, *p*-hydroxybenzoic, gallic, and benzoic acids have all been identified in the genus *Gracilaria* belong to red seaweeds [6]. Green and red seaweeds had higher benzoic and cinnamic acid content than brown seaweeds [7]. Also, gallic acid levels in green and red seaweeds contain high content [8]. The catechin concentration is higher in red macroalgae followed by green and brown macroalgae. In brown seaweeds, phloroglucinol derivatives, and flavonoids make up 35% of the total PCs, with caffeic, gallic, ferulic, and chlorogenic acids being the most prevalent [9].

18.2.2 Bromophenols

In macroalgae, the predominant bromophenolic compounds are polybrominated dibenzo-p-dioxins, methoxylated and hydroxylated bromodiphenyl ethers, bromophenols and their byproducts such as bromoanisoles [7]. The chemical structures of some common bromophenolic compounds present in marine organisms are shown in Figure 18.3. Additionally, brominated sesquiterpenes and other brominated compounds have been discovered in macroalgae [10]. In particular, 2,4,6-tribromophenol is extensively dispersed and produced by marine species to protect them from predators and biofouling. Dibromoethane, bromochloroacetic

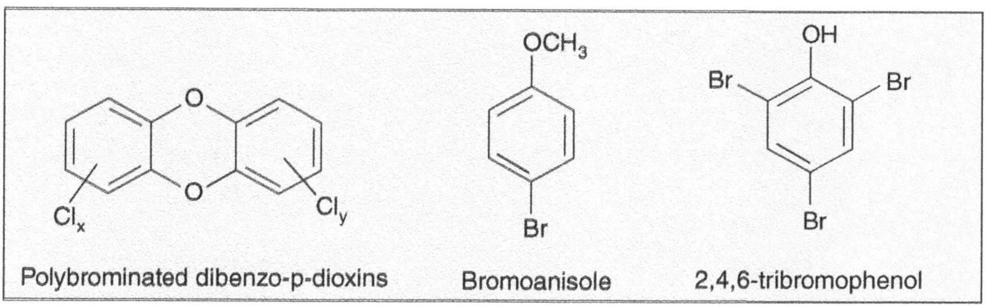

Figure 18.3 Chemical structures of some common bromophenolic compounds present in marine organisms.

acid, dibromochloromethane, and bromoform were found in *Asparagopsis taxformis* [11]. Bromoanisols were found to be present in species such as *Laminaria digitata, Ascophyllum nodosum, Ceramium virgatum, Fucus serratus, Saccharina latissima, and Spongomorpha* sp. [12]. In the diet, some bromophenols are found in prawn species derived from marine algae and bryozoans. Methanol or methanol-dichloromethane can be used to extract them, however, yields may vary depending on other conditions [13].

18.2.3 Phlorotannins

In brown algae, PT makes up the cell wall's structural framework. This structurally heterogeneous group's biological characteristics are governed by its complex chemical makeup, linkage locations, and level of polymerization (125 Da–651 kDa). The chemical structures of some common PT present in marine organisms are shown in Figure 18.4. These PCs, like the tannins of terrestrial plants, have a polymer structure, are strongly linked to polysaccharides, proteins, and other biopolymers, and are extremely soluble in water. Phloroglucinol, eckol, and dieckol are the primary components of the PT from brown algae with antibacterial action. Phloroglucinol (1,3,5-trihydroxybenzene) is a complex polymer only present in

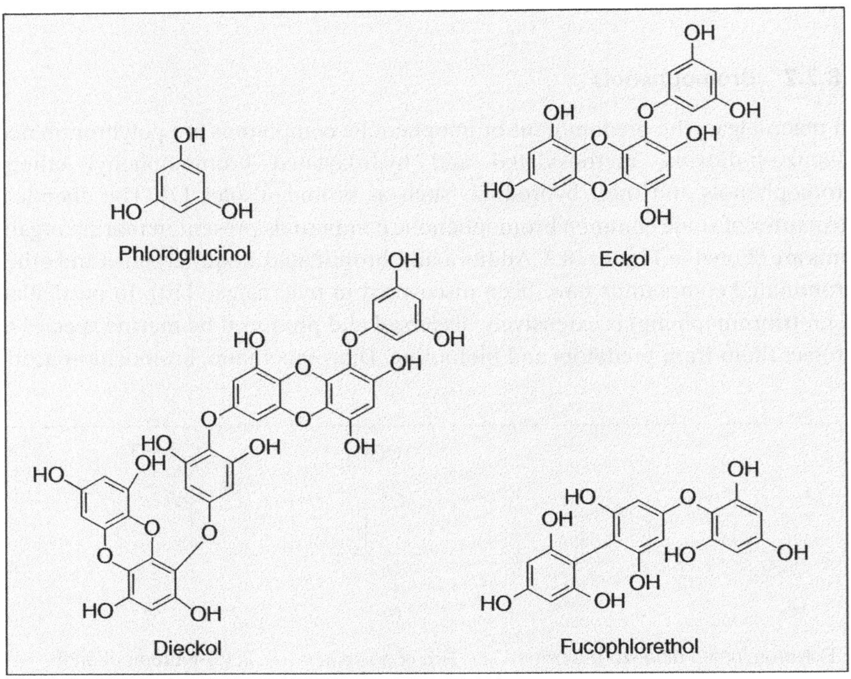

Figure 18.4 Chemical structures of some common phlorotannins present in marine organisms.

brown seaweeds [14]. Algae also contain some PTs that have been sulfated or halogenated. Algal PTs have upto eight phenolic rings, whereas terrestrial plants only make tannins with 3–4 phenolic rings. Their classification is determined by the connections between monomeric units. Fucols are characterized by having aryl-aryl linkages, whereas phlorethols contain aryl-ether linkages. Fuhalols have exclusively ether linkages, along with the presence of extra hydroxyl groups in every third ring. Carmalols are derived from phlorethols and fucophlorethols exhibit both aryl-aryl and aryl-ether units.

The algae, *Fucus* sp. have the highest concentrations of PTs, which can reach 3–12% of their dry weight. While they make up as much as 20% of the weight in some species of algae. About 90% of the total amount of PTs in a free form is present in physodes, membrane-bound vesicles. Phloroglucinol oligomers with five and six rings were identified in *Fucus vesiculosus*, and trimers and tetramers were identified in *Fucus serratus* and *Fucus spiralis*. *Fucus* sp. predominantly contains fucophlorethols, consisting of 3–6 phloroglucinol units (PGUs). The identified isomers include fucofuroeckol, trifucophlorethol, fucotriphlorethol, fucotetraphlorethol, fucophlorethol, dioxinodehydroeckol, difucophlorethol, fucodiphlorethol, and bisfucophlorethol.

The low molecular weight PTs mostly contain phloroglucinol monomers. Different macroalgal species have varying degrees of isomerization, with *F. vesiculosus* exhibiting up to 61 compounds with 12 PGUs. In *F. vesiculosus*, PTs with molecular weights between 498 and 994 Da and 4–8 PGUs constituted the majority of the low molecular weight fractions, while 9–14 PGUs found in *Pelvetia canaliculata* and 8–13 PGUs in *Himanthalia elongata* [7].

18.2.4 Flavonoids

The structure of flavonoids is characterized by heterocyclic oxygen bonded to two aromatic rings. The aromatic rings' degree of hydrogenation varies. Catechins and other flavonoids can be found in abundance in seaweeds. Chemical structures of some common flavonoids present in marine organisms are shown in Figure 18.5. Numerous Chlorophyta, Rhodophyta, and Phaeophyceae species have been shown to contain flavonoids, including quercetin, hesperidin, and rutin. Brown macroalgae, such as *Sargassum muticum*, and red macroalgae such as *Porphyra/Pyropia* spp., and *Chondrus crispus* contain isoflavones, such as daidzein and genistein. The brown macroalgae such as *Durvillae antarctica,* and *Lessonia spicata*, have been reported to have a significant amount of flavonoid glycosides [15]. The green macroalgae, *Nitella hookeri,* contains flavonoids C-glycosides. The flavonoids are biosynthesized from a precursor molecule, naringenin [16]. The flavanone 3-hydroxylase then converts naringenin into dihydrokaempferol, and then subsequent enzymatic mechanism leads to the formation of different flavonoid compounds. *Ecklonia bicyclis, Acetabularia ryukyuensis, Gelidium elegans, Padina arborescens, Portieria hornemannii* (or *Chondrococcus hornemannii*), and *Neopyropia*

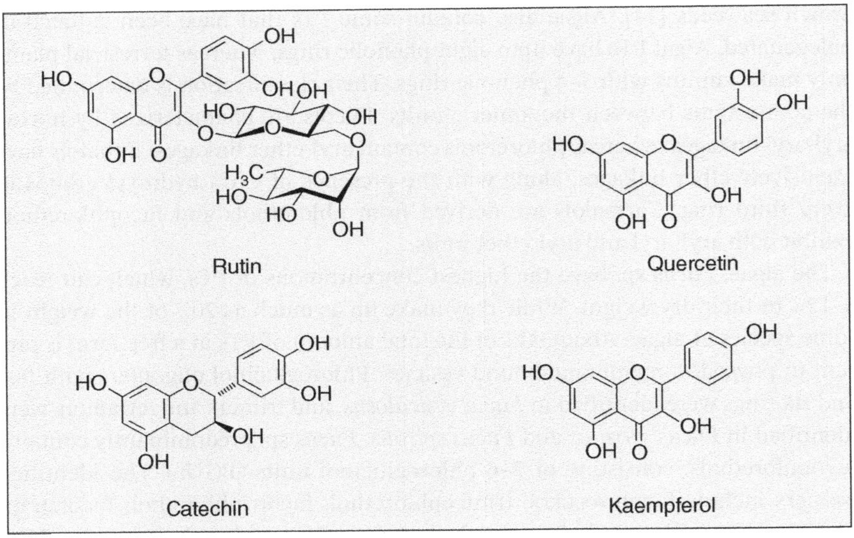

Figure 18.5 Chemical structures of some common flavonoids present in marine organisms.

yezoensis (or *Porphyra yezoensis*) contain flavonoid compounds, including catechin, catechin gallate, epicatechin, epicatechin gallate, epigallocatechin gallate, epigallocatechin.

18.2.5 Phenolic Terpenoids

Red and brown macroalgae have phenolic terpenoids that have been identified and characterized. Phenolic terpenoids in brown macroalgae are known as meroditerpenoids, classified into chromenes, chromanols and plastoquinones. The chemical structures of some common phenolic terpenoids present in marine organisms are shown in Figure 18.6. These compounds are mostly found in the Sargassaceae family and are made up of a hydroquinone ring moiety attached to a polyprenyl chain.

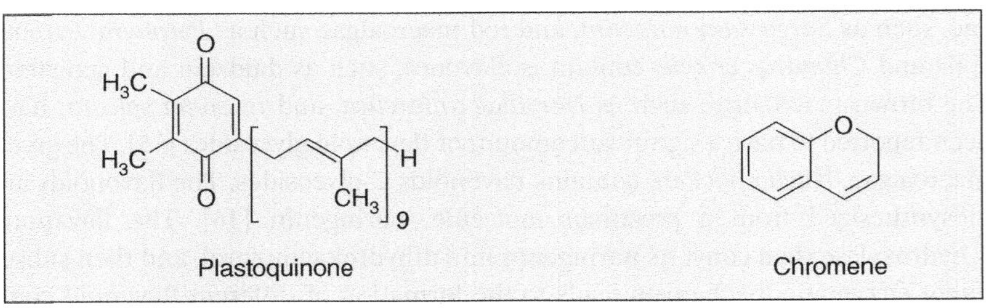

Figure 18.6 Chemical structures of some common phenolic terpenoids present in marine organisms.

Diterpenes and sesquiterpenes were isolated from Rhodomelaceae, and the red seaweed *Callophycus serratus* was reported to possess a macrolide synthesis under secondary cyclization [6].

18.2.6 Mycosporine-like Aminoacids

Mycosporine-like aminoacids (MAA) are the class of ultraviolet (UV)-absorbing compounds found in a wide variety of aquatic animals. These substances were initially discovered in fungi, where they play an important part in UV-induced sporulation [17]. Following that, different MAAs were found in a wide range of aquatic species, such as microalgae, macroalgae, and cyanobacteria. These substances were also found in Rhodophyta species. These substances have 400 Da molecular weights and are water soluble. The chemical structure varies based on an amino acid substituent filled in cyclohexanone or cyclohexenone ring. Depending on the chemical structure's substituents, the conjugated bonds inside the molecule cause broad wavelength absorptions. The Shikimate pathway is involved in the synthesis of majority of MAAs [6].

18.2.7 Non-typical Phenolic Compounds

Non-typical PCs were identified in certain seaweeds. *Colpomenia sinuosa*, a brown seaweed, contains a PC alled colpol. The red macroalga *Tichocarpus crinitus* was found to contain phenylpropanoid derivatives such as tichocarpols. *Calliarthron cheilosporioides*, a red seaweed, was also shown to contain lignin, a polymerized hydroxycinnamyl alcohol believed to be exclusively found in vascular plants.

18.3 Sources

18.3.1 Microalgae

Microalgae, one of the most promising sources for biotechnological applications, must be used to develop an eco-sustainable production of naturally occurring bioactive metabolites. They are photosynthetic organisms that live in aquatic settings and exhibit great degrees of biological and functional diversity. They harbor a wide range of PCs and other families of bioactive molecules. Two essential lignin precursors found in marine microalgae are ferulic and caffeic acids. Microalgae also contain methylated flavonoid, rhamnosythexosyl-methyl-quercetin. The green microalgae, *Haematococcus* sp., and *Haptophyta diacronema* contain naringenin, a crucial precursor of flavones and numerous flavonoids. Rutin, quercetin, and kaempferol are flavonoids extensively dispersed throughout the various microalgal groups. Some green algae, cyanobacteria, and *Diacronema* sp. contain apigenin and luteolin (LT), the derivatives of catechin and epicatechin, which are the building blocks for the synthesis of anthocyanidins [16]. Marennine is another unique example, a PP-classified blue-green pigment that has only ever been reported in the diatom *Hasleaostrearia* sp. [18]. PCs can be synthesized from a variety of microalgae, including cyanobacteria. PCs and phenol derivatives are oxidized by phenol oxidases, such as laccase that found in algae [19]. PCs in seawater may be attributed to the exudates released by diatoms. These exudates have been identified to contain various PCs, including apigenin, catechin, epicatechin, quercitrin, sinapic acid, kaempferol, syringic, gentisic, vanillic, chlorogenic, ferulic, and caffeic acids. These compounds are known to be released by diatoms as part of their biological processes.

The main groups of phenolic acids found in microalgae include HCAs, in particular [20]. The microalgae source of PP are *Nannochloropsis* sp., *Ankistrodesmus* sp., *Euglena cantabrica*, *Spirulina* sp., *Isochrysis galbana*, *Spirogyra* sp., and *Scenedesmus* sp. [21] (Table 18.1).

18.3.2 Macroalgae

Due to chloride and bromide ions in seawater, brominated secondary metabolites are more frequently produced by macroalgae or seaweeds than those containing iodine and fluorine. Iodination can be observed in brown algae, although red and green seaweeds have higher concentrations of bromine or chlorine metabolites [24]. Phenolic acids, bromophenols, and flavonoids are present in large amounts in red and green algae [25]. Apigenin, kaempferol, quercetin, genistein, and catechins were the flavonoids found in algae. PP compounds other than phlorotannins are found in the cell walls of algae, where they do function as a structural element that controls osmotic pressure. They are joined to alginic acid in a complex by covalent ester and hemiacetal linkages [14].

Table 18.1 Phenolic compounds present in marine microalgae.

Phenolic compounds	Microalgae	References
Protocatechuic acid	*Ankistrodesmus* sp.	[22]
Protocatechuic acid; chlorogenic acid; (+) catechin; (−)epicatechin; gallic acid	*Euglena cantabrica*	[22]
Phenolic acids: Vanillic; ferulic; gallic; chlorogenic; syringic; caffeic; protocatechuic and hydroxybenzoic acids.	*Nannochloropsis* sp.	[23]
Protocatechuic; hydroxybenzoic; syringic; vanillic; gallic; chlorogenic acids	*Spirulina* sp.	[23]
Gallic acid	*Spirogyra* sp.	[22]
Naringenin	*Haematococcus* sp., *Haptophyta diacronema*	[16]

In red macroalgae such as *Vidalia* sp. and *Polysiphonia urceolata,* several halogenated PCs have been found. The red algae *Chondrus* sp. produces a compound biochanin A. The green algae (such as *Ulva lactuca*) exhibit a high amount of bromophenols. The green alga, *Cladophora socialis* produces cladophorols, and oligomeric phenols. The red and green seaweed sources of PCs are *Chara* sp., *Cladophora* sp., *Chaetomorpha crassa*, *Chlorodesmis* sp., *Halymenia* sp., *Jania* sp., and *Kappaphycus alvarezii*. The brown algae are well recognized for synthesizing halogenated phlorotannins. Brown algae and diatoms produce phlorotannins in higher concentration comparable to the tannins found in terrestrial plants. Phlorotannins are predominantly present in brown algae, with a lesser occurrence in red algae. In comparison, green algae have a lower concentration of PP [25]. The macroalgae sources are *Acrosiphonia/Spongomorpha* sp., *Ascophyllum nodosum*, *Asparagopsis taxiformis*, *Ceramium tenuicorne*, *Ceramium virgatum*, *Ecklonia cava*, *Ecklonia stolonifera*, *Fucus serratus*, *Laminaria digitata*, *Fucus vesiculosus*, *Saccharina latissima*, and *Fucus radicans* (Tables 18.2–18.4) [49, 50].

18.3.3 Seagrasses

Seagrasses are rich in PP, such as condensed tannins, simple, and sulfated PAs like zosteric acid. The two main PCs identified in eelgrass (*Zostera marina* L.) leaves and roots-rhizomes were caffeic acid and rosmarinic acid, with larger quantities observed in the younger leaves and roots-rhizomes throughout the spring season. *Ruppia* sp. contained 30 mg/g of chicoric acid, which is twice as much as all other flavonoids combined. LT, it's disulfated derivative, 7,3′-disulfate luteolin (DSL), and rosmarinic acid (RA), which are the PP complex (PPC) were isolated from *Zostera marina* and *Z. Asiatica* [7, 51]. The seagrasses sources of PP are *Halodule wrightii*, *Thalassia testudinum*, *Posidonia oceanica* (L.), *Zostera asiatica*, and *Z. Marina*, and *Cymodocea nodosa* (Table 18.5).

Table 18.2 Phenolic compounds present in marine green seaweeds.

Phenolic compounds	Seaweeds	References
Coumarins	*Dasycladus vermicularis*	[4]
Coumarins, vanilic acid	*Cladophora socialis*	[4]
Flavonoids C-glycosides	*Nitella hookeri*	[26]
Epicatechin, epicatechin gallate, catechin, catechin gallate, epigallocatechin gallate, epigallocatechin	*Acetabularia ryukyuensis*	[27]
4-(Sulfooxy)phenylacetic acid; 4-(sulfooxy)benzoic acid	*Dasycladus vermiculari*	[28]

Table 18.3 Phenolic compounds present in marine brown seaweeds.

Phenolic compounds	Seaweeds	References
Plastoquinones, chromanols, chromenes	*Sargassum* sp.	[29, 30]
HBAs, rosmarinic acid, quinic acid derivatives	*Bifurcaria bifurcata, Fucus vesiculosus, Ascophyllum nodosum*	[5, 31]
Bromoanisols	*Acrosiphonia/Spongomorpha* sp., *Fucus radicans, Ceramium virgatum, A. nodosum, Fucus serratus, Laminaria digitata, Saccharina latissima, F. vesiculosus*	[32]
Bisfucophlorethol, difucophlorethol, dioxinodehydroeckol, fucodiphlorethol, fucophlorethol, fucofuroeckol, fucotetraphlorethol, fucotriphlorethol, trifucophlorethol	*Fucus* sp.	[33]
Daidzein, genistein	*Sargassum muticum, Sargassum vulgare*	[34]
Flavonoid glycosides	*Durvillaea antarctica, Macrocystis pyrifera (or Macrocystis integrifolia), Lessonia spicata*	[34, 35]
Epicatechin, epicatechin gallate, epigallocatechin, epigallocatechin gallate, catechin, catechin gallate	*Ecklonia bicyclis, Padina arborescens, Pedinomonas minor*	[6]
Colpol	*Colpomenia sinuosa*	[6]
Fucophlorethol, eckol derivatives	*Cystoseira barbata*	[36]
Bieckol, dieckol, 7-phloroeckol, fucodiphloroethol, fucophloroethol, fucotriphloroethol, phlorofucofuroeckol	*F. spiralis, Cystoseira usneoides, Cystoseira nodicaulis, Cystoseira tamariscifolia*	[37, 38]
Phlorotannin	*Durvillaea antarctica, Lessonia spicata*	[39]
Eckol, dieckol, eckstolonol, phloroglucinol, triphlorethol-A, phlorofucofuroeckol-A	*Ecklonia cava*	[40]
Fucols, fuhalols, fucophlorethols, fucofurodiphlorethol, fucofurotriphlorethol, fucofuropentaphlorethol, phlorotannin derivatives	*F. vesiculosus*	[41]
Trifuhalol, tetrafuhalol, diphlorethol, triphlorethol	*Halidrys siliquosa*	[7]

Table 18.3 (Continued)

Phenolic compounds	Seaweeds	References
Caffeic acid, gallic acid, chlorogenic acid, ferulic acid, quercetin, hydroxybenzaldehyde, myricetin, kaempferol, phloroglucinol	*H. elongata*	[9]
Fucophlorethol (2–18 PGU)	*Padina tetrastromatica*	[42]
Bifuhalol, hydroxytrifuhalol, trifuhalol, tetrafuhalol	*Carpophyllum flexuosum, Carpophyllump lumosum, Ecklonia radiata*	[43]
Catechin-catechin-*O*-gallate, phlorofucofuroeckol, phloroglucinol, quercetin, 3-O-rutinosyl-kaempferol, rutin	*Cystoseira barbata*	[44]
Gallocatechin derivative, p-coumaric acid derivative, phlorotannins derivatives (2–3 PGU)	*Lessonia trabeculate*	[45]
Dieckol, dioxinodehydroeckol, eckols, fuhalols, phlorethols, fucols, fucophlorethols	*Sargassum fusiforme*	[46]

Table 18.4 Phenolic compounds present in marine red seaweeds.

Phenolic compounds	Seaweeds	References
Benzoic, gallic, gentisic, *p*-hydroxybenzoic, protocatechuic, salicylic, syringic, vanillic acids	*Gracilaria* sp.	[6]
Daidzein, genistein	*Chondrus crispus, Porphyra/ Pyropia* sp.	[6]
Catechin, catechin gallate, epicatechin, epicatechin gallate, epigallocatechin, epigallocatechin gallate	*Gelidium elegans, Portieria hornemannii* (*Chondrococcus hornemannii*), *Neopyropia yezoensis* (*Porphyra yezoensis*)	[27]
Phenylpropanoid derivatives—tichocarpols	*Tichocarpus crinitus*	[47]
Lignin	*Calliarthron cheilosporioides*	[48]
Bromochloroacetic acid, dibromoacetic acid, dibromochloromethane, bromoform	*Asparagopsis taxiformis*	[11]

Table 18.5 Phenolic compounds present in marine seagrasses.

Phenolic compounds	Seagrass	References
Diosmetin 7-sulfate, coutaric acid, caftaric acid	*Cymodocea nodosa*	[52]
Gallic acid, syringic acid, ferulic acid, p-hydroxybenzoic acid, p-coumaric acid	*Halodulewrightii*, *Thalassia testudinum*	[53]
Luteolin, rosmarinic acid	*Zostera asiatica*, *Z. marina*	[54]
Apigenin, caffeic acid, diosmetin and its sulfated derivatives, ferulic acid, 3-hydroxyhexanoic acid, Deoxycymodienol, isotedarene A, kaempferol-7,40-dimethylether-3-*O*-sulfate, luteolin, *p*-coumaric, rosmarinic and zosteric acids	*Z. marina*	[7]
Acacetin, apigenin, caffeic acid, diosmetin, luteolin, luteolin-7-sulfate and its sulfated derivatives, rosmarinic acid, zosteric acid	*Zostera noltii*	[7]
Apigenin 7-sulphate, luteolin and its sulfated derivatives, diosmetin, rosmarinic acid	*Z. noltii*, *Z. marina*	[55]
Gallic acid, proanthocyanidins, rosmarinic acid	*Zostera muelleri*	[7]

18.3.4 Sponges

Sponges are rich in certain PCs. The member of (S)-(+)-curcuphenol family is commonly present in sponges and exhibits several biological activities. Bisabolenes are particular polyphenolic compounds present in sponges [56]. The main sources are *Didiscus aceratus, Hyrtios erectus,* and *Myrmekioderma* sp. (Table 18.6).

Table 18.6 Phenolic compounds present in marine sponges.

Phenolic compounds	Sponges	References
(S)-(+)-Curcuphenol, dicurcuphenols A–E, 10β-hydroxycurcudiol, dicurcuphenol ether F	*Didiscus aceratus*	[56]
Phenolic alkenes, erectuseneols A – F	*Hyrtios erectus*	[7]
(S)-Biscurcudiol, (R)-Biscurcudiol, (+)-curcudiol, (+)-dehydrocurcudiol, abolene, myrmekiodermaral, myrmekiodermaral, myrmekioperoxide A, myrmekioperoxide B, (+)-oxoabolene, (+)-curcuphenol, $5^0\alpha$-hydroxycurcudiol, $5^0\beta$- hydroxycurcudiol, curcuepoxide A, curcuepoxide B	*Myrmekioderma* sp.	[7]

Table 18.7 Phenolic compounds present in marine cyanobacteria.

Phenolic compounds	Cyanobacteria	References
Quinic acid, catechin	*Anabaena*	[16, 57]
Catechin	*Arthrospira*	[16, 57]
Phloroglucinol, rutin, protocatechuic acid, orcinol	*Calothrix* sp., *Leptolyngbya* sp.	[16, 58]
Apiin, catechin, chlorogenic acid, epicatechin, gallic acid, kaempferol, quinic acid, rutin	*Nostoc* sp.	[16, 57]

18.3.5 Cyanobacteria

Two cyanobacteria, *Scenedesmus*, and *Phaeodactylum* have been found to contain salicylic acid, a well-known signaling chemical in plants. *Scenedesmus* and *Phormidium* have also been found to contain rosmarinic and gentisic acids [16]. Cyanobacteria contain apigenin and LT. Interesting flavonoids, including biochanin (*Synechocystis* and *Phormidium*), ononin (*Synechocystis*), esculetin (*Nostoc*), sissotrin have also been identified in cyanobacteria (Table 18.7) [16].

18.3.6 Other Marine Sources

Natural marine bromophenols are mostly produced by red algae, but they are also consumed by fish, prawns, and other crustaceans as part of the food chain. The finfish contains more polybrominated diphenyl ethers (PBDEs) than shellfish. Bivalves (clams and mussels) often have greater concentrations of hydroxylated and methoxylated PBDEs than other fishes. Some levels of tribromophenols are identified in certain crustaceans, mollusks, and fishes [7, 59].

18.4 Influence of Environmental Factors on PCs

Marine organisms naturally produce PCs, however, these compounds have been linked to environmental stressors like UV radiation, temperature, salinity, desiccation, and nutrition availability. The amount of phenols in photosynthetic organisms varies seasonally [60]. The growing condition of macroalgae, seagrass, and microalgae exhibits variation in their PP content and shows dependency on species, seasonality, and environmental factors. Algae can produce a variety of PCs, such as terpenoids, MAAs, or other uncommon PCs, to adapt to environmental changes. Algal PC content is significantly influenced by various environmental factors such as temperature and salinity. The biosynthesis of phenolics and flavonoids in diatoms is modulated by the light's spectrum characteristics, intensity, and frequency change [61].

In *Skeletonema marinoi*, PC concentrations were shown to be lower in low light or complete darkness, whereas higher light intensities encouraged the synthesis of these compounds, following exposure to moderate-high blue light intensity [62]. Flavonoid content also changed in response to the light environment [63]. When exposed to UV-C radiation, the green microalga *Chlamydomonas nivalis* boosted the synthesis of PCs. On the other hand, *Scenedesmus quadricauda* exposed to UV-A and UV-C radiations significantly reduced the synthesis of quercetin and kaempferol and altered the overall phenolic profile [64, 65]. However, PC formation in algae can be influenced by the chemical composition of water. It has been demonstrated that ocean acidification-related high CO_2 levels in marine water increase the PC concentration in microalgae. Due to their ability to act as antioxidants, high salinity also caused the accumulation of PP and flavonoids in green macroalgae and cyanobacteria [16, 66]. Additionally, when subjected to rising metal ion concentrations, the green algae *Dunaliella salina, Scenedesmus* sp., and *Chlorella* sp., as well as the diatom *Phaeodactylum* sp., produced more PCs. CuO nanoparticles (non-natural factors) found to stimulate the formation of phenolics in *Nannochloropsis oculata*, which lowers growth rates and the amount of chlorophyll and carotenoids content [67, 68]. Variations in nutrient concentration potentially have an impact on PC synthesis in algae. In contrast, limiting nitrogen or phosphate circumstances in microalgae decreased PC production in freshwater alga *Chlorella vulgaris* and other algae such as *Tetraselmis suecica* and *Phaeodactylum tricornutum* [69].

In Fucus, there was an increase in phenolic production in a low-light environment, which differs from studies conducted on other brown algae that demonstrated higher PP production under high-light conditions. The addition of nutrients did not affect the production of phlorotannins in *Fucus vesciculosus*. However, nitrogen limitation led to an increased accumulation of total PCs, especially phlorotannins, in brown algae. In macroalgae, biotic interactions are reported to stimulate the phenolic production [16].

18.5 Pharmaceutical and Biomedical Applications

The structural importance of PPs, particularly the level of polymerization, is related to their pharmacological value. The majority of PCs have bioactive properties that are important for human health, including the ability to control oxidative and inflammatory stress, regulate the digestion of macronutrients, and influence gut microbiota in a manner similar to prebiotics. Physiologically active substances also have anti-inflammatory, antiallergic, antiviral, anticancer, antioxidant, antidiabetic, and radio-protective properties. Diets with high PP content show strong preventive benefits in age-related chronic diseases. Due to their diverse bioactivities, the phenols found in seaweeds have a lot of potential uses in the biomedical and pharmaceutical industries. Phenols, especially phlorotannins, have exceptional antioxidant capacity because they can chelate reactive oxygen species (ROS), avoiding oxidative stress and cell damage [7]. Therefore, scavenging oxidants is crucial for controlling a variety of ailments, making the PCs in seaweeds a highly valuable source of antioxidants.

Phenolic chemicals are being studied for their potential use in improving or curing some of the most prevalent diseases in the modern world, including cardiovascular, diabetic, neurodegenerative, and psychiatric disorders. The damage that free radicals do to the cellular macromolecules is a key factor in a number of illnesses, including diabetes, cancer, inflammatory diseases, and neurodegenerative diseases. Antioxidants may, therefore, improve human health by reducing the risk of free radical damage.

18.5.1 Antioxidant Activity

The antioxidant activity of PCs is primarily responsible for their positive impacts on human health. Phlorotannin with eight rings traps electrons from free radicals. PT can produce more hydrogen peroxide due to the abundance of hydroxyl groups, which is toxic to bacteria when produced in an aerobic environment. Phenolic acids, phlorotannins, flavonoids, and stilbenoids found in marine organisms have the capability to eliminate and decrease the levels of ROS within cells [70]. This, in turn, improves the antioxidant capacity of the cells by producing enzymes responsible for oxygen metabolism and detoxification of harmful substances. Additionally, these compounds inhibit signaling pathways associated with inflammatory responses [71]. The amount of hydroxyl groups that form the molecule's structure and are strengthened by steric hindrance determines the antioxidant activity of PCs [72–75].

18.5.2 Anticancer Property

PCs and their by-products can act as anticancer metabolites during different cancer stages, encompassing proliferative signaling, metastasis, cell cycle regulation, resistance to cell death, evasion mechanisms, angiogenesis, and evasion of growth suppressors. Cytochrome P450 family 1B1 (CYP1B1), an enzyme overexpressed in a variety of tumors and linked to angiogenesis, is inhibited by quercetin, kaempferol, apigenin, and genistein. The methylated flavonoids have higher intestinal absorption and more resistance to metabolic processes in the liver and intestine, due to which they have more anticancer activity than their corresponding hydroxylated derivatives. However, because of their limited selectivity, PCs can adversely affect normal cell lines [16, 76]. In this instance, chemical modification may be required to improve the selectivity of the molecules f.

18.5.3 Antiviral Property

The ability of PPs to restrict ROS propagation in cell systems plays a crucial role in the reduction of viral replication, particularly in the case of a ssRNA-positive virus. Angiotensin-converting enzymes (ACE) are naturally down-regulated by apigenin and kaempferol. The production of the ACE2 enzyme by kidney is known to be increased by apigenin, which leads to low blood pressure. This property makes apigenin potentially useful for the treatment of viral diseases, such as COVID-19. Kaempferol and apigenin, however, block the renin-angiotensin-aldosterone system, which helps coronavirus infections to enter lung cells. Haslea diatoms can synthesize marennine, which also has antiviral effects [16]. It has been demonstrated that a PT from the brown alga *Ecklonia cava* has the ability to prevent HIV viral multiplication.

18.5.4 Antibacterial Property

Gram-positive and Gram-negative bacteria react differently to PTs, where Gram-negative bacteria are less sensitive. Enzymes and proteins from the bacterial cell membrane can bind to PTs and lyse the cells. The aromatic rings and hydroxyl groups present in PGUs form hydrogen bonds and hydrophobic interactions with the -NH-groups of bacterial proteins. The cell membranes and cell walls of *Vibrio parahaemolyticus* are damaged by low-molecular-weight PTs from the alga *Sargassum thunbergii*, disrupting membrane permeability and allowing intracellular bacteria to flow out. The cell loses phosphates, nucleic acids, and intracellular potassium. The *Salmonella* sp. is sensitive to the acetone extract of the green alga *Cladophora socialis*, which contains PCs as its primary component. The antibacterial action of PCs is increased by an increase in their lipophilic characteristics [14]. Both aerobic and anaerobic bacteria are harmed and killed by algal PTs. Phlorotannins, a broad category of natural polymers, have particularly alluring antibacterial properties. In addition to being non-toxic to eukaryotic cells, extracts from brown algae containing these polyphenolic compounds have a strong antibacterial action on a variety of pathogenic bacteria, including Methicillin-resistant *Staphylococcus aureus*. The human dermatophytic nail infection-causing fungus *Trichophyton rubrum* is susceptible to the dieckol's fungicidal action. PP found in brown algae can be used to develop new antimicrobial medications for the treatment and prevention of infectious disorders [77].

18.5.5 Antibiofilm Property

Seaweed is a particularly alluring, limitless, reservoir of PPs with significant potential as antibiofilm and antibacterial agent. Biofilm formation can be prevented using PTs derived from marine algae. PT prevents the biofilm formation in human skin keratinocytes infected with *Psuedomonas acnes* and *Escherichia coli* [14].

18.5.6 Antidiabetic Property

Seaweeds that produce phlorotannins have demonstrated therapeutic benefits and a variety of distinct health-improving effects. In addition to their antioxidant activity, flavonoids also exhibit additional intriguing health benefits as a result of their chemical structure. The majority of brown seaweeds in the family *Lessoniaceae* and genus *Ecklonia* have been found to have antidiabetic properties. Phlorotannins from *E. cava*, such as fucodiphloroethol G, with a concentration of 19.5 μM, dieckol with 10.8 μM, 6,6–bieckol with the value of 22.2 μM, 7-phloroeckol with the value of 49.5 μM, and phlorofucofuroeckol-A with the value of 19.7 μM, have exhibited a marked inhibition of α-glucosidase [78, 79]. Phlorotannins obtained from *Fucus distichus, Alaria marginata,* and *Ascophyllum nodosum* are also capable of inhibiting α-glucosidase and α-amylase, whereas PTs from *F. vesiculosus* inhibits glucosidase activity. *L. trabeculate* inhibits α-glucosidase and lipase activity [79, 80]. Protein tyrosine phosphatase 1B (PTP1B) is a significant inhibitor of insulin signaling, which can be inhibited by bromophenol (IC 50 = 0.84 μM). Red algal 3-bromo-4,5-bis(2,3-dibromo-4,5-dihydroxybenzyl)-1,2-benzenediol have the ability to stimulate insulin signaling and stop palmitate-induced insulin resistance by inhibiting PTP1B intrinsically [7].

18.5.7 Cardiovascular Disease

It is evidenced that a diet high in flavonoids enhances endothelial function and decreases blood pressure. PTs have a beneficial impact on lowering blood pressure. The blood pressure-lowering compound Sargachromenol D, which was isolated from the *Saragassum siliquastrum*, has the potential to replace existing combination therapies in the treatment of severe hypertension. Phlorotannin supplements for cardiovascular disorders primarily aim to reduce arteriosclerosis and raise levels of HDL-C, a beneficial cholesterol. The products HealSea™, IdAlg™, and Seanol™ approved for cardioprotection contain phlorotannins in its formulation. Angiotensin-converting enzyme (ACE) inhibition by PCs, along with the inhibition of zinc metalloproteinases, appears to depend heavily on the presence of many hydroxyl groups [81]. An unsaturated 2–3 bond coupled with a 4-oxo-function as well as a catechol B-ring in rutin, quercetin, and LT are extremely useful in treating cardiovascular disorders [82].

18.5.8 Neurodegenerative and Mental Disorders

Bromophenols in macroalgae have been employed in the treatment of schizophrenia and Parkinson's disease. Phlorotannins also serve as neuroprotectants and are used in curing neurodegenerative illnesses. *Eisenia bicyclis* extracts reduced the ability of the β-amyloid cleavage enzyme. Phlorotannins from other brown seaweeds like *Ishige foliacea* and *Ecklonia maxima* were also used in the treatment of Alzheimer's disease [83].

18.5.9 Tissue and Bone Regeneration

Phlorotannins were added to the fish collagen/alginate biocomposite, and the result was better in terms of cell proliferation. This new mixture performed better in the *in vitro* assay for calcium deposition and osteogenesis, indicating that it has potential as a biomaterial for the formation of bone tissue. Commercially, *E. cava* phenolic extract, known as Seanol®, was examined as a component of hydrogels for bone regeneration with the potential to boost antibacterial activity and improve bone mineralization [6].

18.5.10 Food Applications

PCs act as a natural, safe food stabilizer, and preservative, and their antioxidant potential may be of interest to the food industry. Many seaweed PP components possess antibacterial effects against major food spoilage and infectious microbes. To improve food's oxidative stability and maintain or improve its natural quality and nutritional content, extracts with seaweed phenolic antioxidants have been employed. In the food industry, their antioxidant and antibacterial potential is useful as nutraceutical ingredients in functional food products or food supplements, and also to increase the shelf-life of processed foods.

Bromophenols from *Pterocladiella capillacea* and *U. lactuca* have been investigated as marine flavor agents and as feed in farmed fish and other aquatic organisms. The European Food Safety Authority has approved the food supplement Seapolynol™ (Botamedi Inc., Seoul, Korea), based on dieckol and other PP from *E. Cava*, which has been investigated as a cardioprotective and antihyperlipidemic agent against doxorubicin-induced cardiotoxicity. The sensitivity of insulin in type 2 diabetes was also impacted by Seapolynol, which may be crucial in preventing metabolic disorders [6].

18.5.11 Cosmetic Applications

Researchers are working to create novel, natural, pharmaceutical, and cosmeceutical formulations. Certain seaweed extracts with the corresponding phenolic chemicals, called phlorotannins, exist in cosmetics, including skincare and anti-aging treatments. Phlorotannins are less toxic than other natural antioxidants and prevent skin elasticity loss in older skin. For instance, the extract from the seaweed *Saccharina japonica* (previously known as *Laminaria japonica*) is used in facial masks because of its anti-inflammatory, antioxidant, and antimicrobial properties. These substances offer a basis for the creation of cosmetics and medications. As a result, phlorotannins from seaweed are potential substitutes for natural antibiotics and candidates for acne-treating drugs and nutritional supplements.

Ecklonia cava has anti-inflammatory, antioxidant, antibacterial, and UV protection properties. Its antibacterial activity acts against the acne-causing bacteria *Propionibacterium acnes*. As a result, these phlorotannins are utilized in many cosmetic compositions. One of the key sectors driving the use of seaweed PCs in commercial applications is the cosmetic industry [6]. Examples include natural UV protection Helioguard® 365, produced by Mibelle Biochemistry in Buchs, Switzerland; AethicSôvée® created with photamin, an anti-aging agent extracted from seaweeds. The MAA-containing extract from *Asparagopsis armata* with anti-aging effects is used in lotions. Tannins, flavonoids, and acrylic acid are the active ingredients in *Ulva compressa* (or *Enteromorpha compressa*) extract employed for cosmetic applications, which have biofunctional properties include antioxidant, antimicrobial, and antiallergic effects. The active ingredients of *F. vesiculosus* extract loaded cosmetic product containing fucoidan and phlorotannins, which have antioxidant, anti-aging, and antimicrobial properties [84, 85].

18.5.12 Feed and Animal Health

It has been proven that the polyphenolic chemicals in seaweed are accessible to animals through the colon. These PCs can be absorbed in two different ways: directly in the upper digestive tract in their unchanged state or indirectly in the lower intestine in their altered form. Phlorotannins from *Ecklonia kurome* were shown to have potential use as antibacterial medications that could serve as a natural alternative to the recently banned feed antibiotics. Phlorotannins are used as an ingredient in pig and poultry feed, but only at very low concentrations typically up to 5–6% in growing animals and never more than 10% to avoid overdosing, which can have detrimental effects on the health of the animals. Phlorotannins improve the quality (better meat quality), productivity (increased animal development and milk production), and safety (reduced shedding of harmful microbes) of animal products when taken in the approved dosage.

18.5.13 Agriculture

Due to their inherent antimicrobial activity, the seaweed PCs in some commercial seaweed extracts provide protection against plant diseases. The defense of plant organisms against harm from free radicals and other oxidants is another putative function of PP. The development of a PC-containing solution for commercial use as a stimulant for mycorrhizal and rhizobial symbiosis, which can occur during fertilization. The PCs used for the treatment are mostly found in the *Ascophyllum* and *Fucus* genus. Similarly, a liquid fertilizer containing PCs from marine algae that actively protects against plant diseases is also found.

18.6 Conclusion

PCs, one of the most active families of bioactive compounds, are widely distributed in marine organisms, which provide a rich and sustainable supply of bioactive, including several unique compounds exclusive to this ecosystem. A sufficient intake of PP is reported to provide health benefits, particularly in relation to the prevention of non-communicable illnesses. Previous studies have shown that marine phenolics have multiple protective effects against common disease conditions such as obesity, diabetes mellitus, Alzheimer's disease, cancer, and other infectious diseases. Due to their enzyme-inhibitory impact and their antidiabetic, antiobesity, antihypertensive, anti-inflammatory, anticancer, antimicrobial, or antiviral activity, PP are involved in numerous multifactorial pathways for many disorders. The potential of marine PCs to intervene in many pathways involved in biological processes is a noteworthy distinction from the currently available drugs. Thus, they might be used as drug alternatives/additives in the pharmaceutical and cosmetic industries. In many types of PP, bioactivity of phlorotannins, and their derivatives are considerably more explored than other PP. A considerable number of studies have been demonstrated the health benefits of bromophenols specifically from marine organisms. In conclusion, marine species are a significant and promising source of PP that could be potentially utilized in the therapeutic management of cancer, obesity, diabetes, and neurodegenerative disorders.

Acknowledgement

The authors thank the Management of Tamil Nadu Dr. J. Jayalalithaa Fisheries University, and the Director, CSIR-Central Leather Research Institute (CLRI) for their support to carry out this work.

Conflict of Interest

The authors of this chapter do not have any conflict of interest.

References

1 Pietta, P., Minoggio, M., and Bramati, L.J. (2003). Plant polyphenols: structure, occurrence and bioactivity. *Studies in Natural Products Chemistry*. 28: 257–312.
2 Liwa, A., Barton, E., Cole, W., and Nwokocha, C. (2017). Bioactive plant molecules, sources and mechanism of action in the treatment of cardiovascular disease, *Pharmacognosy*. In: , 315–336. Elsevier.

3 Luna-Guevara, M.L., Luna-Guevara, J.J., Hernández-Carranza, P. et al. (2018). Phenolic compounds: a good choice against chronic degenerative diseases. *Studies in Natural Products Chemistry* 59: 79–108.

4 Feng, Y., Carroll, A.R., Addepalli, R. et al. (2007). Vanillic acid derivatives from the green algae *Cladophora socialis* as potent protein tyrosine phosphatase 1B inhibitors. *Journal of Natural Products* 70 (11): 1790–1792.

5 Agregan, R., Munekata, P.E., Franco, D. et al. (2017). Phenolic compounds from three brown seaweed species using LC-DAD–ESI-MS/MS. *Food Research International* 99: 979–985.

6 Cotas, J., Leandro, A., Monteiro, P. et al. (2020). Seaweed phenolics: from extraction to applications. *Marine Drugs* 18 (8): 384.

7 Mateos, R., Pérez-Correa, J.R., and Domínguez, H. (2020). Bioactive properties of marine phenolics. *Marine Drug* 18 (10): 501.

8 Tanna, B., Brahmbhatt, H.R., and Mishra, A. (2019). Phenolic, flavonoid, and amino acid compositions reveal that selected tropical seaweeds have the potential to be functional food ingredients. *Journal of Food Processing and Preservation* 43 (12): e14266.

9 Rajauria, G. (2018). Optimization and validation of reverse phase HPLC method for qualitative and quantitative assessment of polyphenols in seaweed. *Journal of Pharmaceutical and Biomedical Analysis* 148: 230–237.

10 Topcu, G., Aydogmus, Z., Imre, S. et al. (2003). Brominated Sesquiterpenes from the Red Alga Laurencia o btusa. *Journal of Natural Products* 66 (11): 1505–1508.

11 Machado, L., Magnusson, M., Paul, N.A. et al. (2016). Identification of bioactives from the red seaweed *Asparagopsis taxiformis* that promote antimethanogenic activity in vitro. *Journal of Applied Phycology* 28: 3117–3126.

12 Bidleman, T.F., Andersson, A., Brugel, S. et al. (2019). Bromoanisoles and methoxylated bromodiphenyl ethers in macroalgae from Nordic coastal regions. *Environmental Science: Processes & Impacts* 21 (5): 881–892.

13 Koch, C. and Sures, B. (2018). Environmental concentrations and toxicology of 2,4,6-tribromophenol (TBP). *Environmental Pollution* 233, 706–213.

14 Besednova, N.N., Andryukov, B.G., Zaporozhets, T.S. et al. (2020). Algae polyphenolic compounds and modern antibacterial strategies: current achievements and immediate prospects. *Biomedicines* 8 (9): 342.

15 Klejdus, B., Lojková, L., Plaza, M. et al. (2010). Hyphenated technique for the extraction and determination of isoflavones in algae: ultrasound-assisted supercritical fluid extraction followed by fast chromatography with tandem mass spectrometry. *Journal of Chromatography A* 1217 (51): 7956–7965.

16 Del Mondo, A., Smerilli, A., Ambrosino, L. et al. (2021). Insights into phenolic compounds from microalgae: structural variety and complex beneficial activities from health to nutraceutics. *Critical Reviews in Biotechnology* 41 (2): 155–171.

17 Llewellyn, C.A. and Airs, R.L. (2010). Distribution and abundance of MAAs in 33 species of microalgae across 13 classes. *Marine Drugs* 8 (4): 1273–1291.

18 Al-Mola, H.F. (2009). Antibacterial activity of crude extracts and phlorotannin isolated from the diatom *Cymbella* spp. *Journal of Pharmacy Research* 2: 304–308.

19 Otto, B. and Schlosser, D. (2014). First laccase in green algae: purification and characterization of an extracellular phenol oxidase from *Tetracystis aeria*. *Planta* 240: 1225–1236.

20 Santana-Casiano, J., González-Dávila, M., González, A. et al. (2014). Characterization of phenolic exudates from *Phaeodactylum tricornutum* and their effects on the chemistry of Fe (II)–Fe (III). *Marine Chemistry* 158: 10–16.

21 Mishra, N., Prasad, S.M., and Mishra, N. (2019). Influence of high light intensity and nitrate deprivation on growth and biochemical composition of the marine microalgae *Isochrysis galbana*. *Brazilian Archives of Biology and Technology* 62.

22 Jerez-Martel, I., García-Poza, S., Rodríguez-Martel, G. et al. (2017). Phenolic profile and antioxidant activity of crude extracts from microalgae and cyanobacteria strains. *Polyphenols and Food Quality* 2017.

23 Scaglioni, P.T., Pagnussatt, F.A., Lemos, A.C. et al. (2019). *Nannochloropsis* sp. and *Spirulina* sp. as a source of antifungal compounds to mitigate contamination by *Fusarium graminearum* species complex. *Current Microbiology* 76: 930–938.

24 Cabrita, M.T., Vale, C., and Rauter, A.P. (2010). Halogenated compounds from marine algae. *Marine Drugs* 8 (8): 2301–2317.

25 Machu, L., Misurcova, L., Vavra Ambrozova, J. et al. (2015). Phenolic content and antioxidant capacity in algal food products. *Molecules* 20 (1): 1118–1133.

26 Markham, K. and Porter, L. (1969). Flavonoids in the green algae (chlorophyta). *Phytochemistry* 8 (9): 1777–1781.

27 Yoshie, Y., Wang, W., Petillo, D., and Suzuki, T. (2000). Distribution of catechins in Japanese seaweeds. *Fisheries Science* 66 (5): 998–1000.

28 Kurth, C., Welling, M., and Pohnert, G. (2015). Sulfated phenolic acids from *Dasycladales siphonous* green algae. *Phytochemistry* 117: 417–423.

29 Reddy, P. and Urban, S. (2009). Meroditerpenoids from the southern Australian marine brown alga *Sargassum fallax*. *Phytochemistry* 70 (2): 250–255.

30 Powers, L.C., Hertkorn, N., McDonald, N. et al. (2019). *Sargassum* sp. act as a large regional source of marine dissolved organic carbon and polyphenols. *Global Biogeochemical Cycles* 33 (11): 1423–1439.

31 Hemmi, A., Mäkinen, A., Jormalainen, V., and Honkanen, T. (2005). Responses of growth and phlorotannins in *Fucus vesiculosus* to nutrient enrichment and herbivory. *Aquatic Ecology* 39: 201–211.

32 Bidleman, T.F., Andersson, A., Jantunen, L.M. et al. (2019). A review of halogenated natural products in Arctic, Subarctic and Nordic ecosystems. *Emerging Contaminants* 5: 89–115.

33 Heffernan, N., Brunton, N.P., FitzGerald, R.J., and Smyth, T.J. (2015). Profiling of the molecular weight and structural isomer abundance of macroalgae-derived phlorotannins. *Marine Drugs* 13 (1): 509–528.

34 Santos, S.A., Félix, R., Pais, A.C. et al. (2019). The quest for phenolic compounds from macroalgae: a review of extraction and identification methodologies. *Biomolecules* 9 (12): 847.

35 Leyton, A., Pezoa-Conte, R., Barriga, A. et al. (2016). Identification and efficient extraction method of phlorotannins from the brown seaweed *Macrocystis pyrifera* using an orthogonal experimental design. *Algal Research* 16: 201–208.

36 Trifan, A., Vasincu, A., Luca, S.V. et al. (2019). Unravelling the potential of seaweeds from the Black Sea coast of Romania as bioactive compounds sources. Part I: *Cystoseira barbata* (Stackhouse) *C. Agardh*. 134: 110820.

37 Ferreres, F., Lopes, G., Gil-Izquierdo, A. et al. (2012). Phlorotannin extracts from fucales characterized by HPLC-DAD-ESI-MS n: approaches to hyaluronidase inhibitory capacity and antioxidant properties. *Marine Drugs* 10 (12): 2766–2781.

38 Freile-Pelegrín, Y. and Robledo, D. (2013). *Bioactive Compounds from Marine Foods: Plant and Animal Sources*, 113–129.

39 Olate-Gallegos, C., Barriga, A., Vergara, C. et al. (2019). Identification of polyphenols from chilean brown seaweeds extracts by LC-DAD-ESI-MS/MS. *Journal of Aquatic Food Product Technology* 28 (4): 375–391.

40 Cho, S., Yang, H., Jeon, Y.-J. et al. (2012). Phlorotannins of the edible brown seaweed Ecklonia cava Kjellman induce sleep via positive allosteric modulation of gamma-aminobutyric acid type A–benzodiazepine receptor: a novel neurological activity of seaweed polyphenols. *Food Chemistry* 132 (3): 1133–1142.

41 Catarino, M.D., Silva, A.M., Mateus, N., and Cardoso, S.M. (2019). Optimization of phlorotannins extraction from *Fucus vesiculosus* and evaluation of their potential to prevent metabolic disorders. *Marine Drugs* 17 (3): 162.

42 Nair, D., Vanuopadath, M., Balasubramanian, A. et al. (2019). Phlorotannins from *Padina tetrastromatica*: structural characterisation and functional studies. *Journal of Applied Phycology* 31 (5): 3131–3141.

43 Zhang, R., Yuen, A.K., Magnusson, M. et al. (2018). A comparative assessment of the activity and structure of phlorotannins from the brown seaweed *Carpophyllum flexuosum*. *Algal Research* 29: 130–141.

44 Sellimi, S., Benslima, A., Barragan-Montero, V. et al. (2017). Polyphenolic-protein-polysaccharide ternary conjugates from *Cystoseira barbata* Tunisian seaweed as potential biopreservatives: chemical, antioxidant and antimicrobial properties. *International Journal of Biological Macromolecules* 105: 1375–1383.

45 Yuan, Y., Zheng, Y., Zhou, J. et al. (2019). Polyphenol-rich extracts from brown macroalgae *Lessonia trabeculate* attenuate hyperglycemia and modulate gut microbiota in high-fat diet and streptozotocin-induced diabetic rats. *Journal of Agricultural and Food Chemistry* 67 (45): 12472–12480.

46 Li, Y., Fu, X., Duan, D. et al. (2017). Extraction and identification of phlorotannins from the brown alga, *Sargassum fusiforme* (Harvey), Setchell. *Marine Drugs* 15 (2): 49.

47 Ishii, T., Okino, T., Suzuki, M., and Machiguchi, Y. (2004). Tichocarpols A and B, two novel phenylpropanoids with feeding-deterrent activity from the red alga *Tichocarpus crinitus*. *Journal of Natural Products* 67 (10): 1764–1766.

48 Martone, P.T., Estevez, J.M., Lu, F. et al. (2009). Discovery of lignin in seaweed reveals convergent evolution of cell-wall architecture. *Current Biology* 19 (2): 169–175.

49 López, J.A., Gómez, A.L., Rodríguez, A. et al. (2016). Production of phenolic compounds by four species of marine microalgae under different light conditions. *Latin American Journal of Aquatic Research* 44 (1): 137–143.

50 Dahlgren, E., Enhus, C., Lindqvist, D. et al. (2015). Induced production of brominated aromatic compounds in the alga *Ceramium tenuicorne*. *Environmental Science and Pollution Research* 22: 18107–18114.

51 Ravn, H., Pedersen, M.F., Borum, J. et al. (1994). Seasonal variation and distribution of two phenolic compounds, rosmarinic acid and caffeic acid, in leaves and roots-rhizomes of eelgrass (*Zostera marina* L.). *Ophelia* 40 (1): 51–61.

52 Milović, S., Stanković, I., Nikolić, D. et al. (2019). Chemical analysis of selected seaweeds and seagrass from the Adriatic Coast of Montenegro. *Chemistry & Biodiversity* 16 (10): e1900327.

53 Steele, L. and Valentine, J.F. (2012). Idiosyncratic responses of seagrass phenolic production following sea urchin grazing. *Marine Ecology Progress Series* 466: 81–92.

54 Styshova, O.N., Popov, A.M., Artyukov, A.A., and Klimovich, A.A. (2017). Main constituents of polyphenol complex from seagrasses of the genus Zostera, their antidiabetic properties and mechanisms of action. *Experimental and Therapeutic Medicine* 13 (5): 1651–1659.

55 Hasle Enerstvedt, K., Lundberg, A., and Jordheim, M. (2017). Characterization of polyphenolic content in the aquatic plants *Ruppia cirrhosa* and *Ruppia maritima* – A source of nutritional natural products. *Molecules* 23 (1): 16.

56 Cichewicz, R.H., Clifford, L.J., Lassen, P.R. et al. (2005). Stereochemical determination and bioactivity assessment of (S)-(+)-curcuphenol dimers isolated from the marine sponge Didiscus aceratus and synthesized through laccase biocatalysis. *Marine Drugs* 13 (19): 5600–5612.

57 Blagojević, D., Babić, O., Rašeta, M. et al. (2018). Antioxidant activity and phenolic profile in filamentous cyanobacteria: the impact of nitrogen. *Journal of Applied Phycology* 30: 2337–2346.

58 Ijaz, S. and Hasnain, S.J. (2016). Antioxidant potential of indigenous cyanobacterial strains in relation with their phenolic and flavonoid contents. *Natural Product Research* 30 (11): 1297–1300.

59 Cade, S.E., Kuo, L.-J., and Schultz, I.R. (2018). Polybrominated diphenyl ethers and their hydroxylated and methoxylated derivatives in seafood obtained from Puget Sound, WA. *Science of the Total Environment* 630: 1149–1154.

60 Fairhead, V.A., Amsler, C.D., McClintock, J.B., and Baker, B. (2006). Lack of defense or phlorotannin induction by uv radiation or mesograzers in *Desmarestia anceps* and *D. menziesii* (phaeophyceae) 1. *Journal of Phycology* 42 (6): 1174–1183.

61 Smerilli, A., Orefice, I., Corato, F. et al. (2017). Photoprotective and antioxidant responses to light spectrum and intensity variations in the coastal diatom S keletonema marinoi. *Environmental Microbiology* 19 (2): 611–627.

62 Smerilli, A., Balzano, S., Maselli, M. et al. (2019). Antioxidant and photoprotection networking in the coastal diatom *Skeletonema marinoi*. *Antioxidants* 8 (6): 154.

63 Dillon, J.G., Tatsumi, C.M., Tandingan, P.G., and Castenholz, R. (2002). Effect of environmental factors on the synthesis of scytonemin, a UV-screening pigment, in a cyanobacterium (*Chroococcidiopsis* sp.). *Archives of Microbiology* 177: 322–331.

64 Kováčik, J., Klejdus, B., and Bačkor, M.J.P. (2010). Physiological responses of *Scenedesmus quadricauda* (Chlorophyceae) to UV-A and UV-C light. *Photochemistry and Photobiology* 86 (3): 612–616.

65 Duval, B., Shetty, K., and Thomas, W.H. (1999). Phenolic compounds and antioxidant properties in the snow alga *Chlamydomonas nivalis* after exposure to UV light. *Journal of Applied Phycology* 11: 559–566.

66 Jin, P., Wang, T., Liu, N. et al. (2015). Ocean acidification increases the accumulation of toxic phenolic compounds across trophic levels. *Nature Communications* 6 (1): 8714.

67 Fazelian, N., Movafeghi, A., Yousefzadi, M., and Rahimzadeh, M. (2019). Cytotoxic impacts of CuO nanoparticles on the marine microalga *Nannochloropsis oculata*. *Environmental Science and Pollution Research* 26: 17499–17511.

68 Belghith, T., Athmouni, K., Bellassoued, K. et al. (2016). Physiological and biochemical response of *Dunaliella salina* to cadmium pollution. *Journal of Applied Phycology* 28: 991–999.

69 Goiris, K., Van Colen, W., Wilches, I. et al. (2015). Impact of nutrient stress on antioxidant production in three species of microalgae. *Algal Research* 7: 51–57.

70 Singh, D.P., Prabha, R., Meena, K.K. et al. (2014). Induced accumulation of polyphenolics and flavonoids in cyanobacteria under salt stress protects organisms through enhanced antioxidant activity. *American Journal of Plant Sciences* 2014.

71 Boo, Y.C. (2019). Can plant phenolic compounds protect the skin from airborne particulate matter? *Antioxidants* 8 (9): 379.

72 Strejckova, A., Dvorak, M., Klejdus, B. et al. (2019). The strong reaction of simple phenolic acids during oxidative stress caused by nickel, cadmium and copper in the microalga *Scenedesmus quadricauda*. *New Biotechnology* 48: 66–75.

73 Safafar, H., Van Wagenen, J., Møller, P., and Jacobsen, C. (2015). Carotenoids, phenolic compounds and tocopherols contribute to the antioxidative properties of some microalgae species grown on industrial wastewater. *Marine Drugs* 13 (12): 7339–7356.

74 Mukherjee, P., Gorain, P.C., Paul, I. et al. (2020). Investigation on the effects of nitrate and salinity stress on the antioxidant properties of green algae with special reference to the use of processed biomass as potent fish feed ingredient. *Aquaculture International* 28: 211–234.

75 Hamed, S.M., Selim, S., Klöck, G., and AbdElgawad, H. (2017). Sensitivity of two green microalgae to copper stress: growth, oxidative and antioxidants analyses. *Ecotoxicology and Environmental Safety* 144: 19–25.

76 Pádua, D., Rocha, E., Gargiulo, D., and Ramos, A. (2015). Bioactive compounds from brown seaweeds: phloroglucinol, fucoxanthin and fucoidan as promising therapeutic agents against breast cancer. *Phytochemistry Letters* 14: 91–98.

77 Kannan, R.R.R., Arumugam, R., Iyapparaj, P. et al. (2013). In vitro antibacterial, cytotoxicity and haemolytic activities and phytochemical analysis of seagrasses from the Gulf of Mannar, South India. *Food Chemistry* 136 (3–4): 1484–1489.

78 Lee, S.-H. and Jeon, Y. (2013). Anti-diabetic effects of brown algae derived phlorotannins, marine polyphenols through diverse mechanisms. *Fitoterapia* 86: 129–136.

79 Lee, S.H., Karadeniz, F., Kim, M.M., and Kim, S.K. (2009). Agriculture. α-Glucosidase and α-amylase inhibitory activities of phloroglucinal derivatives from edible marine brown alga, *Ecklonia cava*. *Journal of the Science of Food and Agriculture* 89 (9): 1552–1558.

80 Kellogg, J., Grace, M.H., and Lila, M.A. (2014). Phlorotannins from Alaskan seaweed inhibit carbolytic enzyme activity. *Marine Drugs* 12 (10): 5277–5294.

81 Ende, C. and Gebhardt, R. (2004). Inhibition of matrix metalloproteinase-2 and-9 activities by selected flavonoids. *Planta Medica* 70 (10): 1006–1008.

82 Fraga, C.G., Croft, K.D., Kennedy, D.O., and Tomás-Barberán, F.A. (2019). The effects of polyphenols and other bioactives on human health. *Food and Function* 10 (2): 514–528.

83 Lomartire, S. and Gonçalves, A.M. (2023). Marine macroalgae polyphenols as potential neuroprotective antioxidants in neurodegenerative diseases. *Marine Drugs* 21 (5): 261.

84 Peckol, P., Krane, J., and Yates, J. (1996). Interactive effects of inducible defense and resource availability on phlorotannins in the North Atlantic brown alga *Fucus vesiculosus*. *Marine Ecology Progress Series* 138: 209–217.

85 Koivikko, R., Loponen, J., Honkanen, T., and Jormalainen, V.J. (2005). Contents of soluble, cell-wall-bound and exuded phlorotannins in the brown alga *Fucus vesiculosus*, with implications on their ecological functions. *Journal of Chemical Ecology* 31: 195–212.

19

Polyphenols of Triberry in Dietary Supplements and Health Care

Palmiro Poltronieri[1], *Sekar Sasikala*[2], *and Srinivasan Narasimhan*[2]

[1]*Department of Agriculture, Institute of Sciences of Food Productions, National Research Council of Italy, CNR-ISPA, Lecce, Italy*
[2]*Department of Biochemistry, Asthagiri Herbal Research Foundation, Chennai, Tamil Nadu, India*

19.1 Introduction

Berries are small, pulpy, and usually edible fruits that are typically brightly colored and have a sweet or tart taste. They are a source of antioxidant compounds in vitamins and minerals, making them a popular addition to a healthy diet. Common berries include strawberries, blueberries, raspberries, blackberries, cranberries, and grapes [1]. Berries have been a part of human diets from the beginning of humanity and have been a medicine source for several ailments. They have also been the subject of scientific research thanks to their beneficial effects on human health. Studies have suggested that regular consumption of berries may contribute to the reduction of the risk of chronic ailments, such as cardiovascular diseases (CVD) and type 2 diabetes, may delay the onset of cancer, and also may slow the loss of cognitive function, and may regulate the functionality of the gut [2, 3].

Several bioactive compounds in berries, such as anthocyanins, flavonols, chalcones, and the non-flavonoid polyphenol resveratrol, belong to the flavonoid class. These compounds have been shown to have anti-inflammatory and antioxidant properties, which may protect against inflammation and oxidative stress, both of which are implicated in developing chronic diseases. Berries also contain dietary fiber, which may promote satiety and improve digestive health [4]. The objective of this review is to examine the existing literature dealing with the potential benefits on health from various berries and components present in them, with a focus on how they can potentially lower the risk of chronic diseases

Science and Engineering of Polyphenols: Fundamentals and Industrial Scale Applications, First Edition. Edited by Chandrabhan Verma.
© 2024 John Wiley & Sons Inc. Published 2024 by John Wiley & Sons Inc.

that are commonly seen in the modern day such as CVD, obesity, diabetes, and cognitive decline. There are many different berries, each with unique flavor, nutritional composition, and health benefits. Some of the most common types of berries are listed in Table 19.1.

Table 19.1 Examples of berries and their phytocompounds.

Serial number	Scientific name	Common name	Description	Reported phytocompounds
1	*Fragaria ananassa*	Strawberries	Pink-colored berry with a sweet and tangy flavor	Vitamin C, folate, potassium, anthocyanins, phenolic compounds, quercetin, ellagic acid, and antioxidants [5]
2	*Vaccinium spp.*	Blueberries	Blue-colored with a sweet and slightly tart taste	High antioxidant contents, anthocyanins, myricetin, resveratrol, ellagic acid, quercetin, flavonoids, phenolic acids, Vitamin C, Vitamin K, and dietary fiber [6]
3	*Rubus idaeus*	Raspberries	Colored with a sweet and tart taste	High in dietary fiber, Vitamin C, and manganese, as well as various antioxidants, anthocyanins, ellagitannins, ellagic acid, sanguiin, and anthocyanins [7]
4	*Rubus spp.*	Blackberries	Colored with a sweet and tart taste	Contain various antioxidants, including anthocyanins, ellagic acid, flavonoids, phenolic compounds, and quercetin [8]
5	*Vaccinium macrocarpon*	Cranberries	A tart and slightly bitter taste	Vitamin C and dietary fiber, as well as various antioxidants, flavanols, proanthocyanidins, anthocyanins, and phenolic acids [9]
6	*Vitis spp.*	Grapes	A sweet and juicy	Good source of Vitamin C, Vitamin K, and potassium, as well as various antioxidants, phenolic acids, stilbenes, anthocyanins, flavonoids, resveratrol, and quercetin [10]

Table 19.1 (Continued)

Serial number	Scientific name	Common name	Description	Reported phytocompounds
7	*Emblica officinalis* Gaertn. or *Phyllanthus emblica* Linn	Indian gooseberry or amla	Light green colored with a sour and tart taste	Vitamin C, phenolic compounds, gallic acids, amino acids, tannins, quercetin, ellagic acid, phyllembelic acid, rutin, curcuminoids, kaempferol, quercetin, mucic acid, and emblicol [11]
8	*Terminalia bellerica* Roxb	Vibhitaki, Tanni, Bahera, Belleric myrobalan	Fruits are yellowish-brown, honed, not winged, softly tomentose, and ellipsoid. It has a sour taste	Flavonoid glycosides, tannins, ellagic acid, alkaloids, belleric acid, flavonoids, phenols, saponins, carbohydrates, proteins [12]
9	*Terminalia chebula* Retz	Haritaki, Kadukkai, Chebulic myrobalan	Fruit is green when unripe and yellowish grey when ripe. It doesn't taste enjoyable	Tannins, chebulinic acid, ellagic acid and anthraquinones, polyphenols, triterpenoid glycosides, phenolic, and fatty acids [13]

Polyphenols are a group of naturally occurring compounds in many plant foods, including berries. They are known for their anti-inflammatory and antioxidant properties and have been shown to possess health benefits, such as reduction of the risk of chronic diseases [14].

19.2 Types of Polyphenols Found in Berries

Anthocyanins: They are hydrophilic pigments conferring to berries the characteristic red, blue, or purple color. They are potent antioxidants linked to a reduced risk of chronic diseases, CVD, and cancer. Examples of berries rich in anthocyanins include blueberries, blackberries, raspberries, and strawberries [14, 15]. Figure 19.1 represents the basic structure of anthocyanin.

Flavonols: They are a type of flavonoid that is found in many plant foods, including berries. Examples of berries rich in flavonols include cranberries, lingonberries, and bilberries [14, 16]. Figure 19.2 represents the basic structure of flavonols.

Flavanols: This is another type of flavonoid found in berries. Flavanols have been linked to reduced CVD risk and improved cognitive function. Examples of

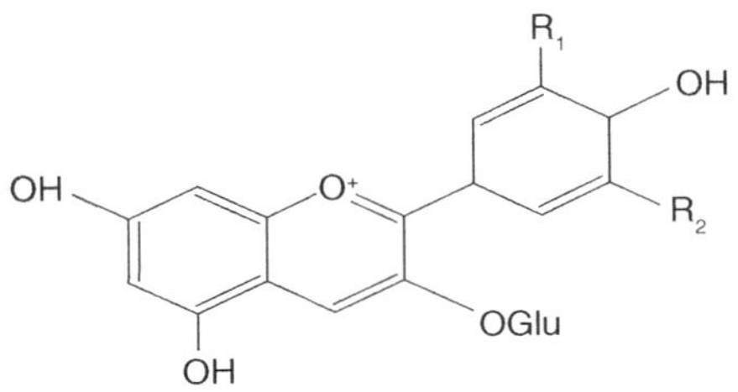

Figure 19.1 Chemical structure of anthocyanin ($C_{78}H_{89}O_{44}$).

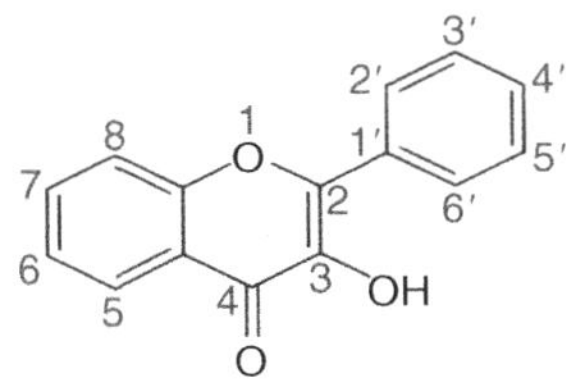

Figure 19.2 Chemical structure of flavonols ($C_{15}H_{10}O_3$).

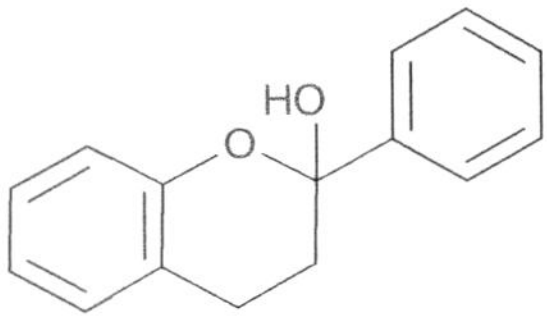

Figure 19.3 Chemical structure of flavanols ($C_{15}H_{14}O_2$).

berries rich in flavanols include strawberries, blackberries, and raspberries [14, 16]. Figure 19.3 represents the basic structure of flavanols.

Ellagitannins: These are a type of tannin found in berries, particularly in raspberries, blackberries, and strawberries. They are potent antioxidants linked to a reduced risk of obesity, diabetes, CVD, and cancer [14, 17]. Figure 19.4 represents the basic structure of *ellagitannins*.

19.3 Potential Health Benefits of Polyphenols from Berries

Cardiovascular Health: The consumption of berries rich in polyphenols has been linked to a reduced risk of CVD. Polyphenols have improved blood pressure, reduced inflammation and CVD risk, and improved endothelial function [17].

Cancer Prevention: Polyphenols have shown potential anti-cancer properties, particularly in reducing the risk of colon, prostate, and breast cancer. Anthocyanins

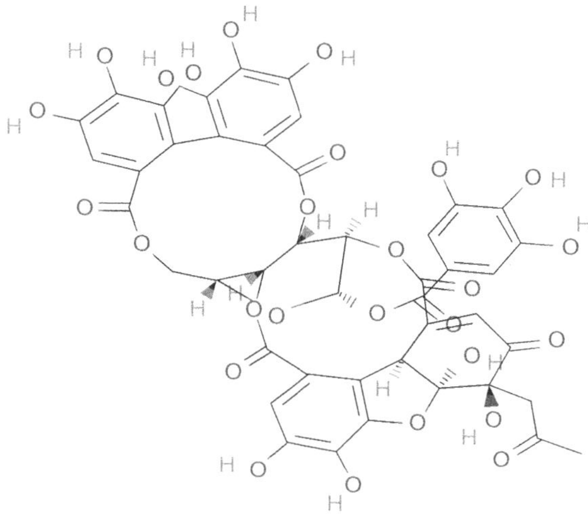

Figure 19.4 Chemical structure of ellagitannins ($C_{44}H_{32}O_{27}$).

and ellagitannins, in particular, have been associated with a reduced risk of cancer [17, 18].

Cognitive Function: Polyphenols improve cognitive function, particularly memory and learning. Flavanols found in berries have been particularly effective in improving cognitive function [16].

Immune System Support: The consumption of berries rich in polyphenols has also been associated with improved immune function, particularly reducing inflammation [19].

Inflammation: This has been linked to the risk of chronic diseases, such as CVD and type 2 diabetes, and cancer. Polyphenols have effectively reduced inflammation by inhibiting the release of pro-inflammatory molecules [20].

Gut Health: Polyphenols found in berries have also been shown to promote and maintain a healthy gut microbiome, essential for overall health and well-being. Polyphenols serve as a good feed for beneficial bacteria in the gut, and this has been associated with a reduced risk of inflammatory bowel disease and improved digestive function [21, 22].

Skin Health: Polyphenols also promote skin health by reducing oxidative stress and inflammation. This may help to reduce the signs of aging and to sustain skin health [23].

Weight Management: Polyphenols have effectively reduced body fat accumulation by inhibiting fat cell production and promoting the breakdown of fats [24].

Berries have been used in scientific research for various purposes, including investigating their health benefits, studying their chemical composition, and exploring their potential use in multiple industries. Here are some examples of scientific services of berries with references:

Investigating the health benefits of berries: Researchers have conducted numerous studies to investigate the health benefits of various berries, including blueberries, strawberries, raspberries, and cranberries. For example, a report found that the antioxidants in blueberries could help protect against neurodegenerative diseases by reducing inflammation in the brain and protecting the membranes from oxidative stress [25]. Another study showed that consuming cranberry extract reduced the risk of urinary tract infections in women [26]. Herbal Teas are used for their therapeutic and energizing properties, with the potential to induce relaxation and protect the stomach from digestive problems; for these beneficial effects, herbal teas can provide cleansing properties to the body and strengthen the immune system.

Studying the chemical composition of berries: Berries are complex mixtures of various chemicals, including phenolic compounds, organic acids, sugars, and pigments. Scientists have studied the chemistry and design of berries better to understand their flavor, aroma, and nutritional properties. For example, a study analyzed blackberries' chemical components and found high levels of anthocyanins responsible for their dark color and antioxidant activity [27]. Another report analyzed the volatile compounds in strawberries and found they were responsible for their characteristic aroma [28].

Exploring the potential use of berries in various industries: Berries are used in a wide range of industries, including food, pharmaceuticals, and cosmetics. For example, a study investigated the potential use of raspberry ketones as a natural food preservative [29]. Another study investigated the use of cranberry extract in skin care products due to its anti-inflammatory and antioxidant properties [30].

Berries have been used in various scientific studies for their potential health benefits. Here are some examples of scientific uses of berries:

- *Antioxidant activity*—Berries are rich in antioxidants, and these compounds can help to protect against oxidative damage caused by free radicals. Many studies have investigated the antioxidant activity of different types of berries, such as strawberries, blueberries, raspberries, and blackberries [5, 7, 31].
- *Cardiovascular health*—Berries have been shown to benefit cardiovascular health, improving blood pressure, reducing inflammation, and reducing oxidative stress. Reports have studied the effects of various types of berries, including blueberries, strawberries, and cranberries [32–34].
- *Cancer prevention*—Some studies have suggested that berries may have potential cancer-preventive properties due to their high antioxidant content and inflammatory effects. Scientists have studied the impact of various types of berries, including strawberries, blueberries, and raspberries [7, 35].
- *Cognitive function*—Some reports proposed that berries have potential cognitive benefits, including improving memory and reducing age-related cognitive decline. Scientists have studied the effects of various types of berries, including blueberries and strawberries [36, 37].

19.4 Conclusion

In recent decades, the chemical molecules called polyphenols, widely present in plants, have attracted increasing attention in nutrition. A rising amount of studies suggests that consuming polyphenols may be essential for maintaining good health through controlling metabolism, weight, chronic disease, and cell proliferation. In conclusion, berry polyphenols have numerous potential health-beneficial effects, such as protecting from CVD, reducing cancer risk, improving cognitive function, supporting the immune system, reducing inflammation, promoting gut health, improving skin health, and promoting weight management. While more research is needed in this area, consuming a variety of berries rich in polyphenols may be a simple and delicious way to reduce the risk of chronic diseases and improve overall health.

References

1 Skrovankova, S., Sumczynski, D., Mlcek, J. et al. (2015). Bioactive compounds and antioxidant activity in different types of berries. *International Journal of Molecular Sciences* 16 (10): 24673–24706. https://doi.org/10.3390/ijms161024673.

2 Gopalan, A., Reuben, S.C., Ahmed, S. et al. (2016). The health benefits of berries. *Critical Reviews in Food Science and Nutrition* 56 (3): 430–446. https://doi.org/10.1080/10408398.2012.741660.

3 Basu, A., Rhone, M., and Lyons, T.J. (2010). Berries: emerging impact on cardiovascular health. *Nutrition Reviews* 68 (3): 168–177. https://doi.org/10.1111/j.1753-4887.2010.00273.x.

4 Devi, B.A., Devi, S.G., and Kumar, M.S. (2015). Health benefits of berries: a review. *Journal of Food Science and Technology* 52 (8): 4550–4560. https://doi.org/10.1007/s13197-014-1393-4.

5 Giampieri, F. et al. (2012). The strawberry: composition, nutritional quality, and impact on human health. *Nutrition* 28 (1): 9–19.

6 Prior, R.L. et al. (2015). Blueberries improve biomarkers of cardiometabolic function in participants with metabolic syndrome—results from a 6-month, double-blind, randomized controlled trial. *The American Journal of Clinical Nutrition* 102 (3): 649–660.

7 Seeram, N.P. (2008). Berry fruits for cancer prevention: current status and future prospects. *Journal of Agricultural and Food Chemistry* 56 (3): 630–635.

8 Puupponen-Pimia, R. et al. (2001). Antimicrobial properties of phenolic compounds from berries. *Journal of Applied Microbiology* 90 (4): 494–507.

9 Koo, H. et al. (2013). Cranberries and their bioactive constituents in human health. *Advances in Nutrition* 4 (6): 618–632.

10 Tundis, R. et al. (2020). Grape berries: ripe for valorization. *Foods* 9 (5): 571.
11 Krishnaveni, M. and Mirunalini, S. (2010). Therapeutic potential of Phyllanthus emblica (amla): the ayurvedic wonder. *Journal of Basic and Clinical Physiology and Pharmacology 21* (1): 93–105. https://doi.org/10.1515/jbcpp.2010.21.1.93.
12 Kumar, N. and Khurana, S.M.P. (2018). Phytochemistry and medicinal potential of the Terminalia bellirica Roxb (Bahera). *Indian Journal of Natural Products and Resources* 9 (2): 97–107.
13 Bag, A., Bhattacharyya, S.K., and Chattopadhyay, R.R. (2013). The development of Terminalia chebula Retz. (Combretaceae) in clinical research. *Asian Pacific Journal of Tropical Biomedicine 3* (3): 244–252. https://doi.org/10.1016/S2221-1691(13)60059-3.
14 Pap, N. et al. (2021). Berry polyphenols and human health: evidence of antioxidant, anti-inflammatory, microbiota modulation, and cell-protecting effects. *Current Opinion in Food Science* 42: 167–186. https://doi.org/10.1016/j.cofs.2021.06.003.
15 Bilawal, A. et al. (2021). A review of the bioactive ingredients of berries and their applications in curing diseases. *Food Bioscience* 44 (A): 101407. https://doi.org/10.1016/j.fbio.2021.101407.
16 Wang, X., Ouyang, Y.Y., Liu, J., and Zhao, G. (2014). Flavonoid intake and risk of CVD: a systematic review and meta-analysis of prospective cohort studies. *The British Journal of Nutrition* 111 (1): 1–11. https://doi.org/10.1017/S000711451300278X.
17 Burton-Freeman, B.M., Sandhu, A.K., and Edirisinghe, I. (2016). Red raspberries and their bioactive polyphenols: cardiometabolic and neuronal health links. *Advances in Nutrition (Bethesda, Md.)* 7 (1): 44–65. https://doi.org/10.3945/an.115.009639.
18 Ding, M., Feng, R., Wang, S.Y. et al. (2006). Cyanidin-3-glucoside, a natural product derived from blackberry, exhibits chemopreventive and chemotherapeutic activity. *Journal of Biological Chemistry* 281 (25): 17359–17368. https://doi.org/10.1074/jbc.M600861200.
19 Amagase, H. et al. (2009 Oct). Immunomodulatory effects of a standardized Lycium barbarum fruit juice in Chinese older healthy human subjects. *Journal of Medicinal Food* 12 (5): 1159–1165.
20 González, R., Ballester, I., López-Posadas, R. et al. (2011). Effects of flavonoids and other polyphenols on inflammation. *Critical Reviews in Food Science and Nutrition* 51 (4): 331–362. https://doi.org/10.1080/10408390903584094.
21 Govers, C. et al. (2018). Review of the health effects of berries and their phytochemicals on the digestive and immune systems. *Nutrition Reviews* 76 (1): 29–46.
22 Williamson, G. (2017). The role of polyphenols in modern nutrition. *Nutrition Bulletin* 42 (3): 226–235. https://doi.org/10.1111/nbu.12278.

23 Yenny, S.W. and Suryani, Y.E. (2020). Polyphenols as natural antioxidants in skin aging. *Sumatera Medical Journal* 3 (3): https://doi.org/10.32734/sumej.v3i3.4047.
24 Lovisa Heyman, Ulrika Axling, Narda Blanco, Olov Sterner, Cecilia Holm, Karin Berger, Evaluation of beneficial metabolic effects of berries in high-fat fed C57BL/6J mice, *Journal of Nutrition and Metabolism*, 2014, 403041, 12, 2014. https://doi.org/10.1155/2014/403041.
25 Kalt, W., Blumberg, J.B., McDonald, J.E. et al. (2010). Identification of anthocyanins in the liver, eye, and brain of blueberry-fed pigs. *Journal of Agricultural and Food Chemistry* 58 (10): 6606–6615.
26 Howell, A.B., Vorsa, N., Marderosian, A.D. et al. (2010). Inhibition of the adherence of P-fimbriated Escherichia coli to uroepithelial-cell surfaces by proanthocyanidin extracts from cranberries. *New England Journal of Medicine* 339 (15): 1085–1086.
27 Wu, X., Beecher, G.R., Holden, J.M. et al. (2004). Lipophilic and hydrophilic antioxidant capacities of common foods in the United States. *Journal of Agricultural and Food Chemistry* 52 (12): 4026–4037.
28 González-Ferrero, C., Salvador, A., and Sanz, J. (2018). Characterization of the volatile profile of strawberry wine using headspace solid-phase microextraction and gas chromatography-mass spectrometry. *Journal of Food Science* 83 (7): 1881–1890. https://doi.org/10.1111/1750-3841.14223.
29 Tiwari, B.K., Muthukumarappan, K., and O'Donnell, C.P. (2017). Raspberry ketone: an overview of its chemistry, health effects and potential as a natural preservative. *Journal of Food Science and Technology* 54 (8): 2231–2240. https://doi.org/10.1007/s13197-017-2680-0.
30 Choi, S.Y., Ko, E.J., Lee, Y.H. et al. (2016). The effect of cranberry extract on the skin: Using the Korean standard model of skin aging. *Journal of Cosmetic Science* 67 (2): 97–104. PMID: 27006185.
31 Curtis, P.J., van der Velpen, V., Berends, L. et al. (2019). Blueberries improve biomarkers of cardiometabolic function in participants with metabolic syndrome-results from a 6-month, double-blind, randomized controlled trial. *The American Journal of Clinical Nutrition* 109 (6): 1535–1545. https://doi.org/10.1093/ajcn/nqy380.
32 Basu, A. et al. (2010). Blueberries decrease cardiovascular risk factors in obese men and women with metabolic syndrome. *The Journal of Nutrition* 140 (9): 1582–1587. https://doi.org/10.3945/jn.110.124701.
33 Erlund, I. et al. (2008). Favorable effects of berry consumption on platelet function, blood pressure, and HDL cholesterol. *The American Journal of Clinical Nutrition* 87 (2): 323–331. https://doi.org/10.1093/ajcn/87.2.323.
34 Ruel, G. et al. (2013). Evidence that cranberry juice may improve augmentation index in overweight men. *Nutrition Research* 33 (1): 41–49. https://doi.org/10.1016/j.nutres.2012.11.002.

35 Aiyer, H.S. and Gupta, R.C. (2010). Berries and ellagic acid prevent estrogen-induced mammary tumorigenesis by modulating enzymes of estrogen metabolism. *Cancer Prevention Research (Philadelphia, PA)* 1 (4): 227–235. https://doi.org/10.1158/1940-6207.capr-09-0260.

36 Krikorian, R. et al. (2010). Blueberry supplementation improves memory in older adults. *Journal of Agricultural and Food Chemistry* 58 (7): 3996–4000. https://doi.org/10.1021/jf9029332.

37 Devore, E.E., Kang, J.H., Breteler, M.M., and Grodstein, F. (2012). Dietary intakes of berries and flavonoids in relation to cognitive decline. *Annals of Neurology* 72 (1): 135–143. https://doi.org/10.1002/ana.23594.

20

Dietary Polyphenols: Recent Trends in Its Biological Significance

P. Indra Arulselvi, K.R. Kavya, and M. Kesavapriya

Department of Biotechnology, Periyar University, Salem, Tamil Nadu, India

20.1 Introduction

Dietary polyphenols encompass various biologically active compounds in plant-based foods, including fruits, vegetables, whole grains, and beverages like tea, coffee, and wine. These compounds can be categorized into various subclasses based on their chemical structure, such as flavonoids, phenolic acids, stilbenes, and lignans. Studies have shown that the regular consumption of polyphenol-rich foods is associated with a lower risk of chronic diseases like cardiovascular disease, cancer, and neurodegenerative disorders. The health benefits of polyphenols are believed to stem from their antioxidant, anti-inflammatory, and antimicrobial properties. Moreover, polyphenols have been found to influence important cellular signaling pathways involved in cell survival, proliferation, and differentiation.

20.1.1 Definition of Polyphenols

Polyphenols are naturally occurring organic compounds found in plants. They are characterized by one or more hydroxyl groups attached to an aromatic ring. These compounds are classified as secondary metabolites and are synthesized by plants in response to various environmental stressors such as ultraviolet radiation, pathogens, and herbivores.

Science and Engineering of Polyphenols: Fundamentals and Industrial Scale Applications,
First Edition. Edited by Chandrabhan Verma.
© 2024 John Wiley & Sons Inc. Published 2024 by John Wiley & Sons Inc.

20.1.2 Sources of Dietary Polyphenols

Dietary polyphenols can be obtained from various plant-based sources, including fruits, vegetables, grains, nuts, seeds, and beverages like tea, coffee, and wine. The main classes of polyphenols include flavonoids, phenolic acids, stilbenes, lignans, and tannins. Table 20.1 provides an overview of these polyphenol classes and their food sources.

Table 20.1 An overview of these polyphenol classes and their food sources.

Class and subclass of dietary polyphenols	Source	References
Flavonoids and anthocyanidins consist of compounds such as cyanidin 3-galactoside, cyanidin 3-glucoside, cyanidin 3-arabinoside, cyanidin 3-xyloside, malvidin, delphinidin, and pelargonidin **Anthoxanthins and flavonols** encompass myricetin, fisetin, quercetin, kaempferol, and isorhamnetin **Flavanones** include naringenin, eriodictyol, and hesperetin **Flavones** consist of apigenin and luteolin **Flavanols** (flavan-3-ols) comprise catechin, epicatechin, epicatechin 3-gallate, morin, epigallocatechin, epigallocatechin-3-gallate, gallocatechin, procyanidins, and prodelphinidins **Isoflavones** (flavans) include genistein, daidzein, and equol **Other subclasses and compounds** encompass rutin, hesperidin, naringin, caffeic acid, chlorogenic acid, ferulic acid, neochlorogenic acid, p-coumaric acid, sinapic acid, ellagic acid, gallic acid, corilagin, resveratrol, trans-resveratrol, catechin polymers, epicatechin polymers, ellagitannins, proanthocyanidins, casuarictin, sanguiin H6, tannic acids, and curcumin	**Fruits**: Various fruits are rich in polyphenols, including blackberries, black currants, blueberries, black grapes, elderberries, strawberries, cherries, plums, cranberries, raspberries, apples, apricots, grapes, peaches, nectarines, pears, raisins, and citrus fruits like lemon, lime, orange, grapefruit, and tangerine	[1–4] [5–9]
	Vegetables: Polyphenol-rich vegetables include capers, celery, chives, onions, red onions, dock leaves, fennel, hot peppers, cherry tomatoes, spinach, sweet potato leaves, lettuce, celery, broccoli, kale, celery hearts, fresh parsley, and potatoes	[10–13]
	Cereal and legumes: Buckwheat, green and yellow beans, chickpeas, black-eyed peas, and lentils are polyphenol-containing cereals and legumes	[14–16]
	Others: Red wine, green and black tea, cocoa powder, and spices like dill weed, peppermint, oregano, rosemary, and thyme are additional polyphenols	[5–9, 14, 17–19]
	Soybean and soy products (soy nuts, soy flour/bread, tofu, miso, soy milk, tofu yogurt, soy cheese/sauce/hot dog), walnuts, peanuts, oak-aged red wine, white wine, chocolate (dark and milk), coffee beans, cider, and immature fruits also contain polyphenols	[20–22]
	Herbal remedy and spice: Turmeric is a dietary spice and herbal remedy that contains polyphenols	[23–26]

20.1.3 Biological Significance of Polyphenols

Polyphenols exhibit diverse biological activities, encompassing antioxidant, anti-inflammatory, anticancer, cardioprotective, neuroprotective, and antimicrobial effects. These beneficial properties are associated with polyphenols' capacity to counteract free radicals, mitigate oxidative stress, regulate signaling pathways, and modulate gene expression.

20.1.4 Classification of Dietary Polyphenols

a) Flavonoids
   1) Flavanols
   2) Flavonols
   3) Isoflavones
b) Non-flavonoids
   1) Hydroxycinnamic acids
   2) Stilbenes
   3) Lignans

Polyphenols are a diverse class of natural compounds found throughout the plant kingdom. They are characterized by their multiple phenol functional groups that contribute to their wide range of chemical and biological properties. Dietary polyphenols are derived from our foods and have been extensively studied for their various health benefits. They can be classified into two main groups: flavonoids and non-flavonoids. Flavonoids are the most abundant and extensively researched subgroup, including flavanols, flavonols, and isoflavones. Flavanols, such as catechins, are highly concentrated in tea, cocoa, apples, and grapes and have been associated with positive effects on cardiovascular health, cognition, and insulin sensitivity. Flavonols, such as quercetin and kaempferol, are commonly found in onions, broccoli, berries, and tea and have been linked to antioxidant, anti-inflammatory, and potential protective effects against chronic diseases. Isoflavones, primarily present in soybeans and soy products and legumes like chickpeas and lentils, have been studied for their potential roles in reducing the risk of certain cancers, improving bone health, and alleviating menopausal symptoms. Non-flavonoid polyphenols, such as hydroxycinnamic acids, are widely distributed in plants and often occur as esters, including caffeic acid, ferulic acid, and coumaric acid. These compounds have demonstrated antioxidant, anti-inflammatory, and anticancer properties.

Stilbenes are a group of non-flavonoid dietary polyphenols found in grapes, peanuts, and berries. Resveratrol, a well-known stilbene, has been extensively studied for its potential health benefits. It exhibits antioxidant, anti-inflammatory, and anticancer properties and positively affects cardiovascular health and blood sugar regulation. Lignans, another type of polyphenol, are present in flaxseeds, sesame seeds, and whole grains. When consumed, they are converted by gut bacteria into enterolignans, which have estrogenic and antiestrogenic effects. Lignans also possess antioxidant and anti-inflammatory properties and have been associated with improved cardiovascular health and blood sugar regulation. Hydroxycinnamic acids, such as ferulic and caffeic acids, are highly concentrated in coffee and whole grains. Stilbenes, like resveratrol, can be found in red wine and grapes. Lignans, such as secoisolariciresinol and matairesinol, are present in flaxseed, sesame seeds, and whole grains. It is crucial to comprehend the categorization of dietary polyphenols because various subgroups within polyphenols exhibit distinct health advantages. For instance, flavonoids have been linked to a decreased likelihood of cardiovascular disease and specific types of cancer, whereas lignans have been associated with a lower risk of breast cancer. By studying the diverse subgroups of polyphenols, we can enhance our understanding of their potential health benefits and formulate dietary approaches to maximize their consumption [27–31].

20.2 Biological Activities of Dietary Polyphenols

20.2.1 Antioxidant Activity

One of dietary polyphenols' most extensively studied and recognized biological properties is their ability to act as antioxidants. Polyphenols function as antioxidants by counteracting reactive oxygen species (ROS) and preventing oxidative damage to cells and tissues. Several in vitro and in vivo investigations have demonstrated the antioxidant effects of different polyphenols, including flavonoids, hydroxycinnamic acids, stilbenes, and lignans. For instance, quercetin, a flavonoid commonly present in fruits and vegetables, has been shown to neutralize free radicals and inhibit lipid peroxidation in laboratory settings and living organisms. Similarly, resveratrol, a stilbene found in red wine, has demonstrated protective effects against oxidative stress in various cellular and animal models. The antioxidant activity of polyphenols is believed to contribute to their potential health benefits, which include reducing the risk of chronic diseases like cardiovascular disease, cancer, and neurodegenerative disorders. However, it is essential to acknowledge that polyphenols' in vivo antioxidant activity is intricate and influenced by factors such as bioavailability, metabolism, and interactions with other components in the diet.

20.2.1.1 Mechanisms of Antioxidant Action

Antioxidants counteract or minimize the harmful impacts of ROS and reactive nitrogen species (RNS) within the body. They achieve this through various mechanisms, such as scavenging ROS and RNS, hindering enzymes that generate ROS, and activating antioxidant enzymes. By donating an electron or hydrogen atom, antioxidants can stabilize free radicals, thereby preventing additional harm to cells and tissues.

20.2.1.2 Role in Disease Prevention

Antioxidants significantly impact the prevention and management of chronic illnesses such as cardiovascular disease, cancer, neurodegenerative diseases, and diabetes. Here is an example of how antioxidants contribute to disease prevention: Cardiovascular disease: Multiple studies have demonstrated that antioxidants can inhibit the oxidation of low density lipoproteins (LDL) cholesterol, which is a crucial factor in the development of atherosclerosis. Additionally, antioxidants possess anti-inflammatory properties that help mitigate the underlying inflammation associated with cardiovascular disease [32].

Cancer: Cancer development is influenced by oxidative stress, which can lead to DNA damage. Antioxidants play a crucial role in neutralizing ROS and RNS, thereby reducing DNA damage and preventing the initiation and progression of cancer cells [33].

Neurodegenerative diseases: Neurodegenerative diseases like Alzheimer's and Parkinson's are associated with oxidative stress and inflammation in the brain. Antioxidants can protect neurons from oxidative damage and reduce inflammation, potentially mitigating the development and progression of these diseases [34].

Diabetes: Oxidative stress is known to play a role in developing insulin resistance and the progression of diabetes. Antioxidants have the potential to enhance insulin sensitivity and reduce oxidative stress, thereby lowering the risk of diabetes [35]. A diet rich in antioxidants derived from fruits, vegetables, whole grains, and nuts has been linked to a decreased likelihood of developing chronic diseases. Additionally, certain antioxidant supplements, including vitamins C and E, beta-carotene, and selenium, have shown some positive effects in preventing and managing chronic diseases [36].

20.2.2 Anti-inflammatory Activity

The anti-inflammatory activity of dietary polyphenols refers to their ability to reduce inflammation in the body. These polyphenolic compounds are commonly found in fruits, vegetables, and tea. Inflammation is a natural response of the immune system to injury or infection. Still, chronic inflammation can contribute to the development of various chronic diseases, including cardiovascular disease, cancer, and neurodegenerative diseases. Polyphenols have demonstrated anti-inflammatory effects by inhibiting the production of molecules that promote inflammation and reduce oxidative stress. For instance, resveratrol, a polyphenol in red wine and grapes, has been observed to possess anti-inflammatory properties by decreasing the expression of pro-inflammatory cytokines and chemokines. Curcumin in turmeric and quercetin in onions and apples are additional examples of polyphenolic compounds with anti-inflammatory properties. These compounds have the potential to mitigate inflammation in the body and contribute to the prevention of chronic diseases. In summary, the anti-inflammatory properties of dietary polyphenols hold significant implications for promoting health and preventing infections. By including polyphenol-rich foods in our diet, we can reduce inflammation and safeguard against the onset of chronic diseases.

20.2.2.1 Mechanisms of Anti-inflammatory Action

Chronic inflammation, which can contribute to the development of various diseases, is a prolonged and detrimental response to injury or infection. Anti-inflammatory compounds are crucial in reducing inflammation and managing or preventing chronic diseases. They achieve this by employing several mechanisms, including:

- Inhibiting the production of pro-inflammatory cytokines and chemokines, which are molecules that promote inflammation.
- Reducing the activity of enzymes involved in inflammation, such as cyclooxygenase-2 and lipoxygenase.
- Suppressing the activation of immune cells like macrophages and T cells, which play a role in initiating and perpetuating inflammation.
- Enhancing the activity of anti-inflammatory enzymes like heme oxygenase-1 and glutathione peroxidase, which help regulate and counteract inflammation.

These mechanisms collectively contribute to the anti-inflammatory effects of these compounds and their potential to manage chronic diseases associated with inflammation.

20.2.2.2 Role in Disease Prevention

Anti-inflammatory compounds can prevent or manage several chronic diseases, including the following:

Cardiovascular disease: Atherosclerosis, a leading cause of cardiovascular disease, is influenced by chronic inflammation. Anti-inflammatory compounds can reduce inflammation and impede the progression of atherosclerosis [37].

Diabetes: Inflammation plays a role in insulin resistance and the advancement of diabetes. Anti-inflammatory compounds can enhance insulin sensitivity and lower the risk of diabetes [35].

Cancer: Chronic inflammation contributes to the development of cancer. Anti-inflammatory compounds can suppress the production of pro-inflammatory cytokines and chemokines, thereby inhibiting the growth and spread of cancer cells [38].

Neurodegenerative diseases: Neurodegenerative diseases, including Alzheimer's and Parkinson's, are linked to chronic inflammation and oxidative stress. Anti-inflammatory compounds can alleviate inflammation and shield neurons from oxidative damage [34].

In general, consuming foods with anti-inflammatory properties, such as fruits, vegetables, whole grains, and fatty fish, has been associated with a decreased risk of chronic diseases. Additionally, supplements with anti-inflammatory properties, such as omega-3 fatty acids, curcumin, and resveratrol, have demonstrated beneficial effects in preventing and managing chronic diseases [39].

20.2.3 Anti-cancer Activity

20.2.3.1 Mechanisms of Anti-cancer Action

Cancer is a complex and multifactorial disease that arises from the uncontrolled growth and division of abnormal cells. Polyphenols have undergone extensive research due to their potential anti-cancer effects, and their mechanisms of action include:

1) Inducing cell cycle arrest: Polyphenols can halt the cell cycle at different stages, preventing the uncontrolled proliferation of cancer cells.
2) Promoting apoptosis: Polyphenols can trigger programmed cell death in cancer cells, a natural defense mechanism against cancer.
3) Inhibiting angiogenesis: Polyphenols can impede the formation of new blood vessels that supply nutrients to cancer cells, starving them and reducing their growth.
4) Suppressing inflammation: Polyphenols can diminish inflammation, a key driver of cancer development and progression.
5) Modulating signaling pathways: Polyphenols can regulate various signaling pathways involved in cancer development and progression, such as the PI3K/Akt/mTOR pathway, the NF-κB pathway, and the Wnt/β-catenin pathway.

Regarding the role of polyphenols in cancer prevention and treatment, several examples demonstrate their anti-cancer effects:

Breast cancer: Polyphenols like epigallocatechin-3-gallate (EGCG) in green tea and turmeric curcumin inhibit breast cancer cells' growth and metastasis [40]. Prostate cancer: Polyphenols such as resveratrol in red wine and green tea catechins inhibit the growth and progression of prostate cancer cells [41]. Colorectal cancer: Polyphenols such as quercetin in onions and apples, and EGCG in green tea, hinder the growth and invasion of colorectal cancer cells [42]. Lung cancer: Polyphenols including resveratrol and EGCG inhibit the growth and metastasis of lung cancer cells [43].

Overall, consuming polyphenol-rich foods and beverages like fruits, vegetables, tea, and red wine has been linked to a decreased cancer risk. Furthermore, preclinical and clinical studies have demonstrated anti-cancer effects in polyphenol supplements such as curcumin, resveratrol, and EGCG [44].

20.2.3.2 Cardiovascular Health Benefits

Polyphenols have been shown to have various cardiovascular health benefits, including:

Reducing oxidative stress: Polyphenols have potent antioxidant properties, which can reduce oxidative stress and prevent oxidative damage to lipids, proteins, and DNA that contribute to cardiovascular disease.

Improving endothelial function: Polyphenols can improve the function of the endothelial cells that line the blood vessels, which is a key factor in preventing atherosclerosis and hypertension.

Lowering blood pressure: Polyphenols can lower blood pressure by promoting the production of nitric oxide, a vasodilator that relaxes the blood vessels and reduces resistance to blood flow.

Inhibiting platelet aggregation: Polyphenols can inhibit the aggregation of platelets, which is a key step in forming blood clots that can cause heart attacks and strokes.

Reducing inflammation: Polyphenols can reduce inflammation, a critical cardiovascular disease driver.

Role in disease prevention: Consuming foods and beverages rich in polyphenols has been linked to a decreased risk of cardiovascular disease. The following examples highlight the cardiovascular health benefits associated with polyphenols:

1) *Lowering the risk of coronary heart disease*: Polyphenol-rich foods like fruits, vegetables, cocoa, and red wine have been associated with a reduced risk of coronary heart disease [45].
2) *Lowering blood pressure*: Polyphenol-rich foods such as tea, cocoa, and berries have demonstrated the ability to lower blood pressure in healthy individuals and those with hypertension [46].
3) *Improving endothelial function*: Polyphenol-rich foods, including dark chocolate, have been shown to enhance endothelial function and decrease the risk of atherosclerosis [47].
4) *Reducing inflammation*: Polyphenol-rich foods like extra-virgin olive oil and green tea have exhibited anti-inflammatory properties, reducing inflammation and oxidative stress, key cardiovascular disease factors [48].

Overall, incorporating a diet rich in polyphenol-rich foods and beverages may contribute to preventing cardiovascular disease and promoting cardiovascular health.

The consumption of whole grains has been linked to improved gut health and a reduced risk of cardiovascular disease. This may be attributed to their impact on inflammation, fasting blood glucose levels, and lipid metabolism. Oats, in particular, have been recognized for their cardiovascular protective properties due to their high dietary fiber content and the presence of phenolic compounds, including avenanthramides. When broken down by gut bacteria, these compounds yield bioactive metabolites that can influence the composition and activity of the gut microbiota, potentially leading to the prevention of chronic diseases. Healthy gut microbiota plays a role in various aspects of human health, including vitamin synthesis and immune defense. The cardiovascular benefits associated with oat consumption are believed to involve anti-inflammatory effects, maintenance of endothelial function, and reductions in plasma cholesterol levels. Moreover, oat fiber and phenolics can directly interact with the gut microbiota, resulting in secondary changes in cholesterol and bile metabolism and the production of important metabolites. Similarly, polyphenols present in vegetables, fruits, and coffee, such as resveratrol, EGCG, quercetin, and chlorogenic acid, have demonstrated cardioprotective effects by inhibiting ROS-induced endothelial dysfunction, enhancing nitric oxide production and availability, and exhibiting anti-inflammatory properties. These protective effects on vascular endothelial cells may contribute to the prevention of cardiovascular diseases; however, the precise molecular mechanisms are still being investigated.

20.3 Recent Trends in Polyphenol Research

Polyphenols have gained considerable attention in scientific research due to their diverse biological activities, encompassing antioxidant, anti-inflammatory, and antimicrobial properties. These compounds hold immense potential across various industries, including medicine, cosmetics, dietary supplements, and food. This chapter presents an overview of recent scientific discoveries in polyphenol research, covering areas such as their interactions with the microbiome and mitochondria, their metabolism and associated health benefits, their involvement in cognitive diseases and obesity, advancements in their analysis and extraction methods, the development of cosmetic formulations, and the creation of functional foods. The chapter also explores the future implications of polyphenols in disease prevention and their potential use as preventive measures. Polyphenols can significantly influence the future of food and nutrition by serving as valuable tools in preventing numerous diseases.

20.3.1 Bioavailability of Polyphenols

Factors affecting bioavailability: The bioavailability of polyphenols refers to the amount of polyphenols that are absorbed and available for use by the body. Several factors can affect the bioavailability of polyphenols, including:

Chemical structure: The chemical structure of polyphenols affects their bioavailability. For example, polyphenols with a low molecular weight are more easily absorbed than those with a high molecular weight.

Food matrix: The food matrix, or how polyphenols are consumed, can affect their bioavailability. For example, fat can increase the absorption of certain polyphenols.

Gut microbiota: The gut microbiota can affect the metabolism and absorption of polyphenols. Some gut bacteria can metabolize polyphenols, which can affect their bioavailability.

Processing and cooking: Food processing and cooking can affect polyphenols' bioavailability. For example, cooking can destroy some polyphenols, while processing can make them more accessible for absorption.

20.3.2 Strategies to Enhance Bioavailability

Several strategies can be used to enhance the bioavailability of polyphenols, including:

Food processing: Processing techniques such as fermentation, sprouting, and malting can increase the bioavailability of polyphenols by breaking down complex structures and making them more accessible for absorption.

Co-ingestion with other nutrients: Co-ingestion of polyphenols with other nutrients, such as fats, fiber, or vitamin C, can increase their bioavailability by enhancing absorption or protecting them from degradation in the gut.

Nanoparticle delivery: Nanoparticles can encapsulate polyphenols, protecting them from degradation and increasing their bioavailability.

Formulation into supplements: Polyphenols can be formulated into supplements, such as capsules or tablets, to increase their bioavailability and provide a consistent dose.

Overall, enhancing the bioavailability of polyphenols can increase their efficacy in promoting health and preventing disease [49–52].

In a systematic review conducted by Speer et al. [52], the relationship between habitual polyphenol consumption, iron status, and biomarkers of cardiovascular disease was assessed. The review included seven randomized controlled trials involving 133 participants who consumed polyphenol-rich foods and supplements such as olive oil, cherries, and catechin-enriched drinks. The studies ranged from 56 to 145 days, with varying polyphenol content (45–1015 mg per 100 g). The results indicated that polyphenols did not impact iron status, and most studies showed improvements in inflammatory markers and lipid profiles. However, the limited number of studies and intervention differences highlight the need for more extensive trials to evaluate the relationship between polyphenol intake, iron bioavailability, and cardiovascular disease risk.

The effectiveness of dietary polyphenols in providing health benefits relies on their bioavailability, which varies depending on their chemical structures. Polyphenols undergo metabolism in the gut through conjugation and must be hydrolyzed before absorption. Metabolism primarily occurs in the liver for high doses, while the intestinal mucosa may metabolize in small amounts. Unabsorbed polyphenols reach the hindgut, where they can be transformed into more bioavailable metabolites by the gut microbiota or influence the gut microbial community as prebiotics. However, the limited oral bioavailability of polyphenolic substances remains a significant challenge in their use as chemopreventive agents. Various factors impact bioavailability, including endogenous factors such as microbiota, digestive enzymes, and the food matrix. Dietary fiber, divalent minerals, and viscous and protein-rich meals may decrease bioaccessibility, while digestible carbohydrates, dietary lipids, and additional antioxidants may enhance polyphenol availability. A comprehensive understanding of the factors affecting polyphenol bioavailability is crucial for maximizing their health benefits [53–55].

20.3.3 Combination of Polyphenols With Other Compounds

Polyphenols can exhibit synergistic effects when interacting with other bioactive compounds in the body, such as vitamins, minerals, amino acids, and even other polyphenols. These interactions can lead to a combined effect that surpasses the individual effects of each compound. One example is the synergistic enhancement of antioxidant capacity when polyphenols are combined with vitamin C. Numerous studies have demonstrated the synergistic effects of polyphenols with other bioactive compounds. For instance, the combination of resveratrol and quercetin has shown greater anti-inflammatory effects than each compound alone. Similarly, the combination of green tea catechins and vitamin E has exhibited enhanced antioxidant effects compared to individual compounds.

Furthermore, mixtures of phytochemicals have also been found to display synergistic effects. In such cases, the combined action of different phytochemicals results in a higher biological effect than the sum of the outcomes each phytochemical achieves. Phytochemicals, known for their intense pigments, contribute to the characteristic color of food products. Additionally, there have been findings indicating that the collaboration between Ag plasmonic photocatalysis and conventional AgBr semiconductor photocatalysis can significantly enhance the overall daylight activity of AgBr@Ag through an intriguing synergistic effect [56].

20.3.4 Health Implications of Combination Therapies

The combination of polyphenols with other bioactive compounds has important health implications. For example, combination therapies may be more effective than single therapies for preventing and treating diseases. The synergistic effects of combination therapies can also result in lower doses of individual compounds being needed, reducing the risk of side effects. Combination therapies may also have specific health benefits for certain populations. For example, combining polyphenols with vitamin D may benefit bone health in postmenopausal women. Combining polyphenols with omega-3 fatty acids may offer advantages for cardiovascular health [57–59]. A ROS-enhanced combination chemotherapy platform has been developed using a biocompatible metal-polyphenol network self-assembly process. This platform involves encapsulating doxorubicin and platinum prodrugs within nanoparticles. Studies have shown that polyphenolic compounds, except rutin, can induce apoptosis characterized by mitochondrial membrane depolarization and cytochrome c release, followed by caspase-3 activation. The release of cytochrome c was found to be inhibited by mitochondrial permeability transition inhibitors such as cyclosporine A and aristolochic acid, which suggests their involvement in the apoptotic effects of polyphenols.

In summary, combining polyphenols with other bioactive compounds holds significant potential for promoting health and preventing diseases [60–62].

20.3.5 Qualitative and Quantitative Analysis of Dietary Polyphenols

Polyphenols are diverse plant secondary metabolites widely distributed in the plant kingdom. These compounds are known for their numerous beneficial health effects, including antioxidant and anti-inflammatory properties. Polyphenols are found in various plant-based foods, such as fruits, vegetables, cereals, and beverages, and are responsible for their distinctive color, taste, and aroma. The qualitative and quantitative analysis of dietary polyphenols is crucial for understanding their bioactivity and potential health benefits.

Qualitative analysis of polyphenols involves identifying and characterizing these compounds present in plant extracts or food matrices. Various analytical techniques are used for the qualitative analysis of polyphenols, including high-performance liquid chromatography (HPLC), gas chromatography (GC), mass spectrometry (MS), and nuclear magnetic resonance (NMR) spectroscopy. These techniques allow for the identification of specific polyphenols based on their characteristic retention times, mass spectra, or chemical shifts. Quantitative analysis of polyphenols involves determining their concentration or content in food matrices. Several methods are available for quantitatively analyzing polyphenols, including colorimetric assays, HPLC, and MS-based methods. The most commonly used way for quantifying polyphenols is the Folin–Ciocalteu assay, a colorimetric method that measures the total phenolic content of a sample. HPLC is a widely used method for quantifying specific polyphenols in plant extracts or food matrices. MS-based techniques, such as liquid chromatography–mass spectrometry and gas chromatography–mass spectrometry, provide high sensitivity and specificity for quantifying polyphenols.

Several factors can affect polyphenols' qualitative and quantitative analysis, including sample preparation, extraction, and analysis conditions. Sample preparation is a critical step in the study of polyphenols as it affects the accuracy and precision of the results. Extraction methods, such as solvent extraction, sonication, and microwave-assisted extraction, can also affect the yield and quality of polyphenols extracted from plant matrices. Analysis conditions, such as the mobile phase composition, pH, and temperature, can also affect the separation and detection of polyphenols by HPLC or MS. The qualitative and quantitative analysis of dietary polyphenols has significant implications for human health and nutrition. Identifying and characterizing polyphenols present in plant-based foods can provide insights into their potential health benefits and contribute to the development of functional foods and nutraceuticals. The quantification of polyphenols in food matrices can also provide information on the dietary intake of these compounds and their bioavailability. Moreover, the analysis of polyphenols in food matrices can provide information on the effects of processing and storage on the content and bioactivity of these compounds.

In conclusion, the qualitative and quantitative analysis of dietary polyphenols is crucial for understanding their bioactivity and potential health benefits. Various analytical techniques are available to analyze polyphenols, including HPLC, GC, MS, and NMR spectroscopy. Sample preparation, extraction, and analysis conditions are critical factors that can affect the accuracy and precision of the results. The analysis of polyphenols in food matrices has significant implications for human health and nutrition and can contribute to developing functional foods and nutraceuticals.

20.4 Conclusion and Outlook

To summarize, this book chapter underscores the significance of dietary polyphenols and their associated health advantages. Polyphenols are naturally occurring compounds in plant-based foods like fruits, vegetables, grains, tea, and wine. They possess various biological properties, including antioxidant, anti-inflammatory, anticancer, and cardioprotective effects. The chapter outlines different techniques employed for qualitative and quantitative analysis of polyphenols, such as chromatography, spectrophotometry, and MS. These methods have facilitated identifying and quantifying diverse polyphenols, enhancing our understanding of their health-promoting characteristics. Additionally, the chapter explores the application of polyphenols in the food industry, where they are utilized as natural antioxidants, colorants, and flavor enhancers. Incorporating polyphenols into food products has demonstrated improvements in nutritional value and shelf-life extension.

The chapter emphasizes the importance of integrating polyphenols into our diets and encourages further research into their health benefits. Polyphenol-rich foods can enhance overall well-being and decrease the risk of chronic diseases. Therefore, it is crucial to continue exploring the potential of polyphenols as functional food components and develop innovative and sustainable approaches for their extraction and integration into food products.

The outlook for this book chapter is promising, as it provides a comprehensive overview of the current state of research on dietary polyphenols. The chapter covers various topics related to polyphenols, including their health benefits, sources, extraction methods, and applications in multiple industries. One of the key strengths of this chapter is its emphasis on recent research findings. The authors have incorporated many studies published within the last few years, which gives the chapter a current and up-to-date feel. This is important in a field like polyphenol research, where new studies are constantly being published and our understanding of these compounds rapidly evolves. Another strength of this chapter is its focus on the practical applications of polyphenols. The authors have provided numerous examples of how polyphenols can be incorporated into food products, supplements, and other consumer goods. This makes the chapter highly relevant for researchers and industry professionals interested in developing new polyphenol products.

The chapter also highlights some of the challenges associated with polyphenol research. For example, the authors note that more consensus exists regarding the best methods for quantifying polyphenols in food and other samples. Additionally, they point out that some polyphenols may adversely affect health in specific contexts, underscoring the need for more research in this area. Overall, the outlook for this book chapter is positive. It provides a thorough and up-to-date overview of the current research on dietary polyphenols and highlights some of the practical applications and challenges associated with these compounds. As research in this field continues to evolve, this chapter will serve as a valuable resource for researchers, industry professionals, and anyone interested in dietary polyphenols' health benefits and applications.

References

1 Hartman, R.E., Shah, A., Fagan, A.M. et al. (2006). Pomegranate juice decreases amyloid load and improves behavior in a mouse model of Alzheimer's disease. *Neurobiology of Disease* 24: 506–515.

2 Hollman, P.C., van Trijp, J.M., Buysman, M.N. et al. (1997). Relative bioavailability of the antioxidant flavonoid quercetin from various foods in man. *FEBS Letters* 418: 152–156.

3 Shen S.Q, Zhang Y., Xiang J.J., Xiong C.L. Protective effect of curcumin against liver warm ischemia/reperfusion injury in rat model is associated with regulation of heat shock protein and antioxidant enzymes. *World Journal of Gastroenterology* 2007;13:1953–1961.

4 Molina, M.F., Sanchez-Reus, I., Iglesias, I., and Benedi, J. (2003). Quercetin, a flavonoid antioxidant, prevents and protects against ethanol-induced oxidative stress in mouse liver. *Biological & Pharmaceutical Bulletin* 26: 1398–1402.

5 Chen, C., Yu, R., Owuor, E.D., and Kong, A.N. (2000). Activation of antioxidant response element (ARE), mitogen-activated protein kinases (MAPKs) and caspases by major green tea polyphenol components during cell survival and death. *Archives of Pharmacal Research* 23: 605–612.

6 Butterfield, D.A., Castegna, A., Pocernich, C.B. et al. (2002). Nutritional approaches to combat oxidative stress in Alzheimer's disease. *The Journal of Nutritional Biochemistry* 13: 444–461.

7 Schaffer, S., Podstawa, M., Visioli, F. et al. (2007). Hydroxytyrosol-rich olive mill wastewater extract protects brain cells in vitro and ex vivo. *Journal of Agricultural and Food Chemistry* 55: 5043–5049.

8 Tuck, K.L., Freeman, M.P., Hayball, P.J. et al. (2001). The in vivo fate of hydroxytyrosol and tyrosol, antioxidant phenolic constituents of olive oil, after intravenous and oral dosing of labeled compounds to rats. *The Journal of Nutrition* 131: 1993–1996.

9 Rice-Evans, C.A., Mmiller, N.J., and Paganga, G. (1997). Antioxidant properties of phenolic compounds. *Trends in Plant Science* 2: 152–159.

10 Nielsen, I.L., Dragsted, L.O., Ravn-Haren, G. et al. (2003). Absorption and excretion of black currant anthocyanins in humans and watanabe heritable hyperlipidemic rabbits. *Journal of Agricultural and Food Chemistry* 51: 2813–2820.

11 Bub, A., Watzl, B., Heeb, D. et al. (2001). Malvidin-3-glucoside bioavailability in humans after ingestion of red wine, dealcoholized red wine and red grape juice. *European Journal of Nutrition* 40: 113–120.

12 McAnlis, G.T., McEneny, J., Pearce, J., and Young, I.S. (1999). Absorption and antioxidant effects of quercetin from onions, in man. *European Journal of Clinical Nutrition* 53: 92–96.

13 Manach, C., Morand, C., Gil-Izquierdo, A. et al. (2003). Bioavailability in humans of the flavanones hesperidin and narirutin after the ingestion of two doses of orange juice. *European Journal of Clinical Nutrition* 57: 235–242.

14 Lotito, S.B. and Frei, B. (2006). Consumption of flavonoid-rich foods and increased plasma antioxidant capacity in humans: cause, consequence, or epiphenomenon? *Free Radical Biology & Medicine* 41: 1727–1746.

15 Erlund, I., Meririnne, E., Alfthan, G., and Aro, A. (2001). Plasma kinetics and urinary excretion of the flavanones naringenin and hesperetin in humans after ingestion of orange juice and grapefruit juice. *The Journal of Nutrition* 131: 235–241.

16 Henning, S.M., Niu, Y., Liu, Y. et al. (2005). Bioavailability and antioxidant effect of epigallocatechin gallate administered in purified form versus as green tea extract in healthy individuals. *The Journal of Nutritional Biochemistry* 16: 610–616.

17 Widlansky, M.E., Duffy, S.J., Hamburg, N.M. et al. (2005). Effects of black tea consumption on plasma catechins and markers of oxidative stress and inflammation in patients with coronary artery disease. *Free Radical Biology & Medicine* 38: 499–506.

18 Bell, J.R., Donovan, J.L., Wong, R. et al. (2000). (+)-Catechin in human plasma after ingestion of a single serving of reconstituted red wine. *The American Journal of Clinical Nutrition* 71: 103–108.

19 Holt, R.R., Lazarus, S.A., Sullards, M.C. et al. (2002). Procyanidin dimer B2 [epicatechin-(4beta-8)-epicatechin] in human plasma after the consumption of a flavanol-rich cocoa. *The American Journal of Clinical Nutrition* 76: 798–804.

20 Schwarz, D. and Roots, I. (2003). In vitro assessment of inhibition by natural polyphenols of metabolic activation of procarcinogens by human CYP1A1. *Biochemical and Biophysical Research Communications* 303: 902–907.

21 Gonthier, M.P., Remesy, C., Scalbert, A. et al. (2006). Microbial metabolism of caffeic acid and its esters chlorogenic and caftaric acids by human faecal microbiota in vitro. *Biomedicine & Pharmacotherapy* 60: 536–540.

22 Seeram, N.P., Lee, R., and Heber, D. (2004). Bioavailability of ellagic acid in human plasma after consumption of ellagitannins from pomegranate (Punica granatum L.) juice. *Clinica Chimica Acta* 348: 63–68.

23 Rangkadilok, N., Sitthimonchai, S., Worasuttayangkurn, L. et al. (2007). Evaluation of free radical scavenging and antityrosinase activities of standardized longan fruit extract. *Food and Chemical Toxicology* 45: 328–336.

24 Ray, P.S., Maulik, G., Cordis, G.A. et al. (1999). The red wine antioxidant resveratrol protects isolated rat hearts from ischemia reperfusion injury. *Free Radical Biology & Medicine* 27: 160–169.

25 Zhang, Y., Liu, Y., Wang, T. et al. (2006). Resveratrol, a natural ingredient of grape skin: antiarrhythmic efficacy and ionic mechanisms. *Biochemical and Biophysical Research Communications* 340: 1192–1199.

26 Chung, K.T., Wong, T.Y., Wei, C.I. et al. (1998). Tannins and human health: a review. *Critical Reviews in Food Science and Nutrition* 38: 421–464.

27 Li, Y., Li, S., Lin, C. et al. (2017). Polyphenols in food industry: extraction, purification, and utilization. *Journal of Food Quality* 2017: 9850107.

28 Williamson, G. and Kay, C.D. (2010). The bioavailability, transport, and bioactivity of dietary flavonoids: a review from a historical perspective. *Comprehensive Reviews in Food Science and Food Safety* 9 (2): 104–122.

29 Bhagwat, S., Haytowitz, D.B., Holden, J.M., and USDA Database for the Flavonoid Content of Selected Foods, Release 3.1 (2014). *USDA Database for the flavonoid content of selected foods, Release 3.1*. US Department of Agriculture, Agricultural Research Service.

30 Heiss, C. and Kelm, M. (2010). Flavanols and cardiovascular disease prevention. *European Heart Journal Supplements* 12 (Suppl. D): D34–D41.

31 Mastroiacovo, D., Kwik-Uribe, C., Grassi, D. et al. (2015). Cocoa flavanol consumption improves cognitive function, blood pressure control, and metabolic profile in elderly subjects: The Cocoa, Cognition, and Aging (CoCoA) Study—a randomized controlled trial. *The American Journal of Clinical Nutrition* 101 (3): 538–548.

32 Liguori, I., Russo, G., Curcio, F. et al. (2018). Oxidative stress, aging, and diseases. *Clinical Interventions in Aging* 13: 757–772.

33 Stagos, D., Amoutzias, G.D., Matakos, A. et al. (2015). Chemoprevention of liver cancer by plant polyphenols. *Food and Chemical Toxicology* 80: 216–224.

34 Mecocci, P., Polidori, M.C., and Troiano, L. (2018). Dietary antioxidants in aging and age-related diseases. *Current Pharmaceutical Design* 24 (2): 2–10.

35 Anderson, E.J. and Lustig, M.E. (2018). Diet-gut microbiota interactions and their implications for insulin resistance and diabetes. *Nutrients* 10 (10): 1410.

36 Bjelakovic, G., Nikolova, D., Gluud, L.L. et al. (2014). Antioxidant supplements for prevention of mortality in healthy participants and patients with various diseases. *Cochrane Database of Systematic Reviews* 3.

37 Ridker, P.M., Everett, B.M., Thuren, T. et al. (2017). Antiinflammatory therapy with canakinumab for atherosclerotic disease. *New England Journal of Medicine* 377 (12): 1119–1131.

38 Nagaraju, G.P., Aliya, S., and Alese, O.B. (2017). Role of inflammation in gastrointestinal cancer. *Gastroenterology & Hepatology* 13 (6): 357–369.

39 Calder, P.C., Bosco, N., Bourdet-Sicard, R. et al. (2017). Health relevance of the modification of low grade inflammation in ageing (inflammageing) and the role of nutrition. *Ageing Research Reviews* 40: 95–119. https://doi.org/10.1016/j.arr.2017.09.001.

40 Khan, N. and Mukhtar, H. (2018). Tea polyphenols in promotion of human health. *Nutrients* 11 (1): 39.

41 Salehi, B., Mishra, A.P., Nigam, M. et al. (2019). Resveratrol: a double-edged sword in health benefits. *Biomedicine* 7 (3): 6.

42 Yang, J., Liu, R.H., and Halim, L. (2019). Antioxidant and antiproliferative activities of common edible plants on human colon cancer cells. *Food Science & Nutrition* 7 (8): 2566–2575.

43 Lee, C.H. (2019). Reversal of epithelial-mesenchymal transition by natural anti-inflammatory and pro-resolving lipids. *Cancers (Basel)* 11 (12): 1841. https://doi.org/10.3390/cancers11121841.

44 Guo K, Feng Y, Zheng X, Sun L, Wasan HS, Ruan S, Shen M. (2021) Resveratrol and its analogs: potent agents to reverse epithelial-to-mesenchymal transition in tumors *Frontiers in Oncology*. 11:644134. doi: https://doi.org/10.3389/fonc.2021.644134.

45 Bondonno, C.P., Dalgaard, F., Kyrø, C. et al. (2020). Flavonoid intake is associated with lower mortality in the Danish Diet Cancer and Health Cohort. *Nature Communications* 11 (1): 1–10.

46 Jumar, A., Schröder, H., Kelm, M. et al. (2020). Polyphenol-rich cocoa improves blood pressure in patients with diabetes and hypertension. *European Journal of Clinical Investigation* 50 (2): e13186.

47 Monagas, M., Khan, N., Andrés-Lacueva, C. et al. (2009). Dose-response relationship between cocoa flavanols and vascular endothelial function in humans: a systematic review and meta-analysis of randomized controlled trials. *The American Journal of Clinical Nutrition* 93 (1): 62–72.

48 Schwingshackl, L., Christoph, M., Hoffmann, G., and Boeing, H. (2015). Polyphenol intake and risk of cardiovascular disease: a systematic review and meta-analysis of prospective cohort studies. *The American Journal of Clinical Nutrition* 102 (4): 924–933.

49 Williamson, G. and Clifford, M.N. (2017). Colonic metabolites of berry polyphenols: the missing link to biological activity? *British Journal of Nutrition* 114 (6): 849–864.

50 Goni, I., Serrano, J., Saura-Calixto, F., and Vasquez-Caicedo, A.L. (2019). Factors affecting the bioavailability and bioefficacy of polyphenols in humans. *Food & Function* 10 (10): 6228–6249.

51 Castro-Barquero, S., Ruiz-Roso, B., and Beltrán-de-Miguel, B. (2020). Enhanced bioavailability of phenolic compounds by food processing. *Current Opinion in Food Science* 32: 56–62.

52 Lobo, V., Patil, A., Phatak, A., and Chandra, N. (2010). Free radicals, antioxidants and functional foods: impact on human health. *Pharmacognosy Reviews* 4 (8): 118–126.

53 Etxeberria, U. and Martínez, J.A. (2019). Factors that modify the bioavailability of polyphenols. *Journal of Physiology and Biochemistry* 75 (3): 219–231.

54 Speer, H., D'Cunha, N.M., Botek, M. et al. (2019). The effects of dietary polyphenols on circulating cardiovascular disease biomarkers and iron status: a systematic review. *Nutrition and Metabolic Insights* 21 (12): 1178638819882739. https://doi.org/10.1177/1178638819882739.

55 D'Archivio, M., Filesi, C., Varì, R. et al. (2010). Bioavailability of the polyphenols: status and controversies. *International Journal of Molecular Sciences* 11 (4): 1321–1342. https://doi.org/10.3390/ijms11041321.

56 Aguilera-Carbo, A., Augur, C., Prado-Barragan, L.A. et al. (2020). Polyphenols: factors influencing their bioavailability and health-promoting properties. *Nutrients* 12 (7): 2143. https://doi.org/10.3390/nu12072143.

57 Halvorsen, B.L., Carlsen, M.H., Phillips, K.M. et al. (2006). Content of redox-active compounds (i.e., antioxidants) in foods consumed in the United States. *The American Journal of Clinical Nutrition* 84 (1): 95–135. https://doi.org/10.1093/ajcn/84.1.95.

58 González-Sarrías, A., Larrosa, M., and Tomás-Barberán, F.A. (2017). The synergistic interaction between polyphenols and gut microbiota in the modulation of host health. *Molecules* 22 (1): 1–18.

59 Kuo, Y.C., Lin, C.Y., Tsai, W.J., and Chu, C.Y. (2015). Antioxidant and anti-inflammatory activity of a combination of resveratrol, β-sitosterol, and β-sitosterol glucoside in vitro. *Food & Nutrition Research* 59 (1): 29263.

60 Serafini, M., Peluso, I., and Raguzzini, A. (2010). Flavonoids as anti-inflammatory agents. *Proceedings of the Nutrition Society* 69 (3): 273–278.

61 Jia, X., Liu, J., Shi, Y., and Guo, W. (2020). Synergistic effects of green tea catechins with other bioactive compounds. *Journal of Food and Drug Analysis* 28 (1): 1–12.

62 Abbasi, B., Kimiagar, M., Sadeghniiat, K. et al. (2012). The effect of magnesium supplementation on primary insomnia in elderly: a double-blind placebo-controlled clinical trial. *Journal of Research in Medical Sciences* 17 (12): 1161–1169.

21

Polyphenols for Pharmaceutical (Diagnosis, Therapies, and Prevention of Diseases) Applications

*Nastaran Hashemzadeh, Sanaz Sajedi-Amin, Abolghasem Jouyban, and Elaheh Rahimpour**

Pharmaceutical Analysis Research Center and Faculty of Pharmacy, Tabriz University of Medical Sciences, Tabriz, Iran

21.1 Introduction

Phytopharmaceuticals are a potential field in the pharmaceutical business that is expected to attract more attention. Among the phytochemicals, polyphenols are human diets' most abundant bioactive compounds. They are phytochemicals with at least one phenolic ring and hydroxyl group, and due to their health benefits, they are a key source for policymakers and health professionals. Over 8000 plant polyphenols include berries, herbs and spices, cocoa powder, nuts, flax seeds, vegetables, olives, coffee, and tea. Studies have shown that a diet rich in polyphenols should be 650 mg daily to improve overall health [1]. Plant-based medicines are readily available, less priced, efficient, and seldom cause adverse effects. The World Health Organization believes medicinal plants are the most effective sources of various medications [2]. So, the current chapter aims to review polyphenol's applications in life science.

21.2 Classification of Polyphenols

As biologically active plant-derived substances, polyphenols are classified as flavonoids and non-flavonoids. Figure 21.1 summarizes the classification of polyphenols along with their subgroups. The principle for distinguishing these

* Corresponding author

Science and Engineering of Polyphenols: Fundamentals and Industrial Scale Applications, First Edition. Edited by Chandrabhan Verma.
© 2024 John Wiley & Sons Inc. Published 2024 by John Wiley & Sons Inc.

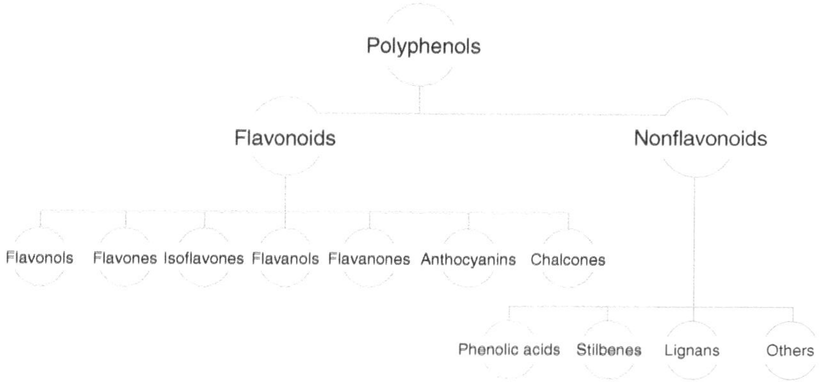

Figure 21.1 Classification of polyphenols families.

two classes is the number of phenol rings, the type and position of functional groups, and the saturation degree.

21.2.1 Flavonoids

Flavonoids are the most popular group of polyphenols. The flavonoid nuclei characterize the molecular structures of flavonoids: two phenyl rings (A and B) linked to the heterocyclic pyran ring (C), commonly called the C6–C3–C6 carbon skeleton. The general structure of flavonoids is shown in Figure 21.2. The fundamental subunits of flavonoids are flavonols, flavones, isoflavones, flavanols, flavanones, anthocyanins, and chalcones, which their structures are shown in Figure 21.2. In each subunit of flavonoids, the patterns of methylation and hydroxylation of the A and B rings and the keto group of the C ring are different. Flavonoids can scavenge free radicals and have strong antioxidant properties [3]. Flavonoids also have metal-chelating properties that prevent the formation of radicals catalyzed by metal ions. The arrangement of functional groups on the flavonoid backbone is important to their antioxidant behavior. The hydroxyl group (—OH) has become predominant for this effect among the functional groups. A significant effect of protection against reactive oxygen species (ROS) can be seen with C ring hydroxyl configuration (position 3), whereas substitution of the rings A and B has little impact [4]. Four primary antioxidant mechanisms—hydrogen atom transfer, electron transfer, single electron transfer followed by proton transfer, and sequential proton loss electron transfer—were used. Various techniques involving 2,2-diphenyl-1-picrylhydrazyl (DPPH), oxygen radical absorbance capacity, trolox equivalent antioxidant capacity, and ferric reducing ability of plasma assays have been used to characterize a flavonoid's antioxidant properties [5].

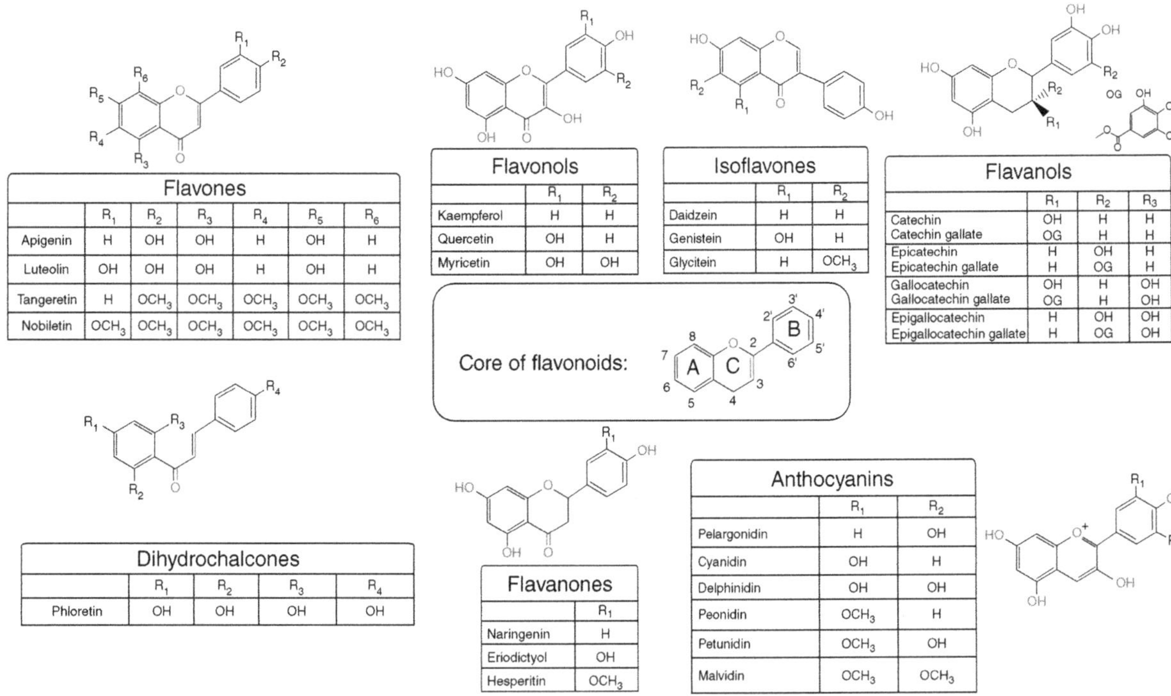

Flavones

	R_1	R_2	R_3	R_4	R_5	R_6
Apigenin	H	OH	OH	H	OH	H
Luteolin	OH	OH	OH	H	OH	H
Tangeretin	H	OCH_3	OCH_3	OCH_3	OCH_3	OCH_3
Nobiletin	OCH_3	OCH_3	OCH_3	OCH_3	OCH_3	OCH_3

Flavonols

	R_1	R_2
Kaempferol	H	H
Quercetin	OH	H
Myricetin	OH	OH

Isoflavones

	R_1	R_2
Daidzein	H	H
Genistein	OH	H
Glycitein	H	OCH_3

Flavanols

	R_1	R_2	R_3
Catechin	OH	H	H
Catechin gallate	OG	H	H
Epicatechin	H	OH	H
Epicatechin gallate	H	OG	H
Gallocatechin	OH	H	OH
Gallocatechin gallate	OG	H	OH
Epigallocatechin	H	OH	OH
Epigallocatechin gallate	H	OG	OH

Dihydrochalcones

	R_1	R_2	R_3	R_4
Phloretin	OH	OH	OH	OH

Flavanones

	R_1
Naringenin	H
Eriodictyol	OH
Hesperitin	OCH_3

Anthocyanins

	R_1	R_2
Pelargonidin	H	OH
Cyanidin	OH	H
Delphinidin	OH	OH
Peonidin	OCH_3	H
Petunidin	OCH_3	OH
Malvidin	OCH_3	OCH_3

Figure 21.2 The flavonoid subclass structures.

Phytochemicals must be metabolized to be bioavailable. The following are some potential mechanisms by that polyphenols might generate bioactivity: (1) in the mouth, through the activity of local microflora or salivary enzymes secreted, (2) in the gut because of the environment's acidity, (3) by the activity of pancreatic, brush border, or microbial enzymes within the small or large intestinal lumen, (4) as the endogenous phase I and phase II enzymes move through the enterocytes, (5) by hepatic phase I and phase II enzymes in the liver, and (6) by phase I and phase II enzymes at different body regions [6]. The metabolism of polyphenols can occur in the entire digestive system; however, the stomach (via endogenous and microbial enzymes) and liver are the most significant sites for phytochemical metabolism.

21.2.2 Non-flavonoids

Another category of polyphenols is nonflavonoids, which have various structures. Non-flavonoids, from the simplest to the most complex structures, are phenolic acids (*e.g.* salicylic acid, protocatechic acid, vanillic acid, gallic acid, syringic acid, *p*-comaric acid, caffeic acid, ferulic acid, and sinapic acid), stilbenes (*e.g.* resveratrol), lignans (*e.g.* furan, furofuran, dibenzylbutane lignans, etc.), and others (*e.g.* xanthones, secoiridoids (oleuropein, demethyloleuropein, ligstroside, etc.) and coumarins (warfarin, psoralen, bergapten, etc.), which the structures of some compounds are given in Figure 21.3.

21.3 Polyphenols Applications as Potential Pharmaceutical Agents

21.3.1 Polyphenols in Diagnosis

Various articles considered the use of polyphenols in diagnostic applications. For example, with the aim of noninvasive liquid biopsy-based cancer diagnosis, Jang et al. conducted a study on the existing difficulties in isolating and detecting extracellular vesicles (EV), including exosomes, in body fluids. They recognized ineffective EV isolation as a significant obstacle. They presented an effective, resource-free pre-enrichment method using the polyphenolic biomolecule (−)-epigallocatechin-3-gallate (EGCG) to isolate and identify exosomal microRNAs (miRNAs) in samples of human blood plasma. Their method consists of three steps: EV aggregation by EGCG, EV isolation via filters, and EV target miRNA identification via molecular beacons. The levels of target miRNAs in EVs were closely associated with those in the freely accessible gene expression omnibus database, according to tests of the methodology using blood samples of gastric cancer or hepatocellular carcinoma cases. The study emphasizes

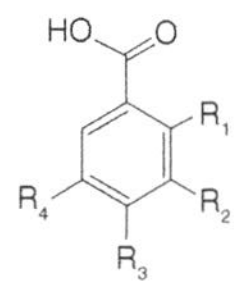

Simple phenolic acids				
	R_1	R_2	R_3	R_4
Salicylic acid	OH	H	H	H
Vanillic acid	H	OH_3	OH	H
Gallic acid	H	OH	OH	OH

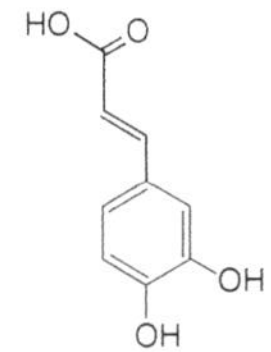

Caffeic acid

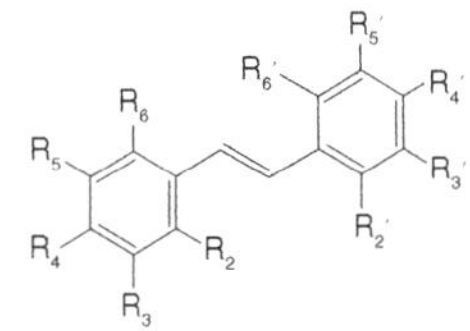

Stilbenes										
	R_2	R_3	R_4	R_5	R_6	$R_{2'}$	$R_{3'}$	$R_{4'}$	$R_{5'}$	$R_{6'}$
Rresveratrol	H	OH	H	OH	H	H	H	OH	H	H

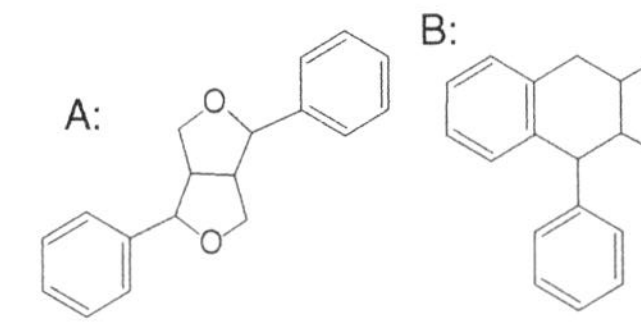

Lignans	
furofuran	A
arylnaphthalene	B

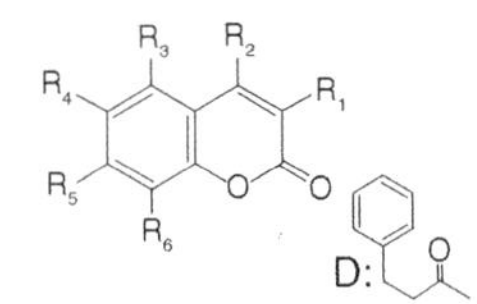

Coumarins						
	R_1	R_2	R_3	R_4	R_5	R_6
warfarin	D	OH	H	H	H	H

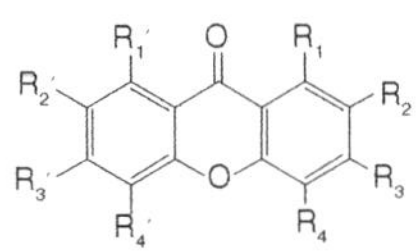

Xanthones								
	R_1	R_2	R_3	R_4	$R_{1'}$	$R_{2'}$	$R_{3'}$	$R_{4'}$
1,3,5-Trihydroxyxanthone	OH	H	OH	H	H	H	H	OH
1,8-Dihydroxy-3,5,6-trimethoxyxanthone	H	H	OCH_3	OCH_3	OH	H	OCH_3	H

C:

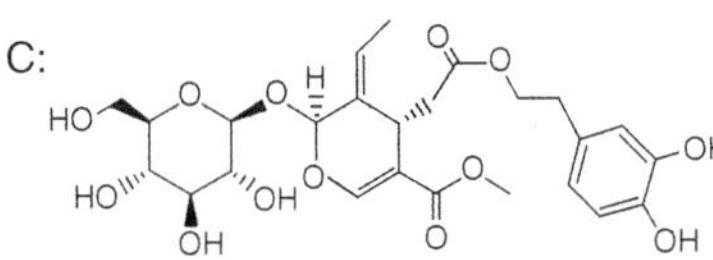

Secoiridoids	
Oleuropein	C

Figure 21.3 The structure of some non-flavonoids.

employing a bio-adhesive-mediated preparation approach in clinical samples to successfully achieve sample isolation and enrichment for exosome-based molecular diagnostics [7]. In another work, Dai et al. discuss the potential biomedical uses of polyphenol-based particles, such as polydopamine particles, metal-phenolic network-based particles, and polymer–phenol particles, in various diagnostic and therapeutic procedures. It describes the creation of colloidal particles using a variety of covalent and noncovalent interactions using a variety of polymeric or biomolecular substances derived from polyphenols or their derivatives. These particles can have particular characteristics or functions, such as the ability to target, be multiresponsive, and scavenge radicals. The benefits of cargo-loaded carriers in the diagnosis and treatment of diseases, particularly cancer, are also highlighted in the article. Because of their excellent biocompatibility and widespread occurrence in nature, polyphenols have drawn growing attention in the engineering of functional materials. Polyphenols or their derivatives have been used to create colloidal particles such as nanogels, hollow capsules, mesoporous particles, and core–shell structures by various covalent and noncovalent interactions. Moreover, various cargos, particularly those that have demonstrated their benefits in detecting and treating diseases, including cancer, can be integrated into these particles. The report emphasizes the potential medicinal applications of polyphenol-based particles and the significance of these particles in creating novel therapeutic and diagnostic approaches [8].

Moreover, small extracellular vesicles (sEVs) have been acknowledged as circulating biomarkers and prognosticators for disease diagnosis, according to Wang et al. [9]. Unfortunately, the absence of a reliable and simple scale-up isolation technique severely restricts the therapeutic applications of sEVs. Here, the viability of a three-dimensional (3D) polyphenol-metal network for label-free sEV separation was investigated. Tannic acid (TA) was used as a proof-of-concept in the design and fabrication of a 3D network covering mesoporous silicon beads (SiO_2@BSA@Fe-TA6) using Fe(III) as the coordinated metal and TA as the polyphenolic ligand. SiO_2@BSA@Fe-TA6 demonstrated universal capture toward sEVs in various cells and plasma samples using the inexpensive TA (as low as US$ 0.18/g) as the probe. The capture efficiency was $85.4\% \pm 1.5\%$, much higher than the ultracentrifugation approach and equivalent to antibody-based capture techniques. In comparison to the ultracentrifugation approach, the purity of the sEVs separated by SiO_2@BSA@Fe-TA6 from the H1299 cell culture supernatant was measured to be (1.07 ± 0.14) 10^{11} particles/μg, which is 3.1 times higher. Another significant advantage of SiO_2@BSA@Fe-TA6 is the straightforward self-assembly method, which can produce a yield of up to grams and enable processing more than 500 plasma samples at once. With an accuracy of 87.1%, the SiO_2@BSA@Fe-TA6-based method was also successfully used to distinguish between nonsmall cell lung cancer (NSCLC) and small cell lung cancer. A label-free, universal, inexpensive, and simple scale-up method for sEV-based liquid biopsy in lung cancer diagnosis and type has been devised using SiO_2@BSA@Fe-TA6 [9]. In another study performed by Corrêa et al. the connection between miRNA expression, polyphenols, and inflammatory biomarkers related to obesity was investigated. The research explores how circulating miRNAs may be employed as possible biomarkers for inflammatory and metabolic illnesses and the potential of polyphenols to affect obesity and its metabolic effects. The research also draws attention to the paucity of information on the relationship between dietary consumption of polyphenols, miRNA expression, and inflammatory biomarkers associated with obesity. Polyphenols can regulate the expression of miRNA, which may have implications for the identification and management of diseases related to obesity [10].

In a study conducted by Gamage et al. [11], evidence examining the mechanisms by which polyphenols lower the likelihood of depression diagnosis or depressive symptoms have been provided. Metabolites of polyphenols are thought to affect several processes involved in inflammation, oxidative stress, the hypothalamic–pituitary–adrenal axis, neurogenesis, and the kynurenine pathway, all factors contributing to depression. The combination of animal studies implies that polyphenol therapies lessen depressive-like behaviors and influence these pathophysiological pathways by lowering proinflammatory mediator concentrations in the blood and brain; acting as an indirect antidepressant and antioxidants; decreasing oxidative stress; reducing circulating levels of the hormones that control the hypothalamic–pituitary–adrenal axis; and glucocorticoids; encouraging hippocampus regeneration; boosting brain-derived neurotrophic factor; lowering the kynurenine to tryptophan ratio; and, switching production from neurotoxic to neuroprotective metabolites along the metabolic pathway.

21.3.2 Polyphenols in the Therapies and Prevention of Diseases

21.3.2.1 Polyphenols in the Treatment and Prevention of Osteoporosis

Osteoporosis is a metabolic bone disease that gradually reduces bone density and raises the risk of fractures. Rao et al. [12] reported the various risk factors for osteoporosis, such as oxidative stress, and the function of osteoblasts and osteoclasts in developing the disease. The study discusses several medications and nutritional solutions for osteoporosis treatment and prevention. The utilization of antioxidants from organic foods, like fruits and vegetables, reduces ROS and wards off osteoporosis. The study's findings suggest that the natural antioxidant lycopene, which is present in tomatoes, may help decrease the incidence of osteoporosis. The paper presents evidence of the potential use of lycopene and polyphenols as alternative or complementary agents with other recognized medications for the prevention or treatment of osteoporosis in postmenopausal women. This evidence comes from *in vitro* studies, animal intervention studies, epidemiological studies, and clinical intervention studies. Comprehensive details on non-flavonoid polyphenolic chemicals and their potential application, in the treatment, and prevention of osteoporosis were provided by Bellavia et al. [13] The study focuses on the pathways involved in these compounds frequent epigenetically active actions. The imbalance between bone resorption and bone production, which defines reduced bone resistance and large risks of fractures, with subsequent pain and functional restrictions, is discussed concerning the development of progressive osteopenia and osteoporosis. The study's findings suggest that non-flavonoid polyphenolic compounds may one day be employed to treat and prevent osteoporosis. These substances have the ability to balance bone resorption and bone formation, which is crucial for maintaining bone resistance and lowering the risk of fractures [13]. The role of polyphenols in the prevention and treatment of osteoporosis is also discussed by Udari et al. [14]. It gives a general overview of the risk factors for osteoporosis and stresses the value of diet and lifestyle choices in lowering the likelihood of occurrence. The study also emphasizes how polyphenols protect bone health by lowering oxidative stress and inflammation and controlling osteoblastogenesis, osteoclastogenesis, and osteoimmunological activity. However, there are few trustworthy human studies linking the consumption of polyphenols to precise advice for osteoporosis prevention and bone health [14].

Furthermore, a nutritional supplement called greens + bone builder may improve bone health, according to studies of Kang et al. The 8-week trial revealed that the supplement, which contains micronutrients and polyphenols, increased antioxidant capacity while lowering oxidative stress indicators and the bone resorption marker of collagen type I cross-linked C-telopeptide (CTX). According to the study, consuming polyphenols along with other micronutrients may lower the risk of developing osteoporosis. The study's findings suggest that postmenopausal women may benefit from taking greens + bone builder daily supplements to help lower oxidative damage and maybe lower their risk of developing osteoporosis. The combination of polyphenols and other micronutrients in the supplement reduces oxidative stress parameters and bone resorption through their antioxidant activity. The marker for bone growth, however, showed no discernible change following treatment [15]. In another work, the correlation between increased levels of tumor necrosis factor-α (TNF-α) and intracellular ROS and osteoporosis was studied by Hasan et al. [16]. The ability of the polyphenol EGCG, a component of green tea, can decline the negative effects of ROS and TNF-α on bone health. As a result, it may be possible to develop new osteoporosis treatments and gain a better understanding of the disease's pathogenesis [16]. According to the study findings conducted by Hou et al., ferulic acid can prevent osteoporosis in newborn rats. Therapy with 20 or 30 mg/kg body weight ferulic acid improved bone mechanical characteristics, decreased alkaline phosphatase (ALP) and osteocalcin levels, and increased bone mineral density (BMD). Moreover, ferulic acid decreased nuclear factor kappa-light-chain-enhancer of activated B cells (NF-κB) levels and elevated sirtuin 1 (SIRT1) miRNA and protein levels. The results of the histological examination revealed that ferulic acid raised

BMD. Overall, the results point to ferulic acid as a possible therapeutic agent for osteoporosis prevention and treatment [17]. Meng et al. talk about how areca nut seed polyphenol (ACP) helps rats with ovariectomies gain more bone mass. The study examines the interplay between serum metabolism, bone metabolism, and ACP. The researchers examined several variables, including bone alkaline phosphatase, osteocalcin, Runt-related transcription factor 2, osteoprotegerin (OPG), receptor activator of nuclear factor NF-κB ligand (RANKL), 5-hydroxytryptophan receptor protein, bone morphogenetic protein-2 (BMP2), and bone expression of β-catenin. The findings demonstrated that ACP could inhibit 5-hydroxytryptamine (5-HT), which was linked to serum metabolism and supported bone mass by regulating bone production and resorption. So, ACP can increase bone mass in rats with ovariectomies by regulating bone production and resorption. It increases the ratio of OPG/RANKL, Lrp5, BMP2, and β-catenin while downregulating 5-HT, which is associated with serum metabolism. According to the study, ACP may one day be used to treat osteoporosis [18]. Other similar studies are the connection between eating foods high in polyphenols (*e.g.* olive oil, fruits and vegetables, tea, and soy) and preventing osteoporosis and the discussion of the impact of lycopene and polyphenols on the risk of osteoporosis and the crucial role in the development of the disease [19].

21.3.2.2 Polyphenols in the Treatment and Prevention of Inflammatory Diseases

The plant metabolites, particularly polyphenols, are reported to be good pancreatic lipase inhibitors. Rahim et al. examined the inhibitory potential and method of action of particular flavonoid and non-flavonoid polyphenols and discovered that epicatechins containing gallic acid and galloyl moiety had better inhibitory power. The inhibition was a competitive process, and the galloyl moiety was important for the binding of the enzyme inhibitor. The authors note that there are relatively few reports on lipase inhibition by flavonoids and non-flavonoids, and those that do exist have wide inconsistencies in assay methods and outcomes. However, to fully comprehend the structure–activity relationship of lipase inhibitors from dietary extracts, the authors contend that additional study is necessary [20].

21.3.2.3 Polyphenols in the Treatment and Prevention of Diabetes

The potential of natural polyphenols in the treatment of diabetes mellitus (DM) is discussed by Solayman et al. It gives an overview of contemporary diabetes care techniques as well as the shortcomings of traditional antidiabetic medications. The remainder of the essay examines how polyphenols, which are present in many plants, fruits, and vegetables, may help to reduce insulin resistance and regulate carbohydrate metabolism. The study's findings suggest that natural polyphenols can be used as complementary or alternative treatments for DM. Via several methods, including improved insulin release activity, increased cellular glucose absorption, and preservation and restoration of beta-cell integrity, these polyphenols can regulate carbohydrate metabolism which proves that they may one day be useful for the treatment of diabetes [21]. Type 2 diabetes mellitus (T2DM) may be also treated with natural products high in polyphenols, such as sea buckthorn berries, red grapes, bilberries, chokeberries, chocolate, coffee, and green tea, according to Dragan et al. [22]. The mechanism is that polyphenols influence lipid and carbohydrate metabolism, lower blood sugar and reduce insulin resistance, boost oxidative stress and inflammation, and increase lipid metabolism. The potential effectiveness of polyphenols in the treatment of diabetes is backed up by studies on both humans and animals [22].

Moreover, a study of complementary and alternative medicine therapies for enhancing glycemic control in T2DM was undertaken by Sharma et al. [23] in which only human clinical studies were chosen. While fiber, green tea, and fenugreek have various advantages, but there is little evidence that they significantly improve glycemic control, the research that analyzed the evidence for each therapy came to the conclusion that chromium and maybe gymnema appears to improve glycemic control. It is necessary to conduct more research on cinnamon and bitter melon. The study's findings suggest that chromium and perhaps gymnema can help T2DM patients achieve better glycemic control. There is, however, little proof that other complementary and alternative medicine treatments, like fenugreek, green tea, and fiber, significantly enhance glycemic control [23]. The impact of several forms of polyphenols, such as phenolic acids, flavonoids, stilbenes, lignans, and polymeric lignans, on metabolic problems and complications brought on by diabetes are also discussed by Bahadoran et al. The study was based on numerous *in vitro* research, animal models, and a few human trials. According to the results, polyphenolic compounds may be special nutraceuticals and adjunctive therapies for different symptoms of T2DM. Plant polyphenols like phenolic acids, flavonoids, stilbenes, lignans, and polymeric lignans alter lipid and carbohydrate metabolism, lessen insulin resistance, hyperglycemia, and dyslipidemia, enhance adipose tissue metabolism, and reduce oxidative stress, stress-sensitive signaling pathways, and inflammatory processes. The development of long-term DM consequences, such as cardiovascular disease, neuropathy, nephropathy, and retinopathy, may also be halted by polyphenolic substances [24]. Activation of insulin receptors and glucose uptake in insulin-sensitive tissues, reduction in gastrointestinal glucose absorption, inhibition of carbohydrate digestion, stimulation of insulin secretion, modulation of liver glucose release, alteration of intracellular signaling pathways, and gene expression are some of the potential mechanisms for the role of foods high in polyphenols in treatment of DM [25]. In another report, polyphenols have been suggested by Kabir et al. as potential therapeutic agents for the management of diabetic neuropathy. Kolaviron, resveratrol, naringenin, quercetin, kaempferol, and curcumin were highlighted as the particular polyphenols to block diabetic neuropathy-related deteriorating pathways. These polyphenols can control calcium entry into neurons to counteract glutamate neurotoxicity and the involvement of the glutamate pathway in diabetic neuropathy [26].

The antidiabetic, anti-inflammatory, and antioxidant properties of *Coriandrum sativum* seeds (PCS) were studied by Mechchate et al. The purpose of the study was to examine the effects of PCS on blood glucose levels, general health, and lipid profile in diabetic mice after 28 days of PCS administration. In addition, two assays (-carotene discoloration and DPPH) and carrageenan-induced paw edema in Wistar rats were used to evaluate PCS's anti-inflammatory and antioxidant activities. They used liquid chromatography-tandem mass spectrometry to confirm the presence of nine polyphenols in PCS. According to the results, PCS significantly reduced high fasting blood glucose levels, enhanced general health, and significantly reduced hyperlipidemia in diabetic mice. The PCS also demonstrated good anti-inflammatory and antioxidant properties, as well as strong antihyperglycemic action in an oral glucose tolerance test which implies PCS may be effective in treating diabetes and its associated consequences [27]. There are also some evidence for the effect of flavonoids in herbal teas and coffee on metabolic and microvascular activity, lowering cholesterol and fasting glucose, and reducing inflammation and oxidative stress to prevent and manage T2DM [28], and the effect of polyphenols on the preservation of vascular homeostasis by the endothelium controls in diabetes using reduce oxidative stress and act on cellular signaling pathways like nitric oxide signaling, vascular endothelial growth factor-mediated angiogenesis, estrogen receptor stress, and the nuclear factor erythroid 2-related factor 2 (Nrf2)-antioxidant pathway [29].

The manufacture of nano polyphenols and their potential as therapeutic agents for the treatment of diabetes are covered in a study conducted by Rambaran et al. They assessed 33 studies that looked at the effectiveness of various types of polyphenols, such as phenolic acids, flavonoids, non-flavonoids, and extracts high in polyphenols, as antidiabetic medicines. Curcumin and quercetin were the polyphenols that had received the greatest research. The most common type of nanocarrier was a nanoparticle, and the size of the nano polyphenols was between 15 and 333 nm. The drawbacks of using polyphenols in these studies include their quick evaporation, poor water solubility, and instability at low pH, which nanotechnology can assist solve these problems. The study's findings support the possibility of developing potent therapeutic agents for the management of diabetes through the manufacture of nano polyphenols employing the appropriate nanocarriers, polymers, and encapsulation procedures. Many investigations found that the effects of nano polyphenol were substantially greater than those of its free form [30].

21.3.2.4 Polyphenols in the Treatment and Prevention of Neurodegenerative Disease (NDDs)

Azam et al. investigated the polyphenol's role in the control of Toll-like receptors (TLRs) signaling and treat neurodegeneration. The release of endotoxins, autoimmune disturbances, and an excess of mitochondrial ROS accompanies neurodegeneration. Some age-related neurological conditions have been linked to the pathogenesis of TLRs, which are important in neuroinflammation. Polyphenols have the capacity to modify TLR signaling and offer therapeutic advantages for neurodegeneration. Various classes of polyphenols can target TLR signaling in various pathways and stop the progression of inflammation by activating TLR signaling modulation [31]. Moreover, Ramassamy et al. discuss the polyphenolic compound's role in preventing the death of neurons in conditions like Alzheimer's and Parkinson's diseases. They emphasize that polyphenols from diverse sources, such as curcumin, blueberries, wine, Ginkgo biloba extract, green tea, and wine, with the aim of their antioxidant potential and modulation of multiple pathways, have the ability to protect neurons from various hazardous molecules and have positive impacts on cells [32]. Mandel et al. pointed out the green tea catechin polyphenols' role in NDDs like Parkinson's and Alzheimer's, where the loss of neurons results from a complicated series of toxic responses. The putative neuroprotective capabilities of these polyphenols activate a variety of cellular modes of action relevant to their neuroprotective effect [33]. With the same mechanism, curcumin, resveratrol, and EGCG were introduced as potential treatment for Alzheimer's disease, due to their low frequency of side effects. However, the mainstay of treatment for Alzheimer's disease continues to be the existing traditional pharmacological medications [34]. Also, there are some evidence for the polyphenols' effect on the treatment of age-related neurological conditions like Alzheimer's disease, Parkinson's disease, multiple sclerosis, and Huntington's disease by the modulation of immune response and signaling pathways like peroxisome proliferator-activated receptor, Nrf2, signal transducer and activator of transcription, hypoxia-inducible factor, and mitogen-activated protein kinase [35]. Moreover, Nargeh et al. explain that the protein aggregations caused by the ubiquitin-proteasome system's (UPS) inefficiency and changes in the gut microbiome are the two primary pathogenic hallmarks, and the polyphenols enhance the UPS and slow the progression of NDDs [36].

The ability of tea polyphenols to control disorders connected to circadian rhythms by altering the functionality of gut microbiota was also reported in the literature. Intestinal microbes can communicate with the brain thanks to the brain–gut–microbiome axis, and the strong relationship between intestinal microbiota and circadian rhythm offers tea polyphenols a new chance to control metabolic illnesses. In order to advance the idea of the prevention of metabolic illnesses, the paper primarily focuses on the interaction between tea polyphenols and intestinal microbiota from the perspective of the gut–brain axis [37]. In another study, the protective properties of stilbenes, phenolic acid, and flavonols against ischemic stroke and cardiovascular disorders was investigated [38]. According to the paper's findings, polyphenols in particular exhibit protective properties against ischemic stroke and cardiovascular disorders. When it comes to lowering vascular risk factors for ischemic strokes, like atrial fibrillation, polyphenol consumption over time is clinically significant. With the activation of various pathways and processes, such as the induction of an antithrombotic action, such as Honokiol, various types of polyphenols protect the brain. The study recommends using polyphenols appropriately and holistically as a potential treatment approach for stroke prevention [38].

21.3.2.5 Polyphenols in the Treatment and Prevention of Cardiovascular Diseases

The effect of polyphenols in preventing endothelial dysfunction, which can result in cardiovascular disease, was reported in the literature [39-44]. These natural substances support endothelium health and have the capacity to operate as radical scavengers. However, more clinical studies are required to establish the right dosages and duration of exposure to these substances to actually improve vascular function. Moreover, methods are required to increase the bioavailability and absorption of these substances [39]. In another study, the bioactivities of grape phenolics are discussed by Xia et al. It provides an overview of the current understanding of grape polyphenol extraction, separation, and identification techniques as well as their bioavailability and potential toxicity. The bioactivities of grape phenolics, such as their anti-inflammatory, anti-cancer, anti-inflammation, anti-aging, and antibacterial effects, are the main topic of discussion. According to the study's findings, eating grapes and grape products is linked to a lower prevalence of degenerative illnesses like cardiovascular disease and some malignancies. The bioactivities of grapes' polyphenols, such as anthocyanins, flavonoids, and resveratrol, are believed to be the main source of their health benefits in the cardiovascular disease [40]. An investigation by Victoria-Montesinos et al. examined the impact of commercial product of Citrolive™, a nutraceutical supplement, on plasma atherogenic capability and lipid metabolism. Citrus fruit flavonoids and olive leaf extracts make up the product. There were 23 untreated participants in the research who had high cholesterol and cardiovascular risk factors. Citrolive® was eaten by participants in the intervention group, while controls were given a placebo. According to the study, consuming Citrolive™ for three months was linked to lower levels of oxidized low-density lipoprotein (LDL)-cholesterol and the LDL-oxidase/LDL-cholesterol ratio when compared to controls. The intervention group also experienced a significant rise in serum paraoxonase activity (PON1). According to the study, using Citrolive™ for 3 months reduced oxidized LDL-cholesterol and the LDL-oxidase/LDL-cholesterol ratio in treatment-naive participants with intermediate atherosclerosis risk compared to controls. The intervention group also experienced a significant rise in serum PON1 [41].

Resveratrol was mentioned as a polyphenol that can decrease the course of diabetic cardiomyopathy (DCM). DCM is a form of cardiac disease that affects patients with diabetes. Many mitochondrial functions have been linked to the onset of DCM, such as lipid and glucose metabolism, the formation of ROS, apoptosis, autophagy, and mitochondrial biogenesis. Polyphenols are molecules that are involved in reducing oxidative stress, promoting mitochondrial biogenesis, and enhancing autophagy, such as B-cell lymphoma 2, uncoupling protein 2, SIRT-1, adenosine 5′ monophosphate-activated protein kinase, and N-terminal kinase1; so, resveratrol has an effective role in the slowness of DCM development by reducing oxidative stress, increasing mitochondrial biogenesis, and decreasing apoptosis [42]. The possible benefit of polyphenols in preventing heart failure is suggested by Najjar et al. It examines research on the preventive effects of polyphenols in heart failure both *in vitro* and *in vivo*. The main ways that polyphenols prevent heart failure are decreased cardiac oxidative stress and inflammation decreased mitochondrial dysfunction, improved Ca^{2+} homeostasis, increased survival signaling, and increased SIRT1 activity. So, eating a diet high in polyphenols may be able to prevent or treat heart failure in people [43]. Moreover, the effect of polyphenols in preventing cardiovascular illnesses is discussed by Gormaz et al. The anti-inflammatory and antioxidant characteristics of polyphenols as well as their potential advantages in lowering oxidative stress, inflammation, and platelet aggregation may lessen the occurrence of cardiovascular disease [44].

21.3.2.6 Polyphenols in the Treatment and Prevention of Viral and Bacterial Infections

Polyphenolic compounds have been demonstrated to have antiviral activity against a number of viruses, including the influenza virus, herpes simplex virus, and HIV. For example, a study explain about the effect of polyphenolic compounds on rabies as a viral disease with an acute, progressive, and lethal course. Rabies is still a serious public health concern in developing nations, particularly in Asia, Africa, and Latin America. The paper seeks to assess the potential *in vitro* antirabies activity of 24 synthesized phenolic compounds including gallic acid, methyl gallate, ethyl gallate, propyl gallate, butyl gallate, pentyl gallate, hexyl gallate, heptyl gallate, octyl gallate, decyl gallate, dodecyl gallate, tetradecyl gallate, cetyl gallate, octadecyl gallate, catechin, epicatechin, epigallocatechin, quercetin, and 1,3,5-trihydroxy benzene 3,4,5-trimethyl. As a result, the authors contend that the compounds' anti-rabies activity was affected by the presence of free hydroxyl and ether groups [45]. Moreover, a study on several natural compounds' possible antiviral, antibacterial, and antifungal properties is provided by Zçelik et al. Alkaloids like yohimbine, scopolamine, and atropine, flavonoid derivatives like quercetin and apigenin, and phenolic acids like gallic acid and caffeic acid are among the substances studied. They reported that all flavonoids screened herein had a notable occurrence of anti-herpes simplex virus (HSV)-1 activity, whereas the investigated alkaloids demonstrated an impressive inhibitory action against HSV-1. Only genistein, gallic, chlorogenic, and quinic acids had any sort of anti-RNA virus parainfluenza type-3 impact among the phenolics. In Madin–Darby bovine kidney cells, the majority of the substances demonstrated less cytotoxicity than acyclovir [46].

21.3.2.7 Polyphenols in the Treatment and Prevention of Cancer

The administration of polyphenols and their ability to prevent or treat cancer is the subject of a critical evaluation by the authors of clinical and preclinical investigations. The chemotherapeutic effects of polyphenols, in particular their capacity to control several stages of cancer growth, were reported by Etienne-Selloum et al. They concentrated on resveratrol, curcumin, and EGCG, three important polyphenolic substances, and their capacity to trigger apoptosis in cancer cells through controlling the p53 tumor suppressor pathway. The other methods involving the p73 tumor suppressor and the potential modulation of p53 expression by polyphenols at the transcriptional and post-translational levels were also studied. They also emphasized that oxidative stress plays in polyphenol-induced cancer cell death. The evidence regarding the function of p53 in polyphenol-mediated apoptosis is still debatable, and additional pathways involving the tumor suppressor p73 may be crucial. Even when p53 is altered or absent, as well as when cells are resistant to modern treatments, polyphenols preserve their selective anticancer capabilities [47]. A study also reported the antiproliferative effects of *Iris pseudopumila* flower and rhizome extracts and the fourteen compounds extracted from them. The purpose of the study was to assess the cytotoxic activity of these extracts and compounds against the large lung carcinoma (CORL-23) and amelanotic melanoma cell lines of humans (C32). The methanol extract from flowers was the most antiproliferative, while kaempferol-3-*O*-beta-D-glucopyranoside and irisolidone were the most antiproliferative chemicals, according to the data. The research sheds important light on the possible application of *Iris pseudopumila* as a source of organic chemicals with antiproliferative capabilities. The CORL-23 and C32 cell lines used in the study were found to be resistant to the proliferative effects of *Iris pseudopumila* flowers and rhizomes extracts and their separated components. These findings imply that *Iris pseudopumila* might be a source of natural substances for the creation of anticancer medications [48]. In another study, the relationship between green tea polyphenol consumption and a decreased cancer risk, specifically in connection to prostate cancer, was investigated. The research was done on cell cultures, animal models, epidemiological data on people, and more. The possible use of tea polyphenols as a conservative treatment for patients with less advanced prostate cancer was also covered. Herein, the possibility of green tea polyphenols, in particular EGCG, in preventing prostate cancer is related to the molecular effects of EGCG on DNA methylation. So, it can be proposed a diet-based combinatorial strategy as a viable treatment for the cancer [49].

Moreover, in the report by Afsar et al., bioassay-guided separation of cytotoxic chemicals from *Acacia hydaspica* and their impacts on prostate and breast cancer cell lines proliferation and signaling activity are discussed. According to the study, catechin, 7-*O*-galloyl catechin, methyl gallate, and catechin-3-*O*-gallate are the four active polyphenolic chemicals. The substances caused apoptosis in both cell lines and hindered cell proliferation in both cell lines. Besides, the therapy decreased signaling pathways and suppressed the expression of antiapoptotic molecules in both cell lines. While methyl gallate and catechin-3-*O*-gallate suppressed the proliferation of MDA-MB-231 breast cancer cells, all four substances did so for PC-3 prostate cancer cells. There was evidence of chromatin condensation, cell shrinkage, and apoptotic bodies, and the chemicals also prevented cell survival and colony development in both cell lines. The results imply that these polyphenolic substances might make promising chemopreventive/therapeutic cancer options. In both cell lines, the compounds prevented colony formation, cell growth, and survival. They also partially induced apoptosis by downregulating several signaling pathways [50]. Also, Little et al. reported that dietary polyphenols can lower the incidence of colorectal cancer by modifying inflammatory pathways. The authors discovered strong epidemiological evidence of a decreased risk of colorectal cancer after conducting a study of case–control and cohort studies evaluating polyphenol intake. The relationship between inflammation and the progression of the disease is also covered in the article, along with important signaling pathways mediated by NF-κB and STAT, phosphatidylinositol 3-kinase, and cyclooxygenase [51]. Hasima et al. with a focus on five substances like resveratrol, rottlerin, genistein, quercetin, and curcumin conduct a study on their effects on autophagy and cell death in cancer. The study comes to the conclusion that autophagy, a lysosomal pathway for the destruction of cellular components and organelles, is a crucial and adaptable activity needed for cellular homeostasis. In response to cellular stresses such as food or growth factor deprivation, autophagy can work as a survival mechanism, but it can also result in a nonapoptotic form of programmed cell death (PCD) called autophagy-induced cell death or autophagy-associated cell death (type II PCD). The study hypothesizes that naturally occurring polyphenols in foods including rottlerin, genistein, quercetin, curcumin, and resveratrol, as well as noncanonical (Beclin-1 independent) autophagy pathways, can cause type II PCD by a variety of methods. So, these substances may offer a way to kill cancer cells that boosts the effects of conventional treatments and should be taken into account when developing new therapeutic approaches [52].

In another study, the potential of polyphenolic chemicals as anticancer adjuvants to decrease tyrosine kinase inhibitors (TKI)-resistant lung cancer is discussed. The TKI resistance is a significant problem in cancer treatment, particularly for NSCLC, and that lung cancer is a major cause of cancer-related mortality globally. The authors found that the majority of the tested polyphenolic compounds (equol, kaempferol, resveratrol, and ellagic acid) displayed a strong anticancer effect in the HCC827 panel and were highly selective for TKI-resistant lung cancer cell line H1993 while sparing the TKI-sensitive one H2073. These polyphenolic chemicals may be turned into medications to combat TKI resistance in advanced lung cancer [53]. Furthermore, a research on the anticancer effects of the main polyphenol classes—flavonoids, phenolic acids, lignans, and stilbenes—was conducted. Epidemiological, *in vitro*, *in vivo*, and clinical investigations completed within the last five years served as the foundation for the survey. In the paper, the probable modes of action of these polyphenols are discussed, including antioxidant and anti-inflammatory properties as well as modification of various molecular processes implicated in carcinogenesis. The findings imply that the antioxidant, anti-inflammatory, and molecular modulating characteristics of natural polyphenols present in fruits and vegetables could be exploited to prevent and treat cancer. The bulk of laboratory investigations have supported the anticancer effects of natural polyphenols, the report says, even if epidemiological studies on the association between dietary polyphenol consumption and cancer risks have produced conflicting results. Molecular events and signaling pathways linked to cell survival, proliferation, differentiation, migration, angiogenesis, hormone activities, detoxifying enzymes, immunological responses, etc. are primarily modulated as part of the mechanisms of action. With different cancer kinds, cell lines, and doses, polyphenols have different anticancer effects. The article examines a number of phenolic compounds, including the phenolic acids, hydroxybenzoic acid, and hydroxycinnamic acid. The article also covers naturally occurring stilbenes, including resveratrol, a significant member of this class [54].

To fully comprehend the underlying molecular mechanisms and possible clinical uses of polyphenols in the treatment and prevention of cancer, more research is required, although some study stated that a distinction between cancer prevention and cancer treatment must be defined clearly and that experimental methods for each should be unique [55].

21.4 Conclusion and Future Trends

Dietary polyphenols, such as flavonoids, phenolic acids, catechins, tannins, lignans, stilbenes, coumarins, and anthocyanidins, are abundant in grains, cereals, pulses, vegetables, spices, fruits, chocolates, and alcoholic beverages. They have drawn a lot of attention due to their potential chemopreventive/protective roles in the upkeep of human health and diseases. Dietary polyphenols are thought to have potent antioxidant effects that protect cells from ROS and oxidative stress, hence preventing OS-related pathological diseases. Preclinical and clinical data, overwhelmingly, support the notion that long-term consumption of diets high in polyphenols offers protection against the emergence of a variety of chronic diseases, including cancer, diabetes, inflammatory disorders, cardiovascular diseases, NDDs, and infectious diseases. Due to their potential, the use of polyphenols as phytopharmaceuticals is expected to grow over time and take on a variety of new forms. The development of dietary polyphenol compounds as potential therapeutic possibilities against a variety of chronic human diseases, however, could result from potential future research with sufficient preclinical and clinical examinations.

Abbreviations

3D	Three-dimensional
5-HT	5-Hydroxytryptamine
ACP	Nut seed polyphenol
BMD	Bone mineral density
BMP2	5-Hydroxytryptophan receptor protein, bone Morphogenetic morphogenetic protein-2
C32	Amelanotic melanoma cell lines in humans
CORL-23	Large lung carcinoma
DCM	Diabetic cardiomyopathy
DM	Diabetes mellitus
DPPH	2,2-Diphenyl-1-picrylhydrazyl
EGCG	Epigallocatechin-3-gallate
EV	Extracellular vesicles
HSV	Herpes simplex virus
LDL	Low-density lipoprotein
miRNAs	microRNAs
NDDs	Neurodegenerative disease disease
NF-κB	Nuclear factor kappa-light-chain-enhancer of activated B cells
Nrf2	Nuclear factor erythroid 2-related factor 2
NSCLC	Non-small cell lung cancer
—OH	Hydroxyl group
OPG	Osteoprotegerin
PCD	Programmed cell death
PCS	*Coriandrum sativum* seeds
PON1	Paraoxonase activity
RANKL	Receptor activator of nuclear factor NF-κB ligand

ROS	Reactive oxygen species
sEVs	Small extracellular vesicles
SIRT1	Sirtuin 1
T2DM	Type 2 diabetes mellitus
TA	Tannic acid
TKI	Tyrosine kinase inhibitors
TLRs	Toll-like receptors
TNF-α	Tumor necrosis factor-α
UPS	Ubiquitin-proteasome system

References

1 Laura, A., Alvarez-Parrilla, E., and González-Aguilar, G.A. (2009). *Fruit and Vegetable Phytochemicals: Chemistry, Nutritional Value and Stability*. John Wiley & Sons.

2 Yadav, R. and Agarwala, M. (2011). Phytochemical analysis of some medicinal plants. *Journal of Phytology* 3 (12).

3 Ciumărnean, L., Milaciu, M.V., Runcan, O. et al. (2020). The effects of flavonoids in cardiovascular diseases. *Molecules* 25 (18): 4320.

4 Procházková, D., Boušová, I., and Wilhelmová, N. (2011). Antioxidant and prooxidant properties of flavonoids. *Fitoterapia* 82 (4): 513–523.

5 Santos-Sánchez, N.F., Salas-Coronado, R., Villanueva-Cañongo, C., and Hernández-Carlos, B. (2019). Antioxidant compounds and their antioxidant mechanism. *Antioxidants* 10: 1–29.

6 Bao, Y. and Fenwick, R. (2004). *Phytochemicals in Health and Disease*. CRC Press.

7 Jang, M., Choi, G., Choi, Y.Y. et al. (2019). Extracellular vesicle (EV)-polyphenol nanoaggregates for microRNA-based cancer diagnosis. *NPG Asia Materials* 11 (1): 79.

8 Dai, Q., Geng, H., Yu, Q. et al. (2019). Polyphenol-based particles for theranostics. *Theranostics* 9 (11): 3170.

9 Wang, H., Liu, D., Zhang, L. et al. (2022). Label-free small extracellular vesicles capturing strategy for lung cancer diagnosis and typing based on a natural polyphenol-metal three-dimensional network. *Analytical Chemistry* 94 (46): 16103–16112.

10 Corrêa, T.A. and Rogero, M.M. (2019). Polyphenols regulating microRNAs and inflammation biomarkers in obesity. *Nutrition* 59: 150–157.

11 Gamage, E., Lane, M., Travica, N. et al. (2023). Polyphenols and depression: exploring the potential mechanisms of action. *Proceedings of the Nutrition Society* 82 (OCE2): E118.

12 Rao, L. and Rao, A. (2013). Oxidative stress and antioxidants in the risk of osteoporosis—role of the antioxidants lycopene and polyphenols. *Topics in Osteoporosis* 5: 117–144.

13 Bellavia, D., Caradonna, F., Dimarco, E. et al. (2021). Non-flavonoid polyphenols in osteoporosis: preclinical evidence. *Trends in Endocrinology & Metabolism* 32 (7): 515–529.

14 Đudarić, L., Fužinac-Smojver, A., Muhvić, D., and Giacometti, J. (2015). The role of polyphenols on bone metabolism in osteoporosis. *Food Research International* 77: 290–298.

15 Kang, N.N. (2012). *Oxidative Stress and the Risk of Osteoporosis: The Role of Dietary Polyphenols and Nutritional Supplements in Postmenopausal Women.* University of Toronto (Canada).

16 Hasan, W., Ahmad, S., Thakur, H. et al. (2014). In vitro regulation of osteoclast generation: a cost-effective strategy to combat osteoporosis with natural antioxidants and polyphenols like EGCG. *European Journal of Academic Research* 2: 2286–4822.

17 Hou, T., Zhang, L., and Yang, X. (2019). Ferulic acid, a natural polyphenol, protects against osteoporosis by activating SIRT1 and NF-κB in neonatal rats with glucocorticoid-induced osteoporosis. *Biomedicine & Pharmacotherapy* 120: 109205.

18 Meng, K., Mei, F., Zhu, L. et al. (2021). Arecanut (Areca catechu L.) seed polyphenol improves osteoporosis via gut-serotonin mediated Wnt/β-catenin pathway in ovariectomized rats. *Journal of Functional Foods* 84: 104598.

19 Rao, L. and Rao, A. (2015). *Oxidative Stress and Antioxidants in the Risk of Osteoporosis—Role of Phytochemical Antioxidants Lycopene and Polyphenol-Containing Nutritional Supplements*, Phytochemicals—Isolation, Characterisation and Role in Human Health (ed. L.R. Venketeshwer Rao), 247–260. Croatia: Intechopen https://doi.org/10.5772/60446.

20 Rahim, A.T.M., Takahashi, Y., and Yamaki, K. (2015). Mode of pancreatic lipase inhibition activity in vitro by some flavonoids and non-flavonoid polyphenols. *Food Research International* 75: 289–294.

21 Solayman, M., Ali, Y., Alam, F. et al. (2016). Polyphenols: potential future arsenals in the treatment of diabetes. *Current Pharmaceutical Design* 22 (5): 549–565.

22 Dragan, S., Andrica, F., Serban, M.-C., and Timar, R. (2015). Polyphenols-rich natural products for treatment of diabetes. *Current Medicinal Chemistry* 22 (1): 14–22.

23 Sharma, P., Hajam, Y.A., Kumar, R., and Rai, S. (2022). Complementary and alternative medicine for the treatment of diabetes and associated complications: a review on therapeutic role of polyphenols. *Phytomedicine Plus* 2 (1): 100188.

24 Bahadoran, Z., Mirmiran, P., and Azizi, F. (2013). Dietary polyphenols as potential nutraceuticals in management of diabetes: a review. *Journal of Diabetes & Metabolic Disorders* 12: 1–9.
25 Aryaeian, N., Sedehi, S.K., and Arablou, T. (2017). Polyphenols and their effects on diabetes management: a review. *Medical Journal of the Islamic Republic of Iran* 31: 134.
26 Kabir, M.T., Tabassum, N., Uddin, M.S. et al. (2021). Therapeutic potential of polyphenols in the management of diabetic neuropathy. *Evidence-Based Complementary and Alternative Medicine* 2021: 1–20.
27 Mechchate, H., Es-Safi, I., Amaghnouje, A. et al. (2021). Antioxidant, anti-inflammatory and antidiabetic proprieties of LC-MS/MS identified polyphenols from coriander seeds. *Molecules* 26 (2): 487.
28 Alkhatib, A., Tsang, C., Tiss, A. et al. (2017). Functional foods and lifestyle approaches for diabetes prevention and management. *Nutrients* 9 (12): 1310.
29 Suganya, N., Bhakkiyalakshmi, E., Sarada, D., and Ramkumar, K. (2016). Reversibility of endothelial dysfunction in diabetes: role of polyphenols. *British Journal of Nutrition* 116 (2): 223–246.
30 Rambaran, T.F. (2020). Nanopolyphenols: a review of their encapsulation and anti-diabetic effects. *SN Applied Sciences* 2 (8): 1335.
31 Azam, S., Jakaria, M., Kim, I.-S. et al. (2019). Regulation of toll-like receptor (TLR) signaling pathway by polyphenols in the treatment of age-linked neurodegenerative diseases: focus on TLR4 signaling. *Frontiers in Immunology* 10: 1000.
32 Ramassamy, C. (2006). Emerging role of polyphenolic compounds in the treatment of neurodegenerative diseases: a review of their intracellular targets. *European Journal of Pharmacology* 545 (1): 51–64.
33 Mandel, S. and Youdim, M.B. (2004). Catechin polyphenols: neurodegeneration and neuroprotection in neurodegenerative diseases. *Free Radical Biology and Medicine* 37 (3): 304–317.
34 Syarifah-Noratiqah, S.-B., Naina-Mohamed, I., Zulfarina, M.S., and Qodriyah, H. (2018). Natural polyphenols in the treatment of Alzheimer's disease. *Current Drug Targets* 19 (8): 927–937.
35 Bhullar, K.S. and Rupasinghe, H. (2013). Polyphenols: multipotent therapeutic agents in neurodegenerative diseases. *Oxidative Medicine and Cellular Longevity* 2013.
36 Nargeh, H., Aliabadi, F., Ajami, M., and Pazoki-Toroudi, H. (2021). Role of polyphenols on gut microbiota and the ubiquitin-proteasome system in neurodegenerative diseases. *Journal of Agricultural and Food Chemistry* 69 (22): 6119–6144.
37 Yan, R., Ho, C.T., and Zhang, X. (2020). Interaction between tea polyphenols and intestinal microbiota in host metabolic diseases from the perspective of the gut–brain axis. *Molecular Nutrition & Food Research* 64 (14): 2000187.

38 Pacifici, F., Rovella, V., Pastore, D. et al. (2021). Polyphenols and ischemic stroke: insight into one of the best strategies for prevention and treatment. *Nutrients* 13 (6): 1967.

39 Di Pietro, N., Baldassarre, M.P.A., Cichelli, A. et al. (2020). Role of polyphenols and carotenoids in endothelial dysfunction: an overview from classic to innovative biomarkers. *Oxidative Medicine and Cellular Longevity* 2020: 1–19.

40 Xia, E.-Q., Deng, G.-F., Guo, Y.-J., and Li, H.-B. (2010). Biological activities of polyphenols from grapes. *International Journal of Molecular Sciences* 11 (2): 622–646.

41 Victoria-Montesinos, D., Abellán Ruiz, M.S., Luque Rubia, A.J. et al. (2021). Effectiveness of consumption of a combination of citrus fruit flavonoids and olive leaf polyphenols to reduce oxidation of low-density lipoprotein in treatment-naive cardiovascular risk subjects: a randomized double-blind controlled study. *Antioxidants* 10 (4): 589.

42 Bhagani, H., Nasser, S.A., Dakroub, A. et al. (2020). The mitochondria: a target of polyphenols in the treatment of diabetic cardiomyopathy. *International Journal of Molecular Sciences* 21 (14): 4962.

43 Najjar, R.S. and Feresin, R.G. (2021). Protective role of polyphenols in heart failure: molecular targets and cellular mechanisms underlying their therapeutic potential. *International Journal of Molecular Sciences* 22 (4): 1668.

44 Gormaz, J.G., Valls, N., Sotomayor, C. et al. (2016). Potential role of polyphenols in the prevention of cardiovascular diseases: molecular bases. *Current Medicinal Chemistry* 23 (2): 115–128.

45 Chávez, J.H., Leal, P.C., Yunes, R.A. et al. (2006). Evaluation of antiviral activity of phenolic compounds and derivatives against rabies virus. *Veterinary Microbiology* 116 (1-3): 53–59.

46 Özçelik, B., Kartal, M., and Orhan, I. (2011). Cytotoxicity, antiviral and antimicrobial activities of alkaloids, flavonoids, and phenolic acids. *Pharmaceutical Biology* 49 (4): 396–402.

47 Etienne-Selloum, N., Dandache, I., Sharif, T. et al. (2013). Polyphenolic compounds targeting p53-family tumor suppressors: current progress and challenges. *Future Aspects of Tumor Suppressor Gene* 129–167.

48 Conforti, F., Menichini, F., Rigano, D., and Senatore, F. (2009). Antiproliferative activity on human cancer cell lines after treatment with polyphenolic compounds isolated from Iris pseudopumila flowers and rhizomes. *Zeitschrift für Naturforschung C* 64 (7–8): 490–494.

49 Davalli, P., Rizzi, F., Caporali, A. et al. (2012). Anticancer activity of green tea polyphenols in prostate gland. *Oxidative Medicine and Cellular Longevity* 2012.

50 Afsar, T., Trembley, J.H., Salomon, C.E. et al. (2016). Growth inhibition and apoptosis in cancer cells induced by polyphenolic compounds of Acacia hydaspica: involvement of multiple signal transduction pathways. *Scientific Reports* 6 (1): 23077.

51 Little, C., Combet, E., McMillan, D. et al. (2017). The role of dietary polyphenols in the moderation of the inflammatory response in early stage colorectal cancer. *Critical Reviews in Food Science and Nutrition* 57 (11): 2310–2320.

52 Hasima, N. and Ozpolat, B. (2014). Regulation of autophagy by polyphenolic compounds as a potential therapeutic strategy for cancer. *Cell death & disease.* 5(11):e1509-e.

53 Jeong, H., Phan, A.N., and Choi, J.-W. (2017). Anti-cancer effects of polyphenolic compounds in epidermal growth factor receptor tyrosine kinase inhibitor-resistant non-small cell lung cancer. *Pharmacognosy Magazine* 13 (52): 595.

54 Zhou, Y., Zheng, J., Li, Y. et al. (2016). Natural polyphenols for prevention and treatment of cancer. *Nutrients* 8 (8): 515.

55 Russo GL, Tedesco I, Spagnuolo C, Russo M, Antioxidant polyphenols in cancer treatment: friend, foe or foil? *Seminars in Cancer Biology*; 2017: Elsevier.

22

Food Additive Applications of Polyphenols

Richika Ganjoo[1], *Shveta Sharma*[1,3], *Humira Assad*[1,4], *Praveen Kumar Sharma*[1], *and Ashish Kumar*[2]

[1] *Department of Chemistry, School of Chemical Engineering and Physical Sciences, Lovely Professional University, Punjab, India*
[2] *Nalanda College of Engineering, Bihar Engineering University, Department of Science, Technology and Technical Education, Government of Bihar, India*
[3] *Department of Chemistry, Government College Una, Affiliated to Himachal Pradesh University, Una, India*
[4] *Department of Chemistry, Faculty of Technology and Science, Lovely Professional University, Phagwara, Punjab, India*

22.1 Introduction

Food additives are compounds added to food to preserve or enhance its quality in terms of taste, texture, freshness, safety, or appearance. Some food additives, like salt (found in meats like bacon or dried fish) or sugar (found in marmalade) or sulfur dioxide (found in wine), have been used to preserve foods for millennia. Since producing food on a large scale is considerably different from producing it on a small scale at home, to meet the demands of food production, a wide variety of food additives have been developed throughout time. Additives are necessary to keep processed food safe and in excellent condition along its journey from factories or kitchens in industrial settings to warehouses and retailers and ultimately to customers [1–3]. Food additives should only be used when they are technologically necessary, do not deceive customers, and have a specific technological objective, like maintaining nutritional value or improving food stability [4–7].

Food additives may be developed synthetically or obtained from plants, animals, or minerals. Customers are purposefully introduced to food to accomplish specific technological functions they frequently take for granted [8–10]. It is common to utilize several thousand food additives, and every one of them aims to improve the appeal or safety of food in a certain way. The WHO and FAO classify

Science and Engineering of Polyphenols: Fundamentals and Industrial Scale Applications,
First Edition. Edited by Chandrabhan Verma.
© 2024 John Wiley & Sons Inc. Published 2024 by John Wiley & Sons Inc.

food additives into three broad groups depending on how they behave. Due to their medicinal benefits and possible technological uses, polyphenols are naturally occurring substances in fruits and vegetables. They are receiving increasing amounts of attention. They may be employed as natural colors, prebiotic components, hydrogels, and nanocomplexes to enhance the physicochemical characteristics of starch and food preservation.

A kind of complex secondary metabolite found in abundance in a plant's foliage, bark, roots, leaves, stems, and fruits is called a polyphenol. Due to their distinctive chemical makeup, plant-derived polyphenols have been proven in several studies to have a wide range of bioactivities, including antioxidant and antibacterial properties as well as the capability to prevent chronic illnesses including cancer, osteoporosis, cardiovascular disease, and neurodegeneration [11–13]. Many human disorders, including neurodegenerative diseases, might be prevented and perhaps treated using polyphenols, which are promising chemicals. However, like any pharmacological medicine, they may also exhibit side effects and/or toxicity, especially if large quantities accumulate in the body [14].

Plant polyphenols, which are widely distributed in plants and have a variety of physiological roles, are one of the study hotspots in food science as natural product development has continuously increased. Due to their outstanding antioxidant and antibacterial activity, plant-derived polyphenols offer a substitute for chemical additives in the food industry, including oil, meat, fish, beverages, and food packaging materials. This chapter describes plant polyphenols' potential antioxidant and antimicrobial properties and their use in food storage.

22.2 Polyphenols as Flavoring Agents

Most food additives are flavoring compounds added to meals to enhance their flavor or scent. Flavorings come in hundreds of meals, including cereal, cake, yogurt, soft drinks, and confections. Natural flavoring ingredients include combinations of nuts, fruits, spices, and those made from vegetables and wine. Other flavorings replicate the tastes of natural foods. In recent years, beverages have evolved into a product that is consumed regularly. However, due to their high sugar levels and preservatives, human quality and possible health issues have gained significant consideration lately. Since we consume many polyphenols in our everyday diet, comprising vegetables, fruits, and whole grain products, we usually consume them directly as meals or drinks, enhancing the beverages' health advantages and helping us address health issues. By stabilizing different substances, it is also utilized to increase the sense of smell and shelf life of goods. Tea polyphenols have a beneficial impact on the preservation of beverage flavor and the lengthening of shelf-life.

Nutritional depletion, the development of off-flavors, and product discoloration occur throughout the production and storage of drinks, negatively impacting the good's customer acceptance and shelf life. Light, temperature, oxygen, pH, microbes, enzymes, and component interactions throughout manufacturing and packing are often implicated in nutrient deterioration, pigments, fragrance, and flavor molecules. Based on these considerations, several research studies have investigated the use of polyphenols to improve the stability of beverage components and ensure a minimal loss of beverage quality. Due to the high quantities of proteins and reducing sugars in ultra-high temperature (UHT) processed drinks, particularly lactose-hydrolyzed milk, beverages made with milk, and beverages made with plant proteins, Maillard reactions are visible. Furfurals, aldehydes, and dicarbonyls are examples of Maillard reaction products that, as a consequence, have a detrimental impact on nutritional value, color, fragrance, flavor, and shelf life while being stored [15].

Polyphenols are diverse molecules with over 8000 different species, including flavonoids (such as catechins in tea and cocoa, isoflavones in beans, quercetin in onions, and anthocyanins in fruit) and non-flavonoids (such as chlorogenic acids in coffee), which are present in all plants for protection against ultraviolet light and reactive oxygen species and also occasionally provide color. Humans absorb a significant portion of the non-nutritive dietary components known as polyphenols. Evidence shows that consuming foods and drinks high in polyphenols lowers the risk of developing non-communicable illnesses. *Streptococcus mutans* and pediatric dental plaque development and biofilm formation were examined by Wu et al. in commercially available plant polyphenol-containing drinks [16]. *S. mutans* was shown to be inhibited in its ability to proliferate and produce biofilms when consumed in commercially available, ready-to-drink tea and cranberry juice beverages. The development, biofilm formation, and matrix mechanical characteristics of the bacteria that cause children's supragingival plaque were similarly inhibited by infusions made from prepackaged teas and cranberry juice beverages. These drinks include plant-derived polyphenols, which target the same cariogenic virulence factors as conventional antibacterial drugs but without side effects. These drinks show potential dental health advantages and are superior to regularly used sugar-sweetened beverages when eaten in the correct order between meals and with adequate mouth hygiene. A fresh, practical, and creative idea for promoting children's dental health is managing caries by consuming natural foods and drinks [16].

Yogurt contains starter cultures, which have probiotic properties, and polyphenols (PPs), which are an easy method to give both. Sun-Waterhouse et al. showed that drinking yogurts containing pectins (low or high methoxyl) were exposed to physical, chemical, and microbiological tests. Apple PPs were incorporated pre- or post-fermentation. The TEPC, Viscosity, topology, PP profiles, viscosity, storage modulus, and colony number of starting cultures were all impacted by the kind of PPs and pectin, according to the results. More polar PPs and a higher TEPC were formed in yogurts when APE was added before to fermentation as opposed to after. Either method may produce drinkable yogurt that has been improved with apple PPs. Improved TEPC and starting culture growth may make the pre-fermentation process even more favorable, providing a bioprocess for optimizing PP profiles in drinkable yogurt [17].

To improve flavor and boost antioxidant effectiveness, berry fruit juice or an extract from berry materials with a higher PP content may be included in yogurt items before or after fermentation. Sun-Waterhouse et al. additionally examined the impact of adding pure PP (cyanidin 3-*o*-glucopyranoside chloride, Cyanidin) or a blackcurrant PP extract (BPE) before or following fermentation on the microbiological, chemical, and rheological characteristics of consuming yogurt produced with high or low methoxy (HM or LM) pectin levels [18]. Both the control and PP-enhanced yogurts underwent rheological study, PP profiling by HPLC, TEPC analysis, and microbiological examination for the endurance and expansion of starter cultures. The outcomes demonstrate that incorporating BPE before fermentation led to tiny phenolic compounds (such as phenolic acids) in the yogurts, and a TEPC was 3.5–3.9 times higher than BPE applied after fermentation. Yogurt's PP levels were impacted by fermentation.

Before fermentation, the starter cultures *Streptococcus thermophilus* and *Lactobacillus delbrueckii* subsp. *bulgaricus*, in addition to the colony number and appearance of these cultures, as well as the elastic quality and viscosity of the resulting yogurts, were all altered differently by adding BPE and Cyanidin. When pectins were added, the yogurt's rheological characteristics changed, with the HM pectin yogurts exhibiting a more robust gel framework than the LM pectin yogurts. Adding PPs had a variety of impacts on the starting cultures, depending on the yogurt formulation (such as the pectin stabilizer type), the starter culture type, the kind of PP, and the method used to add the PPs. The probiotic impacts of starter cultures and the known health advantages of blackcurrant PPs may be provided by specific outcomes, which can be tuned to increase the TEPC and favorable PPs as metabolites of originally assimilated PPs. Although pre- and post-fermentation methods for PP addition to consumable yogurts are technically possible, they each have unique difficulties regarding product processing and functional benefits.

Ahmad et al. researched how apple peel polyphenol extract (APPE) affected probiotic yogurthurt's microbiological and physicochemical characteristics [19]. Probiotic yogurt had five different APPE concentrations added to it: (i) CTL, a sample without APPE; (ii) AE1, 1% APPE addition; (iii) AE2, 2% APPE addition; (iv) AE3, 3% APPE addition; (v) AE4, 4% APPE addition; and (vi) AE5, 5% APPE addition. The prepared probiotic yogurt was kept at 4 °C for a 21-day days before its physical, chemical, and biological properties were evaluated. All yogurt samples had a comparable preliminary sustainable number of *Strreptococcus thermophilus*, *Lactobacillus acidophilus*, *Bifidobacterium lactis*, and *Lactobacillus bulgaricus* on day 1. Probiotics' most significant viability decrease was seen in CTL ($p < 0.05$). The AE5 samples showed the probiotic's lowest viability decrease ($p < 0.05$). Including APPE improved the viscosity, acidity, and water retention capacity. The inclusion of APPE had no discernible influence on the amount of milk fat or total solids in the probiotic yogurt. In the AE1, AE2, AE3, AE4, and AE5 samples, the overall phenolic levels of probiotic yogurt rose considerably to 0.59, 0.71, 0.97, 1.18, and 1.35, respectively. The AE3 and AE4 samples had improved taste, aroma, color, and decent texture. Including APPE, which improved the probiotic's survival ability and the yogurtshurts' antioxidant activity. In conclusion, apple peels might function well as a prebiotic in yogurt with higher viable probiotic numbers.

22.3 Polyphenols in Enzyme Preparations

Proteins called enzymes are found in nature and speed up biological operations by breaking larger molecules into their component elements. They serve as substitutes for chemical-based technologies and may be acquired through extraction from animals, plants, or microorganisms such as bacteria. They are mainly used in baking, fruit juice, wine and beer, and cheese production (to enhance curd formation).

Bread, muffins, crumpets, cakes, cookies, bagels, pancakes, doughnuts, waffles, rolls, and other bakery items have long been a staple of the human diet. Polyphenols, particularly polyphenol-enriched goods, are employed extensively in bakery foods as nutraceuticals with various biological properties. Most of the fruits and vegetables used to make the polyphenol-augmented goods are vegetables and fruits, comprising whole fruits, puree, juice, jam, dried fruit, pomace, and peel powder. These ingredients, in addition to providing polyphenols, can contribute additional nutrients, particularly dietary fibers for bread items. The incorporation of foods high in polyphenols to bakery items, like the products of fruits and vegetables, is one of two main ways used to include polyphenols in baked foods. The other approach involves incorporating polyphenol standards directly into the recipe. Fruits and vegetables are used to make a variety of products, such as jams, purees or pulp, marmalade, pieces of dried fruit, canned goods, fruit powder, fruit seed oil, fruit peel, and fruit pomace. Polyphenols and polyphenol-enriched waste products have been frequently employed in baked goods for neutraceutical and other applications. By delaying the digestion of starch, the addition of polyphenols may boost antioxidant and antibacterial activity, reduce the presence of advanced glycation end products (AGEs) and other foodborne toxins, and lower postprandial blood glucose.

Byproducts with high levels of polyphenols boost cheese's antioxidant and antibacterial activity, extending its shelf life. According to the byproducts used and the method used for extract extraction, contrasting outcomes have been found regarding the color and sensory qualities of enriched cheeses. Furthermore, the prebiotic properties of functional cheese components and the enhancement of human health are brought about by their bioavailability. Overall, the sustainability of the food supply chain and consumer health may benefit from the use of agri-food byproducts in cheese production [20]. To create cheese with an elevated polyphenol content, grape extracts may be added to milk. Silva et al. characterized and added four commercial preparations from whole grapes, grape seeds, and grape peel to milk at doses of 0, 0.1, 0.2, and 0.3% (wt/vol) [21]. Model cheeses were made to test the impact of grape extracts on the syneresis, milk gel texture, and milk clotting kinetics. Similar amounts of polyphenolic chemicals were present in grape seed and whole grape extracts, which were nearly twice as abundant as those in grape skin extracts. The amount of phenolic compounds closely correlated with the action of radical scavengers. Grape extracts lengthened the time required for rennet-induced coagulation and slowed its pace when added to milk. The concentration introduced to milk was the key element determining the clotting characteristics despite variances being seen across the extracts. Milk gels indicated increased stiffness and decreased stiffness as grape extract concentrations rose. Additionally, the syneresis of milk gels reduced as grape extract concentrations rose, resulting in cheeses with increased moisture content. Grape extracts in milk marginally accelerated the recovery of proteins in cheese but had no impact on the recovery of fat. The recovery coefficient for polyphenols with whole grape or grape seed extracts incorporated into milk at 0.1% (wt/vol) was around 0.63, and it fell with rising extract dose in milk. With no concentration impact, grape seed extracts (0.87) showed a better polyphenol recovery. Cheese may be made with possible health advantages while maintaining cheese output by adding industrial extracts from whole grapes, grape seeds, or grape peel to milk in the range of 0.1 to 0.3% (wt/vol).

Red wine is healthier than beer, spirits, and even white wine due to the high quantities of polyphenols it contains. Red wine in particular has been said to be particularly helpful in preventing cancer, as well as other chronic diseases like heart disease. Despite the existence of risk variables such as excessive intake of saturated fats, higher smoking, and poor physical activity, regular and moderate consumption has been demonstrated in multiple epidemiological studies to correlate inversely with vascular disease and death. Figure 22.1 shows the primary polyphenols found in red wine and the health advantages they have been shown to have in in vitro and in vivo investigations [22].

22.4 Polyphenols in Preservation and Coloring

There are many more uses for food additives, including preservation, coloring, and sweetening. They are added to food when it is being prepared, packed, transported, or stored, and ultimately, they are absorbed into the food itself. Mold, air, bacteria, and yeast-related decomposition may all be slowed down by preservatives. Preservatives assist in preventing contamination that may lead to foodborne

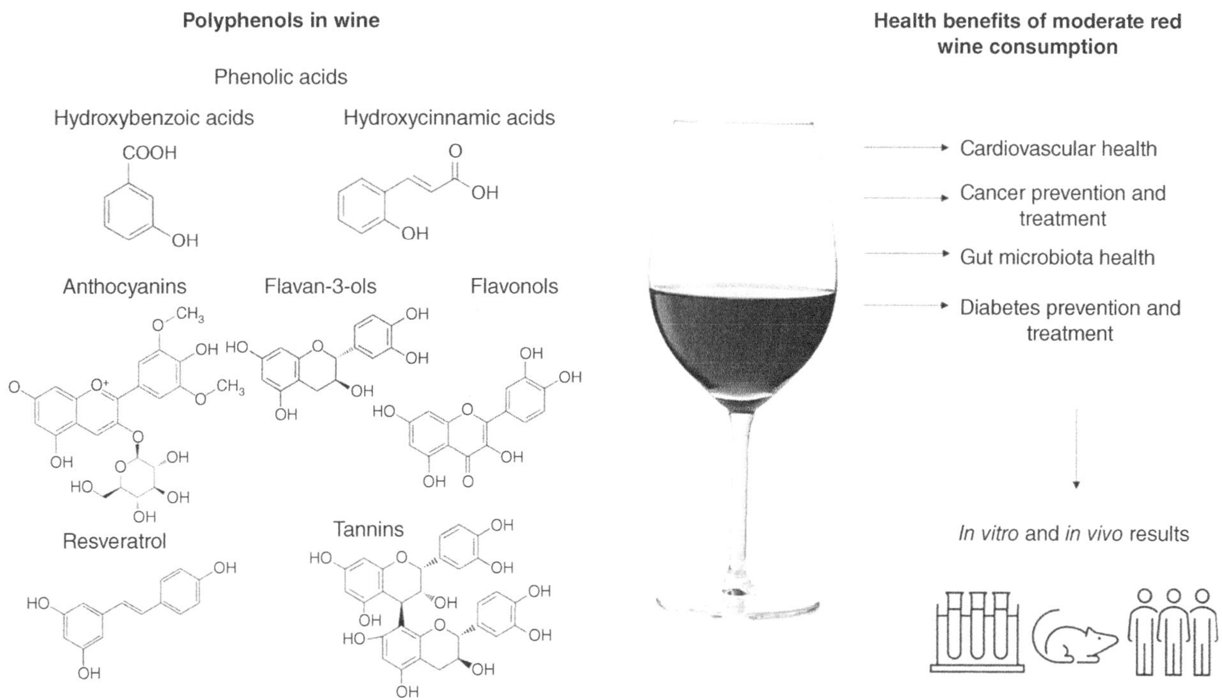

Figure 22.1 Red wine's polyphenol content and the advantages of moderate red wine intake for health. Source: Buljeta et al. [22]/ from MDPI/CC BY 4.0.

illnesses, including potentially fatal botulism, in addition to preserving food quality. Food and drink colors affect both their look and agreeability, which is correlated with both taste and flavor. Intact colors from raw materials are present in foods and drinks, as are synthetic pigments produced during preparation and preservation (assuming food additives are not used) via chemical and enzymatic processes. Natural pigments like polyphenols, carotenoids, and chlorophyll are produced from raw ingredients. The complicated chemical processes that take place during processing and storage change a number of other substances, including proteins, amino acids, lipids, carbohydrates, and others; these reactions' byproducts may also affect the color of meals and drinks.

Food colorings may be classified as synthetic or natural. However, according to the FDA, the word "natural colorants" has a pretty narrow meaning. A strawberry fruit preparation colored with anthocyanins, for instance, would be considered to have a natural color. The United States and other governing organizations classify all other colorants as artificial. According to rules established by the US Food and Drug Administration (FDA), natural colors, or colorants produced from natural sources, are free from certification. Different fruit and vegetable-based sources—which are widely recognized for their coloring properties—are often employed to color food. Another widely used example that gives applications a yellow color while maintaining strong thermal stability is turmeric extract.

Food packaging is a necessary but optional component of the contemporary food business system that primarily safeguards food from chemical, biological, and physical harm as it travels from the farm or factory to the customer while also making food consumption easier. The primary purpose of packaging is to protect the nutritional value and shelf-life of more stable food items while also prolonging their sensory appeal. People are paying increasing attention to bioactive, renewable, functional, and edible containers as living standards rise and awareness of the quality and safety of food increases. This is done to prevent the influx of chemicals or preservatives from conventional packaging materials. More than 8000 polyphenolic chemicals have been separated and identified, and they are abundantly found in many different portions of fruits, vegetables, herbs, and other plants. There are three primary categories that it falls under: stilbenoids, phenolic acids, and flavonoids.

Polyphenols have been used in substances for wrapping food to enhance the physical and functional qualities due to their remarkable antibacterial and antioxidant activity and availability. In the most recent decades, several researchers have tried to combine synthetic polymers or biopolymers with natural polyphenols or extracts high in phenols to enhance the packing qualities of materials, substitute chemical additives, and lower the risk of food migration. In general, there are various stages involved in the creation of packaging products high in polyphenol, like adding polyphenols or phenolic-rich extracts into polymers, structure characterization using a variety of modern equipments, evaluation of a material's bioactivities, mechanical, and barrier properties, followed by food packaging applicability, and lastly, food assessment by physical, chemical, sensory, and biological indexes. Several studies have examined the use of polyphenols in conjunction with polymers for package. The majority of these research overlooked the uses of these coatings in food packing and instead concentrated only on the manufacture and characterization of the film along with a few in vitro antioxidant and antibacterial tests. As a result, their use in food packing has to be given further consideration.

22.4.1 Polyphenols Incorporated with Synthetic and Biopolymers for Food Packaging

Sustainable, renewable, and ecological packaging materials are receiving more and more attention as a solution to the problems of plastic waste and chemical smuggling. As a result, adding natural polyphenols or extracts high in polyphenols to bio-based products has gained popularity. Some researchers have sought to incorporate organic polyphenols or phenolic-rich compounds onto traditional synthetic polymers to substitute chemical ingredients and minimize possible migration in food in an effort to replace synthetic plastic antioxidants. As can be shown in Table 22.1, when chemical synthetic and biological polymers were grafted onto polyphenols or polyphenol-rich extracts, the resulting films had excellent packing characteristics and bioactivities [23–28].

Table 22.1 The uses of polyphenols in food packaging.

Name of synthetic/ biopolymer	Polyphenols	Application
PVA-chitosan, Starch-PVA	Roselle anthocyanins	The film, and observe the freshness of the meat
PET-PP	Mango polyphenols	The film increases the longevity of lettuce and tangerines
PLA-chitosan	Tea polyphenols	The film increases the longevity of cherries
Gelatin	Tea polyphenols	The film prevents rancidity and oxidation of sunflower oil
Gelatin-carrageenan	Curcumin, gallic acid, quercetin	Film, increase the raw grill meat's shelf life
Zein	Curcumin	Coating, minimizing apple coating post-harvest degradation

22.4.2 Polyphenols Incorporated Polysaccharides for Food Packaging

Active and innovative packaging may be created using polysaccharide films and coatings. The physical characteristics of polysaccharide films are improved by adding phenolic acids, increasing their potential commercial usefulness as a food preservation substance. One method is to include bioactive substances like polyphenols or extracts high in polyphenols while creating multifunctional films and coatings. Numerous plants, including fruits, vegetables, and cereal grains, contain polyphenols. They are varied in both function and structure. Polyphenols come in many various forms, such as lignans, polyphenolic amides, flavonoids, phenolic acids, and resveratrol. Polyphenols and plant extract high in polyphenols are added to food items for increased health benefits and functionality. Polyphenols may be used in a variety of edible coatings, such as food packing using polysaccharide structures. As illustrated in Table 22.2, different films and coatings were created by formulating various polysaccharides and polyphenols or extracts into them using compression molding, extrusion blowing, microfluidic spinning, or plate casting [29–34]. The coatings may be employed for active and/or innovative packing depending on the polyphenol content. The compositions, processing histories, and various polysaccharide and polyphenolic extract types have an impact on the film's physical and functional properties. While overloading the coatings with polyphenols might have undesirable effects, such

Table 22.2 Polysaccharide coatings incorporated with phenolic extracts.

Polysaccharide	Polyphenol	Benefits
Carboxymethylcellulose	*Allium tuberosum* extract	Reduced opacity, improved film thickness, and enhanced transmittance.
Chitosan	Thinned young apple polyphenols	Increasing the amount of polyphenol increased film density, expansion, solubility, and thickness (from 0.07 to 0.11 mm).
Starch (corn)	Mango peel powder	The coatings become darker and more orange and yellow. The film surface becomes more supple.
Tremella fuciformis	Roasted peanut skin extract	The films thickened, darkened and became redder, and their surface became rougher.
Konjac glucomannan	Grape peel extract	The extract and nanocrystals were evenly distributed across the films. The film's surface texture and transparency were both improved.

as the creation of aggregations within the polysaccharide matrix, the addition of the proper quantities of polyphenol extract leads to the production of films with superior barrier and mechanical performance. Different meals of both plant and animal sources were packaged using coatings and films with antioxidant and antibacterial properties, significantly extending the shelf life of these goods.

22.5 Polyphenols in Preservation of Meat and Aquatic Items

The majority of the high-quality protein required by the human body comes from aquatic and meat items, which are rich sources of lipids, proteins, and other micronutrients. Fresh meat and aquatic goods, however, are very susceptible to rotting during transportation, trade, manufacturing, and preservation, which results in sensory quality degradation, nutritional loss, shelf-life reduction, and even the development of possible toxicants [35, 36]. For the most part, cold chain storage, the inclusion of preservatives or antioxidants, modified environment packing, irradiation, sterilization, and other contemporary measures were used to solve these difficulties [37, 38]. The most efficient and practical method among them is the addition of preservatives. As a result, chemical preservatives like tert-butylhydroquinone (TBHQ), butylhydroxytoluene (BHT), butylhydroxyanisole (BHA), benzoic acid, sorbic acid, and its salts were widely used to prevent lipids and proteins from oxidizing, to prevent discoloration, and to inhibit the growth of harmful microorganisms. Nevertheless, extended usage of chemical preservatives may be harmful to human health. Much emphasis has been given to the discovery and use of natural stabilizers and antioxidants in response to the increasing request for nutritious and functional meat. Especially plant-derived polyphenols, which have the benefits of being non-toxic, biodegradable, and having specific nutrient content and health functions, are used in food preservation [39, 40].

Natural polyphenols can be used to protect food packaging from potential risks from physical and biochemical influences as well as to preserve or improve the sensory value, nutrient content, and quality of meat and aquatic products. They may also be employed as a cryoprotectant, a colorant, an antibacterial, and an antioxidant. Additionally, plant polyphenols enhance the color stability of red meat and specific fish [41–43]. Animals contain high amounts of water activity levels, lipid content, and protein content, and are easily oxidized to generate acidity, off flavors, and discoloration, which results in loss of edible and economic value [44]. The breakdown of the self-antioxidant defense system, oxygen, metal ions, light, heat, and other variables are the main causes of protein, lipid, and myoglobin oxidation, which is the second-largest cause of sensory and nutrient degradation after microbial infection.

The primary mechanism of lipid oxidation, particularly for fatty acids, is nonenzymatic or autoxidation. The primary byproduct of automatic lipid oxidation is peroxide, which is highly unstable and easily decomposes to alkoxy radicals. The leading causes of the off-odor and bad flavors of oxidized meat and aquatic goods are peroxides, which may also be additionally broken down into compounds with low molecular weights comprising alcohols, acids, dienes, and carbonyls (aldehydes and ketones). Protein oxidation frequently results in lipid oxidation, which can change the physical, chemical, and dietary characteristics of proteins like their dissolution, digestion, emulsification, bioavailability, and gelation. Lipid oxidation is caused by free radicals, peroxides, and their breakdown products. The primary heme protein, myoglobin, which is made up of globin and heme, is what gives meat its red color. The color of the meat depends on the condition of the heme group. The meat's color naturally changes from brilliant red to brown when myoglobin and oxymyoglobin organically oxidize into metmyoglobin.

Foods high in protein such as meat and aquatic goods are prone to spoiling, which is primarily characterized by protein breakdown. In order to preserve meat and aquatic goods, standard chemical antioxidants have been replaced or reduced with tea polyphenols (TP), the most extensively researched and utilized plant polyphenols. In comparison to the TP-free (distilled water) group, silver carp treated with a 0.2% TP solution and wrapped in plastic containers with polyvinyl dichloride film were able to considerably obstruct TVC, reduce TVB-N, pH, TBARS, and K value (nucleotide breakdown), while maintaining flavor characteristics for longer [45]. In the latter investigation, silver carp fillets were cut into slices and submerged in larger amounts (0.5% and 1%) of TP medium for 30 minutes at 4 °C. This experiment revealed that TP treatment also altered the microbiota of the fillets, altered the pattern and process of fish fillet spoilage, and decreased the development of Cadaverine, hypoxanthine, and putrescine [46].

The goal of Zhou et al. was to create films from curdlan and tea polyphenols and ascertain how their concentrations affected the films' mechanical qualities (tensile strength and extension at break) and water vapor permeability (WVP) [47]. The curdlan-based films' characteristics were enhanced by adding various ratios of tea polyphenols. The findings demonstrated that adding tea polyphenols, particularly at a content of 0.6%, which led to a 50% fall in elongation at break, was likely to severely reduce the tensile strength and elongation at breakage of films. In the meanwhile, the films' WVP and moisture levels both fell. Nevertheless, a low WVP may stop food from losing moisture. Investigated were other film characteristics, such as antioxidant effectiveness. According to the findings, tea polyphenols may enhance the film's antioxidant capacity. The shelf life of chilled meat was also prolonged by roughly 3–5 days thanks to the use of composite films in preservation. Some characteristics, such as the composite film's ability to resist water and its DPPH (1,1-diphenyl-2-picrylhydrazyl) capability to scavenge free radicals, were enhanced.

Al-Hijazeen et al. also investigated how tannic acid (2.5–10 mg/kg) improved the quality and had antioxidant effects on raw and cooked chicken breast meat while it was being stored in the refrigerator. Compared to BHA at 5 mg/kg, tannic acid at 5 and 10 mg/kg may effectively suppress lipid and protein oxidation, reduce the production of off-aroma volatiles, and preserve color stability in meat [48]. To test the antioxidant properties on chicken breast and thigh meats, they also continuously mixed tannic acid with oregano oil. According to the findings, the combination outperformed the individuals in lowering generation of TBARS, total carbonyls, and off-aroma volatiles, and it was a great alternative for the chemical antioxidant [49].

According to pH, electrical conductivity, K value, TVB-N, TBARS, complete plate count, and sensory assessment, the combined impact of super chilling and tea polyphenols on maintaining the quality of hairtails was evaluated by Luan et al. [50]. Hairtail blocks underwent three different treatments: freezing at 18 °C, super chilling at 3 °C, and super chilling at 3 °C after being submerged in a solution containing 6.0 g/L tea polyphenols, respectively. In the specimens that were super chilled at 3 °C, frozen at 18 °C, and super chilled at 3 °C with tea polyphenols, respectively, the TBARS value of hairtails rapidly rose from 0.41 mg/100 g to 1.83, 1.38, and 1.14 mg/100 g during storage of 30 days (p 0.05). While the stability of sensory scores dropped (p 0.05), the TVB-N, microbiological growth, electrical conductivity, and K values all showed an increasing tendency (p 0.05), similar to that seen with the TBARS results. All of these quality indicators, in combination with treated samples, have increased compared to super chilling samples. The findings of indicators including pH, TVB-N, and TBARS in combined treatment specimens were even better than those in frozen samples. These findings revealed that super chilling with tea polyphenols was a useful strategy for preventing lipid oxidation and microbiological development, delaying degradation, and preserving the natural freshness of hairtails.

22.6 Conclusion

Polyphenols are seen as a good substitute for artificial preservatives owing to their antioxidant and antimicrobial properties. Polyphenols are gaining popularity as a useful ingredient for enhancing the degradable food packaging film's antibacterial and antioxidant capabilities. Beyond their usage as additives, polyphenols play a crucial function in the preservation of meat and meat products. Aquatic and meat products that have been treated with polyphenols keep their quality and maintain freshness more efficiently. Furthermore, it is anticipated that the use of polyphenols as natural preservatives for food would outweigh the negative consequences of synthetic counterparts.

References

1 Martinengo, P., Arunachalam, K., and Shi, C. (2021). Polyphenolic antibacterials for food preservation: review, challenges, and current applications. *Food* 10 (10): 2469.

2 Hu, Q. and Luo, Y. (2016). Polyphenol-chitosan conjugates: Synthesis, characterization, and applications. *Carbohydrate Polymers* 151: 624–639.

3 Martillanes, S., Rocha-Pimienta, J., Cabrera-Bañegil, M. et al. (2017). Application of phenolic compounds for food preservation: food additive and active packaging. *Phenolic Compounds-Biological Activity* 3 (8): 39–58.

4 Zeng, T., Zhang, X., Guo, Y. et al. (2014). Enhanced catalytic application of Au@ polyphenol-metal nanocomposites synthesized by a facile and green method. *Journal of Materials Chemistry A* 2 (36): 14807–14811.

5 Kyriakoudi, A. and Mourtzinos, I. (2022). Green extraction technology of polyphenols from food by-products. *MDPI* 1109.

6 Milinčić, D.D., Popović, D.A., Lević, S.M. et al. (2019). Application of polyphenol-loaded nanoparticles in food industry. *Nanomaterials* 9 (11): 1629.

7 Deng, J., Yang, H., Capanoglu, E. et al. (2018). Technological aspects and stability of polyphenols. In: *Polyphenols: Properties, Recovery, and Applications* (ed. C.M. Galanakis), 295–323. Elsevier.

8 Nazzaro, F., Fratianni, F., De Feo, V. et al. (2022). *Polyphenols applications in food industry sector* (ed. M.A. Saraiva and E.M. da Cruz Alexandre), 301–336. Technologies to Recover Polyphenols from AgroFood By-products and Wastes: Elsevier.

9 Wang, S., Du, Q., Meng, X., and Zhang, Y. (2022). Natural polyphenols: a potential prevention and treatment strategy for metabolic syndrome. *Food & Function* 13 (19): 9734–9753.

10 Panda, L. and Duarte-Sierra, A. (2022). Recent advancements in enhancing antimicrobial activity of plant-derived polyphenols by biochemical means. *Horticulturae* 8 (5): 401.

11 Mitra, S., Tareq, A.M., Das, R. et al. (2022). Polyphenols: a first evidence in the synergism and bioactivities. *Food Reviews International* 1–23.

12 Wang, L., Pan, X., Jiang, L. et al. (2022). The biological activity mechanism of chlorogenic acid and its applications in food industry: a review. *Frontiers in Nutrition* 9: 943911.

13 Deng, S., Shi, S., and Xia, X. (2022). Effect of plant polyphenols on the physicochemical properties, residual nitrites, and N-nitrosamine formation in dry-fried bacon. *Meat Science* 191: 108872.

14 Brglez Mojzer, E., Knez Hrnčič, M., Škerget, M. et al. (2016). Polyphenols: extraction methods, antioxidative action, bioavailability and anticarcinogenic effects. *Molecules* 21 (7): 901.

15 Jansson, T., Rauh, V., Danielsen, B.P. et al. (2017). Green tea polyphenols decrease strecker aldehydes and bind to proteins in lactose-hydrolyzed UHT milk. *Journal of Agricultural and Food Chemistry* 65 (48): 10550–10561.
16 Wu, C.D., Huang, E., Li, W. et al. (2021). Beverages containing plant-derived polyphenols inhibit growth and biofilm formation of streptococcus mutans and children's supragingival plaque bacteria. *Beverages* 7 (3): 43.
17 Sun-Waterhouse, D., Zhou, J., and Wadhwa, S.S. (2012). Effects of adding apple polyphenols before and after fermentation on the properties of drinking yoghurt. *Food and Bioprocess Technology* 5 (7): 2674–2686.
18 Sun-Waterhouse, D., Zhou, J., and Wadhwa, S.S. (2013). Drinking yoghurts with berry polyphenols added before and after fermentation. *Food Control* 32 (2): 450–460.
19 Ahmad, I., Khalique, A., Shahid, M.Q. et al. (2020). Studying the influence of apple peel polyphenol extract fortification on the characteristics of probiotic yoghurt. *Plants* 9 (1): 77.
20 Difonzo, G., Antonino, C., Squeo, G. et al. (2023). Application of agri-food by-products in cheesemaking. *Antioxidants.* 12 (3): 660.
21 Felix da Silva, D., Matumoto-Pintro, P.T., Bazinet, L. et al. (2015). Effect of commercial grape extracts on the cheese-making properties of milk. *Journal of Dairy Science* 98 (3): 1552–1562.
22 Buljeta, I., Pichler, A., Šimunović, J., and Kopjar, M. (2023). Beneficial effects of red wine polyphenols on human health: comprehensive review. *Current Issues in Molecular Biology* 45 (2): 782–798.
23 Zhang, J., Zou, X., Zhai, X. et al. (2019). Preparation of an intelligent pH film based on biodegradable polymers and roselle anthocyanins for monitoring pork freshness. *Food Chemistry* 272: 306–312.
24 Belizón, M., Fernández-Ponce, M.T., Casas, L. et al. (2018). Supercritical impregnation of antioxidant mango polyphenols into a multilayer PET/PP food-grade film. *Journal of CO_2 Utilization* 25: 56–67.
25 Ye, J., Wang, S., Lan, W. et al. (2018). Preparation and properties of polylactic acid-tea polyphenol-chitosan composite membranes. *International Journal of Biological Macromolecules* 117: 632–639.
26 Liu, F., Antoniou, J., Li, Y. et al. (2015). Preparation of gelatin films incorporated with tea polyphenol nanoparticles for enhancing controlled-release antioxidant properties. *Journal of Agricultural and Food Chemistry* 63 (15): 3987–3995.
27 Khan, M.R., Sadiq, M.B., and Mehmood, Z. (2020). Development of edible gelatin composite films enriched with polyphenol loaded nanoemulsions as chicken meat packaging material. *CyTA Journal of Food* 18 (1): 137–146.
28 Yilmaz, A., Bozkurt, F., Cicek, P.K. et al. (2016). A novel antifungal surface-coating application to limit postharvest decay on coated apples: molecular, thermal and morphological properties of electrospun zein–nanofiber mats loaded with curcumin. *Innovative Food Science & Emerging Technologies* 37: 74–83.

29 Riaz, A., Lagnika, C., Luo, H. et al. (2020). Effect of Chinese chives (*Allium tuberosum*) addition to carboxymethyl cellulose based food packaging films. *Carbohydrate Polymers* 235: 115944.

30 Sun, L., Sun, J., Chen, L. et al. (2017). Preparation and characterization of chitosan film incorporated with thinned young apple polyphenols as an active packaging material. *Carbohydrate Polymers* 163: 81–91.

31 Sun, L., Sun, J., Liu, D. et al. (2018). The preservative effects of chitosan film incorporated with thinned young apple polyphenols on the quality of grass carp (*Ctenopharyngodon idellus*) fillets during cold storage: correlation between the preservative effects and the active properties of the film. *Food Packaging and Shelf Life* 17: 1–10.

32 Rojas-Bravo, M., Rojas-Zenteno, E., Hernández-Carranza, P. et al. (2019). A potential application of mango (*Mangifera indica* L. cv Manila) peel powder to increase the total phenolic compounds and antioxidant capacity of edible films and coatings. *Food and Bioprocess Technology* 12: 1584–1592.

33 Ju, A. and Song, K.B. (2020). Active biodegradable films based on water soluble polysaccharides from white jelly mushroom (Tremella fuciformis) containing roasted peanut skin extract. *LWT* 126: 109293.

34 Tong, C., Wu, Z., Sun, J. et al. (2020). Effect of carboxylation cellulose nanocrystal and grape peel extracts on the physical, mechanical and antioxidant properties of konjac glucomannan films. *International Journal of Biological Macromolecules* 156: 874–884.

35 Xie, W., Huang, Y., Xiang, Y. et al. (2020). Insights into the binding mechanism of polyphenols and fish myofibrillar proteins explored using multi-spectroscopic methods. *Food and Bioprocess Technology* 13: 797–806.

36 Wu, J., Li, C., Li, L. et al. (2022). Improved physicochemical properties and product characteristics of tilapia surimi by tea polyphenols during chilled storage. *LWT* 167: 113822.

37 Huang, Z., Wang, Q., Cao, J. et al. (2023). Mechanisms of polyphenols on quality control of aquatic products in storage: a review. *Critical Reviews in Food Science and Nutrition* 1–20.

38 Nowak, A., Czyzowska, A., Efenberger, M., and Krala, L. (2016). Polyphenolic extracts of cherry (*Prunus cerasus* L.) and blackcurrant (*Ribes nigrum* L.) leaves as natural preservatives in meat products. *Food Microbiology* 59: 142–149.

39 Dong, L., Zhu, J., Li, X., and Li, J. (2013). Effect of tea polyphenols on the physical and chemical characteristics of dried-seasoned squid (*Dosidicus gigas*) during storage. *Food Control* 31 (2): 586–592.

40 Maqsood, S., Benjakul, S., and Shahidi, F. (2013). Emerging role of phenolic compounds as natural food additives in fish and fish products. *Critical Reviews in Food Science and Nutrition* 53 (2): 162–179.

41 Zhu, W., Han, M., Bu, Y. et al. (2022). Plant polyphenols regulating myoglobin oxidation and color stability in red meat and certain fish: a review. *Critical Reviews in Food Science and Nutrition* 1–13.

42 Li, Y., Lei, Y., Tan, Y. et al. (2022). Efficacy of freeze-chilled storage combined with tea polyphenol for controlling melanosis, quality deterioration, and spoilage bacterial growth of Pacific white shrimp (*Litopenaeus vannamei*). *Food Chemistry* 370: 130924.

43 Lan, W., Liu, J., Wang, M., and Xie, J. (2021). Effects of apple polyphenols and chitosan-based coatings on quality and shelf life of large yellow croaker (*Pseudosciaena crocea*) as determined by low field nuclear magnetic resonance and fluorescence spectroscopy. *Journal of Food Safety* 41 (3): e12887.

44 Ramanathan, L. and Das, N. (1992). Studies on the control of lipid oxidation in ground fish by some polyphenolic natural products. *Journal of Agricultural and Food Chemistry* 40 (1): 17–21.

45 Fan, W., Chi, Y., and Zhang, S. (2008). The use of a tea polyphenol dip to extend the shelf life of silver carp (*Hypophthalmicthys molitrix*) during storage in ice. *Food Chemistry* 108 (1): 148–153.

46 Jia, S., Huang, Z., Lei, Y. et al. (2018). Application of Illumina-MiSeq high throughput sequencing and culture-dependent techniques for the identification of microbiota of silver carp (*Hypophthalmichthys molitrix*) treated by tea polyphenols. *Food Microbiology* 76: 52–61.

47 Zhou, Y., Xu, T., Zhang, Y. et al. (2019). Effect of tea polyphenols on curdlan/chitosan blending film properties and its application to chilled meat preservation. *Coatings* 9 (4): 262.

48 Al-Hijazeen, M., Lee, E.J., Mendonca, A., and Ahn, D.U. (2016). Effects of tannic acid on lipid and protein oxidation, color, and volatiles of raw and cooked chicken breast meat during storage. *Antioxidants* 5 (2): 19.

49 Al-Hijazeen, M., Mendonca, A., Lee, E.J., and Ahn, D.U. (2018). Effect of oregano oil and tannic acid combinations on the quality and sensory characteristics of cooked chicken meat. *Poultry Science* 97 (2): 676–683.

50 Luan, L., Fu, S., Yuan, C. et al. (2017). Combined effect of superchilling and tea polyphenols on the preservation quality of hairtail (*Trichiurus haumela*). *International Journal of Food Properties* 20 (sup1): S992–S1001.

23

Biological Significance of Polyphenols as Functional Molecules in Biomaterial Preparations

Kalaiselvi Vinayagamurthy[1,2], Deena Praveena Kumar[1,2], Karthika Yalavarthi[1,2], Kamini Numbi Ramudu[1,2], Jayasri Ravichandran[3], Nimish Mol Stephen[3], and Ganesan Ponesakki[1,2]

[1] *Department of Biochemistry and Biotechnology, CSIR-Central Leather Research Institute (CLRI), Adyar, Chennai, Tamil Nadu, India*
[2] *Academy of Scientific and Innovative Research (AcSIR), Ghaziabad, India*
[3] *Department of Fish Processing Technology, Dr. M.G.R. Fisheries College and Research Institute, Tamil Nadu Dr. J. Jayalalithaa Fisheries University, Ponneri, Tamil Nadu, India*

23.1 Introduction

Polyphenols are a class of ubiquitous chemical compounds that are biosynthesized naturally from plant and marine sources as secondary metabolites with a varied, bioactive class of plant secondary metabolites that are an important component of the human diet and are of significant importance because of their biological characteristics [1]. It consists of multiple phenol units (C_6H_6O) containing catechol and pyrogallol, which allow for its strong non-covalent and covalent interactions with plenty of molecules/ions [2]. Approximately 8000 distinct types of polyphenols have been identified to date. Based on their chemical structures, they are classified into flavonoids such as flavones, isoflavones, neoflavonoids, chalcones, flavonols, anthocyanidins, and proanthocyanidins and non-flavonoids, such as phenolic acids (hydroxybenzoic acids and hydroxycinnamic acids), stilbenoids, lignans, and tannins (Figure 23.1). It has been established that polyphenols contain to have in inherent biological properties and health-promoting activity like antioxidant, antibacterial, antiallergic, antitumor, immune regulation, antidiabetic, antiradiation, antiviral, and various other aspects [3]. Meanwhile, polyphenols have a high safety profile in the human body, so they have been widely utilized in medicine,

Science and Engineering of Polyphenols: Fundamentals and Industrial Scale Applications,
First Edition. Edited by Chandrabhan Verma.
© 2024 John Wiley & Sons Inc. Published 2024 by John Wiley & Sons Inc.

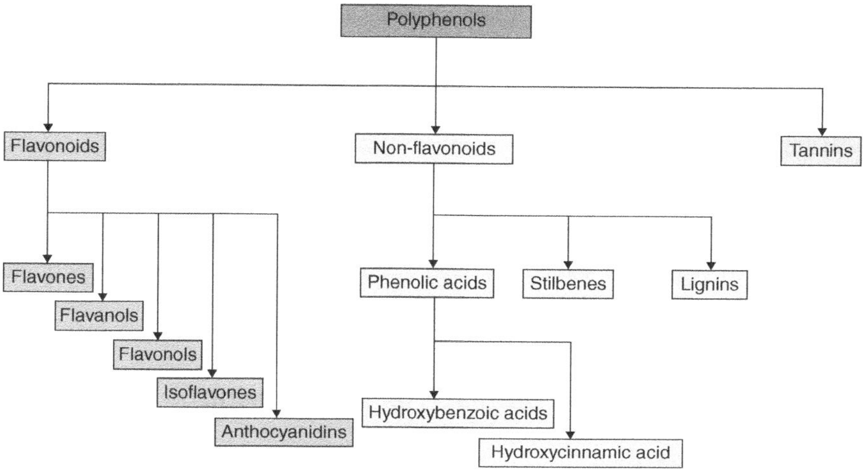

Figure 23.1 Classification of polyphenols based on chemical structure.

cosmetics, and food applications [4]. Rapid advances have been witnessed in exploiting the therapeutic efficiency of polyphenol conjugates for various diseases and treatment for human health through material preparation ([5]). With these health benefits, polyphenols in biomaterials engineering are ever-increasing attention mainly due to their unique structural and chemical features with biodegradable, biocompatible, and capable of assembly through many mechanisms, which are widely distributed and abundant in nature.

23.2 Functional Properties of Polyphenols for Biomaterial Engineering

Polyphenols are gaining scientific attention due to their potential beneficial impact and play a vital function in human health with regard to various chronic diseases including, cardiovascular disease, cancer, chronic inflammation, and many degenerative disorders [6, 7]. Recent research has reported that many of these diseases are linked to oxidative stress caused by reactive oxygen species (ROS) and nitrogen species. Phytochemicals, particularly polyphenols, contribute significantly to the total antioxidant activity, rather than vitamin C's inherent amphiphilicity, which results from the combination of antioxidants [8]. So far, polyphenols have primarily been recognized and used as micronutrients acquired through food or a dietary supplement. Numerous attempts have been made throughout the last three decades that have resulted in substantial advancements in the development of innovative technology and ways to solve intractable medical problems and improve human health care. Biomaterials have been a key component of this advancement. Polyphenols have been researched for their involvement in producing new biomaterials, regulating biomaterial functions, and as new agents in biomaterials, processing for various medicinal and biomedical applications, including tissue engineering and drug delivery [9].

Recent biomaterial development advances are centered on polyphenols in two basic categories: nanomaterials and hydrogels. In the simple, economical, and environmentally friendly synthesis of materials with minimal use of hazardous chemicals, polyphenols serve as valuable building blocks to create functional nanomaterials, films, scaffolds, and gels.

23.3 Polyphenols in Biomaterial Engineering

23.3.1 Nanomaterials

Nanomaterials are materials with sizes in the range of 1–100 nm with unique physical and chemical properties. Even with their excellent biomedical uses, there are serious health concerns with their aftereffects and to overcome this, a new type of nanomaterials with great biocompatibility and good performance is needed. In recent years, polyphenol-based nanoparticles have attracted a substantial amount of research attention due to their bioadhesive, biocompatible, antioxidant property, anticancer activity, and antibacterial properties. Thus, these materials with their significant potential can be used in the preparation, stabilization, and modification of multifunctional nano-assemblies for various biomedical applications [10]. Özen and Wang [11] summarized the developments of nanofibrous structures for hormonal therapies, wound dressings, transdermal patches, and tissue scaffolds. Different types of nanomaterials like nanoparticles, nanotubes, nanofibers, and nanowires have been developed using polyphenols for various biomedical applications such as cancer therapy, cardiovascular diseases, tissue engineering, neurodegenerative diseases, and bone regeneration. Methods for preparing these nanomaterials include encapsulation, complex coacervation, in-situ polymerization, interfacial polymerization, inclusion complexion, liposomes, spray coating, and extrusion [12]. As polyphenols have the ability to interact with plenty of molecules owing to their chemical properties, it is advantageous in producing diverse hybrid polyphenol-containing nanomaterials [10].

23.3.2 Hydrogels

Hydrogel is becoming increasingly important in material science and biomedicine because of its high research value as an essential component for creating the transition from conventional to smart hydrogel systems due to their flexibility in fabrication, good biocompatibility, and desirable physiochemical characteristics. Hydrogel with its three-dimensional polymer network has the ability to mimic the physicochemical properties of native extracellular matrix (ECM). In addition to their ability to absorb water or biological fluids, their intrinsic biocompatibility and hydrophilicity enable them to provide biomimetic environments to cells and proteins *in vitro*, allowing for biomolecule encapsulation and cell attachment, proliferation, and migration [13]. Furthermore, hydrogels have been widely exploited as bioactive drug carriers capable of entrapping and delivering a variety of therapeutic agents such as small molecules, nanoparticles, proteins, enzymes, nucleic acids, and others [14]. Nonetheless, they have significant limitations, such as restricted mechanical properties and fast drug release. Thus, by introducing polyphenols with multiple phenolic moieties that could provide various covalent and non-covalent interactions with diverse biomacromolecules and bioactive components, the hydrogels have significantly improved their structural and chemical properties by enhancing the biological activities of materials that have been recognized from crosslinking structures to derived properties. Their applications give hydrogel systems strong adhesion, antioxidation, antibacterial capabilities, and self-healing abilities that can endure the challenging environment *in vivo*, establishing great vitality in biomedicine that is needed for modern intelligent biomedical hydrogels [13]. Hydrogel with polyphenols is being explored as it offers a promising platform for biomedical applications due to its biocompatibility, bioactivity, and tunable properties.

23.4 The Biological Significance of Polyphenols as Biomaterials

23.4.1 Flavonoid-Based Biomaterial

Flavonoids are widely distributed naturally occurring polyphenolic compounds present in almost all parts of higher plants. It can be further divided into flavones, flavanols, isoflavones, anthocyanidin, anthocyanin, and other flavonoids (Figure 23.2). The health-promoting properties such as antioxidant, anti-inflammatory, and anticancerous properties of flavonoids have advantageous effects on human health. Research has suggested that these beneficial effects may help to reduce the risk of several chronic diseases including cardiovascular diseases, cancer, and neurodegenerative diseases, and aid in tissue engineering and bone regeneration. Epigallocatechin-3-gallate (EGCG), tannins, genistein, quercetin, chrysin, delphinidin, naringenin, butein, luteolin, and taxifolin are some examples of flavonoids effective against several chronic diseases. In spite of its beneficial effects, flavonoids are poorly soluble in water, have rapid metabolism, and have low bioavailability, meaning that they are prevented from being effectively absorbed and distributed

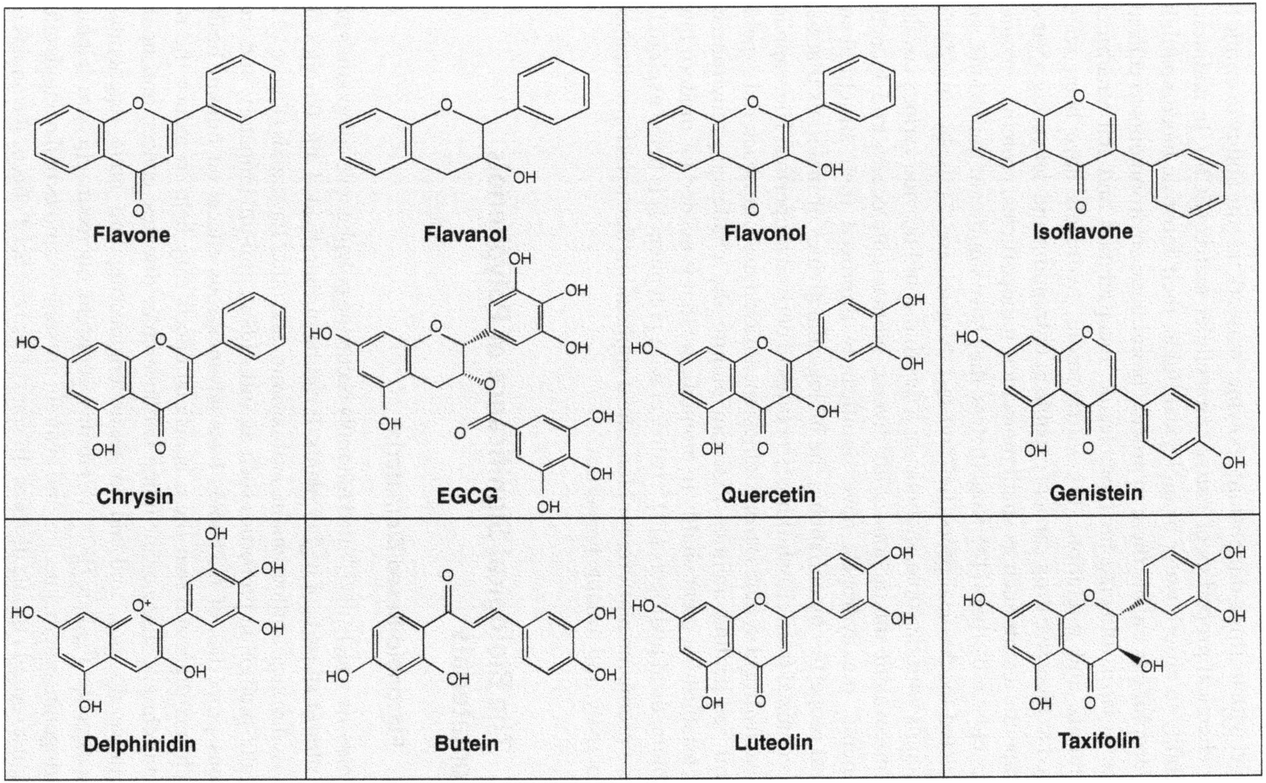

Figure 23.2 Chemical structures of flavonoids with examples.

throughout the body. Flavonoid-based biomaterials exhibit improved bioavailability and solubility [15]. There are several forms of flavonoid-based biomaterials such as hydrogel, polymeric nanoparticles, nanotubes, solid lipid nanocarriers, nanofibers, nanocapsules, and metal polyphenol nanoformulations applied in biomedical applications.

23.4.1.1 Flavones: Chrysin-Based Biomaterial

Chrysin (5,7-dihydroxyflavone), a flavonoid found in a variety of natural products and plants. It is a component of honey, passion fruit, Indian trumpet flower, and European propolis. Chrysin is composed of two benzene rings and one heterocyclic ring containing oxygen. It has 2–3 double-bound carbons and a carbonyl group linked to the fourth carbon, but no 3-carbon hydroxyl group [16]. It has been extensively investigated for its wide range of biological properties like anti-inflammatory, pro-apoptotic, antioxidant, antiviral, and immunomodulatory potential with pharmacological effects in preventing coronary heart disease, stroke, neuro, renal, and cancer [17]. Further, it also acts as an effective stimulator for the differentiation of keratinocytes by accelerating the natural wound-healing process. It effectively inhibits the expression of the key pro-inflammatory enzymes, including inducible nitric oxide synthase and cyclooxygenase-2, and suggested that it can protect neurons from oxidative stress and apoptosis [18]. Despite its diverse biological applications, chrysin faces significant challenges in clinical application due to its weak solubility and low biological stability [19]. So, on the other hand, it can be explored in the preparation of biomaterial. Whereas, in hydrogel, it acts as a hydrophobic crosslinker in a temperature-responsive hydrogel system rather than being investigated as a biological agent. Still, it also has the potential to promote osteogenesis with controlled release and bioavailability in conjunction with other polymeric scaffolds as a stem cell-inducing factor in bone tissue engineering and also upregulated the expression of osteogenic proteins and induced osteogenic differentiation of dental pulp stem cells [20, 21]. To exert their maximum beneficial activity, it is encapsulated as nanoparticle to enhance the efficacy of chemotherapeutic agents, which significantly showed potent anticancer activities *in vitro* that cause the induction of apoptosis and a decrease in the colony formation ability of breast cancer cells [22]. Additionally, as a natural immunomodulatory agent, incorporation with polymeric nanoparticles may provide a novel platform to create smart multifunctional structures for application in regenerative medicine purposes. One of the best ways to obtain a therapeutic drug delivery platform in order to treat oral recurrent aphthous ulcers and increase the bioavailability of chrysin is to entrap it in liposomal or mucoadhesive films. Because of its hydrophobic nature, it can be loaded as solid lipid-based nanoparticles using a novel sensing mechanism for delivering anticancer components that inhibit pancreatic cancer cells by triggering apoptosis and decreasing angiogenesis [23].

23.4.1.2 Flavanols: Tea Polyphenols-Based Biomaterial

EGCG ((2R,3R)-3′,4′,5,5′,7-Pentahydroxyflavan-3-yl 3,4,5-trihydroxybenzoate) is a major polyphenolic molecule present in green tea. Antioxidant properties and therapeutic applications of EGCG have gained extreme attention in the biomedical field. EGCG-based hydrogels and nanomaterials have been developed for numerous applications such as immuno-oncotherapy [24], anti-osteosarcoma [25], drug delivery [26], and several other cancer therapies. EGCG-based metal-polyphenol nanoformulations have also been developed with copper as a metal ion to alleviate chondrocyte inflammation by scavenging excessive intracellular ROS and modulating synovial macrophages [27]. In another study, Liu et al. [28] constructed EGCG nanocapsules incorporated with copper, which was further engineered as a controlled-release antibacterial composite hydrogel system that promotes wound angiogenesis and inhibits inflammation. Reducing properties of tea polyphenols are used in fabricating metallic nanoparticles like silver and gold nanoparticles and non-metallic nanoparticles like single-walled carbon nanotubes (SWNTs). In light of this, Wakayama et al. [29] developed silver nanoparticles using EGCG as a reducing agent for suppressing tumor accumulation, and Dong et al. [30] fabricated gold nanoparticles for periodontitis and bone regeneration. Chen et al. [31] dispersed SWNTs in water using EGCG and developed SWNT/EGCG nanocomposite that exhibited sensitivity towards hydrogen peroxide. Studies show that EGCG inhibits the expression of smooth muscle cell actin by inhibiting TGF-β—Smad2/3 pathway and thus in turn ameliorates pulmonary hypertension. In a study by Haddad et al. [32], an inhalable EGCG nano-liposome was constructed and optimized using high-phase transition phospholipids as a potential treatment for pulmonary arterial hypertension. EGCG-laden hyaluronic acid hydrogel depot fabricated by He et al. [33] reduced oxidative stress and inflammation, which are the common pathological mechanisms for the onset and progression of tissue regeneration. ROS-scavenging EGCG hydrogel acts as a protective cell carrier of bone marrow-derived mesenchymal cells (BMSCs) from ROS-mediated death and osteogenesis inhibition. From further *in vivo* investigations, BMSCs encapsulated EGCG hydrogel promotes osteoporosis by osteointegration of 3D-printed microporous titanium alloy prosthesis [34].

23.4.1.3 Flavonols: Quercetin-Based Biomaterial

Quercetin (2-(3,4-dihydroxyphenyl)-3,5,7-trihydroxy-4Hchromen-4-one) is a flavonoid compound of plant metabolite that is majorly found in vegetables and fruits. A potent flavonoid possesses substantial antioxidative, antiviral, antithrombin, anti-inflammatory properties and prevents cancer and chronic infections, increases endurance, improves athletic performance, and also helps in inhibiting the activity of certain enzymes such as acetylcholinesterase and cyclo-oxygenase. For the treatment of cancer, heart illness, and UV-induced damage, it has been widely employed as a therapeutic drug molecule. Its therapeutic potential and effectiveness are decreased by poor solubility due to the polar hydroxyl groups, environmental factors like light, temperature, pH instability, and short biological half-life with disintegration during local administration inside the host body [35]. Thus, to improve the bioavailability, drug delivery systems for quercetin cutaneous transport have been developed using nanocarrier systems such as liposomes, self-polymerization of quercetin, mannosylated liposomes, nanogels, deformable liposomes, microemulsions, thermo-responsive mesoporous silica, polymeric nanoparticles, and hydrogels with biocompatibility, controlled drug release with targeted drug delivery are feasible. Hence, quercetin is effectively loaded in various polymeric systems and its impact on the therapeutic potential is investigated. The most promising methods include a hydrogel containing quercetin, which increases bioavailability and can work directly on the skin's surface to give it adhesiveness and moisture properties. By encouraging cartilage development, it can also lower the risk of oxidative stress-related chronic diseases including osteoarthritis and rheumatoid arthritis. It may also help with subchondral bone restoration by providing enough oxygen and nutrients for cell proliferation [36]. These systems can support the immune system while a wound is healing and shield the skin from oxidation, inflammation, and photoaging. While the nanoemulsion method successfully increased the bio-accessibility of the drug, coaxial electrospinning techniques were employed to create microparticles infused with quercetin to improve the solubility and infusion qualities for oral or sublingual administration [37]. Quercetin-loaded nanoliposomes are used to research malignant tissue's apoptotic cell death at low concentrations by improving its solubility and increasing its bioactivity in preventing tumor growth [38]. Through several cell signaling pathways, quercetin nanoparticles were used to accomplish the beneficial effects on liver cancer suppression. Quercetin nanoparticles exhibited the potential to decrease liver cancer cell proliferation and induce apoptosis. The inhibitory impact of quercetin nanoparticles in liver cancer was strengthened [39].

23.4.1.4 Isoflavones: Genistein-Based Biomaterial

A natural polyphenolic isoflavone compound, genistein (4′,5,7-Trihydroxyisoflavone) is a phytoestrogen that mimics the effects of estrogen in the body. It is found in plant-based sources such as soy products, chickpeas, and other legumes. Along with its broad range of pharmacological properties like tyrosine and topoisomerase inhibition, it also inhibits cancer cell proliferation inhibition and induces apoptosis in cancer cells [40]. To improve the efficiency of genistein, genistein-based biomaterials like hydrogels, scaffolds, and nanomaterials have been constructed for diverse applications like cancer therapy [22], postmenopausal symptoms [41], intestinal toxicity [42], and drug delivery systems [43]. As reported in a study, carbon dots synthesized from genistein exhibited enhanced biocompatibility, low cytotoxicity and showed a potential against inflammatory conditions [44]. The anti-inflammatory and antioxidant activities of genistein are also used to treat local respiratory diseases. Given this, He et al. [45] designed lipid-coated genistein nanocrystals for pulmonary delivery to reduce systemic absorption and macrophage clearance in local respiratory diseases. Encapsulated genistein with chicory polysaccharide increased the stability of genistein twice that of unencapsulated genistein, which is a critically important determinant of chemopreventive effects in cancer therapy [46]. Wang et al. [47] developed a mesoporous bioactive glass loaded with genistein, enhanced the osteoblast differentiation and matrix mineralization for the treatment of bone defect as it has anti-osteoporotic activity. Genistein with its estrogen-like properties was selected as a bioactive molecule and encapsulated in 1,4-butanediol diglycidyl ether cross-linked chitosan-thioglycolic acid mucoadhesive hydrogel for long-term treatment to ease vascular atrophy [48].

23.4.1.5 Anthocyanidins and Anthocyanins-Based Biomaterial

Anthocyanidins and anthocyanins are plant-derived water-soluble flavonoids mostly rich in fruits and vegetables. These plant pigments are characterized as unsaturated and unoxidized flavonoid derivatives. Anthocyanins have a flavylium cation (C6–C3–C6) as its fundamental structure, which can be connected to different sugars, hydroxyls, and methoxyl groups. Nature has so far identified over 700 structurally different anthocyanins and 27 recognized anthocyanidin compounds. Cyanidins, pelargonidins, delphinidins, petunidins, peonidins, and malvidins are the most frequent anthocyanidins, accounting for more than 90% of the known varieties. Whereas anthocyanins are glycosides of anthocyanidins that give many plants their red, purple, and blue colors [49, 50]. These colored pigments are red in acidic solutions and blue in alkaline solutions. These flavonoids have recently acquired popularity in studies due to their exceptional antioxidant, anticancer, eyesight protecting, and anti-aging characteristics. Furthermore, they have a variety of effects on platelets, blood vessels, and lipoproteins that can lower the risk of coronary heart disease. Irrespective of their properties, the primary challenge posed by these polyphenolic compounds is their molecular instability and easily degradable property when exposed to pH, heat, and light, which necessitates the need to fabricate them in the form of biomaterials [51, 52]. Hanafy [53] reported that anthocyanins can get absorbed through the small intestine and into blood completely in their active form when encapsulated and incorporated into hydrogel made using corn starch and albumin. As reported in a study by Lotfinia et al. [54], the antibacterial, antioxidant, and pH sensitivity properties of anthocyanin meet the requirements of wound dressing when loaded along with honey in alginate hydrogel. Anthocyanins nanocomposite developed using silver as the metallic ion reduction has protective effects on hepatotoxicity and oxidative stress [55]. The gastro-protective effect of encapsulated anthocyanin-loaded chitosan nanoparticles (ACNPs) was evaluated by Perumcherry Raman et al. [56] who reported that ACNPs were able to protect and alleviate gastric ulceration.

As for absorption and stability in the simulated gastrointestinal fluid, ACNPs perform better than free anthocyanins [57]. Similar to this, biodegradable nanoparticle formulations, such as PLGA with PEG nanoparticles in SH-SY5Y cell lines [58], and PEG-AuNPs in *in vivo* and *in vitro* Alzheimer disease models [59] have been used to increase the bioavailability and biostability of anthocyanins due to their hydrophilic unstable nature, improving the neurotherapeutic potential of anthocyanins. According to Ali et al. [60], anthocyanin-loaded polyethylene glycol-gold nanoparticles had a stronger anti-neurotoxic impact on Alzheimer's patients.

23.4.1.6 Other Flavonoids

Some of the other flavonoids present in nature are taxifolin, butein, daidzein, naringenin, and luteolin. These flavonoids are more bioavailable, stable, and have longer half-lives in the body when they are in biomaterial forms. Members of the flavonoid family, such as taxifolin ((2R,3R)-3,3′,4′,5,7-Pentahydroxyflavan-4-one), have been found to have antioxidant, anti-inflammatory, hepatoprotective, anti-Alzheimer's disease, and anti-angiogenic characteristics. Taxifolin's anticancer capabilities have recently received attention. The key components of taxifolin's anticancer mechanism are the inhibition of angiogenesis, cytochrome P450 enzymes, P-glycoprotein, ROS, and cell cycle regulators, as well as the stimulation of apoptosis. Because of these different activities in a single molecule, taxifolin is being evaluated as a potential therapeutic drug or adjuvant in cancer therapy. Taxifolin inhibits lung cancer viability, stemness, and EMT both *in vitro* and *in vivo* [47, 61]. Taxifolin-loaded zinc oxide nanoparticles are effective antiproliferative agents in MCF-7 cells [62]. It acts as a potential drug for osteoporosis by inducing osteogenic differentiation in human bone marrow-derived mesenchymal stem cells when adsorbed by a gelatin nanoparticle coated with polystyrene sodium sulfonate/polyacrylamide hydrochloride [63]. Electrospun nanocomposite membrane constructed using polyvinyl alcohol (PVA) and chitosan mixed with taxifolin accelerates chronic diabetic wound healing [34]. To enhance the gastroprotective property of taxifolin by increasing the residence time and release in the stomach, engineered gastro mucoadhesive microparticles composed of Syloid® AL-1 mesoporous silica, and chitosan hydroxypropylmethylcellulose incorporated with taxifolin were used [64].

Luteolin (3′,4′,5,7-Tetrahydroxyflavone), a naturally occurring flavone co-loaded with TGF-β1, and co-delivered to enhance cellular uptake of nucleus pulposus cells and prolonged release of drugs in intravertebral discs for the treatment of intravertebral disc degeneration [65]. Miao et al. [66] formulated luteolin-loaded methoxy poly (ethylene glycol)-poly(lactide) micelles as an effective therapy for lung infectious diseases. Luteolin micelles substantially increased bacterial elimination and reduced inflammatory infiltration in the lung infectious model.

A basic chalcone derivate, butein (I-1-(2′,4′-dihydroxyphenyl)- 3-(3,4-dihydroxy phenyl) prop-2-en-1-one), exhibits anti-fibrotic properties. In order to address the poor solubility of butein, it was encapsulated inside vitamin A-modified solid-lipid nanoparticle (SLN) [67]. Similarly, Md et al. [68] developed a smart oral pH-sensitive dual-layer nano-hydrogel loaded with naringenin, a natural anticancer polyphenol for colorectal cancer. In another study, naringenin was loaded into gelatin microspheres, which were further encapsulated in the nano-hydroxyapatite/silk fibroin scaffolds for bone tissue engineering by exhibiting larger osteogenic differentiation [69].

Daidzein (7-hydroxy-3-(4-hydroxyphenyl)-4H-chromen-4-one), a significant soy isoflavones was formulated into nanoemulsion and nanoemulsion-based gels due to water-insoluble for topical application against melanoma [70]. Similarly, Nemitz et al. [71] fabricated an isoflavone aglycone-rich (expressed as sum of genistein, daidzein, and glycitein) nanoemulsions and semisolid hydrogels composed of hyaluronic acid for dermal applications. Daidzein could lower low-density lipoprotein (LDL) cholesterol, inhibit pro-inflammatory cytokines, cell adhesion proteins, and platelet aggregation, increase nitric oxide production, and improve vascular reactivity by using a SLN as an injectable delivery system for daidzein, and loading SLNs with daidzein for the treatment of cardio-cerebrovascular disease [72]. The flavanoid-based biomaterials and their biological significance are presented in Table 23.1.

23.4.2 Tannic Acid-Based Biomaterial

Tannins, which are polyphenolic phytochemicals found in plants, are considered safe and environmentally friendly. They are classified as condensed tannins and hydrolysable tannins. As a polyphenol, it possesses a broad spectrum of health-beneficial properties, including anticancer, antioxidant, anti-inflammatory, antiviral, antimicrobial, and homeostatic effects, making them potentially useful in addressing various medical conditions. Among various types of tannins, tannic acid (TA) is being investigated as an organic molecule that can improve the biofunctionalities of host materials [82]. TA (penta-m-digalloyl glucose) classified as a gallotannin is a plant-derived polyphenol molecule with a high molecular weight that protects plants. Depending on the plant source, TA has a central glucose core with five digalloyl ester groups covalently bonded to the center or numerous gallate moieties esterified to a core phenol moiety per molecule ranging from 2 to 12 [83] (Figure 23.3).

The use of TA as a component of biomaterial has spread significantly in biomedical applications ranging from wound healing to wearable sensors utilizing a wide range of natural and synthetic polymers. The potential benefits of using TA as a mediator agent and biochemical additive for tissue engineering and regenerative medicine are beginning to emerge, as well as in the future of biomaterials engineering, particularly with regard to biomedical hydrogels with multiple functions. The range of TA-based biomaterials is fabricated, which include films,

Table 23.1 Flavanoid-based biomaterials and their biological significance.

Type	Compound name	Biomaterial form/preparation technique	Application/functionality	Reference
Flavones	Chrysin	Chrysin-loaded chitosan-folic acid-coated solid-lipid nanoparticles—by hot homogenization and probe-sonication method	Against pancreatic cancer/ antioxidant effects	[23]
		Chrysin-loaded chitosan/carboxymethyl cellulose/hydroxyapatite scaffolds—using simple ionic gelation	Bone tissue engineering/ osteogenesis	[21]
		Polyethylene glycol conjugate chrysin nanoparticle—by two-step reaction	Against human breast cancer/ anticancer activity	[22]
Flavanols	Epigallocatechin gallate	Epigallocatechin gallate-loaded chitosan-sodium alginate nanoparticles	Male infertility/antihypertensive	[73]
		Detachable H_2O_2-responsive gel-based microneedle patches containing methotrexate and epigallocatechin gallate as cross-linkers by two-step micromolding method	Psoriasis management/ anti-oxidative	[74]
Flavonols	Quercetin	Quercetin-loaded polyvinylalcohol and Eudragit nanoparticles by nanoprecipitation technique	Drug delivery/antioxidant activities and anti-lipid peroxidation	[75]
		Gold-coated quercetin conjugated poly (DL-lactide-co-glycolide) nanoparticles	Inhibit liver cancer cell growth/ apoptosis induction	[39]
		Quercetin-loaded liposomes incorporated in gelatin and carbopol hydrogel	Prevent connective tissue disorder/ accelerated wound healing	[76]
		Thermosensitive chitosan/collagen-quercetin hydrogels	Bone tissue engineering/ROS scavenging activity	[77]

(*Continued*)

Table 23.1 (Continued)

Type	Compound name	Biomaterial form/preparation technique	Application/functionality	Reference
Isoflavone	Genistein	Lipid-coated genistein nanocrystals by ball wet milling method	Pulmonary drug delivery/anti-inflammatory and antioxidant	[78]
		Genistein zein chicory polysaccharide nanoparticles	Cancer therapy/increases stability and anticancer activity	[46]
		Genistein carbon dots nanoparticle	Drug delivery vehicle/antioxidant and anti-inflammatory	[44]
Anthocyanidins and anthocyanins		Red cabbage anthocyanins and honey incorporated into sodium alginate hydrogel film	Wound dressing/antioxidative and antibacterial	[54]
		Anthocyanins loaded poly (lactide-*co*-glycolide) and polyethylene glycol nanoparticles by emulsification-solvent evaporation technique	Neuroprotective effect/antioxidant, anti-apoptotic, anti-inflammatory, and anti-Alzheimer's effects	[58]
Other flavonoids	Taxifolin	Fabrication of flavonoid (taxifolin and quercetin) modified surfaces	Osteopromotive/anti-inflammatory and anti-fibrotic potential	[79]
		Taxifolin-loaded zinc oxide nanoparticles	Breast cancer therapy/anticancer activity	[62]
	Butein	Chitosan conjugated butein graft polymerization	Biosensing/better immobilization	[80]
	Daidzein	Daidzein loaded solid-lipid nanoparticles with polyethylene glycol by hot homogenization technique	Cardio-cerebrovascular disease/drug delivery and inhibit platelet aggregation	[72]
	Naringenin and baicalein	Naringenin and baicalein in polymeric nano-micelles	Pharmaceutical applications/antioxidant activity	[81]

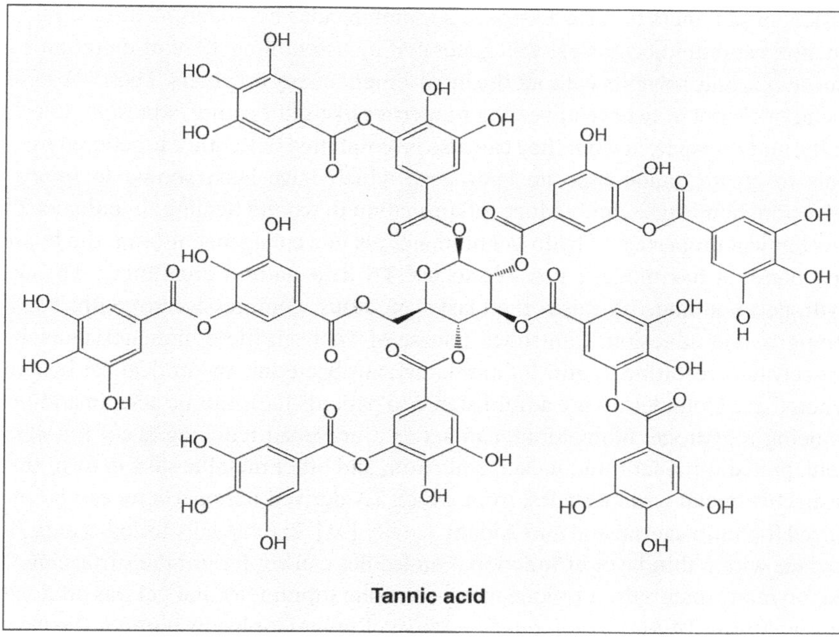

Figure 23.3 Chemical structure of tannic acid.

nanoparticles, and hydrogel materials. Due to interactions between blood and the phenolic hydroxyl groups, TA functions as a hemostatic glue, causing blood to clot quickly [84]. TA is an amphiphilic molecule that can interact with hydrophilic and hydrophobic materials by hydrogen bonding, electrostatic, coordinative bonding, and hydrophobic interactions [85, 86]. This is an essential criterion for hydrogel fabrication with the most prevalent crosslinking approach, making it one of the research areas of polyphenol-based hydrogels for several years. And further, hydrogel's mechanical and self-healing properties are improved with the incorporation of metal ions and boronate that made them more advantageous [87, 88]. Because of its biocompatibility nature, and with outstanding mechanical capabilities, rod-shaped cellulose nanocrystals are commonly employed as nanofillers to improve the mechanical performance of these TA-based hydrogel [89]. Besides that, it also possesses valuable qualities such as antioxidant, antibacterial, and biodegradability for producing new innovative biomaterials. An organohydrogel was created by using hydrogen bonds between PVA and TA, and the chelation of TA and Al_3. This organohydrogel has a superior capacity to sustain high conductivity and mechanical qualities. Aside from its good visibility in visible light, the TA-based organohydrogel possesses a UV-filtering feature that can protect skin from ultraviolet radiation, making it more suitable for use as artificial skin. TA can react with a variety of polymers to form TA-based supramolecular hydrogels. A unique metal-phenolic supramolecular gel was generated by assembling TA and metal ions in various organic solvents without the involvement of any polymers. These TA-based metallogels not only have appealing properties like self-healing, adhesion, injectability, and transparency but they can also be combined with other functional materials to create nanocomposite hydrogels, which have been shown to improve adhesion, hemostasis, and reduce inflammation in wound healing. To enhance the mechanical properties of hydrogel biomaterials like collagen, chitosan, the hydrogen bonding has made it possible to use TA as a natural crosslinker. Through hydrogen bonding, TA raises the elastic modulus, compressive strength, tensile strength, and adhesion. Controlled release of TA from these biomaterials is also conceivable, resulting in anti-inflammatory, antibacterial, and anticancer benefits when these biomaterials are administered to patients [90]. Tannic acid, in addition to being a hydrogel biomaterial, can act as a universal reducing agent for silver, gold, palladium, platinum, nickel, zinc, iron, and other metallic salts to form their respective metal nanoparticles, from which TA-derived nanoparticles can be produced for antibacterial and antioxidant activity [91]. The capacity to coat a material surface with a thin layer of functional molecules can confer unique surface qualities on many materials. A unique metal-phenolic supramolecular gel was produced by combining TA and metal ions in a range of organic solvents without the usage of polymers. On a variety of surfaces, TA spontaneously adsorbs to produce hydrophilic coatings. TA quickly creates monolayers on polycrystalline gold, silica, and silver. Because of TA's capacity to chelate proteins and metals, it has been utilized to encapsulate hydrophobic medicines for better drug delivery applications as well as to stabilize nanoparticles made of gold, other metals, and polymers [92]. TA can be incredibly important in the assembly of nanoparticles because of its capacity to adsorb and bind to a variety of surfaces and molecules. TA can potentially interfere with proteins necessary for bacterial and viral function at specific doses and membrane proteins in some malignancies, which can result in cell-specific apoptosis [93]. TA adsorbs in a monolayer on numerous surfaces, and when paired with its capacity to bind proteins and chelate metal ions, it is possible to build multifunctional thin film coatings. Overall, TA has a staggering array of prospective applications that could make it important in the field of biomedical and biomaterial research. Tannin-based biomaterials and their biological significance are presented in Table 23.2.

23.4.3 Lignin-Based Biomaterial

Lignin is widely known for being an affordable and sustainable organic resource in the plant's xylem for its rigidity [97]. It possesses a polyphenol-like structure (Figure 23.4) with phenolic hydroxyl and methoxyl groups, containing many aromatic rings. The connections produced between the phenylpropanoid lignin

Table 23.2 Tannin-based biomaterials and their biological significance.

Type	Compound name	Biomaterial form/ preparation technique	Application/functionality	Reference
Tannins	Tannic acid	Tannic acid particles by water-in-oil microemulsion system	Potential in biomedical application/antioxidant, antibacterial and fluorescent properties	[94]
		Tannic acid-based conducting polymer hydrogel	Spinal cord injury repair/ neurogenesis	[88]
		Tannic acid-paclitaxel nanoparticles by nanoformulation	Breast cancer therapy/ anticancer activity	[95]
		Sodium alginate and tannic acid hydrogels	Wound dressings/ antibacterial effect	[96]

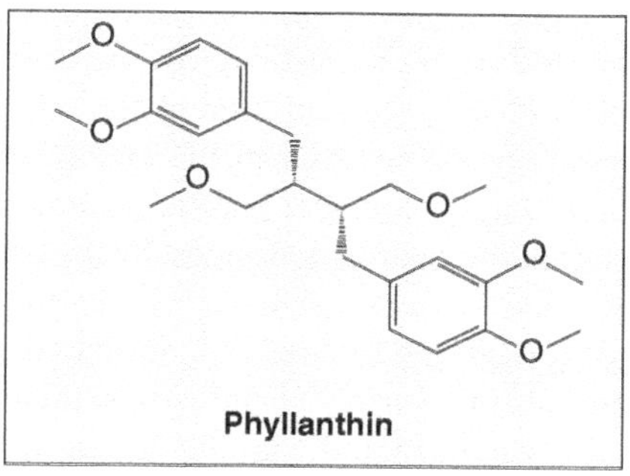

Figure 23.4 Chemical structure of lignin.

precursors determine the variations in lignin structures in different plants. P-coumaryl, coniferyl, and sinapyl are the three aromatic alcohols. Still, it is vital to note its antioxidant, non-toxicity, antifungal, and antibacterial properties when considering its potential in biological applications. Early research that uses nanotechnology in a wide range of domains confirms the potential of lignin-derived composites to be used as scaffolding and also by various nano-oriented approaches to optimize the technological features of pharmaceuticals in traditional biomedical applications. Self-assembly of enzymatic hydrolysis of lignin nanospheres was achieved by adding a non-solvent to the lignin solution. This group of molecules is used to prepare spheres with good stability after 30 days of storage in the 190–590 nm range. Despite indicating a probable carrier function, no trials on drug release or biocompatibility were conducted in this work [98]. Colloidal lignin spheres are also being researched as cosmetic sunscreen components. Sunscreen lotions containing lignin spheres of roughly 50 nm were developed and attained a sun protection factor (SPF) of 15 for UVA radiation, with the size and extraction process of the lignin spheres influencing the SPF [99]. Lignin has been demonstrated to be a viable resource for biodegradable, bioactive materials, either in colloidal form as a "green" alternative for metallic nanoparticles in pharmaceutical drug delivery or in agricultural formulation [100]. In addition to spheres, lignin-based films were synthesized and tested for biomedical applications. Thin films of a lignin-polypropylene blend were discovered to exhibit enhanced antioxidant properties against thermos-oxidative degradation. Furthermore, due to lignin's scavenging properties, the lignin as a stabilizer has been proven to protect mouse DNA from oxidative damage [101]. Thin films were also created utilizing polyethylene and lignin nanoparticles. The antibacterial activity of lignin was shown to be comparable to that of other bactericides such as bronopol or chlorohexidine against *E. coli* and *S. aureus*.

Furthermore, the inclusion of lignin had no effect on the mechanical properties of the films [102]. Lignin has been reported to impact hydrogel mechanical properties with high tensile strength. Particularly, the crosslinking copolymerization, crosslinking of grafted lignin and monomers, and permeative polymer network and polymerization are involved in the fabrication of hydrogels from lignin. Also, it is found to have a compact porous surface with a minimum percentage of lignin content in the hydrogel. Besides, by the availability, biocompatibility, biodegradability, and low cost of lignin, the hydrogels can be prepared with a potential for large-scale production. Lignin-based hydrogels have been widely used as biomaterials in a variety of industries, including those that treat water for industry and agriculture and can also act as an absorbent for heavy metal ions, controlled release agents for controlled delivery and water retention, smart materials for stimuli response, and biosensors and electrodes [103]. Furthermore, electrospinning provides a variety of techniques for immobilizing bioactive compounds, such as chemical or physical conjugation, initial mixing, emulsion electrospinning, and coaxial electrospinning to build a nanofibrous network. This allows for the incorporation of a wide range of therapeutic compounds, regardless of miscibility, into the fibrous mesh [104]. In spite of these benefits, electrospinning pure lignin is challenging due to its chemical makeup, molecular structure, and its low molecular weight. As a result, it has been combined with polymers to form the nanofibrous structure. Thus, adding lignin to nanofiber mixed with polycaprolactone (PCL) confers outstanding antioxidant characteristics; improves the survivability, proliferation, and differentiating the Schwann cells; and eventually facilitates nerve tissue creation [105]. Furthermore,

Table 23.3 Lignin-based biomaterials and their biological significance.

Type	Compound name	Biomaterial form/preparation technique	Application/ functionality	Reference
Lignin	Lignin	Lignin was grafted with polylactic acid nanofibrous scaffolds by electrospinning.	Cartilage tissue engineering/ antioxidant activity	[106]
		Lignin-incorporated nanogel by one-step method	Wound dressing/ antioxidant	[107]
		Polycaprolactone coated chitin lignin gel-based core–shell fibers by coaxial electrospinning.	Wound dressing and healing/drug delivery	[104]
		Lignin-polycaprolactone nanofiber by solvent-free ring-opening polymerization	Nerve regeneration/ antioxidant	[105]

the efficacy of PCL-coated chitin-lignin nanofibrous scaffolds in drug delivery-based wound dressing was revealed. To prevent medicines from being released directly within the chitin-lignin electrospun scaffolds, they adopted a coaxial electrospinning method to wrap the chitin-lignin core layer with a hydrophobic PCL outer layer [104]. Furthermore, electrospinning method is associated with certain limitations like high voltage requirement. In this context, combination of electrospinning with 3D bio-printing could develop nanofibers with specific characteristic features for various applications. Lignin-based biomaterials and their biological significance are presented in Table 23.3

23.4.4 Stilbene-Based Biomaterial

Stilbene is a non-flavanoid natural polyphenol (Figure 23.5) commonly found in grapes, sugarcane, tomato, bilberry, and cranberry [108]. Stilbene synthase, an important enzyme, plays a major role in the synthesis of stilbene, which is not universally expressed. These are produced by plants in stress conditions [109]. Among various types of stilbenes, resveratrol is considered to be one of the important stilbenes with its excellent bioactivities.

23.4.4.1 Resveratrol

Resveratrol, a natural stilbene polyphenol synthesized by the plant to fight against stress and microbes [110]. Resveratrol has a pair of phenolic rings linked by a double bond, which aids in generating cis and trans isomers (Figure 23.6). The

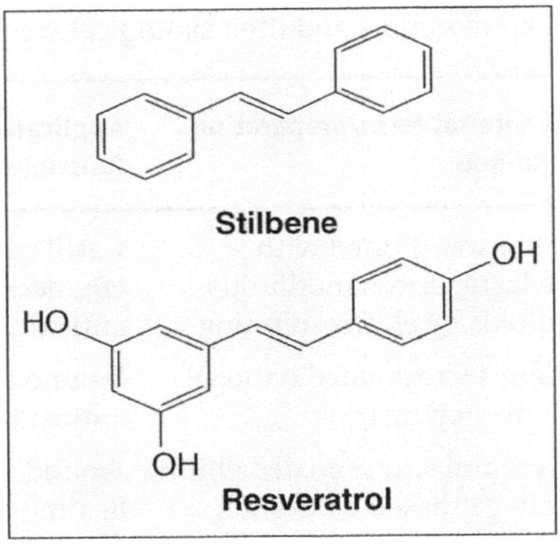

Figure 23.5 Chemical structure of stilbene with example.

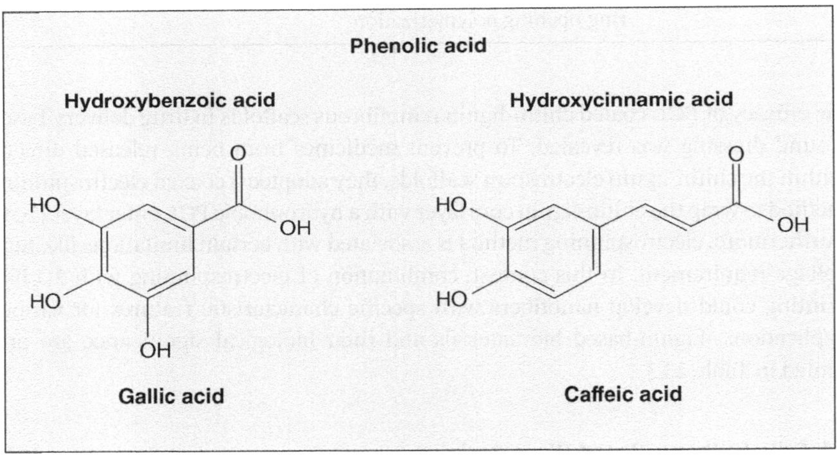

Figure 23.6 Chemical structures of phenolic acids with examples.

trans isomer is more stable compared to the cis isomer if it is protected from light [111]. It is found in grapes, wine, peanuts, and berries, legumes, blueberries [112, 113]. It has various biological activities, such as antioxidants, helps in scavenging ROS in stress condition, anticancer [114], anti-inflammatory by inhibiting NF-κB signaling, antibacterial [115], antiviral, antifungal [113], neuroprotective, anti-aging, cardioprotective and promotes apoptosis in cancer cells. Though it has many advantages, there are some limitations such as photosensitivity, low water soluble, low chemical stability, less bioavailability, and absorption that limit their application in biomedical field [112, 116]. These limitations can be overcome by fabricating resveratrol into biomaterial forms like of nanoparticles, nanofibers, hydrogels. Nanoformulations can be in the form of liposomes, polymeric nanoparticles, SLNs, alginate nanoparticles, mesoporous silica nanoparticles, and co-loaded nanoparticles [112, 116]. The tiny size of 100–200 nm and the high surface area to volume ratio of nanoparticle helps in targeted drug delivery [117], which helps in sustained drug release, and prevents fast metabolism [116]. Curcumin and resveratrol-loaded injectable hydrogel was found to show anticancer effects against hepatocellular carcinoma [116]]. Nanoparticles prepared with resveratrol and Au-Se reported to exhibit growth inhibitory effects in breast, lung, prostate and skin cancer [28, 117]. Electrospinning produces nanofibers, a simple, scalable, and reproducible system by using natural and synthetic polymers. Higher surface-to-volume ratio and small diameter of the nanofiber help in sustained drug delivery [118, 119]. A previous study reported that the use of resveratrol-loaded nanofibers helped in treating periodontal disease [120], in modulating osteoblasts and osteoclast [118], for efficient drug delivery to the gastrointestinal system [121], enhancing osteogenic and antioxidant activity [122]. Hydrogels contain a hydrophilic 3D polymer network that helps in biocompatible, water-holding capacity, and sustained drug release [123]. Resveratrol-loaded hydrogels have various applications, such as tissue engineering of cartilage [124], reducing scar formation [125], antibacterial activity in wound healing [123], and treating burn wounds [126]. Stilbene-based biomaterials and their biological significance are presented in Table 23.4.

23.4.5 Phenolic Acid-Based Biomaterials

Phenolic acids are naturally occurring bioactive compounds commonly found in plant sources such as fruits, vegetables, and seeds [127]. The shikimic acid pathway produces phenolic acid starting with two aromatic amino acids, L-tyrosine and L-phenyl alanine [128]. Hydroxybenzoic and hydroxycinnamic acids are the two classes of phenolic acids [127]. Phenolic acids have various

Table 23.4 Stilbene-based biomaterials and their biological significance.

Type	Compound name	Biomaterial form/preparation technique	Application/ functionality	Reference
Stilbenes	Resveratrol	Resveratrol, Au-Se, gold nanoparticles	Breast cancer therapy/ anticancer activity	[28]
		Resveratrol-loaded Fmoc-FFGGRGD peptide hydrogels	Wound dressing/ anti-inflammation	[125]
		Resveratrol-loaded polycaprolactone nanofibers by electrospinning	Treatment of periodontal disease/ anti-inflammation	[120]

applications, such as antioxidants, anti-inflammatory, anticancer, antiviral, and antimicrobial activities.

23.4.5.1 Hydroxybenzoic Acid-Based Biomaterials

Hydroxybenzoic acid and its derivatives contain a standard C6–C1 seven-carbon framework derived from benzoic acid, available in conjugated form with sugars or organic acids [129]. Hydroxybenzoic acid is more available in potato skin, onion, and red fruits [127]. Gallic acid, vanillic acid, syringic acid, p-hydroxybenzoic acid, and protocatechuic acids are derivatives of hydroxybenzoic acids [130]

23.4.5.1.1 Gallic Acid

Gallic acid is a natural endogenous polyphenol compound known as 3,4,5-trihy droxybenzoic acid, which is found in tea leaves, citrus fruits, berries, wine, and herbs [131, 132]. It has the anticancer, neuroprotective, antihyperglycemic, anti-inflammatory, antimicrobial, and antioxidant ability to scavenge ROS such as hydroxyl, peroxide, and superoxide radicals. Though gallic acid has many advantages, its less bioavailability, fast elimination, and low absorption are the limitations. These limitations are overcome with the help of drug delivery systems such as hydrogels, nanoparticles, and nanofibers. A study reported that gallic acid-loaded nanofibers have sustained release and help treat wounds [133]. Gallic acid-loaded nanoparticles help in treating hepatic and breast cancers [134]. Gallic acid-loaded hydrogels show antibacterial activity and help in treating wounds [135, 136]. Diammonium glycyrrhizinate and gallic acid co-assembled hydrogels helped to cure *S. aureus*-infected wounds [136]. Similarly, antibacterial nanocomposite was constructed using chitosan, copper, and gallic acid as a multifunctional wound dressing in tissue engineering applications. Li et al. [137] and Sun et al. [138] developed a hybrid hydrogel composed of poly (citrate-gallic acid) and encapsulated stem cells from human exfoliated deciduous teeth-derived exosomes (SHED-Exo) for treating traumatic brain injury. Gallic acid-coated copper oxide nanoparticles are used to treat pancreatic cancer [139].

23.4.5.2 Hydroxycinnamic Acid-Based Biomaterials

Hydroxycinnamic acid contains nine carbon C6–C3 structures identified by the hydroxyl group on the benzene ring and the carboxyl group in the lateral chain [140]. It occurs in the form of conjugated forms esterified with quinic and tartaric acid and rarely occurs in free forms [130, 140]. They are commonly found in fruits, seeds, and vegetables [141]. Sinapic acid, ferulic acid, p-coumaric acid, and caffeic acid are the most common derivatives of hydroxycinnamic acid [129].

23.4.5.2.1 Caffeic Acid

Caffeic acid belongs to polyphenol commonly found in fruits, coffee, wine. It has antioxidant, anti-inflammatory, anticancer, anticarcinogenic, antibacterial, anti-amylodoigenic activities [142, 143]. Though it has many beneficial activities, the limitations such as low solubility, stability, bioavailability restrict its use [142]. The use of nanoparticle-based drug delivery systems helps to overcome the limitations. Studies reported a high loading capacity, a surface area containing graphene oxide and caffeic acid nanoparticles that act as a chemotherapeutic agent [143], and caffeic acid-loaded liposomes in Alzheimer's therapy [142], caffeic acid and chitosan nanotubes in treating breast cancer [144]. Hydrogels have properties such as good swelling behavior, porosity, and biodegradability. Injectable hydrogel conjugated with caffeic acid and gelatin is used for transplantation [145], in repairing wounds [146], caffeic acid bioconjugated hydrogel transplants replace damaged cells, initiate axonal remyelination, and recover the damaged tissue [145]. Caffeic acid-loaded nanofibers have antibacterial activity, antitumor activity against human HeLa cervical cells [147]. Stilbene-based biomaterials and their biological significance are presented in Table 23.5.

23.5 Marine Polyphenol-Based Biomaterials

One of the most significant sources of diverse chemically active metabolites is marine organisms. Among others, marine polyphenols exhibit diverse biological significance like enzyme inhibitory effect, antioxidant and UV protective activity, anticancer, anti-HIV, anti-allergic, antimicrobial, antiviral, antidiabetic, and anti-inflammatory activities as a result of their structural complexity and uniqueness [148]. Bromophenolic compounds, simple phenolic acids, phlorotannins as well as flavonoids are commonly found phenolics in marine organisms. The bioactive properties of marine polyphenols manifest their potential in health-beneficial biological effects. However, polyphenols from marine sources are a relatively new field and possess various unique characteristics when compared to polyphenols from terrestrial sources [149]. Biomaterial forms of marine polyphenols enhance their solubility, stability, sensitivity to environmental factors, and availability in the human body. Phlorotannin, a proven therapeutic drug was encapsulated in a chitosan-tripolyphosphate carrier to exhibit enhanced antioxidant activity compared to non-encapsulated for targeted delivery [150]. Similarly, nano-encapsulated phlorotannin in polyvinylpyrrolidone showed improved antioxidant activity and sustained release in the gastrointestinal tract. According to the study conducted by Kim et al. [151], diphlorethohydroxycarmalol (DPHC), a phlorotannin isolated from *Ishige okamurae* brown algae has antibacterial effects against *Staphylococcus aureus* and *Pseudomonas aeruginosa* when loaded into PVA hydrogel. DPHC/PVA hydrogel has considerable potential for its application in wound dressing in order to assist in the healing process and provide antibacterial properties [151]. The scientific exploration of marine polyphenol-based biomaterials is still in its early stages. Limited research studies have been conducted thus far on marine polyphenol-based biomaterials for biomedical applications.

Table 23.5 Phenolic acid-based biomaterials and their biological significance.

Type	Compound name	Biomaterial form/preparation technique	Application/functionality	Reference
Hydroxybenzoic acid	Gallic acid	Gallic acid-loaded nanoparticles by freeze-drying technique	Hepatic and breast cancer therapy/anticancer activity	[134]
		Gallic acid and double agarose network hydrogel	Wound healing/anti-inflammation and antibacterial	[135]
		Gallic acid-loaded cellulose acetate nanofibers by electrospinning	Wound healing/antibacterial	[133]
Hydroxycinnamic acid	Caffeic acid	Caffeic acid-loaded chitosan carbon nanotubes	Breast cancer therapy/anticancer activity	[144]
		Injectable hydrogel conjugated with grafted caffeic acid/ε-polylysine and grafted phenyl boronic acid oxidized dextrin by Schiff base reaction	Diabetic wound healing/antibacterial, antioxidant, and anti-inflammation activity	[146]
		Caffeic acid-loaded poly(3-hydroxybutyrate) and poly (ethylene glycol) nanofiber by one-pot electrospinning	Wound dressing/antibacterial and cervical cancer therapy/anticancer activity	[147]

23.6 Conclusion

In conclusion, polyphenol-based biomaterial scaffolds have received a great deal of attention due to their numerous reactive groups and good functional features such as adhesiveness, antioxidant properties, antibacterial capabilities, and biocompatibility. They could be used as potential tools for various applications with appropriate design and fabrication methodologies. Despite several advances, these polyphenols are generally safe and effective and have a wide range of advantageous biological effects on both plants and human. According to the current findings, polyphenols have similarities in their molecular mechanism of action and are capable of reducing endogenous antioxidants, lowering cell proliferation and angiogenesis, changing transcription factors and their associated kinases, and inhibiting a variety of signaling cascades. Clinical-based research is required to establish polyphenols' concentration, bioavailability, effectiveness, safety, and mechanism of action. There are still many issues that must be resolved in the future in order to advance this crucial topic related to the biological activity of polyphenols in the human body as well as their absorption, distribution, metabolism, and excretion. The utilization of cutting-edge and contemporary techniques (such as nanotechnology, nanofiber, hydrogel biomaterial) to improve polyphenols bioavailability and stability in the human body is constantly expanding. The findings of functional particles have received more focus lately, yet hybrid assemblies and films may provide new avenues for the advancement of polyphenol-based biomaterials and the impressive body of literature on polyphenols shows the therapeutic benefits of these substances and suggests that polyphenols are easily oxidized in air due to the high chemical reactivity of phenolic moieties. Thus, investigating the sustainability over time and/or stability of polyphenol-based scaffolds could be crucial. Second, most polyphenols are still chosen blindly for scaffold design due to inaccurate regulation and an unknown theoretical basis for structure-function performances. The research on utilizing marine polyphenols as biomaterials is comparatively limited, and a comprehensive understanding of their potential applications and functionalities is still being developed. While there is growing interest in marine polyphenols as a promising source of biomaterials, further investigation and research efforts are needed to fully explore and harness their potential benefits. Although there has been a wide range of studies on polyphenols, it appears that current information about these compounds is at the lowest level imaginable, and further research with an emphasis on *in vivo* techniques is required to identify the health advantages of these compounds for humans and to develop a better understanding of these compounds. To determine the precise involvement of these polyphenols in the prevention of major human diseases, genomic and proteomic studies are required.

Acknowledgments

The corresponding author acknowledges the Indian Council of Medical Research, New Delhi (GAP-2010) for the financial support. The authors express gratitude to the Management of the respective organization for their support.

Conflict of Interest

The authors declare that there is no conflict of interest.

References

1 Rasouli, H., Farzaei, M.H., and Khodarahmi, R. (2017). Polyphenols and their benefits: a review. *International Journal of Food Properties 20* (sup2): 1700–1741.

2 Faure, E., Falentin-Daudré, C., Jérôme, C. et al. (2013). Catechols as versatile platforms in polymer chemistry. *Progress in Polymer Science 38* (1): 236–270.

3 Quideau, S., Deffieux, D., Douat-Casassus, C., and Pouységu, L. (2011). Plant polyphenols: chemical properties, biological activities, and synthesis. *Angewandte Chemie International Edition 50* (3): 586–621.

4 Rahim, M.A., Kristufek, S.L., Pan, S. et al. (2019). Phenolic building blocks for the assembly of functional materials. *Angewandte Chemie International Edition 58* (7): 1904–1927.

5 Wu, D., Zhou, J., Creyer, M.N. et al. (2021). Phenolic-enabled nanotechnology: versatile particle engineering for biomedicine. *Chemical Society Reviews 50* (7): 4432–4483.

6 Duthie, G.G. and Brown, K.M. (1994). Reducing the risk of cardiovascular disease. In: *Functional Foods: Designer Foods, Pharmafoods, Nutraceuticals* (ed. I. Goldberg), 19–38.

7 Goldberga, I., Li, R., and Duer, M.J. (2018). Collagen structure–function relationships from solid-state NMR spectroscopy. *Accounts of Chemical Research 51* (7): 1621–1629. https://doi.org/10.1021/acs.accounts.8b00092.

8 Wang, H., Cao, G., and Prior, R.L. (1996). Total antioxidant capacity of fruits. *Journal of Agricultural and Food Chemistry 44* (3): 701–705.

9 Shavandi, A., Bekhit, A.E.-D.A., Saeedi, P. et al. (2018). Polyphenol uses in biomaterials engineering. *Biomaterials 167*: 91–106. https://doi.org/10.1016/j.biomaterials.2018.03.018.

10 Guo, Y., Sun, Q., Wu, F.-G. et al. (2021). Polyphenol-containing nanoparticles: synthesis, properties, and therapeutic delivery. *Advanced Materials 33* (22): 2007356.

11 Özen, İ. and Wang, X. (2023). Biomedicine: electrospun nanofibrous hormonal therapies through skin/tissue—a review. *International Journal of Polymeric Materials and Polymeric Biomaterials 72* (1): 21–39.

12 Thuan, P.T., Van Thinh, P., Thao, P.T.T., and Minh, T.N. (2021). Study about nano polyphenol: biomaterial in functional food and its preparation. *Current Aspects in Pharmaceutical Research and Development* 112.

13 Lee, K.Y. and Mooney, D.J. (2001). Hydrogels for tissue engineering. *Chemical Reviews 101* (7): 1869–1880.

14 Kesharwani, P., Bisht, A., Alexander, A. et al. (2021). Biomedical applications of hydrogels in drug delivery system: an update. *Journal of Drug Delivery Science and Technology 66*: 102914.

15 Dobrzynska, M., Napierala, M., and Florek, E. (2020). Flavonoid nanoparticles: a promising approach for cancer therapy. *Biomolecules 10* (9): 1268.

16 Balta, C., Herman, H., Boldura, O.M. et al. (2015). Chrysin attenuates liver fibrosis and hepatic stellate cell activation through TGF-β/Smad signaling pathway. *Chemico-Biological Interactions 240*: 94–101.

17 Tang, S., Floy, M., Bhandari, R. et al. (2017). Development of novel N-isopropylacrylamide (NIPAAm) based hydrogels with varying content of chrysin multiacrylate. *Gels 3* (4): 40.

18 Aishwarya, V. and Sumathi, T. (2015). Chrysin, a natural flavonoid attenuates cognitive dysfunction and neuronal loss associated with amyloid β (25-35)-induced oxidative stress: an experimental model of Alzheimer's disease. *International Journal of Pharmacognosy and Phytochemical Research 7*: 224–236.

19 Maleki, S.J., Crespo, J.F., and Cabanillas, B. (2019). Anti-inflammatory effects of flavonoids. *Food Chemistry 299*: 125124.

20 Huo, J.F., Zhang, M.L., Wang, X.X., and Zou, D.H. (2021). Chrysin induces osteogenic differentiation of human dental pulp stem cells. *Experimental Cell Research 400* (2): 112466.

21 Menon, A.H., Soundarya, S.P., Sanjay, V. et al. (2018). Sustained release of chrysin from chitosan-based scaffolds promotes mesenchymal stem cell proliferation and osteoblast differentiation. *Carbohydrate Polymers 195*: 356–367.

22 Jangid, A.K., Solanki, R., Patel, S. et al. (2022). Improving anticancer activity of chrysin using tumor microenvironment pH-responsive and self-assembled nanoparticles. *ACS Omega 7* (18): 15919–15928. https://doi.org/10.1021/acsomega.2c01041.

23 Farhadi, A., Homayouni Tabrizi, M., Sadeghi, S. et al. (2023). Targeted delivery and anticancer effects of Chrysin-loaded chitosan-folic acid coated solid lipid nanoparticles in pancreatic malignant cells. *Journal of Biomaterials Science, Polymer Edition 34* (3): 315–333.

24 Jannatun, N., Zhang, Y., Wu, B. et al. (2023). Tea polyphenol coordinated with nanoparticles of ZIF-8 and coated with polydopamine and PEG for immuno-oncotherapy. *ACS Applied Nano Materials 6* (6): 4379–4389.

25 Ren, J., Sun, L., Xiao, C. et al. (2023). Chemical bonding of Epigallocatechin-3-gallate to the surface of nano-hydroxyapatite to enhance its biological activity for anti-osteosarcoma. *Smart Materials in Medicine 4*: 396–406.

26 Yang, R., Ma, J., Hu, J. et al. (2023). Formation of ferritin-agaro oligosaccharide-epigallocatechin gallate nanoparticle induced by CHAPS and partitioned by the ferritin shell with enhanced delivery efficiency. *Food Hydrocolloids 137*: 108396.

27 Wei, H., Qin, J., Huang, Q. et al. (2023). Epigallocatechin-3-gallate (EGCG) based metal-polyphenol nanoformulations alleviates chondrocytes inflammation by modulating synovial macrophages polarization. *Biomedicine & Pharmacotherapy 161*: 114366.

28 Liu, X., Liu, J., Xu, S. et al. (2023). Gold nanoparticles functionalized with Au–Se-bonded peptides used as gatekeepers for the off-target release of resveratrol in the treatment of triple-negative breast cancer. *ACS Applied Materials & Interfaces 15* (2): 2529–2537. https://doi.org/10.1021/acsami.2c10221.

29 Wakayama, C., Inubushi, S., Kunihisa, T. et al. (2023). Protein corona formation on epigallocatechin gallate-Au nanoparticles suppressed tumor accumulation. *JCIS Open* 100074.

30 Dong, Z., Lin, Y., Xu, S. et al. (2023). NIR-triggered tea polyphenol-modified gold nanoparticles-loaded hydrogel treats periodontitis by inhibiting bacteria and inducing bone regeneration. *Materials & Design 225*: 111487.

31 Chen, Y., Wang, L., Huang, S. et al. (2021). Lutein attenuates angiotensin II- induced cardiac remodeling by inhibiting AP-1/IL-11 signaling. *Redox Biology 44*: 102020. https://doi.org/10.1016/j.redox.2021.102020.

32 Haddad, F., Mohammed, N., Gopalan, R.C. et al. (2023). Development and optimisation of inhalable EGCG nano-liposomes as a potential treatment for pulmonary arterial hypertension by implementation of the design of experiments approach. *Pharmaceutics 15* (2): 539.

33 He, Z., Luo, H., Wang, Z. et al. (2023). Injectable and tissue adhesive EGCG-laden hyaluronic acid hydrogel depot for treating oxidative stress and inflammation. *Carbohydrate Polymers 299*: 120180.

34 Ding, Q., Ding, C., Liu, X. et al. (2023). Preparation of nanocomposite membranes loaded with taxifolin liposome and its mechanism of wound healing in diabetic mice. *International Journal of Biological Macromolecules* 124537.

35 Dhasmana, A., Malik, S., Ranjan, A. et al. (2022). A bioengineered quercetin-loaded 3D bio-polymeric graft for tissue regeneration and repair. *Biomedicine 10* (12): 3157.

36 Yu, W., Zhu, Y., Li, H., and He, Y. (2019). Injectable quercetin-loaded hydrogel with cartilage-protection and immunomodulatory properties for articular cartilage repair. *ACS Applied Bio Materials 3* (2): 761–771.

37 Li, C., Yu, D.-G., Williams, G.R., and Wang, Z.-H. (2014). Fast-dissolving core-shell composite microparticles of quercetin fabricated using a coaxial electrospray process. *PLoS One 9* (3): e92106.

38 Wang, G., Wang, J.J., Chen, X.L. et al. (2013). The JAK2/STAT3 and mitochondrial pathways are essential for quercetin nanoliposome-induced C6 glioma cell death. *Cell Death & Disease 4* (8): e746–e746.

39 Ren, K.-W., Li, Y.-H., Wu, G. et al. (2017). Quercetin nanoparticles display antitumor activity via proliferation inhibition and apoptosis induction in liver cancer cells. *International Journal of Oncology 50* (4): 1299–1311.

40 Jaiswal, N., Akhtar, J., Singh, S.P., and Ahsan, F. (2019). An overview on genistein and its various formulations. *Drug Research 69* (06): 305–313.

41 Na Takuathung, M., Klinjan, P., Sakuludomkan, W. et al. (2023). Efficacy and safety of the genistein nutraceutical product containing vitamin E, vitamin B3, and ceramide on skin health in postmenopausal women: a randomized, double-blind, placebo-controlled clinical trial. *Journal of Clinical Medicine 12* (4): 1326.

42 Fan, W., Liu, S., Wu, Y. et al. (2022). Genistein-based reactive oxygen species-responsive nanomaterial site-specifically relieves the intestinal toxicity of endocrine-disrupting chemicals. *International Journal of Pharmaceutics 615*: 121478.

43 Li, C., Chen, R., Xu, M. et al. (2018). Hyaluronic acid modified MPEG-b-PAE block copolymer aqueous micelles for efficient ophthalmic drug delivery of hydrophobic genistein. *Drug Delivery* 25 (1): 1258–1265.

44 Jaiswal, K.S., Malka, O., Shauloff, N. et al. (2023). Genistein carbon dots exhibit antioxidant and anti-inflammatory effects in vitro. *Colloids and Surfaces B: Biointerfaces 223*: 113173.

45 He, Y., Liu, C., Han, R. et al. (2023). Reducing systemic absorption and macrophages clearance of genistein by lipid-coated nanocrystals for pulmonary delivery. *Chinese Chemical Letters 34* (1): 107484.

46 Shi, L., Lin, Z., Hou, J. et al. (2023). Purification and characterization of chicory polysaccharide and its application in stabling genistein for cancer therapy. *International Journal of Biological Macromolecules* 124635.

47 Wang, X., Chen, W., Liu, Q., and Liu, L. (2020). Genistein adsorbed mesoporous bioactive glass with enhanced osteogenesis properties. *Biotechnology Letters 42*: 321–328.

48 Yang, Y., Campbell Ritchie, A., and Everitt, N.M. (2021). Recombinant human collagen/chitosan-based soft hydrogels as biomaterials for soft tissue engineering. *Materials Science and Engineering: C 121*: 111846. https://doi.org/10.1016/j.msec.2020.111846.

49 Mottaghipisheh, J., Doustimotlagh, A.H., Irajie, C. et al. (2022). The promising therapeutic and preventive properties of anthocyanidins/anthocyanins on prostate cancer. *Cell 11* (7): 1070.

50 Wallace, T.C. (2011). Anthocyanins in cardiovascular disease. *Advances in Nutrition 2* (1): 1–7.

51 Rosales, T.K.O. and Fabi, J.P. (2022). Nanoencapsulated anthocyanin as a functional ingredient: technological application and future perspectives. *Colloids and Surfaces B: Biointerfaces* 112707.

52 Wallace, T.C. and Giusti, M.M. (2013). *Anthocyanins in Health and Disease*. CRC Press.

53 Hanafy, N.A. (2021). Starch based hydrogel NPs loaded by anthocyanins might treat glycogen storage at cardiomyopathy in animal fibrotic model. *International Journal of Biological Macromolecules 183*: 171–181.

54 Lotfinia, F., Norouzi, M.-R., Ghasemi-Mobarakeh, L., and Naeimirad, M. (2023). Anthocyanin/honey-incorporated alginate hydrogel as a bio-based pH-responsive/antibacterial/antioxidant wound dressing. *Journal of Functional Biomaterials 14* (2): 72.

55 Sheikha, M.A., Soheir, N.A., and SyragEldin, F.M. (2022). Synthesis, characterization and protection effect of black rice anthocyanins nano-composite against hepatotoxicity induced by methotrexate in rats. *Brazilian Journal of Biology 84*.

56 Perumcherry Raman, S., Dara, P.K., Vijayan, D.K. et al. (2022). Anti-ulcerogenic potential of anthocyanin-loaded chitosan nanoparticles against alcohol-HCl induced gastric ulcer in rats. *Natural Product Research 36* (5): 1306–1310.

57 He, B., Ge, J., Yue, P. et al. (2017). Loading of anthocyanins on chitosan nanoparticles influences anthocyanin degradation in gastrointestinal fluids and stability in a beverage. *Food Chemistry 221*: 1671–1677.

58 Amin, F.U., Shah, S.A., Badshah, H. et al. (2017). Anthocyanins encapsulated by PLGA@ PEG nanoparticles potentially improved its free radical scavenging capabilities via p38/JNK pathway against Aβ 1–42-induced oxidative stress. *Journal of Nanobiotechnology 15*: 1–16.

59 Kim, M.J., Rehman, S.U., Amin, F.U., and Kim, M.O. (2017). Enhanced neuroprotection of anthocyanin-loaded PEG-gold nanoparticles against Aβ1-42-induced neuroinflammation and neurodegeneration via the NF-KB/JNK/GSK3β signaling pathway. *Nanomedicine: Nanotechnology, Biology and Medicine 13* (8): 2533–2544.

60 Ali, T., Kim, M.J., Rehman, S.U. et al. (2017). Anthocyanin-loaded PEG-gold nanoparticles enhanced the neuroprotection of anthocyanins in an Aβ 1–42 mouse model of Alzheimer's disease. *Molecular Neurobiology 54*: 6490–6506.

61 Sunil, C. and Xu, B. (2019). An insight into the health-promoting effects of taxifolin (dihydroquercetin). *Phytochemistry 166*: 112066.

62 Sundraraman, G. and Jayakumari, L.S. (2020). Meticulous taxifolin releasing performance by the zinc oxide nanoparticles: as a short road to drug delivery system for cancer therapeutics. *Journal of Cluster Science 31*: 241–255.

63 Liu, Y. (2020). Preparation of taxifolin-loaded gelatin-polystyrene sodium sulfonate/polyacrylamide hydrochloride nanocomposite and its induction on bone marrow mesenchymal stem cells in vitro. *Nanoscience and Nanotechnology Letters 12* (6): 729–737.

64 Moura, F.S., Perioli, L., Pagano, C. et al. (2019). Chitosan composite microparticles: a promising gastroadhesive system for taxifolin. *Carbohydrate Polymers 218*: 343–354.

65 Ding, Y., Wang, H., Wang, Y. et al. (2022). Co-delivery of luteolin and TGF-β1 plasmids with ROS-responsive virus-inspired nanoparticles for microenvironment regulation and chemo-gene therapy of intervertebral disc degeneration. *Nano Research 15* (9): 8214–8227.

66 Miao, J., Lin, F., Huang, N., and Teng, Y. (2021). Improving anti-inflammatory effect of luteolin with nano-micelles in the bacteria-induced lung infection. *Journal of Biomedical Nanotechnology 17* (6): 1229–1241.

67 Mahdinloo, S., Hemmati, S., Valizadeh, H. et al. (2022). Synthesis and preparation of vitamin A coupled butein-loaded solid lipid nanoparticles for liver fibrosis therapy in rats. *International Journal of Pharmaceutics* 625: 122063.

68 Md, S., Abdullah, S., Awan, Z.A., and Alhakamy, N.A. (2022). Smart oral pH-responsive dual layer nano-hydrogel for dissolution enhancement and targeted delivery of naringenin using protein-polysaccharides complexation against colorectal cancer. *Journal of Pharmaceutical Sciences 111* (11): 3155–3164.

69 Yu, X., Shen, G., Shang, Q. et al. (2021). A naringin-loaded gelatin-microsphere/nano-hydroxyapatite/silk fibroin composite scaffold promoted healing of critical-size vertebral defects in ovariectomised rat. *International Journal of Biological Macromolecules 193*: 510–518.

70 Kaplan, A.B.U., Cetin, M., Orgul, D. et al. (2019). Formulation and in vitro evaluation of topical nanoemulsion and nanoemulsion-based gels containing daidzein. *Journal of Drug Delivery Science and Technology 52*: 189–203.

71 Nemitz, M.C., von Poser, G.L., and Teixeira, H.F. (2019). In vitro skin permeation/retention of daidzein, genistein and glycitein from a soybean isoflavone rich fraction-loaded nanoemulsions and derived hydrogels. *Journal of Drug Delivery Science and Technology 51*: 63–69.

72 Gao, Y., Gu, W., Chen, L. et al. (2008). The role of daidzein-loaded sterically stabilized solid lipid nanoparticles in therapy for cardio-cerebrovascular diseases. *Biomaterials 29* (30): 4129–4136.

73 El-Shahawy, A. and A. G., Khaled, E., Ali El-Nesr, K., Gamal, A., Mustafa, F. M., & Abo El-Ela, F. I. (2023). Resolving male infertility that is induced by β-and ca channel antagonist drugs: propranolol/verapamil using an optimized nanohybrid formula: experimental and computational studies. *European Polymer Journal 183*: 111653. https://doi.org/10.1016/j.eurpolymj.2022.111653.

74 Bi, D., Qu, F., Xiao, W. et al. (2023). Reactive oxygen species-responsive gel-based microneedle patches for prolonged and intelligent psoriasis management. *ACS Nano 17* (5): 4346–4357. https://doi.org/10.1021/acsnano.2c08979.

75 Wu, T.-H., Yen, F.-L., Lin, L.-T. et al. (2008). Preparation, physicochemical characterization, and antioxidant effects of quercetin nanoparticles. *International Journal of Pharmaceutics 346* (1): 160–168. https://doi.org/10.1016/j.ijpharm.2007.06.036.

76 Jangde, R., Srivastava, S., Singh, M.R., and Singh, D. (2018). In vitro and in vivo characterization of quercetin loaded multiphase hydrogel for wound healing application. *International Journal of Biological Macromolecules 115*: 1211–1217.

77 Arpornmaeklong, P., Sareethammanuwat, M., Apinyauppatham, K., and Boonyuen, S. (2021). Characteristics and biologic effects of thermosensitive quercetin-chitosan/collagen hydrogel on human periodontal ligament stem cells. *Journal of Biomedical Materials Research Part B: Applied Biomaterials 109* (10): 1656–1670.

78 Yang, Z., McClements, D.J., Peng, X. et al. (2023). Fabrication of zein–carboxymethyl cellulose nanoparticles for co-delivery of quercetin and resveratrol. *Journal of Food Engineering 341*: 111322. https://doi.org/10.1016/j.jfoodeng.2022.111322.

79 Cordoba, A., Satué, M., Gómez-Florit, M. et al. (2015). Flavonoid-modified surfaces: multifunctional bioactive biomaterials with osteopromotive, anti-inflammatory, and anti-fibrotic potential. *Advanced Healthcare Materials 4* (4): 540–549.

80 Krishnan, V., Pandey, G.R., Babu, K.A. et al. (2021). Chitosan grafted butein: a metal-free transducer for electrochemical genosensing of exosomal CD24. *Carbohydrate Polymers 269*: 118333.

81 Singla, P., Parokie, G., Garg, S. et al. (2023). Enhancing encapsulation of hydrophobic phyto-drugs naringenin and baicalein in polymeric nano-mics. *Journal of Drug Delivery Science and Technology* https://doi.org/10.1016/j.jddst.2023.104403.

82 Bigham, A., Rahimkhoei, V., Abasian, P. et al. (2022). Advances in tannic acid-incorporated biomaterials: infection treatment, regenerative medicine, cancer therapy, and biosensing. *Chemical Engineering Journal 432*: 134146. https://doi.org/10.1016/j.cej.2021.134146.

83 Jafari, H., Ghaffari-Bohlouli, P., Vahid Niknezhad, S. et al. (2022). Tannic acid: a versatile polyphenol for design of biomedical hydrogels. *Journal of Materials Chemistry B 10* (31): 5873–5912. https://doi.org/10.1039/D2TB01056A.

84 Geng, H., Dai, Q., Sun, H. et al. (2020). Injectable and sprayable polyphenol-based hydrogels for controlling hemostasis. *ACS Applied Bio Materials 3* (2): 1258–1266.

85 Erel-Unal, I. and Sukhishvili, S.A. (2008). Hydrogen-bonded multilayers of a neutral polymer and a polyphenol. *Macromolecules 41* (11): 3962–3970.

86 Fan, H., Wang, L., Feng, X. et al. (2017). Supramolecular hydrogel formation based on tannic acid. *Macromolecules 50* (2): 666–676.

87 Lin, F., Wang, Z., Shen, Y. et al. (2019). Natural skin-inspired versatile cellulose biomimetic hydrogels. *Journal of Materials Chemistry A 7* (46): 26442–26455.

88 Zhou, L., Fan, L., Yi, X. et al. (2018). Soft conducting polymer hydrogels cross-linked and doped by tannic acid for spinal cord injury repair. *ACS Nano 12* (11): 10957–10967.

89 Shao, C., Wang, M., Meng, L. et al. (2018). Mussel-inspired cellulose nanocomposite tough hydrogels with synergistic self-healing, adhesive, and strain-sensitive properties. *Chemistry of Materials 30* (9): 3110–3121.

90 Zhang, C., Wu, B., Zhou, Y. et al. (2020). Mussel-inspired hydrogels: from design principles to promising applications. *Chemical Society Reviews 49* (11): 3605–3637.

91 Sahiner, N., Sagbas, S., and Aktas, N. (2016). Preparation of macro-, micro-, and nano-sized poly (Tannic acid) particles with controllable degradability and multiple biomedical uses. *Polymer Degradation and Stability 129*: 96–105.

92 Ball, V. and Meyer, F. (2016). Deposition kinetics and electrochemical properties of tannic acid on gold and silica. *Colloids and Surfaces A: Physicochemical and Engineering Aspects 491*: 12–17.

93 Baldwin, A. and Booth, B.W. (2022). Biomedical applications of tannic acid. *Journal of Biomaterials Applications 36* (8): 1503–1523.

94 Sahiner, N., Sagbas, S., and Aktas, N. (2015). Single step natural poly (tannic acid) particle preparation as multitalented biomaterial. *Materials Science and Engineering: C 49*: 824–834.

95 Chowdhury, P., Nagesh, P.K., Hatami, E. et al. (2019). Tannic acid-inspired paclitaxel nanoparticles for enhanced anticancer effects in breast cancer cells. *Journal of Colloid and Interface Science 535*: 133–148.

96 Kaczmarek, B., Milek, O., Michalska-Sionkowska, M. et al. (2020). Novel eco-friendly tannic acid-enriched hydrogels-preparation and characterization for biomedical application. *Materials 13* (20): 4572.

97 Sugiarto, S., Leow, Y., Tan, C.L. et al. (2022). How far is Lignin from being a biomedical material? *Bioactive Materials 8*: 71–94.

98 Xiong, F., Han, Y., Wang, S. et al. (2017). Preparation and formation mechanism of size-controlled lignin nanospheres by self-assembly. *Industrial Crops and Products 100*: 146–152.

99 Qian, Y., Zhong, X., Li, Y., and Qiu, X. (2017). Fabrication of uniform lignin colloidal spheres for developing natural broad-spectrum sunscreens with high sun protection factor. *Industrial Crops and Products 101*: 54–60.

100 Chowdhury, M.A. (2014). The controlled release of bioactive compounds from lignin and lignin-based biopolymer matrices. *International Journal of Biological Macromolecules 65*: 136–147.

101 Košíková, B. and Lábaj, J. (2009). Lignin-stimulated protection of polypropylene films and DNA in cells of mice against oxidation damage. *BioResources 4* (2).

102 Košíková, B. and Sláviková, E. (2011). Use of lignin and pla as renewable additives of polypropylene films. *Wood Research 56* (3): 393–402.

103 Meng, Y., Lu, J., Cheng, Y. et al. (2019). Lignin-based hydrogels: a review of preparation, properties, and application. *International Journal of Biological Macromolecules 135*: 1006–1019.

104 Abdullah, T., Gauthaman, K., Mostafavi, A. et al. (2020). Sustainable drug release from polycaprolactone coated chitin-lignin gel fibrous scaffolds. *Scientific Reports 10* (1): 20428.

105 Wang, J., Tian, L., Luo, B. et al. (2018). Engineering PCL/lignin nanofibers as an antioxidant scaffold for the growth of neuron and Schwann cell. *Colloids and Surfaces B: Biointerfaces 169*: 356–365.

106 Liang, R., Yang, X., Yew, P.Y.M. et al. (2022). PLA-lignin nanofibers as antioxidant biomaterials for cartilage regeneration and osteoarthritis treatment. *Journal of Nanobiotechnology 20* (1): 327.

107 Xu, Z., Yuan, L., Liu, Q. et al. (2022). Crosslinking effect of dialdehyde cholesterol modified starch nanoparticles on collagen hydrogel. *Carbohydrate Polymers 285*: 119237. https://doi.org/10.1016/j.carbpol.2022.119237.

108 El Khawand, T., Courtois, A., Valls, J. et al. (2018). A review of dietary stilbenes: sources and bioavailability. *Phytochemistry Reviews 17* (5): 1007–1029. https://doi.org/10.1007/s11101-018-9578-9.

109 Reinisalo, M., Kårlund, A., Koskela, A. et al. (2015). Polyphenol stilbenes: molecular mechanisms of defence against oxidative stress and aging-related diseases. *Oxidative Medicine and Cellular Longevity 2015*: 340520. https://doi.org/10.1155/2015/340520.

110 Lin, Y.-C., Hu, S.C.-S., Huang, P.-H. et al. (2020). Electrospun resveratrol-loaded polyvinylpyrrolidone/cyclodextrin nanofibers and their biomedical applications. *Pharmaceutics 12* (6): 552. https://doi.org/10.3390/pharmaceutics12060552.

111 Figueiras, T.S., Neves-Petersen, M.T., and Petersen, S.B. (2011). Activation energy of light induced isomerization of resveratrol. *Journal of Fluorescence 21* (5): 1897–1906. https://doi.org/10.1007/s10895-011-0886-3.

112 Annaji, M., Poudel, I., Boddu, S.H.S. et al. (2021). Resveratrol-loaded nanomedicines for cancer applications. *Cancer Reports 4* (3): e1353. https://doi.org/10.1002/cnr2.1353.

113 Xiao, Q., Zhu, W., Feng, W. et al. (2019). A review of resveratrol as a potent chemoprotective and synergistic agent in cancer hemotherapy. *Frontiers in Pharmacology 9*: https://www.frontiersin.org/articles/10.3389/fphar.2018.01534.

114 Alavi, M., Farkhondeh, T., Aschner, M., and Samarghandian, S. (2021). Resveratrol mediates its anti-cancer effects by Nrf2 signaling pathway activation. *Cancer Cell International 21* (1): 579. https://doi.org/10.1186/s12935-021-02280-5.

115 Salehi, B., Mishra, A.P., Nigam, M. et al. (2018). Resveratrol: a double-edged sword in health benefits. *Biomedicine 6* (3): 91. https://doi.org/10.3390/biomedicines6030091.

116 Zheng, Y., Jia, R., Li, J. et al. (2022). Curcumin- and resveratrol-co-loaded nanoparticles in synergistic treatment of hepatocellular carcinoma. *Journal of Nanobiotechnology 20* (1): 339. https://doi.org/10.1186/s12951-022-01554-y.

117 Sharifi-Rad, J., Quispe, C., Mukazhanova, Z. et al. (2021). Resveratrol-based nanoformulations as an emerging therapeutic strategy for cancer. *Frontiers in Molecular Biosciences 8*: https://www.frontiersin.org/articles/10.3389/fmolb.2021.649395.

118 Riccitiello, F., De Luise, A., Conte, R. et al. (2018). Effect of resveratrol release kinetic from electrospun nanofibers on osteoblast and osteoclast differentiation. *European Polymer Journal 99*: 289–297. https://doi.org/10.1016/j.eurpolymj.2017.12.035.

119 Snetkov, P., Morozkina, S., Olekhnovich, R., and Uspenskaya, M. (2022). Electrospun curcumin-loaded polymer nanofibers: solution recipes, process parameters, properties, and biological activities. *Materials Advances 3* (11): 4402–4420. https://doi.org/10.1039/D2MA00440B.

120 Zupančič, Š., Baumgartner, S., Lavrič, Z. et al. (2015). Local delivery of resveratrol using polycaprolactone nanofibers for treatment of periodontal disease. *Journal of Drug Delivery Science and Technology 30*: 408–416. https://doi.org/10.1016/j.jddst.2015.07.009.

121 Rostami, M., Ghorbani, M., Aman Mohammadi, M. et al. (2019). Development of resveratrol loaded chitosan-gellan nanofiber as a novel gastrointestinal delivery system. *International Journal of Biological Macromolecules 135*: 698–705. https://doi.org/10.1016/j.ijbiomac.2019.05.187.

122 Karimi-Soflou, R., Mohseni-Vadeghani, E., and Karkhaneh, A. (2022). Controlled release of resveratrol from a composite nanofibrous scaffold: effect of resveratrol on antioxidant activity and osteogenic differentiation. *Journal of Biomedical Materials Research Part A 110* (1): 21–30. https://doi.org/10.1002/jbm.a.37262.

123 Wang, J., Hu, J., Yuan, X. et al. (2022). Recombinant collagen hydrogels induced by disulfide bonds. *Journal of Biomedical Materials Research Part A 110* (11): 1774–1785. https://doi.org/10.1002/jbm.a.37427.

124 Sheu, S.-Y., Chen, W.-S., Sun, J.-S. et al. (2013). Biological characterization of oxidized hyaluronic acid/resveratrol hydrogel for cartilage tissue engineering. *Journal of Biomedical Materials Research Part A 101* (12): 3457–3466. https://doi.org/10.1002/jbm.a.34653.

125 Zhao, C.-C., Zhu, L., Wu, Z. et al. (2020). Resveratrol-loaded peptide-hydrogels inhibit scar formation in wound healing through suppressing inflammation. *Regenerative Biomaterials 7* (1): 99–107. https://doi.org/10.1093/rb/rbz041.

126 Zheng, Y., Yuan, W., Liu, H. et al. (2020). Injectable supramolecular gelatin hydrogel loading of resveratrol and histatin-1 for burn wound therapy. *Biomaterials Science 8* (17): 4810–4820. https://doi.org/10.1039/D0BM00391C.

127 Lafay, S. and Gil-Izquierdo, A. (2008). Bioavailability of phenolic acids. *Phytochemistry Reviews 7* (2): 301–311. https://doi.org/10.1007/s11101-007-9077-x.

128 Kumar, N. and Pruthi, V. (2014). Potential applications of ferulic acid from natural sources. *Biotechnology Reports 4*: 86–93. https://doi.org/10.1016/j.btre.2014.09.002.

129 Kumar, N. and Goel, N. (2019). Phenolic acids: natural versatile molecules with promising therapeutic applications. *Biotechnology Reports 24*: e00370. https://doi.org/10.1016/j.btre.2019.e00370.

130 Rashmi, H.B. and Negi, P.S. (2020). Phenolic acids from vegetables: a review on processing stability and health benefits. *Food Research International 136*: 109298. https://doi.org/10.1016/j.foodres.2020.109298.

131 Ashrafizadeh, M., Zarrabi, A., Mirzaei, S. et al. (2021). Gallic acid for cancer therapy: molecular mechanisms and boosting efficacy by nanoscopical delivery. *Food and Chemical Toxicology 157*: 112576. https://doi.org/10.1016/j.fct.2021.112576.

132 Badhani, B., Sharma, N., and Kakkar, R. (2015). Gallic acid: a versatile antioxidant with promising therapeutic and industrial applications. *RSC Advances 5* (35): 27540–27557. https://doi.org/10.1039/C5RA01911G.

133 Wutticharoenmongkol, P., Hannirojram, P., and Nuthong, P. (2019). Gallic acid-loaded electrospun cellulose acetate nanofibers as potential wound dressing materials. *Polymers for Advanced Technologies 30* (4): 1135–1147. https://doi.org/10.1002/pat.4547.

134 Hassani, A., Azarian, M.M.S., Ibrahim, W.N., and Hussain, S.A. (2020). Preparation, characterization and therapeutic properties of gum arabic-stabilized gallic acid nanoparticles. *Scientific Reports 10* (1), Article 1. https://doi.org/10.1038/s41598-020-71175-8.

135 Gong, W., Wang, R., Huang, H. et al. (2023). Construction of double network hydrogels using agarose and gallic acid with antibacterial and anti-inflammatory properties for wound healing. *International Journal of Biological Macromolecules 227*: 698–710. https://doi.org/10.1016/j.ijbiomac.2022.12.085.

136 He, W., Wang, X., Gong, W. et al. (2023). Construction of an antibacterial hydrogel based on diammonium glycyrrhizinate and gallic acid for bacterial-infected wound healing. *Colloids and Surfaces B: Biointerfaces 222*: 112975. https://doi.org/10.1016/j.colsurfb.2022.112975.

137 Li, Y., Wang, M., Sun, M. et al. (2022). Engineering antioxidant poly (citrate-gallic acid)-Exosome hybrid hydrogel with microglia immunoregulation for traumatic brain Injury-post neuro-restoration. *Composites Part B: Engineering 242*: 110034. https://doi.org/10.1016/j.compositesb.2022.110034.

138 Sun, X., Dong, M., Guo, Z. et al. (2021). Multifunctional chitosan-copper-gallic acid based antibacterial nanocomposite wound dressing. *International Journal of Biological Macromolecules 167*: 10–22.

139 Majeed, S., Shamsunazatul, N.I.N.B., Danish, M. et al. (2023). Mitochondrial membrane depolarization, oxidative stress induced cell death in human pancreatic adenocarcinoma cells (PANC-1) treated with biologically engineered gallic acid coated copper oxide nanoparticles. *Journal of Cluster Science* https://doi.org/10.1007/s10876-023-02420-z.

140 Sova, M. and Saso, L. (2020). Natural sources, pharmacokinetics, biological activities and health benefits of hydroxycinnamic acids and their metabolites. *Nutrients 12* (8): 2190. https://doi.org/10.3390/nu12082190.

141 Alam, M.A., Subhan, N., Hossain, H. et al. (2016). Hydroxycinnamic acid derivatives: a potential class of natural compounds for the management of lipid metabolism and obesity. *Nutrition & Metabolism 13* (1): 27. https://doi.org/10.1186/s12986-016-0080-3.

142 Andrade, S., Pereira, M.C., and Loureiro, J.A. (2023). Caffeic acid loaded into engineered lipid nanoparticles for Alzheimer's disease therapy. *Colloids and Surfaces B: Biointerfaces 225*: 113270. https://doi.org/10.1016/j.colsurfb.2023.113270.

143 Tiwari, S., Sontakke, A.D., Baruah, K., and Purkait, M.K. (2023). Development of graphene oxide-based nano-delivery system for natural chemotherapeutic agent (Caffeic Acid). *Materials Today Proceedings 76*: 325–335. https://doi.org/10.1016/j.matpr.2022.11.373.

144 Rahimi, H., Mohammadgholi, A., and Koosha, M. (2023). Cytotoxic effect of caffeic acid loaded in chitosan-grafted carbon nanotubes on MDA-MB-231 breast cancer cell lines. *Iran Journal of Science* https://doi.org/10.1007/s40995-023-01465-z.

145 Subbarayan, R., Girija, D.M., and Rao, S.R. (2019). Human umbilical cord tissue stem cells and neuronal lineages in an injectable caffeic acid–bioconjugated gelatin hydrogel for transplantation. *Journal of Cellular Physiology 234* (3): 1967–1977. https://doi.org/10.1002/jcp.26834.

146 Wu, Y., Wang, Y., Long, L. et al. (2022). A spatiotemporal release platform based on pH/ROS stimuli-responsive hydrogel in wound repairing. *Journal of Controlled Release 341*: 147–165. https://doi.org/10.1016/j.jconrel.2021.11.027.

147 Ignatova, M.G., Manolova, N.E., Rashkov, I.B. et al. (2016). Poly(3-hydroxybutyrate)/caffeic acid electrospun fibrous materials coated with polyelectrolyte complex and their antibacterial activity and in vitro antitumor effect against HeLa cells. *Materials Science and Engineering: C 65*: 379–392. https://doi.org/10.1016/j.msec.2016.04.060.

148 Sang, V.T., Dai Hung, N., and Se-Kwon, K. (2019). Pharmaceutical properties of marine polyphenols: an overview. *ACTA Pharmaceutica Sciencia 57* (2).

149 Mateos, R., Pérez-Correa, J.R., and Domínguez, H. (2020). Bioactive properties of marine phenolics. *Marine Drugs 18* (10): 501.

150 Kaushalya, K.G.D. and Gunathilake, K. (2022). Encapsulation of phlorotannins from edible brown seaweed in chitosan: effect of fortification on bioactivity and stability in functional foods. *Food Chemistry 377*: 132012.

151 Kim, S.-H., Kim, K., Kim, B.S. et al. (2020). Fabrication of polyphenol-incorporated anti-inflammatory hydrogel via high-affinity enzymatic crosslinking for wet tissue adhesion. *Biomaterials 242*: 119905. https://doi.org/10.1016/j.biomaterials.2020.119905.

Index

Science and Engineering of Polyphenols: Fundamentals and Industrial Scale Applications, First Edition. Edited by Chandrabhan Verma.
© 2024 John Wiley & Sons Inc. Published 2024 by John Wiley & Sons Inc.

Printed and bound by CPI Group (UK) Ltd, Croydon, CR0 4YY

17/09/2026

14964309-0007